Subcellular Biochemistry

Volume 114

The book series SUBCELLULAR BIOCHEMISTRY is a renowned and well recognized forum for disseminating advances of emerging topics in Cell Biology and related subjects. All volumes are edited by established scientists and the individual chapters are written by experts on the relevant topic. The individual chapters of each volume are fully citable and indexed in Medline/Pubmed to ensure maximum visibility of the work.

Nadia M. Hamdy • Ahmed S. Sultan
John Frederick Kennedy
Editors

From (Epi)genetics-to-Precision (E2P)

Non-protein Coding RNAs (ncRNAs) a Step-toward Precision Medicine

Springer

Editors
Nadia M. Hamdy
Biochemistry and Molecular Biology Department, Faculty of Pharmacy
Ain Shams University
Cairo, Egypt

Ahmed S. Sultan
Georgetown University Medical Center
Washington DC, WA, USA

John Frederick Kennedy
Chembiotech Laboratories
Worcestershire, UK

ISSN 0306-0225 ISSN 2542-8810 (electronic)
Subcellular Biochemistry
ISBN 978-3-032-08529-0 ISBN 978-3-032-08530-6 (eBook)
https://doi.org/10.1007/978-3-032-08530-6

This Springer imprint is published by the registered company Springer Nature Switzerland AG
The registered company address is: Gewerbestrasse 11, 6330 Cham, Switzerland

Contents

About the Authors

Nadia M. Hamdy is a renowned researcher with significant contributions to the scientific community since 2004. She has over 120 publications, 5 book chapters, and a Scopus h-index of 43. She has supervised 50 masters' and PhD students and has received numerous awards, including the life-time acheivement award 2025, Obada International Award for Distinguished Researchers 2025, ASU Appreciation Award in Technological Sciences 2024, ASRT-Egypt State Award Dr. Shawky el Hadad Award in cancer 2023, and the International "Wendy Harvan" Top Honorary Award in February 2023, and several international Honorary Awards in 2025 and 2021. Her research interests include pharmaceutical clinical biochemistry, molecular biology, genetic studies, biomarkers, immuno-oncology, biotechnology, and nano-biotechnology. She focuses on understanding genetic and epigenetic mechanisms of cancers and diseases, developing biomarkers for disease prognosis, and exploring innovative therapeutic approaches. She is also a member of the editorial board of 9 Q1 & Q2 Scopus journals and a reviewer for international and regional grants, awards, and scientific journals.

Ahmed S. Sultan is a distinguished Professor of Stem Cell and Molecular Oncology at Georgetown University Medical Center. He received his MD/PhD from Osaka University School of Medicine in 1999 and an MPH from Georgetown University in 2005. His research interests include clinical biochemistry, molecular biology, gene therapy for breast and liver cancer, and the role of stem cells in cancer models. His innovative research focuses on discovering new methods and treatment strategies for metastatic cancers, isolating and manipulating cancer stem cells versus human cancer cells. He aims to enhance the quality of life for cancer patients through innovative research.

John Frederick Kennedy is a distinguished scientist with four degrees from The University of Birmingham. He has received prestigious awards such as the Meldola Medal for Chemistry and the Carbohydrate Chemistry Award by the Royal Society of Chemistry. He has held various professional appointments, including Director of the Birmingham Carbohydrate and Protein Technology Group and Professor of

Applied Chemistry at Glyndwr University. He has published over 800 papers and authored or edited around 20 books. His research interests include carbohydrate chemistry, enzyme technology, and the environmental impact of industrial processes. He has been involved in developing innovative solutions for medical and environmental challenges, demonstrating a commitment to advancing science for the benefit of society.

Chapter 1
Epigenetic Modulation of Immunity: Mechanisms, Implications, and Emerging Therapeutic Horizons; a Step Toward Epigenetics to Precision

Dalia O. Saleh, Mohamed L. Salem, Sophie Van Linthout, Queran Lin, and Nadia M. Hamdy

Abstract Epigenetic regulation plays a central role in immune cell development, specialization, and memory formation by dynamically modifying DNA, histones, and RNA. These processes enable adaptation to environmental cues, precise pathogen responses, and maintenance of immune tolerance, while their disruption contributes to autoimmune, inflammatory, and cancer pathogenesis. DNA methylation, histone modifications, and noncoding RNA regulation shape the lineage and activation states of T cells, B cells, macrophages, and natural killer (NK) cells, with specific alterations linked to diseases, such as systemic lupus erythematosus (SLE) and rheumatoid arthritis (RA). Emerging insights also highlight roles for RNA modifications and exosome-mediated RNA transfer in immune activation, trained immu-

D. O. Saleh (✉)
Pharmacology Department, Institute of Medical Research and Clinical Studies, National Research Centre, Giza, Egypt
e-mail: do.saleh@nrc.sci.eg

M. L. Salem
Immunology and Biotechnology Unit, Department of Zoology, Faculty of Science, Tanta University, Tanta, Egypt

Center of Excellence in Cancer Research, Tanta University, Tanta, Egypt

S. Van Linthout
Berlin Institute of Health (BIH) at Charité—Universitätmedizin Berlin, BIH Center for Regenerative Therapies (BCRT), Berlin, Germany

Q. Lin
Clinical Research Centre & Breast Tumor Centre, Sun Yat-sen Memorial Hospital, Sun Yat-sen University, Faculty of Medicine, WHOCC, Imperial College London, London, UK

N. M. Hamdy
Biochemistry and Molecular Biology Department, Faculty of Pharmacy, Ain Shams University, Cairo, Egypt
e-mail: Nadia_hamdy@pharma.asu.edu.eg

N. M. Hamdy et al. (eds.), *From (Epi)genetics-to-Precision (E2P)*, Subcellular Biochemistry 114, https://doi.org/10.1007/978-3-032-08530-6_1

nity, and antigen presentation. Advances in single-cell epigenomics, CRISPR-based editing, and RNA sequencing are driving the development of targeted therapies—such as DNA methyltransferase (DNMT) inhibitors, histone deacetylase (HDAC) inhibitors, RNA-based interventions, and exosome delivery systems—that aim to reprogram immune responses. Understanding immune cell epigenetics paves the way for precision immunotherapies tailored to patient-specific profiles, offering highly specific, effective treatments with minimal immune suppression.

Graphical Abstract

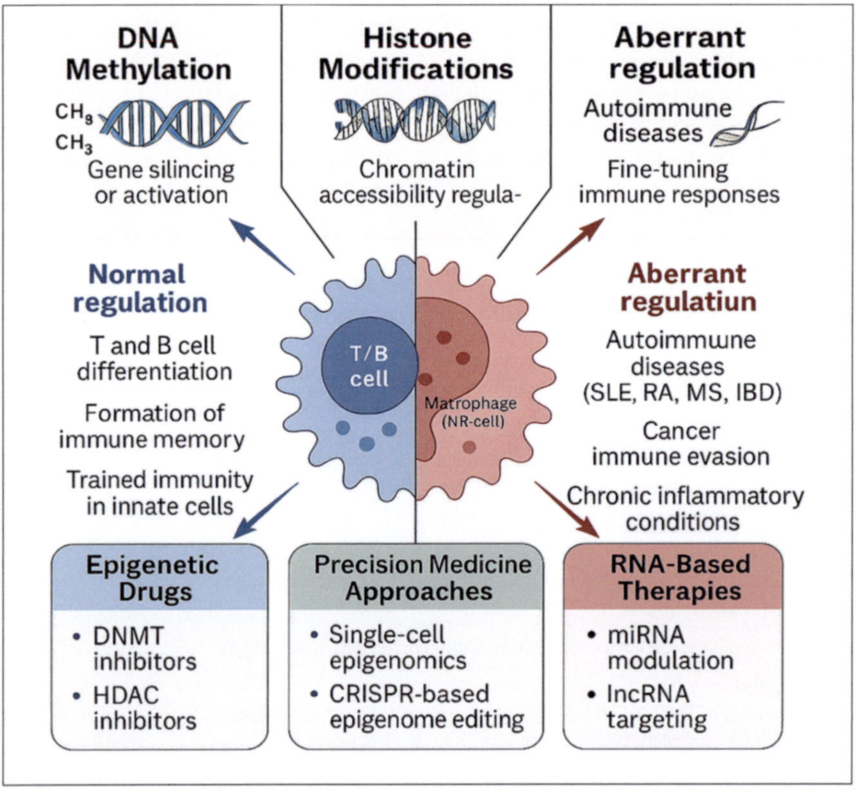

This scientific diagram illustrates how epigenetic mechanisms—DNA methylation, histone modifications, and noncoding RNAs (miRNAs, lncRNAs)—regulate immune cell function under normal and pathological conditions. On the left, normal regulation supports T and B cell differentiation, immune memory formation, and trained immunity in innate cells like macrophages and NK cells. On the right, aberrant regulation contributes to autoimmune diseases [e.g., SLE, RA, multiple sclerosis (MS), and inflammatory bowel disease (IBD)], cancer immune evasion, and chronic inflammation. At the bottom, therapeutic strategies are mapped: epigenetic drugs (e.g., DNMT and HDAC inhibitors), precision medicine approaches (e.g., single-cell epigenomics and CRISPR editing), and RNA-based therapies (e.g., miRNA modulation and lncRNA targeting) aim to restore healthy immune function

Keywords Epigenetics · Immune cells · Autoimmunity · RNA changes · Targeted immunotherapy

Abbreviations

ATAC-seq	Assay for Transposase-Accessible Chromatin with high-throughput sequencing
Cas9	CRISPR-associated Protein 9
ChIP	Chromatin Immunoprecipitation
CNS	Central nervous system
CpG	Cytosine-phosphate-Guanine
CRISPR	Clustered Regularly Interspaced Short Palindromic Repeats
DNA	Deoxyribonucleic acid
DNMT	DNA methyltransferase
FOXP3	Forkhead Box Protein P3
H3K27ac	Acetylation at Lysine 27 of Histone H3
HDAC	Histone deacetylase
IBD	Inflammatory bowel disease
IFN	Interferon
IgH	Immunoglobulin heavy chain
IL	Interleukin
lncRNA	long noncoding RNA
miRNA	microRNA
MS	Multiple sclerosis
ncRNA	noncoding RNA
NK	Natural killer (cell)
PAMPs	Pathogen-associated molecular patterns
RA	Rheumatoid arthritis
RNA	Ribonucleic acid
RNA-seq	RNA sequencing
scRNA-seq	Single-cell RNA sequencing
SLE	Systemic lupus erythematosus
Th	T helper cell
TLR	Toll-like receptor
TNF-α	Tumor necrosis factor-alpha
Treg	Regulatory T cell
Ψ	Pseudouridine

Introduction

Epigenetic mechanisms critically modulate immune function, governing the specificity, adaptability, and maintenance of physiological homeostasis required for effective host defense (Morandini et al. 2016). While genetic information provides the basic blueprint, it is the dynamic epigenetic landscape—including reversible yet heritable modifications, such as DNA methylation, histone modifications, chromatin remodeling, and noncoding RNA regulation—that refines immune responses without changing the underlying DNA sequence (Handy et al. 2011). These epigenetic processes enable immune cells to swiftly respond and adapt to environmental stimuli, pathogens, and inflammatory signals, thus ensuring timely and appropriate immune reactions (Dai et al. 2021).

This chapter provides a comprehensive overview of the pivotal role of epigenetic regulation in immune homeostasis, the pathogenesis of autoimmune and inflammatory diseases, and the development of novel therapeutic strategies, underscoring the dynamic interplay between the epigenome and immune function (Caldwell and Li 2024).

Recent advances have deepened our understanding of both innate and adaptive immune responses to cancer, where epigenetic regulation plays a critical role (Yang et al. 2023). Innate immune cells—such as natural killer (NK) cells, dendritic cells (DCs), and macrophages—serve as first-line defenders against infections and malignancies. Special attention has been given to the intricate pathways by which Toll-like receptors (TLRs) mediate interactions between the innate immune system and tumors (Stojanovic et al. 2018). TLRs, a subset of pattern recognition receptors (PRRs), recognize self and nonself antigens, bridge innate and adaptive immunity, regulate cytokine production, and control cell proliferation and survival (Li and Wu 2021). While mammals possess 13 TLRs (TLR1–13), humans express only 10 (TLR1–10). Although the roles of TLR1–9 and TLR11 are well characterized, TLR10, TLR12, and TLR13 remain less understood (Tanekhy 2016).

Beyond pathogen recognition, TLRs also detect endogenous danger-associated molecular patterns (DAMPs), such as heat shock proteins and mitochondrial DNA, thereby modulating immune responses in various physiological and pathological contexts. Activation of TLR pathways can exert opposing effects on tumor biology: promoting inflammation and tumor progression or enhancing antitumor immune responses by activating DCs, NK cells, and cytotoxic T lymphocytes (CTLs) (Bezhaeva et al. 2022). Ligand binding—via infection or vaccination—initiates key molecular events that shape the immune response through transcriptional and epigenetic regulation (Verbeke et al. 2022).

Epigenetic mechanisms, including DNA methylation, chromatin remodeling, and RNA-based regulation (miRNA, soRNA, icnRNA, siRNA, and piRNA), modulate gene expression critical to immune function (Pajares et al. 2021). Notably, noncoding RNAs transported via exosomes facilitate intercellular genetic communication (Marima et al. 2024). Tumor-derived exosomes carrying noncoding RNAs can reprogram neighboring cells, influencing immune surveillance and tumor progression. This underscores the importance of investigating exosome-mediated communication within the tumor microenvironment (Xu et al. 2022).

Additionally, the epigenetic modification of RNA, particularly pseudouridylation—the isomerization of uridine to pseudouridine (Ψ)—has emerged as a critical regulatory mechanism (Jia et al. 2025). Dysregulation of pseudouridylation machinery has been implicated in cancer, affecting the structure and function of key RNAs (rRNA, tRNA, and mRNA) involved in translation and gene expression.

Non-mutational epigenetic reprogramming is increasingly recognized as a hallmark of cancer (Añazco-Guenkova et al. 2024). Epigenetic modifications critically regulate oncogenic signaling pathways, tumor immunogenicity, and immune effector cell functionality, thereby influencing cancer progression and the effectiveness of therapeutic interventions (Tien et al. 2023). These findings have major implications for developing immunotherapies and epigenetic-targeted treatments (Wang et al. 2022).

In parallel, substantial progress has been made in cancer vaccine development. Vaccines targeting tumor-associated antigens, particularly neoantigens derived from tumor-specific mutations, aim to elicit sustained T cell-mediated antitumor responses (Blass and Ott 2021). Researchers are also exploring epigenetic markers as novel vaccine targets (Bannister and Messina 2020). However, reports of potential frameshift events during ribosomal translation of mRNA-based vaccines—resulting in truncated, nonfunctional proteins—highlight the need for further investigation to optimize vaccine design and efficacy (Mulroney et al. 2024).

Mechanisms of Epigenetic Regulation

The immune system's ability to mount specific, robust, and regulated responses relies heavily on precise gene expression control. Beyond genetic information, **epigenetic mechanisms**—heritable changes in gene activity without alterations to the DNA sequence—play a central role in governing immune cell development, differentiation, and function. These modifications enable immune cells to respond dynamically to environmental cues, pathogens, and stressors while maintaining tolerance to self-antigens. Among the major epigenetic regulators are **DNA methylation**, **histone modifications**, and **noncoding RNAs**. Each mechanism contributes uniquely to shaping the immune landscape, from guiding T and B cell lineage decisions to modulating inflammatory and tolerogenic responses. A summary of these key epigenetic mechanisms and their immune-specific roles is presented in Table 1.1.

DNA Methylation in Immune Cell Differentiation and Immune Regulation

DNA methylation is a crucial regulatory mechanism in immune system function, influencing the **development**, **activation**, and **differentiation** of immune cells, as illustrated in Fig. 1.1. By dynamically adjusting in response to environmental cues, DNA methylation allows the immune system to **adapt** and respond effectively to threats, such as infections and tissue damage (Qin et al. 2021a). Aberrant methylation patterns are implicated in many **autoimmune disorders**, **chronic inflammation**, and **cancers**, offering promising therapeutic targets for **epigenetic therapies**

Table 1.1 Epigenetic mechanisms and their roles in immune regulation

Mechanism	Description	Key features
DNA methylation	Addition of methyl groups to cytosine bases in DNA, particularly at CpG islands, which typically leads to transcriptional repression	Essential for **lineage-specific gene silencing** and activation **Dynamic and reversible** modification Plays a key role in **immune cell differentiation**, **tolerance**, and **activation**
Histone modifications	Posttranslational chemical modifications of histone proteins, including acetylation, methylation, phosphorylation, and ubiquitination	**Alter chromatin architecture**, making it more or less accessible to transcription machinery Involved in **gene expression regulation**: acetylation usually activates, while methylation can repress gene expression Important for **immune cell activation**, **differentiation**, and **immune response** modulation
Noncoding RNAs	Includes microRNAs (miRNAs) and long noncoding RNAs (lncRNAs) that regulate transcriptional and posttranscriptional gene expression	**Fine-tune immune responses** by modulating the expression of immune-related genes **Control immune cell differentiation** (e.g., Th17, Treg, B cells) and their **effector functions** (e.g., cytokine production) miRNAs regulate posttranscriptional gene silencing by binding to mRNA, while lncRNAs can act as scaffolds to recruit chromatin-modifying complexes

aimed at **restoring immune balance** and **enhancing immune tolerance** (**Mei et al.** 2022).

DNA methylation is a fundamental epigenetic modification that plays a critical role in regulating gene expression, particularly in immune cell development and function (Lanata et al. 2018), as illustrated in Fig. 1.1. It involves the addition of a **methyl group** (CH_3) to the **cytosine** base of DNA, predominantly at **CpG dinucleotides**, leading to the repression of gene expression. This modification is dynamic and reversible, allowing for precise control over gene activation and silencing, which is essential for the proper functioning of the immune system (Breiling and Lyko 2015; Kumar et al. 2018).

DNA Methylation in T Cell Lineage Commitment

T cells, which are essential for adaptive immunity, undergo a series of differentiation events during their development in the thymus. DNA methylation plays a pivotal role in regulating the **commitment** of precursor T cells into specific subsets,

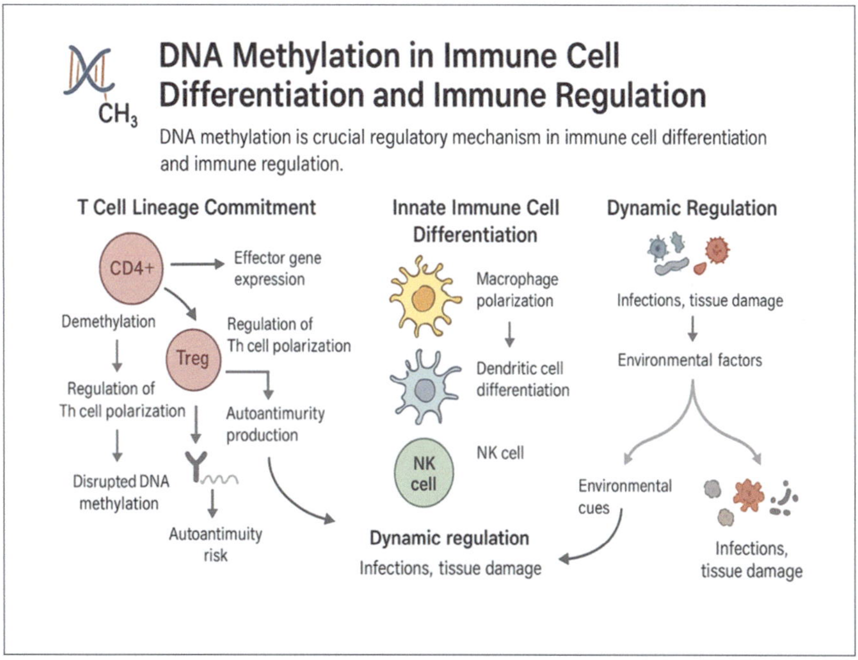

Fig. 1.1 DNA methylation in immune cell differentiation and regulation. This schematic illustrates the pivotal role of DNA methylation in modulating immune cell development and function. In the **T cell lineage**, DNA demethylation promotes effector gene expression and regulates T helper (Th) cell polarization, including differentiation into regulatory T cells (Tregs). Disruption of normal DNA methylation patterns can impair Th cell differentiation and promote autoimmunity. In **innate immune cells**, such as macrophages, dendritic cells, and natural killer (NK) cells, methylation dynamics control macrophage polarization (e.g., M1 vs. M2), dendritic cell differentiation, and NK cell function. Finally, **dynamic regulation** of DNA methylation occurs in response to **infections, tissue injury**, and **environmental factors** (e.g., diet, stress, and toxins), which collectively influence immune responses and homeostasis. Aberrant methylation in any of these processes may contribute to immune dysregulation, autoimmunity, or impaired pathogen defense

such as **CD4+ helper T cells (Th1, Th2, Th17, and Treg)** and **CD8+ cytotoxic T cells** (Correa et al. 2020).

One critical aspect of this regulation involves the expression of effector genes that drive immune functions. Demethylation of specific regions within these genes facilitates their transcriptional activation. For example, the *interferon-gamma (IFN-γ)* gene, essential for Th1 cell responses, becomes actively expressed when its promoter region undergoes demethylation. This allows Th1 cells to produce IFN-γ in response to intracellular pathogens, contributing to robust cell-mediated immunity (Mittelstaedt et al. 2021).

Beyond effector function, DNA methylation also orchestrates the polarization of T helper cells by modulating the expression of key transcription factors. A notable example is the *Foxp3* gene, which encodes a transcription factor indispensable for the development and function of Tregs. The methylation status of *Foxp3* determines not only whether Tregs can emerge from naïve T cells but also affects their stability and suppressive capacity. Proper methylation ensures that Tregs can effectively maintain immune homeostasis and prevent overactive immune responses that could harm the host (Morales-Nebreda et al. 2019).

However, when DNA methylation patterns become dysregulated, the consequences for immune regulation can be profound. Hypomethylation of immune-related genes may lead to their aberrant expression, resulting in hyperactive T cell responses (Liu et al. 2022b). This phenomenon has been implicated in the pathogenesis of several autoimmune disorders, including systemic lupus erythematosus (SLE) (Gorelik et al. 2007) and rheumatoid arthritis (RA) (Cribbs et al. 2014). Thus, the epigenetic regulation of T cell differentiation and function through DNA methylation is not only essential for shaping effective immune responses but also for maintaining immune tolerance and preventing autoimmunity (Bagni et al. 2025).

DNA Methylation in B Cell Maturation

B cells are central to the humoral immune response, responsible for producing antibodies. DNA methylation is also a key regulatory mechanism in **B cell development**, maturation, and **antibody class switching** (Lai et al. 2013).

DNA methylation critically regulates gene expression during B cell development, guiding differentiation from hematopoietic stem cells into mature, antibody-secreting plasma cells (Shi et al. 2024). Throughout B cell maturation, specific patterns of DNA methylation ensure the timely activation or repression of genes critical to each developmental stage. One prominent example is the methylation of the *immunoglobulin heavy chain* (IgH) locus. This locus encodes a key component of the B cell receptor (BCR) and antibodies, and its regulation is essential for the generation of antibody diversity (Oudinet et al. 2019; Demczuk et al. 2012). Methylation of this region in early progenitor stages maintains gene silencing, preventing premature expression. As B cells mature, selective demethylation at this locus occurs, enabling the recombination of gene segments necessary for creating a diverse repertoire of antibodies. This precise control ensures that B cells develop with functional, yet highly variable, antigen receptors capable of recognizing a wide array of pathogens (Manoharan et al. 2015).

In addition to orchestrating antibody production, DNA methylation also governs immune tolerance in mature B cells by repressing genes associated with self-reactivity and autoantibody generation, thus safeguarding against autoimmunity (Wang et al. 2024a). When these regulatory methylation marks are disrupted—for instance, through hypermethylation of genes that typically suppress autoimmune responses—B cells may become autoreactive. This aberrant gene silencing can result in the inappropriate production of autoantibodies, which target and damage host tissues. Therefore, the dynamic and tightly controlled DNA methylation landscape in B cells is essential not only for their normal development and antibody function but also for maintaining immune homeostasis and preventing pathological autoimmunity (Zan and Casali 2015; Celarain and Tomas-Roig 2020).

DNA Methylation in Immune Cell Differentiation

The differentiation of **innate immune cells** (e.g., **macrophages**, **dendritic cells**, and **natural killer cells (NK cells)**) also depends on tightly regulated DNA methylation patterns. These immune cells respond to pathogens and play critical roles in initiating immune responses and maintaining tissue homeostasis. DNA methylation ensures the precise regulation of gene expression required for these innate immune cells to develop appropriately and perform their specific roles effectively (Tuong et al. 2022).

Macrophages, for instance, can differentiate into functionally distinct subtypes in response to environmental signals, such as infection or tissue damage. This polarization process results in either M1 macrophages, which produce pro-inflammatory cytokines and contribute to pathogen clearance, or M2 macrophages, which are involved in resolving inflammation and promoting tissue repair (Chen et al. 2023). DNA methylation plays a central role in regulating the expression of genes associated with these phenotypes. For example, in M2 macrophages, hypermethylation of promoter regions of pro-inflammatory cytokine genes like *TNF-α* and *IL-6* suppresses their expression, thereby limiting the inflammatory response. This epigenetic control is crucial to prevent chronic inflammation and to facilitate recovery following infection or injury (Chen et al. 2020).

Similarly, dendritic cells (DCs) depend on well-regulated DNA methylation during their development from progenitor cells. These antigen-presenting cells are key players in bridging innate and adaptive immunity, as they process and present antigens to naïve T cells, thereby initiating adaptive immune responses (Soto et al. 2020). DNA methylation not only governs their differentiation but also modulates their functional capacity to activate T cells. Aberrant DNA methylation patterns in dendritic cells—such as inappropriate silencing of genes required for antigen processing or costimulatory signaling—can result in impaired immune activation. This dysfunction may lead to weakened immunity against infections or, conversely, to inappropriate immune activation that contributes to autoimmunity or chronic inflammation (Mittelstaedt et al. 2021).

Dynamic Regulation of DNA Methylation in Response to Environmental Cues

DNA methylation is **dynamic** and responds to various **environmental stimuli**, such as infections, inflammation, and tissue damage. The adaptability of the immune system is partly due to the ability of immune cells to alter their **DNA methylation patterns** in response to these signals (Rasmi et al. 2023).

During infections, immune cells rapidly adapt to combat invading pathogens by undergoing extensive transcriptional reprogramming, much of which is driven by dynamic changes in DNA methylation that regulate gene accessibility and expression. Specifically, pro-inflammatory cytokine genes, such as *TNF-α*, *IL-6*, and *IFN-γ*, undergo site-specific demethylation at their promoter regions, enabling their transcription and subsequent protein production. This demethylation is a critical mechanism that ensures immune cells can produce the appropriate signals needed to coordinate and amplify the immune response during infection (Fu and Wang 2024).

In the context of tissue damage, DNA methylation plays a dual role in regulating the immune response. On one hand, hypermethylation of inflammatory gene promoters can suppress excessive inflammation, protecting tissues from further injury (Qin et al. 2021b). On the other hand, demethylation of genes involved in cellular repair and regeneration promotes healing by activating pathways necessary for tissue recovery. This dynamic regulation of methylation enables the immune system to finely balance pro-inflammatory and reparative processes, ensuring that the response to injury is both effective and non-damaging (Das et al. 2021).

Environmental factors also significantly influence DNA methylation and, consequently, immune function. Lifestyle elements, such as diet, psychological stress, exposure to environmental toxins, and the composition of the gut microbiome, can all induce epigenetic changes that persist over time. For example, a high-fat diet has been shown to alter DNA methylation in immune cells, leading to dysregulated immune responses (Breton et al. 2021). Such epigenetic shifts may contribute to chronic low-grade inflammation, which is a hallmark of metabolic disorders, including type 2 diabetes and obesity (El Mesallamy et al. 2013). These findings underscore the importance of environmental and lifestyle factors in shaping immune competence through epigenetic mechanisms like DNA methylation (Wu and Yin 2022).

Implications of Aberrant DNA Methylation in Disease

The **dysregulation** of DNA methylation can lead to various immune-related disorders. **Hypomethylation** of genes can lead to the **activation of pro-inflammatory pathways** and **autoimmune diseases**, while **hypermethylation** can result in the **silencing of immune regulatory genes**, impairing immune tolerance and leading to chronic inflammation (Li et al. 2021a).

Autoimmune disorders, such as multiple sclerosis (MS), rheumatoid arthritis (RA), and type 1 diabetes (T1D), are characterized by a loss of immune tolerance, wherein the immune system aberrantly targets self-antigens. Emerging evidence implicates dysregulated DNA methylation as a critical epigenetic mechanism contributing to the pathogenesis of these diseases (Fakour et al. 2024; El-Mesallamy et al. 2013a). Hypomethylation of genes involved in immune activation may lead to the inappropriate expression of pro-inflammatory mediators and activation of autoreactive lymphocytes, while hypermethylation of regulatory genes can impair mechanisms that normally suppress autoimmunity (Danieli et al. 2024). These epigenetic alterations result in the breakdown of peripheral and central tolerance, promoting chronic inflammation and autoantibody production. Targeting aberrant methylation patterns through epigenetic therapies holds promise for reestablishing immune homeostasis and ameliorating autoimmune responses by restoring the expression of genes essential for immune regulation (Ashwani et al. 2025).

In cancer, DNA methylation is a key mechanism of immune evasion. Hypermethylation of CpG islands within promoter regions of tumor suppressor and immune-related genes leads to their transcriptional silencing, thereby diminishing tumor antigenicity and impairing immune surveillance. This epigenetic repression disrupts antigen processing and presentation, dampens interferon signaling, and hinders the recruitment and activation of immune effector cells (Lakshminarasimhan and Liang 2016). Consequently, tumor cells acquire the ability to escape immunosurveillance. Therapeutic strategies aimed at demethylating these silenced genes—such as the use of DNA methyltransferase inhibitors—are being explored to enhance tumor immunogenicity and improve the efficacy of immunotherapies. By reversing aberrant methylation patterns, it may be possible to potentiate antitumor immune responses and overcome resistance to immune checkpoint blockade (Cai et al. 2025).

Histone Modifications in Epigenetic Regulation of Immune Cells

Histone modifications, such as **acetylation** and **methylation**, play a pivotal role in **epigenetic regulation** of immune cells. These modifications influence **immune cell differentiation**, **activation**, and **immune responses** to infections or autoimmune challenges, as demonstrated in Fig. 1.2. **Histone acetylation** promotes gene expression by opening chromatin, while **histone methylation** can either activate or repress gene expression, depending on the modification (Sawada and Gallo 2021). **Histone-modifying enzymes**, including **HATs**, **HDACs**, and **HMTs**, regulate these processes and play a key role in determining immune cell fate and function. Dysregulation of these **epigenetic mechanisms** can lead to **immune-related diseases,** such as **autoimmunity** and **chronic inflammation**, making **histone-modifying enzymes** potential therapeutic targets for treating immune-mediated disorders (Yang et al. 2022c).

Histone modifications, particularly **acetylation** and **methylation**, are integral to **epigenetic regulation** and play a critical role in **gene expression**, **immune cell differentiation**, and **activation** (Yao et al. 2024). These modifications influence the **structure of chromatin**, thereby regulating access to the **transcriptional**

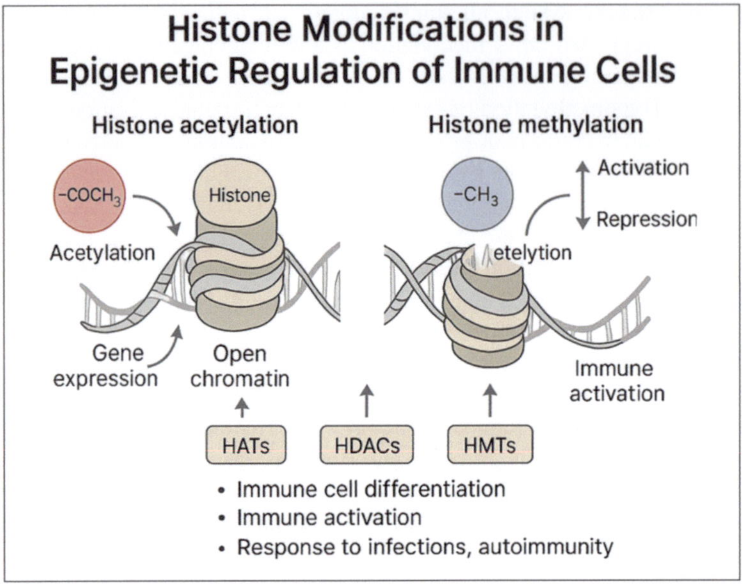

Fig. 1.2 Histone modifications in epigenetic regulation of immune cells. This figure illustrates the pivotal roles of histone modifications—specifically acetylation and methylation—in the epigenetic regulation of immune cell differentiation, activation, and function. Histone acetylation, mediated by histone acetyltransferases (HATs), promotes chromatin relaxation and gene transcription, facilitating immune responses, such as cytokine production and T cell activation. Conversely, histone deacetylases (HDACs) remove acetyl groups, leading to chromatin condensation and gene repression. Histone methylation exhibits dual regulatory effects: H3K4 methylation (e.g., H3K4me3) activates gene expression essential for immune activation, whereas H3K27 methylation (e.g., H3K27me3) represses genes to maintain immune tolerance, particularly in regulatory T cells (Tregs). Histone methyltransferases (HMTs) orchestrate these methylation patterns, influencing immune cell lineage commitment, including Th1, Th2, Th17, and Treg differentiation. The dynamic interplay of these modifications underlies the immune system's ability to mount effective responses against pathogens while preventing autoimmunity

machinery. By controlling the accessibility of DNA, histone modifications guide the **fate** and **function** of immune cells, contributing to proper immune responses and the development of immune tolerance (Ur Rehman et al. 2025).

Histone Acetylation: A Promoter of Gene Expression and Immune Activation

Histone acetylation involves the addition of **acetyl groups** to **lysine residues** on histones, leading to **chromatin relaxation** and increased gene transcription. Acetylation is a central mechanism in **epigenetic regulation** because it allows transcription factors and other regulatory proteins to access DNA and initiate transcription (Su et al. 2022).

Histone acetylation plays a vital role in regulating chromatin accessibility and gene expression in immune cells. The addition of acetyl groups to histone tails reduces their positive charge, thereby weakening the interaction between histones and the **negatively charged DNA** (Kim et al. 2023). This biochemical change results in a relaxed chromatin configuration known as euchromatin, which is more accessible to transcription factors and the transcriptional machinery. The open chromatin structure created by histone acetylation facilitates the expression of genes that are essential for immune activation (Eberharter and Becker 2002).

In immune responses, histone acetylation is especially critical. In **T lymphocytes**, for example, acetylation at the promoter region of genes, such as *IL-2,* is necessary for T cell activation and proliferation in response to antigenic stimulation. Similarly, macrophages and dendritic cells depend on histone acetylation to upregulate the expression of pro-inflammatory cytokines, like TNF-α and IL-6, which are key mediators in the innate immune response to infection (Wu et al. 2023).

Histone acetyltransferases (HATs) are the enzymes responsible for catalyzing the addition of acetyl groups to histones. Within immune cells, HATs play a crucial role in modulating the expression of genes involved in immune regulation and activation. By promoting the transcription of genes encoding cytokines and other immune effectors, HATs enable immune cells to mount coordinated responses to infections, inflammatory stimuli, and environmental challenges (Wu et al. 2023).

Histone Methylation: A Dual Role in Gene Regulation

Histone **methylation** involves the addition of **methyl groups** (-CH_3) to specific lysine or arginine residues on histones. Unlike acetylation, which generally promotes gene expression, **histone methylation** can either **activate** or **repress** gene expression, depending on the specific histone residue that is modified (Jambhekar et al. 2019).

Histone methylation represents a complex and finely tuned epigenetic mechanism that can either activate or repress gene expression, depending on the specific site and degree of methylation. One key activating modification is the methylation of histone H3 at lysine 4 (H3K4), particularly in its **trimethylated form (H3K4me3)** (Jambhekar et al. 2019). This mark is closely associated with active transcription and is typically found at the promoter regions of genes essential for immune function. In T lymphocytes, H3K4me3 facilitates the expression of genes like *IL-2*, which is critical for T cell activation. Similarly, in B cells, this modification enhances the transcription of genes necessary for antibody production, thereby promoting adaptive immune responses (Lamere et al. 2016; Anwar et al. 2022).

Conversely, histone methylation at **lysine 27 on histone H3 (H3K27)** serves as a repressive signal. The trimethylated form, H3K27me3, is a hallmark of transcriptionally silent chromatin and is enriched in genomic regions that are not actively expressed (Abu-Hanna et al. 2022). This modification is particularly important in **regulatory T cells (Tregs)**, where H3K27me3 at the *Foxp3* gene promoter stabilizes Treg identity and function. By repressing the transcription of genes that could promote inappropriate immune activation, H3K27me3 ensures that Tregs maintain immune tolerance and prevent autoimmune pathology (Yue et al. 2024).

The role of **H3K27me3** in immune tolerance is crucial. Tregs depend on the sustained presence of this repressive mark to silence genes that may otherwise lead to uncontrolled immune responses. Disruptions in the normal pattern of H3K27 methylation have been implicated in the pathogenesis of autoimmune diseases, where loss of gene silencing can result in the breakdown of tolerance and the development of chronic inflammation and immune-mediated tissue damage (Peeters et al. 2024).

Role of Histone-Modifying Enzymes in Immune Cell Fate and Function

Histone-modifying enzymes are critical to **epigenetic regulation** by determining the outcome of histone modifications. These enzymes include **histone acetyltransferases (HATs)**, **histone deacetylases (HDACs)**, and **histone methyltransferases (HMTs)**, which catalyze the addition or removal of acetyl and methyl groups on histones. These enzymes orchestrate the **fate** and **differentiation** of immune cells, ensuring that the appropriate immune responses are activated in response to environmental cues (Peeters et al. 2024).

Histone-modifying enzymes play a central role in regulating chromatin structure and, consequently, gene expression in immune cells. **HATs** are responsible for adding acetyl groups to histone tails, which leads to chromatin relaxation and enhanced transcriptional activity (Gujral et al. 2020). In the immune system, HATs are essential for activating genes involved in T cell differentiation, macrophage polarization, and the expression of key pro-inflammatory cytokines. By loosening chromatin structure, HATs facilitate the binding of transcription factors and RNA polymerase to DNA, enabling robust immune gene expression (Parajuli et al. 2025).

On the other hand, **HDACs** remove acetyl groups from histones, resulting in chromatin condensation and gene repression. HDAC activity is associated with the suppression of immune gene transcription, thereby modulating immune responses. Importantly, pharmacological inhibition of HDACs has emerged as a promising strategy in cancer immunotherapy. HDAC inhibitors can enhance T cell activation, increase the production of cytokines, and improve the effectiveness of immune-based treatments by reversing the silencing of immune effector genes (Watson et al. 2024).

HMTs, another class of histone-modifying enzymes, catalyze the addition of methyl groups to specific lysine residues on histones. These enzymes are critical for immune cell fate decisions. For example, the methylation of H3K4 is associated with gene activation, while H3K27 methylation is linked to gene repression (Mao et al. 2024). HMTs regulate the transcriptional programs that control the differentiation of various T cell subsets, including Th1, Th2, Th17, and regulatory T cells (Tregs). Through their site-specific methylation activity, HMTs fine-tune the balance between immune activation and tolerance, influencing immune cell function and polarization in response to different environmental and pathological cues (Zhang and Kuchroo 2019).

Epigenetic Regulation of Immune Responses

Histone modifications are critical in controlling immune responses by ensuring that immune cells respond appropriately to **infections**, **tissue injury**, and **immune tolerance signals** (Mugisha and University Iii 2024).

Histone modifications play a fundamental role in shaping immune responses by controlling the transcriptional programs of immune cells. In macrophages, histone acetylation is essential for initiating the expression of pro-inflammatory cytokines, such as TNF-α, during infections. This acetylation relaxes chromatin structure, allowing transcriptional machinery to access and activate genes that drive the macrophage's inflammatory response, thereby enabling these cells to effectively combat invading pathogens (Xu et al. 2025).

In T cells, histone modifications are key regulators of lineage-specific transcription factors that direct cell polarization. For instance, acetylation and methylation patterns control the expression of T-bet in Th1 cells, GATA3 in Th2 cells, and Foxp3 in regulatory T cells (Tregs). These epigenetic changes ensure that T cells adopt the appropriate functional phenotype in response to infections or immunological challenges, allowing for tailored and effective immune responses (Wan 2014; Tripathi and Lahesmaa 2014).

Moreover, the regulation of immune tolerance is critically dependent on histone methylation in Tregs. Methylation at the promoter region of the Foxp3 gene ensures its stable and sustained expression, which is vital for the suppressive function of Tregs (Lu et al. 2017). This stability underpins their role in maintaining immune homeostasis and preventing harmful autoimmune responses. Disruption in these epigenetic mechanisms can impair Treg function, leading to immune dysregulation and the development of autoimmune diseases (Golzari-Sorkheh and Zúñiga-Pflücker 2023).

Noncoding RNAs and Their Role in Epigenetic Regulation of Immune Cells

Noncoding RNAs (ncRNAs), which include **microRNAs (miRNAs)** and **long noncoding RNAs (lncRNAs)**, have emerged as key players in the **epigenetic regulation** of immune cell differentiation, function, and gene expression. These mechanisms are illustrated in Fig. 1.3. Unlike protein-coding genes, ncRNAs do not translate into proteins, but they regulate a vast array of cellular processes by modulating **gene expression** at both the **transcriptional** and **posttranscriptional** levels. In the immune system, ncRNAs serve as **critical regulators** of **immune responses**,

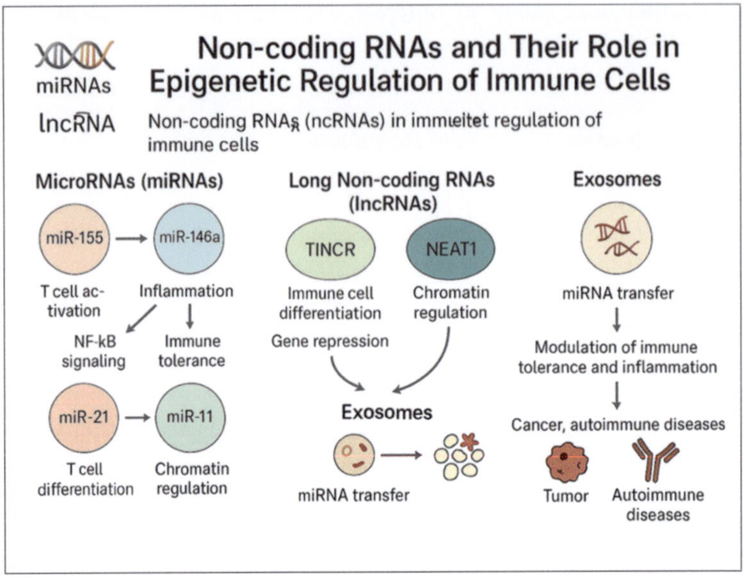

Fig. 1.3 Noncoding RNAs and their role in epigenetic regulation of immune cells. This infographic highlights the epigenetic functions of noncoding RNAs (ncRNAs), including microRNAs (miRNAs) and long noncoding RNAs (lncRNAs), as well as exosomes, in immune regulation. Key microRNAs like miR-155 enhance T cell activation and inflammation via NF-κB signaling, while miR-146a promotes immune tolerance by suppressing inflammatory pathways. Other miRNAs, such as miR-21 and miR-11, regulate T cell differentiation, apoptosis, and chromatin structure. Long noncoding RNAs like TINCR and NEAT1 influence immune cell differentiation and gene expression by interacting with chromatin-modifying complexes and regulating chromatin architecture. Exosomes facilitate the transfer of miRNAs between immune cells, modulating immune tolerance, inflammation, and cell polarization. This intercellular communication plays significant roles in autoimmune diseases and cancer progression by transferring epigenetically active miRNAs that alter gene expression in recipient cells. These ncRNAs and exosome-mediated mechanisms serve as crucial epigenetic regulators, maintaining immune balance and influencing disease outcomes

helping to fine-tune the balance between immune activation and tolerance. Their ability to interact with **chromatin**, **transcription factors**, and other components of the epigenetic machinery makes them pivotal in the regulation of immune cell fate and function (Ratti et al. 2020; Abd El Fattah et al. 2023; Rizk et al. 2024).

MicroRNAs (miRNAs) in Immune System Regulation

MicroRNAs (miRNAs) are small (~22 nucleotide) **noncoding RNAs** that regulate **gene expression** by binding to the **3′ untranslated regions (UTRs)** of target messenger RNAs (mRNAs), leading to mRNA degradation or translational repression. MiRNAs play a central role in modulating **immune responses**, including

Table 1.2 Immunoregulatory roles of miRNAs in immune cell function and disease implications

miRNA	Key functions	Immune cells affected	Implications
miR-155	Promotes Th1 and Th17 differentiation- Regulates macrophage polarization- Controls cytokines like TNF-α, IL-6	T helper cells, macrophages	Supports immune activation; overexpression linked to autoimmunity
miR-146a	Regulates immune tolerance- Suppresses excessive inflammation- Targets NF-κB and cytokine pathways	Macrophages, dendritic cells	Prevents tissue damage; maintains balanced immune response
miR-21	Modulates T cell and B cell function- Promotes Th17 differentiation- Regulates apoptosis	T cells, B cells	Maintains homeostasis; implicated in autoimmunity and inflammation

inflammation, **immune tolerance**, **T cell activation**, and the **differentiation of immune cells** (O'Brien et al. 2018), **as tabulated in** Table 1.2.

Among the microRNAs involved in immune regulation, **miR-155** is one of the most extensively characterized. It plays a pivotal role in promoting the activation and differentiation of T helper cells, particularly the Th1 and Th17 subsets, and is also involved in steering macrophage polarization. miR-155 exerts its effects by modulating the expression of key immune-related genes, including those encoding pro-inflammatory cytokines, such as tumor necrosis factor-alpha (TNF-α) and interleukin-6 (IL-6) (Sufianov et al. 2025). Through this regulation, miR-155 contributes significantly to the epigenetic control of immune responses, supporting effective immune cell activation against pathogens. However, its dysregulation has been associated with pathological immune activation, contributing to the development of chronic inflammation and autoimmune diseases (Hu and Huang 2022).

miR-146a serves as a critical mediator in maintaining immune homeostasis and controlling inflammation. It is upregulated in response to pro-inflammatory stimuli and microbial exposure and acts by targeting components of the NF-κB signaling cascade and inflammatory cytokines. By suppressing these pathways, miR-146a helps temper the activity of innate immune cells, such as macrophages and dendritic cells, thus preventing excessive inflammatory responses that can lead to tissue injury (Poe et al. 2022). This miRNA plays a protective role by fostering immune tolerance and limiting inappropriate immune activation (Xiao et al. 2020).

In contrast, **miR-21** is known for its regulatory functions in both T cell differentiation and B cell activation. It influences immune responses by modulating the expression of genes involved in inflammatory signaling and immune tolerance (Ha 2011). miR-21 has been shown to enhance the proliferation and activation of T cells and is particularly important for the differentiation of Th17 cells—an immune cell subset implicated in several autoimmune disorders (Murugaiyan et al. 2015). Furthermore, miR-21 contributes to the regulation of apoptosis, thereby supporting the maintenance of immune equilibrium and preventing immune-mediated tissue damage (Tang et al. 2022).

Long Noncoding RNAs (lncRNAs) in Immune Epigenetic Regulation

Long noncoding RNAs (lncRNAs) are a diverse class of noncoding RNAs that are longer than 200 nucleotides and are involved in a wide range of biological processes, including **chromatin remodeling**, **gene regulation**, and **immune cell differentiation**. **LncRNAs** can modulate the **epigenetic landscape** by interacting with **chromatin modifiers**, **transcription factors**, and the **transcriptional machinery** (**Hu et al.** 2018**; Basu et al.** 2025).

TINCR (Terminal Differentiation-Induced Noncoding RNA) has emerged as a significant long noncoding RNA (lncRNA) involved in the regulation of immune cell development (Sharma et al. 2020). It exerts its influence by interacting with chromatin-modifying complexes, thereby modulating gene expression patterns that direct the differentiation of immune cells, such as T and B lymphocytes. TINCR promotes the activation of immune-related genes while downregulating those that could potentially lead to immune overactivation or dysregulation (Xie et al. 2025). This fine-tuning of immune responses is essential for maintaining immune equilibrium, and disruptions in TINCR expression have been associated with the pathogenesis of autoimmune disorders and chronic inflammatory diseases (Stergioti et al. 2022).

Similarly, **NEAT1 (Nuclear Enriched Abundant Transcript 1)** plays a crucial role in immune regulation through its involvement in the formation of nuclear paraspeckles—subnuclear structures that modulate gene expression at the epigenetic level (An et al. 2018). In the context of immune function, NEAT1 interacts with both chromatin remodelers and transcription factors to influence the transcription of genes linked to inflammation, immune tolerance, and cellular activation. It has been shown to impact key immunological processes, such as T cell differentiation and macrophage polarization, thereby shaping the immune system's ability to respond to infections while preventing excessive or misdirected inflammatory reactions (Ghahramani Almanghadim et al. 2025).

Exosomes and Their Role in ncRNA-Based Epigenetic Regulation

Exosomes are small, **extracellular vesicles** (30–150 nm) secreted by many types of cells, including immune and nonimmune cells. These vesicles play an essential role in **intercellular communication**, as they carry a variety of molecular cargoes, including **miRNAs**, **proteins**, **lipids**, and **mRNA**. Exosomes are involved in the transfer of **genetic material** between cells, allowing for **epigenetic regulation** across different cell types. This **cell-to-cell communication** is particularly important in **immune responses (Hu et al.** 2024).

Exosomes play a pivotal role in intercellular communication within the immune system, particularly through the transfer of microRNAs (miRNAs). One notable example is the transport of miR-146a via exosomes between immune cells, such as macrophages and dendritic cells. This transfer can modulate inflammatory signaling pathways and contribute to the regulation of immune tolerance, thereby ensuring that immune responses remain balanced and do not escalate into harmful inflammation (Guo et al. 2024; Nail et al. 2023).

Beyond simple miRNA exchange, exosomal miRNAs critically regulate immune processes, including antigen presentation and immune cell polarization. In autoimmune disorders, exosomes can transport miRNAs targeting autoantigens, modulating immune responses by either reinforcing immune tolerance or amplifying pathogenic activation, depending on the immunological context (Lv et al. 2025). Similarly, in cancer, exosomes have been implicated in aiding tumor cells by delivering miRNAs that inhibit immune detection, thereby facilitating immune escape and supporting tumor survival and progression (Kumar et al. 2024).

The involvement of exosomal miRNAs extends to the propagation of pathological signals in both cancer and autoimmune conditions. In cancer, these vesicles can deliver miRNAs that impair the immune system's ability to target tumor cells effectively (Kumar et al. 2024). In autoimmune diseases, exosomal miRNA communication may intensify inflammation by activating autoreactive immune cells, ultimately contributing to tissue damage and disease progression (Saadh et al. 2025).

Noncoding RNAs (ncRNAs), including **miRNAs** and **lncRNAs**, play pivotal roles in the **epigenetic regulation** of immune cell differentiation, activation, and function. **miRNAs** regulate immune responses by fine-tuning gene expression at the posttranscriptional level, influencing **inflammation**, **immune tolerance**, and immune cell activation. **LncRNAs**, on the other hand, modify the **epigenetic landscape** by interacting with chromatin modifiers and transcription factors, thereby influencing **immune cell fate** and **differentiation** (Gareev et al. 2023b). Furthermore, **exosomes** serve as crucial mediators of **intercellular communication**, facilitating the transfer of **ncRNAs** and other molecules that regulate **immune responses** and **immune cell polarization**. Together, these ncRNAs, along with **exosomal transfer**, play a key role in shaping immune system function and maintaining homeostasis, with significant implications for **autoimmune diseases**, **cancer**, and **therapeutic interventions (Liu et al.** 2019).

RNA Modifications: Pseudouridylation and beyond

While **DNA methylation** and **histone modifications** are well-established as critical players in **epigenetic regulation**, recent research has highlighted the significance of **RNA modifications** as essential contributors to the **fine-tuning of gene expression** and the regulation of immune system function. Figure 1.4 represents these

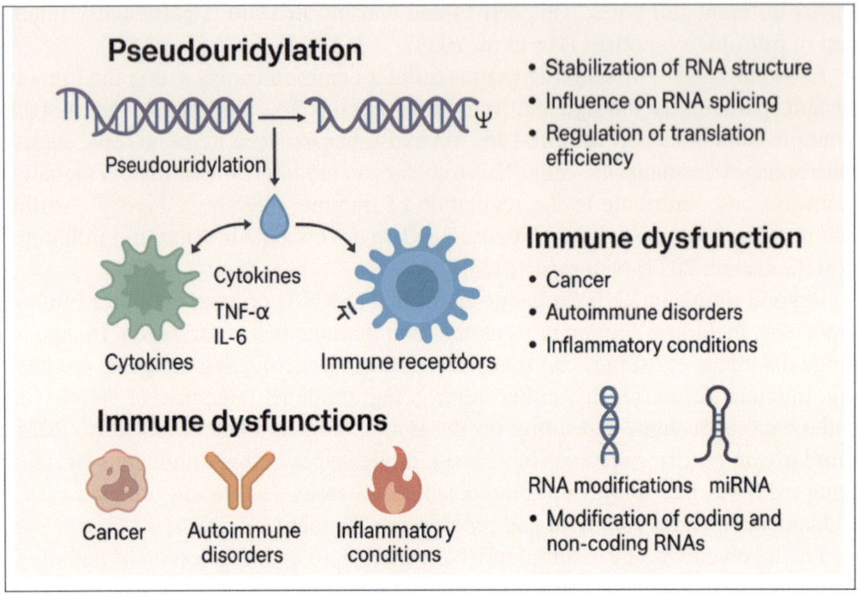

Fig. 1.4 Role of pseudouridylation and other RNA modifications in immune regulation. This infographic illustrates the multifaceted functions of RNA modifications, with a focus on pseudouridylation, in immune system regulation. Uridine is enzymatically converted to pseudouridine (Ψ), enhancing RNA stability and structural integrity. Pseudouridylation influences alternative splicing and increases translation efficiency of immune-related mRNAs, such as those encoding TNF-α and IL-6. Enhanced mRNA stability promotes robust cytokine production and immunity. Aberrant pseudouridylation contributes to immune evasion in cancer, overactivation in autoimmune diseases (e.g., SLE, RA), and chronic inflammation (e.g., IBD, asthma). RNA modifications also regulate the expression and function of miRNAs and lncRNAs, affecting immune cell differentiation and epigenetic programming. These mechanisms highlight the critical role of RNA modifications in maintaining immune homeostasis and their potential as therapeutic targets

modifications with an additional layer of complexity in the regulation of cellular processes, and they are involved in the **transcriptional** and **posttranscriptional** regulation of immune cell activity. One of the key RNA modifications gaining attention in the context of **immune system regulation** is **pseudouridylation** (Kobayashi and Kitagawa 2024).

Pseudouridylation: A Key RNA Modification

Pseudouridylation is a process in which **uridine (U)** residues in RNA molecules are modified to **pseudouridine (Ψ)**. This modification involves the enzymatic conversion of uridine into its isomeric form, pseudouridine, which plays several crucial roles in RNA biology (Lin et al. 2021).

Pseudouridylation, a posttranscriptional RNA modification, contributes significantly to the regulation of immune responses by enhancing RNA structure and function. One of its primary roles is to stabilize RNA molecules by reinforcing their secondary and tertiary structures (Kornienko et al. 2024). This structural integrity is essential for the optimal performance of various RNA types, including mRNAs that encode critical components of the immune system, such as pro-inflammatory cytokines, immune receptors, and transcription factors (Liu et al. 2024).

In addition to structural stabilization, pseudouridylation influences RNA splicing—a vital step in the maturation of pre-mRNA into functional mRNA. By affecting splicing patterns, this modification determines the specific isoforms of immune-related genes that are produced. Consequently, it plays a role in regulating immune cell differentiation, activation, and overall functionality (Jia et al. 2025).

Moreover, pseudouridylation enhances translational efficiency by stabilizing certain mRNAs, thereby promoting their translation into proteins. This is particularly relevant during immune activation, where the efficient production of proteins, such as TNF-α and IL-6, is crucial. By supporting the expression of these inflammatory mediators, pseudouridylation contributes to the fine-tuning of immune signaling and the orchestration of effective immune responses (Lin et al. 2024).

Pseudouridylation in Immune Cell Activation and Function

Recent studies have revealed that **pseudouridylation** regulates the expression of **immune-related genes**, including those involved in **immune cell activation**, **cytokine production**, and **immune receptor function**. This process directly impacts how immune cells respond to external stimuli, such as infections or tissue damage, and regulates their ability to maintain **immune homeostasis** (Zhang et al. 2024a).

Pseudouridylation plays a vital role in modulating immune responses by affecting both cytokine production and immune receptor expression. This RNA modification enhances the stability of mRNAs encoding key pro-inflammatory cytokines, such as TNF-α, IL-6, and IFN-γ. By promoting the rapid and efficient translation of these cytokines, pseudouridylation ensures a swift and effective inflammatory response during immune activation, enabling the body to respond promptly to infections or tissue injury (Ding et al. 2025).

In addition to cytokines, pseudouridylation also regulates the expression of immune receptors, including Toll-like receptors (TLRs) and various cytokine receptors. Modification of the mRNAs encoding these receptors can influence their stability and translation, ultimately affecting the expression levels of receptors on immune cells (Duan et al. 2022). This fine-tuned regulation is critical for maintaining immune cell sensitivity to pathogenic signals and ensuring a balanced immune response that is neither underactive nor excessively aggressive (Cui et al. 2022).

Implications of Dysregulated Pseudouridylation in Immune Dysfunction

As with other RNA modifications, the dysregulation of **pseudouridylation** can have significant consequences for immune system function and contribute to the development of various diseases, including cancer, autoimmune disorders, and inflammatory conditions (Wang et al. 2024b).

Dysregulation of pseudouridylation has been increasingly recognized as a contributing factor in various pathological conditions, including cancer, autoimmune disorders, and chronic inflammatory diseases (Miller 2023). In cancer, aberrant pseudouridylation can facilitate immune evasion by altering the expression of immune checkpoint molecules and inflammatory cytokines. By modifying the stability and translation efficiency of mRNAs encoding key immune-related proteins, tumor cells may escape immune surveillance and promote immune resistance, thereby supporting cancer progression (Ding et al. 2025).

In the context of autoimmune diseases, abnormal pseudouridylation patterns in mRNAs involved in immune regulation can lead to the overactivation of immune cells. This dysregulation may drive the excessive production of inflammatory cytokines, autoantibodies, and the activation of self-reactive T cells, all of which contribute to the pathogenesis of conditions such as rheumatoid arthritis, multiple sclerosis, and systemic lupus erythematosus (SLE) (Sundaresan and Shirafkan 2023).

Moreover, altered pseudouridylation has been associated with chronic inflammatory disorders, including inflammatory bowel disease (IBD) and asthma. These modifications may affect the expression of cytokines and immune receptors, exacerbating the inflammatory response and perpetuating disease activity. Overall, disruptions in pseudouridylation underscore its critical role in maintaining immune homeostasis and preventing disease development (Cui et al. 2022).

RNA Modifications and Noncoding RNAs in Epigenetic Regulation

In addition to pseudouridylation, RNA modifications are also crucial for regulating the expression of **noncoding RNAs** (ncRNAs), which are key players in immune regulation. Noncoding RNAs, such as **microRNAs (miRNAs)** and **long noncoding RNAs (lncRNAs)**, are often subject to **RNA modifications** that influence their stability, function, and interactions with **chromatin-modifying enzymes**. These modifications further add complexity to the regulation of immune responses at the epigenetic level (Leng et al. 2024).

For example, pseudouridylation may regulate the expression of specific **miRNAs** or **lncRNAs** that modulate the epigenetic regulation of immune cells, including their differentiation and functional programming. The modification of **ncRNAs** can influence their ability to interact with **target mRNAs** or **chromatin-associated proteins**, ultimately affecting immune cell fate decisions and the ability to mount an effective immune response (Kumar et al. 2020; Hamdy et al. 2024).

RNA modifications, particularly **pseudouridylation**, represent a vital mechanism in **epigenetic regulation** that affects immune cell function, differentiation, and response to external stimuli. By stabilizing RNA structures, influencing RNA splicing, and regulating translation efficiency, pseudouridylation plays a central role in modulating the expression of immune-related genes, including those involved in **cytokine production** and **immune receptor signaling**. Dysregulation of RNA modification pathways, including pseudouridylation, can lead to immune dysfunction, contributing to diseases such as **cancer**, **autoimmune disorders**, and **chronic inflammation**. Furthermore, RNA modifications are essential for regulating the expression of **noncoding RNAs**, adding another layer of complexity to the **epigenetic regulation** of immune responses. Understanding the role of RNA modifications in the immune system opens new avenues for therapeutic interventions targeting immune-related diseases (Qiu et al. 2023).

Epigenetic Regulation in Immune Cell Development and Function

Epigenetic regulation is a key factor in controlling the **differentiation**, **activation**, and **functional programming** of immune cells, particularly **T lymphocytes**. These modifications shape the immune system's ability to respond to infections, inflammation, and other stimuli, ensuring a coordinated and effective immune response, as shown in Fig. 1.5 (Frias et al. 2021).

T Lymphocytes

T lymphocytes, or T cells, are central players in adaptive immunity. They differentiate into distinct subsets, including **Th1**, **Th2**, **Th17**, and **Treg** cells, each with specialized roles in immune responses. Epigenetic mechanisms, including **DNA methylation**, **histone modifications**, and **noncoding RNAs**, are essential in regulating these processes (Chi et al. 2024).

T Cell Differentiation and Lineage Commitment

T cell differentiation into the various subsets (Th1, Th2, Th17, and Treg) is driven by a complex network of epigenetic modifications that influence gene expression and cell fate. The interplay between **histone acetylation**, **methylation**, and **DNA demethylation** is particularly crucial in orchestrating the development of these T cell subsets (Renaude et al. 2020).

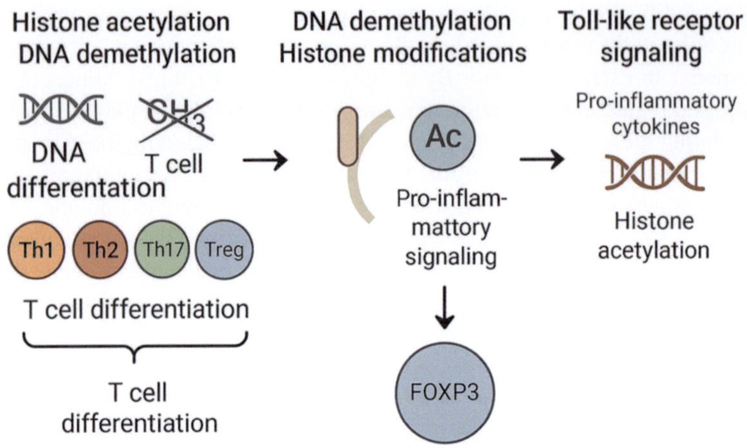

Fig. 1.5 Epigenetic regulation of T lymphocyte differentiation and function. This infographic illustrates the central role of epigenetic mechanisms in the differentiation and function of T lymphocytes. (1) **T cell subset differentiation**: Naïve T cells differentiate into distinct subsets—Th1, Th2, Th17, and Treg—driven by specific cytokines and epigenetic changes. Histone acetylation and DNA demethylation facilitate the expression of key transcription factors and cytokines (e.g., IFN-γ for Th1, IL-4 for Th2). (2) **FOXP3 regulation in Tregs**: Stable FOXP3 expression in Tregs is ensured through DNA demethylation at CpG islands and histone modifications near its promoter, securing long-term suppressive function. (3) **TLR-mediated epigenetic reprogramming**: Toll-like receptor (TLR) signaling upon pathogen recognition induces inflammatory cytokines and recruits chromatin-modifying enzymes (e.g., HATs, HDACs), leading to histone acetylation and chromatin remodeling in T cells. This reprogramming enhances effector gene expression and suppresses regulatory pathways. Collectively, these epigenetic modifications ensure proper immune activation, tolerance, and homeostasis

Histone acetylation promotes a relaxed, transcriptionally active chromatin state, facilitating the expression of genes essential for T cell activation and lineage commitment. Histone acetyltransferases (HATs) play a critical role by acetylating histone tails, thereby enabling the transcription of key cytokine genes, such as **IFN-γ** for Th1 differentiation and **IL-4** for Th2 differentiation (Falvo et al. 2013).

Similarly, **DNA demethylation** serves as a pivotal epigenetic mechanism in T cell programming. This modification can activate previously silenced genes by removing methyl groups from cytosine residues in promoter regions. In Th cell subsets, the **demethylation of the IFN-γ promoter** is essential for Th1 cell development, while **IL-4 promoter demethylation** facilitates Th2 cell differentiation. These modifications ensure that T cells differentiate into the appropriate subset based on environmental cues, such as cytokine signals and antigen recognition (Cribbs et al. 2015; Hammad et al. 2022).

Regulation of FOXP3 in Tregs

Tregs (regulatory T cells) are essential for maintaining **immune tolerance** and preventing autoimmunity. The stability and function of Tregs depend heavily on **epigenetic regulation**, particularly the **FOXP3** gene, a master regulator of Treg differentiation and function (Golzari-Sorkheh and Zúñiga-Pflücker 2023).

The stable expression of **FOXP3**, the master transcription factor of regulatory T cells (Tregs), is tightly controlled by epigenetic mechanisms. One critical regulator is **DNA demethylation**. In Tregs, specific **CpG islands within the FOXP3 promoter region** are demethylated, facilitating sustained transcriptional activation of FOXP3 and ensuring the suppressive function of these cells (Morikawa et al. 2014).

Additionally, **histone modifications** play a complementary role in maintaining FOXP3 expression. **Histone acetylation** near the FOXP3 locus promotes an open chromatin configuration conducive to transcription, while specific **histone methylation patterns** contribute to the stability of FOXP3 expression (Li et al. 2021b). Together, these epigenetic modifications are essential for the long-term functionality and lineage identity of Tregs, enabling their pivotal role in maintaining immune homeostasis and tolerance. This **epigenetic regulation** of FOXP3 is critical for preventing autoimmune diseases, as it ensures that Tregs can effectively suppress the activation of self-reactive T cells (Yue et al. 2024).

Toll-Like Receptor (TLR) Signaling and T Cell Epigenetic Reprogramming

Toll-like receptors (TLRs) are key components of the **innate immune system** that detect **pathogen-associated molecular patterns (PAMPs)** and activate **inflammatory responses**. TLRs also play an essential role in modulating the **epigenetic landscape** of T cells, influencing their differentiation and activation in response to infection (El-Zayat et al. 2019).

Toll-like receptors (TLRs) play a central role in bridging innate and adaptive immunity by recognizing pathogen-associated molecular patterns (PAMPs), such as bacterial lipopolysaccharides (LPS) or viral RNA. Upon activation, TLR signaling initiates intracellular cascades that result in the production of pro-inflammatory cytokines, including **TNF-α, IL-6**, and **IL-1β** (Chen et al. 2024). These cytokines not only activate innate immune cells but also shape the adaptive immune response by directing T cell differentiation. For instance, TLR activation can promote the polarization of naïve $CD4^+$ T cells toward the **Th17 lineage**, which is essential for defense against bacterial and fungal pathogens (Xie et al. 2023).

Beyond cytokine production, **TLR signaling also induces epigenetic remodeling** in T cells. This involves the recruitment of chromatin-modifying enzymes, such as **histone acetyltransferases (HATs)** and **histone deacetylases (HDACs)**, leading to modifications in chromatin structure that regulate gene transcription (Bhat et al. 2024). Specifically, **histone acetylation** induced by TLR activation promotes the expression of pro-inflammatory and effector genes, while concurrently repressing the expression of genes involved in immune regulation and tolerance. This epigenetic reprogramming facilitates the full activation and functional commitment of T cells during immune responses (Rosemblatt et al. 2019).

Overall, **epigenetic modifications**—including **histone modifications**, **DNA demethylation**, and the action of **noncoding RNAs**—are critical for guiding the differentiation and functional specialization of T lymphocytes into subsets, such as **Th1**, **Th2**, **Th17**, and **Treg** cells. The **stable expression of lineage-defining transcription factors** (e.g., FOXP3 in Tregs) is tightly regulated through these mechanisms, ensuring the fidelity and longevity of immune responses (Bai et al. 2022). Moreover, the integration of signals from receptors like TLRs further fine-tunes chromatin landscapes, influencing both immune activation and tolerance. These processes are fundamental to the maintenance of immune homeostasis and the prevention of autoimmunity (Cui et al. 2022).

B Lymphocytes

Epigenetic reprogramming in B lymphocytes plays a fundamental role in their development, activation, and immune function. DNA methylation controls critical processes, such as **antigen receptor diversification** and **class-switch recombination**, while **miRNAs** and **lncRNAs** modulate gene expression during **B cell differentiation** and **antibody production,** as presented in Fig. 1.6 (Ruth and University Ix 2024). Dysregulation of these processes can lead to **autoimmune diseases** or altered immune responses, underlining the importance of precise **epigenetic regulation** in maintaining immune homeostasis. Additionally, **exosomes** mediate intercellular communication, enhancing or suppressing B cell functions and influencing disease outcomes. As such, **epigenetic therapies** targeting B cell regulation hold potential for treating immune-related disorders and improving therapeutic outcomes (Khan et al. 2020).

B lymphocytes (B cells) are critical components of the **adaptive immune system**, responsible for **antibody production** and **immune memory**. The development and function of B cells are tightly regulated by **epigenetic mechanisms**, which govern key processes, such as **antigen receptor diversification**, **class-switch recombination**, and **immune activation** (Khan et al. 2020). These epigenetic modifications include **DNA methylation**, **histone modifications**, and the regulation by **noncoding RNAs** (ncRNAs), including **microRNAs (miRNAs)** and **long noncoding RNAs (lncRNAs)** (Bure et al. 2022).

B Cell Epigenetic Reprogramming in Development

The development of B cells occurs in several stages, from early progenitors to mature, antigen-responsive cells. During this process, **epigenetic regulation** ensures that B cells progress through these stages and are capable of responding

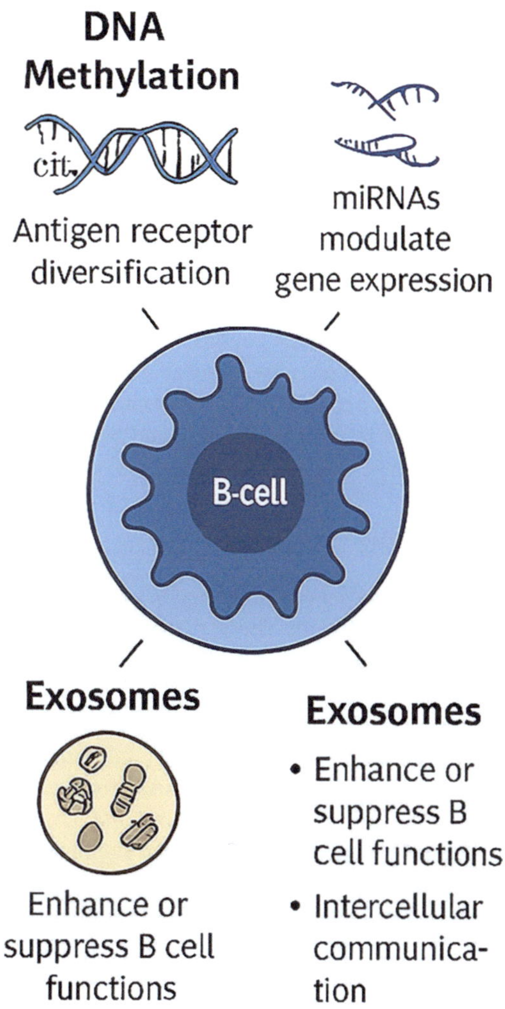

Fig. 1.6 Epigenetic regulation in B lymphocytes. The figure illustrates the complex epigenetic regulation of B lymphocytes, highlighting how **DNA methylation**, **histone modifications**, and **noncoding RNAs** (including **miRNAs** like *miR-155* and **lncRNAs,** such as *NEAT1* and *TINCR*) orchestrate B cell development, differentiation, and antibody production. These mechanisms ensure proper immune function by regulating key processes, such as **class-switch recombination** and **antigen receptor diversification**. Additionally, **exosomes** act as intercellular messengers, transferring regulatory molecules that influence B cell activity and immune responses

appropriately to pathogens. **DNA methylation**, in particular, is critical in regulating B cell activation, differentiation, and the immune response (Zhao et al. 2023).

DNA methylation is a key epigenetic mechanism involved in regulating the immunoglobulin heavy chain (IgH) locus, which is essential for antibody production in B cells. During the process of class-switch recombination (CSR), B cells switch from producing one antibody isotype (e.g., IgM) to another (e.g., IgG, IgA, or IgE) in response to antigenic stimulation. This recombination is tightly regulated by dynamic changes in DNA methylation (Vaidyanathan and Chaudhuri 2015). Demethylation of certain regions within the IgH locus activates switch (S) regions, allowing the rearrangement of heavy chain genes and enabling isotype switching. This epigenetic remodeling is crucial for generating a diverse and effective antibody response to a wide range of pathogens (Ma et al. 2024).

However, aberrant DNA methylation patterns in B cells can disrupt this process and contribute to the development of autoimmune diseases. Hypomethylation in regulatory regions of the IgH locus may lead to uncontrolled or inappropriate class switching, resulting in the production of self-reactive antibodies. Such dysregulation has been implicated in autoimmune conditions like SLE and RA, where B cells generate autoantibodies that target host tissues. These findings underscore the importance of proper epigenetic regulation in preserving immune tolerance and preventing pathological immune responses (Corneth and Neys 2022).

Regulation by Noncoding RNAs (ncRNAs)

Both **miRNAs** and **lncRNAs** have been identified as crucial regulators of B cell differentiation, activation, and function. These noncoding RNAs add an additional layer of regulation, fine-tuning the gene expression programs that control B cell behavior (Heydari et al. 2024).

miRNAs are small, noncoding RNAs that primarily regulate gene expression posttranscriptionally by binding to mRNAs, either inhibiting their translation or promoting their degradation. Among the miRNAs involved in B cell biology, miR-155, miR-21, and miR-146a are particularly important. miR-155, in particular, is one of the most extensively studied miRNAs in the immune system (Gareev et al. 2023a). Its expression is tightly regulated during B cell activation, and it plays a crucial role in promoting B cell proliferation and differentiation in response to antigen stimulation. miR-155 modulates the expression of various genes critical for B cell signaling and antibody production. However, dysregulated expression of miR-155 has been implicated in autoimmune diseases, such as SLE, where excessive B cell activation leads to the production of pathogenic autoantibodies (Heydari et al. 2024).

lncRNAs, which are typically longer than 200 nucleotides, regulate gene expression through mechanisms such as chromatin remodeling, modulation of transcription factor activity, and interaction with epigenetic enzymes. In B cells, lncRNAs are emerging as key players in controlling activation, antigen receptor expression, and the generation of effective antibody responses (Mattick et al. 2023). lncRNAs, such as NEAT1 and TINCR, have been shown to interact with chromatin-modifying enzymes and transcriptional regulators, thereby maintaining chromatin structure and influencing gene expression profiles essential for B cell differentiation. These lncRNAs are also involved in regulating processes, such as class-switch recombination and somatic hypermutation, both of which are fundamental for the generation of a diverse and effective antibody repertoire (Statello et al. 2021).

Exosomes in B Cell Activation

Exosomes are small, **extracellular vesicles** that mediate **intercellular communication** by transporting genetic material, proteins, lipids, and other molecules between cells. These vesicles are involved in various aspects of immune regulation, including the activation and differentiation of B cells (Javdani-Mallak and Salahshoori 2024).

Exosomes, which are secreted by various immune cells—including B cells themselves—serve as potent mediators of intercellular communication within the immune system. These vesicles carry a diverse cargo of bioactive molecules, including regulatory miRNAs, proteins, and lipids, which can be transferred to recipient cells, modulating their behavior (Schwarzenbach and Gahan 2021). In the context of B cell activation, exosomes play a crucial role by delivering functional miRNAs and signaling proteins that influence B cell proliferation, differentiation, and antibody production. For example, exosomes derived from activated T cells or dendritic cells can carry specific miRNAs that modulate the gene expression profiles of B cells, thereby enhancing their activation and promoting effective humoral immune responses (Sadeghi et al. 2025).

Beyond their role in stimulating immune activation, exosomes also contribute significantly to the regulation of immune tolerance and the pathogenesis of autoimmune diseases. By transporting immunoregulatory molecules, exosomes can either enhance or suppress immune responses depending on the context (Essola et al. 2024). In autoimmune diseases, exosomes may facilitate the spread of inflammatory signals or autoantigenic material, thereby exacerbating immune dysregulation. Conversely, in cancer, tumor-derived exosomes can suppress immune surveillance by delivering immunosuppressive factors that impair B cell and T cell functions. Thus, exosomes represent a dual-faceted regulatory mechanism in immunity, with implications for both protective immune responses and pathological conditions, such as autoimmunity and cancer (Saadh et al. 2025).

Innate Immune Cells and Epigenetic Remodeling: The Phenomenon of Trained Immunity

The phenomenon of **trained immunity** represents a significant shift in our understanding of the immune system's ability to "remember" pathogens through **epigenetic reprogramming**. This process was illustrated in Fig. 1.7, which involves a range of **epigenetic modifications**, including **DNA methylation**, **histone modifications**, and the regulation by **noncoding RNAs** (miRNAs and lncRNAs), which together ensure enhanced immune responses upon reexposure to pathogens (Guo et al. 2022). **Exosomes** play a crucial role in mediating trained immunity by transferring epigenetic information between cells, thereby amplifying the immune response. This discovery has profound implications for **vaccine design** and **immune therapies**, as manipulating the **epigenetic landscape** of innate immune cells could

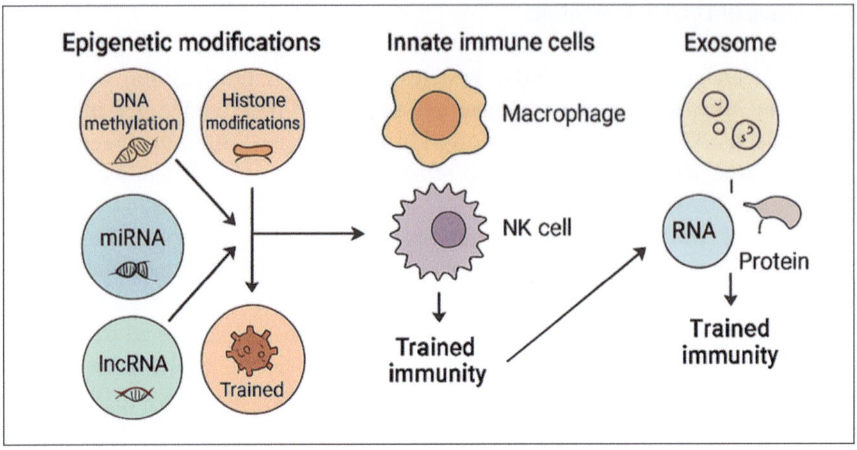

Fig. 1.7 Innate immune cells and epigenetic remodeling (the phenomenon of trained immunity). This figure depicts the concept of *trained immunity*, where innate immune cells like macrophages and NK cells develop memory-like behavior through epigenetic remodeling. Key mechanisms include DNA methylation, histone modifications, and regulation by noncoding RNAs (miRNAs and lncRNAs), which enhance immune responses upon reexposure to pathogens. Exosomes mediate intercellular communication by transferring these regulatory molecules, reinforcing immune readiness across cells. This process has important implications for vaccine design, offering new strategies to boost long-term immunity by targeting the epigenetic landscape of innate immune cells

enhance vaccine efficacy and provide long-term protection against infections (Li et al. 2025; Megdiche and Salerno-Gonçalves 2025).

Innate immune cells, such as **macrophages** and **natural killer (NK) cells**, are crucial components of the immune system, providing the first line of defense against infections. Unlike adaptive immune cells, which develop long-term memory of pathogens through antigen-specific receptors, innate immune cells have traditionally been considered to lack this ability to "remember" previous encounters with pathogens. However, recent research has uncovered a phenomenon known as **trained immunity**, which reveals that innate immune cells can undergo **epigenetic remodeling** in response to infections and other environmental stressors, allowing them to exhibit enhanced responses upon subsequent infections (Marshall et al. 2018).

Epigenetic Remodeling in Innate Immune Cells

Epigenetic changes refer to modifications in the genome that affect gene expression without altering the underlying DNA sequence. In the context of **innate immune cells**, these changes play a pivotal role in their ability to adapt to and "remember" previous encounters with pathogens (Sun et al. 2023).

During trained immunity, **innate immune cells,** such as macrophages and NK cells, undergo epigenetic remodeling, including DNA hypomethylation and histone modifications, that prime them for enhanced responses upon reexposure to pathogens. DNA hypomethylation at inflammatory gene loci and increased histone acetylation promote a more accessible chromatin structure, facilitating the expression of immune-related genes. These changes enable macrophages to produce higher levels of cytokines like TNF-α, IL-1β, and IL-6 in response to previously encountered pathogens (Netea et al. 2020). Similarly, NK cells can enter a memory-like state, demonstrating elevated cytotoxicity and cytokine secretion after prior infections or vaccinations, thus contributing to a more robust innate immune defense (Gabrielli et al. 2016).

Exosomes and the Transfer of Trained Immunity

Exosomes, small **extracellular vesicles**, are emerging as key mediators in the process of **trained immunity**. These vesicles are involved in **intercellular communication** and can carry a variety of molecules, including **proteins**, **lipids**, and **RNA** (including **miRNAs** and **lncRNAs**). Exosomes facilitate the transfer of these bioactive molecules between cells, allowing them to influence the **epigenetic state** of recipient cells (Feng et al. 2023).

RNA and protein transfer via exosomes: During trained immunity, **macrophages** and **NK cells** can release exosomes that contain **modified RNA** (such as **miRNAs**) and **proteins** that have been altered in response to an infection. These exosomes can then be taken up by neighboring cells, leading to changes in gene expression and **epigenetic reprogramming** of the recipient cells (Ye et al. 2024). Exosomes from **trained macrophages** might carry **miRNAs** that promote the expression of inflammatory genes, thus making other immune cells more responsive to pathogens. This transfer of **epigenetic information** via exosomes can contribute to the sustained changes in immune cell function, ensuring a **heightened immune response** upon reexposure to pathogens (Xiao et al. 2024).

Trained Immunity and Vaccine Design

The discovery of **trained immunity** has had significant implications for **vaccine development**. Traditional vaccines primarily focus on generating long-term memory through the activation of **adaptive immunity** (i.e., through **T cells** and **B cells**). However, trained immunity provides an additional layer of protection by enhancing the **innate immune response**, which acts as the first line of defense and plays a crucial role in the early stages of infection (Bhargavi and Subbian 2024).

Manipulating epigenetic landscapes for vaccine efficacy: By understanding how **epigenetic modifications** influence the memory-like behavior of innate immune cells, researchers are now exploring ways to **manipulate** the **epigenetic landscape** of these cells to enhance vaccine efficacy (Wimmers et al. 2021). Specific compounds or adjuvants that promote **epigenetic reprogramming** of innate immune cells could improve their ability to respond to subsequent infections or vaccinations, providing **long-lasting protection** against a wide range of pathogens (Lee et al. 2022).

Moreover, **enhancing innate immune responses** as trained immunity can help enhance vaccine effectiveness by priming the **innate immune system** for rapid and efficient responses upon pathogen reexposure. This could be particularly valuable for combating infections where adaptive immunity takes longer to respond or for vaccines targeting pathogens that evade the adaptive immune system (Wang and Yu 2025).

Role of miRNAs and lncRNAs in Trained Immunity

miRNAs and **lncRNAs** are integral players in the regulation of **trained immunity** by controlling gene expression and modulating key immune pathways (Yang et al. 2022b).

miRNAs are small noncoding RNAs that regulate gene expression by binding to **mRNAs** and either degrading them or preventing their translation. In the context of trained immunity, **miRNAs** like **miR-155** and **miR-21** have been shown to play critical roles in the activation of **pro-inflammatory genes**, the **polarization** of immune cells, and the regulation of **immune responses**. These miRNAs can be transferred via **exosomes** to other immune cells, thereby enhancing the **memory-like function** of these cells (Sprenkle and Serezani 2023).

lncRNAs are involved in the regulation of gene expression at various levels, including chromatin remodeling and transcriptional regulation. In trained immunity, **lncRNAs,** such as **NEAT1** and **TUG1,** are implicated in the regulation of immune-related genes, influencing the ability of innate immune cells to respond more robustly to reexposure to pathogens. LncRNAs also interact with chromatin-modifying complexes to establish **long-term epigenetic changes** that sustain the memory of previous infections (Montano et al. 2024).

Epigenetics in Autoimmune and Inflammatory Diseases

Epigenetic mechanisms play a crucial role in the regulation of immune responses, and their dysregulation is implicated in the pathogenesis of various **autoimmune and inflammatory diseases** presented in Table 1.3. These alterations, which include **DNA methylation**, **histone modifications**, and **noncoding RNA regulation**, can disrupt immune homeostasis, leading to immune system dysfunction and the development of disease. Below is an in-depth elaboration on the epigenetic changes observed in key autoimmune and inflammatory diseases and their consequences (Surace and Hedrich 2019).

Understanding these **epigenetic mechanisms** opens new avenues for **targeted therapies** aimed at **correcting these alterations**. By developing therapies that target **DNA methylation**, **histone modifications**, and **noncoding RNA expression**, it

Table 1.3 Epigenetic alterations in autoimmune and inflammatory diseases

Disease	Epigenetic alterations	Consequences	References
Systemic lupus erythematosus (SLE)	Hypomethylation of immune activation genes	Loss of tolerance, autoantibody production, exacerbated inflammation	Methylation changes in lupus (Ferreté-Bonastre et al. 2022); DNA methylation and autoimmunity (Mazzone et al. 2019)
Rheumatoid arthritis (RA)	Histone modification-induced overexpression of inflammatory cytokines (e.g., IL-6, TNF-α)	Joint destruction, chronic synovitis, synovial inflammation	Histone modification in RA pathogenesis (Nemtsova et al. 2019); epigenetic alterations in RA (Doody et al. 2017)
Multiple sclerosis (MS)	Hypomethylation of neuroantigen-related genes	CNS inflammation, demyelination, neurodegeneration	DNA methylation in MS (Li et al. 2017); epigenetic mechanisms in MS (Huynh and Casaccia 2013)
Inflammatory bowel disease (IBD)	Dysregulated noncoding RNA networks (miRNAs, lncRNAs)	Impaired gut barrier function, chronic gut inflammation, dysbiosis	Noncoding RNAs in IBD (Ghafouri-Fard et al. 2020); RNA expression in IBD and inflammation (Grillo et al. 2021)
Psoriasis	Histone acetylation and methylation patterns influencing inflammatory cytokine genes	Skin inflammation, hyperkeratosis, altered immune responses	Epigenetics in psoriasis (Olejnik-Wojciechowska et al. 2024); histone modification in psoriasis (Liu et al. 2024)
Type 1 diabetes (T1D)	DNA hypomethylation of immune regulator genes (e.g., IL-2, IFN-γ)	Autoimmune destruction of pancreatic beta cells	DNA methylation in T1D (Paul et al. 2016); immune dysregulation in diabetes (Pan et al. 2021)
Hashimoto's thyroiditis	DNA methylation and histone modifications at the thyroid peroxidase (TPO) gene	Thyroid dysfunction, autoimmune attack on thyroid tissue	Epigenetic changes in Hashimoto's disease (Coppedè 2017)
Crohn's disease	Altered histone modification profiles and microRNA deregulation	Chronic inflammation, intestinal fibrosis, dysregulated immune responses	Histone and miRNA changes in Crohn's (Fernández-Ponce et al. 2021); epigenetic mechanisms in Crohn's (Hornschuh et al. 2021)

(continued)

Table 1.3 (continued)

Disease	Epigenetic alterations	Consequences	References
Asthma	DNA methylation of immune regulatory genes (e.g., IL-4, IL-13)	Chronic airway inflammation, airway remodeling	DNA methylation in asthma (Legaki et al. 2022); epigenetics of asthma and inflammation (Gomez 2019)
Sjögren's syndrome	Histone acetylation and methylation modifications at cytokine gene loci	Dry mouth, dry eyes, autoimmune inflammation of exocrine glands	Epigenetic regulation in Sjögren's syndrome (Ibáñez-Cabellos et al. 2019)

may be possible to treat or even prevent the progression of these diseases, providing patients with more effective and personalized treatment options. Ultimately, advancing our understanding of the epigenetic regulation of immune responses will be key to unlocking new therapeutic strategies for autoimmune and inflammatory diseases, helping to mitigate their impact on affected individuals (Castro-Muñoz et al. 2023).

Systemic Lupus Erythematosus (SLE)

Epigenetic alterations, such as **hypomethylation** of genes involved in immune activation, are a hallmark of SLE. These immune activation genes, including those encoding cytokines and costimulatory molecules, are typically regulated by DNA methylation. In SLE, **reduced methylation** at key loci, such as **IFN-γ** and **TNF-α**, leads to **uncontrolled immune activation (Zhang et al.** 2024b).

The **loss of immune tolerance** caused by hypomethylation results in the activation of autoreactive T cells and the production of **autoantibodies** against nuclear antigens, including **double-stranded DNA** (dsDNA). These autoantibodies form immune complexes that deposit in tissues, leading to **inflammation**, tissue damage, and **organ dysfunction**. The breakdown of immune tolerance is a central feature of SLE pathogenesis, and epigenetic dysregulation plays a key role in perpetuating this process. Therefore, targeting DNA methylation pathways could potentially restore immune tolerance in SLE patients and prevent autoantibody production, providing a new avenue for **epigenetic therapies** in autoimmune diseases (Moulton et al. 2017).

Rheumatoid Arthritis (RA)

Another key **epigenetic alteration**, **histone modification**, plays a critical role in driving the overexpression of inflammatory **cytokines,** such as **TNF-α**, **IL-6**, and **IL-1β, which** are implicated in RA. Histone acetylation and methylation at promoters of these cytokine genes have been shown to be **dysregulated**, promoting a **pro-inflammatory environment** within the synovium (Yang et al. 2022a).

The **overexpression of pro-inflammatory cytokines** leads to **chronic synovitis**, which is a key feature of RA. This chronic inflammation induces **joint destruction**, cartilage degradation, and bone resorption, contributing to the progressive **disability** observed in RA patients. Histone modifications also affect the **recruitment and activation** of inflammatory immune cells, amplifying the inflammatory response and disease progression. Interestingly, targeting histone modification enzymes, such as **histone deacetylases (HDACs)**, or using **small molecules** to reverse **histone acetylation** could offer new strategies for treating RA by **reducing inflammation** and preventing tissue damage (Kondo et al. 2021).

Multiple Sclerosis (MS)

In MS, **hypomethylation** of genes involved in **neuroantigen** recognition, such as those encoding myelin proteins, such as **myelin basic protein (MBP)**, contributes to an aberrant immune response targeting the central nervous system (CNS). This hypomethylation results in the **upregulation** of these neuroantigens in **T cells**, allowing them to recognize and attack myelin as foreign (Liu et al. 2022a).

The **hypomethylation** leads to **CNS inflammation** and **demyelination**, which are the hallmark features of MS. The immune system's attack on the myelin sheath disrupts neural communication, causing a wide range of neurological symptoms, including muscle weakness, paralysis, and cognitive dysfunction. This immune-mediated destruction of the myelin sheath is a progressive process in MS, leading to long-term disability. **DNA methyltransferase inhibitors (DNMTis)** or **small molecules** targeting specific **epigenetic modifiers** may help to **restore proper DNA methylation** patterns and reduce the autoimmune attack on the CNS. Restoring the epigenetic landscape could prevent the progression of demyelination and **ameliorate symptoms** (Sarkar et al. 2024).

Inflammatory Bowel Disease (IBD)

In IBD, **dysregulated noncoding RNA networks**, including **microRNAs (miRNAs)** and **long noncoding RNAs (lncRNAs)**, contribute to the pathogenesis of the disease. These noncoding RNAs regulate various aspects of immune cell function, including cytokine production, immune cell differentiation, and barrier integrity in the gut. The **dysregulation** of miRNAs and lncRNAs impairs the function of the **intestinal epithelial barrier**, leading to **intestinal permeability** (leaky gut). This compromised barrier allows **pathogens** and **toxic molecules** to enter the bloodstream, triggering a **chronic inflammatory response** (Ghafouri-Fard et al. 2020).

The failure to regulate this inflammation leads to the development of **ulcers, tissue damage**, and **increased immune cell infiltration** into the gut, characteristic of IBD. Restoring **noncoding RNA regulation** in IBD could help **normalize immune responses** and improve **intestinal barrier function**. Therapeutic interventions targeting specific miRNAs or lncRNAs involved in IBD pathogenesis might offer promising strategies for controlling gut inflammation and preventing disease progression (Kong et al. 2024).

Insights from Recent Research on Epigenetics and Immune Regulation

Recent advancements in epigenetics have provided new and deeper insights into how immune cells function and adapt in response to various environmental and pathogenic challenges. Epigenetic modifications play a crucial role in the dynamic regulation of immune responses, offering mechanisms by which immune cells can "remember" previous encounters with pathogens and maintain homeostasis during periods of stress or injury (Banihashemian and Mirmajlessi 2025).

Recent research in **epigenetics** has unveiled critical insights into the **dynamic regulation** of immune cells, emphasizing how **epigenetic modifications** govern immune cell differentiation, function, and memory. The ability to **rapidly modulate epigenetic landscapes** in response to **infection**, **tissue damage**, and **vaccination** is central to the immune system's ability to adapt and protect the host (Mutchler et al. 2025). The discovery of **cell-specific epigenetic signatures** has enhanced our understanding of immune cell fate decisions, opening the door for **precision immunotherapies** tailored to the individual patient's needs. Furthermore, the application of cutting-edge technologies, such as **single-cell epigenomics** and **CRISPR-based genome editing**, is transforming our ability to manipulate the epigenome in therapeutic settings, providing new avenues for the treatment of autoimmune diseases, cancer, and chronic infections. These insights will continue to shape the future of **immunotherapy**, allowing for the development of more effective and personalized treatments that harness the power of the **immune system** while minimizing side effects and improving patient outcomes (Mutchler et al. 2025).

Dynamic Modulation of Epigenetic Changes

Epigenetic modifications in immune cells are not static but occur rapidly in response to **infections**, **tissue injury**, and **vaccination**. These modifications allow immune cells to **adapt quickly** to environmental cues, enabling an effective and timely immune response. During an infection, the activation of **T lymphocytes** and

macrophages is often accompanied by changes in **DNA methylation** and **histone modification**, which enhance the expression of genes involved in **inflammatory cytokine production** and **immune cell recruitment (Placek et al.** 2019).

One of the key features of epigenetic modulation is its role in **immune memory**. Immune cells can "remember" prior encounters with pathogens or vaccines through lasting epigenetic changes, allowing for a more robust response upon reexposure. This is particularly relevant in the context of **trained immunity**, where innate immune cells like macrophages and NK cells retain epigenetic modifications that enhance their responses to subsequent infections, even without specific antigen recognition (Netea et al. 2019).

These dynamic epigenetic changes are being leveraged in the development of **immunotherapies**, where **epigenetic reprogramming** is used to either boost immune responses, as seen in **vaccines** or **cancer immunotherapies**, or restore immune tolerance, as seen in **autoimmune therapies** (Villanueva et al. 2020).

Cell-Specific Epigenetic Landscapes

Unique Epigenetic Signatures

Each immune cell type—including **T cells**, **B cells**, **macrophages**, and **dendritic cells**—is characterized by a **distinct epigenetic landscape** that governs its **differentiation potential** and **functional identity**. In **T helper cells** (Th1, Th2, Th17, and Treg), the transition from naïve T cells to specialized subtypes is orchestrated by **DNA demethylation** and **histone acetylation**, which facilitate the expression of lineage-defining transcription factors, such as **T-bet** for Th1 and **FOXP3** for Tregs. In **B cells**, activation involves **methylation changes** in loci, such as the **IgH region**, driving **antibody production** and **class-switch recombination**. Aberrant methylation in these regulatory regions can result in **autoimmune pathologies** through the production of **autoantibodies**. Similarly, **macrophages** and **dendritic cells** undergo **epigenetic reprogramming** in response to pathogens or cytokine cues, altering the transcription of **inflammatory genes** and enhancing innate immune responses (Roy et al. 2021).

Functional Programming of Immune Cells

Beyond differentiation, **epigenetic modifications** guide the **functional specialization** of immune cells, enabling them to participate in **inflammation**, **tissue repair**, and **pathogen defense**. For instance, **histone acetylation** in macrophages facilitates the expression of **pro-inflammatory cytokines**, such as TNF-α and IL-6. Conversely, the **silencing of immune tolerance genes** can amplify immune activation during infection. These dynamic and reversible epigenetic events ensure immune cells can rapidly adapt to the microenvironment while retaining plasticity (Placek et al. 2019).

Personalized Immunotherapy

Deciphering **cell-type-specific epigenetic signatures** opens the door to **precision immunotherapy**. A deep understanding of how individual immune cells respond to environmental stimuli through epigenetic regulation allows for the identification of **disease-driving epigenetic changes**. This knowledge can be harnessed to design **personalized immunotherapeutic strategies** that precisely target pathological epigenetic alterations, offering promising avenues for the treatment of **autoimmune diseases**, **chronic inflammation**, and **cancer** (Rocha et al. 2023).

Clinical Implications of Epigenetic Dysregulation

Autoimmune Diseases

Aberrant epigenetic programming is a hallmark of several **autoimmune diseases**, including **SLE**, **RA**, and **MS**. In these conditions, **hypomethylation** of key **immune activation genes** in **T and B cells** can disrupt **immune tolerance**, promoting the **production of autoantibodies** and triggering **chronic inflammation** and **tissue damage**. These epigenetic defects not only contribute to disease onset and progression but also represent a promising therapeutic target. A deeper understanding of these **epigenetic alterations** paves the way for the development of **targeted epigenetic therapies** aimed at restoring immune balance and halting disease progression (Araki and Mimura 2024).

Cancer Immune Evasion

Epigenetic changes also play a key role in **cancer immunoevasion**. **Tumor cells** frequently undergo **epigenetic modifications** that silence tumor suppressor genes and immune checkpoint molecules, allowing them to evade immune detection. **Immune cells**, such as **T cells** and **macrophages**, may also undergo epigenetic reprogramming that impairs their ability to effectively target and eliminate tumor cells. Epigenetic therapies that **reprogram immune cells** or **reactivate tumor suppressors** are a promising avenue in cancer immunotherapy (Panda et al. 2024).

Chronic Infections

Chronic infections, such as **HIV** or **hepatitis,** can induce **persistent epigenetic changes** in immune cells, contributing to **immune exhaustion** and dysfunctional immune responses. These changes often involve **DNA methylation**, **histone modifications**, and alterations in **noncoding RNA expression**, which collectively silence critical immune genes or sustain the expression of inhibitory pathways. Over time, this epigenetic reprogramming impairs the ability of **T cells**, **NK cells**, and **macrophages** to mount effective responses. Targeting these **epigenetic disruptions**—for example, through **epigenetic modulators** or **precision gene editing tools**—holds therapeutic potential by **reinvigorating immune cell function**, restoring **cytokine production**, and enhancing **viral control** in individuals with chronic infections (Żychowska and Ćmil 2024).

Technological Advances Transforming Epigenetics in Immunology

Single-Cell Epigenomic Profiling

Recent advancements in **single-cell sequencing** technologies have enabled researchers to map the **epigenetic landscape** of individual immune cells. This technology has allowed for the identification of **epigenetic signatures** that define immune cell states during infections, autoimmune diseases, and cancer. Single-cell epigenomics also allows for the tracking of **epigenetic changes** over time, providing valuable insights into how immune cells **respond to stimuli** and **retain memory** of past encounters (Ginhoux et al. 2022).

CRISPR-Based Genome Editing

The advent of **CRISPR/Cas9** technology has revolutionized the ability to manipulate the epigenome in a targeted manner. Researchers can now **edit specific epigenetic marks** at precise locations on the genome, enabling the study of **gene function** and **epigenetic regulation** in immune cells. CRISPR-based techniques are being explored to **correct epigenetic defects** in immune cells, potentially leading to new treatments for autoimmune diseases and enhancing the effectiveness of **immunotherapies** (Ansori et al. 2023).

High-Throughput RNA Sequencing

Advances in **RNA sequencing** technologies allow researchers to profile the expression of **noncoding RNAs**, including **miRNAs** and **lncRNAs**, which are key regulators of immune cell function. This has expanded our understanding of how **noncoding RNAs** modulate the **epigenetic regulation** of gene expression, influencing **immune responses** in health and disease. By profiling these RNA molecules, researchers can identify **biomarkers** for disease diagnosis and **predict therapeutic outcomes** (Chodurska and Kunej 2025).

Therapeutic Implications of Epigenetic Targeting

Epigenetic Targeting Therapies

Epigenetic regulation is a promising avenue for therapeutic intervention, especially for diseases where traditional genetic-based treatments fall short. Various **epigenetic drugs** and **genome-editing technologies** target the molecular processes that control gene expression, offering novel strategies for **precision medicine**. These approaches focus on manipulating the **DNA**, **histones**, and **noncoding RNAs** to restore normal immune function or prevent disease progression (Dai et al. 2024). The key classes of epigenetic therapies are mentioned later:

DNA Methyltransferase (DNMT) Inhibitors

DNA methylation involves the addition of a methyl group to the cytosine base in DNA, typically occurring at **CpG dinucleotides**. This modification can lead to gene silencing, especially in tumor suppressor genes, contributing to various cancers and autoimmune diseases. **DNMT inhibitors** work by blocking the activity of **DNA methyltransferases**, which are responsible for adding these methyl groups to DNA. This inhibition prevents the silencing of critical genes and can **reactivate suppressed tumor suppressor genes**, leading to the restoration of normal cellular function (Kim 2025). DNMT inhibitors, such as azacitidine and decitabine, function by disrupting abnormal DNA methylation patterns to restore gene expression and are clinically used in treating myelodysplastic syndromes and acute myeloid leukemia, as summarized in Table 1.4.

These inhibitors hold significant promise for **cancer therapy**, especially in cancers driven by epigenetic silencing, and in **autoimmune diseases** where reactivation of silenced immune-regulatory genes could help restore immune tolerance (Chiappinelli et al. 2016).

Table 1.4 DNA methyltransferase (DNMT) inhibitors

Agent	Mechanism of action	Clinical use
Azacitidine	A nucleoside analog that incorporates into DNA during replication and inhibits DNMTs, leading to hypomethylation and reexpression of silenced genes	Treatment of myelodysplastic syndromes (MDS) and acute myeloid leukemia (AML) (Raj and Mufti 2006)
Decitabine	Similar to azacitidine, it is a demethylating agent that disrupts DNA methylation patterns, reactivating silenced genes	Used in cancer therapy, particularly in MDS and AML (Yang et al. 2021)

Histone Deacetylase (HDAC) Inhibitors

Histone acetylation plays a crucial role in **chromatin relaxation** and gene activation. Histone **acetyltransferases (HATs)** add acetyl groups to histone tails, which reduce the positive charge on histones, leading to an open chromatin structure and allowing gene transcription (Milazzo and Mercatelli 2020). On the other hand, **histone deacetylases (HDACs)** remove these acetyl groups, resulting in chromatin condensation and gene repression. **HDAC inhibitors** work by blocking the activity of HDACs, thereby promoting **histone acetylation**, opening up the chromatin, and **reactivating silenced genes** (Hassell 2019). Several histone deacetylase (HDAC) inhibitors, such as vorinostat, panobinostat, and romidepsin, are already approved or under investigation for cancer treatment, as summarized in Table 1.5.

Bromodomain and Extraterminal (BET) Inhibitors

Bromodomains are protein domains that specifically recognize and bind to **acetylated lysine residues** on histones, leading to the recruitment of **transcriptional machinery** and promoting gene activation. **BET inhibitors** block this interaction, preventing the transcriptional activation of genes involved in **inflammation**, **cell cycle regulation**, and **tumorigenesis**. BET inhibitors are particularly valuable in targeting the **BRD2, BRD3, and BRD4** proteins, which are critical for the regulation of **oncogene expression** (Wu et al. 2024b). BET inhibitors, such as JQ1 and OTX015, have emerged as promising epigenetic drugs targeting transcriptional regulators in various cancers, as shown in Table 1.6.

CRISPR/dCas9-Based Epigenome Editing

The **CRISPR-Cas9** system is a revolutionary genome-editing tool that allows precise modifications of the DNA sequence. However, a **catalytically inactive form of Cas9 (dCas9)** can be used for **epigenome editing** without causing double-strand breaks in DNA. By fusing **dCas9** with epigenetic regulators (e.g., **DNA methyltransferases**, **histone acetyltransferases**), researchers can **directly modify**

Table 1.5 Histone deacetylase (HDAC) inhibitors

Agent	Mechanism/Description	Clinical use/Applications
Vorinostat	Also known as suberoylanilide hydroxamic acid (SAHA), a first-in-class HDAC inhibitor	Approved for cutaneous T cell lymphoma (CTCL); investigated for glioblastoma, non-small cell lung cancer (NSCLC), and other cancers (Sato 2012)
Panobinostat	Potent HDAC inhibitor that alters the expression of genes involved in cell cycle regulation, apoptosis, and angiogenesis	Used in the treatment of multiple myeloma (Hoon Phi et al. 2017)
Romidepsin	HDAC inhibitor that induces cell cycle arrest and promotes apoptosis in tumor cells	Approved for CTCL and peripheral T cell lymphoma (PTCL) (Bates et al. 2015)

Table 1.6 Bromodomain and extraterminal (BET) inhibitors

Agent	Mechanism/Description	Clinical use/Applications
JQ1	Potent, selective BET inhibitor; disrupts transcriptional activation of oncogenes like MYC	Targets cancers, especially acute myeloid leukemia (AML) and solid tumors (Jiang et al. 2020, Radwan et al. 2016)
OTX015	Clinical-stage BET inhibitor; inhibits BRD4, causing cell cycle arrest and apoptosis in cancer cells	Being evaluated for breast cancer, small cell lung cancer, and other malignancies (Riveiro et al. 2016)

chromatin and **gene expression** at specific loci without altering the underlying DNA sequence (Tadić et al. 2019). This technique allows for **reversible and targeted modification** of the epigenome to either **activate or repress genes** in a precise, localized manner. **CRISPR/dCas9-based epigenome editing** holds immense potential in **gene therapy**, **cancer**, and **autoimmune diseases**. It offers the ability to **target gene silencing** or **reactivate genes** involved in immune dysfunction, cancer progression, or disease-associated mutations. The ability to edit the epigenome without permanently altering the genome reduces the risk of off-target effects and long-term complications, making it an attractive option for **personalized medicine (Rahman and Tollefsbol** 2021). Specific applications of CRISPR-based epigenetic editing tools, such as dCas9-p300 and dCas9-DNMT3A, are summarized in Table 1.7: *CRISPR-based epigenetic editing tools and their therapeutic applications*.

These emerging epigenetic therapies, such as **DNMT inhibitors, HDAC inhibitors, BET inhibitors**, and **CRISPR-based epigenome editing**, represent a new frontier in **precision medicine**. By targeting the fundamental processes that regulate gene expression, these therapies offer the potential to **reprogram** immune responses, **reverse epigenetic abnormalities**, and **treat a wide range of diseases**, including **cancers**, **autoimmune disorders**, and **inflammatory diseases**. As research continues to evolve, these tools will likely become key components of therapeutic strategies aimed at **restoring normal immune function** and **treating disease at the epigenetic level (Majchrzak-Celińska et al.** 2021).

Table 1.7 CRISPR-based epigenetic editing tools and their therapeutic applications

System	Mechanism/Description	Applications
CRISPR-dCas9-p300	Fusion of p300 (a histone acetyltransferase) to dCas9, enabling site-specific histone acetylation to activate gene expression	Potential in gene therapy, cancer, and autoimmune diseases; reactivates beneficial genes without altering DNA sequence (Cai and Lv 2023)
CRISPR-dCas9-DNMT3A	Targets DNA methylation at specific loci via fusion of DNMT3A to dCas9, allowing gene silencing or activation to correct epigenetic defects	Used for correcting epigenetic defects and gene silencing/reactivation in personalized medicine without permanent genome changes (Park et al. 2022)

Biomarkers and Personalized Medicine

Epigenetic profiling has emerged as a powerful tool in the realm of **personalized medicine**, particularly in the context of **immune-mediated diseases**. By examining the **epigenetic signatures** of individuals, including changes in **DNA methylation**, **histone modifications**, and **noncoding RNA expression**, clinicians can gain deeper insights into the underlying molecular mechanisms that drive diseases, such as autoimmune disorders, cancer, and chronic infections. This knowledge enables a more **tailored approach** to treatment, optimizing therapeutic strategies based on the unique epigenetic landscape of a patient's disease (Molla and Bitew 2024).

Stratifying Patients by Disease Stage

Epigenetic biomarkers are particularly useful for stratifying patients according to **disease progression**. For example, in autoimmune diseases like **systemic lupus erythematosus (SLE)** or **rheumatoid arthritis (RA)**, **DNA methylation** patterns and **histone modifications** can be used to distinguish between early and advanced stages of the disease. In cancer, **epigenetic profiling** helps to identify the **tumor microenvironment's** unique features, which may vary between early-stage and metastatic disease, guiding treatment decisions (Wu et al. 2018).

Predicting Therapeutic Responses

Epigenetic alterations can provide insights into a patient's likely **response to treatment**. For example, **methylation of immune checkpoint genes** or the expression of specific **noncoding RNAs** in immune cells may indicate how well a patient will respond to **immune checkpoint inhibitors** in cancer immunotherapy. Similarly, **miRNAs** and **lncRNAs** can be used as biomarkers to predict how a patient might respond to **biologic agents** used to treat autoimmune diseases, such as **TNF inhibitors** in rheumatoid arthritis or **B cell depletion therapies** in SLE (Kang et al. 2025).

Guiding Precision Interventions

Epigenetic profiling allows clinicians to customize therapies based on a patient's unique epigenetic landscape. For instance, patients with **specific epigenetic alterations** in their immune cells (e.g., altered **histone acetylation** or **methylation patterns**) might benefit from **epigenetic therapies** that specifically correct those defects. By using **epigenetic drugs,** such as **DNMT inhibitors** or **HDAC inhibitors**, clinicians could directly target and modify the epigenetic regulation of genes that are contributing to disease progression, thereby **personalizing treatment** for maximum efficacy (Ramachandran 2025).

Future Prospects

The future of **epigenetic-based therapies** in immunology and personalized medicine holds incredible promise.

Trained Immunity Engineering

One of the most innovative areas of research is the concept of **trained immunity**, where **innate immune cells** like **macrophages** and **natural killer (NK) cells** retain a memory of previous pathogen encounters, leading to an enhanced response upon reexposure. By **modulating the epigenetic memory** of these cells, researchers hope to improve **vaccine efficacy** and enhance **protection** against infections. For example, **epigenetic reprogramming** of innate immune cells could make vaccines more effective in conferring long-term immunity, providing protection even against variants or emerging infectious agents (Velikova et al. 2024).

Moreover, the development of vaccines that can **stimulate trained immunity** through **epigenetic modifications** might provide an advantage over traditional vaccines. By **targeting epigenetic regulators** involved in innate immune cell memory, vaccines could be designed to induce **stronger, more durable immune responses**, offering greater protection and **long-lasting immunity** (Li et al. 2025; Megdiche and Salerno-Gonçalves 2025).

Combination Immunotherapies

Overcoming cancer immune evasion is a critical challenge in oncology. Cancer cells often develop mechanisms to escape immune detection by **epigenetically modifying immune checkpoint genes** and other regulatory pathways (El-Mesallamy et al. 2013b). Combining **epigenetic drugs,** such as **HDAC inhibitors** or **DNMT inhibitors,** with **immune checkpoint inhibitors** like **anti-PD-1** or **anti-CTLA-4**

offers a promising strategy to counteract this immune suppression. Through **epigenetic modulation**, immune cells' capacity to recognize and eliminate tumor cells can be enhanced, while checkpoint inhibitors block cancer cells' ability to suppress the immune response (Almawash 2025).

Additionally, **targeted epigenetic reprogramming** can fine-tune the immune response in combination therapies. For instance, **HDAC inhibitors** may increase the expression of immune-related genes in **tumor-infiltrating lymphocytes**, while **checkpoint inhibitors** work to reinvigorate **exhausted T cells**. This integrated approach has the potential to improve overall treatment response rates in cancer patients by synergistically enhancing immune function and reducing tumor immune escape (Wu et al. 2024a; El-Mesallamy et al. 2012).

Selective Epigenetic Reprogramming

Precision correction of immune dysfunction is one of the key promises of epigenetic therapy. It is the ability to **correct immune dysfunction** with a high degree of specificity. Unlike traditional immunosuppressive therapies, which affect the immune system globally, **selective epigenetic reprogramming** could be used to correct specific defects in immune cell function without causing broad **immune suppression**. For example, in **autoimmune diseases**, epigenetic drugs could **reverse aberrant gene expression** in immune cells without compromising overall immune function, reducing the risk of infections and other side effects (Pisetsky 2023).

Exosome-mediated therapeutic strategies leverage the unique role of **exosomes**—extracellular vesicles that transport **modified RNA, proteins, and lipids**—as critical mediators of immune regulation. These vesicles can transfer **epigenetic signals**, such as **miRNAs** and **RNA modifications**, between cells, thereby influencing immune responses. By manipulating exosomes, it is possible to deliver **epigenetic modifiers** or correct aberrant RNA modifications in recipient cells. This approach offers a **noninvasive** and highly **targeted method** to modulate immune function, presenting a promising avenue for novel therapies (Payandeh et al. 2024).

Role of RNA Modifications and Pseudouridylation

Therapeutic RNA modifications, particularly **pseudouridylation**, are gaining attention as a potential therapeutic target. Pseudouridylation stabilizes RNA molecules and influences their translation efficiency. In immune cells, this modification can regulate the translation of **pro-inflammatory cytokines** and **immune receptors**. By **modulating RNA modifications**, such as pseudouridylation, **therapeutic strategies** could be developed to **fine-tune immune cell activation** and **inhibit the development of autoimmune diseases** and **chronic inflammation** (Luo et al. 2025).

Conclusion

Epigenetic modulation constitutes a fundamental regulatory layer in the immune system, profoundly influencing immune cell development, activation, differentiation, memory formation, and tolerance. Through dynamic and reversible modifications of DNA methylation, histone posttranslational changes, and noncoding RNAs, such as miRNAs and lncRNAs, the epigenetic landscape orchestrates the transcriptional programs that govern immune cell fate and function. These tightly regulated processes enable the immune system to adapt swiftly to environmental cues—during infections, tissue injury, or vaccination—and to form immunological memory, ensuring a faster and more robust response upon reexposure to pathogens.

However, aberrant epigenetic programming can disrupt immune homeostasis, contributing to the pathogenesis of autoimmune and inflammatory diseases. In SLE, for instance, hypomethylation of immune activation genes promotes autoantibody production and loss of tolerance. In RA, histone modifications lead to the overexpression of inflammatory cytokines, fueling chronic synovitis and joint destruction. Similarly, in MS and IBD, defective DNA methylation and dysregulated noncoding RNA networks exacerbate central nervous system and gut inflammation, respectively. Environmental factors—including infections, diet, and microbiome alterations—can further induce epigenetic reprogramming, often tipping the balance toward pathogenic immune responses.

The growing understanding of immune epigenetics offers an unprecedented opportunity for the development of novel diagnostics and targeted therapeutics. Epigenetic therapies, such as DNA methyltransferase (DNMT) inhibitors and histone deacetylase (HDAC) inhibitors, are emerging as promising strategies to modulate aberrant gene expression without global immune suppression. Furthermore, RNA-based interventions targeting dysregulated miRNAs and lncRNAs hold great potential for fine-tuning immune responses and restoring immune tolerance in autoimmunity and chronic inflammation.

Technological advancements—including single-cell epigenomics, CRISPR/Cas9-mediated epigenome editing, and high-throughput RNA sequencing—are accelerating our ability to dissect the complex epigenetic landscapes of individual immune cell types. These tools facilitate the precision correction of immune dysfunction by selectively modulating epigenetic marks without altering the underlying genome, minimizing off-target effects and paving the way for personalized immunotherapies.

Epigenetic profiling is poised to revolutionize personalized medicine by enabling patient stratification based on distinct epigenetic signatures. Such profiling can predict disease progression, therapeutic response, and relapse risk, allowing the design of individualized treatment regimens that maximize efficacy while minimizing adverse effects. Strategies such as trained immunity engineering—modulating the epigenetic memory of innate immune cells—are under investigation to enhance vaccine efficacy and combat chronic infections and cancer immune evasion. The role of exosomes in the transfer of modified RNAs and epitranscriptomic changes, such as pseudouridylation, also opens new therapeutic avenues for regulating immune responses in autoimmune diseases and cancer. Integrating epigenetic modulators with immune checkpoint inhibitors or other immunotherapies represents a promising frontier to overcome immune resistance mechanisms and improve clinical outcomes.

Therefore, epigenetic modulation is not only pivotal for maintaining immune competence but also represents a critical determinant in the development of immune-mediated diseases. Harnessing the potential of epigenetic therapies—whether through targeted drugs, RNA-based interventions, or precise epigenome editing—offers a transformative path toward highly specific, personalized treatments. This emerging paradigm holds the promise of fundamentally improving outcomes for patients suffering from autoimmune, inflammatory, and cancer-related conditions, ushering in a new era of precision immunology.

References

Abd El Fattah, Y. K., Abulsoud, A. I., Abdelhamid, S. G., Abdelhalim, S. & Hamdy, N. M. 2023. Ccdc144nl-As1/Hsa-Mir-143-3p/Hmga2 Interaction: In-Silico And Clinically Implicated In Crc Progression, Correlated To Tumor Stage And Size In Case-Controlled Study; Step Toward Ncrna Precision. *Int J Biol Macromol,* 253, 126739.

Abu-Hanna, J., Patel, J. A., Anastasakis, E., Cohen, R., Clapp, L. H., Loizidou, M. & Eddama, M. M. R. 2022. Therapeutic Potential Of Inhibiting Histone 3 Lysine 27 Demethylases: A Review Of The Literature. *Clinical Epigenetics,* 14, 98.

Almawash, S. 2025. Revolutionary Cancer Therapy For Personalization And Improved Efficacy: Strategies To Overcome Resistance To Immune Checkpoint Inhibitor Therapy. *Cancers (Basel),* 17.

An, H., Williams, N. G. & Shelkovnikova, T. A. 2018. Neat1 And Paraspeckles In Neurodegenerative Diseases: A Missing Lnc Found? *Noncoding Rna Res,* 3, 243–252.

Añazco-Guenkova, A. M., Miguel-López, B., Monteagudo-García, Ó., García-Vílchez, R. & Blanco, S. 2024. The Impact Of Trna Modifications On Translation In Cancer: Identifying Novel Therapeutic Avenues. 6, Zcae012.

Ansori, A. N., Antonius, Y., Susilo, R. J., Hayaza, S., Kharisma, V. D., Parikesit, A. A., Zainul, R., Jakhmola, V., Saklani, T., Rebezov, M., Ullah, M. E., Maksimiuk, N., Derkho, M. & Burkov, P. 2023. Application Of Crispr-Cas9 Genome Editing Technology In Various Fields: A Review. *Narra J,* 3, E184.

Anwar, M. M., Albanese, C., Hamdy, N. M. & Sultan, A. S. 2022. Rise Of The Natural Red Pigment 'Prodigiosin' As An Immunomodulator In Cancer. *Cancer Cell International,* 22, 419.

Araki, Y. & Mimura, T. 2024. Epigenetic Dysregulation In The Pathogenesis Of Systemic Lupus Erythematosus. 25.

Ashwani, Sharma, A., Choudhary, M. K., Gugulothu, D., Pandita, D., Verma, S., Vora, L. K., Khatri, D. K. & Garabadu, D. 2025. Epigenetic And Mitochondrial Metabolic Dysfunction In Multiple Sclerosis: A Review Of Herbal Drug Approaches And Current Clinical Trials. *Molecular Neurobiology.*

Bagni, G., Biancalana, E., Chiara, E., Costanzo, I., Malandrino, D., Lastraioli, E., Palmerini, M., Silvestri, E., Urban, M. L. & Emmi, G. 2025. Epigenetics In Autoimmune Diseases: Unraveling The Hidden Regulators Of Immune Dysregulation. *Autoimmunity Reviews,* 24, 103784.

Bai, L., Hao, X., Keith, J. & Feng, Y. 2022. Dna Methylation In Regulatory T Cell Differentiation And Function: Challenges And Opportunities. 12.

Banihashemian, S. N. & Mirmajlessi, S. M. 2025. Epigenetic Modifications, Immune Control Processes, And Plant Responses To Nematodes. *Agriculture,* 15, 742.
Bannister, S. & Messina, N. L. 2020. The Emerging Role Of Epigenetics In The Immune Response To Vaccination And Infection: A Systematic Review. 15, 555–593.
Basu, S., Nadhan, R. & Dhanasekaran, D. N. 2025. Long Non-Coding Rnas In Ovarian Cancer: Mechanistic Insights And Clinical Applications. *Cancers,* 17, 472.
Bates, S. E., Eisch, R., Ling, A., Rosing, D., Turner, M., Pittaluga, S., Prince, H. M., Kirschbaum, M. H., Allen, S. L., Zain, J., Geskin, L. J., Joske, D., Popplewell, L., Cowen, E. W., Jaffe, E. S., Nichols, J., Kennedy, S., Steinberg, S. M., Liewehr, D. J., Showe, L. C., Steakley, C., Wright, J., Fojo, T., Litman, T. & Piekarz, R. L. 2015. Romidepsin In Peripheral And Cutaneous T-Cell Lymphoma: Mechanistic Implications From Clinical And Correlative Data. *Br J Haematol,* 170, 96–109.
Bezhaeva, T., Karper, J., Quax, P. H. A. & De Vries, M. R. 2022. The Intriguing Role Of Tlr Accessory Molecules In Cardiovascular Health And Disease. *Front Cardiovasc Med,* 9, 820962.
Bhargavi, G. & Subbian, S. 2024. The Causes And Consequences Of Trained Immunity In Myeloid Cells. *Front Immunol,* 15, 1365127.
Bhat, M. F., Srdanović, S., Sundberg, L.-R., Einarsdóttir, H. K., Marjomäki, V. & Dekker, F. J. 2024. Impact Of Hdac Inhibitors On Macrophage Polarization To Enhance Innate Immunity Against Infections. *Drug Discovery Today,* 29, 104193.
Blass, E. & Ott, P. A. 2021. Advances In The Development Of Personalized Neoantigen-Based Therapeutic Cancer Vaccines. *Nature Reviews Clinical Oncology,* 18, 215–229.
Breiling, A. & Lyko, F. 2015. Epigenetic Regulatory Functions Of Dna Modifications: 5-Methylcytosine And Beyond. *Epigenetics & Chromatin,* 8, 24.
Breton, C. V., Landon, R., Kahn, L. G., Enlow, M. B., Peterson, A. K., Bastain, T., Braun, J., Comstock, S. S., Duarte, C. S., Hipwell, A., Ji, H., Lasalle, J. M., Miller, R. L., Musci, R., Posner, J., Schmidt, R., Suglia, S. F., Tung, I., Weisenberger, D., Zhu, Y. & Fry, R. 2021. Exploring The Evidence For Epigenetic Regulation Of Environmental Influences On Child Health Across Generations. *Communications Biology,* 4, 769.
Bure, I. V., Nemtsova, M. V. & Kuznetsova, E. B. 2022. Histone Modifications And Non-Coding Rnas: Mutual Epigenetic Regulation And Role In Pathogenesis. *Int J Mol Sci,* 23.
Cai, R. & Lv, R. 2023. Crispr/Dcas9 Tools: Epigenetic Mechanism And Application In Gene Transcriptional Regulation. 24.
Cai, W. Y., Cai, X. X., Fei, Y. R., Ye, R., Song, D. M., Hu, D., Zhang, W. W., Xia, M. F. & Yang, X. X. 2025. Dna Methylation And Immune Evasion In Triple-Negative Breast Cancer: Challenges And Therapeutic Opportunities. *Front Oncol,* 15, 1534055.
Caldwell, B. A. & Li, L. 2024. Epigenetic Regulation Of Innate Immune Dynamics During Inflammation. 115, 589–606.
Castro-Muñoz, L. J., Ulloa, E. V., Sahlgren, C., Lizano, M., De La Cruz-Hernández, E. & Contreras-Paredes, A. 2023. Modulating Epigenetic Modifications For Cancer Therapy (Review). *Oncol Rep,* 49.
Celarain, N. & Tomas-Roig, J. 2020. Aberrant Dna Methylation Profile Exacerbates Inflammation And Neurodegeneration In Multiple Sclerosis Patients. *Journal Of Neuroinflammation,* 17, 21.
Chen, S., Saeed, A. F. U. H., Liu, Q., Jiang, Q., Xu, H., Xiao, G. G., Rao, L. & Duo, Y. 2023. Macrophages In Immunoregulation And Therapeutics. *Signal Transduction And Targeted Therapy,* 8, 207.
Chen, S., Yang, J., Wei, Y. & Wei, X. 2020. Epigenetic Regulation Of Macrophages: From Homeostasis Maintenance To Host Defense. *Cellular & Molecular Immunology,* 17, 36–49.
Chen, Y.-H., Wu, K.-H. & Wu, H.-P. 2024. Unraveling The Complexities Of Toll-Like Receptors: From Molecular Mechanisms To Clinical Applications. *International Journal Of Molecular Sciences,* 25, 5037.
Chi, H., Pepper, M. & Thomas, P. G. 2024. Principles And Therapeutic Applications Of Adaptive Immunity. *Cell,* 187, 2052–2078.

Chiappinelli, K. B., Zahnow, C. A., Ahuja, N. & Baylin, S. B. 2016. Combining Epigenetic And Immunotherapy To Combat Cancer. *Cancer Res,* 76, 1683–9.

Chodurska, B. & Kunej, T. 2025. Long Non-Coding Rnas In Humans: Classification, Genomic Organization And Function. *Non-Coding Rna Research,* 11, 313–327.

Coppedè, F. 2017. Epigenetics And Autoimmune Thyroid Diseases. *Front Endocrinol (Lausanne),* 8, 149.

Corneth, O. B. J. & Neys, S. F. H. 2022. Aberrant B Cell Signaling In Autoimmune Diseases. 11.

Correa, L. O., Jordan, M. S. & Carty, S. A. 2020. Dna Methylation In T-Cell Development And Differentiation. *Crit Rev Immunol,* 40, 135–156.

Cribbs, A., Feldmann, M. & Oppermann, U. 2015. Towards An Understanding Of The Role Of Dna Methylation In Rheumatoid Arthritis: Therapeutic And Diagnostic Implications. *Therapeutic Advances In Musculoskeletal Disease,* 7, 206–219.

Cribbs, A. P., Kennedy, A., Penn, H., Read, J. E., Amjadi, P., Green, P., Syed, K., Manka, S. W., Brennan, F. M. & Gregory, B. 2014. Treg Cell Function In Rheumatoid Arthritis Is Compromised By Ctla-4 Promoter Methylation Resulting In A Failure To Activate The Indoleamine 2, 3-Dioxygenase Pathway. *Arthritis & Rheumatology,* 66, 2344–2354.

Cui, L., Ma, R., Cai, J., Guo, C., Chen, Z., Yao, L., Wang, Y., Fan, R., Wang, X. & Shi, Y. 2022. Rna Modifications: Importance In Immune Cell Biology And Related Diseases. *Signal Transduction And Targeted Therapy,* 7, 334.

Dai, E., Zhu, Z., Wahed, S., Qu, Z., Storkus, W. J. & Guo, Z. S. 2021. Epigenetic Modulation Of Antitumor Immunity For Improved Cancer Immunotherapy. *Molecular Cancer,* 20, 171.

Dai, W., Qiao, X., Fang, Y., Guo, R., Bai, P., Liu, S., Li, T., Jiang, Y., Wei, S., Na, Z., Xiao, X. & Li, D. 2024. Epigenetics-Targeted Drugs: Current Paradigms And Future Challenges. *Signal Transduction And Targeted Therapy,* 9, 332.

Danieli, M. G., Casciaro, M., Paladini, A., Bartolucci, M., Sordoni, M., Shoenfeld, Y. & Gangemi, S. 2024. Exposome: Epigenetics And Autoimmune Diseases. *Autoimmunity Reviews,* 23, 103584.

Das, D., Karthik, N. & Taneja, R. 2021. Crosstalk Between Inflammatory Signaling And Methylation In Cancer. *Front Cell Dev Biol,* 9, 756458.

Demczuk, A., Gauthier, M., Veras, I., Kosiyatrakul, S., Schildkraut, C., Busslinger, M., Bechhoefer, J. & Norio, P. 2012. Regulation Of Dna Replication Within The Immunoglobulin Heavy-Chain Locus During B Cell Commitment. *Plos Biology,* 10, E1001360.

Ding, Y. P., Liu, C. C. & Yu, K. D. 2025. Rna Modifications In The Tumor Microenvironment: Insights Into The Cancer-Immunity Cycle And Beyond. *Exp Hematol Oncol,* 14, 48.

Doody, K. M., Bottini, N. & Firestein, G. S. 2017. Epigenetic Alterations In Rheumatoid Arthritis Fibroblast-Like Synoviocytes. *Epigenomics,* 9, 479–492.

Duan, T., Du, Y., Xing, C., Wang, H. Y. & Wang, R. F. 2022. Toll-Like Receptor Signaling And Its Role In Cell-Mediated Immunity. *Front Immunol,* 13, 812774.

Eberharter, A. & Becker, P. 2002. Histone Acetylation: A Switch Between Repressive And Permissive Chromatin. Second In Review Series On Chromatin Dynamics. *Embo Reports,* 3, 224–9.

El-Mesallamy, H. O., Hamdy, N. M. & Sallam, A. A. 2013a. Effect Of Obesity And Glycemic Control On Serum Lipocalins And Insulin-Like Growth Factor Axis In Type 2 Diabetic Patients. *Acta Diabetol,* 50, 679–85.

El-Mesallamy, H. O., Hamdy, N. M., Zaghloul, A. S. & Sallam, A. M. 2012. Serum Retinol Binding Protein-4 And Neutrophil Gelatinase-Associated Lipocalin Are Interrelated In Pancreatic Cancer Patients. *Scand J Clin Lab Invest,* 72, 602–7.

El-Mesallamy, H. O., Hamdy, N. M., Zaghloul, A. S. & Sallam, A. M. 2013b. Clinical Value Of Circulating Lipocalins And Insulin-Like Growth Factor Axis In Pancreatic Cancer Diagnosis. *Pancreas,* 42, 149–54.

El-Zayat, S. R., Sibaii, H. & Mannaa, F. A. 2019. Toll-Like Receptors Activation, Signaling, And Targeting: An Overview. *Bulletin Of The National Research Centre,* 43, 187.

El Mesallamy, H. O., Hamdy, N. M., Mostafa, D. M. & Amin, A. I. 2013. The Serine Protease Granzyme B As An Inflammatory Marker, In Relation To The Insulin Receptor Cleavage In Human Obesity And Type 2 Diabetes Mellitus. *Journal Of Interferon & Cytokine Research,* 34, 179–186.

Essola, J. M., Zhang, M., Yang, H., Li, F., Xia, B., Mavoungou, J. F., Hussain, A. & Huang, Y. 2024. Exosome Regulation Of Immune Response Mechanism: Pros And Cons In Immunotherapy. *Bioactive Materials,* 32, 124–146.

Fakour, P., Shirizadeh, A., Tahamoli-Roudsari, A. & Solgi, G. 2024. Epigenetic Regulation Of Foxp3 Gene Expression In Relation To Impaired Function Of Regulatory T Cells In Systemic Lupus Erythematosus. *Exploration Of Immunology,* 4, 640–657.

Falvo, J. V., Jasenosky, L. D., Kruidenier, L. & Goldfeld, A. E. 2013. Epigenetic Control Of Cytokine Gene Expression: Regulation Of The Tnf/Lt Locus And T Helper Cell Differentiation. *Adv Immunol,* 118, 37–128.

Feng, J., Yao, Y., Wang, Q., Han, X., Deng, X., Cao, Y., Chen, X., Zhou, M. & Zhao, C. 2023. Exosomes: Potential Key Players Towards Novel Therapeutic Options In Diabetic Wounds. *Biomedicine & Pharmacotherapy,* 166, 115297.

Fernández-Ponce, C., Navarro Quiroz, R., Díaz Perez, A., Aroca Martinez, G., Cadena Bonfanti, A., Acosta Hoyos, A., Gómez Escorcia, L., Hernández Agudelo, S., Orozco Sánchez, C., Villarreal Camacho, J., Atencio Ibarra, L., Consuegra Machado, J., Espinoza Garavito, A., García-Cózar, F. & Navarro Quiroz, E. 2021. Micrornas Overexpressed In Crohn's Disease And Their Interactions With Mechanisms Of Epigenetic Regulation Explain Novel Aspects Of Crohn's Disease Pathogenesis. 13, 39.

Ferreté-Bonastre, A. G., Cortés-Hernández, J. & Ballestar, E. 2022. What Can We Learn From Dna Methylation Studies In Lupus? *Clinical Immunology,* 234, 108920.

Frias, A. B., Boi, S. K., Lan, X. & Youngblood, B. 2021. Epigenetic Regulation Of T Cell Adaptive Immunity. 300, 9–21.

Fu, J. & Wang, Y. 2024. Impacts Of Epigenetic Reprogramming On Innate Immunity. *Animal Nutriomics,* 1, 1–58.

Gabrielli, S., Ortolani, C., Del Zotto, G., Luchetti, F., Canonico, B. & Buccella, F. 2016. The Memories Of Nk Cells: Innate-Adaptive Immune Intrinsic Crosstalk. 2016, 1376595.

Gareev, I., De Jesus Encarnacion Ramirez, M. Goncharov, E., Ivliev, D., Shumadalova, A., Ilyasova, T. & Wang, C. 2023a. Mirnas And Lncrnas In The Regulation Of Innate Immune Signaling. *Non-Coding Rna Research,* 8, 534–541.

Gareev, I., De Jesus Encarnacion Ramirez, M. Goncharov, E., Ivliev, D., Shumadalova, A., Ilyasova, T. & Wang, C. 2023b. Mirnas And Lncrnas In The Regulation Of Innate Immune Signaling. *Noncoding Rna Res,* 8, 534–541.

Ghafouri-Fard, S., Eghtedarian, R. & Taheri, M. 2020. The Crucial Role Of Non-Coding Rnas In The Pathophysiology Of Inflammatory Bowel Disease. *Biomedicine & Pharmacotherapy,* 129, 110507.

Ghahramani Almanghadim, H., Karimi, B., Valizadeh, S. & Ghaedi, K. 2025. Biological Functions And Affected Signaling Pathways By Long Non-Coding Rnas In The Immune System. *Non-Coding Rna Research,* 10, 70–90.

Ginhoux, F., Yalin, A., Dutertre, C. A. & Amit, I. 2022. Single-Cell Immunology: Past, Present, And Future. *Immunity,* 55, 393–404.

Golzari-Sorkheh, M. & Zúñiga-Pflücker, J. C. 2023. Development And Function Of Foxp3+ Regulators Of Immune Responses. 213, 13–22.

Gomez, J. L. 2019. Epigenetics In Asthma. *Curr Allergy Asthma Rep,* 19, 56.

Gorelik, G., Fang, J. Y., Wu, A., Sawalha, A. H. & Richardson, B. 2007. Impaired T Cell Protein Kinase Cδ Activation Decreases Erk Pathway Signaling In Idiopathic And Hydralazine-Induced Lupus. *The Journal Of Immunology,* 179, 5553–5563.

Grillo, T. G., Quaglio, A. E. V., Beraldo, R. F., Lima, T. B., Baima, J. P., Di Stasi, L. C. & Sassaki, L. Y. 2021. Microrna Expression In Inflammatory Bowel Disease-Associated Colorectal Cancer. *World J Gastrointest Oncol,* 13, 995–1016.

Gujral, P., Mahajan, V., Lissaman, A. C. & Ponnampalam, A. P. 2020. Histone Acetylation And The Role Of Histone Deacetylases In Normal Cyclic Endometrium. *Reproductive Biology And Endocrinology,* 18, 84.

Guo, D., Yan, J., Yang, Z., Chen, M., Zhong, W., Yuan, X. & Yu, S. 2024. The Immune Regulatory Role Of Exosomal Mirnas And Their Clinical Application Potential In Heart Failure. *Frontiers In Immunology,* Volume 15 - 2024.

Guo, Z., Wang, L., Liu, H. & Xie, Y. 2022. Innate Immune Memory In Monocytes And Macrophages: The Potential Therapeutic Strategies For Atherosclerosis. 11.

Ha, T. Y. 2011. The Role Of Micrornas In Regulatory T Cells And In The Immune Response. *Immune Netw,* 11, 11–41.

Hamdy, N. M., Zaki, M. B., Rizk, N. I., Abdelmaksoud, N. M., Abd-Elmawla, M. A., Ismail, R. A. & Abulsoud, A. I. 2024. Unraveling The Ncrna Landscape That Governs Colorectal Cancer: A Roadmap To Personalized Therapeutics. *Life Sci,* 354, 122946.

Hammad, R., Aglan, R. B., Mohammed, S. A., Awad, E. A., Elsaid, M. A., Bedair, H. M., Khirala, S. K. & Selim, M. A. 2022. Cytotoxic T Cell Expression Of Leukocyte-Associated Immunoglobulin-Like Receptor-1 (Lair-1) In Viral Hepatitis C-Mediated Hepatocellular Carcinoma. 23.

Handy, D. E., Castro, R. & Loscalzo, J. 2011. Epigenetic Modifications: Basic Mechanisms And Role In Cardiovascular Disease. *Circulation,* 123, 2145–56.

Hassell, K. N. 2019. Histone Deacetylases And Their Inhibitors In Cancer Epigenetics. *Diseases,* 7.

Heydari, R., Karimi, P. & Meyfour, A. 2024. Long Non-Coding Rnas As Pathophysiological Regulators, Therapeutic Targets And Novel Extracellular Vesicle Biomarkers For The Diagnosis Of Inflammatory Bowel Disease. *Biomedicine & Pharmacotherapy,* 176, 116868.

Hoon Phi, J., Ah Choi, S., Ae Kwak, P., Yeoun Lee, J., Wang, K.-C., Won Hwang, D. & Kim, S.-K. 2017. Panobinostat, A Histone Deacetylase Inhibitor, Suppresses Leptomeningeal Seeding In A Medulloblastoma Animal Model. *Oncotarget,* 8.

Hornschuh, M., Wirthgen, E., Wolfien, M., Singh, K. P., Wolkenhauer, O. & Däbritz, J. 2021. The Role Of Epigenetic Modifications For The Pathogenesis Of Crohn's Disease. *Clinical Epigenetics,* 13, 108.

Hu, G., Niu, F., Humburg, B. A., Liao, K., Bendi, S., Callen, S., Fox, H. S. & Buch, S. 2018. Molecular Mechanisms Of Long Noncoding Rnas And Their Role In Disease Pathogenesis. *Oncotarget,* 9, 18648–18663.

Hu, J. & Huang, S. 2022. Mir-155: An Important Role In Inflammation Response. 2022, 7437281.

Hu, S., Feng, L., Yang, Z., Fan, X., Gao, H. & Yang, T. 2024. A Recognition Of Exosomes As Regulators Of Epigenetic Mechanisms In Central Nervous System Diseases. *Front Mol Neurosci,* 17, 1370449.

Huynh, J. L. & Casaccia, P. 2013. Epigenetic Mechanisms In Multiple Sclerosis: Implications For Pathogenesis And Treatment. *Lancet Neurol,* 12, 195-206.

Ibáñez-Cabellos, J. S., Seco-Cervera, M., Osca-Verdegal, R., Pallardó, F. V. & García-Giménez, J. L. 2019. Epigenetic Regulation In The Pathogenesis Of Sjögren Syndrome And Rheumatoid Arthritis. *Front Genet,* 10, 1104.

Jambhekar, A., Dhall, A. & Shi, Y. 2019. Roles And Regulation Of Histone Methylation In Animal Development. *Nat Rev Mol Cell Biol,* 20, 625–641.

Javdani-Mallak, A. & Salahshoori, I. 2024. Environmental Pollutants And Exosomes: A New Paradigm In Environmental Health And Disease. *Science Of The Total Environment,* 925, 171774.

Jia, S., Yu, X., Deng, N., Zheng, C., Ju, M., Wang, F., Zhang, Y., Gao, Z., Li, Y., Zhou, H. & Li, K. 2025. Deciphering The Pseudouridine Nucleobase Modification In Human Diseases: From Molecular Mechanisms To Clinical Perspectives. 15, E70190.

Jiang, G., Deng, W., Liu, Y. & Wang, C. 2020. General Mechanism Of Jq1 In Inhibiting Various Types Of Cancer. *Mol Med Rep,* 21, 1021–1034.

Kang, Z., Wang, J., Liu, J., Du, L. & Liu, X. 2025. Epigenetic Modifications In Breast Cancer: From Immune Escape Mechanisms To Therapeutic Target Discovery. *Front Immunol,* 16, 1584087.

Khan, A. Q., Akhtar, S., Prabhu, K. S., Zarif, L., Khan, R., Alam, M., Buddenkotte, J., Ahmad, A., Steinhoff, M. & Uddin, S. 2020. Exosomes: Emerging Diagnostic And Therapeutic Targets In Cutaneous Diseases. *International Journal Of Molecular Sciences,* 21, 9264.

Kim, D. J. 2025. The Role Of The Dna Methyltransferase Family And The Therapeutic Potential Of Dnmt Inhibitors In Tumor Treatment. *Current Oncology,* 32, 88.

Kim, Tae H., Nosella, Michael L., Bolik-Coulon, N., Harkness, Robert W., Huang, Shuya K. & Kay, Lewis E. 2023. Correlating Histone Acetylation With Nucleosome Core Particle Dynamics And Function. *Proceedings Of The National Academy Of Sciences,* 120, E2301063120.

Kobayashi, A. & Kitagawa, Y. 2024. Emerging Roles And Mechanisms Of Rna Modifications In Neurodegenerative Diseases And Glioma. 13.

Kondo, N., Kuroda, T. & Kobayashi, D. 2021. Cytokine Networks In The Pathogenesis Of Rheumatoid Arthritis. 22.

Kong, C., Yang, M., Yue, N. & Zhang, Y. 2024. Restore Intestinal Barrier Integrity: An Approach For Inflammatory Bowel Disease Therapy. 17, 5389–5413.

Kornienko, I. V., Aramova, O. Y., Tishchenko, A. A., Rudoy, D. V. & Chikindas, M. L. 2024. Rna Stability: A Review Of The Role Of Structural Features And Environmental Conditions. *Molecules,* 29, 5978.

Kumar, M. A., Baba, S. K., Sadida, H. Q., Marzooqi, S. A., Jerobin, J., Altemani, F. H., Algehainy, N., Alanazi, M. A., Abou-Samra, A.-B., Kumar, R., Al-Shabeeb Akil, A. S., Macha, M. A., Mir, R. & Bhat, A. A. 2024. Extracellular Vesicles As Tools And Targets In Therapy For Diseases. *Signal Transduction And Targeted Therapy,* 9, 27.

Kumar, S., Chinnusamy, V. & Mohapatra, T. 2018. Epigenetics Of Modified Dna Bases: 5-Methylcytosine And Beyond. *Frontiers In Genetics,* Volume 9 - 2018.

Kumar, S., Gonzalez, E. A., Rameshwar, P. & Etchegaray, J. P. 2020. Non-Coding Rnas As Mediators Of Epigenetic Changes In Malignancies. *Cancers (Basel),* 12.

Lai, A. Y., Mav, D., Shah, R., Grimm, S. A., Phadke, D., Hatzi, K., Melnick, A., Geigerman, C., Sobol, S. E., Jaye, D. L. & Wade, P. A. 2013. Dna Methylation Profiling In Human B Cells Reveals Immune Regulatory Elements And Epigenetic Plasticity At Alu Elements During B-Cell Activation. *Genome Res,* 23, 2030–41.

Lakshminarasimhan, R. & Liang, G. 2016. The Role Of Dna Methylation In Cancer. *Adv Exp Med Biol,* 945, 151–172.

Lamere, S. A., Thompson, R. C., Komori, H. K., Mark, A. & Salomon, D. R. 2016. Promoter H3k4 Methylation Dynamically Reinforces Activation-Induced Pathways In Human Cd4 T Cells. *Genes Immun,* 17, 283–97.

Lanata, C. M., Chung, S. A. & Criswell, L. A. 2018. Dna Methylation 101: What Is Important To Know About Dna Methylation And Its Role In Sle Risk And Disease Heterogeneity. *Lupus Sci Med,* 5, E000285.

Lee, A., Wimmers, F. & Pulendran, B. 2022. Epigenetic Adjuvants: Durable Reprogramming Of The Innate Immune System With Adjuvants. *Curr Opin Immunol,* 77, 102189.

Legaki, E., Arsenis, C., Taka, S. & Papadopoulos, N. G. 2022. Dna Methylation Biomarkers In Asthma And Rhinitis: Are We There Yet? *Clin Transl Allergy,* 12, E12131.

Leng, X., Zhang, M., Xu, Y., Wang, J., Ding, N., Yu, Y., Sun, S., Dai, W., Xue, X., Li, N., Yang, Y. & Shi, Z. 2024. Non-Coding Rnas As Therapeutic Targets In Cancer And Its Clinical Application. *Journal Of Pharmaceutical Analysis,* 14, 100947.

Li, D. & Wu, M. 2021. Pattern Recognition Receptors In Health And Diseases. *Signal Transduction And Targeted Therapy,* 6, 291.

Li, J., Li, L., Wang, Y., Huang, G., Li, X., Xie, Z. & Zhou, Z. 2021a. Insights Into The Role Of Dna Methylation In Immune Cell Development And Autoimmune Disease. *Front Cell Dev Biol,* 9, 757318.

Li, J., Wang, J. & Chen, Z. 2025. Emerging Role Of Exosomes In Cancer Therapy: Progress And Challenges. *Molecular Cancer,* 24, 13.

Li, J., Xu, B., He, M., Zong, X., Cunningham, T., Sha, C., Fan, Y., Cross, R., Hanna, J. H. & Feng, Y. 2021b. Control Of Foxp3 Induction And Maintenance By Sequential Histone Acetylation And Dna Demethylation. *Cell Rep,* 37, 110124.

Li, X., Xiao, B. & Chen, X. S. 2017. Dna Methylation: A New Player In Multiple Sclerosis. *Mol Neurobiol,* 54, 4049-4059.

Lin, T.-Y., Mehta, R. & Glatt, S. 2021. Pseudouridines In Rnas – Switching Atoms Means Shifting Paradigms. *Febs Letters,* 595.

Lin, Y., Lin, P., Lu, Y., Zheng, J., Zheng, Y., Huang, X., Zhao, X. & Cui, L. 2024. Post-Translational Modifications Of Rna-Modifying Proteins In Cellular Dynamics And Disease Progression. *Advanced Science,* 11, 2406318.

Liu, R., Du, S., Zhao, L., Jain, S., Sahay, K., Rizvanov, A., Lezhnyova, V., Khaibullin, T., Martynova, E., Khaiboullina, S. & Baranwal, M. 2022a. Autoreactive Lymphocytes In Multiple Sclerosis: Pathogenesis And Treatment Target. *Front Immunol,* 13, 996469.

Liu, S.-Y., Qu, H.-T., Sun, R.-J., Yuan, D., Sui, X.-H. & Shan, N.-N. 2022b. High-Throughput Dna Methylation Analysis In Itp Confirms Notch1 Hypermethylation Through The Th1 And Th2 Cell Differentiation Pathways. *International Immunopharmacology,* 111, 109105.

Liu, S. L., Sun, P., Li, Y., Liu, S. S. & Lu, Y. 2019. Exosomes As Critical Mediators Of Cell-To-Cell Communication In Cancer Pathogenesis And Their Potential Clinical Application. *Transl Cancer Res,* 8, 298–311.

Liu, W.-W., Zheng, S.-Q., Li, T., Fei, Y.-F., Wang, C., Zhang, S., Wang, F., Jiang, G.-M. & Wang, H. 2024. Rna Modifications In Cellular Metabolism: Implications For Metabolism-Targeted Therapy And Immunotherapy. *Signal Transduction And Targeted Therapy,* 9, 70.

Lu, L., Barbi, J. & Pan, F. 2017. The Regulation Of Immune Tolerance By Foxp3. *Nat Rev Immunol,* 17, 703–717.

Luo, N., Huang, Q., Zhang, M. & Yi, C. 2025. Functions And Therapeutic Applications Of Pseudouridylation.

Lv, X., Liu, W., Zhou, X., Yang, Y., Zhao, W., Meng, L., Mu, F., Zhang, Z., Zhu, S., Zhang, S. & Wang, Y. 2025. Exosomes In Systemic Autoimmune Diseases: Recent Advances In Diagnostic Biomarkers And Therapeutic Applications. 20, 5137–5160.

Ma, F., Braikia, F. Z. & Sen, R. 2024. Lineage- And Stage-Specific Activity Of Antigen Receptor Gene Enhancers During Lymphocyte Development. *Frontiers In Epigenetics And Epigenomics,* Volume 2 - 2024.

Majchrzak-Celińska, A., Warych, A. & Szoszkiewicz, M. 2021. Novel Approaches To Epigenetic Therapies: From Drug Combinations To Epigenetic Editing. *Genes (Basel),* 12.

Manoharan, A., Roure, C., Rolink, A. & Matthias, P. 2015. De Novo Dna Methyltransferases Dnmt3a And Dnmt3b Regulate The Onset Of Igk Light Chain Rearrangement During Early B-Cell Development. *European Journal Of Immunology,* 45.

Mao, W., Wang, B., Huang, R., Sun, Z., Yan, M. & Dong, P. 2024. Histone Modifications In Head And Neck Squamous Cell Carcinoma. *Front Oncol,* 14, 1427725.

Marima, R., Basera, A., Miya, T., Damane, B. P., Kandhavelu, J., Mirza, S., Penny, C. & Dlamini, Z. 2024. Exosomal Long Non-Coding Rnas In Cancer: Interplay, Modulation, And Therapeutic Avenues. *Non-Coding Rna Research,* 9, 887–900.

Marshall, J. S., Warrington, R., Watson, W. & Kim, H. L. 2018. An Introduction To Immunology And Immunopathology. *Allergy, Asthma & Clinical Immunology,* 14, 49.

Mattick, J. S., Amaral, P. P., Carninci, P., Carpenter, S., Chang, H. Y., Chen, L.-L., Chen, R., Dean, C., Dinger, M. E., Fitzgerald, K. A., Gingeras, T. R., Guttman, M., Hirose, T., Huarte, M., Johnson, R., Kanduri, C., Kapranov, P., Lawrence, J. B., Lee, J. T., Mendell, J. T., Mercer, T. R., Moore, K. J., Nakagawa, S., Rinn, J. L., Spector, D. L., Ulitsky, I., Wan, Y., Wilusz, J. E. & Wu, M. 2023. Long Non-Coding Rnas: Definitions, Functions, Challenges And Recommendations. *Nature Reviews Molecular Cell Biology,* 24, 430–447.

Mazzone, R., Zwergel, C., Artico, M., Taurone, S., Ralli, M., Greco, A. & Mai, A. 2019. The Emerging Role Of Epigenetics In Human Autoimmune Disorders. 11, 34.
Megdiche, Y. & Salerno-Gonçalves, R. 2025. Harnessing Adjuvant-Induced Epigenetic Modulation For Enhanced Immunity In Vaccines And Cancer Therapy. *Frontiers In Immunology,* Volume 16 - 2025.
Mei, X., Zhang, B., Zhao, M. & Lu, Q. 2022. An Update On Epigenetic Regulation In Autoimmune Diseases. *Journal Of Translational Autoimmunity,* 5, 100176.
Milazzo, G. & Mercatelli, D. 2020. Histone Deacetylases (Hdacs): Evolution, Specificity, Role In Transcriptional Complexes, And Pharmacological Actionability. 11.
Miller, F. W. 2023. The Increasing Prevalence Of Autoimmunity And Autoimmune Diseases: An Urgent Call To Action For Improved Understanding, Diagnosis, Treatment, And Prevention. *Curr Opin Immunol,* 80, 102266.
Mittelstaedt, N. N., Becker, A. L., De Freitas, D. N., Zanin, R. F., Stein, R. T. & Duarte De Souza, A. P. 2021. Dna Methylation And Immune Memory Response. *Cells,* 10.
Molla, G. & Bitew, M. 2024. Revolutionizing Personalized Medicine: Synergy With Multi-Omics Data Generation, Main Hurdles, And Future Perspectives. 12.
Montano, C., Flores-Arenas, C. & Carpenter, S. 2024. Lncrnas, Nuclear Architecture And The Immune Response. *Nucleus,* 15, 2350182.
Morales-Nebreda, L., Mclafferty, F. S. & Singer, B. D. 2019. Dna Methylation As A Transcriptional Regulator Of The Immune System. *Transl Res,* 204, 1–18.
Morandini, A. C., Santos, C. F. & Yilmaz, Ö. 2016. Role Of Epigenetics In Modulation Of Immune Response At The Junction Of Host-Pathogen Interaction And Danger Molecule Signaling. *Pathog Dis,* 74.
Morikawa, H., Ohkura, N., Vandenbon, A., Itoh, M., Nagao-Sato, S., Kawaji, H., Lassmann, T., Carninci, P., Hayashizaki, Y., Forrest, A. R. R., Standley, D. M., Date, H., Sakaguchi, S., Consortium, T. F., Forrest, A. R. R., Kawaji, H., Rehli, M., Baillie, J. K., De Hoon, M. J. L., Haberle, V., Lassmann, T., Kulakovskiy, I. V., Lizio, M., Itoh, M., Andersson, R., Mungall, C. J., Meehan, T. F., Schmeier, S., Bertin, N., Jørgensen, M., Dimont, E., Arner, E., Schmidl, C., Schaefer, U., Medvedeva, Y. A., Plessy, C., Vitezic, M., Severin, J., Semple, C. A., Ishizu, Y., Francescatto, M., Alam, I., Albanese, D., Altschuler, G. M., Archer, J. A. C., Arner, P., Babina, M., Baker, S., Balwierz, P. J., Beckhouse, A. G., Pradhan-Bhatt, S., Blake, J. A., Blumenthal, A., Bodega, B., Bonetti, A., Briggs, J., Brombacher, F., Burroughs, A. M., Califano, A., Cannistraci, C. V., Carbajo, D., Chen, Y., Chierici, M., Ciani, Y., Clevers, H. C., Dalla, E., Davis, C. A., Deplancke, B., Detmar, M., Diehl, A. D., Dohi, T., Drabløs, F., Edge, A. S. B., Edinger, M., Ekwall, K., Endoh, M., Enomoto, H., Fagiolini, M., Fairbairn, L., Fang, H., Farach-Carson, M. C., Faulkner, G. J., Favorov, A. V., Fisher, M. E., Frith, M. C., Fujita, R., Fukuda, S., Furlanello, C., Furuno, M., Furusawa, J.-I., Geijtenbeek, T. B., Gibson, A., Gingeras, T., Goldowitz, D., Gough, J., Guhl, S., Guler, R., Gustincich, S., Ha, T. J., Hamaguchi, M., Et Al. 2014. Differential Roles Of Epigenetic Changes And Foxp3 Expression In Regulatory T Cell-Specific Transcriptional Regulation. *Proceedings Of The National Academy Of Sciences,* 111, 5289–5294.
Moulton, V. R., Suarez-Fueyo, A., Meidan, E., Li, H., Mizui, M. & Tsokos, G. C. 2017. Pathogenesis Of Human Systemic Lupus Erythematosus: A Cellular Perspective. *Trends Mol Med,* 23, 615–635.
Mugisha, E. & University Iii, K. I. 2024. Epigenetic Regulation Of Immune Cells In Health And Disease: A Comprehensive Review. *Research Invention Journal Of Biological And Applied Sciences,* 4, 1–6.
Mulroney, T. E., Pöyry, T. & Yam-Puc, J. C. 2024. N(1)-Methylpseudouridylation Of Mrna Causes +1 Ribosomal Frameshifting. 625, 189–194.
Murugaiyan, G., Da Cunha, A. P., Ajay, A. K., Joller, N., Garo, L. P., Kumaradevan, S., Yosef, N., Vaidya, V. S. & Weiner, H. L. 2015. Microrna-21 Promotes Th17 Differentiation And Mediates Experimental Autoimmune Encephalomyelitis. *J Clin Invest,* 125, 1069–80.

Mutchler, A. L., Haynes, A. P., Saleem, M., Jamison, S., Khan, M. M., Ertuglu, L. & Kirabo, A. 2025. Epigenetic Regulation Of Innate And Adaptive Immune Cells In Salt-Sensitive Hypertension. *Circulation Research,* 136, 232–254.

Nail, H. M., Chiu, C.-C., Leung, C.-H., Ahmed, M. M. M. & Wang, H.-M. D. 2023. Exosomal Mirna-Mediated Intercellular Communications And Immunomodulatory Effects In Tumor Microenvironments. *Journal Of Biomedical Science,* 30, 69.

Nemtsova, M. V., Zaletaev, D. V., Bure, I. V., Mikhaylenko, D. S., Kuznetsova, E. B., Alekseeva, E. A., Beloukhova, M. I., Deviatkin, A. A., Lukashev, A. N. & Zamyatnin, A. A., Jr. 2019. Epigenetic Changes In The Pathogenesis Of Rheumatoid Arthritis. *Front Genet,* 10, 570.

Netea, M. G., Domínguez-Andrés, J., Barreiro, L. B., Chavakis, T., Divangahi, M., Fuchs, E., Joosten, L. A. B., Van Der Meer, J. W. M., Mhlanga, M. M., Mulder, W. J. M., Riksen, N. P., Schlitzer, A., Schultze, J. L., Stabell Benn, C., Sun, J. C., Xavier, R. J. & Latz, E. 2020. Defining Trained Immunity And Its Role In Health And Disease. *Nat Rev Immunol,* 20, 375–388.

Netea, M. G., Schlitzer, A., Placek, K., Joosten, L. A. B. & Schultze, J. L. 2019. Innate And Adaptive Immune Memory: An Evolutionary Continuum In The Host's Response To Pathogens. *Cell Host & Microbe,* 25, 13–26.

O'Brien, J., Hayder, H., Zayed, Y. & Peng, C. 2018. Overview Of Microrna Biogenesis, Mechanisms Of Actions, And Circulation. *Front Endocrinol (Lausanne),* 9, 402.

Olejnik-Wojciechowska, J., Boboryko, D., Bratborska, A. W., Rusińska, K., Ostrowski, P., Baranowska, M. & Pawlik, A. 2024. The Role Of Epigenetic Factors In The Pathogenesis Of Psoriasis. *International Journal Of Molecular Sciences* [Online], 25.

Oudinet, C., Braikia, F. Z., Dauba, A., Santos, J. M. & Khamlichi, A. A. 2019. Developmental Regulation Of Dna Cytosine Methylation At The Immunoglobulin Heavy Chain Constant Locus. 15, E1007930.

Pajares, M. J., Alemany-Cosme, E., Goñi, S., Bandres, E., Palanca-Ballester, C. & Sandoval, J. 2021. Epigenetic Regulation Of Micrornas In Cancer: Shortening The Distance From Bench To Bedside. *International Journal Of Molecular Sciences,* 22, 7350.

Pan, W. W., Lin, F. & Fort, P. E. 2021. The Innate Immune System In Diabetic Retinopathy. *Prog Retin Eye Res,* 84, 100940.

Panda, R., Mohan, S. & Vellapandian, C. 2024. Harnessing Epigenetic Mechanisms To Overcome Immune Evasion In Cancer: The Current Strategies And Future Directions. *Cureus,* 16, E70631.

Parajuli, N., Subedi, K., Solone, X. K., Jiang, A., Zhou, L. & Mi, Q.-S. 2025. Epigenetic Control Of Alveolar Macrophages: Impact On Lung Health And Disease. *Cells,* 14, 640.

Park, H., Shin, J., Kim, Y., Saito, T., Saido, T. C. & Kim, J. 2022. Crispr/Dcas9-Dnmt3a-Mediated Targeted Dna Methylation Of App Rescues Brain Pathology In A Mouse Model Of Alzheimer's Disease. 11, 41.

Paul, D. S., Teschendorff, A. E., Dang, M. A., Lowe, R., Hawa, M. I., Ecker, S., Beyan, H., Cunningham, S., Fouts, A. R., Ramelius, A., Burden, F., Farrow, S., Rowlston, S., Rehnstrom, K. & Frontini, M. 2016. Increased Dna Methylation Variability In Type 1 Diabetes Across Three Immune Effector Cell Types. 7, 13555.

Payandeh, Z., Tangruksa, B., Synnergren, J., Heydarkhan-Hagvall, S., Nordin, J. Z., Andaloussi, S. E. L., Borén, J., Wiseman, J., Bohlooly-Y, M., Lindfors, L. & Valadi, H. 2024. Extracellular Vesicles Transport Rna Between Cells: Unraveling Their Dual Role In Diagnostics And Therapeutics. *Molecular Aspects Of Medicine,* 99, 101302.

Peeters, J. G. C., Silveria, S., Ozdemir, M., Ramachandran, S. & Dupage, M. 2024. Hyperactivating Ezh2 To Augment H3k27me3 Levels In Regulatory T Cells Enhances Immune Suppression By Driving Early Effector Differentiation. *Cell Rep,* 43, 114724.

Pisetsky, D. S. 2023. Pathogenesis Of Autoimmune Disease. *Nat Rev Nephrol,* 19, 509–524.

Placek, K., Schultze, J. L. & Aschenbrenner, A. C. 2019. Epigenetic Reprogramming Of Immune Cells In Injury, Repair, And Resolution. *J Clin Invest,* 129, 2994–3005.

Poe, A. J., Shah, R., Khare, D., Kulkarni, M., Phan, H., Ghiam, S., Punj, V., Ljubimov, A. V. & Saghizadeh, M. 2022. Regulatory Role Of Mir-146a In Corneal Epithelial Wound Healing Via Its Inflammatory Targets In Human Diabetic Cornea. *The Ocular Surface,* 25, 92–100.

Qin, W., Scicluna, B. P. & Van Der Poll, T. 2021a. The Role Of Host Cell Dna Methylation In The Immune Response To Bacterial Infection. *Frontiers In Immunology,* Volume 12 - 2021.

Qin, W., Scicluna, B. P. & Van Der Poll, T. 2021b. The Role Of Host Cell Dna Methylation In The Immune Response To Bacterial Infection. *Front Immunol,* 12, 696280.

Qiu, L., Jing, Q., Li, Y. & Han, J. 2023. Rna Modification: Mechanisms And Therapeutic Targets. 4, 25.

Radwan, S. M., Hamdy, N. M., Hegab, H. M. & El-Mesallamy, H. O. 2016. Beclin-1 And Hypoxia-Inducible Factor-1α Genes Expression: Potential Biomarkers In Acute Leukemia Patients. *Cancer Biomark,* 16, 619–26.

Rahman, M. M. & Tollefsbol, T. O. 2021. Targeting Cancer Epigenetics With Crispr-Dcas9: Principles And Prospects. *Methods,* 187, 77–91.

Raj, K. & Mufti, G. J. 2006. Azacytidine (Vidaza(R)) In The Treatment Of Myelodysplastic Syndromes. *Ther Clin Risk Manag,* 2, 377–88.

Ramachandran, A. 2025. Epigenetic Frontiers: Integrating Advanced Ai With Emerging Discoveries In Human Health, Disease, And Environmental Adaptation.

Rasmi, Y., Shokati, A., Hassan, A., Aziz, S. G., Bastani, S., Jalali, L., Moradi, F. & Alipour, S. 2023. The Role Of Dna Methylation In Progression Of Neurological Disorders And Neurodegenerative Diseases As Well As The Prospect Of Using Dna Methylation Inhibitors As Therapeutic Agents For Such Disorders. *Ibro Neurosci Rep,* 14, 28–37.

Ratti, M., Lampis, A., Ghidini, M., Salati, M., Mirchev, M. B., Valeri, N. & Hahne, J. C. 2020. Micrornas (Mirnas) And Long Non-Coding Rnas (Lncrnas) As New Tools For Cancer Therapy: First Steps From Bench To Bedside. 15, 261–278.

Renaude, E., Kroemer, M., Loyon, R., Binda, D., Borg, C., Guittaut, M., Hervouet, E. & Peixoto, P. 2020. The Fate Of Th17 Cells Is Shaped By Epigenetic Modifications And Remodeled By The Tumor Microenvironment. *Int J Mol Sci,* 21.

Riveiro, M. E., Astorgues-Xerri, L., Vazquez, R., Frapolli, R., Kwee, I., Rinaldi, A., Odore, E., Rezai, K., Bekradda, M., Inghirami, G., D'incalci, M., Noel, K., Cvitkovic, E., Raymond, E. & Bertoni, F. 2016. Otx015 (Mk-8628), A Novel Bet Inhibitor, Exhibits Antitumor Activity In Non-Small Cell And Small Cell Lung Cancer Models Harboring Different Oncogenic Mutations. *Oncotarget,* 7, 84675–84687.

Rizk, N. I., Kassem, D. H., Abulsoud, A. I., Abdelhalim, S., Yasser, M. B., Kamal, M. M. & Hamdy, N. M. 2024. Revealing The Role Of Serum Exosomal Novel Long Non-Coding Rna Nampt-As As A Promising Diagnostic/Prognostic Biomarker In Colorectal Cancer Patients. *Life Sci,* 352, 122850.

Rocha, G. I. Y., Gomes, J. E. M. & Leite, M. L. 2023. Epigenome-Driven Strategies For Personalized Cancer Immunotherapy. 15, 1351–1367.

Rosemblatt, M., Doñas, C. & Loyola, A. 2019. Exploring Epigenetic Drugs In The Regulation Of Inflammatory Autoimmune Diseases. *In:* Fernandes Nunes, A. C. (Ed.) *Translational Studies On Inflammation.* Rijeka: Intechopen.

Roy, R., Ramamoorthy, S., Shapiro, B. D., Kaileh, M., Hernandez, D., Sarantopoulou, D., Arepalli, S., Boller, S., Singh, A., Bektas, A., Kim, J., Moore, A. Z., Tanaka, T., Mckelvey, J., Zukley, L., Nguyen, C., Wallace, T., Dunn, C., Wersto, R., Wood, W., Piao, Y., Becker, K. G., Coletta, C., De, S., Sen, J. M., Battle, A., Weng, N. P., Grosschedl, R., Ferrucci, L. & Sen, R. 2021. Dna Methylation Signatures Reveal That Distinct Combinations Of Transcription Factors Specify Human Immune Cell Epigenetic Identity. *Immunity,* 54, 2465–2480.E5.

Ruth, S. & University Ix, K. I. 2024. Epigenetic Regulation Of Immune Cells In Health And Disease: A Comprehensive Review. *Idosr Journal Of Science And Technology,* 10, 1–7.

Saadh, M. J., Allela, O. Q. B., Al-Hussainy, A. F., Baldaniya, L., Rekha, M. M., Nathiya, D., Kaur, P., Aminov, Z., Sameer, H. N., Hameed, H. G., Athab, Z. H. & Adil, M. 2025. Exosomal Non-Coding Rnas: Gatekeepers Of Inflammation In Autoimmune Disease. *J Inflamm (Lond),* 22, 18.

Sadeghi, M., Tavakol Afshari, J., Fadaee, A., Dashti, M., Kheradmand, F., Dehnavi, S. & Mohammadi, M. 2025. Exosomal Mirnas Involvement In Pathogenesis, Diagnosis, And Treatment Of Rheumatoid Arthritis. *Heliyon,* 11, E41983.

Sarkar, S. K., Willson, A. M. L. & Jordan, M. A. 2024. The Plasticity Of Immune Cell Response Complicates Dissecting The Underlying Pathology Of Multiple Sclerosis. *Journal Of Immunology Research,* 2024, 5383099.

Sato, A. 2012. Vorinostat Approved In Japan For Treatment Of Cutaneous T-Cell Lymphomas: Status And Prospects. *Onco Targets Ther,* 5, 67–76.

Sawada, Y. & Gallo, R. L. 2021. Role Of Epigenetics In The Regulation Of Immune Functions Of The Skin. *J Invest Dermatol,* 141, 1157–1166.

Schwarzenbach, H. & Gahan, P. B. 2021. Exosomes In Immune Regulation. *Noncoding Rna,* 7.

Sharma, U., Barwal, T. S., Malhotra, A., Pant, N., Vivek, Dey, Gautam, A., Tuli, H. S., Vasquez, K. M. & Jain, A. 2020. Long Non-Coding Rna Tincr As Potential Biomarker And Therapeutic Target For Cancer. *Life Sci,* 257, 118035.

Shi, Y., Zhu, Z., Chen, Q. & Chen, X. 2024. Dna Methylation Regulates B Cell Activation Via Repressing Pax5 Expression In Teleost. *Frontiers In Immunology,* Volume 15 - 2024.

Soto, J. A., Gálvez, N. M. S., Andrade, C. A., Pacheco, G. A., Bohmwald, K., Berrios, R. V., Bueno, S. M. & Kalergis, A. M. 2020. The Role Of Dendritic Cells During Infections Caused By Highly Prevalent Viruses. *Front Immunol,* 11, 1513.

Sprenkle, N. T. & Serezani, C. H. 2023. Micrornas In Macrophages: Regulators Of Activation And Function. 210, 359–368.

Statello, L., Guo, C. J. & Chen, L. L. 2021. Gene Regulation By Long Non-Coding Rnas And Its Biological Functions. 22, 96–118.

Stergioti, E. M., Manolakou, T., Boumpas, D. T. & Banos, A. 2022. Antiviral Innate Immune Responses In Autoimmunity: Receptors, Pathways, And Therapeutic Targeting. 10.

Stojanovic, A., Correia, M. & Cerwenka, A. 2018. The Nkg2d/Nkg2dl Axis In The Crosstalk Between Lymphoid And Myeloid Cells In Health And Disease. *Frontiers In Immunology,* 9, 827.

Su, Y., Lian, J., Chen, S., Zhang, W. & Deng, C. 2022. Epigenetic Histone Acetylation Modulating Prenatal Poly I:C Induced Neuroinflammation In The Prefrontal Cortex Of Rats: A Study In A Maternal Immune Activation Model. *Front Cell Neurosci,* 16, 1037105.

Sufianov, A., Agaverdiev, M., Mashkin, A. & Ilyasova, T. 2025. The Functions Of Immune System-Derived Mirnas In Cardiovascular Diseases. *Non-Coding Rna Research,* 11, 91–103.

Sun, S., Aguirre-Gamboa, R. & Barreiro, L. B. 2023. Transmission Of Stimulus-Induced Epigenetic Changes Through Cell Division Is Coupled To Continuous Transcription Factor Activity. *Front Immunol,* 14, 1129577.

Sundaresan, B. & Shirafkan, F. 2023. The Role Of Viral Infections In The Onset Of Autoimmune Diseases. 15.

Surace, A. E. A. & Hedrich, C. M. 2019. The Role Of Epigenetics In Autoimmune/Inflammatory Disease. *Front Immunol,* 10, 1525.

Tadić, V., Josipović, G., Zoldoš, V. & Vojta, A. 2019. Crispr/Cas9-Based Epigenome Editing: An Overview Of Dcas9-Based Tools With Special Emphasis On Off-Target Activity. *Methods,* 164–165, 109–119.

Tanekhy, M. 2016. The Role Of Toll-Like Receptors In Innate Immunity And Infectious Diseases Of Teleost. *Aquaculture Research,* 47, 1369–1391.

Tang, F., Zhou, Z., Huang, K., Deng, W., Lin, J., Chen, R., Li, M. & Xu, F. 2022. Micrornas In The Regulation Of Th17/Treg Homeostasis And Their Potential Role In Uveitis. *Frontiers In Genetics,* Volume 13 - 2022.

Tien, F.-M., Lu, H.-H., Lin, S.-Y. & Tsai, H.-C. 2023. Epigenetic Remodeling Of The Immune Landscape In Cancer: Therapeutic Hurdles And Opportunities. *Journal Of Biomedical Science,* 30, 3.

Tripathi, S. K. & Lahesmaa, R. 2014. Transcriptional And Epigenetic Regulation Of T-Helper Lineage Specification. *Immunological Reviews,* 261, 62–83.

Tuong, Z. K., Stewart, B. J., Guo, S. A. & Clatworthy, M. R. 2022. Epigenetics And Tissue Immunity—Translating Environmental Cues Into Functional Adaptations. *Immunological Reviews,* 305, 111–136.

Ur Rehman, S., Abdullah, M. & Shurovi, M. 2025. The Role Of Dna Methylation And Histone Modifications In The Pathogenesis Of Hematological Malignancies And Solid Cancers: Mechanisms, Clinical Implications, And Therapeutic Potential. *Asian Journal Of Medical And Biological Research,* 11, 23–36.

Vaidyanathan, B. & Chaudhuri, J. 2015. Epigenetic Codes Programing Class Switch Recombination. *Frontiers In Immunology,* Volume 6 - 2015.

Velikova, T., Kaouri, I. E., Bakopoulou, K., Gulinac, M., Naydenova, K., Dimitrov, M., Peruhova, M. & Lazova, S. 2024. Mucosal Immunity And Trained Innate Immunity Of The Gut. *Gastroenterology Insights,* 15, 661–675.

Verbeke, R., Hogan, M. J., Loré, K. & Pardi, N. 2022. Innate Immune Mechanisms Of Mrna Vaccines. *Immunity,* 55, 1993–2005.

Villanueva, L., Álvarez-Errico, D. & Esteller, M. 2020. The Contribution Of Epigenetics To Cancer Immunotherapy. *Trends In Immunology,* 41, 676–691.

Wan, Y. Y. 2014. Gata3: A Master Of Many Trades In Immune Regulation. *Trends Immunol,* 35, 233–42.

Wang, R., Lan, C., Benlagha, K., Camara, N. O. S., Miller, H., Kubo, M., Heegaard, S., Lee, P., Yang, L., Forsman, H., Li, X., Zhai, Z. & Liu, C. 2024a. The Interaction Of Innate Immune And Adaptive Immune System. *Medcomm,* 5, E714.

Wang, X., Gao, H., Pu, W., Zeng, Z., Xu, N., Luo, X., Tang, D. & Dai, Y. 2024b. Dysregulation Of Pseudouridylation In Small Rnas Contributes To Papillary Thyroid Carcinoma Metastasis. *Cancer Cell International,* 24, 337.

Wang, X. & Yu, G. 2025. Advancing Veterinary Vaccines Design Through Trained Immunity Insights. *Frontiers In Veterinary Science,* Volume 11 - 2024.

Wang, Y., Huang, Z., Li, B., Liu, L. & Huang, C. 2022. The Emerging Roles And Therapeutic Implications Of Epigenetic Modifications In Ovarian Cancer. *Front Endocrinol (Lausanne),* 13, 863541.

Watson, N., Kuppuswamy, S., Ledford, W. L. & Sukumari-Ramesh, S. 2024. The Role Of Hdac3 In Inflammation: Mechanisms And Therapeutic Implications. *Frontiers In Immunology,* Volume 15 - 2024.

Wimmers, F., Donato, M., Kuo, A., Ashuach, T., Gupta, S., Li, C., Dvorak, M., Foecke, M. H., Chang, S. E., Hagan, T., De Jong, S. E., Maecker, H. T., Van Der Most, R., Cheung, P., Cortese, M., Bosinger, S. E., Davis, M., Rouphael, N., Subramaniam, S., Yosef, N., Utz, P. J., Khatri, P. & Pulendran, B. 2021. The Single-Cell Epigenomic And Transcriptional Landscape Of Immunity To Influenza Vaccination. *Cell,* 184, 3915–3935.E21.

Wu, B., Zhang, B., Li, B., Wu, H. & Jiang, M. 2024a. Cold And Hot Tumors: From Molecular Mechanisms To Targeted Therapy. *Signal Transduction And Targeted Therapy,* 9, 274.

Wu, D., Shi, Y., Zhang, H. & Miao, C. 2023. Epigenetic Mechanisms Of Immune Remodeling In Sepsis: Targeting Histone Modification. *Cell Death & Disease,* 14, 112.

Wu, F.-Y. & Yin, R.-X. 2022. Recent Progress In Epigenetics Of Obesity. *Diabetology & Metabolic Syndrome,* 14, 171.

Wu, H., Liao, J., Li, Q., Yang, M., Zhao, M. & Lu, Q. 2018. Epigenetics As Biomarkers In Autoimmune Diseases. *Clin Immunol,* 196, 34–39.

Wu, M., Guan, G., Yin, H. & Niu, Q. 2024b. A Review Of The Bromodomain And Extraterminal Domain Epigenetic Reader Proteins: Function On Virus Infection And Cancer. *Viruses,* 16.

Xiao, C., Nemazee, D. & Gonzalez-Martin, A. 2020. Microrna Control Of B Cell Tolerance, Autoimmunity And Cancer. *Semin Cancer Biol,* 64, 102–107.

Xiao, Y., Yuan, Y., Hu, D. & Wang, H. 2024. Exosome-Derived Microrna: Potential Target For Diagnosis And Treatment Of Sepsis. *Journal Of Immunology Research,* 2024, 4481452.

Xie, L., Fang, J., Yu, J., Zhang, W., He, Z., Ye, L. & Wang, H. 2023. The Role Of Cd4(+) T Cells In Tumor And Chronic Viral Immune Responses. *Medcomm (2020),* 4, E390.

Xie, X., Huang, M., Ma, S., Xin, Q., Wang, Y., Hu, L., Zhao, H., Li, P., Liu, M., Yuan, R., Miao, Y., Zhu, Y. & Cong, W. 2025. The Role Of Long Non-Coding Rnas In Cardiovascular Diseases: A Comprehensive Review. *Non-Coding Rna Research,* 11, 158–187.

Xu, Y., Zhang, H. & Nie, D. 2025. Histone Modifications And Metabolic Reprogramming In Tumor-Associated Macrophages: A Potential Target Of Tumor Immunotherapy. *Front Immunol,* 16, 1521550.

Xu, Z., Chen, Y., Ma, L., Chen, Y., Liu, J., Guo, Y., Yu, T., Zhang, L., Zhu, L. & Shu, Y. 2022. Role Of Exosomal Non-Coding Rnas From Tumor Cells And Tumor-Associated Macrophages In The Tumor Microenvironment. *Mol Ther,* 30, 3133–3154.

Yang, C., Li, D., Teng, D., Zhou, Y., Zhang, L., Zhong, Z. & Yang, G. J. 2022a. Epigenetic Regulation In The Pathogenesis Of Rheumatoid Arthritis. *Front Immunol,* 13, 859400.

Yang, J., Li, Z., Wang, L., Yun, X., Zeng, Y., Ng, J. P. L., Lo, H., Wang, Y., Zhang, K., Law, B. Y. K. & Wong, V. K. W. 2022b. The Role Of Non-Coding Rnas (Mirna And Lncrna) In The Clinical Management Of Rheumatoid Arthritis. *Pharmacological Research,* 186, 106549.

Yang, J., Xu, J., Wang, W., Zhang, B., Yu, X. & Shi, S. 2023. Epigenetic Regulation In The Tumor Microenvironment: Molecular Mechanisms And Therapeutic Targets. *Signal Transduction And Targeted Therapy,* 8, 210.

Yang, Y., Li, J., Geng, Y., Liu, L. & Li, D. 2021. Azacitidine Regulates Dna Methylation Of Gadd45γ In Myelodysplastic Syndromes. 35, E23597.

Yang, Y., Zhang, M. & Wang, Y. 2022c. The Roles Of Histone Modifications In Tumorigenesis And Associated Inhibitors In Cancer Therapy. *J Natl Cancer Cent,* 2, 277–290.

Yao, W., Hu, X. & Wang, X. 2024. Crossing Epigenetic Frontiers: The Intersection Of Novel Histone Modifications And Diseases. *Signal Transduction And Targeted Therapy,* 9, 232.

Ye, J., Li, D., Jie, Y., Luo, H., Zhang, W. & Qiu, C. 2024. Exosome-Based Nanoparticles And Cancer Immunotherapy. *Biomedicine & Pharmacotherapy,* 179, 117296.

Yue, Y., Ren, Y., Lu, C., Li, P. & Zhang, G. 2024. Epigenetic Regulation Of Human Foxp3+ Tregs: From Homeostasis Maintenance To Pathogen Defense. *Frontiers In Immunology,* Volume 15 - 2024.

Zan, H. & Casali, P. 2015. Epigenetics Of Peripheral B-Cell Differentiation And The Antibody Response. *Front Immunol,* 6, 631.

Zhang, D., Zhang, M., Zhang, L., Wang, W., Hua, S., Zhou, C. & Sun, X. 2024a. Long Non-Coding Rnas And Immune Cells: Unveiling The Role In Viral Infections. *Biomedicine & Pharmacotherapy,* 170, 115978.

Zhang, H. & Kuchroo, V. 2019. Epigenetic And Transcriptional Mechanisms For The Regulation Of Il-10. *Semin Immunol,* 44, 101324.

Zhang, Y., Maskan Bermudez, N., Sa, B., Maderal, A. D. & Jimenez, J. J. 2024b. Epigenetic Mechanisms Driving The Pathogenesis Of Systemic Lupus Erythematosus, Systemic Sclerosis And Dermatomyositis. *Experimental Dermatology,* 33, E14986.

Zhao, A., Zhou, H., Yang, J., Li, M. & Niu, T. 2023. Epigenetic Regulation In Hematopoiesis And Its Implications In The Targeted Therapy Of Hematologic Malignancies. *Signal Transduction And Targeted Therapy,* 8, 71.

Żychowska, J. & Ćmil, M. 2024. The Role Of Epigenetic Mechanisms In The Pathogenesis Of Hepatitis C Infection. 14.

Chapter 2
The Intricate Interplay of Noncoding RNAs and the Gut Microbiome in Gastrointestinal and Endocrine-Related Cancers

Sherien M. El-Daly, Bahgat Fayed, Roba M. Talaat, Shaimaa A. Gouhar, Cinderella A. Fahmy, Amira Mohamed Abd El-Jawad, Nadia M. Hamdy, and Zakaria Y. Abd Elmageed

Abstract The human gut microbiome and noncoding RNAs (ncRNAs) represent interconnected regulatory networks that profoundly influence cancer development, particularly in gastrointestinal and endocrine-related malignancies. This chapter

S. M. El-Daly (✉)
Medical Biochemistry Department, Medical Research and Clinical Studies Institute, National Research Centre, Giza, Egypt

Cancer Biology and Genetics Laboratory, Centre of Excellence for Advanced Sciences, National Research Centre, Giza, Egypt
e-mail: sm.el-daly@nrc.sci.eg

B. Fayed
Department of Chemistry of Natural and Microbial Products, Pharmaceutical and Drug Industries Research Institute, National Research Centre, Giza, Egypt

R. M. Talaat
Molecular Biology Department, Genetic Engineering and Biotechnology Research Institute (GEBRI), University of Sadat City, El Sadat City, Egypt

S. A. Gouhar · A. M. A. El-Jawad
Medical Biochemistry Department, Medical Research and Clinical Studies Institute, National Research Centre, Giza, Egypt

C. A. Fahmy
Cancer Biology and Genetics Laboratory, Centre of Excellence for Advanced Sciences, National Research Centre, Giza, Egypt

Biochemistry Department, Biotechnology Research Institute, National Research Centre, Giza, Egypt

N. M. Hamdy
Biochemistry and Molecular Biology Department, Faculty of Pharmacy, Ain Shams University, Cairo, Egypt
e-mail: Nadia_hamdy@pharma.asu.edu.eg

Z. Y. Abd Elmageed
Department of Biomedical Science, Discipline of Pharmacology, Edward Via College of Osteopathic Medicine, Monroe, LA, USA

N. M. Hamdy et al. (eds.), *From (Epi)genetics-to-Precision (E2P)*, Subcellular Biochemistry 114, https://doi.org/10.1007/978-3-032-08530-6_2

delineates the intricate interplay of microbiome-ncRNA crosstalk in the context of gastrointestinal and endocrine-related cancers.

The chapter begins with a comprehensive overview of the taxonomic and functional landscape of the healthy adult gut microbiome. The gut microbiome, comprising trillions of microorganisms, plays a crucial role in endocrine regulation through hormone metabolism, synthesis of bioactive compounds, and modulation of immune responses, thereby establishing a critical crosstalk with the host endocrine system. Dysbiosis, or microbial imbalance, has been linked to endocrine dysfunction and the pathogenesis of various diseases, including gastrointestinal and endocrine-related cancers.

We then elucidate the classifications of noncoding RNAs and their function as key molecular regulators in cellular communication, gene expression, and disease progression. NcRNAs contribute significantly to the development and progression of endocrine-related malignancies. The intricate crosstalk between the gut microbiome and host ncRNAs demonstrates how gut dysbiosis can disrupt host ncRNA expression patterns, thereby affecting oncogenic pathways, immune surveillance, and metabolic reprogramming linked to tumor initiation, progression, and metastasis. Conversely, host-derived ncRNAs, secreted into the gut lumen, can directly shape microbial gene expression. In this section, we explore how dysregulation of this axis contributes to carcinogenesis through the promotion of chronic inflammation, epithelial barrier dysfunction, and oncogenic signaling. Therapeutic strategies targeting this interplay, including probiotics, prebiotics, fecal microbiota transplantation, and dietary interventions, are introduced in the context of restoring microbial balance.

This comprehensive chapter provides crucial insights into the molecular mechanisms governing microbiome-ncRNA interactions and their implications for cancer biology, offering new perspectives for therapeutic interventions in gastrointestinal and endocrine-related malignancies.

Graphical Abstract

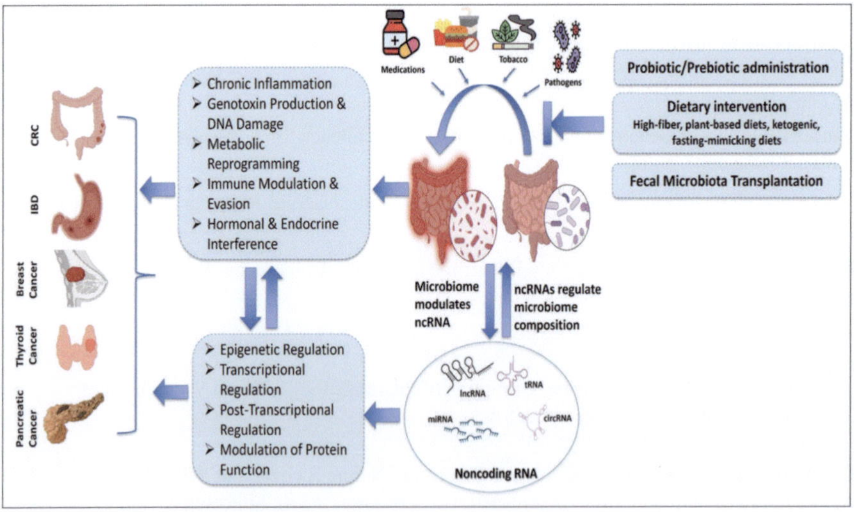

Keywords Microbiome · Dysbiosis · Noncoding RNA · Gastrointestinal cancer · Endocrine cancer · Oncogenesis · Biomarkers · Therapeutics

Abbreviations

ABCC3	ATP-binding cassette transporter
AhR	Aryl hydrocarbon receptor
AIEC	Adherent-invasive *E. coli*
AKT	Protein kinase B
AR	Androgen receptor
ASCVD	Atherosclerotic cardiovascular disease
ASO	Antisense oligonucleotide
BC	Breast cancer
BCL	B-cell lymphoma
BMI	Body mass index
BRI3	Brain-specific angiogenesis inhibitor 3
CA	Carbonic anhydrase
CAD	Coronary artery disease
cAMP	cyclic adenosine monophosphate
CCND	Cyclin D
CDKN1C	Cyclin-dependent kinase inhibitor
ceRNAs	competing endogenous ribonucleic acid
CHEK2	Checkpoint kinase 2
CiRNAs	Intron circular ribonucleic acid
circRNAs	circular ribonucleic acid
c-Myc	Cellular myelocytomatosis oncogene
CNS	Central nervous system
COPD	Chronic obstructive pulmonary disease
CRC	Colorectal cancer
CREB	cAMP response element-binding protein
CRPC	Castration-resistant prostate cancer
DCA	Deoxycholic acid
DNA	Deoxyribonucleic acid
DSS	Dextran sulfate sodium
EciRNAs	Exon circular ribonucleic acid
EIciRNAs	Exon-intron circular ribonucleic acid
EMT	Epithelial-mesenchymal transition
ER	Estrogen receptor
ERK	Extracellular signal-regulated kinases
ESR	Estrogen receptor
ETBF	Enterotoxigenic Bacteroides fragilis
EVs	Electric vehicles
FASN	Fatty acid synthase

FMT	Fecal microbiota transplantation
FOXP3	Forkhead box P3
FXR	Farnesoid X receptor
GAB2	Grb2-associated binding protein 2
GABA	Gamma-aminobutyric acid
GALT	Gut-associated lymphoid tissue
GF	Germ-free
GIT	Gastrointestinal tract
GLP	Glucagon-like peptide
GPCRs	G protein-coupled receptors
GPR	G-protein-coupled receptors
GSK3β	Glycogen synthase kinase 3 beta
HDACs	Histone deacetylases
HER2	Human epidermal growth factor receptor 2
HMGA2	High mobility group AT-hook 2
HNRNP A2B1	Heterogeneous nuclear ribonucleoprotein A2/B1
HPA	Hypothalamic-pituitary-adrenal
IBD	Inflammatory bowel disease
IESC	Intestinal epithelial stem cells
IL-6	Interleukin-6
ILC3	Innate lymphoid cells
IM	Intestinal metaplasia
IPA	Indole-3-propionic acid
JAK	Janus kinase
KSRP	KH-type splicing regulatory protein
LDHA	Lactate dehydrogenase
LITAF	Lipopolysaccharide-induced tumor necrosis factor
lncRNAs	Long noncoding ribonucleic acid
MAPK	Mitogen-activated protein kinase
MAVS	Mitochondrial antiviral signaling protein
MEG3	Maternally expressed gene 3
MMP14	Matrix metallopeptidase 14
mRNA	messenger ribonucleic acid
miRNAs	Microribonucleic acids
MYD88	Myeloid differentiation primary response protein 88
NAFLD	Nonalcoholic fatty liver disease
ncRNAs	noncoding ribonucleic acids
NEN	Neuroendocrine neoplasms
NF-κB	Nuclear factor kappa beta
NOD1	Nucleotide-binding oligomerization domain protein 1
OGT	O-GlcNAc transferases
PCDH9	Protocadherin 9
PCOS	Polycystic ovary syndrome
PDE4D	Phosphodiesterase 4

PI3K	Phosphatidylinositol 3-kinase
PIK3CA	Phosphatidylinositol-4,5-bisphosphate 3-kinase catalytic subunit alpha
piRNAs	PIWI-interacting ribonucleic acid
PIWIL4	Piwi like RNA-mediated gene silencing 4
PKA	Protein kinase A
PKM2	Pyruvate Kinase M2
PPARγ	Peroxisome proliferator-activated receptor gamma activity
PPP1CA	Protein phosphatase 1, catalytic subunit, alpha isoform
PR	Progesterone receptor
PTEN	Phosphatase and TENsin
PTMs	Posttranslational modifications
RAF	Raf-1 proto-oncogene, serine/threonine kinase
RBPs	RNA-binding proteins
RELMβ	Resistin-like molecule beta
rRNAs	ribosomal ribonucleic acids
RS	Resistant starch
RUNX1	Runt-related transcription factor 1
SENP1	Sentrin-specific protease 1
SFB	Segmented filamentous bacteria
siRNAs	short interfering ribonucleic acids
SNAI2	Snail family zinc finger 2
Snhg9	short nucleolar RNA host gene 9
snoRNAs	Small nucleolar ribonucleic acids
snRNAs	Small nuclear ribonucleic acids
SOCS3	Suppressor of cytokine signaling 3
SOX4	SRY-box transcription factor 4
SPF	Specific pathogen-free
SRSF5	Serine/Arginine-rich splicing factor 5
STAT	Signal transducer and activator of transcription
TGFBR2	Transforming growth factor beta receptor 2
TGR5	Takeda G-protein receptor 5
Th17	T helper 17
TJP	Tight junction proteins
TLR4	Toll-like receptor 4
TNF-α	Tumor necrosis factor-alpha
TOR	Target of rapamycin
TRIM72	Tripartite motif containing 72
tRNAs	transfer ribonucleic acid
TRPC1	Transient receptor potential
TSLP	Thymic stromal lymphopoietin
UTR	Untranslated region
YAP	Yes-associated protein 1

Introduction

The human gut microbiome, a vast and dynamic environment comprising trillions of microorganisms residing within the gastrointestinal tract, is essential in maintaining systemic health. While its contributions to digestion, immune defense, and metabolic balance are well established, growing research shows that its influence extends far beyond the gut. In fact, this microbial community can affect a range of systemic processes, including those regulated by the endocrine system. An altered equilibrium within the gut microbiota, known as dysbiosis, can contribute to the progression of various pathologies, particularly malignancies influenced by hormonal signaling (Arnone and Cook 2023; Zhang et al. 2019a; Zhong et al. 2022).

Endocrine-related cancers include malignancies that develop from endocrine glands or tissues, as well as tumors that are significantly influenced by hormonal signaling. These cancers can be divided into two main categories: those that originate directly from classical endocrine organs and those that arise elsewhere but are strongly affected by hormones. The first category includes thyroid cancer, adrenal gland malignancies (such as adrenocortical carcinoma and pheochromocytoma), parathyroid and pituitary tumors (the majority of which are benign), and pancreatic neuroendocrine tumors (NETs). The second category includes hormone-responsive cancers, such as breast cancer (driven by estrogen and/or progesterone), prostate cancer (driven by androgens), endometrial cancer (sensitive to estrogen), ovarian cancer (influenced by gonadotropins and sex steroids), and testicular cancer (linked to sex hormones). Endocrine-related cancers, such as those affecting the breast, prostate, thyroid, and pancreas (specifically pancreatic neuroendocrine tumors), rely heavily on hormonal pathways for tumor initiation, growth, metastasis, and response to treatment. Importantly, even malignancies that do not arise in classical endocrine glands are considered endocrine-related when hormonal exposure significantly contributes to their pathogenesis and therapeutic response (Belfiore and Perks 2013).

Recent studies suggest that the gut microbiome may influence these pathways directly through interactions with host cells (Clemente-Suárez et al. 2024). Indirect contributions may also arise from the synthesis of bioactive metabolites. These underlying mechanisms could influence both oncological progression and diminished responsiveness to treatment (Liu et al. 2022).

At the same time, noncoding RNAs (ncRNAs)—including microRNAs (miRNAs), long noncoding RNAs (lncRNAs), and circular RNAs (circRNAs)—have gained prominence as pivotal modulators of gene expression across various regulatory mechanisms, from epigenetic modifications to posttranscriptional control (El-Daly et al. 2025a, b; Gouhar et al. 2022). These molecules are now recognized as critical players in cancer development, capable of shaping tumor behavior, metastatic potential, and therapeutic outcomes (Hamdy et al. 2024a, b, 2025a, b). Interestingly, recent findings have begun to uncover a complex and reciprocal relationship between the gut microbiome and host ncRNA networks. This interplay may represent a previously underappreciated axis in the pathophysiology of endocrine-related cancers and other endocrine disorders.

In this chapter, we present a comprehensive review of current knowledge regarding the gut microbiome's function in endocrine-related cancers, with a particular focus on the functions and regulation of ncRNAs. We illustrate how microbial imbalances can influence ncRNA expression and activity, and conversely, how ncRNAs may modulate the structural characteristics and physiological roles of the gut microbiome. We further explore how the emerging microbiome-ncRNA–cancer axis may present novel opportunities for advancing therapeutic strategies in endocrine-related cancers and disorders.

The Gut Microbiome: A Key Player in Endocrine Regulation of the Host

The gut microbiome, a complex and dynamic community of microorganisms residing in the gastrointestinal tract, has garnered significant attention for its far-reaching impact on human health. Beyond its roles in digestion and immune function, emerging evidence highlights the gut microbiome's critical influence on endocrine regulation (Rastelli et al. 2019). The composition of the gut microbiome includes bacteria, viruses, fungi, and archaea, all of which interact with the host in intricate ways (Ahmed et al. 2022; Matijašić et al. 2020). Gut microorganisms contribute to hormone metabolism, synthesize bioactive compounds, and modulate immune responses, establishing a direct link between gut health and endocrine function (Hussain et al. 2021). Dysbiosis, or microbial imbalance, has been implicated in endocrine-related cancers, such as breast and prostate cancers, underscoring the significance of gut microbiota in disease prevention and health maintenance (Clemente-Suárez et al. 2024). In the next part, we explore the mechanisms through which the gut microbiome interacts with the endocrine system, emphasizing its pivotal role in maintaining metabolic and hormonal homeostasis and the implications for health and disease. Additionally, we discuss emerging therapeutic approaches targeting microbiota-endocrine crosstalk, including probiotics, prebiotics, fecal microbiota transplantation (FMT), dietary interventions, and pharmacological strategies.

Overview of the Gut Microbiome and Its Composition

The gut microbiome is an extraordinarily diverse and dynamic community of microorganisms that reside within the gastrointestinal tract (GIT). It consists of trillions of microbes whose diversity is shaped by various factors such as age, diet, genetics, and environmental exposures. The majority of these microorganisms are bacteria, with over 1000 different species identified. These bacteria can be classified into major phyla, including Firmicutes, which are primarily responsible for breaking down complex carbohydrates (Flint et al. 2012), and Bacteroidetes, which play a role in digesting dietary fibers and polysaccharides (Magne et al. 2020). Proteobacteria are another group, often associated with pathogenic microbes, but also containing some beneficial species (Moon et al. 2018). Additionally, Actinobacteria, such as *Bifidobacterium*, contribute to gut health by supporting beneficial bacterial populations (Hidalgo-Cantabrana et al. 2017).

While bacteria are the predominant microorganisms, fungi also reside in the gut, though in smaller numbers. These fungi, including species like *Candida* and *Aspergillus*, play important roles in immune regulation (Wheeler et al. 2017). Viruses, particularly bacteriophages, interact with bacteria in the gut, affecting their populations and activities (Shuwen and Kefeng 2022). The precise role of viruses in gut health remains an emerging field of study. Archaea, such as *Methanobrevibacter smithii*, are less common than bacteria but are involved in methane production and other metabolic processes (Mafra et al. 2022). Eukaryotic microbes, such as protozoa, parasites, and algae, are also found in the gut in smaller numbers and may influence gut health by contributing to microbial diversity or, in some cases, by promoting infections and diseases. Overall, in healthy patients, Firmicutes were the most dominant, making up 78% of the relative abundance, followed by Actinobacteria at 16.5%. Bacteroidetes were present at a lower level, with a mean relative abundance of 2.2% (Doumatey et al. 2020).

Factors Affecting Gut Microbiome Composition

Diet plays a critical role in shaping the gut microbiome, as the types of food we consume can either support or disrupt the balance of microbial communities. A diet rich in fiber, prebiotics, and fermented foods promotes the growth of beneficial microbes (Fayed et al. 2018). Fiber serves as a food source for many gut bacteria, particularly those that produce short-chain fatty acids (SCFAs) such as butyrate, which have anti-inflammatory effects and support gut health. Prebiotics, found in foods such as garlic, onions, and bananas, help stimulate the growth of beneficial bacteria like *Bifidobacterium* and *Lactobacillus* (Fayed et al. 2021). Fermented foods, such as yogurt, kefir, and sauerkraut, provide live microorganisms that can help enhance the diversity of the microbiome (Ismail et al. 2020). In contrast, a high-fat or high-sugar diet can lead to the overgrowth of harmful microbes while reducing the abundance of beneficial species (Satokari 2020). This imbalance, known as dysbiosis, is linked to various health issues, including obesity, metabolic syndrome, and inflammation (Breton et al. 2022; El-Mesallamy et al. 2013).

The composition of the microbiome also evolves with age (Bradley and Haran 2024). In infancy, the microbiome is heavily influenced by the mode of delivery (vaginal birth versus cesarean section), breast milk, and early environmental exposures. As children grow into adolescence and adulthood, the microbiome stabilizes and becomes more diverse, reflecting dietary habits, lifestyle, and environmental factors. In old age, however, the microbiome may become less diverse and more prone to dysbiosis, which is associated with a decline in immune function and an increased susceptibility to infections, inflammation, and age-related diseases. Studies show that the microbiome of elderly individuals often contains a lower diversity of beneficial microbes and a higher prevalence of potentially pathogenic microbes (Salazar et al. 2017).

The use of antibiotics is another significant factor that can disrupt the gut microbiota (B. Fayed et al. 2023). Antibiotics, while effective at treating infections, can cause temporary or long-lasting changes in the microbiome by killing off not only harmful bacteria but also beneficial microbes (Fayed et al. 2023). This disruption can lead to dysbiosis, which in turn may increase the risk of secondary infections, such as those caused by *Clostridium difficile*, and may contribute to the development of antibiotic resistance (Seekatz et al. 2022). Long-term or repeated antibiotic use can also make the microbiome less stable, reducing its ability to recover after disturbances (Patangia et al. 2022).

Finally, geography and lifestyle factors have a substantial impact on microbiome composition. Geographic location influences the types of foods available and the exposure to different environmental conditions, which can shape microbial communities. For example, people living in rural areas or regions with less access to sanitation may have microbiomes that are more diverse due to greater exposure to nature and environmental microbes. Conversely, individuals living in urban environments with higher sanitation standards may have less diverse microbiomes, potentially due to reduced exposure to diverse microbial sources (Panthee et al. 2022). Lifestyle factors, such as stress, sleep patterns, and physical activity, also affect the microbiome (Sochacka et al. 2024). For instance, regular physical activity supports a more diverse microbiome, while chronic stress can contribute to dysbiosis and impact gut health. These factors, including diet, age, antibiotic use, and geography, all interact to shape the composition of the gut microbiome, ultimately influencing its function and its role in maintaining overall health.

Functional Roles of the Gut Microbiome

The gut microbiome plays a central role in numerous vital functions, contributing significantly to overall health and well-being. One of the most well-established functions of the gut microbiome is its involvement in the breakdown of complex carbohydrates, fibers, and proteins that the human digestive system cannot process on its own (Mora-Flores et al. 2023). Microbes in the gut ferment undigested carbohydrates and fibers, producing SCFAs such as butyrate, acetate, and propionate (Morrison and Preston 2016). These SCFAs have numerous health benefits, including anti-inflammatory effects, supporting the integrity of the gut lining, and serving as an energy source for colonocytes (Shin et al. 2023). Butyrate, in particular, has been shown to enhance gut barrier function, reduce inflammation, and may even play a role in preventing colorectal cancer (Kang et al. 2023). Moreover, gut microbes are involved in metabolism by influencing the absorption of nutrients and the storage of fats, which affects energy balance and body weight regulation (Wolters et al. 2019).

The gut microbiome is intricately linked to the immune system, playing a crucial role in immune modulation. It helps shape both the innate and adaptive immune responses by interacting with gut-associated lymphoid tissue (GALT) (Bemark et al. 2024). A healthy, balanced microbiome promotes immune tolerance by helping the immune system distinguish between harmful pathogens and harmless antigens, such as food and beneficial microbes, thereby preventing inappropriate immune responses. This tolerance is critical for reducing the risk of autoimmune diseases and allergies. Dysbiosis has been associated with various autoimmune conditions, including inflammatory bowel disease (IBD) and rheumatoid arthritis (Salem et al. 2019). Additionally, the microbiome produces metabolites that influence immune cell differentiation and activity, impacting inflammatory responses and the body's defense against pathogens (C. H. Kim 2018). The microbiome can also regulate the production of antimicrobial peptides, cytokines, and other immune-related factors, thus supporting the body's ability to fight off infections and maintain overall immune homeostasis (Gubatan et al. 2021). Beyond immune modulation, the gut microbiome also plays a pivotal role in nutrient synthesis, contributing to essential physiological processes that sustain overall health. These include essential B vitamins (such as B12, folate, and riboflavin), vitamin K, and various amino acids (Yang et al. 2024). The production of these vitamins plays a critical role in various physiological processes, including red blood cell formation, DNA synthesis, and bone health. Furthermore, the gut microbiome assists in the absorption of other important nutrients by enhancing the bioavailability of minerals like calcium, magnesium, and iron (Hadadi et al. 2021). This synthesis of vitamins and other nutrients is vital for maintaining metabolic processes and supporting overall health.

Finally, the gut microbiome serves as a critical defense system against harmful pathogens. It does so by preventing pathogenic microbes from colonizing the gut through a variety of mechanisms. One key method is competitive exclusion, where beneficial microbes outcompete pathogens for available nutrients and space, effectively limiting their growth (Melo-Bolívar et al. 2022). Additionally, the microbiome produces a wide range of antimicrobial substances, including bacteriocins, hydrogen peroxide, and other metabolites that directly inhibit or kill pathogens (Hou et al. 2022). This protective role is particularly important in preventing infections caused by opportunistic pathogens such as *Clostridium difficile* and *Escherichia coli* (*E. coli*). The microbiome also influences the gut's mucosal immunity, stimulating the production of mucins and other protective molecules that act as a barrier to pathogen invasion (Fang et al. 2021). By maintaining a healthy microbiome, the gut can better resist infection, thereby protecting the host from a variety of gastrointestinal and systemic diseases.

Gut Microbiome and Endocrine Crosstalk

The gut microbiome communicates with the endocrine system through multiple pathways, playing a crucial role in maintaining physiological balance. One key mechanism is the production of bioactive metabolites by gut bacteria, such as

SCFAs, which modulate hormone release and receptor activity. Additionally, certain gut microbes are capable of synthesizing neurotransmitters, such as serotonin and dopamine, thereby influencing the hypothalamic-pituitary-adrenal (HPA) axis and stress responses. The gut microbiota also shapes immune responses, indirectly affecting endocrine signaling and inflammation. Furthermore, microbial enzymes can metabolize hormones, altering their bioavailability and overall function, demonstrating a direct interaction between gut microbiota and the endocrine system.

Hormone Metabolism

Gut microorganisms play a pivotal role in hormone metabolism through the production of enzymes that can activate, deactivate, or modify hormones, influencing various physiological processes and health outcomes. One prominent example is the production of β-glucuronidase by gut bacteria, an enzyme that deconjugates estrogen from its inactive, conjugated form, thereby increasing its bioavailability, as illustrated in Fig. 2.1 (Hu et al. 2023). This process, referred to as the estrobolome, plays a critical role in regulating circulating estrogen levels, which has far-reaching implications for reproductive health, bone density, and the risk of developing

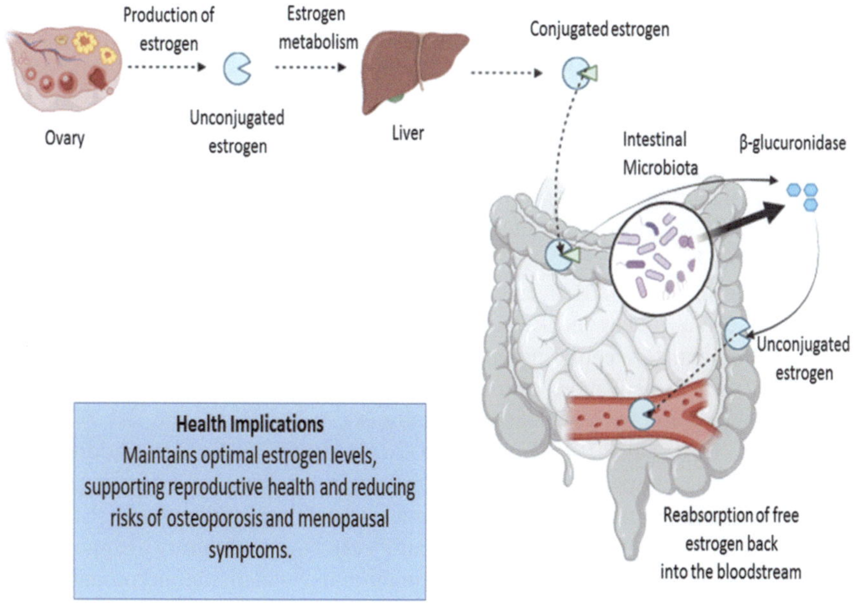

Fig. 2.1 Role of gut microbiota in estrogen metabolism and regulation. Following hepatic conjugation, estrogens pass through the gut and are deconjugated by microbial β-glucuronidase enzymes, allowing for their reabsorption into circulation. This enterohepatic recycling contributes to maintaining estrogen homeostasis, improves reproductive health, and minimizes menopausal complications

hormone-dependent cancers such as breast and endometrial cancers (Kwa et al. 2016).

The estrobolome exemplifies how the gut microbiome can influence systemic hormonal balance by modulating estrogen metabolism and reabsorption in the intestines, thus contributing to overall endocrine homeostasis. Dysbiosis can disrupt this process, potentially leading to conditions of estrogen level disturbances (Rishabh et al. 2024). For instance, a reduced population of β-glucuronidase-producing bacteria may lower estrogen bioavailability, contributing to symptoms of menopause or osteoporosis, while an overactive estrobolome may elevate circulating estrogen levels, increasing susceptibility to estrogen-driven cancers (Hu et al. 2023).

Beyond estrogen metabolism, gut microbes also interact with other hormones, including cortisol, thyroid hormones, and androgens, through similar enzymatic activities. The microbial metabolism of cortisol, for example, can influence the body's stress response and immune function by modulating glucocorticoid activity (Ly et al. 2021). As research advances, a deeper understanding of the microbiome's role in hormone metabolism could pave the way for more personalized and microbiome-targeted treatments, enhancing overall endocrine and metabolic health.

Synthesis of Bioactive Compounds

Gut microbes produce a diverse array of bioactive compounds that play a fundamental role in regulating hormone levels, metabolic processes, and overall endocrine function (Table 2.1). Among these compounds, SCFAs, secondary bile acids, and neurotransmitters are particularly influential in shaping hormonal pathways, underscoring the intricate relationship between the gut microbiome and endocrine health.

SCFAs, including butyrate, propionate, and acetate, are key metabolic byproducts of microbial fermentation of dietary fibers. These molecules exert a significant influence over hormone regulation by interacting with endocrine tissues and receptors (den Besten et al. 2013). For instance, SCFAs stimulate the secretion of gut hormones such as glucagon-like peptide-1 (GLP-1) and peptide YY (PYY), both of which are crucial for regulating appetite, satiety, and insulin sensitivity (Tolhurst et al. 2012). GLP-1 enhances insulin secretion from pancreatic β-cells in response to glucose, thus improving glucose homeostasis and reducing the risk of type 2 diabetes (Carlessi et al. 2017; Mostafa et al. 2016; Mostafa et al. 2015). Additionally, SCFAs modulate adipocyte function, promoting the release of leptin, a hormone that signals energy sufficiency and regulates food intake (Tang and Li 2021). Butyrate, in particular, exerts anti-inflammatory effects by inhibiting histone deacetylases (HDACs), reducing systemic inflammation that can disrupt hormonal signaling (Davie 2003).

Secondary bile acids, generated by gut microbial metabolism of primary bile acids produced by the liver, also play a pivotal role in endocrine regulation (Wahlström et al. 2016). These metabolites activate nuclear receptors such as the farnesoid X receptor (FXR) and the Takeda G-protein receptor 5 (TGR5), which are

Table 2.1 Key gut microbial metabolites and their endocrine effects

Metabolite	Microbial source	Endocrine effects	References
Acetate	Fermentation of fibers	Modulates lipid metabolism, promotes leptin secretion, and controls insulin sensitivity	Hernández et al. (2019)
β-glucuronidase	Gut bacteria (estrobolome)	Deconjugates estrogen, increasing its bioavailability	Hu et al. (2023) and Hu et al. (2023)
Butyrate	Fermentation of dietary fibers	Enhances gut barrier function, releases gut hormones such as GLP-1 and PYY, reduces inflammation, and reduces secretion of the "hunger hormone" ghrelin	Bridgeman et al. (2020)
Dopamine	Gut bacterial synthesis	Modulate stress response, influence cortisol secretion	Hamamah et al. (2022)
Gamma-aminobutyric acid	Gut microbial synthesis	Control insulin metabolism	Braga et al. (2024)
Indole derivatives	Tryptophan metabolism	Activates AhR, promotes Treg differentiation, reduces inflammation, and enters the liver to control glucose and lipid metabolism	Li et al. (2021)
Propionate	Fermentation of fibers	Stimulates GLP-1 and PYY secretion, regulates appetite and insulin levels	Zhang et al. (2022)
Secondary bile acids	Microbial metabolism of primary bile	Activates FXR and TGR5, influencing glucose and lipid metabolism	Chiang and Ferrell (2020)
Serotonin	Microbial influence on enterochromaffin cells	Modulate endocrine organs via the gut-brain axis	Hills et al. (2019)

involved in lipid metabolism, glucose homeostasis, and energy balance. By modulating FXR and TGR5 activity, secondary bile acids influence the synthesis and secretion of hormones like insulin, glucagon, and adiponectin, thus contributing to metabolic health (Kim and Fang 2018). Dysregulation of bile acid metabolism has been implicated in metabolic disorders such as obesity, nonalcoholic fatty liver disease (NAFLD) (El-Mesallamy et al. 2011; Hussein et al. 2024), and insulin resistance (Wang et al. 2024c).

Furthermore, gut microbes synthesize neurotransmitters, including serotonin, dopamine, gamma-aminobutyric acid (GABA), and norepinephrine, which bridge the gut-brain axis and impact endocrine function (Strandwitz 2018). Approximately 95% of the body's serotonin is produced in the gut by enterochromaffin cells, facilitated by microbial metabolites that enhance serotonin biosynthesis (Wei et al. 2022). Serotonin not only influences mood and behavior but also regulates intestinal motility and modulates the hypothalamic-pituitary-adrenal (HPA) axis, a central player in the body's stress response (Rusch et al. 2023). Further, dopamine produced by gut bacteria can similarly affect the HPA axis and contribute to the regulation of cortisol, the primary stress hormone. Through these mechanisms, the microbiome indirectly shapes the body's ability to respond to stress, with implications for mental health and endocrine disorders such as adrenal insufficiency and Cushing's syndrome.

Moreover, microbial-derived bioactive compounds influence reproductive hormones. SCFAs and secondary bile acids can modulate estrogen and androgen metabolism by interacting with nuclear receptors and affecting enzyme activity (Qi et al. 2021). This has significant implications for conditions such as polycystic ovary syndrome (PCOS), endometriosis, and infertility, where hormone imbalances are central to pathophysiology (Insenser et al. 2018). By regulating the bioavailability and activity of sex hormones, gut microbes contribute to menstrual regularity, ovulation, and overall reproductive health.

Immune Responses

The gut microbiota plays a central role in shaping immune responses, acting as a critical mediator between the immune and endocrine systems. This intricate relationship influences the balance between pro- and anti-inflammatory signals, directly impacting the body's ability to maintain hormonal equilibrium and prevent endocrine-related autoimmune and inflammatory conditions. By producing a wide range of metabolites, gut microbes communicate with immune cells, orchestrating the fine-tuning of immune reactions that are essential for endocrine health. SCFAs, such as butyrate, propionate, and acetate, exemplify the immunomodulatory power of the microbiome (Zhang et al. 2023). These metabolites bind to G-protein-coupled receptors on immune cells (GPR41, GPR43, and GPR109A), leading to the suppression of pro-inflammatory cytokines such as interleukin-6 (IL-6) and tumor necrosis factor-alpha (TNF-α) (Mohan et al. 2017) and induce ERK1/2 and p38 MAPK phosphorylations (Liu et al. 2023). At the same time, SCFAs promote the production of anti-inflammatory cytokines by inhibiting the NF-κB (Liu et al. 2023). Further, gut microbiota metabolites play a crucial role in the differentiation and expansion of regulatory T cells (Tregs) (He et al. 2024). Tregs play a pivotal role in preventing autoimmunity by suppressing excessive immune responses and maintaining immune tolerance. This process is particularly critical for protecting endocrine organs, such as the thyroid, pancreas, and adrenal glands, from autoimmune attacks, thereby reducing the risk of disorders such as Hashimoto's thyroiditis (Guo et al. 2024).

Beyond SCFAs, other microbial metabolites, such as indole derivatives from tryptophan metabolism, exert similar anti-inflammatory effects by activating the aryl hydrocarbon receptor (AhR) (Dong and Perdew 2020). Activation of AhR in immune cells enhances $Foxp3^{+}$ Treg cell differentiation and inhibits the development of pathogenic Th17 cells, which are often implicated in autoimmune disorders. By limiting Th17 expansion, the gut microbiome can mitigate inflammation-driven damage to endocrine tissues, thereby preserving hormonal function.

The gut microbiota also contributes to the development and training of the GALT, a key component of the immune system. Through constant interaction with gut microbes, GALT learns to distinguish between harmful pathogens and beneficial commensals, promoting immune tolerance and preventing chronic inflammation (Bemark et al. 2024). This immune tolerance extends systemically, impacting distant endocrine organs and ensuring that the immune system does not overreact to endogenous hormones or hormone-producing cells.

In addition to systemic effects, immune responses shaped by the microbiome influence local endocrine environments. For example, immune cells residing in the pancreas are heavily influenced by gut microbial metabolites, which affect insulin secretion and β-cell function (Wu et al. 2023). In the thyroid gland, microbiome-driven immune regulation prevents excessive antibody production that could otherwise lead to conditions such as Graves' disease or hypothyroidism (Chen et al. 2024).

The bidirectional nature of gut-immune-endocrine communication underscores the potential for microbiome-targeted therapies to manage autoimmune and inflammatory endocrine disorders. Probiotics, prebiotics, and dietary interventions aimed at restoring microbial diversity have shown promise in mitigating inflammation, enhancing immune tolerance, and supporting endocrine health. As the understanding of the gut microbiome's role in immune-endocrine crosstalk deepens, novel therapeutic strategies may emerge to prevent and treat conditions driven by immune dysregulation, reinforcing the importance of gut health as a cornerstone of overall endocrine balance.

Dysbiosis and Its Disruptive Impact on Endocrine Functions

Dysbiosis, defined as an imbalance in the composition and function of the gut microbiota, has profound consequences for endocrine regulation (DeGruttola et al. 2016). A healthy gut microbiome maintains hormonal homeostasis through metabolite production, immune modulation, and hormone metabolism. However, when dysbiosis occurs—due to factors like poor diet, antibiotics, stress, or disease—this equilibrium is disturbed, leading to widespread endocrine dysfunction.

One of the key consequences of dysbiosis is the alteration of microbial enzyme activity, which directly affects hormone metabolism. For example, a reduction in β-glucuronidase-producing bacteria can impair the estrobolome's ability to deconjugate estrogens, thereby reducing their bioavailability and contributing to estrogen deficiency states such as osteoporosis or irregular menstruation. Conversely, excessive β-glucuronidase activity may elevate active estrogen levels, heightening the risk for hormone-dependent cancers, including breast and endometrial cancer (Hu et al. 2023).

Dysbiosis also disrupts the production of SCFAs, which are critical for stimulating the secretion of metabolic hormones like GLP-1 and PYY. A reduction in SCFA levels contributes to insulin resistance, impaired glucose metabolism, and increased appetite, ultimately predisposing individuals to metabolic syndrome and type 2 diabetes. For instance, Gueddouri and his colleagues observed a link between insulin resistance and increased gut permeability that appeared independent of obesity. Their findings suggest that a compromised gut barrier function may serve as a potential risk factor for insulin resistance in young adults (Gueddouri et al. 2022). Furthermore, the study by Tan et al. (2021) found that a shortage of SCFAs, stemming from dysbiosis, was associated with worse functional outcomes in patients following acute ischemic stroke, highlighting the importance of SCFAs in neurological recovery and overall health (Tan et al. 2021).

The inflammatory environment disrupted by dysbiosis further exacerbates endocrine disruption, amplifying the cascade of hormonal imbalances. Overgrowth of pathogenic bacteria elevates systemic levels of pro-inflammatory cytokines (e.g., IL-6, TNF-α), which interfere with insulin signaling and disrupt reproductive hormones. Of interest, a recent study investigated the effects of continuous antibiotic treatment on gut microbiota, serum cytokines, and intestinal inflammatory responses in mice. The study found that antibiotic-induced dysbiosis led to significant changes in gut microbiota composition and increased levels of pro-inflammatory cytokines. These findings suggest that antibiotic-induced alterations in gut microbiota can exacerbate systemic inflammation, potentially disrupting endocrine function and contributing to metabolic disorders (Gao et al. 2024). Additionally, a study utilizing a mouse model of inflammatory bowel disease demonstrated that persistent intestinal inflammation leads to gut dysbiosis, which may in turn contribute to negative reproductive outcomes by disrupting endocrine homeostasis (Martell et al. 2025). Figure 2.2 illustrates the multifaceted impact of gut microbiota dysbiosis, highlighting its triggers, the underlying pathophysiological mechanisms, and the downstream consequences on endocrine function and related disorders.

Approaches Targeting Microbiota-Endocrine Crosstalk (Fig. 2.3)

Probiotics and Prebiotics

Probiotics, defined as live beneficial bacteria, and prebiotics, which are fibers that nourish gut bacteria, are key players in modulating the gut microbiome and influencing systemic metabolic health (Fayed et al. 2018). Numerous studies have highlighted the role of probiotics in modifying the growth of beneficial microbial communities that can outcompete pathogenic bacteria (Fayed et al. 2019), leading to improved gut integrity and reduced intestinal permeability. This is particularly significant for preventing the translocation of endotoxins that contribute to systemic inflammation and insulin resistance (Di Vincenzo et al. 2024).

Specific strains such as *Lactobacillus rhamnosus, Lactobacillus plantarum,* and *Bifidobacterium longum* have been shown to enhance glucose metabolism, alleviate oxidative stress, and reduce inflammation (Virk et al. 2024). Additionally, probiotics can modulate the HPA axis, regulating cortisol levels and lowering the adverse effects of chronic stress on endocrine function (Sabit et al. 2023).

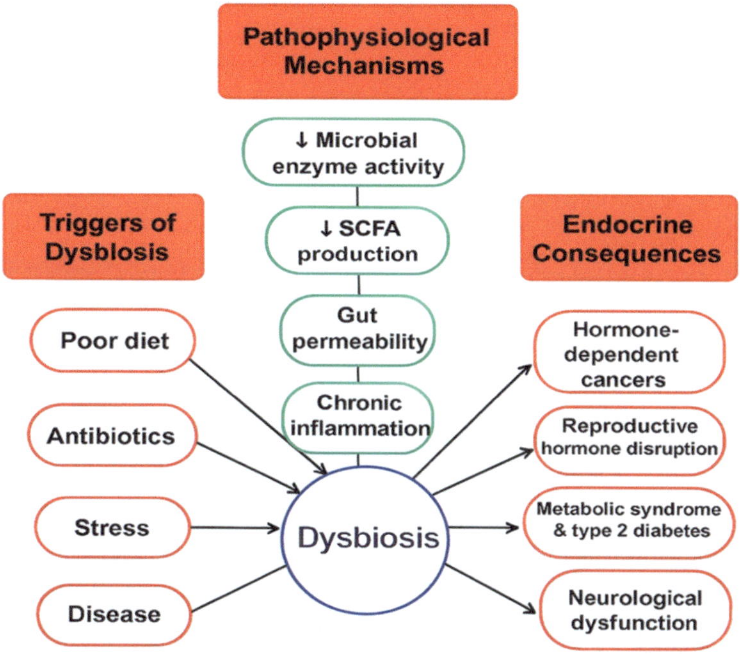

Fig. 2.2 Impact of dysbiosis on endocrine-related pathophysiological outcomes. The diagram illustrates how triggers such as poor diet, antibiotics, stress, and disease lead to gut dysbiosis. Dysbiosis disrupts microbial activity and SCFA production, increasing gut permeability and inflammation. These changes contribute to endocrine-related consequences, including cancers, hormone disruption, metabolic disorders, and neurological dysfunction

Prebiotics, including inulin, fructooligosaccharides, chitooligosaccharides, and galactooligosaccharides, selectively stimulate the growth of probiotic bacteria (Nour et al. 2024). This contributes to improved gut barrier function, reduced low-grade inflammation, and enhanced bile acid metabolism, which collectively improve endocrine health. Moreover, prebiotics influence the production of incretin gut peptide, promoting glucose homeostasis and appetite regulation (Cani et al. 2009). Emerging research also highlights the synergistic effects of synbiotics—combinations of probiotics and prebiotics—which offer enhanced efficacy in modulating microbiota composition and restoring dysbiotic conditions (Zhang et al. 2024a). Synbiotic interventions have shown promise in reducing adiposity, improving lipid profiles, and enhancing insulin sensitivity, particularly in individuals with metabolic syndrome or type 2 diabetes (Markowiak and Śliżewska 2017). Beyond metabolic health, probiotic and prebiotic interventions have demonstrated the potential to influence reproductive and thyroid health by modulating systemic inflammatory responses and improving nutrient absorption, essential for hormonal balance (Shu et al. 2024). The expanding understanding of the gut-endocrine axis continues to drive innovations in probiotic and prebiotic formulations, paving the way for targeted therapies to address a spectrum of endocrine and metabolic disorders.

Approach	Probiotics/ Prebiotics	Postbiotics	Fecal Microbiota Transplantation	Dietary Interventions
MECHANISM	Modulate gut microbiome, improve gut integrity	Modulate metabolic pathways, influence immune responses	Restore gut microbial diversity, re-establish normal communities	Shape gut microbiome, promote beneficial bacteria growth
BENEFITS	Enhance glucose metabolism, reduce inflammation, modulate HPA-axis	Modulate microbiota-endocrine interactions, manage metabolic disorders	Improve metabolic activities, endocrine functions, treat metabolic syndrome	Improve insulin sensitivity, hormone regulation, reduce disease risks
EXAMPLES	Lactobacillus, inulin, synbiotics	Non-living microbial products, metabolic byproducts	Transfer of fecal microbiota from healthy donor	Fiber-rich diets, fermented foods, Mediterranean diet

Fig. 2.3 Summary of the approaches for modulating the communication between gut microbiota and the endocrine system

Postbiotic

Postbiotics, which are nonliving microbial products or metabolic byproducts, are gaining recognition as valuable agents in modulating the complex communication between gut microbiota and the endocrine system (Fang et al. 2024). This microbiota-endocrine crosstalk plays a crucial role in regulating host metabolism, immunity, and overall health.

One of the primary mechanisms through which postbiotics exert their effects is by modulating host metabolic pathways and influencing endocrine control of metabolism. Postbiotics can mimic host-derived hormones or affect glucose metabolism and lipid storage through immunomodulation (Fang et al. 2024). Additionally, they interact with the host's immune system, helping to modulate inflammatory responses that are closely linked to metabolic and endocrine functions (Anwar et al. 2022; Hamdy et al. 2024b; Ozma et al. 2022). This immune modulation can lead to enhanced insulin sensitivity and a reduction in adiposity. The gut microbiota, through the production of metabolites, engages in bidirectional communication with host hormones, impacting a range of physiological processes.

The therapeutic potential of postbiotics lies in their ability to modulate microbiota-endocrine interactions without the risks associated with live microbial therapies. This precision makes them attractive for managing metabolic disorders such as obesity and diabetes. Recent studies have provided insights into how postbiotics influence host metabolism and immunity, reinforcing their potential in therapeutic interventions (Fang et al. 2024).

Fecal Microbiota Transplantation

Fecal microbiota transplantation (FMT) is an emerging therapeutic approach that involves transferring fecal microbiota from a healthy donor into a recipient suffering from microbiome dysbiosis (Zikou et al. 2024). FMT aims to restore gut microbial diversity and reestablish normal microbial communities, which has shown potential in improving the body's metabolic activities and endocrine functions. It has been successfully used in treating conditions like metabolic syndrome, insulin resistance, inflammatory bowel disease (IBD), and autoimmune disorders, which are often associated with altered gut microbiota (Boicean et al. 2023). By influencing the gut-liver axis, FMT may modulate hepatic metabolism and improve insulin sensitivity, further contributing to improved endocrine regulation (Zhao et al. 2023). However, more research is essential to optimize FMT protocols and to elucidate the long-term effects of these interventions on metabolic regulation and hormonal balance.

Dietary Interventions

Dietary habits play a central role in shaping the gut microbiome and influencing metabolic and endocrine health. Diets rich in fiber, polyphenols, and fermented foods have been shown to promote gut microbiota diversity, supporting the growth of beneficial bacteria that produce metabolites beneficial for metabolic and endocrine function (Leeuwendaal et al. 2022). Fiber-rich diets, such as those abundant in fruits, vegetables, whole grains, and legumes, provide substrates for the production of SCFAs, which contribute to improved insulin sensitivity and hormone regulation. Fermented foods, such as functional yogurt, kefir, and kimchi, contain probiotics that can directly enhance microbial diversity and support gut health (Ismail et al. 2020). Moreover, dietary patterns like the Mediterranean diet, which emphasizes plant-based foods, healthy fats, and moderate consumption of fish, have been associated with improved gut health and reduced risks of metabolic and endocrine diseases, including obesity, type 2 diabetes, and cardiovascular disease (Nagpal et al. 2019).

Noncoding RNAs (ncRNA): Types and Mechanisms

NcRNAs comprise a diverse class of RNA molecules that, by definition, do not code for proteins. The identification of ncRNA is regarded as a significant advancement in the understanding of the "genetic central dogma" (DNA/RNA/protein) and the "junk RNA" hypothesis, which claims that junk RNA covers all those RNA molecules that are not translated (70% of the genome undergoes transcription), whereas only roughly 2% are subjected to translation, and, as a result, they are claimed to be nonfunctional (Maia et al. 2014).

Since their initial identification, ncRNAs have been recognized as key regulators of various biological processes across different cell types and tissues. Their altered expression has been increasingly associated with the development and progression of numerous diseases. While miRNAs have been extensively studied in the context of human cancers, emerging evidence highlights the growing importance of other subclasses, such as lncRNAs and circRNAs, in various pathological processes (Nemeth et al. 2024).

Classification of ncRNAs (Fig. 2.4)

NcRNAs can be classified based on their function into housekeeping and regulatory types and based on their size into small ncRNAs and long ncRNAs. Housekeeping ncRNAs are constitutively expressed and are essential for the basic maintenance and functioning of cellular processes. Regulatory ncRNAs influence gene expression at multiple levels, including transcription, post-transcription, and epigenetics. In contrast to housekeeping ncRNAs, they are frequently expressed in a cell-type or condition-specific manner (Röther and Meister 2011).

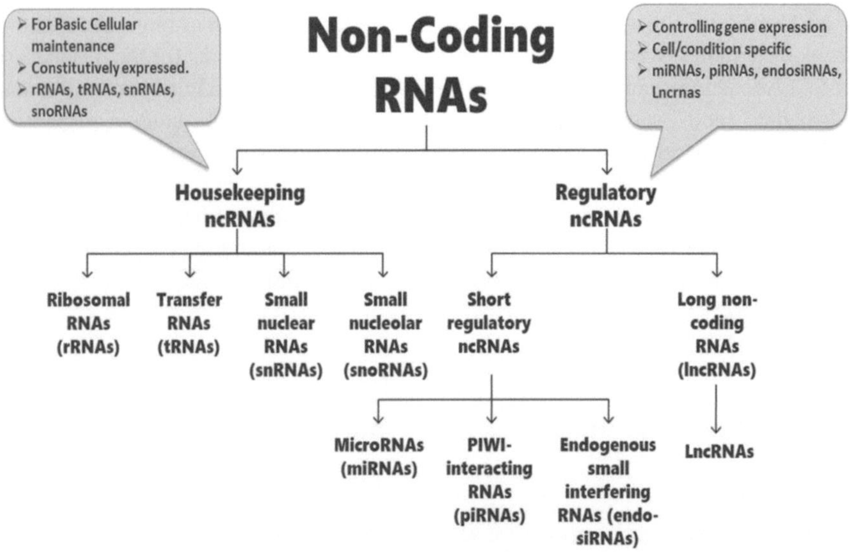

Fig. 2.4 Functional classification of ncRNAs as housekeeping and regulatory. NcRNAs are also classified based on their size into small ncRNAs and long ncRNAs (lncRNAs). The small ncRNAs (typically <200 nucleotides) include miRNAs, short interfering RNAs (siRNAs), Endogenous small interfering RNAs (endo-siRNAs), PIWI-interacting RNAs (piRNAs), Small nuclear RNAs (snRNAs), and small nucleolar RNAs (snoRNAs). On the other hand, long ncRNAs are typically >200 nt

Small ncRNA

MicroRNA (also termed miRNA or miR), an endogenous small ncRNA molecule, contains around 18 to 25 nucleotides and controls the protein-coding expression of genes via binding to the 3′ untranslated region (UTR) of messenger RNA (mRNA), therefore inhibiting translation and promoting the destruction of target mRNAs. A single miRNA can target multiple mRNAs, and conversely, a single mRNA can be targeted by multiple miRNAs, allowing for intricate regulatory networks. miRNAs are crucial in various cellular functions, including proliferation, apoptosis, differentiation, and metabolism. Additionally, miRNAs are involved in intercellular communications, particularly through their incorporation into extracellular vesicles like small extracellular vesicles, "exosomes," which mediate their transport between cells. This allows miRNAs to influence the behavior of neighboring or distant cells, thereby contributing to tissue homeostasis and disease progression (El-Daly et al. 2025a, b; Gouhar et al. 2025).

Circular RNAs (circRNAs) are a unique class of single-stranded, covalently closed RNA molecules that lack free 5′ or 3′ ends. Unlike linear RNAs, circRNAs form a continuous loop, making them more stable and resistant to exonucleases. They can be divided into three categories according to their composition: exon circRNAs (EciRNAs), intron circRNAs (CiRNAs), and exon-intron circRNAs (EIciRNAs). CircRNAs can act as miRNA sponges, effectively sequestering miRNAs and preventing them from interacting with their mRNA targets. Certain circRNAs engage with RNA-binding proteins (RBPs), influencing their function or positioning. Furthermore, certain circRNAs can be translated into functional peptides, although this occurs infrequently. CircRNAs are also found in extracellular vesicles, suggesting a role in cell-cell communication, particularly under pathological conditions such as cancer, where they may contribute to tumor progression and metastasis (Hwang and Kim 2024).

Long Noncoding RNAs (lncRNAs)

Long noncoding RNAs (lncRNAs) are generally defined as RNA transcripts longer than 200 nucleotides that lack protein-coding potential. They exhibit diverse modes of action and can localize to both the nucleus and cytoplasm. In the nucleus, lncRNAs regulate gene expression by recruiting chromatin-modifying complexes, guiding transcription factors, or modulating enhancer-promoter interactions. In the cytoplasm, they can function as competing endogenous RNAs (ceRNAs), sponging miRNAs and thereby relieving repression on target mRNAs. LncRNAs also serve as scaffolds, bringing together RNA, DNA, and proteins to form functional regulatory complexes. Moreover, lncRNAs can be secreted into extracellular vesicles, contributing to intercellular communication and the modulation of gene expression in recipient cells (Xiao et al. 2024).

Mechanisms of Action of Noncoding RNAs in Cellular Regulation and Communication

Noncoding RNAs (ncRNAs) perform diverse and essential functions that can be broadly grouped into three categories:

(1) Regulatory functions: Small ncRNAs and lncRNAs modulate chromatin architecture, transcription, RNA stability, translation efficiency, and protein activity through sequence-specific base pairing or by acting as guides, scaffolds, decoys, or sponges for DNA, RNA, and protein partners.(2) Structural and catalytic functions: Ribosomal RNAs (rRNAs) fold with ribosomal proteins to build the small and large ribosomal subunits and provide the peptidyl-transferase activity required for peptide-bond formation.(3) Adaptor/transport functions: Transfer RNAs (tRNAs) are aminoacylated with specific amino acids and decode mRNA codons within the ribosome, thereby delivering their cargo to the growing polypeptide chain during translation (Polikanov et al. 2015; Winkle et al. 2021).

MiRNAs bind to complementary sequences in target mRNAs, leading to mRNA degradation or translation inhibition. Approximately 30% of protein-encoding genes are regulated by miRNAs, which can regulate various normal and pathophysiological pathways such as cell differentiation, cell proliferation, and tumor development. MiRNAs are increasingly recognized as valuable biomarkers for a wide range of pathophysiological conditions, including diabetes, cancers, and other complex disorders (El-Daly et al. 2023b; El-Toukhy et al. 2023; Zhang et al. 2024b).

Noncoding RNAs Modulate Intestinal Function

MiRNAs can affect the epithelial and endothelial permeability through the regulation of tight junction protein (TJP) expression. MiR-122a, miR-144, and miR-200C-3p can enhance the permeability of intestinal tight junctions by directly targeting and degrading occludin mRNA. Occludin is a crucial transmembrane protein that plays a significant role in the formation and maintenance of tight junctions within epithelial and endothelial cells. Tight junctions are critical for regulating paracellular permeability and maintaining barrier integrity in tissues such as the intestinal epithelium, blood-brain barrier, and other epithelial linings (Hou et al. 2017; Ye et al. 2011). Another miRNA is miR-29, which has the ability to enhance the permeability of intestinal epithelial cells by directly targeting and reducing claudin-1 mRNA. (Zhou et al. 2015). Claudin-1 is a critical tight junction protein that plays a major role in regulating paracellular permeability in the intestinal epithelium (Neurath et al. 2025). By modulating epithelial barrier function, these miRNAs can influence interactions between the host and gut microbiota, thereby contributing to gut microbiome modulation.

Dysregulation of the intestinal epithelial barrier contributes significantly to the pathogenesis of autoimmune and inflammatory diseases, often accompanied by alterations in gut microbiome composition and function. Epithelial miRNAs serve as important mediators in the dialogue between epithelial cells, immune cells, and the gut microbiota. For example, studies using a mouse model with gut epithelial-specific deletion of Dicer1, the enzyme essential for miRNA maturation, demonstrate that epithelial miRNAs are critical for mounting effective mucosal immune responses, including the initiation of antiparasitic T helper type 2 (Th2) responses. In particular, miR-375 expressed in colonic epithelial cells promotes the production of cytokines such as resistin-like molecule beta (RELMβ) and thymic stromal lymphopoietin (TSLP), which orchestrate local Th2 immunity (Biton et al. 2011). By modulating these cytokines and immune pathways, epithelial miRNAs influence gut mucosal homeostasis and help shape the gut microbial ecosystem, thereby contributing to gut microbiome modulation and the maintenance of intestinal health.

As stated earlier, ncRNA mediates key communications between gut epithelial cells, immune cells, and endothelial cells, thereby influencing immune responses crucial for maintaining intestinal homeostasis and modulating the gut microbiome (Hamdy et al. 2025a, 2025b). These ncRNAs regulate gene expression at transcriptional and posttranscriptional levels during innate immune activation, impacting the production of inflammatory mediators and cytokines that shape host–microbe interactions. For instance, lncRNAs have emerging roles in macrophage function, where they can influence polarization states that are critical in tissue homeostasis and inflammation (Zhao et al. 2021a).

LncRNAs, such as linc00514, have been shown to promote an immunosuppressive microenvironment by driving macrophage M2 polarization via pathways like STAT3/Jagged1/Notch signaling, resulting in the secretion of cytokines IL-4 and IL-6, which affect local immune regulation. These mechanisms suggest that lncRNAs participate integrally in host-pathogen and host-microbiota communication networks polarization (Rossetto et al. 2013; Su et al. 2025). Although the precise molecular pathways remain incompletely understood, ncRNAs represent vital regulators of gut immunity and microbiome composition, providing promising avenues for therapeutic manipulation of gut-related diseases characterized by dysregulated immunity and microbial imbalance.

The Role of Noncoding RNAs in Endocrine-Related Cancers: Molecular Mechanisms and Clinical Implications

Endocrine-related cancers are malignancies arising in organs and tissues that produce hormones. These include the breast, prostate, thyroid, pancreas, and adrenal glands. Hormonal regulation and receptor signaling pathways are central to their initiation and progression (Chen et al. 2022). A growing body of evidence highlights RNA biology as a central player in the molecular pathology of endocrine-related cancers (Bradley and Anczuków 2023; Richard et al. 2025). Aberrant mRNA translation and altered RNA stability further exacerbate malignancy by disrupting the expression balance of essential regulatory proteins (Fard and Holz 2023). Dysregulation of noncoding RNAs—including miRNAs, lncRNAs, and circRNAs—can disrupt gene regulatory networks, facilitating hallmark features of cancer such as immune evasion, enhanced survival, and metastatic dissemination (El-Daly et al. 2023a, 2025a, b). These insights not only deepen our understanding of cancer biology but also open new avenues for diagnostic/prognostic biomarkers and targeted therapies, positioning ncRNA-based mechanisms at the forefront of endocrine oncology research.

The crucial molecular pathways that are targeted and modulated by various types of ncRNAs in different endocrine-related cancers are presented in Table 2.2.

Noncoding RNA and Gut Microbiome Interplay

It has been proposed that host genetics, diet, geography, and social contact patterns all influence the composition of the gut microbiota. Despite being relatively stable, the richness and composition of mature microbiota can vary depending on both internal and external factors, such as age, lifestyle, drug usage, and food choices (Schmidt et al. 2018). The host's metabolism, including body mass index (BMI), insulin secretion, and fat synthesis, can be regulated by several metabolites produced by the gut microbiota, such as butyrate and bile acids. Methyl donors that influence host DNA methylation are provided by dietary intake of nutrients (such as folic acid, methionine, and vitamin B12), which may regulate the inflammatory condition of the digestive tract. Important mediators in the interaction between microbiota and epigenetics include microbial metabolites. Certain microbial metabolites can alter histone acetylation, DNA methylation, and ncRNAs, as well as impair the homeostasis of the intestinal environment (Wu et al. 2021).

Microbiome Regulation of Host ncRNAs

Maintaining epigenetic modification requires the gut microbiota's interaction with the host through ncRNA (Ramos-Molina et al. 2019). Overall, epigenetic regulation has been generally recognized as a potential mechanism by which bacteria may modify the host's physiology. According to D'Aquila et al. (2020), biomolecules generated by the host's food metabolism, such as SCFAs, polyphenols, vitamins, tryptophan catabolites, and polyamines, can be used to connect gut microbiota with host epigenetics. Both the gut microbiota and chemicals derived from it can trigger particular epigenetic modifications that control a variety of host physiological processes. Epigenetic changes in the host may result from the abrupt rise or fall in

Table 2.2 Key ncRNA-target gene regulatory axes in endocrine-related cancer progression and therapy resistance

Type of ncRNA	Target gene(s)	Cancer type	Affected pathway	Clinical outcome	References
lncRNA MIR497HG/ miR-497/miR-195	MAP2K1, AKT3, BCL2, RAF1, CCND1	ER^+ breast cancer	PI3K–AKT	Loss of MIR497HG → tamoxifen resistance; PI3K–AKT repression restores tamoxifen sensitivity	Tian et al. (2023)
miR-203	SOCS3	ER^+ breast cancer	JAK/STAT3, MAPK, c-Myc, stemness	miR-203 inhibition → ↓ proliferation, ↓ tumor growth *in vivo*	Muhammad et al. (2016)
miR-222, miR-378a, miR-22, miR-181a, miR-19b	Various targets	$HER2^+$ and ER/ PR^- breast cancer	Proliferation (Ki-67), metastasis	High Ki-67, lymph node metastasis	Kalinina et al. (2021)
lncRNA LINC00152	PDE4D → TRPC1	Tamoxifen-resistant ER^+ BC	cAMP/PKA/CREB → Ca^{2+}–ferroptosis	↑ LINC00152 → tamoxifen resistance, poor survival; inhibition restores ferroptosis sensitivity	Saatci et al. (2024)
lncRNA LINC00173	ERα; LITAF	ER^+ breast cancer	ERα signaling & TNFα transcription; apoptosis	LINC00173 knockdown + estrone → ↑ TNFα, apoptosis, ↓ tumor growth	Xie et al. (2024)
lncRNA LINC02568	ESR1; CA12	ER^+ breast cancer	ERα-driven transcription; pH regulation	↑ LINC02568 → proliferation and endocrine resistance; ASO targeting sensitizes to therapy	Chen et al. (2023a)
circ_0086722	miR-339-5p → STAT5A	Prostate cancer	STAT5A activation	↑ circ_0086722 → promotes proliferation & progression	Deng et al. (2022)
circ-SLC19A1 (exosomal)	miR-497 → SEPTIN2	Prostate cancer	Cell proliferation & invasion	electric vehicles (EV)-circ-SLC19A1 → ↑ proliferation & invasiveness	Zheng et al. (2020)
lncRNA SNHG4	let-7 miRNAs → RRM2	CRPC (enzalutamide-resistant)	Cell cycle, senescence, DNA repair	SNHG4 overexpression → enzalutamide resistance; knockdown → cell cycle arrest and senescence	Dong et al. (2023)
miR-145	DNA repair factors	Prostate cancer	DNA repair/ radiosensitivity	↑ miR-145 → enhanced radiosensitivity	Gong et al. (2015)

(continued)

Table 2.2 (continued)

Type of ncRNA	Target gene(s)	Cancer type	Affected pathway	Clinical outcome	References
circ-ITCH	miR-17	Prostate cancer	PI3K/AKT/mTOR (PAM) attenuation	↑ circ-ITCH → inhibits oncogenic PAM signaling	Li et al. (2020b)
circMBOAT2	miR-1271-5p	Prostate cancer	PAM signaling activation	↑ circMBOAT2 → enhances oncogenic PAM signaling	Shi et al. (2020)
circSMARCC1	miR-1322	Prostate cancer	PAM signaling; M2 macrophage polarization	↑ circSMARCC1 → activates PAM, recruits M2 macrophages; diagnostic biomarker	Xie et al. (2022)
piR-001773, piR-017184	PIWIL4 → PCDH9	Prostate cancer	PAM signaling via PCDH9 suppression	↑ piRNAs → advanced stage and higher Gleason scores	Zhang et al. (2020)
miR-218	GAB2/PI3K/AKT/GSK-3β	Metastatic prostate cancer	EMT inhibition	↑ miR-218 → ↓ EMT, migration, metastasis	Guo et al. (2025 and Tian et al. (2020)
miR-21, miR-92a, miR-4534, PlncRNA-1	PTEN	Prostate cancer	Cell cycle & apoptosis bypass	↑ these ncRNAs → cell cycle progression, chemoresistance	Cui et al. (2021, Yanshen et al. (2021), and Zhao et al. (2021b)
lncRNA ZFAS1	p53 network; miR-135b-5p/193a-3p/34b	Papillary thyroid carcinoma	p53-mediated proliferation	↑ ZFAS1 → oncogenic; p53 repression; therapeutic target	Wang et al. (2024a)
lncRNA LINC00092	miR-4741	Papillary thyroid carcinoma	Feed-forward loop (MYB–LINC00092–miR-4741)	↓ LINC00092 → tumor progression; tumor suppressor	Cheng et al. (2024)
lncRNA HAGLROS	miR-206 → HMGA2	Papillary thyroid carcinoma	Proliferation, migration, anti-apoptosis	↑ HAGLROS → tumor growth; silencing → growth suppression	Zeng et al. (2023)

lncRNA HCG22	miR-425-5p	Papillary thyroid carcinoma	Proliferation & metastasis	↑ HCG22 → proliferation & metastasis; silencing → ↓ proliferation	Cao et al. (2024)
miR-1246	PIK3AP1	Thyroid cancer	PI3K/AKT	↑ miR-1246 → tumor-suppressive effect: ↓ proliferation, migration, invasion	Li et al. (2022a)
LINC02082, LOC105369812	PI3K/AKT	Papillary thyroid carcinoma	Diagnostic discrimination	LOC105369812 → distinguishes malignant vs. benign nodules	Xu et al. (2023a)
miR-204-5p	HMGA2, SNAI2, TGFBR2, SOX4	Papillary thyroid carcinoma	EMT & MAPK	↑ miR-204-5p → ↓ invasion	Van Branteghem et al. (2023)
lncRNA LNCPTCTS	Snail (via eEF1A2); E-cadherin, PEBP1	Thyroid cancer	EMT & MAPK	↓ LNCPTCTS → ↑ EMT, metastasis; tumor suppressor	Ma et al. (2023)
miR-183	MEG3, BRI3	Pancreatic neuroendocrine tumors (PanNETs)	Proliferation, invasion, angiogenesis, EMT	↑ miR-183 → oncogenic *in vitro*	Michael et al. (2019)
miR-223	NLRP3 inflammasome; CgA processing	Neuroendocrine neoplasms (NEN)	Inflammasome activation; immune modulation	↓ miR-223 → altered CgA fragmentation, neutrophil behavior	Geisler et al. (2025)
lnc-HOXB8-1:2 (exosomal)	miR-6825-5p → CXCR3	NED-CRC	TAM recruitment; M2 polarization	↑ lnc-HOXB8-1:2 → tumor progression	Li et al. (2022b)
circ_0009092	miR-665 → NLK	Colorectal cancer	Wnt/β-catenin, STAT3; TME	↓ circ_0009092 → ↑ proliferation, invasion; low levels → poor prognosis	Song et al. (2023)
LINC00982 (PRDM16-DT peptide)	HNRNPA2B1 → CHEK2 isoform	Colorectal cancer	Metastasis suppression; E-cadherin secretion	↓ LINC00982 → ↑ metastasis; PRDM16-DT restores CHEK2 L-isoform, suppresses metastasis	Hu et al. (2024)
circGPRC5A	PPP1CA → YAP	Colorectal cancer	YAP activation	↑ circGPRC5A → poor prognosis, ↑ proliferation & metastasis	Chen et al. (2023b)
circ_0007379	KSRP → miR-320a → RUNX1	Colorectal cancer	RUNX1 suppression; TME modulation	↓ circ_0007379 → ↑ proliferation & invasion; overexpression → tumor suppression	Long et al. (2023)

(continued)

Table 2.2 (continued)

Type of ncRNA	Target gene(s)	Cancer type	Affected pathway	Clinical outcome	References
LINC01852	TRIM72 → SRSF5 → PKM2 isoforms	Colorectal cancer	Metabolic shift; oxaliplatin sensitivity	↓ LINC01852 → glycolysis, chemoresistance; overexpression → ↓ growth, ↑ chemosensitivity	Bian et al. (2024)
circRERE	miR-6837-3p → MAVS	Colorectal cancer	Type I IFN activation; $CD8^+$ T-cell infiltration	↑ circRERE → enhances anti-PD-1 efficacy	Ding et al. (2023)
lncRNA H19	miR-22-3p → MMP14	Metastatic colorectal cancer	EMT & metastasis	↓ HDAC2 → ↑ H19 → ↑ EMT & metastasis; HDAC2 acts as a metastasis suppressor	Hu et al. (2020)
circMYH9	hnRNPA2B1 → p53; HIF1α regulation	Colorectal cancer	Redox homeostasis; serine/glycine metabolism	↑ circMYH9 → p53 downregulation, ↑ proliferation; xenograft growth acceleration	Liu et al. (2021)

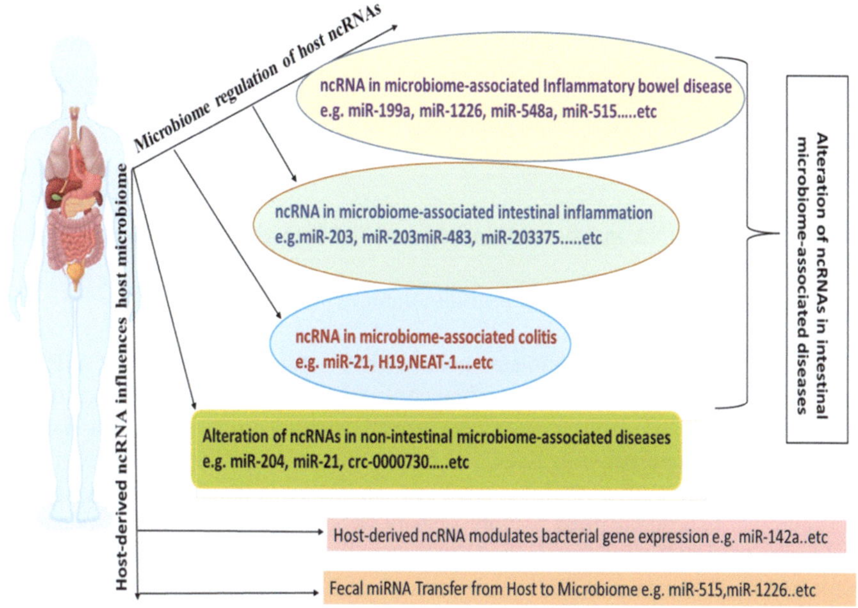

Fig. 2.5 Bidirectional crosstalk between gut microbiome and host ncRNAs. This figure illustrates how host miRNAs are dysregulated in response to gut microbiome alterations in intestinal diseases such as IBD, intestinal inflammation, and colitis, with key examples indicated. It also shows microbiome-associated changes in ncRNAs linked to non-intestinal diseases. Additionally, host-derived ncRNAs modulate bacterial gene expression, while fecal miRNAs are transferred to gut microbes, influencing microbial composition and function

SCFAs brought on by dietary food intake or environmental changes. For example, through several signaling pathways, butyrate can maintain homeostasis and encourage intestinal growth (Hamer et al. 2008; Hu et al. 2011). A growing body of evidence indicates that miRNAs play a critical role in host-microbiota interactions that regulate gut health (Fig. 2.5). Microbiota has been demonstrated to have the ability to alter host miRNA expression patterns (Dalmasso et al. 2011). Accordingly, several microbial taxa are thought to be linked to inflammation and tumors, such as *Fusobacterium nucleatum* (*F. nucleatum*) and *Faecalibacterium prausnitzii* (*F. prausnitzii*), which have a significant impact on miRNA expression in the intestinal epithelium. Downstream targets may subsequently play a role in the development of colorectal cancer (CRC) (Nikolaieva et al. 2022). *Lactobacilli* pretreatment may reverse the downregulation of certain miRNAs (miR-200b, miR-215, and miR-192) that is usually brought on by an infection with *Listeria monocytogenes* (*L. monocytogenes*). According to Wieërs et al. (2020), probiotics that control the expression of miRNA may have positive impacts on intestinal homeostasis and the development of the host's defenses against infection. An investigation revealed that intestinal epithelial stem cells' microbiota dramatically inhibited miR-375-3p. Additionally, they offer proof that the microbiota-sensitive miR-375-3p of intestinal epithelial stem cells (IESC) affects intestinal epithelial cells (IEC) proliferation, most likely by maintaining physiologically active IESC (Peck et al. 2017). Moreover, gut microbiota, together with nearby protein-coding genes, can control the expression of lncRNAs in the local intestine, which in turn controls the tissue-specific route (Dempsey et al. 2018).

Alteration of ncRNA in Microbiome-Associated Intestinal Diseases

ncRNA in Microbiome-Associated Inflammatory Bowel Disease

Gut microbiota dysbiosis modulates host ncRNA networks—principally miRNAs and lncRNAs—which, in turn, influence local intestinal immunity, epithelial homeostasis, and IBD, colitis, and CRC, as well as remote organ functions via gut–heart, gut–brain, and gut–lung axes.

Dysbiosis of the gut microbiota has been implicated in the pathogenesis of several chronic disorders—most notably obesity, type 2 diabetes mellitus, metabolic syndrome, and IBD (DeGruttola et al. 2016; Louis et al. 2014). The proliferation of both beneficial, segmented filamentous bacteria (SFB) and pathogenic *F. nucleatum* and *E. coli* bacteria is influenced by a number of miRNAs that are differentially expressed in IBD, including miR-199a, miR-1226, miR-548a, and miR-515-5p. IBD develops as a result of alterations in the gut microbiota (Ji et al. 2018). Additionally, miR-193a-3p can control bacterial product absorption to influence the host's intestinal disease (Dai et al. 2015).

In Crohn's disease, colonization of the intestinal mucosa by adherent-invasive *E. coli* (AIEC) has been shown to inhibit the expression of autophagy-related genes such as *ATG5* and *ATG16L1*. This effect is mediated through the induction of microRNAs miR-30c and miR-130a. Furthermore, the gut microbiota suppresses the production of the lncRNA short nucleolar RNA host gene 9 (Snhg9) through an immunological signaling cascade that involves group 3 innate lymphoid cells (ILC3) and myeloid cells, according to a recent study by Wang et al. (2023). Small intestine epithelial cells' lipid metabolism is changed because of this regulatory axis's enhancement of PPARγ activity (Y. Wang et al. 2023).

NcRNA in Microbiome-Associated Intestinal Inflammation

Enzymatic hydrolysis of the human food results in indigestible resistant starch (RS), the substrate for the metabolism of gut microbiota, which has been shown to lower intestinal inflammation (Myhrstad et al. 2020). Through the modulation of miRNA expression, the microbiome was found to have an impact on the regulation of intestinal barrier function and intestinal homeostasis (Singh et al. 2011). The primary cells that interact with intestinal bacteria and help to maintain intestinal homeostasis are intestinal epithelial cells (IECs) (Peck et al. 2017; Schroeder and Backhed 2016). For instance, miR-21 has high expression in many diseases such as cancer and IBD (Shi et al. 2016). According to Veltman et al. (2012), enteropathogenic *E. coli* led to the disruption of the tight junction by inhibition of many miRNAs such as miR-203, miR-483–3p, and miR-595, while *E. coli Nissle 1917* caused an increase in barrier integrity via the regulation of specific miRNAs. Therefore, one of the main functional mechanisms of probiotics for the maintenance of homeostasis is miRNA expression modulation.

Additionally, a range of metabolites from the gut microbiota can control intestinal inflammation and homeostasis through epigenetic pathways. By blocking histone deacetylases and particular ligands for G protein-coupled receptors (GPCRs), SCFAs reduce inflammation. Brown et al. (2003) found that SCFAs, particularly microbial-derived butyrate, enhance the development of the intestinal epithelial barrier by downregulating claudin-2 expression. This effect is mediated through a mechanism dependent on the interleukin-10 receptor alpha (IL-10RA). Claudin-2 is a tight junction protein that increases intestinal permeability, so its inhibition helps strengthen the barrier function of the gut lining. By activating IL-10RA signaling, butyrate supports anti-inflammatory pathways and promotes epithelial integrity, which is critical for maintaining gut homeostasis and preventing inflammation. The downregulation of some bacterial transporters via miR-193a-3p led to a decrease in colonic inflammation (Dai et al. 2015; Li et al. 2009; Xie et al. 2013). MiR-375-3p is a microRNA that plays a role in regulating the behavior of IESCs. Researchers found that miR-375 responds specifically to signals from the gut microbiota that are in close contact with IESCs. When these microbes are present, they suppress the expression of miR-375-3p. This suppression stops the inhibitory effect that miR-375-3p normally has on stem cell growth. As a result, inhibiting miR-375-3p enhances the stem cells' proliferation, helping to maintain and repair the intestinal lining (Peck et al. 2017).

The gut microbiota can influence T-cell differentiation by modifying histones, which in turn affects the level of inflammation in the intestine. Most Tregs are generated in the thymus (tTregs), while a smaller portion is produced in peripheral tissues (pTregs). These cells express the transcription factor Foxp3 and are essential for suppressing intestinal inflammation (Josefowicz et al. 2012). SCFAs help to reduce inflammation by acting as histone deacetylase (HDAC) inhibitors. They support Treg cell differentiation, increase FOXP3 protein acetylation and gene expression in $CD4^+$ T cells, and encourage histone acetylation (Park et al. 2015; Smith et al. 2013). The activity of the myeloid differentiation factor 88 (MyD88) and NFκ-B pathways is impacted by gut microbiota's downregulation of miR-107 expression in dendritic cells and macrophages. This, in turn, affects immune homeostasis and the expression of the target gene IL-23p19 (Xue et al. 2014). On the other hand, macrophages can control inflammation, eliminate invasive pathogens, and preserve intestinal homeostasis.

NcRNA in Microbiome-Associated Colitis

Intestinal inflammation may arise from either an imbalance in the gut microbiota or an uncontrolled host defense against the invasive microbiota. Changing the gut microbiota can produce epigenetic factors that influence the development and occurrence of colitis (Lee et al. 2015). The physiological processes involved in preserving intestinal homeostasis can be influenced by host epigenetics, particularly miRNAs, which control the gut microbiota. For example, miR-21-5p expression in IECs can influence intestinal barrier function by targeting ADP ribosylation factor 4 (ARF4), a protein involved in vesicle trafficking and tight junction regulation. By downregulating ARF4, miR-21-5p can increase epithelial permeability, which may contribute to barrier dysfunction and inflammation if not properly controlled (Nakata et al. 2017).

MiR-21 contributes to the pathophysiology of intestinal inflammation by upsetting the gut flora. Mice deficient in miR-21 are less susceptible to colitis. MiR-21 has been demonstrated to impact the differentiation and activation of different T-cell subsets by modulating key genes such as IL12p35 and Stat3. For instance, in various inflammatory conditions, miR-21 promotes the development of Th17 cells, suppresses regulatory T cells (Tregs), and enhances the activation of memory T cells (Johnston et al. 2018). These effects suggest that miR-21 plays a crucial role in shaping immune responses and may contribute to immune imbalance in inflammatory diseases. The gut microbiota can affect the expression of H19, an lncRNA that plays a key role in maintaining the intestinal barrier and mucosal homeostasis (Dey et al. 2014). Another ncRNA, NEAT1, becomes upregulated during microbial-induced inflammation and helps promote inflammation by supporting the assembly of inflammasomes, which are protein complexes involved in the immune response (Zhang et al. 2019b).

In addition, exosomal ncRNAs obtained from dietary sources can influence the composition and activity of the gut microbiota, thereby potentially affecting host physiology in both humans and mice. For example, ginger-derived exosome-like nanoparticles contain miRNAs such as *Monodelphis domestica* mdo-miR7267-3p, which are selectively picked up by *Lactobacillus rhamnosus* and target the *ycnE* monooxygenase gene (Teng et al. 2018). *ycnE* monooxygenase contributes to the bacteria's ability to adapt to environmental conditions. Similarly, garlic-derived exosomes enriched with peu-miR2916-p3 have been shown to promote the growth of *Bacteroides thetaiotaomicron*, a bacterium with anti-inflammatory properties, thereby alleviating colitis symptoms in mice (Wang et al. 2024b).

Alteration of ncRNAs in Non-Intestinal Microbiome-Associated Diseases

Qin and Wade (2018) indicated that gut bacteria not only have local effects in the gut but also control distant organs, including the heart, brain, and lungs. A disruption in the host-gut microbiota balance will lead to an imbalance in the gut microbiota and ultimately contribute to diseases like intestinal (Rodriguez-Nogales et al.

2017; Yu et al. 2017), neurological (Bercik et al. 2011; Stilling et al. 2015), cardiovascular (Wang et al. 2012; Wang et al. 2011), and immune diseases (Kamada et al. 2013; Wilks et al. 2013).

The main cause of increased sympathetic overdrive to the heart and its arteries, which leads to hypertension, is gut dysbiosis, which is caused by the hypothalamus's reduction of miR-204. Furthermore, it suggests that parasympathetic activity is reduced when hypothalamic miR-204 is decreased. Thus, it is thought that hypothalamic miR-204 plays a key role in the interaction between the gut, brain, microbiota, and hypertension brought on by heart damage (Kassan et al. 2021). Additionally, endothelial dysfunction linked to diabetes is influenced by the microbiota-miR-204 axis. According to their research, vascular miR-204 is sensitive to microbiota, and aortic miR-204 is reversibly decreased while endothelial activity is increased when microbes are suppressed. Mice lacking miR-204, on the other hand, did not exhibit any alteration in endothelial function in response to microbial manipulation. Thus, miR-204 plays a crucial role in the effects of microbes. Moreover, gut dysbiosis may influence miR-204, which is essential for controlling the central nervous system (CNS) and the onset of obesity. miR-204 could therefore establish a direct connection between the brain and gut microbiota (Gaddam et al. 2020; Kassan et al. 2022). Mice lacking miR-21 were protected against small intestinal injury, showed less gut microbiota dysbiosis, and had a lower risk of liver damage after bile duct ligation surgery. Mice lacking miR-21 (KO) showed an overrepresentation of *Lactobacillus* in their microbiota profile. When synthetic miR-21 was incubated in vitro, a notable reduction in the *Lactobacillus* load was noted, indicating that miR-21 primarily causes liver dysfunction by directly influencing the gut microbiota (Santos et al. 2020). Through an increase in IL-17a production by miR-122, microbiome imbalance aids in the advancement of chronic obstructive pulmonary disease (COPD) (Zhu et al. 2020). During TB infection, dysbiosis of the gut microbiota led to the downregulation of LncRNA-CGB, an LncRNA regulated by commensal gut bacteria. It has been demonstrated that patients with type 2 diabetes have unique gut flora and miRNA characteristics. There is a reverse relationship between the gut bacteria *Bacteroides uniformis* and *Phascolarcto bacterium* and miR-122-5p. Furthermore, they found that the gut bacteria *Bacteroides uniformis* and *Phascolarcto bacterium* were negatively correlated with miR-122-5p. Faecium was also found, suggesting that interaction between miRNA and gut microbiota may influence crucial glucose metabolism genes during the development of diabetes mellitus type 2 (T2DM), hence regulating insulin production and signal transmission (Li et al. 2020a).

In response to the gut microbiota, circRNA–miRNA–mRNA axis (circ0000730–miR466i-3p–SOX9 axis) suppresses lung tumors (Zhu et al. 2020). This concept adds another layer to the complexity of the host–microbe crosstalk by revealing the interactions among the regulatory circRNAs. Furthermore, Diling et al. (2020) used the overexpressed circNF1-419 adeno-associated virus animal system to show that the overexpression of circNF1-419 in the brain not only influenced the brain's cholinergic system but also changed the gut microbiota composition, intestinal homeostasis, and physiology, and the gut microbiota trajectory in newborn mice. Their findings demonstrate a link between circRNAs and the gut microbiome. Gut microbiota impacts circRNA expression to regulate levels of corresponding miRNAs. Specifically, gut microbiota inhibits mmu_circ_0000730 expression in an IL-11-dependent manner.

Another example is mmu_circ_0000730 which plays a crucial role in microbiota-mediated lung cancer (Lewis lung carcinoma, LLC) metastasis. Additionally, both mmu-miR-466i-3p and mmu-miR-466f-3p inhibit several genes involved in the epithelial-mesenchymal transition (Zhu et al. 2020).

Host-Derived ncRNA Influences Host Microbiome

Host ncRNAs play a significant role in modulating the gut microbiota by regulating microbial gene expression, metabolism, and host–microbe interactions (Fig. 2.5). Host cells, particularly intestinal epithelial cells, secrete extracellular vesicles such as exosomes that carry ncRNA cargo, including miRNAs and circRNAs (Rizk et al. 2024). These vesicle-encapsulated ncRNAs can be taken up by gut bacteria, thereby influencing the composition and function of the microbiota and contributing to host health and disease (Malmuthuge and Guan 2021).

Host-Derived ncRNA Modulates Bacterial Gene Expression

Epigenetic mechanisms of the host can influence the gut microbial community and its metabolite profile, with host miRNAs playing a key role in shaping microbiota composition (Sato-Kuwabara et al. 2015). Secretory intestinal epithelial cells, including goblet cells, Paneth cells, and other IECs, produce miRNAs that are packaged into exosomes and released into the intestinal lumen. These miRNAs can precisely modulate microbial gene expression, thereby impacting microbiota growth and community structure (Liu et al. 2016; Yuan et al. 2019).

Host miRNAs can act as direct regulators of bacterial genes, influencing microbial composition and function (Santos et al. 2020; Yang et al. 2017). For example, certain miRNAs target bacterial ribosomal RNA genes, such as the bacterial genomic rRNA (16S rRNA), and catalytic RNAs like ribozyme RNA (RNaseP), thereby affecting bacterial growth dynamics. IEC-specific miRNA deficiency in mice results in dysregulated intestinal microbiota and exacerbated colitis, highlighting the importance of host miRNAs in maintaining microbiota homeostasis (Liu et al. 2016).

MiRNAs found in host-derived exosomes may serve as a mediator to facilitate cross-border communication between eukaryotic host cells and prokaryotic microbes. These miRNAs may destabilize bacterial mRNAs or inhibit translation by interacting with posttranscriptional regulatory mechanisms, thus modulating bacterial gene expression (Filip et al. 2016). Upon entry into bacteria such as *F.*

nucleatum and *E. coli* —likely via exosome membrane fusion—host miRNAs selectively regulate bacterial transcriptomes. Notably, host-derived miRNAs regulate not only pathogenic bacteria but also beneficial symbionts. For instance, dextran sulfate sodium (DSS) treatment in colitis-tolerant mice induces intestinal epithelial cells to secrete miR-142a-3p, which targets bacterial genes, including locus tag LREU_RS06530 (polA) and LREU_RS03575, thereby promoting the growth of the commensal bacterium *Lactobacillus reuteri* (*L. reuteri*) (He et al. 2022).

Fecal miRNA Transfer from Host to Microbiome

Fecal miRNAs exhibit enhanced stability compared to other RNA species and are frequently encapsulated within extracellular vesicles, such as exosomes, in the intestinal lumen. The gut microbiota is capable of internalizing these fecal miRNAs, which in turn modulate bacterial gene expression and community composition (Liu et al. 2016). Intestinal epithelial cells and specific host cells expressing homeodomain-only protein homeobox (Hopx) are primary sources of fecal miRNAs (Liu et al. 2016). Upon uptake by bacteria, these miRNAs can regulate bacterial gene expression and influence microbial growth dynamics. For example, human miR-515-5p and miR-1226-5p have been demonstrated to promote the proliferation of *F. nucleatum* and *E. coli*, respectively.

According to Tomkovich et al. (2020), the fecal miRNA profile altered as a result of bacterial activities linked to the biofilm state. In CRC, they also discovered novel miRNAs, including miR-2137, miR-5126, miR-6239, miR-6240, and miR-6538. Most of these miRNAs are expected to have redundant targets in bacteria, such as genes that control motility, secretion, outer membrane proteins, stress response, iron uptake, and carbohydrate utilization/transport.

Mice deficient in intestinal miRNAs exhibit dysregulated gut microbiota and exacerbated colitis, phenotypes that can be ameliorated by fecal miRNA transplantation, underscoring the functional significance of fecal miRNAs in maintaining microbial and intestinal homeostasis (Liu et al. 2016). Furthermore, miRNAs detected in tissue and fecal samples serve as biomarkers for active ulcerative colitis and sporadic human colon cancer (Ahmed et al. 2009). Transplantation of fecal miRNAs from wild-type donors can restore microbial balance and mitigate colitis symptoms, suggesting potential immunotherapeutic applications of fecal miRNA transplantation (Tarallo et al. 2019). Consistently, fecal miRNAs represent promising noninvasive biomarkers for monitoring microbial community shifts and intestinal pathology (Moloney et al. 2018).

Unlike the traditional role of miRNAs, which typically suppress gene expression in host cells by binding to mRNA and reducing its translation, fecal miRNAs can regulate gut bacteria in a different way. These miRNAs interact with bacterial components like 16S rRNA and ribozyme RNA, and instead of repressing gene expression, they can actually enhance bacterial gene transcription (Bartel 2009; Fabian et al. 2010). For example, miR-515-5p was shown to increase the expression of

16S/23S rRNA transcripts in *F. nucleatum*, while miR-1226-5p elevates the levels of yegH mRNA in *E. coli*. Both miRNAs promote the growth of these bacteria, which are known contributors to CRC (Liu et al. 2016). This suggests that intestinal miRNAs present in feces can directly influence the composition and activity of the gut microbiota, thereby affecting intestinal health and potentially contributing to disease.

Molecular Pathways Involved in the Microbiome-ncRNA Axis

Even though extensive research has demonstrated the reciprocal relationship between the microbiome and ncRNAs, it is still unknown what molecular mechanisms the gut microbiota uses to control host ncRNA expression and how dietary or fecal exosomal ncRNAs influence the gut microbiota. There are two possible mechanisms: metabolic pathways and immune pathways.

Functions of Metabolites in Microbiome-ncRNA Axis

SCFAs, indole derivatives, polyamines, secondary bile acids, essential vitamins, and tryptophan are among the many metabolites that the gut microbiota produces from food substrates. Research indicates that gut microbiota metabolites may control the activities of noncoding RNAs. SCFA treatment altered the expression of miR-10a-5p, which binds to the 3′UTR of PIK3CA and inhibits the activation of the PI3K-Akt pathway, hence reducing colitis (Liu et al. 2024). Acetate and butyrate, two important SCFAs, have been demonstrated to have a major impact on DNA methylation at the host miR-378a gene's promoter region, helping to avoid obesity and glucose intolerance. However, these protective effects are partially counteracted by Yin Yang 1 (YY1), a common target gene of the miR-378a family. Overall, the findings suggest that betaine alleviates obesity and related metabolic syndrome through the gut microbiota–regulated miR-378a/YY1 axis, unveiling a novel mechanism by which the gut microbiota contributes to host metabolic health (Du et al. 2021).

Indole-3-propionic acid (IPA), a tryptophan metabolite produced exclusively by gut microbiota, was identified as the most significantly downregulated metabolite in patients with coronary artery disease (CAD). In an independent cohort, reduced circulating IPA levels were associated with both the presence and severity of atherosclerotic cardiovascular disease (ASCVD). Dietary supplementation with IPA was found to attenuate atherosclerotic plaque formation in ApoE−/− mice. Mechanistically, IPA administration in both murine and human-derived macrophages enhanced cholesterol efflux to apolipoprotein A-I (ApoA-I) via a previously uncharacterized miR-142-5p/ABCA1 (ATP-binding cassette transporter A1) signaling pathway. In vivo experiments further confirmed that IPA promotes macrophage-mediated reverse cholesterol transport by modulating this pathway. Conversely, reduced IPA levels led to aberrant overexpression of miR-142-5p in macrophages, impairing ABCA1 expression and accelerating atherosclerosis. Notably, dysregulation of the miR-142-5p/ABCA1/reverse cholesterol transport axis was observed in CAD patients and was closely linked to altered circulating IPA levels (Xue et al. 2022). These findings suggest a connection between a specific gut microbiota-derived tryptophan metabolite and ASCVD. Notably, dysregulation of the miR-142-5p/ABCA1/reverse cholesterol transport axis was observed in CAD patients and was closely linked to altered circulating IPA levels (Xue et al. 2022). These findings suggest a connection between a specific gut microbiota-derived tryptophan metabolite and ASCVD.

Li et al. (2019) discovered a regulatory pathway induced by bile acids in gastric cells, which includes the upregulation of miR-92a-1-5p and the downregulation of its target gene, FOXD1. Their analysis revealed that miR-92a-1-5p is significantly elevated in intestinal metaplasia (IM) tissues and is inducible by bile acids. Mechanistically, miR-92a-1-5p promotes the expression of CDX2 and downstream intestinal markers by targeting the FOXD1/FOXJ1 axis and activating the NF-κB signaling pathway. Clinically, elevated levels of miR-92a-1-5p were associated with reduced FOXD1 expression and increased CDX2 expression in IM tissue samples.

Apart from the advantages of nutritional digestion, the gut microbiota also generates genotoxic substances that could potentially control the expression of ncRNA. Colibactin-producing (*pks*$^+$) *E. coli* triggers senescence by disrupting the SUMOylation of p53, a critical posttranslational modification in eukaryotic cells. This disruption is mediated by the upregulation of miR-20a-5p, which targets sentrin-specific protease 1 (SENP1), a key regulator of the SUMOylation process. These molecular alterations—specifically the changes in SENP1, miR-20a-5p, and associated growth factors—were also observed in the CRC mouse model and in human CRC biopsy samples colonized by *pks*$^+$ *E. coli* (Dalmasso et al. 2014).

Serotonin is a key regulator of intestinal homeostasis, and its dysregulation is closely linked to IBD severity. Olivo-Martínez et al. (2024) In 2024, a study examined the effects of electric vehicles (EVs) originating from the probiotic *E. coli* Nissle 1917 (EcN) and a commensal *E. coli* strain on serotonin metabolism under inflammatory conditions, employing an IL-1β-induced Caco-2 cell model. Their findings revealed that EcN-derived EVs significantly reduced free serotonin levels by enhancing serotonin transporter (SERT) expression. This effect was mediated through the downregulation of miR-24 and miR-200a, as well as TLR4 and NOD1 signaling pathways. Furthermore, EcN EVs attenuated IL-1β-induced disruptions in tight junction proteins and oxidative stress markers, contributing to the maintenance of intestinal barrier integrity.

Functions of Immunity in the Microbiome-ncRNA Axis

Although both ncRNAs and the gut microbiota are known to play crucial roles in regulating gastrointestinal immune activity, direct interactions between the gut immune system and the microbiota–ncRNA axis have only recently gained significant research interest. The role of ncRNAs in gastrointestinal tract immunity has been shown in numerous studies. For instance, miR-375 controls the growth of IESCs and the formation of the intestinal mucus layer. miR-375 also controls the growth of intestinal epithelial stem cells and suppresses the development of enteroendocrine lineages (Biton et al. 2011). MiR-146a plays a significant role in regulating host immunity against *L. monocytogenes* infection by modulating the gut microbiota. Du et al. (2018) showed that miR-146a expression increases during infection, but mice lacking miR-146a were more resistant to the bacteria, with macrophages showing reduced bacterial invasion and survival. Differences in gut microbiota composition were observed between miR-146a-deficient and wild-type mice, notably an increase in beneficial short-chain fatty acid–producing bacteria in the deficient mice. These findings reveal a critical interplay between miR-146a, gut microbiota, and immune defense, suggesting potential therapeutic targets for infection control.

By preserving a barrier to enhance permeability selectivity for commensal microorganisms and their metabolites, the gut microbiota has also been shown to play important roles in controlling gut immune activities in addition to ncRNAs. A study demonstrated that adherent-invasive *E. coli* (AIEC) promotes intestinal fibrosis in Crohn's disease by suppressing the microRNA let-7b, particularly its exosomal form released from intestinal epithelial cells. Decreased levels of let-7b resulted in the polarization of intestinal macrophages toward a fibrotic phenotype by regulating TGFβR1. TGFβR1 plays a pivotal role in immunity by mediating anti-inflammatory signals and regulating immune cell differentiation, particularly maintaining immune tolerance and controlling inflammation. Restoring let-7b or targeting AIEC effectively alleviates fibrosis, highlighting a key interaction between microbiota, miRNA, and immune cells in driving intestinal fibrogenesis and offering potential therapeutic avenues (Xu et al. 2023b). Ultimately, the disruption of gut microbiota caused by antibiotics negatively affects the host's antiviral immunity by influencing the expression levels of miR-146b and miR-29c in the lungs. Ultimately, the disruption of gut microbiota caused by antibiotics negatively affects the host's antiviral immunity by influencing the expression levels of miR-146b and miR-29c in the lungs (Pang et al. 2018).

These studies emphasize that the interaction between the gut microbiota and noncoding RNAs (ncRNAs) is intricately linked to both microbial metabolites and immune regulation. Crucially, these pathways do not operate independently; instead, they are significantly interlinked. For instance, immune responses influence gut homeostasis and microbiota composition, thereby altering metabolite production, while microbiota-derived metabolites, in turn, modulate immune function. Therefore, further investigation into the molecular mechanisms underlying these complex interactions is both necessary and promising for advancing our understanding of host-microbiota-immune system dynamics.

Gut Microbiome-ncRNA Crosstalk in the Regulation of Carcinogenesis of Endocrine-Related Cancers

As stated earlier, most evidence underscores the complex, bidirectional relationship between ncRNAs and the microbiota in the context of cancer, particularly in pancreatic and colorectal malignancies. Data from the competitive endogenous RNA (ceRNA) network revealed that lncRNAs, together with specific miRNAs and mRNAs, constitute the majority of the ceRNA regulatory network influenced by probiotics or pathogenic microbes. These molecular components are strongly associated with the initiation and progression of microbiota-driven cancers (Mortazavi-Jahromi et al. 2020), highlighting the importance of deciphering this crosstalk for disease prevention and health maintenance.

The Interplay of Microbiome and ncRNAs in Promoting Cancer Development and Progression (Table 2.3)

In CRC, Yuan et al. (2018) identified 76 miRNAs differentially expressed between tumor and adjacent normal tissues, including well-characterized oncogenic miRNAs such as miR-182, miR-503, and the miR-17 ~ 92 cluster. These miRNAs exhibited significant associations with the relative abundance of key bacterial phyla, including Firmicutes, Bacteroidetes, and Proteobacteria. Importantly, the microbiota linked to these miRNAs was enriched for metabolic pathways, particularly glycan biosynthesis, suggesting that miRNA-regulated modulation of glycosylation may facilitate the recruitment and colonization of pathogenic bacteria within tumors. This supports the hypothesis that miRNAs indirectly shape the tumor-associated microbiome by affecting glucose metabolism and glycosylation processes.

Dysbiosis can promote carcinogenesis by altering miRNA expression profiles. For example, the anaerobic gut bacterium genus Blautia shows a negative correlation with several oncogenic miRNAs, such as miR-20a, miR-21, miR-96, miR-182, miR-183, and miR-7974, that are all elevated in CRC tissues, highlighting a potential protective role of Blautia against tumorigenesis (Yuan et al. 2018). This interaction underscores how the microbiota may influence colorectal inflammation and cancer through miRNA regulation and metabolite production.

In CRC tissues that tested positive for *F. nucleatum*, miR-4474 and miR-4717 were significantly upregulated, correlating with both early and advanced disease stages (Feng et al. 2019). Both *F. nucleatum* and *E. coli* have been implicated in CRC promotion, partly through miRNA-mediated pathways (Liu et al. 2016). Conversely, beneficial bacteria like *F. prausnitzii*, a major butyrate producer, modulate oncogenic signaling by influencing host miRNA expression. Butyrate suppresses the oncogene c-Myc, resulting in downregulation of the miR-17 ~ 92 cluster—known for its roles in angiogenesis, metastasis, and proliferation—in CRC

Table 2.3 Interplay of gut microbiome and non-coding RNAs in cancer development and suppression

Microbiome involved	Key findings	Implications	Study
Firmicutes, Bacteroidetes, Proteobacteria, *F. prausnitzii*	Identified 76 differentially expressed miRNAs in CRC associated with bacterial phyla and glycan biosynthesis. Butyrate suppresses miR-17 ~ 92, reducing c-Myc	miRNAs shape the tumor microbiome via glycosylation; butyrate suppresses oncogenic signaling	Yuan et al. (2018)
Blautia, *F. nucleatum*, *Porphyromonas*, *Bacteroides*	Negative correlation between *Blautia* and oncogenic miRNAs; miRNA-regulated glycans correlate with *F. nucleatum*	*Blautia*'s protective role: miRNAs influence bacterial colonization	Yuan et al. (2018)
F. nucleatum	Upregulation of miR-4474 and miR-4717 in *F. nucleatum*-positive CRC	Links specific miRNAs to bacterial presence and CRC progression	Feng et al. (2019)
F. nucleatum and *E. coli*	*F. nucleatum* and *E. coli* promote CRC via miRNA-mediated pathways	Bacterial influence on cancer through ncRNAs	Liu et al. (2016)
Gut microbiota (butyrate producers, e.g., *F. prausnitzii*)	Butyrate reduces miR-17 ~ 92a and miR-92a, enhancing p57 and apoptosis	Butyrate's tumor-suppressive role via miRNA regulation	Yuan et al. (2018)
Gut microbiota (butyrate producers)	Fiber-rich diets enhance butyrate producers, suppressing oncogenic miRNAs; red meat upregulates miR-17 ~ 92	Diet modulates CRC risk through microbiome-ncRNA interactions	Humphreys et al. (2014) and Schlörmann et al. (2015)
Gut microbiota (bile acid metabolizers)	DCA downregulates miR-199a-5p, promoting CRC progression	Dietary microbial metabolites drive oncogenic ncRNA changes	Kong et al. (2012)
ETBF	Upregulation of lncRNA BFAL1 sequesters tumor-suppressive miRNAs	Microbial regulation of lncRNAs drives cancer progression	Bao et al. (2019)
SPF microbiota	Microbiota colonization upregulates specific miRNAs in gut mucosa	Microbiota directly impacts host miRNA expression	Dalmasso et al. (2011)
F. nucleatum	*F. nucleatum* enhances CRC via TLR4-MyD88-NF-κB signaling, upregulating miR-21	Reinforces the role of bacterial signaling in cancer development	Proenca et al. (2018), Sun et al. (2019, and Yang et al. (2017)

Salaquimonas_sp002400845, *Clostridium_AI_sp002297865*, *Enterocloster_citroniae*	Gut dysbiosis correlates with dysregulated miRNAs impacting inflammation and oncogenesis	Microbiome-ncRNA interactions in endocrine-related cancers	Li et al. (2024)
Gut microbiota (butyrate producers)	Butyrate inhibits miR-92a, reducing c-Myc and promoting apoptosis	Butyrate's anticancer effects via miRNA regulation	Hu et al. (2015)
Probiotics; *L. fermentum*, *L. salivarius*; *F. nucleatum*	Probiotics enhance miR-155 and miR-223; *F. nucleatum* downregulates miR-18a and miR-4802	Probiotics suppress cancer; *F. nucleatum* promotes resistance via ncRNAs	Li et al. (2020a)
F. nucleatum	*F. nucleatum* induces miR-21 and downregulates miR-18a and miR-4802 via TLR4-MyD88	Bacterial influence on treatment resistance through miRNAs	Yu et al. (2017)
Probiotic: *L. mesenteroides*	*L. mesenteroides* downregulates miR-21 and miR-200b, inducing apoptosis.	Probiotic potential in cancer therapy via ncRNA modulation	Zununi Vahed et al. (2017)

cells (Yuan et al. 2018). Supplementation with butyrate reduces miR-17 ~ 92a and miR-92a levels in human colorectal cancer lines, supporting its tumor-suppressive effects through enhancing CDKN1C (p57) expression and promoting cancer cell apoptosis (Yuan et al. 2018).

Dietary influences, the gut microbiome, and host ncRNAs, notably miRNAs and lncRNAs, interact intricately to modulate cancer CRC risk and progression. For example, fiber-rich diets enhance butyrate-producing microbiota and suppress oncogenic miRNAs, whereas high red meat intake upregulates miR-17 ~ 92 clusters and proliferative pathways in colonic mucosa (Humphreys et al. 2014; Schlörmann et al. 2015). This suggests that colonization of diet-derived fiber or butyrate-producing microbiota may prevent CRC.

Dietary patterns, particularly high-fat intake, stimulate hepatic secretion of primary bile acids, which are subsequently transformed by gut microbiota into secondary bile acids such as deoxycholic acid (DCA). DCA is recognized as a genotoxic metabolite that promotes CRC through several mechanisms. For example, DCA has been shown to downregulate the tumor-suppressive miR-199a-5p in CRC, thereby enhancing tumor cell survival and contributing to chemoresistance (Wang et al. 2012). This illustrates how dietary factors, via microbial metabolism, can drive oncogenic ncRNA alterations in the host.

Host miRNAs also regulate microbial composition and metabolism in the tumor microenvironment. For instance, miRNA-mediated regulation of glycan biosynthetic pathways can influence the colonization of bacteria such as *F. nucleatum*, which is frequently enriched in CRC. Studies have shown that increased mucin-type O-glycan biosynthesis positively correlates with the abundance of *F. nucleatum*, while inversely relating with Porphyromonas and Bacteroides (Yuan et al. 2018). Microbial signals also regulate host lncRNA expression with effects on tumorigenesis and immune function. For example, *F. nucleatum* enhances SP1 transcription factor binding at the ENO1-IT1 promoter, increasing lncRNA ENO1-IT1 expression, an event implicated in cancer and inflammation. In Enterotoxigenic *Bacteroides fragilis* (ETBF)-associated CRC, upregulation of BFAL1 (a *B. fragilis*–associated lncRNA) drives tumorigenesis by sequestering miR-155-5p and miR-200a-3p, attenuating their tumor-suppressive effects (Bao et al. 2019).

In a study by Dalmasso et al. (2011), miRNA expression patterns in the ileal and colonic mucosa were analyzed in germ-free (GF) mice compared to GF mice colonized with microbiota from specific pathogen-free (SPF) mice. Using a microarray containing 656 miRNAs, the researchers observed that colonization led to the upregulation of miR-298 in the ileum and miR-128, miR-200c-5p, and miR-342-5p in the colon. Conversely, expression of miR-465c-5p, miR-466d-3p, miR-466d-5p, miR-665, and miR-683 was decreased in the colon following colonization. By integrating computational target prediction with DNA microarray analysis, the study identified several genes modulated by these differentially expressed miRNAs. Notably, the ATP-binding cassette transporter ABCC3 was found to be a target of miR-665, suggesting that the gut microbiota may enhance ABCC3 expression in the colon by suppressing miR-665.

By triggering the Toll-like receptor 4 (TLR4)-MyD88 signaling pathway, *F. nucleatum* increased miR-21 in tumor cells, promoting the growth of tumors and the proliferation of CRC. RAS p21 protein activator 1 (RASA1) and programmed cell death 4 (PDCD4) are two tumor suppressor genes that miR-21 targets and inhibits their expression. It is widely acknowledged that *F. nucleatum* potentiates CRC development through TLR4 signaling, where TLR4 binds to MYD88 (Proenca et al. 2018; Sun et al. 2019; Yang et al. 2017) and subsequently activates the key downstream transcription factor (NF-κB) that can bind to the promoter region of miR-21 and elevate miR-21 expression.

Recent integrative analysis of early Hashimoto's thyroiditis (HT) reveals that gut microbiota dysbiosis may influence host transcriptomic regulation via ncRNAs, contributing to endocrine autoimmunity and cancer-relevant signaling. In the study by Li et al. (2024), blood and fecal samples of 31 early HT patients and 30 healthy individuals underwent RNA and metagenomic sequencing. The results revealed key microbial shifts and a strong association between three characteristic microbial species (*Salaquimonas_sp002400845, Clostridium_AI_sp002297865, and Enterocloster_citroniae*) that correlate strongly with dysregulated miRNA expression, particularly miR-548aq-3p and miR-374a-5p, impacting pathways linked to immune modulation, inflammation, and thyroid oncogenesis. These findings highlight the gut–ncRNA axis that mediates chronic inflammation and tissue-specific transcriptome remodeling, offering a mechanistic bridge between microbial perturbations and endocrine-related cancer development.

In their meta-transcriptomic investigation of prostate cancer tissues, Salachan et al. (2022) reveal microbial dysbiosis marked by an overabundance of Shewanella and reduced Vibrio parahaemolyticus and Staphylococcus saprophyticus in malignant samples as compared to benign prostate tissue samples, alongside significant host transcriptomic shifts. Although ncRNAs were not directly profiled in this study, altered immune signaling—such as suppressed Toll-like receptor pathways and changes in dendritic and macrophage cell enrichment—suggests a microbiome-induced rewiring of the tumor microenvironment, which is likely mediated by key point lncRNAs and miRNAs known to govern immune surveillance, inflammation, and stromal remodeling. These findings suggest a potential bidirectional axis whereby microbial communities may influence ncRNA expression, indirectly modulating tumor progression or suppression through immune and metabolic pathways.

Although the study by Mu et al. (2025), focuses primarily on the acidic microenvironment and the involvement of lncRNA LOC100507424 in the regulation of pancreatic cancer progression via the E2F1/FOXM1 axis, this study offers a framework that can be extrapolated to microbiome-ncRNA crosstalk. The acidic conditions in pancreatic tumors, largely driven by lactate accumulation and metabolic reprogramming as mentioned before, mirror microbiome-influenced environments where bacterial metabolites such as SCFAs and bile acids can alter pH and modulate ncRNA expression. LOC100507424, which regulates FOXM1-mediated glycolysis and stress adaptation, may act as a biochemical pipeline through which microbial signals influence tumor aggressiveness. Though the microbiota were not explicitly examined, the data suggest that lncRNA-mediated transcriptional responses to stress, including those influenced by microbial communities, could reinforce malignant phenotypes and may have therapeutic potential in altering tumor microenvironment dynamics.

The 2024 review by Xu et al. (2024), the authors explored how various ncRNAs serve as master regulators of metabolic reprogramming in breast cancer, particularly impacting glycolysis, lipid metabolism, and glutaminolysis. Although the microbiome is not the central focus in this review, the study reveals several mechanistic overlaps with microbially driven metabolic pathways. Key ncRNAs such as miR-155, circ_0001955, and TSPEAR-AS2 modulate glucose transporters (e.g., GLUT1 and GLUT3), while others (e.g., HOXC-AS3 and circPDCD11) regulate enzymes like lactate dehydrogenase (LDHA) and pyruvate kinase isoenzyme M2 (PKM2), which are responsive to microenvironmental changes and often shaped by microbial metabolites like SCFAs. Lipogenic changes induced by fatty acid synthase (FASN)-targeting circRNAs and tsRNAs interact with immunological signaling cascades that could be driven by gut-derived lipopolysaccharides and bile acids. Additionally, cholesterol and glycogen-modulating ncRNAs (e.g., tDR-4733 and tRF Lys−CTT−010) may be indirectly sensitive to microbial enzymatic activity through gut-liver axis signaling. These ncRNA-mediated metabolic adaptations contribute to treatment resistance, immune escape, and subtype specificity, reflecting a potential axis of microbiome-ncRNA crosstalk in tumor evolution. Clinically, such crosstalk may open new avenues for personalized therapy and biomarker development, positioning metabolism-regulating ncRNAs as integrators of microbial influence and metabolic vulnerabilities in breast cancer pathophysiology.

Another piece of evidence that underscores the intersection of noncoding RNA regulation and progression of endocrine-related cancer is the study by Cai et al., where the authors demonstrated that the androgen-repressed lncRNA LINC01126 drives castration-resistant prostate cancer (CRPC) by modulating posttranslational modifications of the androgen receptor (AR). Specifically, LINC01126 promotes a switch from O-GlcNAcylation at threonine 80 to phosphorylation at serine 81, enhancing AR nuclear translocation and transcriptional activity in the absence of androgens. These findings could also be linked indirectly to microbiome modulation, where the microbial metabolites such as SCFAs and polyamines could influence host epigenetic landscapes, potentially altering lncRNA expression and enzymes that mediate posttranslational modifications (PTMs) (e.g., CDK9, O-GlcNAc transferases (OGT), methyltransferases), thereby contributing to androgen-independent AR reactivation. This implies a possible pathway for gut microbiota-mediated epitranscriptomic alterations to drive lncRNA-dependent processes such as LINC01126 overexpression, supporting a microbiome-noncoding RNA axis in CRPC pathogenesis (Cai et al. 2024).

In summary, these studies collectively illustrate a dynamic regulatory network wherein the gut microbiota modulates host ncRNA expression, which in turn influences microbial ecology and tumor biology, emphasizing the therapeutic potential of targeting microbiota–ncRNA interactions.

Interplay of Microbiome and ncRNAs in Tumor Suppression (Table 2.3)

On the other hand, it has been demonstrated that intestinal microbiota and the metabolites they produce have beneficial tumor-suppression properties. Butyrate can cause apoptosis and differentiation in addition to inhibiting the growth of colorectal cancer cells via dysregulating the expression of miRNAs. According to Hu et al. (2015), butyrate inhibited the overexpression of oncogenic miR-92a in human CRC. They also observed a sharp decline in c-MYC and pri-miR-17-92a levels following butyrate therapy. Two intestinal probiotics, *Lactobacillus fermentum* (*L. fermentum*) and *Lactobacillus salivarius* (*L. salivarius*), have been shown to enhance miR-155 and miR-223 expression, while miR-150 and miR-143 expression were restored by *L. fermentum*. Additionally, the expression of miR-18a* and miR-4802, which were involved in autophagy-related pathways, may be inhibited by *F. nucleatum*, potentially leading to CRC chemoresistance. miR-4802 and miR-18a, which similarly depend on the TLR4 and MYD88 signaling pathway, are significantly downregulated in the *F. nucleatum–mediated* chemoresistance of CRC cells (Li et al. 2020a).

F. nucleatum infection is associated with chemotherapy tolerance, which induces miR-21, among other miRNAs, but downregulates miR-18a and miR-4802, triggered by TLR4-MyD88 activation (Yu et al. 2017). By controlling miRNAs, probiotic supplements can prevent tumor progression. When *Leuconostoc mesenteroides* (*L. mesenteroides*) and HT-29 cells were cocultured, it was demonstrated that this bacteria could efficiently downregulate miRNA-21 and miRNA-200b, which in turn triggered the apoptosis of colon cancer cells (Zununi Vahed et al. 2017).

Clinical Implications and Future Directions of Gut Microbiome-ncRNA Crosstalk in Cancer Development

The complex interaction between noncoding RNAs and the gut microbiome in gastrointestinal and endocrine-related cancers offers many opportunities for clinical application and future research.

1. Diagnostic and prognostic applications: The bidirectional interplay between the gut microbiome and ncRNAs, particularly miRNAs and lncRNAs, presents a promising source of disease biomarkers. Fecal and circulating miRNAs affected by microbiome composition have shown potential as noninvasive biomarkers for early detection of gastrointestinal cancers and inflammatory bowel disease (Ahmed et al. 2009). (Moloney et al. 2018). Moreover, specific microbiome-ncRNA signatures can predict patient responses to cancer immunotherapies and chemotherapies. For example, an abundance of certain bacterial taxa together with upregulated oncogenic miRNAs has been linked to chemoresistance in

colorectal cancer (D. Wang et al. 2012). Integration of patient microbiome and ncRNA profiles could guide precision oncology approaches.
2. Therapeutic applications: Engineered probiotics, which produce beneficial metabolites or even ncRNA therapeutics, are a novel approach to correcting dysbiosis and modulating gene expression. These therapies could target the underlying causes of chronic inflammation and metabolic imbalance (Drago et al. 2025; Murali and Mansell 2024). On the other hand, the development of novel delivery systems for fecal microbiota transplantation could reshape microbial communities and regulate host and microbial ncRNA expression.
3. Multi-omics integration: Future research should prioritize the development of comprehensive multi-omics techniques that profile the microbiome (metagenomics, metatranscriptomics), host ncRNA expression (RNA-seq, microarray), and proteomics all at once. This integrative research could provide a systems-level understanding of the microbiome-ncRNA crosstalk in cancer development and progression, revealing new regulatory networks and possible treatment targets.
4. Research beyond bacteria: Most studies have focused on bacterial components of the microbiome, but future studies should explore other components of the microbiome, viruses (phages), fungi (mycobiomes), and archaea, and their potential interaction with host ncRNAs.
5. Expanding research to additional cancer types: Current studies on the crosstalk between the microbiome and noncoding RNAs primarily focus on gastrointestinal and endocrine-related cancers. Future research should explore the microbiome-ncRNA axis in various cancers, including gynecological, lung, skin, and systemic effects on distant tumor sites.
6. Clinical translation and validation: Large-scale, long-term clinical investigations are required to validate preclinical findings and show the clinical implications of microbiome-ncRNA crosstalk and the efficacy of microbiota-targeted therapies in cancer patients.

Conclusion

The intricate interplay between the gut microbiome and ncRNAs has emerged as a critical factor in the regulation of endocrine function and the pathogenesis of various malignancies, especially the gastrointestinal and endocrine-related disorders. This book chapter highlights how gut microbial communities influence hormone metabolism, immune responses, and the synthesis of bioactive compounds, while host-derived ncRNAs modulate both microbial composition and activity. Conversely, microbiota-derived metabolites and ncRNAs also impact host gene regulation, illustrating a bidirectional communication essential for maintaining endocrine homeostasis.

Furthermore, the involvement of various ncRNA classes, including miRNAs, lncRNAs, and circRNAs, in various types of cancer underscores their significance as both regulators and potential biomarkers. Dysregulation of ncRNAs can alter the gut microbiome's structure and function, potentially exacerbating disease progression.

Collectively, these findings underscore the importance of investigating the microbiome-ncRNA axis to better unravel the molecular mechanisms underlying endocrine health and disease. Future research focused on this crosstalk holds the potential to promote the development of innovative therapeutic strategies that simultaneously target microbial communities and ncRNA regulatory networks, ultimately advancing precision medicine and improving clinical outcomes in various disorders.

References

Ahmed, F. E., et al. (2009). Diagnostic microRNA markers for screening sporadic human colon cancer and active ulcerative colitis in stool and tissue. *Cancer genomics & proteomics, 6*(5), 281–295.

Ahmed, H. M., Shehata, H. H., El-Saeed, G. S. M., Gabal, H. H. A., & El-Daly, S. M. (2022). Ameliorative effect of Lactobacillus rhamnosus GG on acetaminophen-induced hepatotoxicity via PKC/Nrf2/PGC-1α pathway. *J Genet Eng Biotechnol, 20*(1), 142. doi:https://doi.org/10.1186/s43141-022-00422-4

Anwar, M. M., Albanese, C., Hamdy, N. M., & Sultan, A. S. (2022). Rise of the natural red pigment 'prodigiosin' as an immunomodulator in cancer. *Cancer Cell International, 22*(1), 419. doi:https://doi.org/10.1186/s12935-022-02815-4

Arnone, A. A., & Cook, K. L. J. E. (2023). Gut and breast microbiota as endocrine regulators of hormone receptor-positive breast cancer risk and therapy response. *164*(1), bqac177.

Bao, Y., et al. (2019). Long noncoding RNA BFAL1 mediates enterotoxigenic Bacteroides fragilis-related carcinogenesis in colorectal cancer via the RHEB/mTOR pathway. *Cell Death Dis, 10*(9), 675. doi:https://doi.org/10.1038/s41419-019-1925-2

Bartel, D. P. (2009). MicroRNAs: target recognition and regulatory functions. *cell, 136*(2), 215-233.

Belfiore, A., & Perks, C. M. J. F. i. E. (2013). Grand challenges in cancer endocrinology: endocrine related cancers, an expanding concept. In (Vol. 4, pp. 141): Frontiers Media SA.

Bemark, M., Pitcher, M. J., Dionisi, C., & Spencer, J. (2024). Gut-associated lymphoid tissue: a microbiota-driven hub of B cell immunity. *Trends in Immunology, 45*(3), 211–223. doi:https://doi.org/10.1016/j.it.2024.01.006

Bercik, P., et al. (2011). The intestinal microbiota affect central levels of brain-derived neurotropic factor and behavior in mice. *Gastroenterology, 141*(2), 599–609, 609 e591–593. doi:https://doi.org/10.1053/j.gastro.2011.04.052

Bian, Z., et al. (2024). LINC01852 inhibits the tumorigenesis and chemoresistance in colorectal cancer by suppressing SRSF5-mediated alternative splicing of PKM. *Mol Cancer, 23*(1), 23. doi:https://doi.org/10.1186/s12943-024-01939-7

Biton, M., et al. (2011). Epithelial microRNAs regulate gut mucosal immunity via epithelium–T cell crosstalk. *12*(3), 239–246.

Boicean, A., Birlutiu, V., Ichim, C., Anderco, P., & Birsan, S. (2023). Fecal Microbiota Transplantation in Inflammatory Bowel Disease. *Biomedicines, 11*(4). doi:https://doi.org/10.3390/biomedicines11041016

Bradley, E., & Haran, J. (2024). The human gut microbiome and aging. *Gut Microbes, 16*(1), 2359677. doi:https://doi.org/10.1080/19490976.2024.2359677

Bradley, R. K., & Anczuków, O. (2023). RNA splicing dysregulation and the hallmarks of cancer. *Nat Rev Cancer, 23*(3), 135–155. doi:https://doi.org/10.1038/s41568-022-00541-7

Braga, J. D., Thongngam, M., & Kumrungsee, T. (2024). Gamma-aminobutyric acid as a potential postbiotic mediator in the gut-brain axis. *NPJ Sci Food, 8*(1), 16. doi:https://doi.org/10.1038/s41538-024-00253-2

Breton, J., Galmiche, M., & Déchelotte, P. (2022). Dysbiotic Gut Bacteria in Obesity: An Overview of the Metabolic Mechanisms and Therapeutic Perspectives of Next-Generation Probiotics. *Microorganisms, 10*(2). doi:https://doi.org/10.3390/microorganisms10020452

Bridgeman, S. C., Northrop, W., Melton, P. E., Ellison, G. C., Newsholme, P., & Mamotte, C. D. S. (2020). Butyrate generated by gut microbiota and its therapeutic role in metabolic syndrome. *Pharmacological Research, 160*, 105174. doi:https://doi.org/10.1016/j.phrs.2020.105174

Brown, A. J., et al. (2003). The Orphan G protein-coupled receptors GPR41 and GPR43 are activated by propionate and other short chain carboxylic acids. *J Biol Chem, 278*(13), 11312–11319. doi:https://doi.org/10.1074/jbc.M211609200

Cai, Y., et al. (2024). Androgen-repressed lncRNA LINC01126 drives castration-resistant prostate cancer by regulating the switch between O-GlcNAcylation and phosphorylation of androgen receptor. *14*(1), e1531. https://doi.org/10.1002/ctm2.1531

Cani, P. D., et al. (2009). Gut microbiota fermentation of prebiotics increases satietogenic and incretin gut peptide production with consequences for appetite sensation and glucose response after a meal. *American Journal of Clinical Nutrition, 90*(5), 1236–1243. doi:https://doi.org/10.3945/ajcn.2009.28095

Cao, X., Ma, C., Wu, Y., & Huang, J. (2024). lncRNA HCG22 regulated cell growth and metastasis of papillary thyroid cancer via negatively modulating miR-425-5p. *Endokrynol Pol, 75*(1), 20–26. doi:https://doi.org/10.5603/ep.97425

Carlessi, R., et al. (2017). GLP-1 receptor signalling promotes β-cell glucose metabolism via mTOR-dependent HIF-1α activation. *Scientific Reports, 7*(1), 2661. doi:https://doi.org/10.1038/s41598-017-02838-2

Chen, C., et al. (2022). Tumor-suppressive circRHOBTB3 is excreted out of cells via exosome to sustain colorectal cancer cell fitness. *21*(1), 46.

Chen, H., Cao, J., Zhang, F., & Xiong, W. (2024). Significance of Gut Microbiota on Graves' Disease. *International Journal of General Medicine, 17*, 3967–3974. doi:https://doi.org/10.2147/ijgm.S467888

Chen, X., et al. (2023a). Estrogen-Induced LncRNA, LINC02568, Promotes Estrogen Receptor-Positive Breast Cancer Development and Drug Resistance Through Both In Trans and In Cis Mechanisms. *Adv Sci (Weinh), 10*(25), e2206663. doi:https://doi.org/10.1002/advs.202206663

Chen, Z., Li, Y., He, K., Yang, J., Deng, Q., Chen, Y., & Fu, Z. (2023b). CircGPRC5A enhances colorectal cancer progress by stabilizing PPP1CA and inducing YAP dephosphorylation. *J Exp Clin Cancer Res, 42*(1), 334. doi:https://doi.org/10.1186/s13046-023-02915-7

Cheng, L., Deng, X., & Le, Y. (2024). MYB/LINC00092 regulatory axis promotes the progression of papillary thyroid carcinoma. *Endokrynol Pol, 75*(1), 27–34. doi:https://doi.org/10.5603/ep.98120

Chiang, J. Y. L., & Ferrell, J. M. (2020). Bile acid receptors FXR and TGR5 signaling in fatty liver diseases and therapy. *American Journal of Physiology: Gastrointestinal and Liver Physiology, 318*(3), G554-g573. doi:https://doi.org/10.1152/ajpgi.00223.2019

Clemente-Suárez, V. J., Redondo-Flórez, L., Rubio-Zarapuz, A., Martín-Rodríguez, A., & Tornero-Aguilera, J. F. (2024). Microbiota Implications in Endocrine-Related Diseases: From Development to Novel Therapeutic Approaches. *Biomedicines, 12*(1). doi:https://doi.org/10.3390/biomedicines12010221

Cui, Z., et al. (2021). LncRNA PlncRNA-1 accelerates the progression of prostate cancer by regulating PTEN/Akt axis. *13*(8), 12113.

D'Aquila, P., Carelli, L. L., De Rango, F., Passarino, G., & Bellizzi, D. (2020). Gut Microbiota as Important Mediator Between Diet and DNA Methylation and Histone Modifications in the Host. *Nutrients, 12*(3). doi:https://doi.org/10.3390/nu12030597

Dai, X., et al. (2015). MicroRNA-193a-3p Reduces Intestinal Inflammation in Response to Microbiota via Down-regulation of Colonic PepT1. *J Biol Chem, 290*(26), 16099–16115. doi:https://doi.org/10.1074/jbc.M115.659318

Dalmasso, G., Antony, C., Julien, D., Arlette, D.-M., & Bonnet, R. (2014). The bacterial genotoxin colibactin promotes colon tumor growth by modifying the tumor microenvironment. *Gut Microbes, 5*(5), 675–680. doi:https://doi.org/10.4161/19490976.2014.969989

Dalmasso, G., et al. (2011). Microbiota modulate host gene expression via microRNAs. *PloS one, 6*(4), e19293. doi:https://doi.org/10.1371/journal.pone.0019293

Davie, J. R. (2003). Inhibition of histone deacetylase activity by butyrate. *Journal of Nutrition, 133*(7 Suppl), 2485s-2493s. doi:https://doi.org/10.1093/jn/133.7.2485S

DeGruttola, A. K., Low, D., Mizoguchi, A., & Mizoguchi, E. (2016). Current Understanding of Dysbiosis in Disease in Human and Animal Models. *Inflamm Bowel Dis, 22*(5), 1137–1150. doi:https://doi.org/10.1097/MIB.0000000000000750

Dempsey, J., Zhang, A., & Cui, J. Y. (2018). Coordinate regulation of long non-coding RNAs and protein-coding genes in germ-free mice. *BMC genomics, 19*(1), 834. doi:https://doi.org/10.1186/s12864-018-5235-3

den Besten, G., van Eunen, K., Groen, A. K., Venema, K., Reijngoud, D. J., & Bakker, B. M. (2013). The role of short-chain fatty acids in the interplay between diet, gut microbiota, and host energy metabolism. *Journal of Lipid Research, 54*(9), 2325–2340. doi:https://doi.org/10.1194/jlr.R036012

Deng, W., et al. (2022). Novel circular RNA circ_0086722 drives tumor progression by regulating the miR-339-5p/STAT5A axis in prostate cancer. *533*, 215606.

Dey, B. K., Pfeifer, K., & Dutta, A. (2014). The H19 long noncoding RNA gives rise to microRNAs miR-675-3p and miR-675-5p to promote skeletal muscle differentiation and regeneration. *Genes Dev, 28*(5), 491–501. doi:https://doi.org/10.1101/gad.234419.113

Di Vincenzo, F., Del Gaudio, A., Petito, V., Lopetuso, L. R., & Scaldaferri, F. (2024). Gut microbiota, intestinal permeability, and systemic inflammation: a narrative review. *Internal and Emergency Medicine, 19*(2), 275–293. doi:https://doi.org/10.1007/s11739-023-03374-w

Diling, C., et al. (2020). CircNF1-419 improves the gut microbiome structure and function in AD-like mice. *Aging (Albany NY), 12*(1), 260–287. doi:https://doi.org/10.18632/aging.102614

Ding, N., You, A. B., Yang, H., Hu, G. S., Lai, C. P., Liu, W., & Ye, F. (2023). A Tumor-suppressive Molecular Axis EP300/circRERE/miR-6837-3p/MAVS Activates Type I IFN Pathway and Antitumor Immunity to Suppress Colorectal Cancer. *Clin Cancer Res, 29*(11), 2095–2109. doi:https://doi.org/10.1158/1078-0432.Ccr-22-3836

Dong, F., & Perdew, G. H. (2020). The aryl hydrocarbon receptor as a mediator of host-microbiota interplay. *Gut Microbes, 12*(1), 1859812. doi:https://doi.org/10.1080/19490976.2020.1859812

Dong, Q., Qiu, H., Piao, C., Li, Z., & Cui, X. (2023). LncRNA SNHG4 promotes prostate cancer cell survival and resistance to enzalutamide through a let-7a/RREB1 positive feedback loop and a ceRNA network. *J Exp Clin Cancer Res, 42*(1), 209. doi:https://doi.org/10.1186/s13046-023-02774-2

Doumatey, A. P., Adeyemo, A., Zhou, J., Lei, L., Adebamowo, S. N., Adebamowo, C., & Rotimi, C. N. (2020). Gut Microbiome Profiles Are Associated With Type 2 Diabetes in Urban Africans. *Front Cell Infect Microbiol, 10*, 63. doi:https://doi.org/10.3389/fcimb.2020.00063

Drago, L., et al. (2025). Systematic review of bidirectional interaction between gut microbiome, miRNAs, and human pathologies. *16*, 1540943.

Du, C.-T., et al. (2018). MicroRNA-146a Deficiency Protects against Listeria monocytogenes Infection by Modulating the Gut Microbiota. *19*(4), 993.

Du, J., et al. (2021). Dietary betaine prevents obesity through gut microbiota-drived microRNA-378a family. *Gut Microbes, 13*(1), 1862612. doi:https://doi.org/10.1080/19490976.2020.1862612

El-Daly, S. M., Abdelrahman, S. S., Abd El-Jawad, A. M., Abdel-Monem, M. A., & El-Saeed, G. S. M. (2025a). Diagnostic Potential of Exosomal and Non-Exosomal Biomarkers in Lung Cancer: A Comparative Analysis Using a Rat Model of Lung Carcinogenesis. *11*(3), 47.

El-Daly, S. M., Abdelrahman, S. S., El-Bana, M. A., Abdel-Latif, Y., Medhat, D., Morsy, S. M., & Wafay, H. A. (2025b). Deciphering the Interplay of the PD-L1/MALT1/miR-200a Axis During Lung Cancer Development. *Biotechnol Appl Biochem*. doi:https://doi.org/10.1002/bab.2724

El-Daly, S. M., El-Bana, M. A., Abd El-Rahman, S. S., Latif, Y. A., & Medhat, D. (2023a). Dynamic expression of H19 and MALAT1 and their correlation with tumor progression biomarkers in a multistage hepatocarcinogenesis model. *Cell Biochem Funct, 41*(3), 331–343. doi:https://doi.org/10.1002/cbf.3785

El-Daly, S. M., Talaat, R. M., Braoudaki, M., Youness, R. A., & Cho, W. C. (2023b). Editorial: Recent breakthroughs in the decoding of circulating nucleic acids and their applications to human diseases. *Front Mol Biosci, 10*, 1203495. doi:https://doi.org/10.3389/fmolb.2023.1203495

El-Mesallamy, H. O., Hamdy, N. M., Rizk, H. H., & El-Zayadi, A. R. (2011). Apelin serum level in Egyptian patients with chronic hepatitis C. *Mediators Inflamm, 2011*, 703031. doi:https://doi.org/10.1155/2011/703031

El-Mesallamy, H. O., Hamdy, N. M., & Sallam, A. A. (2013). Effect of obesity and glycemic control on serum lipocalins and insulin-like growth factor axis in type 2 diabetic patients. *Acta Diabetol, 50*(5), 679–685. doi:https://doi.org/10.1007/s00592-012-0373-6

El-Toukhy, S. E., El-Daly, S. M., Kamel, M. M., & Nabih, H. K. (2023). The diagnostic significance of circulating miRNAs and metabolite profiling in early prediction of breast cancer in Egyptian women. *J Cancer Res Clin Oncol, 149*(8), 5437–5451. doi:https://doi.org/10.1007/s00432-022-04492-2

Fabian, M. R., Sonenberg, N., & Filipowicz, W. (2010). Regulation of mRNA translation and stability by microRNAs. *Annu Rev Biochem, 79*, 351–379. doi:https://doi.org/10.1146/annurev-biochem-060308-103103

Fang, H., Rodrigues, E. L. R., Barra, N. G., Kukje Zada, D., Robin, N., Mehra, A., & Schertzer, J. D. (2024). Postbiotic impact on host metabolism and immunity provides therapeutic potential in metabolic disease. *Endocrine Reviews*. doi:https://doi.org/10.1210/endrev/bnae025

Fang, J., Wang, H., Zhou, Y., Zhang, H., Zhou, H., & Zhang, X. (2021). Slimy partners: the mucus barrier and gut microbiome in ulcerative colitis. *Experimental and Molecular Medicine, 53*(5), 772–787. doi:https://doi.org/10.1038/s12276-021-00617-8

Fard, S. S., & Holz, M. K. J. S. (2023). Regulation of mRNA translation by estrogen receptor in breast cancer. *200*, 109316.

Fayed, B., Abood, A., El-Sayed, H. S., Hashem, A. M., & Mehanna, N. S. J. C. p. (2018). A synbiotic multiparticulate microcapsule for enhancing inulin intestinal release and Bifidobacterium gastro-intestinal survivability. *193*, 137–143.

Fayed, B., El-Sayed, H., & Ismail, S. (2021). Novel assessment of synergistic stimulatory effect of prebiotic chitooligosaccharide and some commercial prebiotics on the probiotic growth: a preliminary study. *J. microbiol., biotechnol. food sci., 11*(1), e3341.

Fayed, B., El-Sayed, H. S., Abood, A., Hashem, A. M., Mehanna, N. S. J. B., & Biotechnology, A. (2019). The application of multi-particulate microcapsule containing probiotic bacteria and inulin nanoparticles in enhancing the probiotic survivability in yoghurt. *22*, 101391.

Fayed, B., et al. (2023). Co-administration of amoxicillin-loaded chitosan nanoparticles and inulin: A novel strategy for mitigating antibiotic resistance and preserving microbiota balance in Helicobacter pylori treatment. *International Journal of Biological Macromolecules, 253*(Pt 2), 126706. doi:https://doi.org/10.1016/j.ijbiomac.2023.126706

Feng, Y. Y., et al. (2019). Alteration of microRNA-4474/4717 expression and CREB-binding protein in human colorectal cancer tissues infected with Fusobacterium nucleatum. *PLoS One, 14*(4), e0215088. doi:https://doi.org/10.1371/journal.pone.0215088

Filip, A. T., Balacescu, O., Marian, C., & Anghel, A. (2016). Microbiota Small RNAs in Inflammatory Bowel Disease. *J Gastrointestin Liver Dis, 25*(4), 509–516. doi:https://doi.org/10.15403/jgld.2014.1121.254.lip

Flint, H. J., Scott, K. P., Louis, P., & Duncan, S. H. (2012). The role of the gut microbiota in nutrition and health. *Nature Reviews: Gastroenterology & Hepatology, 9*(10), 577–589. doi:https://doi.org/10.1038/nrgastro.2012.156

Gaddam, R. R., et al. (2020). Microbiota-governed microRNA-204 impairs endothelial function and blood pressure decline during inactivity in db/db mice. *Sci Rep, 10*(1), 10065. doi:https://doi.org/10.1038/s41598-020-66786-0

Gao, W., et al. (2024). Alterations in gut microbiota and inflammatory cytokines after administration of antibiotics in mice. *Microbiol Spectr, 12*(8), e0309523. doi:https://doi.org/10.1128/spectrum.03095-23

Geisler, L., et al. (2025). miR-223 and Chromogranin A Affect Inflammatory Immune Cell Activation in Liver Metastasis of Neuroendocrine Neoplasms. *Cells, 14*(2). doi:https://doi.org/10.3390/cells14020111

Gong, P., Zhang, T., He, D., & Hsieh, J.-T. J. R. (2015). MicroRNA-145 modulates tumor sensitivity to radiation in prostate cancer. *184*(6), 630–638.

Gouhar, S. A., Abd Elmageed, Z. Y., & El-Daly, S. M. (2025). Extracellular vesicles as dynamic biomarkers: Monitoring therapeutic responses and drug resistance in cancer. In *International Review of Cell and Molecular Biology*: Academic Press.

Gouhar, S. A., Abo-elfadl, M. T., Gamal-Eldeen, A. M., & El-Daly, S. M. J. E. T. (2022). Involvement of miRNAs in response to oxidative stress induced by the steroidal glycoalkaloid α-solanine in hepatocellular carcinoma cells. *37*(2), 212–223.

Gubatan, J., Holman, D. R., Puntasecca, C. J., Polevoi, D., Rubin, S. J., & Rogalla, S. (2021). Antimicrobial peptides and the gut microbiome in inflammatory bowel disease. *World Journal of Gastroenterology, 27*(43), 7402–7422. doi:https://doi.org/10.3748/wjg.v27.i43.7402

Gueddouri, D., et al. (2022). Insulin resistance per se drives early and reversible dysbiosis-mediated gut barrier impairment and bactericidal dysfunction. *Mol Metab, 57*, 101438. doi:https://doi.org/10.1016/j.molmet.2022.101438

Guo, J., Si, G., & Si, F. (2024). Treg cells as a protective factor for Hashimoto's thyroiditis: a Mendelian randomization study. *Frontiers in Endocrinology, 15*, 1347695. doi:https://doi.org/10.3389/fendo.2024.1347695

Guo, R., Shi, L., Chen, Y., Lin, C., & Yin, W. (2025). Exploring the roles of ncRNAs in prostate cancer via the PI3K/AKT/mTOR signaling pathway. *Front Immunol, 16*, 1525741. doi:https://doi.org/10.3389/fimmu.2025.1525741

Hadadi, N., Berweiler, V., Wang, H., & Trajkovski, M. (2021). Intestinal microbiota as a route for micronutrient bioavailability. *Curr Opin Endocr Metab Res, 20*, 100285. doi:https://doi.org/10.1016/j.coemr.2021.100285

Hamamah, S., Aghazarian, A., Nazaryan, A., Hajnal, A., & Covasa, M. (2022). Role of Microbiota-Gut-Brain Axis in Regulating Dopaminergic Signaling. *Biomedicines, 10*(2). doi:https://doi.org/10.3390/biomedicines10020436

Hamdy, N. M., et al. (2024a). In silico analysis and comprehensive review of circular-RNA regulatory roles in breast diseases; a step-toward non-coding RNA precision. *Pathol Res Pract, 263*, 155651. doi:https://doi.org/10.1016/j.prp.2024.155651

Hamdy, N. M., et al. (2025a). LincRNA-miR interactions in hepatocellular carcinoma: comprehensive review and in silico analysis: a step toward ncRNA precision. *Naunyn Schmiedebergs Arch Pharmacol*. doi:https://doi.org/10.1007/s00210-025-04285-7

Hamdy, N. M., et al. (2025b). Insights into the genetic and epigenetic mechanisms governing X-chromosome-linked-miRNAs expression in cancer; a step-toward ncRNA precision. *289*, 138773.

Hamdy, N. M., Zaki, M. B., Rizk, N. I., Abdelmaksoud, N. M., Abd-Elmawla, M. A., Ismail, R. A., & Abulsoud, A. I. (2024b). Unraveling the ncRNA landscape that governs colorectal cancer: A roadmap to personalized therapeutics. *Life Sci, 354*, 122946. doi:https://doi.org/10.1016/j.lfs.2024.122946

Hamer, H. M., Jonkers, D., Venema, K., Vanhoutvin, S., Troost, F. J., & Brummer, R. J. (2008). Review article: the role of butyrate on colonic function. *Aliment Pharmacol Ther, 27*(2), 104–119. doi:https://doi.org/10.1111/j.1365-2036.2007.03562.x

He, L., Zhong, Z., Wen, S., Li, P., Jiang, Q., & Liu, F. (2024). Gut microbiota-derived butyrate restores impaired regulatory T cells in patients with AChR myasthenia gravis via mTOR-mediated autophagy. *Cell Commun Signal, 22*(1), 215. doi:https://doi.org/10.1186/s12964-024-01588-9

He, L., Zhou, X., Liu, Y., Zhou, L., & Li, F. (2022). Fecal miR-142a-3p from dextran sulfate sodium-challenge recovered mice prevents colitis by promoting the growth of Lactobacillus reuteri. *Mol Ther, 30*(1), 388–399. doi:https://doi.org/10.1016/j.ymthe.2021.08.025

Hernández, M. A. G., Canfora, E. E., Jocken, J. W. E., & Blaak, E. E. (2019). The Short-Chain Fatty Acid Acetate in Body Weight Control and Insulin Sensitivity. *Nutrients, 11*(8). doi:https://doi.org/10.3390/nu11081943

Hidalgo-Cantabrana, C., Delgado, S., Ruiz, L., Ruas-Madiedo, P., Sánchez, B., & Margolles, A. (2017). Bifidobacteria and Their Health-Promoting Effects. *Microbiol Spectr, 5*(3). doi:https://doi.org/10.1128/microbiolspec.BAD-0010-2016

Hills, R. D., Pontefract, B. A., Mishcon, H. R., Black, C. A., Sutton, S. C., & Theberge, C. R. (2019). Gut Microbiome: Profound Implications for Diet and Disease. *Nutrients, 11*(7). doi:https://doi.org/10.3390/nu11071613

Hou, K., et al. (2022). Microbiota in health and diseases. *Signal Transduct Target Ther, 7*(1), 135. doi:https://doi.org/10.1038/s41392-022-00974-4

Hou, Q., Huang, Y., Zhu, S., Li, P., Chen, X., Hou, Z., & Liu, F. (2017). MiR-144 Increases Intestinal Permeability in IBS-D Rats by Targeting OCLN and ZO1. *Cell Physiol Biochem, 44*(6), 2256–2268. doi:https://doi.org/10.1159/000486059

Hu, H. F., et al. (2024). LINC00982-encoded protein PRDM16-DT regulates CHEK2 splicing to suppress colorectal cancer metastasis and chemoresistance. *Theranostics, 14*(8), 3317–3338. doi:https://doi.org/10.7150/thno.95485

Hu, S., Ding, Q., Zhang, W., Kang, M., Ma, J., & Zhao, L. (2023). Gut microbial beta-glucuronidase: a vital regulator in female estrogen metabolism. *Gut Microbes, 15*(1), 2236749. doi:https://doi.org/10.1080/19490976.2023.2236749

Hu, S., Dong, T. S., Dalal, S. R., Wu, F., Bissonnette, M., Kwon, J. H., & Chang, E. B. (2011). The microbe-derived short chain fatty acid butyrate targets miRNA-dependent p21 gene expression in human colon cancer. *PloS one, 6*(1), e16221. doi:https://doi.org/10.1371/journal.pone.0016221

Hu, S., Liu, L., Chang, E. B., Wang, J. Y., & Raufman, J. P. (2015). Butyrate inhibits pro-proliferative miR-92a by diminishing c-Myc-induced miR-17-92a cluster transcription in human colon cancer cells. *Mol Cancer, 14*, 180. doi:https://doi.org/10.1186/s12943-015-0450-x

Hu, X. T., et al. (2020). HDAC2 inhibits EMT-mediated cancer metastasis by downregulating the long noncoding RNA H19 in colorectal cancer. *J Exp Clin Cancer Res, 39*(1), 270. doi:https://doi.org/10.1186/s13046-020-01783-9

Humphreys, K. J., et al. (2014). Dietary manipulation of oncogenic microRNA expression in human rectal mucosa: a randomized trial. *Cancer Prev Res (Phila), 7*(8), 786–795. doi:https://doi.org/10.1158/1940-6207.Capr-14-0053

Hussain, T., et al. (2021). Relationship between gut microbiota and host-metabolism: Emphasis on hormones related to reproductive function. *Anim Nutr, 7*(1), 1–10. doi:https://doi.org/10.1016/j.aninu.2020.11.005

Hussein, J., Medhat, D., El Bana, M., El-Naggar, M., Omara, E., & El-Daly, S. M. (2024). Delivery of Eicosapentaenoic acid-loaded cellulose nanocrystal alleviates liver fibrosis via modulating phospholipids and apoptotic regulators. *Journal of Drug Delivery Science and Technology, 97*, 105796. https://doi.org/10.1016/j.jddst.2024.105796

Hwang, H. J., & Kim, Y. K. (2024). Molecular mechanisms of circular RNA translation. *Experimental & Molecular Medicine, 56*(6), 1272–1280. doi:https://doi.org/10.1038/s12276-024-01220-3

Insenser, M., Murri, M., Del Campo, R., Martínez-García, M., Fernández-Durán, E., & Escobar-Morreale, H. F. (2018). Gut Microbiota and the Polycystic Ovary Syndrome: Influence of Sex, Sex Hormones, and Obesity. *Journal of Clinical Endocrinology and Metabolism, 103*(7), 2552–2562. doi:https://doi.org/10.1210/jc.2017-02799

Ismail, S. A., El-Sayed, H. S., & Fayed, B. (2020). Production of prebiotic chitooligosaccharide and its nano/microencapsulation for the production of functional yoghurt. *Carbohydrate Polymers, 234*, 115941. doi:https://doi.org/10.1016/j.carbpol.2020.115941

Ji, Y., Li, X., Zhu, Y., Li, N., Zhang, N., & Niu, M. (2018). Faecal microRNA as a biomarker of the activity and prognosis of inflammatory bowel diseases. *Biochem Biophys Res Commun, 503*(4), 2443-2450. doi:https://doi.org/10.1016/j.bbrc.2018.06.174

Johnston, D. G., et al. (2018). Loss of microRNA-21 influences the gut microbiota, causing reduced susceptibility in a murine model of colitis. *Journal of Crohn's and Colitis, 12*(7), 835–848.

Josefowicz, S. Z., et al. (2012). Extrathymically generated regulatory T cells control mucosal TH2 inflammation. *Nature, 482*(7385), 395–399. doi:https://doi.org/10.1038/nature10772

Kalinina, T., Kononchuk, V., Alekseenok, E., Obukhova, D., Sidorov, S., Strunkin, D., & Gulyaeva, L. (2021). Expression of Estrogen Receptor- and Progesterone Receptor-Regulating MicroRNAs in Breast Cancer. *Genes (Basel), 12*(4). doi:https://doi.org/10.3390/genes12040582

Kamada, N., Seo, S. U., Chen, G. Y., & Nunez, G. (2013). Role of the gut microbiota in immunity and inflammatory disease. *Nat Rev Immunol, 13*(5), 321–335. doi:https://doi.org/10.1038/nri3430

Kang, J., Sun, M., Chang, Y., Chen, H., Zhang, J., Liang, X., & Xiao, T. (2023). Butyrate ameliorates colorectal cancer through regulating intestinal microecological disorders. *Anti-Cancer Drugs, 34*(2), 227–237. doi:https://doi.org/10.1097/cad.0000000000001413

Kassan, A., Ait-Aissa, K., & Kassan, M. (2021). Hypothalamic miR-204 Induces Alteration of Heart Electrophysiology and Neurogenic Hypertension by Regulating the Sympathetic Nerve Activity: Potential Role of Microbiota. *Cureus, 13*(10), e18783. doi:https://doi.org/10.7759/cureus.18783

Kassan, A., Ait-Aissa, K., & Kassan, M. (2022). Gut Microbiota Regulates the Sympathetic Nerve Activity and Peripheral Serotonin Through Hypothalamic MicroRNA-204 in Order to Increase the Browning of White Adipose Tissue in Obesity. *Cureus, 14*(2), e21913. doi:https://doi.org/10.7759/cureus.21913

Kim, C. H. (2018). Immune regulation by microbiome metabolites. *Immunology, 154*(2), 220–229. doi:https://doi.org/10.1111/imm.12930

Kim, H., & Fang, S. (2018). Crosstalk between FXR and TGR5 controls glucagon-like peptide 1 secretion to maintain glycemic homeostasis. *Lab Anim Res, 34*(4), 140–146. doi:https://doi.org/10.5625/lar.2018.34.4.140

Kong, Y., Bai, P. S., Sun, H., Nan, K. J., Chen, N. Z., & Qi, X. G. (2012). The deoxycholic acid targets miRNA-dependent CAC1 gene expression in multidrug resistance of human colorectal cancer. *Int J Biochem Cell Biol, 44*(12), 2321–2332. doi:https://doi.org/10.1016/j.biocel.2012.08.006

Kwa, M., Plottel, C. S., Blaser, M. J., & Adams, S. (2016). The Intestinal Microbiome and Estrogen Receptor-Positive Female Breast Cancer. *Journal of the National Cancer Institute, 108*(8). doi:https://doi.org/10.1093/jnci/djw029

Lee, J., et al. (2015). Profiles of microRNA networks in intestinal epithelial cells in a mouse model of colitis. *Scientific Reports, 5*(1), 18174.

Leeuwendaal, N. K., Stanton, C., O'Toole, P. W., & Beresford, T. P. (2022). Fermented Foods, Health and the Gut Microbiome. *Nutrients, 14*(7). doi:https://doi.org/10.3390/nu14071527

Li, J., Zhang, Z., Hu, J., Wan, X., Huang, W., Zhang, H., & Jiang, N. (2022a). MiR-1246 regulates the PI3K/AKT signaling pathway by targeting PIK3AP1 and inhibits thyroid cancer cell proliferation and tumor growth. *Mol Cell Biochem, 477*(3), 649–661. doi:https://doi.org/10.1007/s11010-021-04290-3

Li, L., Li, C., Lv, M., Hu, Q., Guo, L., & Xiong, D. (2020a). Correlation between alterations of gut microbiota and miR-122-5p expression in patients with type 2 diabetes mellitus. *Ann Transl Med, 8*(22), 1481. doi:https://doi.org/10.21037/atm-20-6717

Li, M., et al. (2024). Integrative analysis of gut microbiome and host transcriptome reveal novel molecular signatures in Hashimoto's thyroiditis. *Journal of Translational Medicine, 22*(1), 1045. doi:https://doi.org/10.1186/s12967-024-05876-3

Li, S., et al. (2020b). Circular RNA cir-ITCH is a potential therapeutic target for the treatment of castration-resistant prostate cancer. *2020*(1), 7586521.

Li, T., et al. (2019). MicroRNA-92a-1–5p increases CDX2 by targeting FOXD1 in bile acids-induced gastric intestinal metaplasia. *Gut, 68*(10), 1751. doi:https://doi.org/10.1136/gutjnl-2017-315318

Li, X., et al. (2022b). Exosome-derived lnc-HOXB8-1:2 induces tumor-associated macrophage infiltration to promote neuroendocrine differentiated colorectal cancer progression by sponging hsa-miR-6825-5p. *BMC Cancer, 22*(1), 928. doi:https://doi.org/10.1186/s12885-022-09926-1

Li, X., Zhang, B., Hu, Y., & Zhao, Y. (2021). New Insights Into Gut-Bacteria-Derived Indole and Its Derivatives in Intestinal and Liver Diseases. *Frontiers in Pharmacology, 12*, 769501. doi:https://doi.org/10.3389/fphar.2021.769501

Li, Y., Tan, W., Neo, T. W., Aung, M. O., Wasser, S., Lim, S. G., & Tan, T. M. (2009). Role of the miR-106b-25 microRNA cluster in hepatocellular carcinoma. *Cancer Sci, 100*(7), 1234–1242. doi:https://doi.org/10.1111/j.1349-7006.2009.01164.x

Liu, Q., Sun, W., & Zhang, H. (2022). Interaction of Gut Microbiota with Endocrine Homeostasis and Thyroid Cancer. *Cancers (Basel), 14*(11). doi:https://doi.org/10.3390/cancers14112656

Liu, S., et al. (2016). The Host Shapes the Gut Microbiota via Fecal MicroRNA. *Cell Host Microbe, 19*(1), 32–43. doi:https://doi.org/10.1016/j.chom.2015.12.005

Liu, T., et al. (2024). Dietary Dihydroquercetin Alleviated Colitis via the Short-Chain Fatty Acids/miR-10a-5p/PI3K-Akt Signaling Pathway. *Journal of Agricultural and Food Chemistry, 72*(42), 23211–23223. doi:https://doi.org/10.1021/acs.jafc.4c03278

Liu, X., et al. (2021). CircMYH9 drives colorectal cancer growth by regulating serine metabolism and redox homeostasis in a p53-dependent manner. *Mol Cancer, 20*(1), 114. doi:https://doi.org/10.1186/s12943-021-01412-9

Liu, X. F., et al. (2023). Regulation of short-chain fatty acids in the immune system. *Frontiers in Immunology, 14*, 1186892. doi:https://doi.org/10.3389/fimmu.2023.1186892

Long, F., et al. (2023). Intergenic CircRNA Circ_0007379 Inhibits Colorectal Cancer Progression by Modulating miR-320a Biogenesis in a KSRP-Dependent Manner. *Int J Biol Sci, 19*(12), 3781–3803. doi:https://doi.org/10.7150/ijbs.85063

Louis, P., Hold, G. L., & Flint, H. J. (2014). The gut microbiota, bacterial metabolites and colorectal cancer. *Nat Rev Microbiol, 12*(10), 661–672. doi:https://doi.org/10.1038/nrmicro3344

Ly, L. K., Doden, H. L., & Ridlon, J. M. (2021). Gut feelings about bacterial steroid-17,20-desmolase. *Molecular and Cellular Endocrinology, 525*, 111174. doi:https://doi.org/10.1016/j.mce.2021.111174

Ma, C., et al. (2023). Inflammatory cytokine-regulated LNCPTCTS suppresses thyroid cancer progression via enhancing Snail nuclear export. *Cancer Lett, 575*, 216402. doi:https://doi.org/10.1016/j.canlet.2023.216402

Mafra, D., et al. (2022). Archaea from the gut microbiota of humans: Could be linked to chronic diseases? *Anaerobe, 77*, 102629.

Magne, F., Gotteland, M., Gauthier, L., Zazueta, A., Pesoa, S., Navarrete, P., & Balamurugan, R. (2020). The Firmicutes/Bacteroidetes Ratio: A Relevant Marker of Gut Dysbiosis in Obese Patients? *Nutrients, 12*(5). doi:https://doi.org/10.3390/nu12051474

Maia, B. M., Rocha, R. M., & Calin, G. A. (2014). Clinical significance of the interaction between non-coding RNAs and the epigenetics machinery: challenges and opportunities in oncology. *Epigenetics, 9*(1), 75–80. doi:https://doi.org/10.4161/epi.26488

Malmuthuge, N., & Guan, L. L. (2021). Noncoding RNAs: Regulatory Molecules of Host-Microbiome Crosstalk. *Trends Microbiol, 29*(8), 713–724. doi:https://doi.org/10.1016/j.tim.2020.12.003

Markowiak, P., & Śliżewska, K. (2017). Effects of Probiotics, Prebiotics, and Synbiotics on Human Health. *Nutrients, 9*(9). doi:https://doi.org/10.3390/nu9091021

Martell, M., Quarnstrom, C. F., Khoruts, A., Vezys, V., Staley, C., & Shmidt, E. (2025). Chronic Intestinal Inflammation and Microbial Dysbiosis are Associated with Female Reproductive Outcomes in a Mouse Model of Inflammatory Bowel Disease. *Gastro Hep Advances*, 100670.

Matijašić, M., Meštrović, T., Paljetak, H., Perić, M., Barešić, A., & Verbanac, D. (2020). Gut Microbiota beyond Bacteria-Mycobiome, Virome, Archaeome, and Eukaryotic Parasites in IBD. *International Journal of Molecular Sciences, 21*(8). doi:https://doi.org/10.3390/ijms21082668

Melo-Bolívar, J. F., Ruiz Pardo, R. Y., Junca, H., Sidjabat, H. E., Cano-Lozano, J. A., & Villamil Díaz, L. M. (2022). Competitive Exclusion Bacterial Culture Derived from the Gut Microbiome of Nile Tilapia (Oreochromis niloticus) as a Resource to Efficiently Recover Probiotic Strains: Taxonomic, Genomic, and Functional Proof of Concept. *Microorganisms, 10*(7). doi:https://doi.org/10.3390/microorganisms10071376

Michael, I. P., Saghafinia, S., & Hanahan, D. J. P. o. t. N. A. o. S. (2019). A set of microRNAs coordinately controls tumorigenesis, invasion, and metastasis. *116*(48), 24184–24195.

Mohan, M. L., Vasudevan, N. T., & Naga Prasad, S. V. (2017). Proinflammatory Cytokines Mediate GPCR Dysfunction. *Journal of Cardiovascular Pharmacology, 70*(2), 61–73. doi:https://doi.org/10.1097/fjc.0000000000000456

Moloney, G. M., Viola, M. F., Hoban, A. E., Dinan, T. G., & Cryan, J. F. (2018). Faecal microRNAs: indicators of imbalance at the host-microbe interface? *Benef Microbes, 9*(2), 175–183. doi:https://doi.org/10.3920/BM2017.0013

Moon, C. D., Young, W., Maclean, P. H., Cookson, A. L., & Bermingham, E. N. (2018). Metagenomic insights into the roles of Proteobacteria in the gastrointestinal microbiomes of healthy dogs and cats. *Microbiologyopen, 7*(5), e00677. doi:https://doi.org/10.1002/mbo3.677

Mora-Flores, L. P., Moreno-Terrazas Casildo, R., Fuentes-Cabrera, J., Pérez-Vicente, H. A., de Anda-Jáuregui, G., & Neri-Torres, E. E. (2023). The Role of Carbohydrate Intake on the Gut Microbiome: A Weight of Evidence Systematic Review. *Microorganisms, 11*(7). doi:https://doi.org/10.3390/microorganisms11071728

Morrison, D. J., & Preston, T. (2016). Formation of short chain fatty acids by the gut microbiota and their impact on human metabolism. *Gut Microbes, 7*(3), 189–200. doi:https://doi.org/10.1080/19490976.2015.1134082

Mortazavi-Jahromi, S. S., Aslani, M., & Mirshafiey, A. J. I. L. (2020). A comprehensive review on miR-146a molecular mechanisms in a wide spectrum of immune and non-immune inflammatory diseases. *227*, 8–27.

Mostafa, A. M., Hamdy, N. M., Abdel-Rahman, S. Z., & El-Mesallamy, H. O. (2016). Effect of vildagliptin and pravastatin combination on cholesterol efflux in adipocytes. *IUBMB Life, 68*(7), 535–543. doi:https://doi.org/10.1002/iub.1510

Mostafa, A. M., Hamdy, N. M., El-Mesallamy, H. O., & Abdel-Rahman, S. Z. (2015). Glucagon-like peptide 1 (GLP-1)-based therapy upregulates LXR-ABCA1/ABCG1 cascade in adipocytes. *Biochem Biophys Res Commun, 468*(4), 900–905. doi:https://doi.org/10.1016/j.bbrc.2015.11.054

Mu, D., Shi, Y., Sun, R., Han, B., Zhong, K., Ye, Y., & Zhang, J. (2025). The acidic microenvironment promotes pancreatic cancer progression via the lncRNA-LOC100507424/E2F1/FOXM1 axis. *BMC Cancer, 25*(1), 655. doi:https://doi.org/10.1186/s12885-025-14073-4

Muhammad, N., Bhattacharya, S., Steele, R., & Ray, R. B. (2016). Anti-miR-203 suppresses ER-positive breast cancer growth and stemness by targeting SOCS3. *Oncotarget, 7*(36), 58595–58605. doi:https://doi.org/10.18632/oncotarget.11193

Murali, S. K., & Mansell, T. J. (2024). Next generation probiotics: Engineering live biotherapeutics. *Biotechnol Adv, 72*, 108336. doi:https://doi.org/10.1016/j.biotechadv.2024.108336

Myhrstad, M. C., Tunsjø, H., Charnock, C., & Telle-Hansen, V. H. (2020). Dietary fiber, gut microbiota, and metabolic regulation—Current status in human randomized trials. *Nutrients, 12*(3), 859.

Nagpal, R., Shively, C. A., Register, T. C., Craft, S., & Yadav, H. (2019). Gut microbiome-Mediterranean diet interactions in improving host health. *F1000Res, 8*, 699. doi:https://doi.org/10.12688/f1000research.18992.1

Nakata, K., et al. (2017). Commensal microbiota-induced microRNA modulates intestinal epithelial permeability through the small GTPase ARF4. *J Biol Chem, 292*(37), 15426–15433. doi:https://doi.org/10.1074/jbc.M117.788596

Nemeth, K., Bayraktar, R., Ferracin, M., & Calin, G. A. (2024). Non-coding RNAs in disease: from mechanisms to therapeutics. *Nature Reviews Genetics, 25*(3), 211–232. doi:https://doi.org/10.1038/s41576-023-00662-1

Neurath, M. F., Artis, D., & Becker, C. (2025). The intestinal barrier: a pivotal role in health, inflammation, and cancer. *The Lancet Gastroenterology & Hepatology, 10*(6), 573–592. doi:https://doi.org/10.1016/S2468-1253(24)00390-X

Nikolaieva, N., Sevcikova, A., Omelka, R., Martiniakova, M., Mego, M., & Ciernikova, S. (2022). Gut Microbiota-MicroRNA Interactions in Intestinal Homeostasis and Cancer Development. *Microorganisms, 11*(1). doi:https://doi.org/10.3390/microorganisms11010107

Nour, S. A., Foda, D. S., Elsehemy, I. A., & Hassan, M. E. (2024). Co-administration of xylo-oligosaccharides produced by immobilized Aspergillus terreus xylanase with carbimazole to mitigate its adverse effects on the adrenal gland. *Scientific Reports, 14*(1), 17481. doi:https://doi.org/10.1038/s41598-024-67310-4

Olivo-Martínez, Y., Martínez-Ruiz, S., Cordero-Alday, C., Bosch, M., Badia, J., & Baldoma, L. (2024). Modulation of Serotonin-Related Genes by Extracellular Vesicles of the Probiotic Escherichia coli Nissle 1917 in the Interleukin-1β-Induced Inflammation Model of Intestinal Epithelial Cells. *25*(10), 5338.

Ozma, M. A., et al. (2022). Postbiotics as the key mediators of the gut microbiota-host interactions. *Le Infezioni in Medicina, 30*(2), 180–193. doi:https://doi.org/10.53854/liim-3002-3

Pang, P., Yu, B., Shi, Y., Deng, L., Xu, H., Wu, S., & Chen, X. (2018). Alteration of Intestinal Flora Stimulates Pulmonary microRNAs to Interfere with Host Antiviral Immunity in Influenza. *23*(12), 3151.

Panthee, B., Gyawali, S., Panthee, P., & Techato, K. (2022). Environmental and Human Microbiome for Health. *Life (Basel), 12*(3). doi:https://doi.org/10.3390/life12030456

Park, J., Kim, M., Kang, S. G., Jannasch, A. H., Cooper, B., Patterson, J., & Kim, C. H. (2015). Short-chain fatty acids induce both effector and regulatory T cells by suppression of histone deacetylases and regulation of the mTOR-S6K pathway. *Mucosal Immunol, 8*(1), 80–93. doi:https://doi.org/10.1038/mi.2014.44

Patangia, D. V., Anthony Ryan, C., Dempsey, E., Paul Ross, R., & Stanton, C. (2022). Impact of antibiotics on the human microbiome and consequences for host health. *Microbiologyopen, 11*(1), e1260. doi:https://doi.org/10.1002/mbo3.1260

Peck, B. C., Mah, A. T., Pitman, W. A., Ding, S., Lund, P. K., & Sethupathy, P. (2017). Functional Transcriptomics in Diverse Intestinal Epithelial Cell Types Reveals Robust MicroRNA Sensitivity in Intestinal Stem Cells to Microbial Status. *J Biol Chem, 292*(7), 2586–2600. doi:https://doi.org/10.1074/jbc.M116.770099

Polikanov, Y. S., Melnikov, S. V., Söll, D., & Steitz, T. A. (2015). Structural insights into the role of rRNA modifications in protein synthesis and ribosome assembly. *Nat Struct Mol Biol, 22*(4), 342–344. doi:https://doi.org/10.1038/nsmb.2992

Proenca, M. A., et al. (2018). Relationship between Fusobacterium nucleatum, inflammatory mediators and microRNAs in colorectal carcinogenesis. *World J Gastroenterol, 24*(47), 5351–5365. doi:https://doi.org/10.3748/wjg.v24.i47.5351

Qi, X., Yun, C., Pang, Y., & Qiao, J. (2021). The impact of the gut microbiota on the reproductive and metabolic endocrine system. *Gut Microbes, 13*(1), 1–21. doi:https://doi.org/10.1080/19490976.2021.1894070

Qin, Y., & Wade, P. A. (2018). Crosstalk between the microbiome and epigenome: messages from bugs. *J Biochem, 163*(2), 105–112. doi:https://doi.org/10.1093/jb/mvx080

Ramos-Molina, B., et al. (2019). Gut Microbiota Composition Is Associated With the Global DNA Methylation Pattern in Obesity. *Front Genet, 10*, 613. doi:https://doi.org/10.3389/fgene.2019.00613

Rastelli, M., Cani, P. D., & Knauf, C. (2019). The Gut Microbiome Influences Host Endocrine Functions. *Endocrine Reviews, 40*(5), 1271–1284. doi:https://doi.org/10.1210/er.2018-00280
Richard, V., Lee, K., & Kerin, M. J. (2025). MicroRNAs as Endocrine Modulators of Breast Cancer. *26*(7), 3449.
Rishabh, Bansal, S., Goel, A., Gupta, S., Malik, D., & Bansal, N. (2024). Unravelling the Crosstalk between Estrogen Deficiency and Gut-biotaDysbiosis in the Development of Diabetes Mellitus. *Current Diabetes Reviews, 20*(10), e240124226067. doi:https://doi.org/10.2174/0115733998275953231129094057
Rizk, N. I., Kassem, D. H., Abulsoud, A. I., AbdelHalim, S., Yasser, M. B., Kamal, M. M., & Hamdy, N. M. (2024). Revealing the role of serum exosomal novel long non-coding RNA NAMPT-AS as a promising diagnostic/prognostic biomarker in colorectal cancer patients. *Life Sci, 352*, 122850. doi:https://doi.org/10.1016/j.lfs.2024.122850
Rodriguez-Nogales, A., et al. (2017). Differential intestinal anti-inflammatory effects of Lactobacillus fermentum and Lactobacillus salivarius in DSS mouse colitis: impact on microRNAs expression and microbiota composition. *Mol Nutr Food Res, 61*(11). doi:https://doi.org/10.1002/mnfr.201700144
Rossetto, C. C., Tarrant-Elorza, M., Verma, S., Purushothaman, P., & Pari, G. S. (2013). Regulation of viral and cellular gene expression by Kaposi's sarcoma-associated herpesvirus polyadenylated nuclear RNA. *J Virol, 87*(10), 5540–5553. doi:https://doi.org/10.1128/jvi.03111-12
Röther, S., & Meister, G. (2011). Small RNAs derived from longer non-coding RNAs. *Biochimie, 93*(11), 1905–1915. doi:https://doi.org/10.1016/j.biochi.2011.07.032
Rusch, J. A., Layden, B. T., & Dugas, L. R. (2023). Signalling cognition: the gut microbiota and hypothalamic-pituitary-adrenal axis. *Frontiers in Endocrinology, 14*, 1130689. doi:https://doi.org/10.3389/fendo.2023.1130689
Saatci, O., et al. (2024). Targeting LINC00152 activates cAMP/Ca(2+)/ferroptosis axis and overcomes tamoxifen resistance in ER+ breast cancer. *Cell Death Dis, 15*(6), 418. doi:https://doi.org/10.1038/s41419-024-06814-3
Sabit, H., et al. (2023). The Effect of Probiotic Supplementation on the Gut-Brain Axis in Psychiatric Patients. *Current Issues in Molecular Biology, 45*(5), 4080–4099. doi:https://doi.org/10.3390/cimb45050260
Salachan, P. V., Rasmussen, M., Fredsøe, J., Ulhøi, B., Borre, M., & Sørensen, K. D. (2022). Microbiota of the prostate tumor environment investigated by whole-transcriptome profiling. *Genome Medicine, 14*(1), 9. doi:https://doi.org/10.1186/s13073-022-01011-3
Salazar, N., Valdés-Varela, L., González, S., Gueimonde, M., & de Los Reyes-Gavilán, C. G. (2017). Nutrition and the gut microbiome in the elderly. *Gut Microbes, 8*(2), 82–97. doi:https://doi.org/10.1080/19490976.2016.1256525
Salem, F., et al. (2019). Gut microbiome in chronic rheumatic and inflammatory bowel diseases: Similarities and differences. *United European Gastroenterol J, 7*(8), 1008–1032. doi:https://doi.org/10.1177/2050640619867555
Santos, A. A., Afonso, M. B., Ramiro, R. S., Pires, D., Pimentel, M., Castro, R. E., & Rodrigues, C. M. P. (2020). Host miRNA-21 promotes liver dysfunction by targeting small intestinal Lactobacillus in mice. *Gut Microbes, 12*(1), 1–18. doi:https://doi.org/10.1080/19490976.2020.1840766
Sato-Kuwabara, Y., Melo, S. A., Soares, F. A., & Calin, G. A. (2015). The fusion of two worlds: non-coding RNAs and extracellular vesicles--diagnostic and therapeutic implications (Review). *Int J Oncol, 46*(1), 17–27. doi:https://doi.org/10.3892/ijo.2014.2712
Satokari, R. (2020). High Intake of Sugar and the Balance between Pro- and Anti-Inflammatory Gut Bacteria. *Nutrients, 12*(5). doi:https://doi.org/10.3390/nu12051348
Schlörmann, W., Naumann, S., Renner, C., & Glei, M. (2015). Influence of miRNA-106b and miRNA-135a on butyrate-regulated expression of p21 and Cyclin D2 in human colon adenoma cells. *Genes Nutr, 10*(6), 50. doi:https://doi.org/10.1007/s12263-015-0500-4
Schmidt, T. S. B., Raes, J., & Bork, P. (2018). The Human Gut Microbiome: From Association to Modulation. *Cell, 172*(6), 1198–1215. doi:https://doi.org/10.1016/j.cell.2018.02.044

Schroeder, B. O., & Backhed, F. (2016). Signals from the gut microbiota to distant organs in physiology and disease. *Nat Med, 22*(10), 1079–1089. doi:https://doi.org/10.1038/nm.4185

Seekatz, A. M., Safdar, N., & Khanna, S. (2022). The role of the gut microbiome in colonization resistance and recurrent Clostridioides difficile infection. *Therapeutic Advances in Gastroenterology, 15*, 17562848221134396. doi:https://doi.org/10.1177/17562848221134396

Shi, C., et al. (2016). Novel evidence for an oncogenic role of microRNA-21 in colitis-associated colorectal cancer. *Gut, 65*(9), 1470–1481. doi:https://doi.org/10.1136/gutjnl-2014-308455

Shi, J., et al. (2020). Circular RNA circMBOAT2 promotes prostate cancer progression via a miR-1271-5p/mTOR axis. *12*(13), 13255.

Shin, Y., Han, S., Kwon, J., Ju, S., Choi, T. G., Kang, I., & Kim, S. S. (2023). Roles of Short-Chain Fatty Acids in Inflammatory Bowel Disease. *Nutrients, 15*(20). doi:https://doi.org/10.3390/nu15204466

Shu, Q., Kang, C., Li, J., Hou, Z., Xiong, M., Wang, X., & Peng, H. (2024). Effect of probiotics or prebiotics on thyroid function: A meta-analysis of eight randomized controlled trials. *PloS One, 19*(1), e0296733. doi:https://doi.org/10.1371/journal.pone.0296733

Shuwen, H., & Kefeng, D. (2022). Intestinal phages interact with bacteria and are involved in human diseases. *Gut Microbes, 14*(1), 2113717. doi:https://doi.org/10.1080/19490976.2022.2113717

Singh, N., Shirdel, E. A., Waldron, L., Zhang, R.-H., Jurisica, I., & Comelli, E. M. (2011). The murine caecal microRNA signature depends on the presence of the endogenous microbiota. *International journal of biological sciences, 8*(2), 171.

Smith, P. M., et al. (2013). The microbial metabolites, short-chain fatty acids, regulate colonic Treg cell homeostasis. *Science, 341*(6145), 569–573. doi:https://doi.org/10.1126/science.1241165

Sochacka, K., Kotowska, A., & Lachowicz-Wiśniewska, S. (2024). The Role of Gut Microbiota, Nutrition, and Physical Activity in Depression and Obesity-Interdependent Mechanisms/Co-Occurrence. *Nutrients, 16*(7). doi:https://doi.org/10.3390/nu16071039

Song, J., et al. (2023). Hsa_circ_0009092/miR-665/NLK signaling axis suppresses colorectal cancer progression via recruiting TAMs in the tumor microenvironment. *J Exp Clin Cancer Res, 42*(1), 319. doi:https://doi.org/10.1186/s13046-023-02887-8

Stilling, R. M., et al. (2015). Microbes & neurodevelopment--Absence of microbiota during early life increases activity-related transcriptional pathways in the amygdala. *Brain Behav Immun, 50*, 209–220. doi:https://doi.org/10.1016/j.bbi.2015.07.009

Strandwitz, P. (2018). Neurotransmitter modulation by the gut microbiota. *Brain Research, 1693*(Pt B), 128–133. doi:https://doi.org/10.1016/j.brainres.2018.03.015

Su, Y., Bai, Q., Zhang, W., Xu, B., & Hu, T. (2025). The Role of Long Non-Coding RNAs in Modulating the Immune Microenvironment of Triple-Negative Breast Cancer: Mechanistic Insights and Therapeutic Potential. *15*(3), 454.

Sun, C. H., et al. (2019). The role of Fusobacterium nucleatum in colorectal cancer: from carcinogenesis to clinical management. *Chronic Dis Transl Med, 5*(3), 178–187. doi:https://doi.org/10.1016/j.cdtm.2019.09.001

Tan, C., et al. (2021). Dysbiosis of Gut Microbiota and Short-Chain Fatty Acids in Acute Ischemic Stroke and the Subsequent Risk for Poor Functional Outcomes. *JPEN: Journal of Parenteral and Enteral Nutrition, 45*(3), 518–529. doi:https://doi.org/10.1002/jpen.1861

Tang, R., & Li, L. (2021). Modulation of Short-Chain Fatty Acids as Potential Therapy Method for Type 2 Diabetes Mellitus. *Canadian Journal of Infectious Diseases & Medical Microbiology, 2021*, 6632266. doi:https://doi.org/10.1155/2021/6632266

Tarallo, S., et al. (2019). Altered Fecal Small RNA Profiles in Colorectal Cancer Reflect Gut Microbiome Composition in Stool Samples. *mSystems, 4*(5). doi:https://doi.org/10.1128/mSystems.00289-19

Teng, Y., et al. (2018). Plant-Derived Exosomal MicroRNAs Shape the Gut Microbiota. *Cell Host Microbe, 24*(5), 637–652.e638. doi:https://doi.org/10.1016/j.chom.2018.10.001

Tian, J., et al. (2020). The miR-218/GAB2 axis regulates proliferation, invasion and EMT via the PI3K/AKT/GSK-3β pathway in prostate cancer. *394*(1), 112128.

Tian, Y., et al. (2023). MIR497HG-Derived miR-195 and miR-497 Mediate Tamoxifen Resistance via PI3K/AKT Signaling in Breast Cancer. *Adv Sci (Weinh), 10*(12), e2204819. doi:https://doi.org/10.1002/advs.202204819

Tolhurst, G., et al. (2012). Short-chain fatty acids stimulate glucagon-like peptide-1 secretion via the G-protein-coupled receptor FFAR2. *Diabetes, 61*(2), 364–371. doi:https://doi.org/10.2337/db11-1019

Tomkovich, S., et al. (2020). Human Colon Mucosal Biofilms and Murine Host Communicate via Altered mRNA and microRNA Expression during Cancer. *mSystems, 5*(1). doi:https://doi.org/10.1128/mSystems.00451-19

Van Branteghem, C., Augenlicht, A., Demetter, P., Craciun, L., & Maenhaut, C. (2023). Unraveling the Roles of miR-204-5p and HMGA2 in Papillary Thyroid Cancer Tumorigenesis. *Int J Mol Sci, 24*(13). doi:https://doi.org/10.3390/ijms241310764

Veltman, K., Hummel, S., Cichon, C., Sonnenborn, U., & Schmidt, M. A. (2012). Identification of specific miRNAs targeting proteins of the apical junctional complex that simulate the probiotic effect of E. coli Nissle 1917 on T84 epithelial cells. *Int J Biochem Cell Biol, 44*(2), 341–349. doi:https://doi.org/10.1016/j.biocel.2011.11.006

Virk, M. S., et al. (2024). The Anti-Inflammatory and Curative Exponent of Probiotics: A Comprehensive and Authentic Ingredient for the Sustained Functioning of Major Human Organs. *Nutrients, 16*(4). doi:https://doi.org/10.3390/nu16040546

Wahlström, A., Sayin, S. I., Marschall, H. U., & Bäckhed, F. (2016). Intestinal Crosstalk between Bile Acids and Microbiota and Its Impact on Host Metabolism. *Cell Metabolism, 24*(1), 41–50. doi:https://doi.org/10.1016/j.cmet.2016.05.005

Wang, D., et al. (2012). Gut microbiota metabolism of anthocyanin promotes reverse cholesterol transport in mice via repressing miRNA-10b. *Circ Res, 111*(8), 967–981. doi:https://doi.org/10.1161/CIRCRESAHA.112.266502

Wang, G., Wei, L., & Yang, H. (2024a). p53-associated miRNAs repress lncRNA ZFAS1 to retard the proliferation of papillary thyroid carcinoma. *Endokrynol Pol, 75*(1), 12–19. doi:https://doi.org/10.5603/ep.97567

Wang, X., et al. (2024b). peu-MIR2916-p3-enriched garlic exosomes ameliorate murine colitis by reshaping gut microbiota, especially by boosting the anti-colitic Bacteroides thetaiotaomicron. *Pharmacol Res, 200*, 107071. doi:https://doi.org/10.1016/j.phrs.2024.107071

Wang, Y., et al. (2023). The gut microbiota reprograms intestinal lipid metabolism through long noncoding RNA Snhg9. *Science, 381*(6660), 851–857. doi:https://doi.org/10.1126/science.ade0522

Wang, Y., Xu, H., Zhou, X., Chen, W., & Zhou, H. (2024c). Dysregulated bile acid homeostasis: unveiling its role in metabolic diseases. *Med Rev (2021), 4*(4), 262–283. doi:https://doi.org/10.1515/mr-2024-0020

Wang, Z., et al. (2011). Gut flora metabolism of phosphatidylcholine promotes cardiovascular disease. *Nature, 472*(7341), 57–63. doi:https://doi.org/10.1038/nature09922

Wei, L., Singh, R., & Ghoshal, U. C. (2022). Enterochromaffin Cells-Gut Microbiota Crosstalk: Underpinning the Symptoms, Pathogenesis, and Pharmacotherapy in Disorders of Gut-Brain Interaction. *Journal of Neurogastroenterology and Motility, 28*(3), 357–375. doi:https://doi.org/10.5056/jnm22008

Wheeler, M. L., Limon, J. J., & Underhill, D. M. (2017). Immunity to Commensal Fungi: Detente and Disease. *Annual Review of Pathology, 12*, 359–385. doi:https://doi.org/10.1146/annurev-pathol-052016-100342

Wieërs, G., et al. (2020). How probiotics affect the microbiota. *Frontiers in cellular and infection microbiology, 9*, 454.

Wilks, J., Beilinson, H., & Golovkina, T. V. (2013). Dual role of commensal bacteria in viral infections. *Immunological reviews, 255*(1), 222–229.

Winkle, M., El-Daly, S. M., Fabbri, M., & Calin, G. A. (2021). Noncoding RNA therapeutics - challenges and potential solutions. *Nat Rev Drug Discov, 20*(8), 629–651. doi:https://doi.org/10.1038/s41573-021-00219-z

Wolters, M., et al. (2019). Dietary fat, the gut microbiota, and metabolic health - A systematic review conducted within the MyNewGut project. *Clinical Nutrition, 38*(6), 2504–2520. doi:https://doi.org/10.1016/j.clnu.2018.12.024

Wu, J., Yang, K., Fan, H., Wei, M., & Xiong, Q. (2023). Targeting the gut microbiota and its metabolites for type 2 diabetes mellitus. *Frontiers in Endocrinology, 14*, 1114424. doi:https://doi.org/10.3389/fendo.2023.1114424

Wu, Y., Wang, C. Z., Wan, J. Y., Yao, H., & Yuan, C. S. (2021). Dissecting the Interplay Mechanism between Epigenetics and Gut Microbiota: Health Maintenance and Disease Prevention. *Int J Mol Sci, 22*(13). doi:https://doi.org/10.3390/ijms22136933

Xiao, Y., et al. (2024). Long non-coding RNA-encoded micropeptides: functions, mechanisms and implications. *Cell Death Discovery, 10*(1), 450. doi:https://doi.org/10.1038/s41420-024-02175-0

Xie, T., et al. (2022). CircSMARCC1 facilitates tumor progression by disrupting the crosstalk between prostate cancer cells and tumor-associated macrophages via miR-1322/CCL20/CCR6 signaling. *21*(1), 173.

Xie, Y., et al. (2024). LINC00173 silence and estrone supply suppress ER(+) breast cancer by estrogen receptor α degradation and LITAF activation. *Cancer Sci, 115*(7), 2318–2332. doi:https://doi.org/10.1111/cas.16201

Xie, Y., et al. (2013). MicroRNA-24 regulates XIAP to reduce the apoptosis threshold in cancer cells. *Oncogene, 32*(19), 2442–2451.

Xu, J., Liu, W., Li, R., & Wang, Y. (2023a). Significance of LINC02082 and LOC105369812 in differentiating papillary thyroid cancer from benign nodules. *Medicine (Baltimore), 102*(44), e35542. doi:https://doi.org/10.1097/md.0000000000035542

Xu, S., Wang, L., Zhao, Y., Mo, T., Wang, B., Lin, J., & Yang, H. (2024). Metabolism-regulating non-coding RNAs in breast cancer: roles, mechanisms and clinical applications. *Journal of Biomedical Science, 31*(1), 25. doi:https://doi.org/10.1186/s12929-024-01013-w

Xu, Y., et al. (2023b). Crohn's disease-associated AIEC inhibiting intestinal epithelial cell-derived exosomal let-7b expression regulates macrophage polarization to exacerbate intestinal fibrosis. *Gut Microbes, 15*(1), 2193115. doi:https://doi.org/10.1080/19490976.2023.2193115

Xue, H., et al. (2022). Gut Microbially Produced Indole-3-Propionic Acid Inhibits Atherosclerosis by Promoting Reverse Cholesterol Transport and Its Deficiency Is Causally Related to Atherosclerotic Cardiovascular Disease. *Circulation Research, 131*(5), 404–420. doi:https://doi.org/10.1161/CIRCRESAHA.122.321253

Xue, X., et al. (2014). Downregulation of micro RNA-107 in intestinal CD 11c+ myeloid cells in response to microbiota and proinflammatory cytokines increases IL-23p19 expression. *European journal of immunology, 44*(3), 673–682.

Yang, Y., et al. (2024). Navigating the B vitamins: Dietary diversity, microbial synthesis, and human health. *Cell Host & Microbe, 32*(1), 12–18.

Yang, Y., et al. (2017). Fusobacterium nucleatum increases proliferation of colorectal cancer cells and tumor development in mice by activating toll-like receptor 4 signaling to nuclear factor–κB, and up-regulating expression of microRNA-21. *Gastroenterology, 152*(4), 851–866. e824.

Yanshen, Z., Lifen, Y., Xilian, W., Zhong, D., & Huihong, M. J. L. J. o. M. (2021). miR-92a promotes proliferation and inhibits apoptosis of prostate cancer cells through the PTEN/Akt signaling pathway. *16*(1), 1971837.

Ye, D., Guo, S., Al-Sadi, R., & Ma, T. Y. (2011). MicroRNA regulation of intestinal epithelial tight junction permeability. *Gastroenterology, 141*(4), 1323–1333. doi:https://doi.org/10.1053/j.gastro.2011.07.005

Yu, T., et al. (2017). Fusobacterium nucleatum promotes chemoresistance to colorectal cancer by modulating autophagy. *cell, 170*(3), 548–563. e516.

Yuan, C., Burns Michael, B., Subramanian, S., & Blekhman, R. (2018). Interaction between Host MicroRNAs and the Gut Microbiota in Colorectal Cancer. *mSystems, 3*(3), https://doi.org/10.1128/msystems.00205-00217. doi:10.1128/msystems.00205-17

Yuan, C., Steer, C. J., & Subramanian, S. (2019). Host(-)MicroRNA(-)Microbiota Interactions in Colorectal Cancer. *Genes (Basel), 10*(4). doi:https://doi.org/10.3390/genes10040270
Zeng, Z., Tang, S., Chen, L., Hou, H., Liu, Y., & Li, J. (2023). LncRNA HAGLROS contribute to papillary thyroid cancer progression by modulating miR-206/HMGA2 expression. *Aging (Albany NY), 15*(24), 14930–14944. doi:https://doi.org/10.18632/aging.205321
Zhang, C., Fang, T., Shi, L., Wang, Y., Deng, X., Wang, J., & Zhou, Y. (2024a). The synbiotic combination of probiotics and inulin improves NAFLD though modulating gut microbiota. *The Journal of Nutritional Biochemistry, 125*, 109546. doi:https://doi.org/10.1016/j.jnutbio.2023.109546
Zhang, J., et al. (2019a). Dysbiosis of the gut microbiome is associated with thyroid cancer and thyroid nodules and correlated with clinical index of thyroid function. *64*(3), 564–574.
Zhang, L., Meng, X., Li, D., & Han, X. (2020). piR-001773 and piR-017184 promote prostate cancer progression by interacting with PCDH9. *Cellular Signalling, 76*, 109780. https://doi.org/10.1016/j.cellsig.2020.109780
Zhang, M., et al. (2024b). The role of circRNAs and miRNAs in drug resistance and targeted therapy responses in breast cancer. *Cancer Drug Resist, 7*, 30. doi:https://doi.org/10.20517/cdr.2024.62
Zhang, P., Cao, L., Zhou, R., Yang, X., & Wu, M. (2019b). The lncRNA Neat1 promotes activation of inflammasomes in macrophages. *Nat Commun, 10*(1), 1495. doi:https://doi.org/10.1038/s41467-019-09482-6
Zhang, W., Mackay, C. R., & Gershwin, M. E. (2023). Immunomodulatory Effects of Microbiota-Derived Short-Chain Fatty Acids in Autoimmune Liver Diseases. *Journal of Immunology, 210*(11), 1629–1639. doi:https://doi.org/10.4049/jimmunol.2300016
Zhang, Y., et al. (2022). Propionate stimulates the secretion of satiety hormones and reduces acute appetite in a cecal fistula pig model. *Anim Nutr, 10*, 390–398. doi:https://doi.org/10.1016/j.aninu.2022.06.003
Zhao, Q., Pang, G., Yang, L., Chen, S., Xu, R., & Shao, W. (2021a). Long Noncoding RNAs Regulate the Inflammatory Responses of Macrophages. *Cells, 11*(1). doi:https://doi.org/10.3390/cells11010005
Zhao, W., Ning, L., Wang, L., Ouyang, T., Qi, L., Yang, R., & Wu, Y. J. A. (2021b). miR-21 inhibition reverses doxorubicin-resistance and inhibits PC3 human prostate cancer cells proliferation. *53*(5), e14016.
Zhao, Y., Gong, C., Xu, J., Chen, D., Yang, B., Chen, Z., & Wei, L. (2023). Research Progress of Fecal Microbiota Transplantation in Liver Diseases. *J Clin Med, 12*(4). doi:https://doi.org/10.3390/jcm12041683
Zheng, Y., Li, J. X., Chen, C. J., Lin, Z. Y., Liu, J. X., & Lin, F. J. (2020). Extracellular vesicle-derived circ_SLC19A1 promotes prostate cancer cell growth and invasion through the miR-497/septin 2 pathway. *Cell Biol Int, 44*(4), 1037–1045. doi:https://doi.org/10.1002/cbin.11303
Zhong, W., et al. (2022). Gut dysbiosis promotes prostate cancer progression and docetaxel resistance via activating NF-κB-IL6-STAT3 axis. *10*(1), 94.
Zhou, Q., et al. (2015). MicroRNA 29 targets nuclear factor-κB-repressing factor and Claudin 1 to increase intestinal permeability. *Gastroenterology, 148*(1), 158–169.e158. doi:https://doi.org/10.1053/j.gastro.2014.09.037
Zhu, K., et al. (2020). Microbiota Imbalance Contributes to COPD Deterioration by Enhancing IL-17a Production via miR-122 and miR-30a. *Mol Ther Nucleic Acids, 22*, 520–529. doi:https://doi.org/10.1016/j.omtn.2020.09.017
Zikou, E., Koliaki, C., & Makrilakis, K. (2024). The Role of Fecal Microbiota Transplantation (FMT) in the Management of Metabolic Diseases in Humans: A Narrative Review. *Biomedicines, 12*(8). doi:https://doi.org/10.3390/biomedicines12081871
Zununi Vahed, S., Barzegari, A., Rahbar Saadat, Y., Goreyshi, A., & Omidi, Y. (2017). Leuconostoc mesenteroides-derived anticancer pharmaceuticals hinder inflammation and cell survival in colon cancer cells by modulating NF-κB/AKT/PTEN/MAPK pathways. *Biomedicine & Pharmacotherapy, 94*, 1094–1100. https://doi.org/10.1016/j.biopha.2017.08.033

Chapter 3
Mapping Non-Coding Epimutations in Breast Cancer: Advancing Epigenetics Towards Precision Medicine

Mohamed E. Abdel Aziz, Heba Momtaz, Basel Mohamed, Mariam Amr, Alhassan Soliman Wadan, Shaimaa Abdel-Ghany, Borros Arneth, and Hussein Sabit

Abstract Breast cancer remains a leading cause of cancer-related morbidity and mortality in women worldwide. While extensive research has focused on mutations in protein-coding genes, emerging evidence underscores the pivotal role of the non-coding genome—including long noncoding RNAs (lncRNAs), microRNAs (miRNAs), circular RNAs (circRNAs), and piwi-interacting RNAs (piRNAs)—in tumorigenesis, progression, and therapeutic resistance. These noncoding RNAs (ncRNAs) are regulated through diverse epigenetic mechanisms such as DNA methylation, histone modifications, and N6-methyladenosine (m6A) RNA methylation. Aberrant epigenetic modifications in noncoding regions can silence tumor suppressors or activate oncogenes, thereby reprogramming cellular behavior and contributing to breast cancer heterogeneity. High-throughput techniques like whole-genome bisulfite sequencing (WGBS), ATAC-seq, and ChIP-seq have facilitated the discovery of noncoding epimutations with clinical significance. Moreover, ncRNA-based epigenetic alterations are increasingly explored as diagnostic biomarkers, prognostic indicators, and therapeutic targets, particularly in subtype-specific contexts such

M. E. Abdel Aziz · H. Sabit (✉)
Department of Medical Biotechnology, College of Biotechnology, Misr University for Science and Technology, Giza, Egypt
e-mail: hussein.sabit@must.edu.eg

H. Momtaz · B. Mohamed · M. Amr · S. Abdel-Ghany
Department of Environmental Biotechnology, College of Biotechnology, Misr University for Science and Technology, Giza, Egypt

A. S. Wadan
Oral Biology Department, Faculty of Dentistry, Galala University, Galala Plateau, Attaka, Suez Governorate, Egypt

B. Arneth
Institute of Laboratory Medicine and Pathobiochemistry, Molecular Diagnostics, Hospital of the Universities of Giessen and Marburg UKGM, Justus Liebig University Giessen, Giessen, Germany

N. M. Hamdy et al. (eds.), *From (Epi)genetics-to-Precision (E2P)*, Subcellular Biochemistry 114, https://doi.org/10.1007/978-3-032-08530-6_3

as triple-negative breast cancer (TNBC) and HER2-positive tumors. Despite advances, challenges remain in interpreting functional noncoding elements and translating findings into clinical interventions. This chapter provides a comprehensive examination of the noncoding epigenome in breast cancer, highlighting current methodologies, molecular mechanisms, and translational potential while also identifying future directions needed to leverage noncoding epigenetics for personalized cancer care.

Graphical Abstract

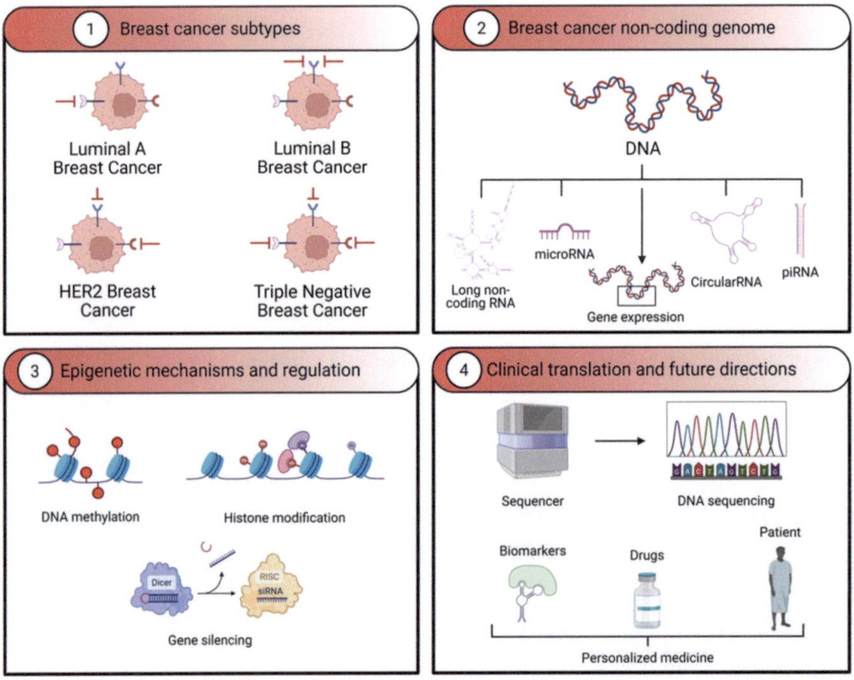

Graphical summary of the role of the noncoding epigenome in breast cancer. Breast cancer subtypes (luminal A, luminal B, HER2-positive, and triple-negative) exhibit distinct molecular features influenced by the noncoding genome. Long noncoding RNAs (lncRNAs), microRNAs (miRNAs), circular RNAs (circRNAs), and piwi-interacting RNAs (piRNAs) are regulated by diverse epigenetic mechanisms, including DNA methylation, histone modifications, and N6-methyladenosine (m6A) RNA methylation. Aberrant epigenetic regulation can silence tumor suppressors or activate oncogenes, driving tumorigenesis, progression, and therapeutic resistance. High-throughput techniques such as whole-genome bisulfite sequencing (WGBS), ATAC-seq, and ChIP-seq enable the identification of clinically relevant noncoding epimutations. Translational applications include their use as a diagnostic

Keywords Noncoding RNAs · Epigenetics · Breast cancer · DNA methylation · Therapeutic resistance

Abbreviations

ATAC-seq	Assay for transposase-accessible chromatin using sequencing
BC	Breast cancer
ceRNA	Competing endogenous RNA
cfDNA	Cell-free DNA;
ChIP-seq	Chromatin immunoprecipitation sequencing
circRNA	Circular RNA
CSC	Cancer stem cell
CTCF	CCCTC-binding factor
DNMT	DNA methyltransferase
eRNA	Enhancer RNA
EMT	Epithelial-mesenchymal transition
ER	Estrogen receptor
3′ UTR	3′ untranslated region
ER+	Estrogen receptor positive.
EZH2	Enhancer of zeste homolog 2
FGFR4	Fibroblast growth factor receptor 4
FOXC1	Forkhead box C1
GATA3	GATA binding protein 3
HDAC	Histone deacetylase
HER2	Human epidermal growth factor receptor 2
HOTAIR	HOX transcript antisense RNA
H3K27ac	Histone H3 lysine 27 acetylation
H3K27me3	Histone H3 lysine 27 trimethylation
H3K4me3	Histone H3 lysine 4 trimethylation
H3K9ac	Histone H3 lysine 9 acetylation
H3K9me2/3	Histone H3 lysine 9 di−/trimethylation
H4K12ac	Histone H4 lysine 12 acetylation
IGF2BP	Insulin-like growth factor 2 mRNA-binding protein
lncRNA	Long noncoding RNA
m6A	N6-methyladenosine
mRNA	Messenger RNA
miRNA	MicroRNA
ncRNA	Noncoding RNA
NF-κB	Nuclear factor kappa-light-chain-enhancer of activated B cells
PIK3CA	Phosphatidylinositol-4,5-bisphosphate 3-kinase catalytic subunit alpha
piRNA	Piwi-interacting RNA
PR	Progesterone receptor

PRC2	Polycomb repressive complex 2
PTEN	Phosphatase and tensin homolog
qPCR	Quantitative polymerase chain reaction
RISC	RNA-induced silencing complex
RRBS	Reduced representation bisulfite sequencing
scATAC-seq	Single-cell ATAC-sequencing
scBS-seq	Single-cell bisulfite sequencing
scChIP-seq	Single-cell chromatin immunoprecipitation sequencing
SE	Super-enhancer
snoRNA	Small nucleolar RNA
TAD	Topologically associating domain
TET	Ten-eleven translocation enzyme
TNBC	Triple-negative breast cancer
TP53	Tumor protein p53
WTAP	Wilms tumor 1-associating protein
YTHDF	YTH N6-methyladenosine RNA-binding protein family
ZNF	Zinc finger protein

Introduction

Breast cancer remains the most frequently diagnosed cancer and a leading cause of cancer mortality in women worldwide, with 2.3 million new cases and 670,000 deaths estimated in 2022 (WHO 2024). Despite advances in early detection and therapy, its incidence has risen in many regions, and outcomes are often suboptimal (Sedeta et al. 2023). Factors such as delayed diagnoses, the biological complexity of diverse breast cancer subtypes, and emerging treatment resistance contribute to this ongoing health challenge (Prabhu et al. 2024). These realities underscore the need for deeper insights into breast cancer biology beyond the classical paradigms of protein-coding gene mutations. Traditionally, breast cancer research focused on mutations in protein-coding oncogenes and tumor suppressors (e.g., *TP53*, *BRCA1*/2, *PIK3CA*) (Santarpia et al. 2016). Nevertheless, it has become increasingly evident that the genetic code alone fails to comprehensively elucidate the complexities associated with the onset and heterogeneity of breast cancer (Kutasovic et al. 2020). Large-scale genome analyses reveal that over 90% of somatic mutations in cancers reside in noncoding regions of the genome (Furuta et al. 2014; Torcivia and Mazumder 2021). These findings highlight the immense "dark matter" of the genome—regulatory DNA sequences, introns, enhancers, promoters, and noncoding RNA loci—as crucial areas of study (Dietlein et al. 2022). Alterations in these domains, particularly epigenetic modifications, can significantly impact gene functionality without modifying the DNA sequence itself. When such epigenetic modifications occur aberrantly in noncoding regulatory regions, they are sometimes termed "noncoding epimutations" (Ibrahim et al. 2024). In breast cancer, these epigenetic alterations can mimic or complement genetic mutations by silencing tumor suppressors or activating oncogenic pathways (Hazra et al. 2021).

Adding another layer of complexity, the human genome is pervasively transcribed into a myriad of noncoding RNAs (ncRNAs) that do not encode proteins but exert diverse regulatory roles (Onoguchi-Mizutani et al. 2023). This class includes small RNAs like microRNAs (miRNAs, ~22 nucleotides in length) and long noncoding RNAs (lncRNAs, >200 nucleotides), among others (e.g., circular RNAs, piRNAs) (Beňačka et al. 2023). MicroRNAs primarily facilitate mRNA degradation or inhibit translation (Karagkouni et al. 2021), whereas lncRNAs act as scaffolds, decoys, or guides for chromatin-modifying enzymes (Tsai et al. 2010). Breast cancer cells exploit these ncRNAs to rewire gene expression networks in favor of malignancy (Al Khaled 2023). Research endeavors have systematically documented numerous lncRNAs and miRNAs exhibiting differential expression in breast tumors relative to normal tissue, signifying the extensive participation of ncRNAs in tumorigenic mechanisms (Wehida et al. 2023) (Fig. 3.1).

Centered within the nucleus, the figure delineates a tightly coordinated transcriptional initiation process beginning at the chromatin level, where enhancer elements, often harboring somatic mutations near oncogenes such as MYC, facilitate chromatin looping to bring distal regulatory regions into proximity with core promoter sequences. This looping is mediated by a complex network of activators and the mediator complex, which act synergistically to recruit and stabilize the assembly of the pre-initiation complex (PIC) at the TATA box, as demonstrated by the sequential recruitment of general transcription factors (GTFs), including TBP, TFIID, IIB, IIA, IIF, IIE, and IIH, culminating in the engagement of RNA polymerase II (Pol II). Notably, the transcriptional machinery is phosphorylated in an ATP-dependent

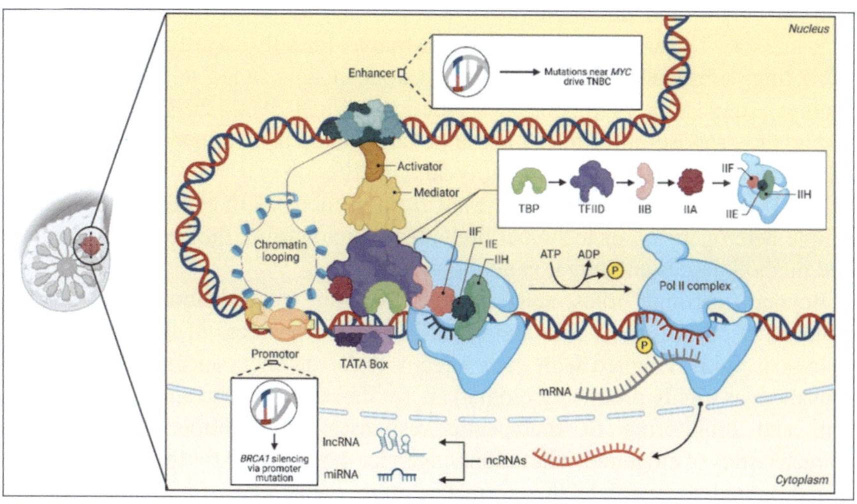

Fig. 3.1 This intricate schematic illustrates the multilayered regulation of gene expression in triple-negative breast cancer (TNBC) through transcriptional activation and epigenetic silencing mechanisms involving both coding and noncoding genomic elements

manner to initiate mRNA synthesis, a process highlighted by the deposition of phosphate groups on Pol II. Intriguingly, the diagram emphasizes how mutations in the promoter region of tumor suppressor genes, such as BRCA1, lead to epigenetic silencing, potentially via disruption of transcription factor binding and loss of promoter accessibility. This process is further influenced by the action of noncoding RNAs (ncRNAs), including long noncoding RNAs (lncRNAs) and microRNAs (miRNAs), which are shown to posttranscriptionally modulate gene expression, possibly via RNA interference or RNA-induced transcriptional silencing complexes, contributing to oncogenic reprogramming in TNBC. The spatial and functional interplay among these elements represents a systems-level regulatory network where aberrant transcriptional control, ncRNA-mediated gene silencing, and mutational landscapes converge to define the transcriptional dysregulation characteristic of aggressive breast cancers. This illustration, therefore, provides a comprehensive, mechanistically dense visualization of transcriptional control, integrating mutational hotspots, chromatin dynamics, transcription factor synergy, and the regulatory role of ncRNAs in cancer epigenetics (adapted from source image).

An emerging body of evidence links epigenetic changes in noncoding regions to key hallmarks of breast cancer, such as initiation, metastasis, and therapeutic resistance (Sarkar et al. 2024). For example, methylation-induced silencing of miR-200 family members promotes epithelial-mesenchymal transition (EMT) (Mongroo and Rustgi 2010; Tellez et al. 2011), while overexpression of miR-221/222 contributes to tamoxifen resistance in estrogen receptor-positive breast cancer (Han et al. 2017). Aberrant DNA methylation patterns and histone modifications in noncoding loci have also been associated with subtype-specific behaviors, such as the aggressive phenotype of basal-like breast cancers (Holm et al. 2010, 2016).

However, the field is not without contradictions. Certain ncRNAs have shown both oncogenic and tumor-suppressive roles depending on the cellular context (Hayat 2024). The case of MALAT1 is illustrative: Initially described as metastasis-promoting, subsequent studies using knockout models failed to reproduce these findings, suggesting more nuanced functions (Chen et al. 2020c). Conversely, despite the proposition of miR-21 as a potential diagnostic biomarker, its dependability exhibits considerable variability among different research studies (Gao et al. 2016). These inconsistencies may result from differences in experimental design, sample heterogeneity, or technical limitations, highlighting the need for standardized methodologies and larger validation cohorts.

Beyond functional roles, ncRNAs and noncoding epimutations are also being explored as clinical tools (Maia et al. 2014). The robustness of their presence in biological fluids, coupled with their selectivity for distinct cancer subtypes, positions them as highly promising biomarkers for the early identification, prognostication, and monitoring of therapeutic responses. In addition, the reversible characteristics of epigenetic alterations engender the potential for therapeutic strategies designed to reinstate the normative expression levels of pivotal ncRNAs (Ferreira and Esteller 2018). Nevertheless, challenges persist, encompassing the establishment of efficacious distribution mechanisms for RNA-oriented treatments and the potential for unintended consequences.

Several critical gaps persist in our understanding. Mechanistic insight into how noncoding epimutations drive cancer remains limited for many ncRNAs (Davalos and Esteller 2013). Furthermore, the noncoding epigenomic frameworks of various breast carcinoma subtypes remain inadequately delineated, constraining our capacity to formulate subtype-specific therapeutic strategies (Xu et al. 2018). Integration of epigenetic and transcriptomic data is needed to resolve these complexities and distinguish between driver and passenger alterations.

This chapter seeks to furnish a thorough and analytical examination of the noncoding genome in the context of breast cancer. It will initially examine epigenetic processes influencing noncoding domains, encompassing DNA methylation and histone alterations. Next, it will present the current landscape of technologies used to map noncoding epimutations, followed by an examination of how these alterations differ across molecular subtypes. The functional consequences of such alterations will be discussed, particularly their influence on cancer hallmarks and therapeutic resistance. Finally, the chapter will address clinical implications, challenges, and emerging therapeutic strategies, offering insights into the future of noncoding epigenomics in breast cancer management. By consolidating existing knowledge and highlighting areas of uncertainty, this work aims to chart a path forward in leveraging the noncoding genome for improved breast cancer outcomes.

The Noncoding Genome Landscape in Breast Cancer: Mechanisms, Biomarkers, and Therapeutic Potential

The Vast and Complex Noncoding Genome in Humans

Only a tiny fraction of the human genome encodes proteins, roughly 1–2%, representing ~20,000 genes, while the vast majority (~98%) is nonprotein coding. This "noncoding genome" was once thought to be genetic "junk" but is now understood to play crucial regulatory roles in the cell biology (Medico-Salsench et al. 2021). The noncoding genome encompasses a bewildering complexity of elements, from regulatory DNA sequences to the growth of noncoding RNA transcripts. Advances in genome annotation have revealed tens of thousands of noncoding RNA genes in humans, far outnumbering protein-coding genes. For example, current GENCODE annotations include over 35,000 long noncoding RNA (lncRNA) genes, compared to ~19,000 protein-coding genes, as well as thousands of small noncoding RNA genes (miRNAs, snoRNAs, etc.) (Harrison et al. 2024).

These noncoding sequences give rise to a diverse "regulatory RNA" network that regulates gene expression programs (Zhang et al. 2023c). Indeed, unlike highly conserved protein-coding genes or microRNAs, many long noncoding RNAs are rapidly gained and lost during evolution, exhibiting poor primary-sequence conservation (Johnsson et al. 2014). The noncoding majority of our genome is a rich landscape of regulatory information, featuring layers of transcripts and elements that fine-tune gene expression in both time and space. Understanding the interactions between noncoding RNAs and their regulatory roles in breast cancer could unveil novel therapeutic targets that improve patient outcomes (Dvorská et al. 2021). Examining these associations might aid in the revelation of unique noncoding RNAs that could play roles as biomarkers or therapeutic targets in the fight against breast cancer (Beňačka et al. 2024).

The Major Noncoding RNA Classes

In recent years, researchers have identified several major classes of noncoding RNAs (ncRNAs), each with its unique mode of production and function within the cell. Among these, the types most closely linked to cancer development include microRNAs (miRNAs), long noncoding RNAs (lncRNAs), circular RNAs (circRNAs), Piwi-interacting RNAs (piRNAs), and small nucleolar RNAs (snoRNAs) (Hamdy et al. 2025). Although these ncRNAs vary widely in size, evolutionary conservation, and how they regulate cellular processes, they all play meaningful roles in the context of breast cancer (Eldash et al. 2023; Hamdy et al. 2024b). Each class has been found to contribute in its way to the complex biology of the disease (Yan and Bu 2021) (Fig. 3.2).

MicroRNAs (miRNAs)

MicroRNAs are short, single-stranded RNA molecules, about 20 to 22 nucleotides long, that help control gene expression after a gene has been transcribed. They derive from hairpin precursor transcripts and integrate into the RNA-induced silencing complex (RISC) to guide sequence-specific silencing of target mRNAs, typically by binding complementary sites in the 3′ untranslated region of mRNAs (Stavast and Erkeland 2019).

In cancer and breast cancer in particular, miRNAs are well-established players. Dysregulation of specific miRNAs can promote tumorigenesis or act as a brake on cancer progression (Subramanian and Sinha 2024). For example, the oncomiR miR-21 is often overexpressed in breast tumors and promotes cell survival and invasion by silencing tumor suppressors like PTEN and PDCD4 (Dan et al. 2021). Conversely, miR-200 family members are frequently downregulated in aggressive breast cancers, relieving inhibition on epithelial-to-mesenchymal transition (EMT) drivers and thus facilitating metastasis (Górecki and Rak 2021).

Due to their small size and conservation, miRNAs were the first noncoding RNAs (ncRNAs) extensively profiled across breast cancer subtypes. Many miRNAs are now known to correlate with specific phenotypes; for example, miR-155 is upregulated in basal-like/triple-negative breast cancers and modulates immune evasion and inflammation, whereas miR-205 is downregulated in basal tumors, contributing to loss of cell polarity in cancer (Angius et al. 2020).

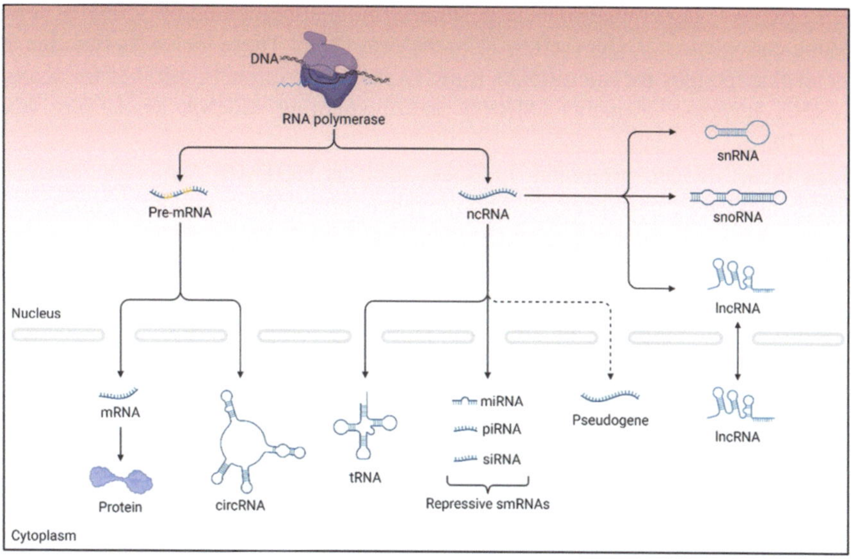

Fig. 3.2 This figure illustrates the transcriptional and posttranscriptional processing pathways that give rise to various coding and noncoding RNA species within the cell. At the top, DNA is transcribed by RNA polymerase to produce primary RNA transcripts. These transcripts diverge into two main categories: pre-messenger RNA (pre-mRNA) and noncoding RNA (ncRNA). Pre-mRNA undergoes processing to form mature mRNA, which exits the nucleus into the cytoplasm, where it is translated into proteins. Alternatively, pre-mRNA can be processed into circular RNA (circRNA) which is functional in the cytoplasm. Noncoding RNAs, on the other hand, include a variety of RNA species that are not translated into proteins. In the nucleus, ncRNAs are processed into small nuclear RNA (snRNA) and small nucleolar RNA (snoRNA), which are involved in splicing and ribosomal RNA modification, respectively. Long noncoding RNAs (lncRNAs) are also generated from ncRNAs and can function both in the nucleus and cytoplasm. Some ncRNAs are processed into repressive small RNAs (smRNAs), such as microRNA (miRNA), PIWI-interacting RNA (piRNA), and small interfering RNA (siRNA), which regulate gene expression posttranscriptionally in the cytoplasm. Additionally, pseudogenes can transcribe ncRNAs that may act as decoys or regulators for other RNAs, often feeding into the lncRNA network. Overall, the figure summarizes the complex landscape of RNA biogenesis, highlighting the intricate regulatory interplay between coding and noncoding RNA species in gene expression and cellular function

Long Noncoding RNAs (lncRNAs)

Long noncoding RNAs (lncRNAs) are typically defined as RNA transcripts longer than 200 nucleotides with little or no protein-coding potential (Hamdy et al. 2024c; Rizk et al. 2024; Zocchi et al. 2021). Thousands of lncRNAs are transcribed in the human genome, generally in a tissue-specific manner and at lower expression levels than mRNAs (Chen et al. 2018). LncRNAs act through a remarkable variety of mechanisms. They can function in the nucleus, for example, as scaffolds assembling chromatin-modifying complexes on DNA, as guides that recruit epigenetic enzymes to specific gene loci, or as decoys sequestering transcription factors and preventing them from binding DNA (Kumar and Bhandari 2022).

LncRNAs play crucial roles in posttranscriptional regulation by acting as competing endogenous RNAs (ceRNAs) in the cytoplasm. These lncRNAs function as molecular sponges for microRNAs (miRNAs), competing with mRNAs for shared miRNA response elements (MREs) and modulating miRNA availability (Oo et al. 2022).

In breast cancer, prominent lncRNAs such as HOTAIR, MALAT1, NEAT1, GAS5, and PVT1 have been shown to regulate tumor cell proliferation, apoptosis, metastasis, and therapy responses (El-Fattah et al. 2021; Yue et al. 2021). For example, the lncRNA HOTAIR was one of the first described oncogenic lncRNAs in breast cancer: it is overexpressed in metastatic breast tumors. It can reprogram chromatin states by recruiting the polycomb repressive complex 2 (PRC2) to hundreds of genes, thereby silencing tumor suppressor genes and promoting metastasis (Ma et al. 2022b).

Circular RNAs (circRNAs)

Circular RNAs are a class of covalently closed RNA molecules found across eukaryotes, with potential roles in posttranscriptional gene regulation (Patop et al. 2019). Back-splicing events typically form them from precursor mRNA transcripts. In back-splicing, a downstream splice donor joins to an upstream splice acceptor, producing a circular transcript with a scrambled exon order that has no free 5′ or 3′ end (Welden and Stamm 2019). In breast cancer, multiple circRNAs have now been identified as dysregulated and functionally essential. One notable example is circFBXW7, which was reported to encode a novel 21-kDa protein that has tumor suppressor activity in breast cancer cells (Ye et al. 2019).

More commonly, breast cancer-associated circRNAs such as circANKS1B and circAGFG1 have been shown to promote metastasis by sponging tumor-suppressive miRNAs, like the miR-148 family, leading to upregulation of pro-metastatic genes (Zeng et al. 2018; Xue et al. 2019). Because circRNAs are unusually stable in blood and tissue, they are being actively explored as cancer biomarkers. For instance, certain circRNAs in plasma have been proposed as diagnostic markers to distinguish breast cancer patients from healthy individuals (De Palma et al. 2022).

Piwi-Interacting RNAs (piRNAs)

Piwi-interacting RNAs (piRNAs) are a distinct class of small ncRNAs, typically 24–32 nucleotides in length (Shaker et al. 2025). PiRNAs are generated from long single-stranded precursor transcripts and processed through a Dicer-independent pathway involving a “ping-pong” amplification cycle in the cytoplasm (Yao et al. 2022). PiRNA expression profiles can distinguish tumor tissues from normal tissues in a cancer-specific manner (Zhou et al. 2023a).

Hundreds of distinct piRNA sequences have been detected in breast cancer samples. Notably, piR-021285 was the first piRNA found to have a tumorigenic role in breast cells. It promotes breast cancer cell invasion by inducing DNA methylation of the pro-apoptotic gene *ARHGAP11A*, thereby silencing that gene and inhibiting apoptosis (Maleki Dana et al. 2020).

Another example is piR-823, which is significantly upregulated in breast tumors and has been shown to support cell proliferation and cancer stem cell properties in estrogen receptor-positive breast cancer cell lines (Zhu et al. 2022). There is also evidence of aberrant expression of Piwi-family proteins in breast tumors. For instance, PIWIL4 is overexpressed in basal-like breast cancer, and its knockdown suppresses breast cancer cell migration and growth by affecting growth factor signaling (Qian et al. 2021).

Small Nucleolar RNAs (snoRNAs)

Small nucleolar RNAs (snoRNAs) are typically 60–150 nucleotide RNAs that localize to the nucleolus and guide the chemical modification of other RNAs, most famously ribosomal RNAs. There are two prominent families, C/D box snoRNAs (guiding 2-O-methylation) and H/ACA box snoRNAs (guiding pseudouridylation), which generally act as part of small nucleolar ribonucleoprotein complexes to modify rRNAs or small nuclear RNAs (Bergeron et al. 2020).

Certain snoRNAs are aberrantly expressed and have an impact on tumor behavior. For example, SNORA71A was recently shown to be upregulated in metastatic breast tumors and to promote epithelial-mesenchymal transition (EMT) and metastasis, which interacts with RNA-binding protein G3BP1 to stabilize *ROCK2* mRNA, enhancing pro-metastatic signaling (Hu et al. 2022). Another notable example is SNORD50A/B, whose deletion or downregulation has been observed in breast cancer. Its loss was associated with resistance to the mTOR inhibitor everolimus drug, as U50 normally regulates mTOR signaling, so its loss confers drug resistance (Li et al. 2024b).

Functional Roles of ncRNAs in Breast Cancer Biology

Noncoding RNAs are now understood to be master regulators of gene expression, operating at virtually every stage of the gene regulation cascade. In breast cancer, a plethora of ncRNAs of all classes described above contribute to hallmark cancer processes by modulating gene expression programs and signaling pathways. We now summarize how ncRNAs function in key regulatory capacities and how this drives breast cancer biology, including their effects on cell proliferation, apoptosis, invasion and metastasis, immune evasion, and therapy resistance. On the other hand, ncRNAs play a significant role in the suppression of breast cancer through various mechanisms. Their suppressive mechanisms are diverse, targeting multiple pathways and cellular functions to inhibit tumor growth and metastasis.

Transcriptional and Epigenetic Regulation

Noncoding RNAs play a crucial role in the transcriptional and epigenetic regulation of gene expression, significantly impacting breast cancer biology. It can act as a signal, guide, and scaffold in transcriptional regulation. They influence gene expression by interacting with chromatin-modifying complexes, thereby affecting chromatin state and transcription factor accessibility (Good 2023).

LncRNAs are involved in establishing and maintaining DNA methylation patterns and histone modifications, which are critical for epigenetic regulation. These modifications can lead to the activation or silencing of genes associated with cancer (Xie et al. 2023). For example, lncRNA HOTAIR recruits the PRC2 complexes to hundreds of genomic sites, leading to repressive histone H3 lysine 27 trimethylation (H3K27me3) at tumor suppressor loci, effectively silencing antimetastatic genes and reprogramming cells toward a metastatic state (Rajagopal et al. 2020).

Another lncRNA, LINC01234, was recently found to bind the chromatin remodeler CHD4 and direct it to the promoters of differentiation genes, maintaining an undifferentiated, proliferative state in breast cancer cells. Through such mechanisms, ncRNAs can control the transcription of oncogenes and tumor suppressors (Bi et al. 2021). Conversely, the lncRNA MEG3 has tumor-suppressive functions through its ability to recruit chromatin-modifying enzymes to silence pro-tumor genes, reestablishing growth control in breast cancer cells (Dong et al. 2021).

Nuclear miRNAs can mediate transcriptional gene silencing (TGS) or activation (TGA) by binding to promoter or enhancer regions of target genes (Billi et al. 2024). Some miRNAs in breast cancer (e.g., miR-320) have been reported to directly impact hormone receptor expression, contributing to the regulation of estrogen receptor (ER) gene transcription in breast cancer cells (Shaath et al. 2021). In addition to piRNAs, it was noted that it can recruit Piwi–Argonaute complexes to chromatin, leading to DNA methylation at complementary sequences (Dabi et al. 2022). In breast cancer, piR-021285's ability to methylate the *ARHGAP11A* promoter is a prime example of an ncRNA enforcing an epigenetic change that suppresses a pro-apoptotic gene, thereby favoring tumor cell survival (Maleki Dana et al. 2020).

Posttranscriptional Gene Regulation

Noncoding RNAs also exert profound control over gene expression posttranscriptionally. They heavily influence the fates of mRNAs once they are produced. MicroRNAs are the quintessential posttranscriptional regulators in breast cancer. This posttranscriptional regulation fine-tunes gene expression, ensuring that proteins involved in key cellular processes are appropriately controlled. Overexpressed oncomiRs like miR-21 and miR-10b directly bind to target mRNAs and trigger their degradation or translational repression, thereby shutting off antiproliferative and antimetastatic brakes (Patutina et al. 2019; El Fatimy et al. 2022). Conversely, loss of tumor suppressor miRNAs leads to aberrant upregulation of proteins that promote cell cycle progression, invasion, or stemness (Otmani and Lewalle 2021).

In breast cancer, the lncRNA NEAT1 serves as a sponge for miR-204, a microRNA that would usually suppress the antiapoptotic protein BCL2, promoting tumor cell survival (Bermúdez et al. 2024). Similarly, circANKS1B sponges the miR-148 family to upregulate *ITGA5* and *PDGFRA* genes that encode proteins involved in cell adhesion, migration, and signaling, key processes in cancer progression, particularly epithelial-mesenchymal transition (EMT) and metastasis (Zeng et al. 2018). However, miRNAs like miR-200 act as tumor suppressors by targeting ZEB1/2, which are transcriptional repressors of E-cadherin. Downregulation of the miR-200 family in breast cancer is associated with epithelial-mesenchymal transition (EMT) and metastasis (Chen et al. 2020a).

Cell Cycle Progression and Proliferation

Many ncRNAs drive breast cancer cell proliferation by modulating cell cycle regulators. Oncogenic miRNAs, such as miR-19a/b, downregulate negative cell cycle regulators and are frequently amplified in breast cancer (O'Bryan et al. 2017). Conversely, tumor-suppressive miRNAs like miR-15a/16-1 can target cyclin-dependent kinases and arrest the cell cycle, inhibiting growth in breast cancer cells (Liu et al. 2019).

LncRNA CCAT1 is overexpressed in breast tumors and promotes G1/S transition by sequestering p15^INK4B^ mRNA away from translation (Liu et al. 2019). However, some ncRNAs constrain proliferation. For example, the lncRNA SCIRT was recently identified as a tumor-suppressive lncRNA that opposes cell cycle gene expression programs in breast cancer, and its loss leads to hyperproliferation (Sideris et al. 2022).

Apoptosis Evasion

Many ncRNAs play roles in apoptosis regulation, either promoting or suppressing programmed cell death. Tumor cells frequently evade apoptosis, a hallmark of cancer, by modulating noncoding RNAs (ncRNAs) that regulate apoptotic pathways. We saw how piR-021285 epigenetically silences *ARHGAP11A* to inhibit apoptosis (Maleki Dana et al. 2020). Additionally, miR-21 targets multiple pro-apoptotic mRNAs, thereby conferring apoptosis resistance in breast cancer cells (Ferreira et al. 2024). LncRNA UCA1 has been reported to bind and sequester pro-apoptotic proteins in the cytosol, blocking their function, as well as to sponge miR-143, which generally targets Bcl-2 (Sideris et al. 2022). The net effect is an increased apoptotic threshold, allowing cancer cells to survive stress and DNA damage that would kill normal cells.

Invasion and Metastasis

Non-coding RNAs regulate various stages of metastasis, from EMT initiation in the primary tumor to invasion through the basement membrane, to survival in circulation, to colonization of distant organs, orchestrating the metastatic process. The miR-200 family plays a crucial role in suppressing epithelial-mesenchymal transition (EMT) and cancer progression by directly targeting and repressing ZEB1 and ZEB2, transcriptional repressors of E-cadherin (Title et al. 2018). Loss of miR-200 expression leads to increased ZEB1/2 levels, promoting EMT, invasion, and metastasis in various cancers (Mongroo and Rustgi 2010). Conversely, overexpression of miR-200 enhances E-cadherin expression, reverses EMT, and reduces cancer cell motility (Hou et al. 2014).

lncRNA ROR induces EMT by sponging miR-205 and activating the ZEB2 pathway (Hou et al. 2014); MALAT1 and NEAT1 both promote migration and invasion in triple-negative breast cancer by regulating EMT-associated gene splicing and transcription (Kim et al. 2018; Su et al. 2025). Recent work on snoRNAs like SNORA71A reinforces that ncRNAs can drive metastasis via novel routes, such as stabilizing mRNAs of metastasis-related kinases (ROCK2) to enhance cell motility (Hu et al. 2022).

Immune Modulation

There is a growing appreciation that ncRNAs help tumors evade or suppress antitumor immunity (Anwar et al. 2022). In breast cancer, miR-155, when expressed in cancer cells, can create an immunosuppressive microenvironment by inducing cytokines that recruit suppressive immune cells and by downregulating antigen presentation machinery (Wang et al. 2022b).

Some lncRNAs in breast cancer cells are induced by interferon or Toll-like receptor signaling and feed back to inhibit immune responses. For example, lncRNA NKILA is upregulated by nuclear factor kappa-light-chain-enhancer of activated B cells (NF-κB) and, in turn, binds to the NF-κB/IκB complex to enhance IκB stability, thus attenuating NF-κB activity. This mechanism in breast tumor-infiltrating T cells causes T cell apoptosis, helping tumors escape immune destruction (Huang et al. 2018).

Therapeutic Resistance

One of the most clinically relevant roles of ncRNAs is in mediating resistance to therapies, including hormonal therapy, chemotherapy, and targeted agents. Breast cancers frequently develop resistance to endocrine therapies (like tamoxifen) or HER2-targeted therapies (like trastuzumab), and ncRNAs are now implicated as key resistance drivers. For instance, miR-221/222 are overexpressed in many endocrine-resistant ER+ tumors and directly downregulate the estrogen receptor (ESR1), promoting cell cycle progression despite tamoxifen (Ouyang et al. 2021).

LncRNAs are heavily involved in drug resistance as well. DSCAM-AS1, an estrogen-regulated lncRNA highly expressed in luminal A/B breast cancers, has been shown to mediate tamoxifen resistance by sequestering miR-137 and thereby upregulating the EPHA7 signaling pathway (Tarighi et al. 2021). Knockdown of *DSCAM-AS1* resensitized cells to tamoxifen, highlighting its potential as a therapeutic target in resistant disease. In HER2+ breast cancer, lncRNA TINCR was reported to promote trastuzumab resistance by sponging miR-125b. High TINCR levels thus sustain HER2 pathway activity despite drug treatment (Dong et al. 2022).

Functional Diversity of ncRNAs Among Breast Cancer Subtypes

Breast cancer is a heterogeneous disease comprising several molecular subtypes with distinct clinical behaviors: mainly luminal A, luminal B (hormone receptor-positive subtypes), HER2-enriched (HER2-positive), and basal-like/triple-negative breast cancer (TNBC), among others. Intriguingly, patterns of ncRNA expression often differ by subtype, and specific ncRNAs play especially prominent roles in one subtype versus another.

Luminal A/B (ER-Positive) Subtypes

Luminal tumors, which express estrogen receptors (ER) and often progesterone receptors and have lower grades in luminal A compared to higher proliferation in luminal B, show distinct noncoding RNA (ncRNA) profiles influenced by estrogen signaling. Several long noncoding RNAs (lncRNAs) are directly regulated by the estrogen receptor (ER) and influence the behavior of luminal tumors. For instance, DSCAM-AS1 is an estrogen-responsive lncRNA that is highly and specifically expressed in luminal A/B breast cancers (Tarighi et al. 2021).

DSCAM-AS1 drives the proliferation of luminal breast cancer cells and mediates tamoxifen resistance by forming a competing endogenous RNA (ceRNA) network with miR-137 and the SRC kinase signaling adaptor EPS8. Knockdown of DSCAM-AS1 in ER+ cell lines reduced tumor growth and restored drug sensitivity (Yue et al. 2020).

From the microRNA side, research showed that enforced miR-221 expression in ER+ cells leads to ER downregulation and antiestrogen resistance, whereas a miR-221 “sponge” could resensitize tamoxifen-resistant cells (Ouyang et al. 2021; Garrido-Cano et al. 2022).

HER2-Positive Subtype

HER2-positive breast cancers, which overexpress the ERBB2/HER2 receptor tyrosine kinase, comprise about 15–20% of cases and historically were aggressive, though HER2-targeted therapies have improved outcomes (Schramm et al. 2015). Several ncRNAs are implicated in modulating HER2-driven oncogenesis and response to therapy. One such lncRNA is HER2OS, also called *ERBB2OS*, which is transcribed from the *ERBB2* locus. HER2OS is often co-amplified with the *HER2* gene and was found to stabilize ERBB2 mRNA and protein levels, thus creating a positive feedback loop sustaining high HER2 signaling (Al-Matouq et al. 2021).

Additionally, lncRNA ZNF649-AS1 was identified as a mediator of trastuzumab resistance; it was found that trastuzumab treatment itself can induce ZNF649-AS1 expression via H3K27 acetylation of its promoter. Once upregulated, this lncRNA represses apoptotic genes and enhances cell survival under therapy. Knockdown of ZNF649-AS1 in HER2+ cells restored sensitivity to trastuzumab, making it a candidate therapeutic target for resistant HER2+ breast cancer (Chauhan et al. 2020).

From the microRNA perspective, miR-125a/b and miR-205 are known to directly target HER2 transcripts; these miRNAs are generally underexpressed in HER2+ tumors, which is consistent with the need for tumor cells to maintain high HER2 levels. Reintroducing miR-205 into HER2+ breast cancer models suppressed HER2 and HER3 levels and curbed tumor growth (Chauhan et al. 2020), illustrating the therapeutic promise of miRNA mimics in this subtype. Another interesting case is miR-4728, an intronic microRNA encoded within the HER2 gene itself; miR-4728-3p is co-expressed with HER2 and has been shown to regulate estrogen receptor signaling, providing a link between HER2 and ER pathways in HER2+/ER+ cancers (Rui et al. 2022).

Basal-Like and Triple-Negative Breast Cancer (TNBC)

Basal-like breast cancers are typically high-grade, aggressive tumors with limited treatment options. This subtype has attracted intense interest in ncRNA research. Several ncRNAs are preferentially associated with triple-negative breast cancer (TNBC). For example, MALAT1 has been reported to be particularly important in TNBC cells, where it facilitates metastatic traits. MALAT1 drives TNBC proliferation and lung metastasis by sponging miR-1, which in turn elevates the expression of the oncogenic transcription factor Slug—inhibiting MALAT1 with antisense oligonucleotides significantly reduced metastasis in TNBC mouse models (Kim et al. 2018).

Another lncRNA, some TNBC studies have found that XIST downregulation correlates with worse outcomes, while others suggest that XIST is context-dependent. Thus, its role remains under investigation and possibly contradictory across studies (Zhang et al. 2022a). lncRNA AC112721.1 that is specifically upregulated in TNBC tissues. They found AC112721.1 directly binds to and stabilizes the pro-metastatic protein THBS1 (Thrombospondin-1) and also sponges miR-491-5p, altogether promoting invasion and lung colonization of TNBC cells (Ma et al. 2024).

From the microRNA angle, TNBCs often exhibit reduced levels of many tumor suppressor miRNAs. For instance, miR-29c is commonly downregulated in TNBC, leading to upregulation of DNMT3A and consequent DNA hypermethylation that silences multiple genes (Bhardwaj et al. 2017). On the other hand, certain oncomiRs are highly expressed in triple-negative breast cancer (TNBC). MiR-155 is one of the highest in basal-like tumors and contributes to their mesenchymal phenotype and immune evasion, as noted earlier. Another, miR-135b, is elevated in claudin-low TNBC and drives invasion by targeting the MID1IP1 (Shao et al. 2025). piRNAs have also been profiled in TNBC cell lines, with some, like piR-016658 and piR-018076, reported at higher levels in triple-negative vs. luminal cells (Guo et al. 2020).

Epigenetic Mechanisms

Essential regulators act without altering the underlying DNA sequence; crucial regulators of cellular identity and function include DNA methylation, histone modifications, genomic imprinting, chromatin accessibility, and random X-chromosome inactivation. In breast cancer (BC), particularly in its aggressive subtypes, epigenetic modifications are significant in regulating noncoding RNAs (ncRNAs), including long noncoding RNAs (lncRNAs). In this cancer, these ncRNAs are being more and more recognized as major controllers of therapeutic response, disease progression, and tumorigenesis (Zolota et al. 2021).

Long noncoding RNAs (lncRNAs) have two opposing roles in cancer: they can either act as tumor suppressors or oncogenes by altering epigenetic landscapes, primarily through changes in DNA methylation. This epigenetic alteration directly affects gene expression by regulating transcriptional activation or suppression (Jin et al. 2021; Mattick et al. 2023).

A set of lncRNAs, including BLAT1, BCLIN25, and H91, has been shown to influence DNA methylation patterns in a manner conducive to tumor growth. For example, the study found that the BLAT1 promoter is hypomethylated, leading to increased BLAT1 expression. Epigenetic activation of BLAT1 is associated with worse prognosis in breast cancer patients and indicates that elevated BLAT1 activity through epigenetic alteration of the gene contributes to the aggressive breast tumor phenotype. Proceeding on the same track of evidence, BCLIN25 is yet another lncRNA that drives breast cancer development through epigenetic alterations. Here, BCLIN25 facilitates oncogene ERBB2 expression through enhanced CpG methylation within the miR-125b promoter. This hypermethylation results in the silencing of the miR-125b.These examples collectively characterize the instructive role of epigenetic modifications—DNA methylation, histone modification, and chromatin remodeling—in regulating the expression of key noncoding RNAs. These ncRNAs, in turn, shape the molecular features of the breast cancer (Jin et al. 2021; Zolota et al. 2021).

DNA Methylations

DNA methylation is a reversible covalent alteration that results when a methyl group is enzymatically transferred to the fifth carbon of the cytosine residues of the genome's CpG dinucleotides. DNA methyltransferases (DNMTs) are a family of enzymes that drive it. DNA methylation plays a critical role in controlling gene expression and regulating genomic stability. Transcriptional repression typically occurs due to the hypermethylation of gene promoter regions, while elevated gene expression results from hypomethylation. Breast cancer (BC) pathophysiology may be impacted due to abnormal patterns of DNA methylation, for example, hyper- and hypomethylation in the target loci that may alter the transcription of tumor suppressor genes and oncogenic factors directly or indirectly (Papanicolau-Sengos and Aldape 2022).

N6-Methyladenosine (m6A)

RNA methylation, particularly m6A, regulates noncoding RNAs in breast cancer. This common mRNA modification in eukaryotes affects miRNAs and lncRNAs through methyltransferases (writers like METTL3, METTL14, METTL16, and WTAP), demethylases (erasers like FTO, ALKBH5, and ALKBH3), and m6A-binding proteins (readers). This reversible modification influences RNA splicing, stability, degradation, and translation, thereby affecting gene expression (Zheng et al. 2021).

Epitranscriptomic regulation influences breast cancer progression by modulating m6A marks. FTO, an m6A demethylase, removes methyl groups from BNIP3 mRNA's 3′ UTR. BNIP3, a tumor suppressor, mediates apoptosis to prevent tumors. FTO's demethylation destabilizes BNIP3 mRNA, reducing expression, inhibiting apoptosis, and promoting cancer progression. Dysregulated RNA methylation suppresses RNA expression, driving tumors. FTO-mediated m6A demethylation silences tumor suppressors like BNIP3, showing epitranscriptomic marks' impact on breast cancer (Liu et al. 2022c).

Histone Modification

In addition to their structural role, histone proteins are highly covalently posttranslationally modified at their tail or side chain, which plays critical roles in the regulation of chromatin dynamics and gene expression. Such modifications are accomplished by adding or deleting certain functional groups. These alterations can significantly impact the charge density at the interface between histones and DNA, which in turn affects chromatin structure and, consequently, transcription (Prabhu et al. 2024).

Epigenetic marks on histone proteins control gene expression. Lysine acetylation (H3K9ac, H3K18ac, H4K12ac), lysine trimethylation (H3K4me3), and arginine dimethylation (H4R3me2) appear in active gene promoters, while lysine methylation (H3K9me2, H3K9me3, H4K20me3) occurs in silenced genes. Loss of H4 histone acetylation (K16) and trimethylation (K20) characterizes cancer, with these modifications predicting cancer prognosis (Feng and Meng 2022).

Histone modifications in breast cancer cells affect regulatory mechanisms of noncoding RNAs (lncRNAs and miRNAs), which influence tumor progression and treatment resistance. Gene expression control through methylation patterns on lncRNA promoters depends on marks determining gene activation or silencing. Breast cancer shows an increased polycomb repressive complex 2 (PRC2), containing histone methyltransferase EZH2, which performs H3K27me3 trimethylation to suppress tumor-suppressive lncRNAs like MEG3. Oncogenic ncRNA levels rise with acetylated histone markers like H3K27ac (Mehrmohamadi et al. 2021).

Histone Acetylation and Methylation

Changes in histone modifications are linked to tumor development and breast cancer progression. About 650 genes contribute to cancer cell transformation. H3K9 alteration affects breast cancer neoplastic transformation. H3K9ac levels increase during tumor development, while H3K9me2 and H3K9me3 levels decline (Prabhu et al. 2024).

LSD1 demethylates histone markers H3K4 and H3K9, linking to early breast cancer development. Cancers show mutations in histone acetyltransferases (HATs), which add acetyl groups to histones, affecting gene expression. Cancer emerges from mutations in acetylation-regulating enzymes. Histone deacetylases (HDACs) remove acetyl groups, and their mutations cause aberrant deacetylation linked to cancer. The imbalance in acetylation processes controls gene expression and promotes tumors (Zhang et al. 2021c).

Beyond acetylation processes, histone tail methylation on lysine and arginine residues introduces an additional layer of intricate chromatin modification. Histone methylation operates through a complex mechanism that differs from the straightforward gene activation or repression seen with acetylation. Instead, it fine-tunes gene expression and chromatin dynamics (Prabhu et al. 2024).

Histone tail methylation and acetylation regulate long noncoding RNAs (lncRNAs). DNA packaging within chromatin, controlled by histone methylation at lysine and arginine sites, affects lncRNA gene activity. While methylation can activate or repress genes, acetylation activates them. Histone methylation affects cancer progression by regulating tumor suppressor or oncogenic lncRNAs (Weinzapfel et al. 2024).

Histone Phosphorylation

Histone phosphorylation, a critical modification in breast cancer noncoding RNAs, involves the addition of phosphate groups to histone proteins. This process alters chromatin structure and influences the expression and function of lncRNAs and miRNAs (Gong et al. 2025; Morgan and Shilatifard 2020).

H3S10 phosphorylation, which occurs during chromatin condensation in cell division, triggers the activation of key oncogenic lncRNAs, like HOTAIR (HOX transcript antisense intergenic RNA), which can be activated by H3S10 phosphorylation. One of the leading causes of breast cancer is HOTAIR, which promotes metastasis (Komar and Juszczynski 2020).

The histone phosphorylation itself may be catalyzed by noncoding RNAs (ncRNAs), miR-941, which is directly in charge of H3S10 phosphorylation, or the phosphorylation of histone H3 at serine 10. Consequently, this leads to breast cancer cells becoming able to divide and migrate. So miR-941 regulates histone modifications, which, in turn, impact the gene expression within the cell (Surapaneni et al. 2020).

H2AX phosphorylation at serine 139 (H2AX-S139ph) affects noncoding RNAs (ncRNAs) in breast cancer and is crucial for DNA damage response and repair. H2AX-S139 phosphorylation interacts with long noncoding RNAs, like the promoter of CDKN1A antisense DNA damage-activated RNA (PANDAR), which controls tumor growth and cell survival. PANDAR's promotion of cell cycle arrest and DNA repair is enhanced by H2AX phosphorylation, enabling cancer cell persistence despite DNA damage (Grohmann et al. 2021).

Chromatin Accessibility in ncRNA

Chromatin accessibility regulates ncRNA expression by controlling promoter availability. In breast cancer, chromatin accessibility changes affect ncRNA transcription and stability, influencing cancer progression. The chromatin structure around ncRNA genes, modified by histones and DNA methylation, determines gene expression. Understanding this relationship is crucial for comprehending breast cancer development (Wang et al. 2021).

HOTAIR, an oncogenic lncRNA, associates with open chromatin and promotes breast cancer metastasis through histone acetylation. Tumor-suppressive lncRNAs like MEG3 are silenced through chromatin compaction and promoter hypermethylation, inhibiting cancer cell growth. These chromatin accessibility changes activate oncogenic lncRNAs while repressing tumor-suppressive lncRNAs, promoting cancer progression (Wang et al. 2021).

miRNAs, like lncRNAs, play a role in regulating gene expression by modulating chromatin accessibility and histone modifications. Specifically, open chromatin and active histone marks, such as H3K27ac, which promote transcription, are associated with miR-200. This miRNA is implicated in the epithelial-to-mesenchymal transition (Bure et al. 2022). Transposable elements and repetitive DNA sequences affect chromatin accessibility and ncRNA expression in gene regulation. These elements can increase ncRNA expression, leading to genomic instability and malignant transformation of breast cancer cells (Chen et al. 2024).

Techniques for Mapping Noncoding Epimutations

Bisulfite-Based Methods

Bisulfite-based methodologies remain the most reliable methods for DNA methylation analysis. They provide single-nucleotide resolution with low DNA input (10–100 pg), are cost-effective, and work across organisms. Sodium bisulfite converts cytosines to uracils, while 5-methylcytosines (5mC) remain unchanged (Farrell et al. 2021).

The resulting material is compatible with various downstream applications, including locus-specific Sanger or pyrosequencing methylation-specific PCR and high-throughput mass spectrometry platforms such as EpiTYPER. For genome-wide profiling (Piao et al. 2021).

Whole-Genome Bisulfite Sequencing (WGBS)

Whole-genome bisulfite sequencing (WGBS) is the most comprehensive method for analyzing DNA methylation status at single-nucleotide resolution across the genome, identifying 5mC changes in coding and noncoding regions like promoters, enhancers, and regulatory positions. The WGBS process involves genomic DNA extraction, bisulfite treatment, library preparation, sequencing, and computational alignment using tools like Bismark or BS-Seeker (Gong et al. 2022).

Being able to measure methylation inside noncoding sequences, which are frequently ignored by targeted or array-based techniques, makes it particularly useful in epigenetic investigations of diseases like cancer (Piao et al. 2021).

High sequencing costs, DNA deterioration during bisulfite treatment, and the computational overhead of processing bisulfite-altered sequences are the main drawbacks of WGBS. Reduced Representation Bisulfite Sequencing (RRBS) has been developed as a targeted alternative to reduce expenses and increase efficiency (Tsutsui and Grossniklaus 2024).

Reduced Representation Bisulfite Sequencing (RRBS)

Reduced Representation Bisulfite Sequencing (RRBS) is a low-cost method for analyzing DNA methylation at the single-nucleotide level in CpG-rich regions like promoters and CpG islands. RRBS enriches functionally significant genomic locations while reducing costs compared to WBGS. The method uses MspI, a methylation-insensitive restriction enzyme that cuts at CCGG sites, to digest genomic DNA targeting CpG-rich regions. DNA fragments of 40–220 base pairs are selected, then ligated using methylated adapters, end-repaired, and A-tailed (Nakabayashi et al. 2022).

After that, sodium bisulfite is added to the DNA. High-throughput sequencing is performed after PCR amplification, and the treated sequences are compared to the reference genome to assess the methylation status (Yang et al. 2024b). It is ideal for studies with limited sample availability, as it requires minimal DNA input (as little as 10–300 ng). Additionally, RRBS was created for single-cell applications, enabling the investigation of DNA methylation at the single-cell level (Nakabayashi et al. 2022).

RRBS has limitations due to the MspI enzyme's CCGG site restriction, excluding genomic areas lacking these sites. Conventional bisulfite sequencing cannot differentiate between 5-methylcytosine (5mC) and 5-hydroxymethylcytosine (5hmC), as both appear as cytosine after bisulfite treatment. Distinguishing these requires additional methods like oxidative bisulfite sequencing (OxBS-seq) and TET-assisted bisulfite sequencing (TAB-seq). The bisulfite treatment method denatures DNA (Shareef et al. 2021).

Amplicon-Based Targeted Bisulfite Sequencing

Amplicon-based targeted bisulfite sequencing is one of the most frequently used methods for determining DNA methylation in specific genomic regions with greater precision and lower cost. With the use of bisulfite-treated DNA-specific primers, PCR is then performed. After PCR, sequencing adapters and sample barcodes, or indices, which are specific to the sample, are then applied (Moser et al. 2020).

Libraries are sequenced on Illumina platforms using barcodes and indices for multiplexing, increasing throughput, and reducing costs for large-scale studies. After sequencing, bioinformatics workflows retrieve methylation data. This approach enables accurate methylation analysis of noncoding regions like lncRNA promoters and enhancers, particularly useful in cancer research (Smith et al. 2022).

Histone Modification Profiling in Noncoding Regions via ChIP-Seq

Epigenetic modifications play a vital role in controlling how genes are expressed. These changes include DNA methylation and various histone modifications like acetylation and methylation (Jin et al. 2020).

Technologies such as chromatin immunoprecipitation followed by sequencing (ChIP-seq) are used to map proteins associated with chromatin and histone modifications, enabling the investigation of epigenetic marks throughout the genome. This method helps identify DNA regions bound by specific proteins or histone modifications. In living cells, the process begins with the crosslinking of proteins to DNA, which is followed by the fragmentation of the chromatin. Chromatin fragments are isolated using antibodies that seek particular histone modifications, such as H3K4me3 (active promoters), H3K27ac (active enhancers), or H3K27me3 (repressed regions). The DNA pieces are sequenced following crosslink elimination and purification. To map the locations of histone modifications and transcription factor binding, the resultant sequences are aligned to a reference genome (Hino et al. 2022).

ChIP-seq has been especially valuable in understanding noncoding regulatory elements, such as enhancers, silencers, and insulators, which regulate gene expression without being translated into proteins. For instance, the histone modification H3K4me3 is typically found in high levels at the start sites of actively transcribed genes, H3K27ac marks enhancers that are actively working, and H3K27me3, which is deposited by Polycomb complexes, indicates regions where genes are silenced. Furthermore, structural proteins like CTCF and cohesin bind to boundary elements that define topologically associating domains (TADs), organizing the genome into three-dimensional structures that are crucial for coordinating the interactions between enhancers and promoters over long distances (Huo et al. 2025).

To enhance the quality and consistency of ChIP-seq data, several normalization approaches are employed, including local regression (LOESS) and the use of spike-in controls, to mitigate inconsistencies resulting from variations in sequencing depth or experimental artifacts. Following data normalization, specialized computational methods such as MACS2 are employed to detect locations with considerable signal enrichment, allowing for the separation of actual epigenetic patterns from random background noise. This degree of analytical rigor is necessary for the proper analysis of ChIP-seq datasets and for revealing the functional importance of chromatin changes in the regulation of genes (Hino et al. 2022).

ATAC-Seq and Tn5 Transposition in Mapping Chromatin Accessibility

The assay for transposase-accessible chromatin using sequencing, the ATAC-seq experiment, identifies the distinct chromatin landscape of cell types and its alterations due to disruption or disease. This method requires few input cells and no prior knowledge of transcription factors or epigenetic markers. Omni-ATAC is an improved version that works with various cell types. The workflow includes sample preparation, transposition, library preparation, sequencing, and data analysis (Grandi et al. 2022).

The ATAC-seq process starts with isolating nuclei from cells to maintain intact chromatin. Tn5 transposase then inserts sequencing adapters into accessible chromatin regions. After PCR amplification and library generation with size selection, high-throughput sequencing produces DNA reads from open chromatin regions. Following genome alignment and peak calling to identify accessible regions, these regions are annotated and analyzed for differential accessibility between conditions. Visualization tools and integration with other omics data help interpret chromatin accessibility and identify regulatory elements, revealing insights into gene regulation (Yan et al. 2020).

Emerging Technologies

Single-cell epigenomics and long-read sequencing technologies enhance mapping of noncoding epimutations in breast cancer. While bulk approaches mask cellular heterogeneity, techniques like scATAC-seq, scBS-seq, and scChIP-seq enable profiling of chromatin accessibility, DNA methylation, and histone modifications in individual cells. This helps identify rare cell populations and track cell lineages while revealing noncoding regulatory element behavior across cellular contexts (Wu and Schmitz 2023; Gupta et al. 2024).

Long-read sequencing platforms, such as Oxford Nanopore and PacBio, enable direct reading of noncoding regulatory regions spanning tens of kilobases. These technologies improve the resolution of complex genomic regions and detect base modifications without chemical conversion. When combined, single-cell and long-read methods reveal cell-type-specific enhancer usage, epigenetic memory, and noncoding drivers of therapy resistance, essential for developing precise diagnostic biomarkers and personalized epigenetic therapies in oncology (Wu and Schmitz 2023) (Table 3.1).

Epimutational Landscape and Clinical Implications of Breast Cancer

Epimutational Landscape of Luminal A Subtype

DNA Methylation The promoter region of the Ca^2+/Mg^2+ channel gene TRPM7 is frequently methylated in luminal A tumors, which correlates with a more favorable prognosis and a decreased incidence of lymph node metastases (Wang et al. 2022b). Conversely, the proteoglycan gene ADAMTSL2 shows hypermethylation in luminal A (and hypomethylation in luminal B) at a specific intragenic CpG (cg14397888). These methylation changes likely alter gene expression and influence tumor biology (Panigoro et al. 2025).

Table 3.1 A summary of the techniques used for mapping noncoding epimutations

Technique	Description	Application in breast cancer	Advantages	Limitations	Citations
Whole-genome bisulfite sequencing (WGBS)	Sequencing of bisulfite-treated DNA to map methylation at single-nucleotide resolution	Comprehensive analysis of DNA methylation across the genome, including noncoding regions	High-resolution, genome-wide coverage	High cost, requires significant computational resources	Gong et al. (2022), Piao et al. (2021), and Tsutsui and Grossniklaus (2024)
Reduced representation bisulfite sequencing (RRBS)	Bisulfite sequencing focused on CpG-rich regions after restriction enzyme digestion	Cost-effective study of methylation in promoter regions and CpG islands	Lower cost than WGBS, suitable for limited samples	Limited to regions with restriction sites, cannot distinguish 5mC from 5hmC	Nakabayashi et al. (2022, Shareef et al. (2021, and Yang et al. (2024b)
Amplicon-based targeted bisulfite sequencing	Sequencing of specific genomic regions after bisulfite treatment	Detailed analysis of methylation in selected noncoding regions, such as lncRNA promoters	High specificity, lower cost for targeted regions	Limited to preselected regions	Moser et al. (2020) and Smith et al. (2022)
ChIP-seq	Immunoprecipitation of chromatin fragments using antibodies against specific histone modifications, followed by sequencing	Mapping histone modifications in noncoding regions, identifying active and repressed regulatory elements	Genome-wide profiling of histone marks	Requires specific antibodies, high sequencing depth	Hino et al. (2022 and Huo et al. (2025)
ATAC-seq	Transposase-based method to map open chromatin regions	Identifying accessible chromatin regions that may contain regulatory elements like enhancers and promoters	Low input requirement, simple protocol	May miss some regulatory regions, resolution limited to fragment size	Grandi et al. (2022) and Yan et al. (2020)
Emerging technologies	Single-cell epigenomics, long-read sequencing	Profiling epigenetic marks at single-cell resolution, direct detection of base modifications	High resolution, ability to study heterogeneity	Still developing, high cost	Gupta et al. (2024) and Wu and Schmitz (2023)

DAB2IP functions as a tumor suppressor by inhibiting cellular pathways. In luminal A tumors, DAB2IP is inactivated by promoter changes, not gene deletion. Low DAB2IP expression correlates with shorter remission and increased proliferation genes typical of luminal B cancers. DAB2IP promoter alterations indicate poor prognosis in luminal A tumors, suggesting its value as a marker and therapeutic target (Mukherjee et al. 2024).

RASSF1A, a tumor suppressor regulating cell cycle and apoptosis, is frequently inactivated in breast cancer through promoter hypermethylation. This methylation is prevalent in luminal A tumors, reducing RASSF1A expression. Early epigenetic alteration enables uncontrolled cell proliferation. RASSF1A hypermethylation correlates with advanced disease stages and poor survival, indicating potential as a diagnostic biomarker and therapeutic target (Park et al. 2023; Quintas-Granados et al. 2021).

Luminal A breast cancers are distinguished by specific hypomethylation patterns, notably involving ESR1, the gene encoding estrogen receptor-α (ERα), which is pivotal in defining the hormone-dependent characteristics of this subtype. A study conducted in 2021 identified a critical 151-base pair CpG island within the ESR1 promoter that is significantly hypomethylated in luminal A tumors, exhibiting approximately 50% less methylation compared to ER-breast cancers or even normal breast tissue. This reduced methylation enhances robust ERα expression, thereby reinforcing the luminal phenotype and the tumor's reliance on estrogen signaling (Quintas-Granados et al. 2021).

Noncoding RNA In luminal A breast cancer, hypermethylation affects protein-coding and lncRNA genes, silencing regulatory transcripts driving tumor progression. MEG3, a tumor-suppressive lncRNA activating p53 and inhibiting cell proliferation, shows hypermethylation in breast tumors, correlating with tumor size and nodal involvement. Hypermethylation of ZNF667-AS1 and SEMA3B-AS1 links to advanced tumors and metastasis. These three lncRNAs represent epigenetically silenced genes that could serve as biomarkers for luminal A progression and targets for epigenetic therapy (Selezneva et al. 2022).

In luminal A tumors, immune activity is lower compared to aggressive subtypes, though specific immune-related genes show hypomethylation. CD40, JAK3, and ICAM3 remain demethylated, affecting tumor microenvironment expression. ICAM3's reduced methylation at enhancer regions, driven by lncRNA RP11-284N8.3, indicates subtype-specific activation. Luminal A tumors show methylation stability in repetitive elements like LINE-1, correlating with lower proliferation. This epigenetic profile balances hypomethylation with methylation preservation, supporting hormone response and treatment efficacy (Zhao et al. 2021; Bediaga et al. 2010).

miRNA genes become active when their promoters exhibit low DNA methylation, thereby enhancing tumor function and hormone sensitivity. The miR-200c/miR-141 cluster, known for inhibiting cell migration and invasion, remains active in luminal A tumors but is inactive in basal-like breast cancer. MiR-29c sustains tumor methylation by reducing DNMT3A activity, contributing to stable methylation in luminal A tumors. Hypomethylated miR-10a and miR-93 affect hormone therapy response. These overexpressed miRNAs contribute to more differentiated, less invasive tumors with improved treatment response, serving as biomarkers and therapeutic targets (Aure et al. 2021; Kudela et al. 2020).

Chromatin Remodeling and Histone Marks The luminal A breast cancer subtype shows distinct epigenetic features, with reduced H3K9me2 levels regulated by G9a/EHMT2 and H3K4 demethylation by LSD1/KDM1A, unlike aggressive subtypes like TNBC. Activating marks H3K4me2 and H3K9ac indicate an active chromatin state. In ER+ luminal A models, estrogen increases G9a and H3K9me2 levels. While G9a expression is lower in luminal A tumors, it promotes estrogen-driven proliferation and EMT, with its elevated levels linked to poor prognosis, making it a potential therapeutic target (Huo et al. 2025).

LSD1 protects against metastasis in luminal A breast cancer, unlike its oncogenic role in other cancers. It demethylates H3K4 and modulates hormone receptors, enhancing GATA3 expression while suppressing TRIM37. Without LSD1, elevated TRIM37 promotes metastatic features despite reduced tumor growth. LSD1 knockout increases lung metastases in ER+ models. High LSD1 levels reduce recurrence risk in ER+ tumors but indicate poor prognosis in basal-like cancers. While LSD1 inhibitors are being studied, caution is needed for luminal A cases, as inhibition may promote metastasis. Patient stratification remains crucial for LSD1-targeted therapies (Hu et al. 2019).

EZH2, a PRC2 complex component, trimethylates H3K27 to repress differentiation and tumor suppressor genes. EZH2 expression is lower in luminal A breast tumors than in luminal B or basal-like tumors, with high levels correlating to poor prognosis and therapy resistance. Elevated EZH2 indicates poor hormone therapy response and silences ERα and tumor suppressor genes through AKT signaling. EZH2 inhibition by Tazemetostat reactivates genes like FOXC1, preventing metastasis. EZH2 functions as both a biomarker and a therapeutic target, with inhibitors potentially complementing hormone therapies (Gan et al. 2022).

Alterations occur in chromatin regulators. In luminal breast cancer, the ARID1A subunit of the SWI/SNF complex shows frequent mutations. Loss-of-function mutations in ARID1A occur in ~10% of estrogen receptor-positive tumors. ARID1A enables chromatin opening at enhancer and promoter regions; its absence causes transcriptional changes affecting survival and therapy resistance. Luminal tumors with ARID1A and PIK3CA mutations form a distinct immunological subgroup, showing increased immune response genes and infiltration according to TCGA/METABRIC analyses (Anabel Sinberger et al. 2023).

The absence of histone methyltransferase KMT2C (MLL3) leads to aggressive behavior in luminal-lineage models. In murine models, Kmt2c-deficient mammary tumors showed accelerated growth with EMT features and a basal-like phenotype. KMT2C loss activated the EMT signature and kinase signaling pathways. In patient data, MLL3/KMT2C inactivating mutations in ER+ cases correlate with endocrine therapy resistance. Histone methylation regulators EZH2, KMT2C, and NSD2 are key epigenetic drivers in luminal A biology, affecting gene expression (Simigdala et al. 2023).

Epitranscriptome RNA methylation regulators show distinct patterns in luminal A breast cancers. METTL3/14 "writers" and YTHDF "readers" modulate estrogen receptor (ER) signaling. METTL14 enhances USP22 and ERα mRNA stability through m6A modifications, promoting tumor progression. METTL14, with YTHDC1/YTHDF1 readers, amplifies USP22/ERα expression. METTL14, or m6A modification inhibition, reduces ER-positive breast cancer cell proliferation. METTL14, USP22, and ERα expression correlate in patient tumors, establishing the METTL14–USP22–ERα axis as an oncogenic pathway in luminal breast cancer (Zhuang et al. 2024).

Transcriptome data show that basal-like m6A regulators can differentiate ER+ tumors. A study classified breast tumors based on methylation and expression profiles of m6A factors. IGF2BP2/3, YTHDC2, and RBM15 levels distinguished basal-like cancers. Half of luminal A/B samples shared these "basal-like m6A" profiles, showing increased immune-related gene expression and poorer disease-free survival. This epitranscriptomic subgroup within luminal A/B suggests a more aggressive phenotype (Yang et al. 2021b).

Epimutational Landscape of Luminal B Subtype

DNA Methylation In luminal B breast cancers, significant changes in DNA methylation are observed in key regulatory genes. Notably, the WNT1 gene is frequently subject to promoter hypermethylation in luminal B cfDNA, resulting in reduced expression and a poorer prognosis (Rodriguez-Casanova et al. 2022).

Luminal B tumors share some methylation traits with luminal A but often show distinct patterns. The same ADAMTSL2 CpG (cg14397888) that is hypermethylated in luminal A is hypomethylated in luminal B. This differential methylation may contribute to subtype-specific gene regulation (Panigoro et al. 2025).

Two subgroups of luminal B breast cancer, Cluster A and Cluster B, were identified through genomic analysis. Cluster A genes relate to cell cycle progression and metabolic signaling. GATA3's hypomethylation enables BRCA1-mediated inhibition of EMT, suggesting tumor suppression. CCND1, E2F1, and SMAD4 show higher expression in Cluster A, enhancing tumor cell proliferation. Cluster B genes associate with EMT and immune response pathways, with hypomethylated YAP1, FN1, SOX11, and COL1A1. YAP1 acts as a Hippo pathway transcriptional activator, promoting metastasis. FN1, SOX11, and COL1A1 regulate cell adhesion and migration. Cluster B shows higher mutation rates in TP53 (54% vs. 25%), PIK3CA (41% vs. 22%), and ERBB2 (18% vs. 7%), contributing to its aggressive characteristics (Wang et al. 2023a).

Noncoding RNA MIAT, a long noncoding RNA, is overexpressed in luminal B breast cancer. MIAT silencing inhibits BT474 cell proliferation and migration, while miR-150-5p levels are reduced. MIAT acts as a competitive endogenous RNA by binding miR-150-5p, relieving PD-L1 inhibition and enhancing its expression to facilitate immune evasion. This establishes MIAT as an oncogene affecting tumor behavior and immune regulation (Mi et al. 2024).

LINC01016, GATA3-AS1, MAPT-IT1, and DSCAM-AS1 form a luminal subtype-specific signature in breast cancer, particularly luminal B (González-Woge et al. 2024). LINC01016 overexpression correlates with lymph node metastasis and poor prognosis (Sun et al. 2023). GATA3-AS1 and MAPT-IT1 are elevated in luminal B cancer and predict nonresponse to neoadjuvant chemotherapy (NAC) (Contreras-Espinosa et al. 2021). DSCAM-AS1, expressed in luminal A, B, and HER2-positive cancers, promotes tumor progression and tamoxifen resistance through miR-204-5p and miR-130a interactions, affecting estrogen receptor signaling. These lncRNAs modulate tumor aggressiveness and treatment response in luminal B breast cancer (Elhasnaoui et al. 2020; Niknafs et al. 2016).

miR-548b-5p was upregulated in luminal B breast cancer compared to luminal A, suggesting an association with a more aggressive cancer subtype. miR-548b-5p shows pro-oncogenic properties in breast cancer, though its precise mechanisms remain incompletely understood (Záveský et al. 2022).

miR-454 shows higher expression in luminal B versus luminal A, exhibiting pro-oncogenic features linked to tumor grade and Ki-67 index. U6 small nuclear RNA upregulation in luminal B correlates with multifocality, lymph node metastasis, and poor survival. Let-7c suppressor miRNA downregulation correlates with Ki-67 and lymphatic spread, while miR-155 upregulation indicates advanced stages. These noncoding RNAs form an epigenetic signature driving luminal B breast cancer aggressiveness (Záveský et al. 2025).

Chromatin Remodeling and Histone Marks While luminal B characteristics are less documented than luminal A, key patterns exist. EZH2, the PRC2 catalytic component that trimethylates H3K27, is overexpressed in luminal B tumors. High EZH2 levels correlate with increased proliferation and poorer survival in ER+ breast cancer. Luminal B shows elevated Ki-67 and EZH2 expression, creating a high-H3K27me3 signature that is more pronounced than in basal-like tumors (Hirukawa et al. 2018).

Epitranscriptomics Luminal B breast cancer shows a similar m6A regulator landscape to luminal A, except more luminal B tumors fall within the "basal-like m6A cluster" with higher IGF2BP2/3, YTHDC2, RBM15, and immune activation. These changes likely drive aggressive tumor behavior, with METTL3 linked to luminal B progression (Yang et al. 2021b).

Furthermore, KIAA1429 (VIRMA), a constituent of the m6A methyltransferase complex, exhibits elevated expression levels in breast tumors compared to normal tissue and is associated with advanced disease stages. KIAA1429 has been demonstrated to facilitate the progression of luminal-type breast cancer by m6A-modifying the oncogenic long noncoding RNA NEAT1, resulting in its overexpression. Although these findings are not specific to luminal B subtypes, they underscore the significance of m6A in luminal tumor biology and imply that luminal B tumors may particularly exploit m6A modifications to sustain their growth (Jing et al. 2022).

Epimutational Landscape of HER2-Enriched Subtype

HER2-enriched breast cancers are characterized by the overexpression or amplification of the HER2 (ERBB2) oncogene, the absence of estrogen receptor (ER) and progesterone receptor (PR), and typically exhibit high proliferation rates. Epigenetically, HER2-enriched tumors share some features with basal-like cancers.

DNA Methylation HER2-enriched tumors exhibit intermediate levels of global DNA methylation, falling between the hypermethylation of luminal A tumors and the hypomethylation of many triple-negative breast cancers (TNBCs). A comprehensive methylation study across various cancer types observed that HER2+ tumors frequently cluster more closely with basal-like tumors in terms of methylation patterns than with luminal A/B tumors (Hohmann et al. 2025).

FGFR4, a receptor tyrosine kinase, is overexpressed in HER2-enriched cancers. Studies show HER2-enriched tumors have reduced methylation near the FGFR4 promoter, correlating with increased FGFR4 mRNA expression. These CpG regions remain methylated in normal breast tissue and non-HER2 tumors. This epigenetic "unlocking" of FGFR4 transcription may enhance cell proliferation and contribute to therapeutic resistance (Hohmann et al. 2025).

PTPRN2 promoter methylation serves as a biomarker for HER2-positive breast cancer. PTPRN2 was hypermethylated in HER2-enriched tumors (cg25910261 in the PTPRN2 promoter) but unmethylated in other subtypes, correlating with reduced PTPRN2 expression. Low PTPRN2 expression predicted poorer relapse-free survival in HER2+ tumors. PTPRN2, involved in secretory vesicle function and insulin signaling, is being studied as a diagnostic tool for identifying HER2+ relapse risk (Panigoro et al. 2025).

Cancer stem cells in breast cancer play a crucial role (Chiang et al. 2024). HOXC8 affects breast cancer stem cells' self-renewal and differentiation. The HOXC8 gene's methylation correlates with tumor progression across cancer types. In HER2-positive breast cancer, elevated HOXC8 expression occurs through DNA methylation and relates to patient pathology. Modulating HOXC8 may inhibit breast cancer development and metastasis (Xie et al. 2024).

HER2-enriched and basal tumors exhibit hypomethylation at repetitive elements and enhancers compared to luminal tumors, showing reduced methylation at heterochromatic regions. This may cause chromatin derepression and activation of proliferation genes, with MYC and CCND1 showing methylation loss and overexpression in HER2+ tumors (Gao et al. 2015; Rønneberg et al. 2011).

Noncoding RNA Numerous noncoding RNAs have been implicated in the HER2+ subtype of breast cancer. Specifically, elevated levels of four key long noncoding RNAs (lncRNAs), namely AC010595.1, AC046168.1, AC069277.1, and AP000904.1, are associated with poorer overall survival. Conversely, reduced levels of LINC00528 and MIR762HG are linked to decreased overall survival in patients with HER2+ breast cancer (Zhang et al. 2021a).

AC009283.1 is an lncRNA that is overexpressed in the HER2-enriched subtype and regulates genes associated with proliferation, cell cycle, and apoptosis in the HER2+ subtype (Cedro-Tanda et al. 2020).

ORLNC1, another lncRNA, functions as a tumor suppressor by modulating the miR-296/PTEN pathway in HER2+ breast cancer. Reduced ORLNC1 levels were associated with a poor prognosis and more advanced stages of cancer (Cheng et al. 2023).

In HER2+ breast cancer, MALAT1 was found to be significantly overexpressed. The inhibition of MALAT1 resulted in a decrease in the proliferation of HER2+ breast cancer cells (Yang et al. 2021a).

LINC00624 acts as an inhibitor of MHC class I antigen presentation, limiting CD8+ T cell infiltration in the cancer microenvironment and causing resistance to anti-HER2 treatment. As a cancer-immunosuppressive long noncoding RNA, targeting LINC00624 through antisense oligonucleotides in tumors with high LINC00624 expression shows therapeutic potential (Zhang et al. 2022b).

Chromatin Remodeling and Histone Marks The overexpression of histone deacetylase sirtuin 6 (SIRT6) is correlated with cancer stem cell (CSC)-like characteristics and tumor dormancy, as well as reduced senescence and oxidative DNA damage in HER2+ breast cancers (Andreani et al. 2023).

In HER2+ breast cancer, H2B monoubiquitination (H2Bub1) is maintained in primary tumors and metastases, unlike its reduction in other subtypes. This RNF40-facilitated modification activates RHO/ROCK/LIMK pathway genes, regulating the cytoskeleton and cell migration. Through H3K4me3 interaction, H2Bub1 regulates gene expression and supports tumor growth and migration, specifically promoting HER2+ breast cancer (Wegwitz et al. 2020).

Epitranscriptomics In HER2+ breast cancers, m6A hypomethylation regulates FGFR4, which phosphorylates GSK-3β and activates β-catenin/TCF4 signaling, driving resistance to anti-HER2 therapies. The suppression of FGFR4 increases reactive oxygen species (ROS) production and labile iron pool accumulation, leading to ferroptosis, an iron-dependent form of oxidative cell death triggered by FGFR4 inhibition (Zou et al. 2022).

An examination of m6A RNA regulators has demonstrated that the expression levels of LRPPRC, METTL16, RBM15B, and ZC3H13 are diminished in HER2+ breast cancer cases compared to HER2- cases. In contrast, YTHDF3 shows increased expression in patients with HER2+ breast cancer relative to those with HER2- breast cancer (Zhang et al. 2020).

Epimutational Landscape of Triple-Negative Subtype

DNA Methylation Basal-like triple-negative breast cancer (TNBC) shows hypermethylation and downregulation of steroid response genes, notably FOXA1, across all basal-like tumors regardless of receptor status. Genes like FOXC1 and ELF5 show hypomethylation and elevated expression in basal-like tumors, suggesting their role in maintaining the basal phenotype. Basal-like TNBC also exhibits differential methylation of immune-related genes, including PD-L1 and PD-L2, potentially contributing to immune evasion (Aine et al. 2025).

DNA methylation patterns contribute to its aggressive phenotype through hypomethylation and hypermethylation of specific genes. ATOX1 and MADCAM1 facilitate cell migration; UCP2 and TXNRD1 regulate oxidative stress; CXCR3 enables invasion; while DHRS4, BEST1, FXYD1, TRAF5, and FUT5 support metabolism and cell extravasation. Tumor suppressors CTXN1, DLC1, CASP3, and DNASE1 are hypermethylated and silenced, affecting apoptosis. Additionally, hypermethylated genes FUNDC2, IDS, TAF13, NKX2-5/NKX2-1, PPT1, FZD6, RECK, and SSBP1 contribute to therapy resistance and metastatic potential, highlighting disrupted signaling pathways in TNBC (Beetch et al. 2020).

The genes FLJ43663 (lncRNA), PBX1 (a member of the PBX homeobox family), and RASA3 (a component of the Ras signaling pathway) exhibited hypomethylation in triple-negative breast cancer (TNBC) compared to nonneoplastic breast tissue and other breast cancer subtypes. This hypomethylation was observed across all stages of TNBC, including the early stages, indicating their potential utility as screening biomarkers (Mendaza et al. 2021).

Noncoding RNA The long noncoding RNA (lncRNA) LINC00299 gene has been identified as hypermethylated in the DNA of patients with triple-negative breast cancer (TNBC) when analyzed through liquid biopsy, compared to healthy individuals. This finding suggests that LINC00299 hypermethylation may serve as a circulating biomarker for TNBC (Manoochehri et al. 2020).

MIR4500HG003 has been shown to promote TNBC metastasis by sponging miR-483-3p, leading to the upregulation of MMP9, a matrix metalloproteinase involved in extracellular matrix degradation and tumor invasion (Lin et al. 2024). Conversely, PTCSC3 acts as a tumor suppressor by downregulating MIR100HG, thereby inhibiting cell proliferation and inducing apoptosis in TNBC cells (Zhang et al. 2023a).

LINC01559 serves as a ceRNA that accelerates TNBC progression by sponging multiple miRNAs, thus promoting cell proliferation and migration (Yang et al. 2022). Similarly, BTG3-7:1, in conjunction with the transcription factor JUND, co-regulates C21ORF91 expression, facilitating TNBC progression (Dan et al. 2020).

MNX1-AS1 promotes TNBC progression by enhancing the phosphorylation of Stat3, a transcription factor involved in cell growth and survival (Fang et al. 2020).

In the context of chemoresistance, ANRIL enhances aerobic glycolysis, contributing to resistance against chemotherapeutic agents in TNBC (Ma et al. 2022a).

LINC-PINT attenuates paclitaxel resistance in TNBC cells by targeting the RNA-binding protein NONO, thereby restoring chemosensitivity (Chen et al. 2020b).

Moreover, SNHG22 facilitates malignant phenotypes in TNBC by sponging miR-324-3p and upregulating SUDS3, a component of the histone deacetylase complex (Fang et al. 2020).

Chromatin Remodeling and Histone Marks Histone lactylation, particularly at H4K12, has emerged as a novel biomarker (Cui et al. 2024) and has been shown to promote malignancy via SLFN5 downregulation (Li et al. 2024a). Furthermore, cancer-associated fibroblasts may enhance ZFP64 lactylation, contributing to chemoresistance by modulating ferroptosis (Huang et al. 2025).

In addition to lactylation, histone phosphorylation plays a pivotal role in TNBC metastasis. NONO has been identified as a regulator of H3 phosphorylation through interactions with PKM2 (Li et al. 2025), while histone H3 proline 16 hydroxylation is implicated in gene expression regulation in mammalian systems (Liu et al. 2022b).

Histone methylation also presents therapeutic implications. The methyltransferase KMT2C, stabilized by Secernin-2, suppresses progression and enhances sensitivity to PARP inhibitors (Huang et al. 2025). CircBRAF was found to recruit KDM4B to H3K9me3, promoting cancer progression (Lan et al. 2024), while repressive trimethylation markers (e.g., H3K27me3) have been correlated with patient survival (Llinàs-Arias et al. 2023b).

Recent research has highlighted the role of histone deacetylase 6 (HDAC6) in forming phase-separated condensates that disrupt chromatin architecture, thereby promoting tumor progression (Lu et al. 2024). Similarly, PML has been shown to reprogram the epigenetic landscape by modulating chromatin composition, driving the expression of pro-metastatic genes in TNBC (Fracassi et al. 2023). These alterations in chromatin state are not merely structural but tightly linked to functional consequences, such as transcriptional reprogramming, immune evasion, and therapeutic resistance.

The dynamic 3D organization of chromatin in TNBC further supports oncogenic behavior. Studies have revealed TNBC-specific changes in chromatin looping and insulation that affect the expression of matrix metalloproteinase clusters, essential for invasion and metastasis (Llinàs-Arias et al. 2023a, b). These structural changes also extend to chromatin accessibility patterns, which vary significantly based on patient ancestry and drive heterogeneity in gene expression (Harris et al. 2023). The use of high-resolution assays has enabled the identification of key chromatin regulators like LOXL2, which maintains chromatin compaction and is critical for sustaining TNBC oncogenic traits (Serra-Bardenys et al. 2024), and ARID1A, which modulates metastatic heterogeneity through its interaction with YAP (Wang et al. 2023c).

Chromatin remodeling also intersects with therapeutic strategies. Pharmacological agents that induce chromatin remodeling have been shown to sensitize TNBC cells to chemotherapy (Bagheri et al. 2024), and novel CRISPR-based screening platforms such as EPIKOL have identified epigenetic vulnerabilities specific to TNBC cells (Yedier-Bayram et al. 2022). Notably, the chromatin-bound factor MUC1-C has been implicated in regulating enhancer accessibility via JUN and BAF complexes, suggesting a role in maintaining cancer stemness (Bhattacharya et al. 2022).

Furthermore, chromatin architectural proteins like SND1 modulate gene-specific methylation, such as CDH1, contributing to metastatic behavior (Zhang et al. 2023b). Alterations in specific chromatin marks, such as H3K4me3 and H3K9me3, also define the epigenetic signatures of TNBC cells (Anjum et al. 2025), and novel players such as BAP18 have been identified as oncogenic transcriptional coactivators (Zhang et al. 2022c).

Epigenetic regulators FOXA1 and GRHL2, identified through accessibility profiling, offer potential lineage-specific therapeutic targets (Yang et al. 2024a). Likewise, synthetic lethality between SETDB1 and CHD4 opens new therapeutic windows in TNBC treatment (Moraes-Almeida et al. 2023).

Epitranscriptomics: METTL3, a major m6A methyltransferase, plays dual roles depending on its regulatory targets. While its reduction facilitates COL3A1 overexpression and TNBC metastasis (Shi et al. 2020), its upregulation in other contexts supports tumorigenesis via PRMT7 methylation (Lu and Yang 2025) and STUB1 suppression, leading to autophagy inhibition (Li et al. 2024d). METTL3 also mediates YAP1 m6A methylation, enhancing endothelial cell activation and metastasis (Du et al. 2025).

Reader proteins such as IGF2BP1-3 have been implicated in promoting oncogenic mRNA stability. IGF2BP2 enhances CDK6 translation by recruiting EIF4A1 (Xia et al. 2024), while IGF2BP3 drives TNBC progression by degrading NF1 (Zhang et al. 2023d) and repressing SLIT2 (Jiang et al. 2022). YTH domain-containing proteins also contribute; YTHDC1 promotes glycolysis and proliferation via the FOXM1/CENPA axis (Shen et al. 2024b), and YTHDF3 stabilizes CENPI to drive tumor growth (Zhang et al. 2025).

Among erasers, ALKBH5 has been shown to regulate FOXO1 mRNA stability and chemoresistance (Liu et al. 2024), while FTO promotes C/EBPβ-LIP translation to exert oncogenic functions (Zuidhof et al. 2025).

Several m6A-regulated transcripts have been identified as key effectors in triple-negative breast cancer (TNBC). POP1 targets CDKN1A for degradation, promoting proliferation (Zhang et al. 2024), while WTAP modulates ferroptosis by regulating LCN2 via m6A-dependent mechanisms (Tan et al. 2024). VIRMA facilitates KIF15 upregulation to enhance malignant progression (Chen et al. 2023). Furthermore, PRMT7, CCL20, and c-MYC have been identified as downstream effectors of m6A-modified transcripts, linking m6A to cell cycle control, migration, and immune modulation (Liu et al. 2022a; Xu et al. 2025).

The manipulation of m6A regulators offers promising therapeutic avenues. For example, inhibition of IGF2BP3 or METTL3 may reverse oncogenic translation programs. In addition, epitranscriptomic modulation shows synergy with chemotherapy and ferroptosis inducers (Tan et al. 2024; Yu et al. 2025) (Table 3.2).

Table 3.2 A summary of the epimutational landscape of the four breast cancer molecular subtypes

Subtype	DNA methylation	Noncoding RNA	Chromatin and histone marks	Epitranscriptomics (m6A)	Citations
Luminal A	Promoter methylation: TRPM7 promoter hypermethylated (favorable prognosis); ADAMTSL2 intragenic hypermethylation (luminal B shows opposite). Tumor suppressors: RASSF1A frequently hypermethylated (silenced). Hormone receptor: ESR1 promoter hypomethylate*d* (↑ ERα expression)	Silenced lncRNAs: MEG3, ZNF667-AS1, and SEMA3B-AS1 lncRNAs are hypermethylated (silenced) in luminal A tumors. Immune genes: CD40, JAK3, and ICAM3 loci remain demethylated (active). Active miRNAs: The miR-200c/141 cluster and miR-29c, miR-10a/93 are hypomethylated/active in luminal A, supporting differentiation and better therapy response	Repressive marks low: Reduced H3K9me2 (via low G9a/EHMT2) and active H3K4me2/H3K9ac marks are characteristic. LSD1 (KDM1A): Acts as an H3K4 demethylase that *protects* against metastasis in luminal A. EZH2: Lower PRC2/EZH2 (H3K27me3) levels than in other subtypes; high EZH2 would correlate with poor outcome. Chromatin regulators: ARID1A (SWI/SNF) loss in ~10% ER+ tumors; KMT2C (MLL3) loss induces aggressive (basal-like) behavior	m6A writers: METTL3/METTL14 and YTHDF readers stabilize ERα/USP22 mRNAs, promoting luminal A tumor growth. m6A profiles: ~50% of luminal A/B tumors share a "basal-like m6A" signature (↑IGF2BP2/3, YTHDC2, RBM15), linked to immune activation and aggressiveness	Anabel Sinberger et al. (2023), Aure et al. (2021), Bediaga et al. (2010), Gan et al. (2022), Hu et al. (2019), Huo et al. (2025), Kudela et al. (2020), Mukherjee et al. (2024), Panigoro et al. (2025), Park et al. (2023), Quintas-Granados et al. (2021), Selezneva et al. (2022), Simigdala et al. (2023, Wang et al. (2022c), Yang et al. (2021b), Zhao et al. (2021), and Zhuang et al. (2024)
Luminal B	Key changes: The WNT1 promoter is hypermethylated in cfDNA. ADAMTSL2 intragenic CpG is hypomethylated (opposite to LumA). Subclusters: Two methylation subgroups—Cluster A (cell cycle genes with hypomethylated GATA3, ↑CCND1/E2F1/SMAD4) and cluster B (EMT/immune genes with hypomethylated YAP1, FN1, SOX11, COL1A1)	Oncogenic lncRNAs: MIAT is overexpressed, sponging miR-150 to derepress PD-L1. Subtype lncRNA signature: LINC01016↑ (lymph node metastasis); GATA3-AS1 and MAPT-IT1↑ (chemoresistance to NAC); DSCAM-AS1↑ (tumor progression and tamoxifen resistance). miRNAs: Upregulated miR-548b-5p and miR-454 (associated with higher grade/Ki-67); others like U6 RNA (poor prognosis) and miR-155 (late-stage) are also elevated	EZH2/H3K27me3: Overexpressed EZH2 (with high Ki-67) creates a strong H3K27me3 signature, driving proliferation and poor survival. (LumB shows higher EZH2/Ki-67 than even basal-like)	m6A regulators: Similar profile to luminal A, but more LumB tumors cluster as "basal-like" (↑IGF2BP2/3, YTHDC2, RBM15). METTL3: Higher METTL3 expression linked to LumB progression. VIRMA (KIAA1429): Upregulated in luminal tumors; m6A-modifies lncRNA NEAT1 (oncogenic)	Contreras-Espinosa et al. (2021), Elhasnaoui et al. (2020), González-Woge et al. (2024), Hirukawa et al. (2018), Jing et al. (2022), Mi et al. (2024), Niknafs et al. (2016), Panigoro et al. (2025), Rodriguez-Casanova et al. (2022), Sun et al. (2023), Wang et al. (2023a), Yang et al. (2021b), Záveský et al. (2022), and Záveský et al. (2025)

(continued)

Table 3.2 (continued)

Subtype	DNA methylation	Noncoding RNA	Chromatin and histone marks	Epitranscriptomics (m6A)	Citations
HER2-enriched	Global methylation: Intermediate between luminal and TNBC; often clusters with basal-like methylation patterns. FGFR4: Promoter hypomethylated (↑FGFR4 expression). PTPRN2: Promoter hypermethylated (HER2-specific marker, low PTPRN2). HOXC8: DNA methylation regulation drives HOXC8 overexpression (stemness). Repeats/enhancers: Hypomethylation at heterochromatic regions (e.g., MYC, CCND1) like basal tumors	Prognostic lncRNAs: High levels of AC010595.1, AC046168.1, AC069277.1, and AP000904.1 correlate with worse survival; low LINC00528 and MIR762HG also predict poor outcome. AC009283.1: Overexpressed lncRNA driving proliferation and cell cycle in HER2+. ORLNC1: Tumor suppressor lncRNA (miR-296/PTEN axis); its downregulation signals advanced HER2+ disease. MALAT1: Upregulated; promotes proliferation (inhibition decreases HER2+ growth). LINC00624: Overexpressed; inhibits MHC I presentation and CD8+ infiltration, causing anti-HER2 resistance	SIRT6: Upregulated histone deacetylase; linked to stemness-like phenotypes and dormancy in HER2+ cells. H2Bub1 (RNF40): Maintained in HER2+ (unlike other subtypes), activating RHO/ROCK/LIMK migration genes (via H3K4me3 interplay)	FGFR4 m6A: Hypomethylation (↓m6A) of FGFR4 mRNA activates β-catenin/TCF4 signaling (anti-HER2 resistance); FGFR4 suppression induces ferroptosis. m6A regulators: ↓LRPPRC, METTL16, RBM15B, and ZC3H13; ↑YTHDF3 in HER2+ vs. HER2– patients	Andreani et al. (2023), Cedro-Tanda et al. (2020), Cheng et al. (2023), Gao et al. (2015), Hohmann et al. (2025), Panigoro et al. (2025), Rønneberg et al. (2011), Wegwitz et al. (2020), Xie et al. (2024), Yang et al. (2021a), Zhang et al. (2020), Zhang et al. (2022b), Zhang et al. (2021a), Zou et al. (2022))

Triple-negative (basal-like)	Steroid genes: FOXA1 promoter hypermethylated (downregulation); basal markers FOXC1 and ELF5 are hypomethylated (↑expression). Immune checkpoints: PD-L1/PD-L2 promoters are often demethylated (immune evasion). Metastasis/invasion: Hypomethylation of genes like ATOX1, MADCAM1, UCP2, TXNRD1, and CXCR3 promotes migration/oxidative stress resistance. Tumor suppressors: Genes such as CTXN1, DLC1, CASP3, and DNASE1 are hypermethylated (silenced). Biomarkers: lncRNA FLJ43663, PBX1, and RASA3 are hypomethylated in TNBC at all stages (potential biomarkers)	Circulating marker: LINC00299 hypermethylated (in ctDNA) in TNBC patients. Metastasis lncRNAs: MIR4500HG003↑ (sponges miR-483-3p, ↑MMP9); PTCSC3↓ (suppresses MIR100HG, inhibits proliferation); LINC01559 (ceRNA) sponges multiple miRNAs to drive progression. Other lncRNAs: BTG3-7:1 with JUND↑ (↑C21ORF91); MNX1-AS1↑ (enhances STAT3 activity). Chemoresistance lncRNAs: ANRIL↑ (promotes glycolysis, drug resistance); LINC-PINT↑ (restores paclitaxel sensitivity); SNHG22↑ (sponges miR-324-3p, ↑SUDS3/HDAC complex)	Histone lactylation: H4K12 lactylation↑ promotes malignancy; CAF-induced ZFP64 lactylation causes chemoresistance. H3 modifications: Enhanced H3 phosphorylation (NONO/PKM2) drives metastasis; changes in H3K9me3 and H3K27me3 correlate with survival; KMT2C stabilization (Secernin-2) suppresses tumor growth; circBRAF recruits KDM4B to H3K9me3 (driving progression). Chromatin remodelers: HDAC6 and PML form condensates that reprogram chromatin (promoting TNBC); LOXL2 maintains compaction, ARID1A loss (through YAP) enhances metastasis; MUC1-C (via JUN/BAF) regulates enhancer accessibility; SND1↑ (methylates CDH1, promoting metastasis). Signature marks: Combined loss of H3K4me3 and gain of H3K9me3/BAP18 coactivator mark TNBC; FOXA1/GRHL2 and SETDB1/CHD4 identified as targetable factors	METTL3: Acts both ways in TNBC—loss leads to ↑COL3A1 and metastasis; overexpression (with PRMT7, STUB1) promotes tumorigenesis. Also mediates m6A on YAP1 (↑endothelial activation/metastasis). Readers: IGF2BP2↑ (stabilizes CDK6 mRNA); IGF2BP3↑ (degrades NF1, represses SLIT2); YTHDC1↑ (↑FOXM1/CENPA glycolysis/proliferation). Erasers: ALKBH5↑ (regulates FOXO1, chemoresistance); FTO↑ (↑C/EBPβ-LIP oncogenic isoform). Targets: POP1 (via m6A) degrades CDKN1A (↑proliferation); WTAP (with m6A) ↑LCN2 (ferroptosis regulation); VIRMA↑ (↑KIF15); downstream effectors PRMT7, CCL20, and c-MYC link m6A to cell cycle/migration	Aine et al. (2025), Anjum et al. (2025), Bagheri et al. (2024), Beetch et al. (2020), Bhattacharya et al. (2022), Chen et al. (2023), Chen et al. (2020b), Cui et al. (2024), Dan et al. (2020), Du et al. (2025), Fang et al. (2020), Fracassi et al. (2023), Harris et al. (2023), Huang et al. (2025), Jiang et al. (2022), Lan et al. (2024), Li et al. (2024a), Li et al. (2025), Li et al. (2024c), Li et al. (2024d), Lin et al. (2024), Liu et al. (2022a), Liu et al. (2024), Liu et al. (2022b), Llinàs-Arias et al. (2023b), Lu et al. (2024), Lu and Yang (2025), Ma et al. (2022a), Manoochehri et al. (2020), Mendaza et al. (2021), Moraes-Almeida et al. (2023), Serra-Bardenys et al. (2024), Shen et al. (2024b), Shi et al. (2020), Tan et al. (2024), Llinàs-Arias et al. (2023a), Xia et al. (2024), Xu et al. (2025), Yang et al. (2024a), Yang et al. (2022), Yedier-Bayram et al. (2022), Yu et al. (2025), Zhang et al. (2024), Zhang et al. (2023a), Zhang et al. (2023b), Zhang et al. (2023d), Zhang et al. (2025), Zhang et al. (2022c), and Zuidhof et al. (2025)

Clinical and Therapeutic Implications

Noncoding epimutations have a great deal of promise as clinical biomarkers to enhance the treatment of breast cancer. NcRNAs, which include miRNAs and lncRNAs, are essential modulators of gene expression and are often dysregulated in breast cancer (Schwarzenbach and Gahan 2023).

The relationship between noncoding RNA expression and DNA methylation implies that methylation marks that impact the regulation of ncRNA are probably also predictive markers (Kumar and Vindal 2025; Gaggi et al. 2025; Dai et al. 2025).

Diagnostic and Prognostic Potential of Noncoding Epimutations

Promoter methylation of noncoding RNAs (ncRNAs), including lncRNAs like H19, MEG3, and GAS5, is increasingly recognized for its diagnostic and prognostic value across cancers, correlating with tumor subtypes and patient outcomes (Sun et al. 2022; Eisa et al. 2025). For instance, studies have shown that variations in methylation levels in ctDNA may predict clinical outcomes, making the detection of methylated ncRNA signatures, such as MEG3, in plasma ctDNA a promising approach for early diagnosis and prognosis. (Thomsen et al. 2020; Zhang et al. 2021b; Panet et al. 2024).

Hypermethylation-induced downregulation of noncoding RNAs, such as LINC00673 and XIST, is associated with aggressive phenotypes and a poor prognosis. Distinct methylation patterns have been observed between breast cancer subtypes, including triple-negative (TNBC) and luminal forms (Zhu et al. 2021; Naciri et al. 2024).

Specific long noncoding RNAs (lncRNAs) exhibit substantial prognostic value in breast cancer. In TNBC, LINC00960 overexpression correlates with poor outcomes (Elango et al. 2024), while LURAP1L-AS1 is similarly linked to poor prognosis (Vishnubalaji et al. 2025). Multi-lncRNA panels, including HOTAIR and MALAT1, are under development to enhance prognostic accuracy and patient stratification (Lewis et al. 2024).

Their broad participation in proliferation, metastasis, and chemoresistance highlights their value (Herzog et al. 2025; Brown et al. 2022; Benacka et al. 2024; Prabhakaran et al. 2024; Hum and Lee 2025b).

Understanding how epimutations affect long noncoding RNAs (lncRNAs) in breast cancer requires precise tools to detect DNA methylation, histone modifications, and RNA methylation, as each influences gene regulation and serves as a diagnostic or prognostic marker.

Hypomethylation of oncogenic lncRNAs like PVT1 can be identified from tumor biopsies or cfDNA using bisulfite pyrosequencing and ELISA-based global methylation assays, aiding early detection and monitoring (Higashimoto et al. 2023; Tarhonska et al. 2025).

Histone marks such as H3K27me3 (repressive) and H3K27ac (activating) regulate lncRNAs like HOTAIR and LINC00511 and are typically detected via ChIP-seq in cultured cells (Hino et al. 2023; Asp 2020; Azadbakht et al. 2022; Wang et al. 2022a).

LncRNAs are used in diagnosis; some lncRNAs, like H19, are explored for diagnostic potential (Elias-Rizk et al. 2020), while others, such as cytoskeletal regulator lncRNAs, may predict patient response to chemotherapy agents like gemcitabine and epirubicin (Asberger et al. 2025).

Promoter hypermethylation in breast cancer silences a large number of tumor suppressor long noncoding RNAs (lncRNAs), including GAS5, MEG3, MORT, LINC00673, and XIST.

Bisulfite conversion, which chemically converts unmethylated cytosines to uracils while leaving methylated ones unaltered, is the most used method for detecting this by taking many types of samples like fresh tumors and plasma cfDNA, and it is detected by methylation-specific PCR and bisulfite sequencing for profiling methylation (Shen et al. 2024a; Zeng et al. 2020; Rodriguez-Casanova et al. 2022).

Similar to lncRNAs, miRNAs are detectable in the bloodstream and have a significant role in the development of breast cancer as well as the response to treatment (Kumar and Ranga 2025). Several miRNAs, including miR-93, have shown promise as prognostic biomarkers as metastasis suppressors by suppressing both invasion ability and cancer stem cell properties (Shibuya et al. 2020).

Enhancers, particularly super-enhancers (SEs), are key regulatory hubs where epigenetic alterations, such as methylation or histone marks, can significantly impact the expression of oncogenes or tumor suppressors. SE activity, inferred through enhancer-derived RNAs or SE-related genes (SERGs), represents a novel biomarker class being developed to predict treatment response and patient prognosis (Wu et al. 2023). Gaining insight into the epigenetic state and activity of SEs that regulate important pathways, such as the estrogen receptor (ER) pathway (Lewis et al. 2024), might help identify possible treatment vulnerabilities and provide valuable prognostic information (Ji et al. 2025).

A noninvasive technique for identifying these epigenetic changes is liquid biopsies, especially for detecting cell-free methylation lncRNA and miRNA promoters in circulating tumor DNA (ctDNA) (Hai et al. 2022).

The limitations of conventional tissue biopsies are overcome by this method, which enables early cancer identification, minimal invasiveness, low risk, multiple repeat sampling, suitability for dynamic monitoring, and the potential to partially mitigate the effect of tumor heterogeneity, longitudinal monitoring of disease development, and treatment response (Lone et al. 2022).

Liquid biopsies are a helpful technique used for prognosis diagnosis of breast cancer. For instance, serum exosomal H19 acts as a novel biomarker for the diagnosis of BC (Zhong et al. 2020).

While liquid biopsy offers advantages over conventional methods, limitations include poor inter-lab standardization, complex sample handling, and the need for improved accuracy, which affect test reliability and clinical adoption.

Furthermore, epigenetic modifications in ncRNAs act as subtype-specific biomarkers, notably in breast cancer. Differences in ncRNA methylation and expression profiles between luminal A/B, HER2+, and TNBC subtypes contribute to their different clinical behaviors and prognosis trajectories (Solaimani et al. 2025; Prabhu et al. 2024). For example, the progression of TNBC and a poor prognosis are associated with some lncRNAs, such as LINC00960 (Elango et al. 2024). Since these subtype-specific ncRNA epimutations are strongly linked to survival, metastasis, and recurrence, understanding them is essential for creating focused treatments and enhancing prognostic accuracy (Kang et al. 2025; Bjorklund et al. 2022).

Therapeutic Potential: Targeting Epigenetic Regulation of Noncoding RNAs

Epigenetic modifiers, including DNA methyltransferases (DNMTs), histone deacetylases (HDACs), enhancers of zeste homolog 2 (EZH2), and ten-eleven translocation (TET) enzymes, regulate the expression of noncoding RNA (ncRNA). EZH2-mediated histone methylation can repress lncRNA transcription, while DNMTs silence tumor-suppressive lncRNAs through promoter hypermethylation (Szczepanek et al. 2023).

Therapeutically, DNMT inhibitors like azacitidine and decitabine demethylate DNA to reactivate gene expression, while HDAC inhibitors such as entinostat alter chromatin structure to modulate transcription (Rahmani et al. 2021; Szczepanek et al. 2023).

Enhancer of zeste homolog 2 (EZH2) is a histone methyltransferase that functions as the catalytic subunit of the polycomb repressive complex 2 (PRC2), which mediates trimethylation of histone H3 on lysine 27 (H3K27me3), leading to transcriptional repression of target genes. In breast cancer, EZH2 is frequently overexpressed and is associated with aggressive tumor phenotypes, including increased proliferation and metastasis. It contributes to oncogenesis by silencing tumor suppressor genes and promoting epithelial-mesenchymal transition (EMT), thus playing a critical role in tumor progression and poor clinical outcomes, and inhibiting EZH2 activity may offer a promising strategy for treating aggressive forms of breast cancer, including triple-negative breast cancer (TNBC) (Li et al. 2024c).

Demethylation strategies can reactivate tumor-suppressive lncRNAs; for instance, MEG3 demethylation leads to its upregulation, which suppresses epithelial-mesenchymal transition (EMT) and Notch1 signaling associated with cancer metastasis (Pan et al. 2022). Reduced expression of MAGI2-AS3 has been linked to poor prognosis in triple-negative breast cancer. Demethylation can restore its expression, suppressing tumor growth and metastasis (Kang et al. 2025).

The effectiveness of epigenetic therapies can be increased by combining them with other therapies. HDAC inhibitors and endocrine therapy have demonstrated synergistic effects in estrogen receptor-positive (ER+) luminal cancers, which may help overcome hormone therapy resistance (Pradel et al. 2024).

By reactivating immune-related lncRNAs, DNMT inhibitors can improve the immune response in triple-negative breast cancer (TNBC), increasing the tumors' susceptibility to immunotherapy (Wang et al. 2023b).

Moreover, drug resistance may be overcome through the epigenetic reprogramming of noncoding RNAs (ncRNAs). Cancer cells may respond better to traditional treatments if the expression of long noncoding RNAs (lncRNAs) implicated in drug resistance pathways is altered. A study published in 2024 demonstrated that the long noncoding RNA (lncRNA) H19 contributes to breast cancer chemoresistance by mediating DNA methylation and regulating the expression of the NAT1 (N-acetyltransferase 1) gene. This epigenetic modification leads to decreased NAT1 expression, which is associated with reduced sensitivity to chemotherapeutic agents. Targeting H19 or its epigenetic regulatory mechanisms could potentially restore drug sensitivity in resistant breast cancer cells. Another study found that the lncRNA UCA1 confers tamoxifen resistance in breast cancer by regulating the EZH2/p21 axis and the PI3K/AKT signaling pathway. UCA1 recruits EZH2 to the p21 promoter, leading to histone methylation and suppression of p21 expression, which promotes cell proliferation and tamoxifen resistance. Targeting UCA1 or its interaction with EZH2 could potentially restore tamoxifen sensitivity in resistant breast cancer cells (Zhou et al. 2023b).

Transitional Challenges and Clinical Opportunities

Tumor heterogeneity in epimutations complicates the discovery of biomarkers and the targeting of treatment. Epigenetic markers from liquid biopsies, like cfDNA methylation (Herzog et al. 2025), need thorough validation for clinical use. Additionally, the development of epigenetic drugs must address off-target effects, given the broad role of epigenetic regulators in normal physiology.

Despite these obstacles, significant opportunities remain. Integrating epigenetic data with other multi-omics platforms (e.g., genomics, transcriptomics) holds promise for refining patient stratification in precision oncology (Kumar and Vindal 2025; Hamdy et al. 2024a). Advances in targeted epigenetic editing tools—such as CRISPR-dCas9 fused with epigenetic modifiers like TET1—enable precise correction of pathogenic epimutations (Lewis et al. 2024). Additionally, the development of companion diagnostics to guide epigenetic therapy selection will be critical for maximizing treatment efficacy and personalizing care for breast cancer patients (Wu et al. 2023).

Challenges and Future Directions

Navigating layers of complexity is necessary to comprehend the functional effects of epigenetic modifications outside of clearly identified protein-coding genes. The size and complexity of the noncoding genome itself are significant obstacles. It encompasses a wide range of regulatory components, including insulators, promoters, enhancers, and silencers of noncoding RNAs, the exact positions, roles, and interactions of which are sometimes unclear. It is technically challenging to map epigenetic alterations (such as DNA methylation or histone marks) throughout these sections of the genome, and it is still difficult to determine the functional significance of changes that are seen. Noncoding epimutations frequently have context-dependent effects, affecting gene expression subtly or indirectly through intricate regulatory networks, in contrast to coding mutations that directly change protein sequences (Ji et al. 2025; Edrei et al. 2023).

An additional element of complexity is introduced by the heterogeneity of cell types within tumors. Breast tumors are composed of various cell types, including immune cells, stromal cells, and cancer cells, each with distinct epigenetic states. Epigenetic signals across various populations are averaged in bulk analysis of tumor tissue, which may obscure important cell-type-specific epimutations. Although they are becoming more popular, single-cell epigenetic technologies are still expensive and computationally demanding (Yost et al. 2025; Regner et al. 2025).

Moreover, the noncoding genome's functional redundancy may complicate interpretation. It can be challenging to identify the key factors causing epigenetic dysregulation since several regulatory elements may influence the expression of a single gene, and the loss or epigenetic silence of one element may be offset by others (Yost et al. 2025).

Despite these obstacles, future studies should concentrate on a few essential topics.

High-resolution, genome-wide epigenetic mapping technologies (such as enhanced bisulfite sequencing, ChIP-seq variants, and ATAC-seq) must continue to be developed and improved. Haplotype-specific epigenetic patterns and intricate structural changes in noncoding areas can be resolved by combining these with long-read sequencing technology (O'neill et al. 2024).

Enhancing the functional annotation of the noncoding genome is one of the primary objectives. Finding noncoding areas and particular epimutations that have a direct influence on the phenotypes of breast cancer cells can be aided by high-throughput functional screens, such as those that use CRISPR-based epigenome editing (Lewis et al. 2024). To create more precise models of noncoding regulatory networks, computational methods combining multi-omics data (epigenomics, transcriptomics, proteomics, and chromatin conformation capture) are required (Kumar and Vindal 2025).

Understanding how genetic variants (SNPs) in noncoding regions influence epigenetic changes, such as enhancer activity or ncRNA function, can help clarify breast cancer etiology. Investigating how genetic background predisposes to epigenetic alterations is crucial. Extending single-cell technologies to simultaneously profile the transcriptome, proteome, and epigenome within individual tumor cells will revolutionize research by revealing cell-type-specific epigenetic heterogeneity and identifying rare cell populations driving therapy resistance or disease progression (Wang et al. 2025).

Translation of promising noncoding epigenetic biomarkers into clinical practice requires validation in large, prospective trials (Hum and Lee 2025a; Herzog et al. 2025; Wu et al. 2023). Developing safer, targeted epigenetic therapies using companion biomarkers remains a key objective in cancer treatment (Brown et al. 2022).

By addressing these challenges and pursuing future directions, researchers can continue to unravel the complexities of the noncoding epigenome, ultimately translating fundamental discoveries into tangible benefits for breast cancer patients.

Conclusion

The noncoding genome, once dismissed as transcriptional "dark matter," has emerged as a central player in breast cancer pathogenesis, influencing gene regulation, tumor progression, and therapeutic resistance. Epigenetic alterations—including DNA methylation, histone modifications, and RNA methylation—profoundly affect the expression and function of noncoding RNAs (ncRNAs) such as lncRNAs, miRNAs, circRNAs, piRNAs, and snoRNAs. These ncRNAs modulate key hallmarks of cancer, including proliferation, metastasis, immune evasion, and response to treatment. Advances in high-throughput technologies, such as WGBS, ATAC-seq, and ChIP-seq, have enabled the fine-resolution mapping of epigenetic landscapes across breast cancer subtypes. Moreover, noncoding epimutations offer significant diagnostic, prognostic, and therapeutic potential, primarily through the use of liquid biopsies and epigenetic drugs. However, challenges remain in functional annotation, cellular heterogeneity, and clinical translation. Continued integration of single-cell and long-read epigenomics, coupled with CRISPR-based editing, will be pivotal in refining our understanding and harnessing noncoding epigenetics for precision oncology in breast cancer.

Author Contribution HS: conceptualized the idea; MEA, HM, BM, and MA: wrote the first draft; ASW: drew the illustrations; SA-G: supervised the work.

Conflict of Interests All authors declare no conflict of interests.

Ethical Approval N/A

Funding This work received no external funding.

References

Aine, M., Nacer, D. F., Arbajian, E., Veerla, S., Karlsson, A., Häkkinen, J., Johansson, H. J., Rosengren, F., Vallon-Christersson, J., Borg, Å. & Staaf, J. 2025. The Dna Methylation Landscape Of Primary Triple-Negative Breast Cancer. *Nat Commun,* 16, 3041.

Al Khaled, J. 2023. Transcriptomic Insights Into The Development And Therapy Of Breast Cancer Via The Ncrna-Mediated Cerna Regulation Network. *Authorea Preprints.*

Al-Matouq, J., Al-Haj, L., Al-Saif, M. & Khabar, K. S. 2021. Post-Transcriptional Screen Of Cancer Amplified Genes Identifies Erbb2/Her2 Signaling As Au-Rich Mrna Stability-Promoting Pathway. *Oncogenesis,* 10, 61.

Anabel Sinberger, L., Zahavi, T., Sonnenblick, A. & Salmon-Divon, M. 2023. Coexistent Arid1a-Pik3ca Mutations Are Associated With Immune-Related Pathways In Luminal Breast Cancer. *Scientific Reports,* 13, 20911.

Andreani, C., Bartolacci, C., Persico, G., Casciaro, F., Amatori, S., Fanelli, M., Giorgio, M., Galié, M., Tomassoni, D., Wang, J., Zhang, X., Bick, G., Coppari, R., Marchini, C. & Amici, A. 2023. Sirt6 Promotes Metastasis And Relapse In Her2-Positive Breast Cancer. *Sci Rep,* 13, 22000.

Angius, A., Cossu-Rocca, P., Arru, C., Muroni, M. R., Rallo, V., Carru, C., Uva, P., Pira, G., Orrù, S. & De Miglio, M. R. 2020. Modulatory Role Of Micrornas In Triple Negative Breast Cancer With Basal-Like Phenotype. *Cancers,* 12, 3298.

Anjum, F., Kaushik, K., Salam, A. & Nandi, C. K. 2025. Chromatin Marks H3k4me3 And H3k9me3 In Triple-Negative Breast Cancer Cell Lines. *Adv Biol (Weinh)*, E2400752.

Anwar, M. M., Albanese, C., Hamdy, N. M. & Sultan, A. S. 2022. Rise Of The Natural Red Pigment 'Prodigiosin' As An Immunomodulator In Cancer. *Cancer Cell International,* 22, 419.

Asberger, J., Ge, I., Schmidt, B., Jager, M., Weiss, D., Berner, K., Erbes, T., Juhasz-Boss, I. & Mayer, S. 2025. Long Non-Coding Rna Signatures In Breast Cancer: Properties As Biomarkers? *Exp Ther Med,* 29, 54.

Asp, P. 2020. Chapter Fourteen - The Chromatin Immunoprecipitation (Chip) Assay And Chip-Qpcr. *In:* Tollefsbol, T. (Ed.) *Epigenetics Methods.* Academic Press.

Aure, M. R., Fleischer, T., Bjørklund, S., Ankill, J., Castro-Mondragon, J. A., Bathen, T. F., Borgen, E., Engebråten, O., Hartman-Johnsen, O.-J., Garred, Ø., Geisler, J., Geitvik, G. A., Hofvind, S., Langerød, A., Lingjærde, O. C., Mælandsmo, G. M., Naume, B., Russnes, H. G., Skjerven, H. K., Schlichting, E., Sørlie, T., Børresen-Dale, A.-L., Tost, J., Sahlberg, K. K., Mathelier, A., Tekpli, X., Kristensen, V. N. & Osbreac 2021. Crosstalk Between Microrna Expression And Dna Methylation Drives The Hormone-Dependent Phenotype Of Breast Cancer. *Genome Medicine,* 13, 72.

Azadbakht, N., Doosti, A. & Jami, M. S. 2022. Crispr/Cas9-Mediated Linc00511 Knockout Strategies, Increased Apoptosis Of Breast Cancer Cells Via Suppressing Antiapoptotic Genes. *Biol Proced Online,* 24, 8.

Bagheri, M., Mohamed, G. A., Mohamed Saleem, M. A., Ognjenovic, N. B., Lu, H., Kolling, F. W., Wilkins, O. M., Das, S., Lacroix, I. S., Nagaraj, S. H., Muller, K. E., Gerber, S. A., Miller, T. W. & Pattabiraman, D. R. 2024. Pharmacological Induction Of Chromatin Remodeling Drives Chemosensitization In Triple-Negative Breast Cancer. *Cell Rep Med,* 5, 101504.

Bediaga, N. G., Acha-Sagredo, A., Guerra, I., Viguri, A., Albaina, C., Ruiz Diaz, I., Rezola, R., Alberdi, M. J., Dopazo, J., Montaner, D., De Renobales, M., Fernández, A. F., Field, J. K., Fraga, M. F., Liloglou, T. & De Pancorbo, M. M. 2010. Dna Methylation Epigenotypes In Breast Cancer Molecular Subtypes. *Breast Cancer Research,* 12, R77.

Beetch, M., Harandi-Zadeh, S., Yang, T., Boycott, C., Chen, Y., Stefanska, B. & Mohammed, S. I. 2020. Dna Methylation Landscape Of Triple-Negative Ductal Carcinoma In Situ (Dcis) Progressing To The Invasive Stage In Canine Breast Cancer. *Sci Rep,* 10, 2415.

Benacka, R., Szaboova, D., Gulasova, Z. & Hertelyova, Z. 2024. Non-Coding Rnas In Breast Cancer: Diagnostic And Therapeutic Implications. *Int J Mol Sci,* 26.

Beňačka, R., Szabóová, D., Guľašová, Z. & Hertelyová, Z. 2024. Non-Coding Rnas In Breast Cancer: Diagnostic And Therapeutic Implications. *International Journal Of Molecular Sciences,* 26, 127.

Beňačka, R., Szabóová, D., Guľašová, Z., Hertelyová, Z. & Radoňak, J. 2023. Non-Coding Rnas In Human Cancer And Other Diseases: Overview Of The Diagnostic Potential. *International Journal Of Molecular Sciences,* 24, 16213.

Bergeron, D., Fafard-Couture, É. & Scott, M. S. 2020. Small Nucleolar Rnas: Continuing Identification Of Novel Members And Increasing Diversity Of Their Molecular Mechanisms Of Action. *Biochemical Society Transactions,* 48, 645–656.

Bermúdez, M., Martínez-Barajas, M. G., Bueno-Urquiza, L. J., López-Gutiérrez, J. A., Villegas-Mercado, C. E. & López-Camarillo, C. 2024. Role Of Microrna-204 In Regulating The Hallmarks Of Breast Cancer: An Update. *Cancers,* 16, 2814.

Bhardwaj, A., Singh, H., Rajapakshe, K., Tachibana, K., Ganesan, N., Pan, Y., Gunaratne, P. H., Coarfa, C. & Bedrosian, I. 2017. Regulation Of Mirna-29c And Its Downstream Pathways In Preneoplastic Progression Of Triple-Negative Breast Cancer. *Oncotarget,* 8, 19645.

Bhattacharya, A., Fushimi, A., Yamashita, N., Hagiwara, M., Morimoto, Y., Rajabi, H., Long, M. D., Abdulla, M., Ahmad, R., Street, K., Liu, S., Liu, T. & Kufe, D. 2022. Muc1-C Dictates Jun And Baf-Mediated Chromatin Remodeling At Enhancer Signatures In Cancer Stem Cells. *Mol Cancer Res,* 20, 556–567.

Bi, M., Zheng, L., Chen, L., He, J., Yuan, C., Ma, P., Zhao, Y., Hu, F., Tang, W. & Sheng, M. 2021. Ln Rna Linc01234 Promotes Triple-Negative Breast Cancer Progression Through Regulating The Mir-429/Synj1 Axis. *American Journal Of Translational Research,* 13, 11399.

Billi, M., De Marinis, E., Gentile, M., Nervi, C. & Grignani, F. 2024. Nuclear Mirnas: Gene Regulation Activities. *International Journal Of Molecular Sciences,* 25, 6066.

Bjorklund, S. S., Aure, M. R., Hakkinen, J., Vallon-Christersson, J., Kumar, S., Evensen, K. B., Fleischer, T., Tost, J., Osbreac, Sahlberg, K. K., Mathelier, A., Bhanot, G., Ganesan, S., Tekpli, X. & Kristensen, V. N. 2022. Subtype And Cell Type Specific Expression Of Lncrnas Provide Insight Into Breast Cancer. *Commun Biol,* 5, 834.

Brown, L. J., Achinger-Kawecka, J., Portman, N., Clark, S., Stirzaker, C. & Lim, E. 2022. Epigenetic Therapies And Biomarkers In Breast Cancer. *Cancers (Basel),* 14.

Bure, I. V., Nemtsova, M. V. & Kuznetsova, E. B. 2022. Histone Modifications And Non-Coding Rnas: Mutual Epigenetic Regulation And Role In Pathogenesis. *International Journal Of Molecular Sciences,* 23, 5801.

Cedro-Tanda, A., Ríos-Romero, M., Romero-Córdoba, S., Cisneros-Villanueva, M., Rebollar-Vega, R. G., Alfaro-Ruiz, L. A., Jiménez-Morales, S., Domínguez-Reyes, C., Villegas-Carlos, F., Tenorio-Torres, A., Bautista-Piña, V., Beltrán-Anaya, F. O. & Hidalgo-Miranda, A. 2020. A Lncrna Landscape In Breast Cancer Reveals A Potential Role For Ac009283.1 In Proliferation And Apoptosis In Her2-Enriched Subtype. *Sci Rep,* 10, 13146.

Chauhan, N., Dhasmana, A., Jaggi, M., Chauhan, S. C. & Yallapu, M. M. 2020. Mir-205: A Potential Biomedicine For Cancer Therapy. *Cells,* 9, 1957.

Chen, C., Wang, Y., Li, Y. & Zhang, C. 2023. Virma Facilitates Triple-Negative Breast Cancer Progression Via Increasing M6a-Dependent Kif15 Expression. *Discov Med,* 35, 787–795.

Chen, H., Li, Z., Zhang, L., Zhang, L., Zhang, Y., Wang, Y., Xu, M. & Zhong, Q. 2020a. Microrna-200c Inhibits The Metastasis Of Triple-Negative Breast Cancer By Targeting Zeb2, An Epithelial-Mesenchymal Transition Regulator. *Annals Of Clinical & Laboratory Science,* 50, 519–527.

Chen, J., Zhu, M., Zou, L., Xia, J., Huang, J., Deng, Q. & Xu, R. 2020b. Long Non-Coding Rna Linc-Pint Attenuates Paclitaxel Resistance In Triple-Negative Breast Cancer Cells Via Targeting The Rna-Binding Protein Nono. *Acta Biochim Biophys Sin (Shanghai),* 52, 801–809.

Chen, L., Zhang, Y.-H., Pan, X., Liu, M., Wang, S., Huang, T. & Cai, Y.-D. 2018. Tissue Expression Difference Between Mrnas And Lncrnas. *International Journal Of Molecular Sciences,* 19, 3416.

Chen, Q., Zhu, C. & Jin, Y. 2020c. The Oncogenic And Tumor Suppressive Functions Of The Long Noncoding Rna Malat1: An Emerging Controversy. *Frontiers In Genetics,* 11, 93.

Chen, Y., Liang, R., Li, Y., Jiang, L., Ma, D., Luo, Q. & Song, G. 2024. Chromatin Accessibility: Biological Functions, Molecular Mechanisms And Therapeutic Application. *Signal Transduction And Targeted Therapy,* 9, 1–39.

Cheng, X., Huang, Z., Pan, A. & Long, D. 2023. Orlnc1 Suppresses Cell Growth In Her2-Positive Breast Cancer Via Mirna-296 Sponging. *Curr Mol Med,* 23, 289–299.

Chiang, Y.-F., Huang, K.-C., Chen, H.-Y., Hamdy, N. M., Huang, T.-C., Chang, H.-Y., Shieh, T.-M., Huang, Y.-J. & Hsia, S.-M. 2024. Hinokitiol Inhibits Breast Cancer Cells In Vitro Stemness-Progression And Self-Renewal With Apoptosis And Autophagy Modulation Via The Cd44/Nanog/Sox2/Oct4 Pathway. *International Journal Of Molecular Sciences,* 25, 3904.

Contreras-Espinosa, L., Alcaraz, N., De La Rosa-Velázquez, I. A., Díaz-Chávez, J., Cabrera-Galeana, P., Rebollar-Vega, R., Reynoso-Noverón, N., Maldonado-Martínez, H. A., González-Barrios, R., Montiel-Manríquez, R., Bautista-Sánchez, D., Castro-Hernández, C., Alvarez-Gomez, R. M., Jiménez-Trejo, F., Tapia-Rodríguez, M., García-Gordillo, J. A., Pérez-Rosas, A., Bargallo-Rocha, E., Arriaga-Canon, C. & Herrera, L. A. 2021. Transcriptome Analysis Identifies Gata3-As1 As A Long Noncoding Rna Associated With Resistance To Neoadjuvant Chemotherapy In Locally Advanced Breast Cancer Patients. *J Mol Diagn,* 23, 1306–1323.

Cui, Z., Li, Y., Lin, Y., Zheng, C., Luo, L., Hu, D., Chen, Y., Xiao, Z. & Sun, Y. 2024. Lactylproteome Analysis Indicates Histone H4k12 Lactylation As A Novel Biomarker In Triple-Negative Breast Cancer. *Front Endocrinol (Lausanne),* 15, 1328679.

Dabi, Y., Bendifallah, S., Suisse, S., Haury, J., Touboul, C., Puchar, A., Favier, A. & Daraï, E. 2022. Overview Of Non-Coding Rnas In Breast Cancers. *Translational Oncology,* 25, 101512.

Dai, C., Qianjiang, H., Fu, R., Yang, H., Shi, A. & Luo, H. 2025. Epigenetic And Epitranscriptomic Role Of Lncrna In Carcinogenesis (Review). *Int J Oncol,* 66.

Dan, T., Shastri, A. A., Palagani, A., Buraschi, S., Neill, T., Savage, J. E., Kapoor, A., Deangelis, T., Addya, S. & Camphausen, K. 2021. Mir-21 Plays A Dual Role In Tumor Formation And Cytotoxic Response In Breast Tumors. *Cancers,* 13, 888.

Dan, Z., Xiujing, H., Ting, L., Xiaorong, Z., Hong, Z., Jiqiao, Y., Yanchu, L. & Jing, J. 2020. Long Non-Coding Rna Btg3-7:1 And Jund Co-Regulate C21orf91 To Promote Triple-Negative Breast Cancer Progress. *Front Mol Biosci,* 7, 605623.

Davalos, V. & Esteller, M. 2013. Unraveling The Complex Network Of Interactions Between Noncoding Rnas And Epigenetics In Cancer. *Non-Coding Rnas And Cancer.* Springer.

De Palma, F. D. E., Salvatore, F., Pol, J. G., Kroemer, G. & Maiuri, M. C. 2022. Circular Rnas As Potential Biomarkers In Breast Cancer. *Biomedicines,* 10, 725.

Dietlein, F., Wang, A. B., Fagre, C., Tang, A., Besselink, N. J., Cuppen, E., Li, C., Sunyaev, S. R., Neal, J. T. & Van Allen, E. M. 2022. Genome-Wide Analysis Of Somatic Noncoding Mutation Patterns In Cancer. *Science,* 376, Eabg5601.

Dong, H., Hu, J., Zou, K., Ye, M., Chen, Y., Wu, C., Chen, X. & Han, M. 2022. Retraction Note To: Activation Of Lncrna Tincr By H3k27 Acetylation Promotes Trastuzumab Resistance And Epithelial-Mesenchymal Transition By Targeting Microrna-125b In Breast Cancer. Springer.

Dong, S., Ma, M., Li, M., Guo, Y., Zuo, X., Gu, X., Zhang, M. & Shi, Y. 2021. Lncrna Meg3 Regulates Breast Cancer Proliferation And Apoptosis Through Mir-141-3p/Rbms3 Axis. *Genomics,* 113, 1689–1704.

Du, S. M., Li, N., Xu, W. J. & Liu, K. 2025. Triple-Negative Breast Cancer Cell-Derived Pir-31115 Promotes The Proliferation And Migration Of Endothelial Cells Via Mettl3-Mediated M6a Modification Of Yap1. *Oncol Rep,* 53.

Dvorská, D., Braný, D., Ňachajová, M., Halašová, E. & Danková, Z. 2021. Breast Cancer And The Other Non-Coding Rnas. *International Journal Of Molecular Sciences,* 22, 3280.

Edrei, Y., Levy, R., Kaye, D., Marom, A., Radlwimmer, B. & Hellman, A. 2023. Methylation-Directed Regulatory Networks Determine Enhancing And Silencing Of Mutation Disease Driver Genes And Explain Inter-Patient Expression Variation. *Genome Biol,* 24, 264.

Eisa, A., Yousif, S., Abo-Alella, M. A., Awad, S., Yousif, H. T. I., Dafaalla, T., Dldom, A. & Alfaki, M. 2025. The Multifaceted Role Of Lncrna Gas5: A Pan-Cancer Analysis Of Its Diagnostic, Prognostic, And Therapeutic Potential. *Cureus,* 17, E77527.

El Fatimy, R., Zhang, Y., Deforzh, E., Ramadas, M., Saravanan, H., Wei, Z., Rabinovsky, R., Teplyuk, N. M., Uhlmann, E. J. & Krichevsky, A. M. 2022. A Nuclear Function For An Oncogenic Microrna As A Modulator Of Snrna And Splicing. *Molecular Cancer,* 21, 17.

El-Fattah, A. A. A., Sadik, N. A. H., Shaker, O. G., Mohamed Kamal, A. & Shahin, N. N. 2021. Serum Long Non-Coding Rnas Pvt1, Hotair, And Neat1 As Potential Biomarkers In Egyptian Women With Breast Cancer. *Biomolecules,* 11, 301.

Elango, R., Radhakrishnan, V., Rashid, S., Al-Sarraf, R., Akhtar, M., Ouararhni, K. & Alajez, N. M. 2024. Long Noncoding Rna Profiling Unveils Linc00960 As Unfavorable Prognostic Biomarker Promoting Triple Negative Breast Cancer Progression. *Cell Death Discov,* 10, 333.

Eldash, S., Sanad, E. F., Nada, D. & Hamdy, N. M. 2023. The Intergenic Type Lncrna (Linc Rna) Faces In Cancer With In Silico Scope And A Directed Lens To Linc00511: A Step Toward Ncrna Precision. *Non-Coding Rna,* 9, 58.

Elhasnaoui, J., Miano, V., Ferrero, G., Doria, E., Leon, A. E., Fabricio, A. S. C., Annaratone, L., Castellano, I., Sapino, A. & De Bortoli, M. 2020. Dscam-As1-Driven Proliferation Of Breast Cancer Cells Involves Regulation Of Alternative Exon Splicing And 3'-End Usage. *Cancers (Basel),* 12.

Elias-Rizk, T., El Hajj, J., Segal-Bendirdjian, E. & Hilal, G. 2020. The Long Non Coding Rna H19 As A Biomarker For Breast Cancer Diagnosis In Lebanese Women. *Sci Rep,* 10, 22228.

Fang, X., Zhang, J., Li, C., Liu, J., Shi, Z. & Zhou, P. 2020. Long Non-Coding Rna Snhg22 Facilitates The Malignant Phenotypes In Triple-Negative Breast Cancer Via Sponging Mir-324-3p And Upregulating Suds3. *Cancer Cell Int,* 20, 252.

Farrell, C., Thompson, M., Tosevska, A., Oyetunde, A. & Pellegrini, M. 2021. Bisulfite Bolt: A Bisulfite Sequencing Analysis Platform. *Gigascience,* 10, Giab033.

Feng, J. & Meng, X. 2022. Histone Modification And Histone Modification-Targeted Anti-Cancer Drugs In Breast Cancer: Fundamentals And Beyond. *Frontiers In Pharmacology,* 13, 946811.

Ferreira, H. J. & Esteller, M. 2018. Non-Coding Rnas, Epigenetics, And Cancer: Tying It All Together. *Cancer And Metastasis Reviews,* 37, 55–73.

Ferreira, T., Da Costa, R. M. G., Dias, F., Gama, A., Gaspar, V. M., Mano, J. F., Oliveira, P. A. & Medeiros, R. 2024. Exploring The Role Of Micrornas As Diagnostic And Prognostic Biomarkers In Canine Mammary Tumors. *Geroscience,* 46, 6641–6657.

Fracassi, C., Ugge, M., Abdelhalim, M., Zapparoli, E., Simoni, M., Magliulo, D., Mazza, D., Lazarevic, D., Morelli, M. J., Collas, P. & Bernardi, R. 2023. Pml Modulates Epigenetic Composition Of Chromatin To Regulate Expression Of Pro-Metastatic Genes In Triple-Negative Breast Cancer. *Nucleic Acids Res,* 51, 11024–11039.

Furuta, M., Fuimoto, A., Shiraishi, Y., Miyano, S., Tsunoda, T. & Nakagawa, H. 2014. Genome-Wide Profiling Of Somatic Mutations In Liver Cancers Revealed Significantly Mutated Genes And Non-Coding Regions In Liver Cancers. *Cancer Research,* 74, 5177–5177.

Gaggi, G., Hausman, C., Cho, S., Badalamenti, B. C., Trinh, B. Q., Di Ruscio, A. & Ummarino, S. 2025. Lncrnas Ride The Storm Of Epigenetic Marks. *Genes (Basel),* 16.

Gan, Y., Lo, Y., Makower, D., Kleer, C., Lu, J. & Fineberg, S. 2022. Ezh2 Protein Expression In Estrogen Receptor Positive Invasive Breast Cancer Treated With Neoadjuvant Endocrine Therapy: An Exploratory Study Of Association With Tumor Response. *Applied Immunohistochemistry & Molecular Morphology,* 30.

Gao, Y., Cai, Q., Huang, Y., Li, S., Yang, H., Sun, L., Chen, K. & Wang, Y. 2016. Microrna-21 As A Potential Diagnostic Biomarker For Breast Cancer Patients: A Pooled Analysis Of Individual Studies. *Oncotarget,* 7, 34498.

Gao, Y., Jones, A., Fasching, P. A., Ruebner, M., Beckmann, M. W., Widschwendter, M. & Teschendorff, A. E. 2015. The Integrative Epigenomic-Transcriptomic Landscape Of Er Positive Breast Cancer. *Clinical Epigenetics,* 7, 126.

Garrido-Cano, I., Pattanayak, B., Adam-Artigues, A., Lameirinhas, A., Torres-Ruiz, S., Tormo, E., Cervera, R. & Eroles, P. 2022. Micrornas As A Clue To Overcome Breast Cancer Treatment Resistance. *Cancer And Metastasis Reviews*, 1–29.

Gong, P., Guo, Z., Wang, S., Gao, S. & Cao, Q. 2025. Histone Phosphorylation In Dna Damage Response. *International Journal Of Molecular Sciences,* 26, 2405.

Gong, T., Borgard, H., Zhang, Z., Chen, S., Gao, Z. & Deng, Y. 2022. Analysis And Performance Assessment Of The Whole Genome Bisulfite Sequencing Data Workflow: Currently Available Tools And A Practical Guide To Advance Dna Methylation Studies. *Small Methods,* 6, 2101251.

González-Woge, M., Contreras-Espinosa, L., García-Gordillo, J. A., Aguilar-Villanueva, S., Bargallo-Rocha, E., Cabrera-Galeana, P., Vasquez-Mata, T., Cervantes-López, X., Vargas-Lías, D. S., Montiel-Manríquez, R., Bautista-Hinojosa, L., Rebollar-Vega, R., Castro-Hernández, C., Álvarez-Gómez, R. M., De La Rosa-Velázquez, I. A., Díaz-Chávez, J., Jiménez-Trejo, F., Arriaga-Canon, C. & Herrera, L. A. 2024. The Expression Profiles Of Lncrnas Are Associated With Neoadjuvant Chemotherapy Resistance In Locally Advanced, Luminal B-Type Breast Cancer. *Int J Mol Sci,* 25.

Good, D. J. 2023. Non-Coding Rnas In Human Health And Diseases. Mdpi.

Górecki, I. & Rak, B. 2021. The Role Of Micrornas In Epithelial To Mesenchymal Transition And Cancers; Focusing On Mir-200 Family. *Cancer Treatment And Research Communications,* 28, 100385.

Grandi, F. C., Modi, H., Kampman, L. & Corces, M. R. 2022. Chromatin Accessibility Profiling By Atac-Seq. *Nature Protocols,* 17, 1518–1552.

Grohmann, C., Walker, F., Devlin, M., Luo, M.-X., Chüeh, A. C., Doherty, J., Vaillant, F., Ho, G.-Y., Wakefield, M. J. & Weeden, C. E. 2021. Preclinical Small Molecule Wehi-7326 Overcomes Drug Resistance And Elicits Response In Patient-Derived Xenograft Models Of Human Treatment-Refractory Tumors. *Cell Death & Disease,* 12, 268.

Guo, B., Li, D., Du, L. & Zhu, X. 2020. Pirnas: Biogenesis And Their Potential Roles In Cancer. *Cancer And Metastasis Reviews,* 39, 567–575.

Gupta, P., O'neill, H., Wolvetang, E. J., Chatterjee, A. & Gupta, I. 2024. Advances In Single-Cell Long-Read Sequencing Technologies. *Nar Genomics And Bioinformatics,* 6, Lqae047.

Hai, L., Li, L., Liu, Z., Tong, Z. & Sun, Y. 2022. Whole-Genome Circulating Tumor Dna Methylation Landscape Reveals Sensitive Biomarkers Of Breast Cancer. *Medcomm (2020),* 3, E134.

Hamdy, N. M., Basalious, E. B., El-Sisi, M. G., Nasr, M., Kabel, A. M., Nossier, E. S. & Abadi, A. H. 2024a. Advancements In Current One-Size-Fits-All Therapies Compared To Future Treatment Innovations For Better Improved Chemotherapeutic Outcomes: A Step-Toward Personalized Medicine. *Current Medical Research And Opinion,* 40, 1943–1961.

Hamdy, N. M., El-Sisi, M. G., Ibrahim, S. M., Elnokoudy, H., Hady, A. A., Abd-Ellatef, G. E. F., Sallam, A.-A. M. & Barakat, B. M. 2024b. In Silico Analysis And Comprehensive Review Of Circular-Rna Regulatory Roles In Breast Diseases; A Step-Toward Non-Coding Rna Precision. *Pathology - Research And Practice,* 263, 155651.

Hamdy, N. M., Zaki, M. B., Abdelmaksoud, N. M., Ismail, R. A., Abd-Elmawla, M. A., Rizk, N. I., Fathi, D. & Abulsoud, A. I. 2025. Insights Into The Genetic And Epigenetic Mechanisms Governing X-Chromosome-Linked-Mirnas Expression In Cancer; A Step-Toward Ncrna Precision. *International Journal Of Biological Macromolecules,* 289, 138773.

Hamdy, N. M., Zaki, M. B., Rizk, N. I., Abdelmaksoud, N. M., Abd-Elmawla, M. A., Ismail, R. A. & Abulsoud, A. I. 2024c. Unraveling The Ncrna Landscape That Governs Colorectal Cancer: A Roadmap To Personalized Therapeutics. *Life Sciences,* 354, 122946.

Han, S.-H., Kim, H. J., Gwak, J. M., Kim, M., Chung, Y. R. & Park, S. Y. 2017. Microrna-222 Expression As A Predictive Marker For Tumor Progression In Hormone Receptor-Positive Breast Cancer. *Journal Of Breast Cancer,* 20, 35–44.

Harris, A. R., Panigrahi, G., Liu, H., Koparde, V. N., Bailey-Whyte, M., Dorsey, T. H., Yates, C. C. & Ambs, S. 2023. Chromatin Accessibility Landscape Of Human Triple-Negative Breast

Cancer Cell Lines Reveals Variation By Patient Donor Ancestry. *Cancer Res Commun,* 3, 2014–2029.

Harrison, P. W., Amode, M. R., Austine-Orimoloye, O., Azov, A. G., Barba, M., Barnes, I., Becker, A., Bennett, R., Berry, A. & Bhai, J. 2024. Ensembl 2024. *Nucleic Acids Research,* 52, D891-D899.

Hayat, M. T. 2024. The Role Of Non-Coding Rnas In Cancer Progression: A Concise Comprehensive Review. *Science Reviews. Biology,* 3, 1.

Hazra, A., Bose, P., Sunita, P. & Pattanayak, S. P. 2021. Molecular Epigenetic Dynamics In Breast Carcinogenesis. *Archives Of Pharmacal Research,* 44, 741–763.

Herzog, C. M. S., Theeuwes, B., Jones, A., Evans, I., Bjorge, L., Zikan, M., Cibula, D., Harbeck, N., Colombo, N., Pashayan, N. & Widschwendter, M. 2025. Systems Epigenetic Approach Towards Non-Invasive Breast Cancer Detection. *Nat Commun,* 16, 3082.

Higashimoto, K., Hara, S. & Soejima, H. 2023. Dna Methylation Analysis Using Bisulfite Pyrosequencing. *In:* Hatada, I. & Horii, T. (Eds.) *Epigenomics: Methods And Protocols.* New York, Ny: Springer Us.

Hino, S., Sato, T. & Nakao, M. 2022. Chromatin Immunoprecipitation Sequencing (Chip-Seq) For Detecting Histone Modifications And Modifiers. *Epigenomics: Methods And Protocols.* Springer.

Hino, S., Sato, T. & Nakao, M. 2023. Chromatin Immunoprecipitation Sequencing (Chip-Seq) For Detecting Histone Modifications And Modifiers. *In:* Hatada, I. & Horii, T. (Eds.) *Epigenomics: Methods And Protocols.* New York, Ny: Springer Us.

Hirukawa, A., Smith, H. W., Zuo, D., Dufour, C. R., Savage, P., Bertos, N., Johnson, R. M., Bui, T., Bourque, G., Basik, M., Giguère, V., Park, M. & Muller, W. J. 2018. Targeting Ezh2 Reactivates A Breast Cancer Subtype-Specific Anti-Metastatic Transcriptional Program. *Nature Communications,* 9, 2547.

Hohmann, L., Sigurjonsdottir, K., Campos, A. B., Nacer, D. F., Veerla, S., Rosengren, F., Reddy, P. T., Häkkinen, J., Nordborg, N., Vallon-Christersson, J., Memari, Y., Black, D., Bowden, R., Davies, H. R., Borg, Å., Nik-Zainal, S. & Staaf, J. 2025. Genomic Characterization Of The Her2-Enriched Intrinsic Molecular Subtype In Primary Er-Positive Her2-Negative Breast Cancer. *Nat Commun,* 16, 2208.

Holm, K., Hegardt, C., Staaf, J., Vallon-Christersson, J., Jönsson, G., Olsson, H., Borg, Å. & Ringnér, M. 2010. Molecular Subtypes Of Breast Cancer Are Associated With Characteristic Dna Methylation Patterns. *Breast Cancer Research,* 12, 1–16.

Holm, K., Staaf, J., Lauss, M., Aine, M., Lindgren, D., Bendahl, P.-O., Vallon-Christersson, J., Barkardottir, R. B., Höglund, M. & Borg, Å. 2016. An Integrated Genomics Analysis Of Epigenetic Subtypes In Human Breast Tumors Links Dna Methylation Patterns To Chromatin States In Normal Mammary Cells. *Breast Cancer Research,* 18, 1–20.

Hou, P., Zhao, Y., Li, Z., Yao, R., Ma, M., Gao, Y., Zhao, L., Zhang, Y., Huang, B. & Lu, J. 2014. Lincrna-Ror Induces Epithelial-To-Mesenchymal Transition And Contributes To Breast Cancer Tumorigenesis And Metastasis. *Cell Death & Disease,* 5, E1287-E1287.

Hu, T., Lu, C., Xia, Y., Wu, L., Song, J., Chen, C. & Wang, Q. 2022. Small Nucleolar Rna Snora71a Promotes Epithelial-Mesenchymal Transition By Maintaining Rock2 Mrna Stability In Breast Cancer. *Molecular Oncology,* 16, 1947–1965.

Hu, X., Xiang, D., Xie, Y., Tao, L., Zhang, Y., Jin, Y., Pinello, L., Wan, Y., Yuan, G. C. & Li, Z. 2019. Lsd1 Suppresses Invasion, Migration And Metastasis Of Luminal Breast Cancer Cells Via Activation Of Gata3 And Repression Of Trim37 Expression. *Oncogene,* 38, 7017–7034.

Huang, D., Chen, J., Yang, L., Ouyang, Q., Li, J., Lao, L., Zhao, J., Liu, J., Lu, Y. & Xing, Y. 2018. Nkila Lncrna Promotes Tumor Immune Evasion By Sensitizing T Cells To Activation-Induced Cell Death. *Nature Immunology,* 19, 1112–1125.

Huang, M. Y., Cai, J. Y., Yang, S. Y., Zhao, Q., Shao, Z. M., Zhang, F. L., Zhang, Y. L., Cao, A. Y. & Li, D. Q. 2025. Secernin-2 Stabilizes Histone Methyltransferase Kmt2c To Suppress Progression And Confer Therapeutic Sensitivity To Parp Inhibition In Triple-Negative Breast Cancer. *Adv Sci (Weinh),* 12, E2413280.

Hum, M. & Lee, A. A.-O. 2025a. Dna Methylation In Breast Cancer: Early Detection And Biomarker Discovery Through Current And Emerging Approaches.

Hum, M. & Lee, A. S. G. 2025b. Dna Methylation In Breast Cancer: Early Detection And Biomarker Discovery Through Current And Emerging Approaches. *J Transl Med,* 23**,** 465.

Huo, Z., Zhang, S., Su, G., Cai, Y., Chen, R., Jiang, M., Yang, D., Zhang, S., Xiong, Y. & Zhang, X. 2025. Immunohistochemical Profiling Of Histone Modification Biomarkers Identifies Subtype-Specific Epigenetic Signatures And Potential Drug Targets In Breast Cancer. *International Journal Of Molecular Sciences,* 26**,** 770.

Ibrahim, M. B., Flanagan, J., Ibrahim, T. & Rouleau, E. 2024. Unraveling Noncoding Dna Variants And Epimutations: A Paradigm Shift In Hereditary Cancer Research. *Future Oncology,* 20**,** 1289–1298.

Ji, Y., Li, B., Lin, R., Yuan, J., Han, Y., Du, Y. & Zhao, Y. A.-O. 2025. Super-Enhancers In Tumors: Unraveling Recent Advances In Their Role In Oncogenesis And The Emergence Of Targeted Therapies.

Jiang, T., He, X., Zhao, Z., Zhang, X., Wang, T. & Jia, L. 2022. Rna M6a Reader Igf2bp3 Promotes Metastasis Of Triple-Negative Breast Cancer Via Slit2 Repression. *Faseb J,* 36**,** E22618.

Jin, H., Du, W., Huang, W., Yan, J., Tang, Q., Chen, Y. & Zou, Z. 2021. Lncrna And Breast Cancer: Progress From Identifying Mechanisms To Challenges And Opportunities Of Clinical Treatment. *Molecular Therapy-Nucleic Acids,* 25**,** 613–637.

Jin, H., Kasper, L. H., Larson, J. D., Wu, G., Baker, S. J., Zhang, J. & Fan, Y. 2020. Chipseqspikeinfree: A Chip-Seq Normalization Approach To Reveal Global Changes In Histone Modifications Without Spike-In. *Bioinformatics,* 36**,** 1270–1272.

Jing, L., Lan, L., Mingxin, Z. & Zhaofeng, Z. 2022. Mettl3/Linc00662/Mir-186-5p Feedback Loop Regulates Docetaxel Resistance In Triple Negative Breast Cancer. *Scientific Reports,* 12**,** 16715.

Johnsson, P., Lipovich, L., Grandér, D. & Morris, K. V. 2014. Evolutionary Conservation Of Long Non-Coding Rnas; Sequence, Structure, Function. *Biochimica Et Biophysica Acta (Bba)-General Subjects,* 1840**,** 1063–1071.

Kang, Z., Wang, J., Liu, J., Du, L. & Liu, X. 2025. Epigenetic Modifications In Breast Cancer: From Immune Escape Mechanisms To Therapeutic Target Discovery. *Frontiers In Immunology,* 16.

Karagkouni, D., Karavangeli, A., Paraskevopoulou, M. D. & Hatzigeorgiou, A. G. 2021. Characterizing Mirna–Lncrna Interplay. *Long Non-Coding Rnas: Methods And Protocols***,** 243–262.

Kim, J., Piao, H.-L., Kim, B.-J., Yao, F., Han, Z., Wang, Y., Xiao, Z., Siverly, A. N., Lawhon, S. E. & Ton, B. N. 2018. Long Noncoding Rna Malat1 Suppresses Breast Cancer Metastasis. *Nature Genetics,* 50**,** 1705–1715.

Komar, D. & Juszczynski, P. 2020. Rebelled Epigenome: Histone H3s10 Phosphorylation And H3s10 Kinases In Cancer Biology And Therapy. *Clinical Epigenetics,* 12**,** 1–14.

Kudela, E., Samec, M., Koklesova, L., Liskova, A., Kubatka, P., Kozubik, E., Rokos, T., Pribulova, T., Gabonova, E., Smolar, M. & Biringer, K. 2020. Mirna Expression Profiles In Luminal A Breast Cancer-Implications In Biology, Prognosis, And Prediction Of Response To Hormonal Treatment. *Int J Mol Sci,* 21.

Kumar, P. & Bhandari, N. 2022. Lncrnas: Role In Regulation Of Gene Expression. *Gene Expression.* Intechopen.

Kumar, S. & Ranga, A. 2025. Role Of Mirnas In Breast Cancer Development And Progression: Current Research. *Biofactors,* 51**,** E2146.

Kumar, S. & Vindal, V. 2025. Decoding Dna Methylation And Non-Coding Rnas Mediated Regulatory Landscape Of Breast Tumor And Adjacent Normal Tissues. *Biorxiv***,** 2025.02.17.638770.

Kutasovic, J. R., Mccart Reed, A. E., Sokolova, A., Lakhani, S. R. & Simpson, P. T. 2020. Morphologic And Genomic Heterogeneity In The Evolution And Progression Of Breast Cancer. *Cancers,* 12**,** 848.

Lan, J., Wang, L., Cao, J., Wan, Y. & Zhou, Y. 2024. Circbraf Promotes The Progression Of Triple-Negative Breast Cancer Through Modulating Methylation By Recruiting Kdm4b To Histone H3k9me3 And Igf2bp3 To Mrna. *Am J Cancer Res,* 14, 2020–2036.

Lewis, M. W., King, C. M., Wisniewska, K., Regner, M. J., Coffey, A., Kelly, M. R., Mendez-Giraldez, R., Davis, E. S., Phanstiel, D. H. & Franco, H. L. 2024. Crispr Screening Of Transcribed Super-Enhancers Identifies Drivers Of Triple-Negative Breast Cancer Progression. *Cancer Res,* 84, 3684–3700.

Li, J., Chen, Z., Jin, M., Gu, X., Wang, Y., Huang, G., Zhao, W. & Lu, C. 2024a. Histone H4k12 Lactylation Promotes Malignancy Progression In Triple-Negative Breast Cancer Through Slfn5 Downregulation. *Cell Signal,* 124, 111468.

Li, J.-N., Loh, Z.-J., Chen, H.-W., Lee, I.-Y., Tsai, J.-H. & Chen, P.-S. 2024b. Snorna U50a Mediates Everolimus Resistance In Breast Cancer Through Mtor Downregulation. *Translational Oncology,* 48, 102062.

Li, Q., Ci, H., Zhao, P., Yang, D., Zou, Y., Chen, P., Wu, D., Shangguan, W., Li, W., Meng, X., Xing, M., Chen, Y., Zhang, M., Chen, B., Kong, L., Zen, K., Huang, D. C. S., Jiang, Z. W. & Zhao, Q. 2025. Nono Interacts With Nuclear Pkm2 And Directs Histone H3 Phosphorylation To Promote Triple-Negative Breast Cancer Metastasis. *J Exp Clin Cancer Res,* 44, 90.

Li, W., Liu, H., Li, W., Zhang, Q., Zhang, Q. & Hu, D. 2024c. Fezf2 Inhibits The Growth Of Triple-Negative Breast Cancer Cells Through Ezh2/Pd-L1 And Enhances Anti-Tumor Immunity In Vivo. *Molecular & Cellular Toxicology,* 20, 553–562.

Li, Y., Luo, B., Lin, X., Bai, D., Li, L., Gao, D., Li, X., Zhong, X., Wei, Y., Yang, L., Zhu, X., Han, L., Tian, H., Zhang, R. & Wang, P. 2024d. 20(R)-Panaxatriol Enhances Mettl3-Mediated M(6)A Modification Of Stub1 To Inhibit Autophagy And Exert Antitumor Effects In Triple-Negative Breast Cancer Cells. *Phytomedicine,* 130, 155537.

Lin, W. D., Chang, C. H., Pan, J. K., Lin, F. C., Chen, Y. C., Chen, Y. J., Wang, P. S., Hong, W. Q., Chen, S. Y., Lin, C. H., Kuo, Y. L., Chung, W. P., Cheng, H. C., Hsiao, M., Yang, C. N. & Lu, P. J. 2024. A Novel Long Non-Coding Rna Mir4500hg003 Promotes Tumor Metastasis Through Mir-483-3p-Mmp9 Axis In Triple-Negative Breast Cancer. *Cell Death Dis,* 15, 310.

Liu, K., Ouyang, Q. Y., Zhan, Y., Yin, H., Liu, B. X., Tan, L. M., Liu, R., Wu, W. & Yin, J. Y. 2022a. Pharmacoepitranscriptomic Landscape Revealing M6a Modification Could Be A Drug-Effect Biomarker For Cancer Treatment. *Mol Ther Nucleic Acids,* 28, 464–476.

Liu, X., Li, P., Huang, Y., Li, H., Liu, X., Du, Y., Lin, X., Chen, D., Liu, H. & Zhou, Y. 2024. M(6)A Demethylase Alkbh5 Regulates Foxo1 Mrna Stability And Chemoresistance In Triple-Negative Breast Cancer. *Redox Biol,* 69, 102993.

Liu, X., Wang, J., Boyer, J. A., Gong, W., Zhao, S., Xie, L., Wu, Q., Zhang, C., Jain, K., Guo, Y., Rodriguez, J., Li, M., Uryu, H., Liao, C., Hu, L., Zhou, J., Shi, X., Tsai, Y. H., Yan, Q., Luo, W., Chen, X., Strahl, B. D., Von Kriegsheim, A., Zhang, Q., Wang, G. G., Baldwin, A. S. & Zhang, Q. 2022b. Histone H3 Proline 16 Hydroxylation Regulates Mammalian Gene Expression. *Nat Genet,* 54, 1721–1735.

Liu, Y., Leng, P., Liu, Y., Guo, J. & Zhou, H. 2022c. Crosstalk Between Methylation And Ncrnas In Breast Cancer: Therapeutic And Diagnostic Implications. *International Journal Of Molecular Sciences,* 23, 15759.

Liu, Z., Chen, Q. & Hann, S. S. 2019. The Functions And Oncogenic Roles Of Ccat1 In Human Cancer. *Biomedicine & Pharmacotherapy,* 115, 108943.

Llinàs-Arias, P., Ensenyat-Mendez, M., Íñiguez-Muñoz, S., Orozco, J. I. J., Valdez, B., Salomon, M. P., Matsuba, C., Solivellas-Pieras, M., Bedoya-López, A. F., Sesé, B., Mezger, A., Ormestad, M., Unzueta, F., Strand, S. H., Boiko, A. D., Hwang, E. S., Cortés, J., Dinome, M. L., Esteller, M., Lupien, M. & Marzese, D. M. 2023a. Chromatin Insulation Orchestrates Matrix Metalloproteinase Gene Cluster Expression Reprogramming In Aggressive Breast Cancer Tumors. *Mol Cancer,* 22, 190.

Llinàs-Arias, P., Ensenyat-Méndez, M., Orozco, J. I. J., Íñiguez-Muñoz, S., Valdez, B., Wang, C., Mezger, A., Choi, E., Tran, Y. Z., Yao, L., Bonath, F., Olsen, R. A., Ormestad, M., Esteller, M.,

Lupien, M. & Marzese, D. M. 2023b. 3-D Chromatin Conformation, Accessibility, And Gene Expression Profiling Of Triple-Negative Breast Cancer. *Bmc Genom Data,* 24, 61.
Lone, S. N., Nisar, S., Masoodi, T., Singh, M., Rizwan, A., Hashem, S., El-Rifai, W., Bedognetti, D., Batra, S. K., Haris, M., Bhat, A. A. & Macha, M. A. 2022. Liquid Biopsy: A Step Closer To Transform Diagnosis, Prognosis And Future Of Cancer Treatments. *Mol Cancer,* 21, 79.
Lu, B., Qiu, R., Wei, J., Wang, L., Zhang, Q., Li, M., Zhan, X., Chen, J., Hsieh, I. Y., Yang, C., Zhang, J., Sun, Z., Zhu, Y., Jiang, T., Zhu, H., Li, J. & Zhao, W. 2024. Phase Separation Of Phospho-Hdac6 Drives Aberrant Chromatin Architecture In Triple-Negative Breast Cancer. *Nat Cancer,* 5, 1622–1640.
Lu, W. & Yang, S. 2025. Mettl3/Igf2bp1 Promotes The Development Of Triple-Negative Breast Cancer By Mediating M6a Methylation Modification Of Prmt7. *Tissue Cell,* 93, 102690.
Ma, J., Zhao, W., Zhang, H., Chu, Z., Liu, H., Fang, X. & Tang, D. 2022a. Long Non-Coding Rna Anril Promotes Chemoresistance In Triple-Negative Breast Cancer Via Enhancing Aerobic Glycolysis. *Life Sci,* 306, 120810.
Ma, P., Kang, S., Li, H., Li, M., Zhao, Y., Yuan, H., Pang, J., Tang, W. & Sheng, M. 2024. A Novel Lncrna Ac112721. 1 Promotes The Progression Of Triple-Negative Breast Cancer By Directly Binding To Thbs1 And Regulating Mir-491-5p/C2cd2l Axis. *Scientific Reports,* 14, 32056.
Ma, Q., Yang, L., Tolentino, K., Wang, G., Zhao, Y., Litzenburger, U. M., Shi, Q., Zhu, L., Yang, C. & Jiao, H. 2022b. Inducible Lncrna Transgenic Mice Reveal Continual Role Of Hotair In Promoting Breast Cancer Metastasis. *Elife,* 11, E79126.
Maia, B. M., Rocha, R. M. & Calin, G. A. 2014. Clinical Significance Of The Interaction Between Non-Coding Rnas And The Epigenetics Machinery: Challenges And Opportunities In Oncology. *Epigenetics,* 9, 75–80.
Maleki Dana, P., Mansournia, M. A. & Mirhashemi, S. M. 2020. Piwi-Interacting Rnas: New Biomarkers For Diagnosis And Treatment Of Breast Cancer. *Cell & Bioscience,* 10, 44.
Manoochehri, M., Jones, M., Tomczyk, K., Fletcher, O., Schoemaker, M. J., Swerdlow, A. J., Borhani, N. & Hamann, U. 2020. Dna Methylation Of The Long Intergenic Noncoding Rna 299 Gene In Triple-Negative Breast Cancer: Results From A Prospective Study. *Sci Rep,* 10, 11762.
Mattick, J. S., Amaral, P. P., Carninci, P., Carpenter, S., Chang, H. Y., Chen, L.-L., Chen, R., Dean, C., Dinger, M. E. & Fitzgerald, K. A. 2023. Long Non-Coding Rnas: Definitions, Functions, Challenges And Recommendations. *Nature Reviews Molecular Cell Biology,* 24, 430–447.
Medico-Salsench, E., Karkala, F., Lanko, K. & Barakat, T. S. 2021. The Non-Coding Genome In Genetic Brain Disorders: New Targets For Therapy? *Essays In Biochemistry,* 65, 671–683.
Mehrmohamadi, M., Sepehri, M. H., Nazer, N. & Norouzi, M. R. 2021. A Comparative Overview Of Epigenomic Profiling Methods. *Frontiers In Cell And Developmental Biology,* 9, 714687.
Mendaza, S., Guerrero-Setas, D., Monreal-Santesteban, I., Ulazia-Garmendia, A., Cordoba Iturriagagoitia, A., De La Cruz, S. & Martín-Sánchez, E. 2021. A Dna Methylation-Based Gene Signature Can Predict Triple-Negative Breast Cancer Diagnosis. *Biomedicines,* 9.
Mi, J., Zhang, H., Jiang, X., Yi, Y., Cao, W., Song, C. & Yuan, C. 2024. Lncrna Miat Promotes Luminal B Breast Cancer Cell Proliferation, Migration, And Invasion In Vitro. *J Appl Genet,* 65, 309–319.
Mongroo, P. S. & Rustgi, A. K. 2010. The Role Of The Mir-200 Family In Epithelial-Mesenchymal Transition. *Cancer Biology & Therapy,* 10, 219–222.
Moraes-Almeida, M. S., Sogayar, M. C. & Demasi, M. A. A. 2023. Evidence Of A Synthetic Lethality Interaction Between Setdb1 Histone Methyltransferase And Chd4 Chromatin Remodeling Protein In A Triple Negative Breast Cancer Cell Line. *Braz J Med Biol Res,* 56, E12854.
Morgan, M. A. & Shilatifard, A. 2020. Reevaluating The Roles Of Histone-Modifying Enzymes And Their Associated Chromatin Modifications In Transcriptional Regulation. *Nature Genetics,* 52, 1271–1281.
Moser, D., Müller, S., Hummel, E., Limberg, A., Dieckmann, L., Frach, L., Pakusch, J., Flasbeck, V., Brüne, M. & Beygo, J. 2020. Targeted Bisulfite Sequencing: A Novel Tool

For The Assessment Of Dna Methylation With High Sensitivity And Increased Coverage. *Psychoneuroendocrinology,* 120, 104784.

Mukherjee, A., Kakati, R. T., Van Alsten, S., Laws, T., Ebbs, A. L., Hollern, D. P., Spanheimer, P. M., Hoadley, K. A., Troester, M. A., Simon, J. M. & Baldwin, A. S. 2024. Dab2ip Loss In Luminal A Breast Cancer Leads To Nf-Kb–Associated Aggressive Oncogenic Phenotypes. *Jci Insight,* 9.

Naciri, I., Liang, M., Yang, Y., Karner, H., Lin, B., De Lourdes Andrade Ludena, M., Hanse, E. A., Lebron, A., Razorenova, O. V., Nicholas, D., Kong, M. & Sun, S. 2024. Loss Of Xist Lncrna Unlocks Stemness And Cellular Plasticity In Ovarian Cancer. *Proceedings Of The National Academy Of Sciences,* 121, E2418096121.

Nakabayashi, K., Yamamura, M., Haseagawa, K. & Hata, K. 2022. Reduced Representation Bisulfite Sequencing (Rrbs). *Epigenomics: Methods And Protocols.* Springer.

Niknafs, Y. S., Han, S., Ma, T., Speers, C., Zhang, C., Wilder-Romans, K., Iyer, M. K., Pitchiaya, S., Malik, R., Hosono, Y., Prensner, J. R., Poliakov, A., Singhal, U., Xiao, L., Kregel, S., Siebenaler, R. F., Zhao, S. G., Uhl, M., Gawronski, A., Hayes, D. F., Pierce, L. J., Cao, X., Collins, C., Backofen, R., Sahinalp, C. S., Rae, J. M., Chinnaiyan, A. M. & Feng, F. Y. 2016. The Lncrna Landscape Of Breast Cancer Reveals A Role For Dscam-As1 In Breast Cancer Progression. *Nature Communications,* 7, 12791.

O'Bryan, S., Dong, S., Mathis, J. M. & Alahari, S. K. 2017. The Roles Of Oncogenic Mirnas And Their Therapeutic Importance In Breast Cancer. *European Journal Of Cancer,* 72, 1–11.

O'neill, K., Pleasance, E., Fan, J., Akbari, V., Chang, G., Dixon, K., Csizmok, V., Maclennan, S., Porter, V., Galbraith, A., Grisdale, C. J., Culibrk, L., Dupuis, J. H., Corbett, R., Hopkins, J., Bowlby, R., Pandoh, P., Smailus, D. E., Cheng, D., Wong, T., Frey, C., Shen, Y., Lewis, E., Paulin, L. F., Sedlazeck, F. J., Nelson, J. M. T., Chuah, E., Mungall, K. L., Moore, R. A., Coope, R., Mungall, A. J., Mcconechy, M. K., Williamson, L. M., Schrader, K. A., Yip, S., Marra, M. A., Laskin, J. & Jones, S. J. M. 2024. Long-Read Sequencing Of An Advanced Cancer Cohort Resolves Rearrangements, Unravels Haplotypes, And Reveals Methylation Landscapes. *Cell Genomics,* 4, 100674.

Onoguchi-Mizutani, R., Taniue, K., Kawata, K., Yamada, T. & Akimitsu, N. 2023. Techniques For Analyzing Genome-Wide Expression Of Non-Coding Rna. *Handbook Of Epigenetics.* Elsevier.

Oo, J. A., Brandes, R. P. & Leisegang, M. S. 2022. Long Non-Coding Rnas: Novel Regulators Of Cellular Physiology And Function. *Pflügers Archiv-European Journal Of Physiology,* 474, 191–204.

Otmani, K. & Lewalle, P. 2021. Tumor Suppressor Mirna In Cancer Cells And The Tumor Microenvironment: Mechanism Of Deregulation And Clinical Implications. *Frontiers In Oncology,* 11, 708765.

Ouyang, Y. X., Feng, J., Wang, Z., Zhang, G. J. & Chen, M. 2021. Mir-221/222 Sponge Abrogates Tamoxifen Resistance In Er-Positive Breast Cancer Cells Through Restoring The Expression Of Erα. *Molecular Biomedicine,* 2, 1–11.

Pan, T., Haiwen, D., Le, J., Shaobo, Z., Delin, W., Wanwan, P., Menghao, D., Xiaopeng, M. & And Chen, Z. 2022. Dnmt1-Mediated Demethylation Of Lncrna Meg3 Promoter Suppressed Breast Cancer Progression By Repressing Notch1 Signaling Pathway. *Cell Cycle,* 21, 2323–2337.

Panet, F., Papakonstantinou, A., Borrell, M., Vivancos, J., Vivancos, A. & Oliveira, M. 2024. Use Of Ctdna In Early Breast Cancer: Analytical Validity And Clinical Potential. *Npj Breast Cancer,* 10, 50.

Panigoro, S. S., Paramita, R. I., Wanandi, S. I., Fadilah, F., Wibisana, I. G. & Sutandyo, N. 2025. Dna Methylation Profiles For Breast Cancer Subtype Classifications: A Translational Study From Microarray To Methylation-Specific Pcr (Msp). *Narra J,* 5, E1364.

Papanicolau-Sengos, A. & Aldape, K. 2022. Dna Methylation Profiling: An Emerging Paradigm For Cancer Diagnosis. *Annual Review Of Pathology: Mechanisms Of Disease,* 17, 295–321.

Park, S. A., Masunaga, N., Kagara, N., Ohi, Y., Gondo, N., Abe, K., Yoshinami, T., Sota, Y., Miyake, T., Tanei, T., Shimoda, M., Sagara, Y. & Shimazu, K. 2023. Evaluation Of <Em>Rassf1a</Em>

Methylation In The Lysate Of Sentinel Lymph Nodes For Detecting Breast Cancer Metastasis: A Diagnostic Accuracy Study. *Oncol Lett,* 26, 475.

Patop, I. L., Wüst, S. & Kadener, S. 2019. Past, Present, And Future Of Circ Rna S. *The Embo Journal,* 38, E100836.

Patutina, O. A., Miroshnichenko, S. K., Mironova, N. L., Sen'kova, A. V., Bichenkova, E. V., Clarke, D. J., Vlassov, V. V. & Zenkova, M. A. 2019. Catalytic Knockdown Of Mir-21 By Artificial Ribonuclease: Biological Performance In Tumor Model. *Frontiers In Pharmacology,* 10, 879.

Piao, Y., Xu, W., Park, K. H., Ryu, K. H. & Xiang, R. 2021. Comprehensive Evaluation Of Differential Methylation Analysis Methods For Bisulfite Sequencing Data. *International Journal Of Environmental Research And Public Health,* 18, 7975.

Prabhakaran, R., Thamarai, R., Sivasamy, S., Dhandayuthapani, S., Batra, J., Kamaraj, C., Karthik, K., Shah, M. A. & Mallik, S. 2024. Epigenetic Frontiers: Mirnas, Long Non-Coding Rnas And Nanomaterials Are Pioneering To Cancer Therapy. *Epigenetics Chromatin,* 17, 31.

Prabhu, K. S., Sadida, H. Q., Kuttikrishnan, S., Junejo, K., Bhat, A. A. & Uddin, S. 2024. Beyond Genetics: Exploring The Role Of Epigenetic Alterations In Breast Cancer. *Pathology-Research And Practice,* 254, 155174.

Pradel, L. S., Ho, Y. L., Gohlke, H. & Kassack, M. U. 2024. The Antioxidant And Hdac-Inhibitor Alpha-Lipoic Acid Is Synergistic With Exemestane In Estrogen Receptor-Positive Breast Cancer Cells. *Int J Mol Sci,* 25.

Qian, L., Xie, H., Zhang, L., Zhao, Q., Lü, J. & Yu, Z. 2021. Piwi-Interacting Rnas: A New Class Of Regulator In Human Breast Cancer. *Frontiers In Oncology,* 11, 695077.

Quintas-Granados, L. I., Cortés, H., Carmen, M. G.-D., Leyva-Gómez, G., Bustamante-Montes, L. P., Rodríguez-Morales, M., Villegas-Vazquez, E. Y., López-Reyes, I., Alcaraz-Estrada, S. L., Sandoval-Basilio, J., Soto-Reyes, E., Sharifi-Rad, J., Figueroa-González, G. & Reyes-Hernández, O. D. 2021. The High Methylation Level Of A Novel 151-Bp Cpg Island In The Esr1 Gene Promoter Is Associated With A Poor Breast Cancer Prognosis. *Cancer Cell International,* 21, 649.

Rahmani, G., Sameri, S., Abbasi, N., Abdi, M. & Najafi, R. 2021. The Clinical Significance Of Histone Deacetylase-8 In Human Breast Cancer. *Pathology - Research And Practice,* 220, 153396.

Rajagopal, T., Talluri, S., Akshaya, R. & Dunna, N. R. 2020. Hotair Lncrna: A Novel Oncogenic Propellant In Human Cancer. *Clinica Chimica Acta,* 503, 1–18.

Regner, M. J., Garcia-Recio, S., Thennavan, A., Wisniewska, K., Mendez-Giraldez, R., Felsheim, B., Spanheimer, P. M., Parker, J. S., Perou, C. M. & Franco, H. L. 2025. Defining The Regulatory Logic Of Breast Cancer Using Single-Cell Epigenetic And Transcriptome Profiling. *Cell Genom,* 5, 100765.

Rizk, N. I., Kassem, D. H., Abulsoud, A. I., Abdelhalim, S., Yasser, M. B., Kamal, M. M. & Hamdy, N. M. 2024. Revealing The Role Of Serum Exosomal Novel Long Non-Coding Rna Nampt-As As A Promising Diagnostic/Prognostic Biomarker In Colorectal Cancer Patients. *Life Sciences,* 352, 122850.

Rodriguez-Casanova, A., Costa-Fraga, N., Castro-Carballeira, C., Gonzalez-Conde, M., Abuin, C., Bao-Caamano, A., Garcia-Caballero, T., Brozos-Vazquez, E., Rodriguez-Lopez, C., Cebey, V., Palacios, P., Cueva, J. F., Lopez-Lopez, R., Costa, C. & Diaz-Lagares, A. 2022. A Genome-Wide Cell-Free Dna Methylation Analysis Identifies An Episignature Associated With Metastatic Luminal B Breast Cancer. *Front Cell Dev Biol,* 10, 1016955.

Rønneberg, J. A., Fleischer, T., Solvang, H. K., Nordgard, S. H., Edvardsen, H., Potapenko, I., Nebdal, D., Daviaud, C., Gut, I., Bukholm, I., Naume, B., Børresen-Dale, A.-L., Tost, J. & Kristensen, V. 2011. Methylation Profiling With A Panel Of Cancer Related Genes: Association With Estrogen Receptor, Tp53 Mutation Status And Expression Subtypes In Sporadic Breast Cancer. *Molecular Oncology,* 5, 61–76.

Rui, T., Xiang, A., Guo, J., Tang, N., Lin, X., Jin, X., Liu, J. & Zhang, X. 2022. Mir-4728 Is A Valuable Biomarker For Diagnostic And Prognostic Assessment Of Her2-Positive Breast Cancer. *Frontiers In Molecular Biosciences,* 9, 818493.

Santarpia, L., Bottai, G., Kelly, C. M., Győrffy, B., Székely, B. & Pusztai, L. 2016. Deciphering And Targeting Oncogenic Mutations And Pathways In Breast Cancer. *The Oncologist,* 21, 1063–1078.

Sarkar, S., Venkatesh, D., Kandasamy, T. & Ghosh, S. S. 2024. Epigenetic Modulations In Breast Cancer: An Emerging Paradigm In Therapeutic Implications. *Frontiers In Bioscience-Landmark,* 29, 287.

Schramm, A., De Gregorio, N., Widschwendter, P., Fink, V. & Huober, J. 2015. Targeted Therapies In Her2-Positive Breast Cancer-A Systematic Review. *Breast Care,* 10, 173–178.

Schwarzenbach, H. & Gahan, P. B. 2023. Interplay Between Lncrnas And Micrornas In Breast Cancer. *Int J Mol Sci,* 24.

Sedeta, E. T., Jobre, B. & Avezbakiyev, B. 2023. Breast Cancer: Global Patterns Of Incidence, Mortality, And Trends. American Society Of Clinical Oncology.

Selezneva, A. D., Filippova, E. A., Selezneva, A. D., Lukina, S. S., Pronina, I. V., Ivanova, N. A., Kazubskaya, T. P., Burdennyy, A. M., Braga, E. A. & Loginov, V. I. 2022. Hypermethylation Of Long Non-Coding Rna Genes Group In The Breast Cancer Development And Progression. *Bulletin Of Experimental Biology And Medicine,* 173, 765–769.

Serra-Bardenys, G., Blanco, E., Escudero-Iriarte, C., Serra-Camprubí, Q., Querol, J., Pascual-Reguant, L., Morancho, B., Escorihuela, M., Tissera, N. S., Sabé, A., Martín, L., Segura-Bayona, S., Verde, G., Aiese Cigliano, R., Millanes-Romero, A., Jerónimo, C., Cebrià-Costa, J. P., Nuciforo, P., Simonetti, S., Viaplana, C., Dienstmann, R., Oliveira, M., Peg, V., Stracker, T. H., Arribas, J., Canals, F., Villanueva, J., Di Croce, L., García De Herreros, A., Tian, T. V. & Peiró, S. 2024. Loxl2-Mediated Chromatin Compaction Is Required To Maintain The Oncogenic Properties Of Triple-Negative Breast Cancer Cells. *Febs J,* 291, 2423–2448.

Shaath, H., Vishnubalaji, R., Elango, R., Khattak, S. & Alajez, N. M. 2021. Single-Cell Long Noncoding Rna (Lncrna) Transcriptome Implicates Malat1 In Triple-Negative Breast Cancer (Tnbc) Resistance To Neoadjuvant Chemotherapy. *Cell Death Discovery,* 7, 23.

Shaker, F. H., Sanad, E. F., Elghazaly, H., Hsia, S. M. & Hamdy, N. M. 2025. Pir-823 Tale As Emerging Cancer-Hallmark Molecular Marker In Different Cancer Types: A Step-Toward Ncrna-Precision. *Naunyn Schmiedebergs Arch Pharmacol,* 398, 47–68.

Shao, Y., Zhang, S., Pan, Y., Peng, Z. & Dong, Y. 2025. Mir-135b: A Key Role In Cancer Biology And Therapeutic Targets. *Non-Coding Rna Research,* 12, 67.

Shareef, S. J., Bevill, S. M., Raman, A. T., Aryee, M. J., Van Galen, P., Hovestadt, V. & Bernstein, B. E. 2021. Extended-Representation Bisulfite Sequencing Of Gene Regulatory Elements In Multiplexed Samples And Single Cells. *Nature Biotechnology,* 39, 1086–1094.

Shen, H. T., Hung, C. A.-O., Davis, C. A.-O. X., Su, C. M., Liao, L. M., Shih, H. M., Lee, K. D., Ansar, M. & Lin, R. A.-O. 2024a. Hypermethylation Of The Gene Body In Srcin1 Is Involved In Breast Cancer Cell Proliferation And Is A Potential Blood-Based Biomarker For Early Detection And A Poor Prognosis. Lid - https://doi.org/10.3390/Biom14050571 [Doi] Lid - 571.

Shen, X., Zhong, J., Yu, P., Liu, F., Peng, H. & Chen, N. 2024b. Ythdc1-Dependent M6a Modification Modulated Foxm1 Promotes Glycolysis And Tumor Progression Through Cenpa In Triple-Negative Breast Cancer. *Cancer Sci,* 115, 1881–1895.

Shi, Y., Zheng, C., Jin, Y., Bao, B., Wang, D., Hou, K., Feng, J., Tang, S., Qu, X., Liu, Y., Che, X. & Teng, Y. 2020. Reduced Expression Of Mettl3 Promotes Metastasis Of Triple-Negative Breast Cancer By M6a Methylation-Mediated Col3a1 Up-Regulation. *Front Oncol,* 10, 1126.

Shibuya, N., Kakeji, Y. & Shimono, Y. A.-O. 2020. Microrna-93 Targets Wasf3 And Functions As A Metastasis Suppressor In Breast Cancer.

Sideris, N., Dama, P., Bayraktar, S., Stiff, T. & Castellano, L. 2022. Lncrnas In Breast Cancer: A Link To Future Approaches. *Cancer Gene Therapy,* 29, 1866–1877.

Simigdala, N., Chalari, A., Sklirou, A. D., Chavdoula, E., Papafotiou, G., Melissa, P., Kafalidou, A., Paschalidis, N., Pateras, I. S., Athanasiadis, E., Konstantopoulos, D., Trougakos, I. P. &

Klinakis, A. 2023. Loss Of Kmt2c In Vivo Leads To Emt, Mitochondrial Dysfunction And Improved Response To Lapatinib In Breast Cancer. *Cell Mol Life Sci,* 80, 100.
Smith, J., Day, R. C. & Weeks, R. J. 2022. Next-Generation Bisulfite Sequencing For Targeted Dna Methylation Analysis. *Chromatin: Methods And Protocols.* Springer.
Solaimani, M., Hosseinzadeh, S. & Abasi, M. 2025. Non-Coding Rnas, A Double-Edged Sword In Breast Cancer Prognosis. *Cancer Cell Int,* 25, 123.
Stavast, C. J. & Erkeland, S. J. 2019. The Non-Canonical Aspects Of Micrornas: Many Roads To Gene Regulation. *Cells,* 8, 1465.
Su, Y., Bai, Q., Zhang, W., Xu, B. & Hu, T. 2025. The Role Of Long Non-Coding Rnas In Modulating The Immune Microenvironment Of Triple-Negative Breast Cancer: Mechanistic Insights And Therapeutic Potential. *Biomolecules,* 15, 454.
Subramanian, K. & Sinha, R. 2024. Functions Of Differentially Regulated Mirnas In Breast Cancer Progression: Potential Markers For Early Detection And Candidates For Therapy. *Biomedicines,* 12, 691.
Sun, H., Wang, G., Cai, J., Wei, X., Zeng, Y., Peng, Y. & Zhuang, J. 2022. Long Non-Coding Rna H19 Mediates N-Acetyltransferase 1 Gene Methylation In The Development Of Tamoxifen Resistance In Breast Cancer. *Exp Ther Med,* 23, 12.
Sun, Y., Zhang, H., Ma, R., Guo, X., Zhang, G., Liu, S., Zhu, W., Liu, H. & Gao, P. 2023. Ets-1-Activated Linc01016 Over-Expression Promotes Tumor Progression Via Suppression Of Rffl-Mediated Dhx9 Ubiquitination Degradation In Breast Cancers. *Cell Death & Disease,* 14, 507.
Surapaneni, S. K., Bhat, Z. R. & Tikoo, K. 2020. Microrna-941 Regulates The Proliferation Of Breast Cancer Cells By Altering Histone H3 Ser 10 Phosphorylation. *Scientific Reports,* 10, 17954.
Szczepanek, J., Skorupa, M., Jarkiewicz-Tretyn, J., Cybulski, C. & Tretyn, A. 2023. Harnessing Epigenetics For Breast Cancer Therapy: The Role Of Dna Methylation, Histone Modifications, And Microrna. *International Journal Of Molecular Sciences,* 24.
Tan, M., He, Y., Yi, J., Chen, J., Guo, Q., Liao, N. & Peng, L. 2024. Wtap Mediates Nupr1 Regulation Of Lcn2 Through M(6)A Modification To Influence Ferroptosis, Thereby Promoting Breast Cancer Proliferation, Migration And Invasion. *Biochem Genet,* 62, 876–891.
Tarhonska, K., Wichtowski, M., Wow, T., Kolacinska-Wow, A., Ploszka, K., Fendler, W., Zawlik, I., Paszek, S., Zuchowska, A. & Jablonska, E. 2025. Dna Methylation And Demethylation In Triple-Negative Breast Cancer: Associations With Clinicopathological Characteristics And The Chemotherapy Response. *Biomedicines,* 13.
Tarighi, M., Khalaj-Kondori, M., Hosseinzadeh, A. & Abtin, M. 2021. Long Non-Coding Rna (Lncrna) Dscam-As1 Is Upregulated In Breast Cancer. *Breast Disease,* 40, 63–68.
Tellez, C. S., Juri, D. E., Do, K., Bernauer, A. M., Thomas, C. L., Damiani, L. A., Tessema, M., Leng, S. & Belinsky, S. A. 2011. Emt And Stem Cell–Like Properties Associated With Mir-205 And Mir-200 Epigenetic Silencing Are Early Manifestations During Carcinogen-Induced Transformation Of Human Lung Epithelial Cells. *Cancer Research,* 71, 3087–3097.
Thomsen, C. A.-O., Hansen, T. F., Andersen, R. F., Lindebjerg, J., Jensen, L. H. & Jakobsen, A. 2020. Early Identification Of Treatment Benefit By Methylated Circulating Tumor Dna In Metastatic Colorectal Cancer.
Title, A. C., Hong, S.-J., Pires, N. D., Hasenöhrl, L., Godbersen, S., Stokar-Regenscheit, N., Bartel, D. P. & Stoffel, M. 2018. Genetic Dissection Of The Mir-200–Zeb1 Axis Reveals Its Importance In Tumor Differentiation And Invasion. *Nature Communications,* 9, 4671.
Torcivia, J. & Mazumder, R. 2021. Scanning Window Analysis Of Non-Coding Regions Within Normal-Tumor Whole-Genome Sequence Samples. *Briefings In Bioinformatics,* 22, Bbaa203.
Tsai, M.-C., Manor, O., Wan, Y., Mosammaparast, N., Wang, J. K., Lan, F., Shi, Y., Segal, E. & Chang, H. Y. 2010. Long Noncoding Rna As Modular Scaffold Of Histone Modification Complexes. *Science,* 329, 689–693.
Tsutsui, H. & Grossniklaus, U. 2024. Whole-Genome Bisulfite Sequencing With A Small Amount Of Dna. *Methods For Plant Nucleus And Chromatin Studies: Methods And Protocols.* Springer.

Vishnubalaji, R., Awata, D. & Alajez, N. M. 2025. Lurap1l-As1 Long Noncoding Rna Promotes Breast Cancer Progression And Associates With Poor Prognosis. *Non-Coding Rna Research,* 12, 1–9.
Wang, F., Bai, X., Wang, Y., Jiang, Y., Ai, B., Zhang, Y., Liu, Y., Xu, M., Wang, Q. & Han, X. 2021. Atacdb: A Comprehensive Human Chromatin Accessibility Database. *Nucleic Acids Research,* 49, D55-D64.
Wang, H., Liu, B., Long, J., Yu, J., Ji, X., Li, J., Zhu, N., Zhuang, X., Li, L., Chen, Y., Liu, Z., Wang, S. & Zhao, S. 2023a. Integrative Analysis Identifies Two Molecular And Clinical Subsets In Luminal B Breast Cancer. *Iscience,* 26, 107466.
Wang, J., Liu, X., Li, P., Wang, J., Shu, Y., Zhong, X., Gao, Z., Yang, J., Jiang, Y., Zhou, X. & Yang, G. 2022a. Long Noncoding Rna Hotair Regulates The Stemness Of Breast Cancer Cells Via Activation Of The Nf-Kappab Signaling Pathway. *J Biol Chem,* 298, 102630.
Wang, J., Wang, Q., Guan, Y., Sun, Y., Wang, X., Lively, K., Wang, Y., Luo, M., Kim, J. A. & Murphy, E. A. 2022b. Breast Cancer Cell–Derived Microrna-155 Suppresses Tumor Progression Via Enhancing Immune Cell Recruitment And Antitumor Function. *The Journal Of Clinical Investigation,* 132.
Wang, Q., Li, G., Ma, X., Liu, L., Liu, J., Yin, Y., Li, H., Chen, Y., Zhang, X., Zhang, L., Sun, L., Ai, J. & Xu, S. 2023b. Lncrna Tincr Impairs The Efficacy Of Immunotherapy Against Breast Cancer By Recruiting Dnmt1 And Downregulating Mir-199a-5p Via The Stat1-Tincr-Usp20-Pd-L1 Axis. *Cell Death Dis,* 14, 76.
Wang, Y., Armendariz, D. A., Wang, L., Zhao, H., Xie, S. & Hon, G. C. 2025. Enhancer Regulatory Networks Globally Connect Non-Coding Breast Cancer Loci To Cancer Genes. *Genome Biol,* 26, 10.
Wang, Y., Chen, X., Qiao, X., Xie, Y., Guo, D., Li, B., Cao, J., Tao, Z. & Hu, X. 2023c. Chromatin Remodelling Molecule Arid1a Determines Metastatic Heterogeneity In Triple-Negative Breast Cancer By Competitively Binding To Yap. *Cancers (Basel),* 15.
Wang, Y., Lu, R., Chen, P., Cui, R., Ji, M., Zhang, X., Hou, P. & Qu, Y. 2022c. Promoter Methylation Of Transient Receptor Potential Melastatin-Related 7 (Trpm7) Predicts A Better Prognosis In Patients With Luminal A Breast Cancers. *Bmc Cancer,* 22, 951.
Wegwitz, F., Prokakis, E., Pejkovska, A., Kosinsky, R. L., Glatzel, M., Pantel, K., Wikman, H. & Johnsen, S. A. 2020. The Histone H2b Ubiquitin Ligase Rnf40 Is Required For Her2-Driven Mammary Tumorigenesis. *Cell Death Dis,* 11, 873.
Wehida, N., Abdel-Rehim, W., El Mansy, H., Karmouty, A. & Kamel, M. A. 2023. A Panel Of Circulating Non-Coding Rnas In The Diagnosis And Monitoring Of Therapy In Egyptian Patients With Breast Cancer. *Biomedicines,* 11, 563.
Weinzapfel, E. N., Fedder-Semmes, K. N., Sun, Z.-W. & Keogh, M.-C. 2024. Beyond The Tail: The Consequence Of Context In Histone Post-Translational Modification And Chromatin Research. *Biochemical Journal,* 481, 219–244.
Welden, J. R. & Stamm, S. 2019. Pre-Mrna Structures Forming Circular Rnas. *Biochimica Et Biophysica Acta (Bba)-Gene Regulatory Mechanisms,* 1862, 194410.
WHO. 2024. *Breast Cancer* [Online]. Available: https://www.Who.Int/News-Room/Fact-Sheets/Detail/Breast-Cancer#:~:Text=In%202022%2c%20there%20were%202,Increasing%20rates%20in%20later%20life13 [Accessed].
Wu, Q., Tao, X., Luo, Y., Zheng, S., Lin, N. & Xie, X. 2023. A Novel Super-Enhancer-Related Gene Signature Predicts Prognosis And Immune Microenvironment For Breast Cancer. *Bmc Cancer,* 23, 776.
Wu, S. & Schmitz, U. 2023. Single-Cell And Long-Read Sequencing To Enhance Modelling Of Splicing And Cell-Fate Determination. *Computational And Structural Biotechnology Journal,* 21, 2373–2380.
Xia, T., Dai, X. Y., Sang, M. Y., Zhang, X., Xu, F., Wu, J., Shi, L., Wei, J. F. & Ding, Q. 2024. Igf2bp2 Drives Cell Cycle Progression In Triple-Negative Breast Cancer By Recruiting Eif4a1 To Promote The M6a-Modified Cdk6 Translation Initiation Process. *Adv Sci (Weinh),* 11, E2305142.

Xie, J., Gan, L., Xue, B., Wang, X. & Pei, X. 2023. Emerging Roles Of Interactions Between Ncrnas And Other Epigenetic Modifications In Breast Cancer. *Frontiers In Oncology,* 13, 1264090.

Xie, X., Tang, S. & Yu, Z. 2024. Dna Methylation-Regulated Hoxc8's Role In Her2-Positive Breast Cancer Function And Its Contribution To Herceptin Resistance. *Discov Med,* 36, 559–570.

Xu, J., Wang, Z., Li, S., Chen, J., Zhang, J., Jiang, C., Zhao, Z., Li, J., Li, Y. & Li, X. 2018. Combinatorial Epigenetic Regulation Of Non-Coding Rnas Has Profound Effects On Oncogenic Pathways In Breast Cancer Subtypes. *Briefings In Bioinformatics,* 19, 52–64.

Xu, X., Liu, Y., Dong, C., Ma, B., Sun, M., Li, H. & Song, X. 2025. Role Of Insulin-Like Growth Factor 2 Mrna Binding Protein 2 In Regulating The Progression Of Triple-Negative Breast Cancer Via Binding With M6a Methylation Modified Downstream Target Genes Ccl20 And C-Myc. *J Physiol Pharmacol,* 76.

Xue, Y. B., Ding, M. Q., Xue, L. & Luo, J. H. 2019. Circagfg1 Sponges Mir-203 To Promote Emt And Metastasis Of Non-Small-Cell Lung Cancer By Upregulating Znf281 Expression. *Thoracic Cancer,* 10, 1692–1701.

Yan, F., Powell, D. R., Curtis, D. J. & Wong, N. C. 2020. From Reads To Insight: A Hitchhiker's Guide To Atac-Seq Data Analysis. *Genome Biology,* 21, 1–16.

Yan, H. & Bu, P. 2021. Non-Coding Rna In Cancer. *Essays In Biochemistry,* 65, 625–639.

Yang, C., Zhu, H., Tan, Y., Zhu, R., Wu, X., Li, Y. & Wang, C. 2021a. Malat1 Promotes Tumorigenesis And Increases Cellular Sensitivity To Herceptin In Her2-Positive Breast Cancer. *Curr Cancer Drug Targets,* 21, 860–869.

Yang, L., Kumegawa, K., Saeki, S., Nakadai, T. & Maruyama, R. 2024a. Identification Of Lineage-Specific Epigenetic Regulators Foxa1 And Grhl2 Through Chromatin Accessibility Profiling In Breast Cancer Cell Lines. *Cancer Gene Ther,* 31, 736–745.

Yang, L., Wu, S., Ma, C., Song, S., Jin, F., Niu, Y. & Tong, W.-M. 2021b. Rna M6a Methylation Regulators Subclassify Luminal Subtype In Breast Cancer. *Frontiers In Oncology,* Volume 10 - 2020.

Yang, Q., Zhu, X., Liu, Y., He, Z., Xu, H., Zheng, H., Huang, Z., Wang, D., Lin, X. & Guo, P. 2024b. Reduced Representative Methylome Profiling Of Cell-Free Dna For Breast Cancer Detection. *Clinical Epigenetics,* 16, 33.

Yang, X., Yang, Y., Qian, X., Xu, X. & Lv, P. 2022. Long Non-Coding Rna Linc01559 Serves As A Competing Endogenous Rna Accelerating Triple-Negative Breast Cancer Progression. *Biomed J,* 45, 512–521.

Yao, J., Xie, M., Ma, X., Song, J., Wang, Y. & Xue, X. 2022. Piwi-Interacting Rnas In Cancer: Biogenesis, Function, And Clinical Significance. *Frontiers In Oncology,* 12, 965684.

Ye, F., Gao, G., Zou, Y., Zheng, S., Zhang, L., Ou, X., Xie, X. & Tang, H. 2019. Circfbxw7 Inhibits Malignant Progression By Sponging Mir-197-3p And Encoding A 185-Aa Protein In Triple-Negative Breast Cancer. *Molecular Therapy Nucleic Acids,* 18, 88–98.

Yedier-Bayram, O., Gokbayrak, B., Kayabolen, A., Aksu, A. C., Cavga, A. D., Cingöz, A., Kala, E. Y., Karabiyik, G., Günsay, R., Esin, B., Morova, T., Uyulur, F., Syed, H., Philpott, M., Cribbs, A. P., Kung, S. H. Y., Lack, N. A., Onder, T. T. & Bagci-Onder, T. 2022. Epikol, A Chromatin-Focused Crispr/Cas9-Based Screening Platform, To Identify Cancer-Specific Epigenetic Vulnerabilities. *Cell Death Dis,* 13, 710.

Yost, K. E., Zhao, Y., Hung, K. L., Zhu, K., Xu, D., Corces, M. R., Shams, S., Louie, B. H., Sarmashghi, S., Sundaram, L., Luebeck, J., Clarke, S., Doane, A. S., Granja, J. M., Choudhry, H., Imielinski, M., Cherniack, A. D., Khurana, E., Bafna, V., Felau, I., Zenklusen, J. C., Laird, P. W., Curtis, C., Cancer Genome Atlas Analysis, N., Greenleaf, W. J. & Chang, H. Y. 2025. Three-Dimensional Genome Landscape Of Primary Human Cancers. *Nat Genet.*

Yu, Y., Hu, J., Wang, W., Lei, H., Xi, Z., Zhang, P., Zhao, E., Lu, C., Chen, H., Liu, C. & Li, L. 2025. Targeting Psmd14 Combined With Arachidonic Acid Induces Synthetic Lethality Via Fads1 M(6)A Modification In Triple-Negative Breast Cancer. *Sci Adv,* 11, Eadr3173.

Yue, X., Wu, W.-Y., Dong, M. & Guo, M. 2021. Lncrna Malat1 Promotes Breast Cancer Progression And Doxorubicin Resistance Via Regulating Mir-570–3p. *Biomedical Journal,* 44, S296-S304.

Yue, Z., Shusheng, J., Hongtao, S., Shu, Z., Lan, H., Qingyuan, Z., Shaoqiang, C. & Yuanxi, H. 2020. Silencing Dscam-As1 Suppresses The Growth And Invasion Of Er-Positive Breast Cancer Cells By Downregulating Both Dctpp1 And Qprt. *Aging (Albany Ny),* 12, 14754.

Záveský, L., Jandáková, E., Weinberger, V., Minář, L., Hanzíková, V., Dušková, D., Faridová, A., Turyna, R., Slanař, O., Hořínek, A. & Kohoutová, M. 2022. Small Non-Coding Rna Profiling In Breast Cancer: Plasma U6 Snrna, Mir-451a And Mir-548b-5p As Novel Diagnostic And Prognostic Biomarkers. *Mol Biol Rep,* 49, 1955–1971.

Záveský, L., Jandáková, E., Weinberger, V., Minář, L., Kohoutová, M., Tefr Faridová, A. & Slanař, O. 2025. The Overexpressed Micrornas Mirs-182, 155, 493, 454, And U6 Snrna And Underexpressed Let-7c, Mir-328, And Mir-451a As Potential Biomarkers In Invasive Breast Cancer And Their Clinicopathological Significance. *Oncology,* 103, 112–127.

Zeng, F., Heng, J., Guo, X., Wang, Y., Wu, W., Tang, L., Chen, M., Wang, S., Deng, H. & Wang, J. 2020. The Novel Tp53 3′-End Methylation Pattern Associated With Its Expression Would Be A Potential Biomarker For Breast Cancer Detection. *Breast Cancer Research And Treatment,* 180, 237–245.

Zeng, K., He, B., Yang, B. B., Xu, T., Chen, X., Xu, M., Liu, X., Sun, H., Pan, Y. & Wang, S. 2018. The Pro-Metastasis Effect Of Circanks1b In Breast Cancer. *Molecular Cancer,* 17, 1–19.

Zhang, B., Gu, Y. & Jiang, G. 2020. Expression And Prognostic Characteristics Of M(6) A Rna Methylation Regulators In Breast Cancer. *Front Genet,* 11, 604597.

Zhang, C., Wang, S., Lu, X., Zhong, W., Tang, Y., Huang, W., Wu, F., Wang, X., Wei, W. & Tang, H. 2024. Pop1 Facilitates Proliferation In Triple-Negative Breast Cancer Via M6a-Dependent Degradation Of Cdkn1a Mrna. *Research (Wash D C),* 7, 0472.

Zhang, G., Gao, L., Zhang, J., Wang, R. & Wei, X. 2023a. Long Non-Coding Rna Ptcsc3 Suppresses Triple-Negative Breast Cancer By Downregulating Long Non-Coding Rna Mir100hg. *Oncol Lett,* 26, 331.

Zhang, H., Gao, M., Zhao, W. & Yu, L. 2023b. The Chromatin Architectural Regulator Snd1 Mediates Metastasis In Triple-Negative Breast Cancer By Promoting Cdh1 Gene Methylation. *Breast Cancer Res,* 25, 129.

Zhang, M.-J., Zhe, Y., Jing, Q., Tian-Hang, L. & Biao, Y. 2022a. Xist As A Valuable Biomarker For Prognosis And Clinical Parameters In Diverse Tumors: A Comprehensive Meta-And Bioinformatics Analysis. *Neoplasma,* 69.

Zhang, P., Zhu, Y., Guo, Q., Li, J., Zhan, X., Yu, H., Xie, N., Tan, H., Lundholm, N. & Garcia-Cuetos, L. 2023c. On The Origin And Evolution Of Rna Editing In Metazoans. *Cell Reports,* 42.

Zhang, Q., Xiu, B., Zhang, L., Chen, M., Chi, W., Li, L., Guo, R., Xue, J., Yang, B., Huang, X., Shao, Z. M., Huang, S., Chi, Y. & Wu, J. 2022b. Immunosuppressive Lncrna Linc00624 Promotes Tumor Progression And Therapy Resistance Through Adar1 Stabilization. *J Immunother Cancer,* 10.

Zhang, X., Shi, L., Sun, H. D., Wang, Z. W., Xu, F., Wei, J. F. & Ding, Q. 2023d. Igf2bp3 Mediates The Mrna Degradation Of Nf1 To Promote Triple-Negative Breast Cancer Progression Via An M6a-Dependent Manner. *Clin Transl Med,* 13, E1427.

Zhang, X., Zhang, H., Li, J., Ma, X., He, Z., Liu, C., Gao, C., Li, H., Wang, X. & Wu, J. 2021a. 6-Lncrna Assessment Model For Monitoring And Prognosis Of Her2-Positive Breast Cancer: Based On Transcriptome Data. *Pathol Oncol Res,* 27, 609083.

Zhang, X., Zhao, D., Yin, Y., Yang, T., You, Z., Li, D., Chen, Y., Jiang, Y., Xu, S., Geng, J., Zhao, Y., Wang, J., Li, H., Tao, J., Lei, S., Jiang, Z., Chen, Z., Yu, S., Fan, J. B. & Pang, D. 2021b. Circulating Cell-Free Dna-Based Methylation Patterns For Breast Cancer Diagnosis. *Npj Breast Cancer,* 7, 106.

Zhang, Y., Chen, S. & Wu, Q. 2025. The M(6)A Reader Ythdf3 Promotes Tnbc Progression By Regulating Cenpi Stabilization. *Front Oncol,* 15, 1546723.

Zhang, Y., Sun, Z., Jia, J., Du, T., Zhang, N., Tang, Y., Fang, Y. & Fang, D. 2021c. Overview Of Histone Modification. *Histone Mutations And Cancer*, 1–16.

Zhang, Y. L., Deng, L., Liao, L., Yang, S. Y., Hu, S. Y., Ning, Y., Zhang, F. L. & Li, D. Q. 2022c. Chromatin Complexes Subunit Bap18 Promotes Triple-Negative Breast Cancer Progression Through Transcriptional Activation Of Oncogene S100a9. *Cell Death Dis,* 13**,** 408.

Zhao, Z., Guo, Y., Liu, Y., Sun, L., Chen, B., Wang, C., Chen, T., Wang, Y., Li, Y., Dong, Q., Ai, L., Wang, R., Gu, Y. & Li, X. 2021. Individualized Lncrna Differential Expression Profile Reveals Heterogeneity Of Breast Cancer. *Oncogene,* 40**,** 4604–4614.

Zheng, F., Du, F., Zhao, J., Wang, X., Si, Y., Jin, P., Qian, H., Xu, B. & Yuan, P. 2021. The Emerging Role Of Rna N6-Methyladenosine Methylation In Breast Cancer. *Biomarker Research,* 9**,** 39.

Zhong, G., Keqiong, W., Jiawei, L., Shuzhe, X., Wei, W. & And Liu, J. 2020. Determination Of Serum Exosomal H19 As A Noninvasive Biomarker For Breast Cancer Diagnosis. *Oncotargets And Therapy,* 13**,** 2563–2571.

Zhou, J., Xie, H., Liu, J., Huang, R., Xiang, Y., Tian, D. & Bian, E. 2023a. Piwi-Interacting Rnas: Critical Roles And Therapeutic Targets In Cancer. *Cancer Letters,* 562**,** 216189.

Zhou, J., Zhu, M., Wang, Q., Deng, Y., Liu, N., Liu, Y. & Liu, Q. 2023b. Serpina3-Ankrd11-Hdac3 Pathway Induced Aromatase Inhibitor Resistance In Breast Cancer Can Be Reversed By Hdac3 Inhibition. *Communications Biology|,* 6**,** 695.

Zhu, D., Yang, J. & Xu, J. 2022. G-Protein-Coupled Estrogen Receptor Enhances The Stemness Of Triple-Negative Breast Cancer Cells And Promotes Malignant Characteristics. *Oncologie (Tech Science Press),* 24.

Zhu, K., Gong, Z., Li, P., Jiang, X., Zeng, Z., Xiong, W. & Yu, J. 2021. A Review Of Linc00673 As A Novel Lncrna For Tumor Regulation.

Zhuang, X., Yin, S., Cheng, J., Sun, W., Fang, Z., Xiang, Y., Peng, E. Y., Yao, Y., Li, Y., He, X., Lu, L., Deng, Y., Huang, H., Cai, G. & Liao, Y. 2024. Mettl14-Mediated M(6)A Modification Enhances Usp22-Erα Axis To Drive Breast Cancer Malignancy. *Pharmacol Res,* 210**,** 107509.

Zocchi, S., David, A., Goodhardt, M. & Garrick, D. 2021. Long Non-Coding Rnas In The Pathophysiology Of Multiple Myeloma New Insights On The Role Of Crnde. *Journal Of Cancer Immunology,* 3**,** 6–12.

Zolota, V., Tzelepi, V., Piperigkou, Z., Kourea, H., Papakonstantinou, E., Argentou, M.-I. & Karamanos, N. K. 2021. Epigenetic Alterations In Triple-Negative Breast Cancer—The Critical Role Of Extracellular Matrix. *Cancers,* 13**,** 713.

Zou, Y., Zheng, S., Xie, X., Ye, F., Hu, X., Tian, Z., Yan, S. M., Yang, L., Kong, Y., Tang, Y., Tian, W., Xie, J., Deng, X., Zeng, Y., Chen, Z. S., Tang, H. & Xie, X. 2022. N6-Methyladenosine Regulated Fgfr4 Attenuates Ferroptotic Cell Death In Recalcitrant Her2-Positive Breast Cancer. *Nat Commun,* 13**,** 2672.

Zuidhof, H. R., Müller, C., Kortman, G., Wardenaar, R., Stepanova, E., Loayza-Puch, F. & Calkhoven, C. F. 2025. The M6a Demethylase Fto Promotes C/Ebpβ-Lip Translation To Perform Oncogenic Functions In Breast Cancer Cells. *Febs J,* 292**,** 2688–2709.

Chapter 4
Unlocking the Epigenetic Landscape of Colorectal Cancer: A Step Toward Epigenetics to Precision

Maha M. Elbrashy, Sally Farouk, Nadia M. Hamdy, Hozaifa Metwally, and Noha G. Bader El Din

Abstract Epigenetic modifications function as central controllers of gene expression in cancer, coordinating crucial cellular activities that trigger the initiation and progression of the tumor, besides their importance in therapeutic response. These modifications can control the gene expression without changing the sequence of DNA. In colorectal cancer (CRC), these alterations involving DNA methylation, histone modifications, chromatin rearrangement, and noncoding ribonucleic acids play a significant role in the pathogenesis of CRC. Abnormal DNA methylation silences the tumor suppressor genes, meanwhile leading to the instability of the genome via reduction of the whole methylation. Specific methylation signatures, such as CpG islands, help in categorizing the subtypes of the tumor and predicting the clinical outcomes. In addition, histone-modifying enzymes, including enhancer of zeste homolog 2 and histone deacetylases, are frequently uncontrolled in cancer, leading to alterations in gene expression. Moreover, small regulatory ribonucleic acids such as microRNA-21 and microRNA-143 contribute to the complex networks that regulate cell survival and growth. Collectively, these epigenetic alterations trigger the transition from benign growth to malignant cancer by continuously suppressing crucial genes. Furthermore, the epigenetic markers can be detected in

---

M. M. Elbrashy (✉)
Biochemistry Department, Biotechnology Research Institute, National Research Centre, Giza, Egypt

S. Farouk · N. G. Bader El Din (✉)
Microbial Biotechnology Department, Biotechnology Research Institute, National Research Centre, Giza, Egypt

N. M. Hamdy
Biochemistry and Molecular Biology Department, Faculty of Pharmacy, Ain Shams University, Cairo, Egypt
e-mail: nadia_hamdy@pharma.asu.edu.eg

H. Metwally
Laboratory of Immune Regulation, The World Premier International Research Center Initiative (WPI) Immunology Frontier Research Center, Osaka University, Osaka, Japan

N. M. Hamdy et al. (eds.), *From (Epi)genetics-to-Precision (E2P)*, Subcellular Biochemistry 114, https://doi.org/10.1007/978-3-032-08530-6_4

blood and stool specimens, offering promising tools for the early detection of cancer. The major obstacle in cancer treatment is the resistance to chemotherapy drugs, which is mainly caused by epigenetic modifications in cancer cells. Therefore, the new therapeutic ways target the modifications that occur in DNA methylation and histone, mostly in conjunction with conventional therapies. As the metabolites produced by the gut microbiome can alter the host epigenetics, they can promote cancer development. Promising technologies help in the concise proofreading of epigenetic marks, and advanced single-cell analysis is paving the way for personalized treatment approaches. This cutting-edge knowledge of epigenetic regulation mechanisms offers new prospects for enhancing diagnosis, prognosis, and targeted therapies in colorectal cancer.

Graphical Abstract

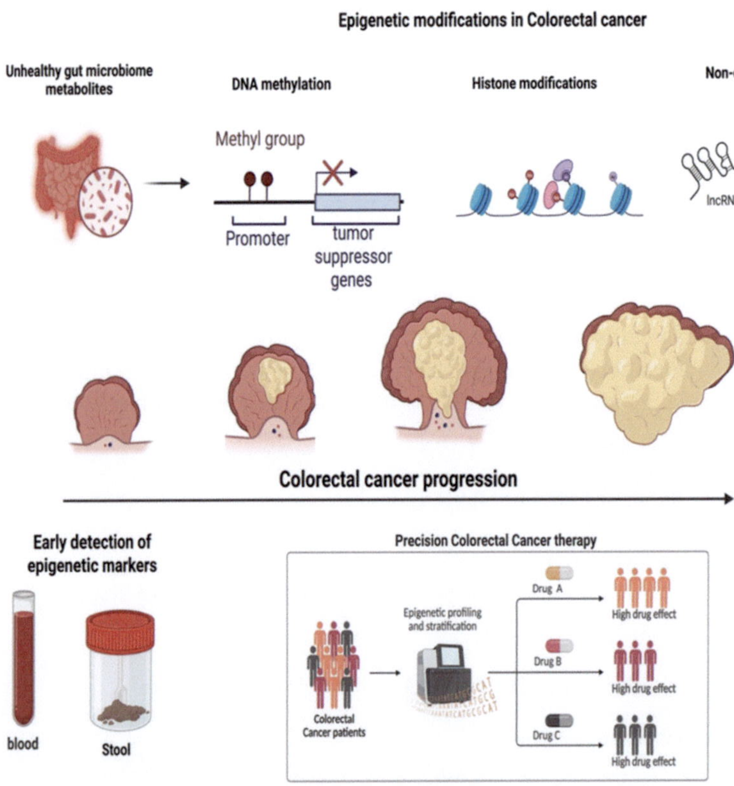

Keywords Colorectal cancer · Epigenetic modifications · Microbiome-epigenome · Epigenetic therapies · Precision treatment

Abbreviations

ADAMTS9-AS2	ADAMTS9 antisense RNA 2
ADAR	Adenosine deaminase RNA
ALDH1A3	Aldehyde dehydrogenase 1 family member A3
ALU	Arthrobacter luteus
APC	Adenomatous polyposis coli
ARF	Alternative reading frame protein
ARID1A	AT-rich interactive domain-containing protein 1A
BACH1	BTB domain and CNC homology 1
BCAT1	Branched chain amino acid transaminase 1
BCL2	B-cell lymphoma-2
BMI1	B lymphoma Mo-MLV insertion region 1 homolog
BMP3	Bone morphogenetic protein 3
BRAF	B-rapidly accelerated fibrosarcoma
BNC1	Basonuclin zinc finger protein 1
BRCA1	Breast cancer type 1
BRG1	Brahma-related gene 1
CA19-9	Carbohydrate antigen 19-9
CACNA1G	Calcium voltage-gated channel subunit alpha1 G
CBP	CREB-binding protein
CCAT	Colon cancer associated transcript
CCAT-1-L	Colon cancer associated transcript 1-long
CCND	Cyclin D
CDH1	Cadherin 1
CDKN2A	Cyclin-dependent kinase inhibitor 2A
CEA	Carcinoembryonic antigen
CHD8	Chromodomain helicase DNA binding protein 8
CHRM3	Cholinergic receptor muscarinic 3
circAGFG1	Circular RNA Arf GAP with FG repeats 1
circ-ERBB2	Circular RNA erb-b2 receptor tyrosine kinase 2
circ-FND3B	Circular RNA fibronectin Type III domain-containing 3B
circPRKCI	Circular RNA protein kinase C iota
CpG	5′@—cytosine (C)—phosphate—guanine (G)—3′
CRNDE	Colorectal neoplasia differentially expressed
CTNNB1	Catenin beta 1
CXCL12	C-X-C motif chemokine ligand 12
DAXX	Death domain-associated protein
DPYD	Dihydropyrimidine dehydrogenase
DNAQ	Epsilon (ε) subunit of DNA polymerase III
DNMT	DNA methyltransferase
DUSP8	Dual-specificity phosphatase 8
EGFR	Epidermal growth factor receptor
eIF4E	Eukaryotic translation initiation factor 4E

EMT	Epithelial-mesenchymal transition
ERK1/2	Extracellular signal-regulated kinase 1/2
EZH2	Enhancer Of zeste 2 polycomb repressive complex 2
Fem1b	Fem-1 homolog B
FEN1	Flap structure-specific endonuclease 1
ELFN1-AS1	ELFN1 antisense RNA 1
FZD5	Frizzled class receptor 5
GADD45B	Growth arrest and DNA damage-inducible protein beta
GAS5	Growth arrest-specific 5
GIHCG	Growth imbalance, hepatocellular carcinoma-associated gene
GNAT	Gcn5-related N-acetyltransferase
GSK-J4	GlaxoSmithKline-J4
GSTP1	Glutathione S-transferase pi 1
H2AX	H2A histone family member X
HDAC	Histone deacetylase
HIF1A-AS2	Hypoxia-inducible factor 1 alpha-antisense RNA 2
HOTAIR	HOX antisense intergenic RNA
HOXD	Homeobox D
IGF2	Insulin-like growth factor 2
IRF	Interferon regulatory factor
INK4	Inhibitors of CDK4
ITGA4	Integrin subunit alpha 4
JMJD3	Jumonji domain-containing protein D3
KCNQ1OT1	KCNQ1 overlapping transcript 1
KDM	Lysine(K) demethylase
KIF18b	Kinesin family member 18B
KRAS	Kirsten rat sarcoma
LINC	long intergenic nonprotein coding RNA
LINE-1	Long interspersed nuclear element-1
LUNAR 1	Lunasin epigenetic regulatory peptide 1
MAL	Myelin and lymphocyte protein
MALAT1	Metastasis-associated lung adenocarcinoma transcript 1
MAPK	Mitogen-activated protein kinase
MEG3	Maternally expressed gene 3
MEIS1	Meis homeobox 1
MET	Mesenchymal epithelial transition
MGMT	Methylated-DNA-protein-cysteine methyltransferase
MIIP-S303	Migration and invasion inhibitory protein Ser303
MINT	Methylated in tumor
miR	MicroRNA
MIR31HG	MIR31 host gene
MLH1	MutL homolog 1
MLL1	Mixed-lineage leukemia 1
MSH	MutS homolog
MSI	Microsatellite instability

MYCN	Myelocytomatosis viral related oncogene, neuroblastoma derived
mTOR	Mechanistic target of rapamycin
MYST	MOZ, Ybf2/Sas3, Sas2, and TIP60 protein family
NEAT1	Nuclear paraspeckle assembly transcript 1
NEUROG1	Neurogenin 1
NDRG4	N-Myc downstream-regulated gene 4
NF-κB	Nuclear factor kappa-light-chain-enhancer of activated B cells
NOTCH	Neurogenic locus notch homolog protein 1
NRF2	Nuclear factor erythroid 2-related factor 2
NY-ESO-1	New York esophageal squamous carcinoma 1
OSMR	Oncostatin M receptor
PARPBP	PARP1 binding protein
PCA3	Prostate cancer antigen 3
PD-1	Programmed cell death protein 1
PDCD4	Programmed cell death 4
PD-L1	Programmed death-ligand 1
PI3K	Phosphatidylinositol 3-kinase
PIK3CA	Phosphatidylinositol-4,5-bisphosphate 3-kinase catalytic subunitalpha
PKCε	Protein kinase C, epsilon
PRUNE2	Prune homolog 2
PTEN	Phosphatase and tensin homolog
PVT1	Plasmacytoma variant translocation 1
RAB3IP	RAB3A interacting protein
RAS	Rat sarcoma
RB1	Retinoblastoma
RNF43	Ring finger protein 43
ROR	RAR-related orphan receptor
RSPO2	R-Spondin 2
RTK	Receptor tyrosine kinase
RUNX3	Runt-related transcription factor3
SCARNA2	Small cajal body-specific RNA 2
SDC2	Syndecan 2
SEDT2	SET domain-containing 2
SFRP	Secreted frizzled-related protein
SIRT1	Sirtuin 1
SLC5A8	Solute carrier family 5 member 8
SLIT	Slit guidance ligand
SMAD	Suppressor of mothers against decapentaplegic
SMARCA2	SWI/SNF related BAF chromatin remodeling complex subunit ATPase 2
SNHG6	Small nucleolar RNA host gene 6
SOCS1	Suppressor of cytokine signaling 1
SOX17	SRY-box transcription factor 17

SST	Somatostatin
SUV39H1	SUV39H1 histone lysine methyltransferase
SWI/SNF	Switch/sucrose non-fermenting
TFPI2	Tissue factor pathway inhibitor 2
TGFβ	Tumor growth factor beta
TIMP3	Tissue inhibitor of metalloproteinases 3
TMEFF2	Transmembrane protein with EGF like and two follistatin like domains 2
TP53	Tumor protein p53
TUG1	Taurine upregulated gene 1
UCA1	Urothelial cancer associated 1
WIF1	Wnt inhibitory factor 1
ZEB1	Zinc finger E-box binding homeobox 1

Introduction: Fundamentals of Epigenetics in Cancer Biology

Epigenetics Overview

Epigenetics is an inherited biological process that alters gene expression without affecting the DNA sequence; rather, it forms covalent and noncovalent changes in the DNA and the chromatin structure (Eldash et al. 2023; Hamdy et al. 2025). The term "epigenetics" originates from the Greek. The prefix "epi-" means "on" or "above," suggesting that these alterations affect the top of the genetic code or are added to it. Epigenetic changes can cause upregulation or downregulation of genes, thus managing the cells' protein production and their functions. Historically, the types of epigenetic modifications have been identified over time, starting from Waddington in 1942, who identified that gene expression could be heritably changed without modifying the DNA sequence. In the 1970s, DNA methylation was discovered with its function in gene silencing. Then, histone modifications were identified in the 1980s and 1990s, connecting epigenetics to cancer and growth. The twenty-first century unraveled the noncoding RNAs and the concept of epigenome (Berger et al. 2009).

Over time, epigenetic mechanisms have been categorized at the molecular level into DNA methylation, histone alterations, chromatin remodeling, and noncoding RNAs, which affect gene expression by altering the chromatin structure and gene availability without modifying the DNA sequence. This complexity highlights the dynamic nature of the genome, where the same genetic code can end with various consequences depending on epigenetic conditions. DNA methylation is the methylation of cytosine residues (at CpG sites) by DNA methyltransferase, causing transcriptional repression by hindering the transcription factor from binding or recruiting silencing complexes. Histone modifications are the posttranslational chemical changes, including acetylation, methylation, and phosphorylation of histone tails, thereby altering the chromatin structure and gene accessibility. Histone acetylation commonly stimulates transcription; however, methylation can either stimulate or suppress, relying on the unique residue and degree of modification. Chromatin remodeling is the ATP-dependent shifting or rearrangement of nucleosomes by specific protein complexes, like the SWI/SNF compound, to dynamically control DNA access. Noncoding RNA regulation includes microRNAs (miRNAs) that bind mRNA to block translation, long noncoding RNAs (lncRNAs) that shield epigenetic regulators to specific genomic loci, and circular RNAs (circRNAs) that make continuous loops, soaking up miRNAs and preventing them from silencing crucial genes (Hamdy et al. 2024a; Liu et al. 2023; Umapathy et al. 2024).

A cell uses flexible epigenetic machinery to activate or suppress specific genes in various cell types; meanwhile, the cells have an identical genotype. This machinery is essential in cellular differentiation, thereby keeping the normal development. Furthermore, environmental parameters, including diet habits, toxins, or lifestyle manners, can cause epigenetic changes, which can sometimes be heritable. Abnormalities in the regulation of the epigenetic machinery can cause different diseases, such as cancer, metabolic syndromes, and neurodegenerative diseases (Hernández-Contreras et al. 2024; Ling 2020).

Cancer as a Genetic and Epigenetic Disease

Cancer is not only considered a genetic disease, but also growing evidence throughout the previous decades suggests that cancer is additionally an epigenetic disease. Epigenetics has recently been considered as a cancer-related factor and one of the hallmarks of cancer. Remarkably, uncontrolled epigenomic machinery has been proposed by previous researchers to possibly provoke all the common cancer characteristics. The development of cancer is caused by interactions between epigenetics and genetics, yet the underlying mechanisms and genetic alterations related to cancer are still under investigation (Hamdy et al. 2024b). However, it is suggested that epigenetic modifications can be initiated by mutations in genes controlling epigenetic processes or cell metabolism (Maleszewska et al. 2018).

Genetic Origin of Cancer

Cancer can be caused by genetic mutations or modifications that disturb the typical regulatory signaling pathways controlling cell division, growth, and death. These abnormalities allow tumor cells to grow irregularly, penetrate nearby tissues, and metastasize to faraway locations in the body. Subsequently, cancer can trigger substantial morbidity and mortality, affecting not only individuals but also families, societies, and whole healthcare organizations. Genetic mutations in the DNA sequence either stimulate oncogenes or inactivate tumor suppressor genes.

Oncogenes can be stimulated by different mechanisms, including gene amplifications, point mutations, or chromosomal reorganizations. To illustrate, gene amplification, like MYCN augmentation in solid tumors, caused an increase in the gene copy counts and translation of protein. Point mutations in RAS or BRAF oncogenes can cause persistent activation of their downstream regulators, leading to abnormal cell proliferation and tumor development (Khan et al. 2019). Chromosomal reorganizations, like translocations, are common in hematopoietic cancers and can stimulate proto-oncogenes. For instance, rearranged during transfection (RET) and tropomyosin receptor kinase (TRK) genes are stimulated by reorganizations in papillary thyroid tumors (Anderson et al. 1992).

However, tumor suppressor genes (TSGs) are considered as genome guardians, as they regulate cell cycle development, DNA repair, and apoptosis. Inhibition or loss of tumor suppressor role, likely rising from genetic mutations or deletions, which can induce hyperproliferation and contribute to tumorigenesis. For instance, genetic mutations in TP53, RB1, and PTEN tumor suppressor genes are usually linked with different cancer types (Hamid et al. 2019). In cancer genomes, it was shown that there are large deletions in the chromosome that can mitigate the activity of many genes simultaneously, resulting in tumorigenesis beyond the loss of a single TSG. For example, deletion of mouse chromosome 11B3 has more effect on lymphoma and leukemia progression than Trp53 deletion only (Y. Liu et al. 2016). The mutual suppression of numerous TSGs in a deleted area can synergistically enhance tumor development and is linked with poor survival in hepatocellular carcinoma (HCC) patients (Xue et al. 2012). Thus, the selective advantage of chromosomal deletions in cancer imitates the accumulative effect of losing many TSGs, rather than the deactivation of a single gene.

Epigenetic Origin of Cancer

DNA Methylation

DNA methylation mechanisms related to cancer include global hypomethylation and regional hypermethylation. Global hypomethylation activates proto-oncogenes and causes chromosomal instability, while regional hypermethylation silences tumor suppressor genes. Of note, the prevalence and site of allele-specific DNA methylation alterations are normally conserved all over various species. In gastrointestinal tumors, hypermethylation disrupts genes involved in cell cycle mediators, DNA repair, and cell death (Suzuki et al. 2006). Brain tumors demonstrate MGMT hypermethylation, whereas prostate cancers exhibit GSTP1 hypermethylation (Esteller 2005). DNA methyltransferases DNMT1 and DNMT3B, which stimulate DNA methylation, are overexpressed in different cancer types. Genome-wide DNA methylation profiles have shown cancer-unique patterns and progressive alterations throughout the tumor evolution (Vidal et al. 2017).

Histone Modifications

Histone modifications can cause cancer development via epigenetic regulation of gene expression. These alterations are regulated by different enzymes and can affect oncogenes and tumor suppressor genes. For example, loss of H3K27 methylation enhances glial cell proliferation and silences tumor suppressor genes in diffuse midline glioma (Day et al. 2022). Whereas hypermethylation of H3K27 promotes tumor vascularization and cell migration in epithelial ovarian cancer (Day et al. 2022). Actually, H3K27 promotes VEGF expression, a key epigenetic mechanism controlling angiogenesis in both physiological and pathological conditions (Salum et al. 2017). Moreover, dysregulation of histone-modifying enzymes is a major cause of triggering and the progression of cancer. As histone modifications are reversible, they are potential targets to improve anticancer drugs, with multitarget drugs affecting numerous molecular targets being among the most effective approaches. Intriguingly, connections between histone demethylase (HDM) and histone methyltransferase (HMT) expression levels have been detected, with unique HDM/HMT gene expression signatures related to normal and cancerous states. Thus, these aligned alterations in histone-modifying enzymes may underscore global epigenetic changes in cancer (Islam et al. 2011).

Chromatin Remodeling

Chromatin remodeling proteins control gene expression and maintain genomic stability. Mutations in these proteins have been involved in different cancer types. Fifty-seven percent of ovarian clear cell carcinoma tumors harbor deactivating mutations in ARID1A, a part of the SWI/SNF chromatin remodeler compound. Likewise, mutations in SWI/SNF complex genes have been recognized in rhabdoid tumors, hepatocellular carcinoma, gastric adenocarcinoma, renal cell carcinoma, lung carcinoma, and malignant sarcoma (Oike et al. 2013). These results suggest that the SWI/SNF complex has a tumor-suppressive role. Chromatin remodeling proteins can change nucleosome architecture through ATP-dependent complexes or histone modifications. Dysregulations in these proteins have been detected in almost all human tumor types, affecting several cellular activities, for example, controlling cell cycle, cell diversity, and genome integrity (C. Zhang et al. 2016).

Noncoding RNAs (ncRNAs)

They have critical roles in cancer development through epigenetic modifications. On one hand, miRNAs can control chromatin structure and gene expression by targeting the main epigenetic enzymes such as DNA methyltransferases (DNMTs) and histone deacetylases (HDACs). MiR-26b, miR-29c, and miR-148b regulate the production of DNMT3B in various breast cancers (Sandhu et al. 2012; Shenawy et al. 2025). Also, deficiency or inhibition of miR-145 is reported to cause abnormal overexpression of HDAC2 and stimulate HCC tumorigenesis (Noh et al. 2013). On the other hand, miRNA dysregulation is caused by DNA methylation and histone alterations in various cancers. EZH2, a histone methyltransferase, is upregulated in HCC and causes metastasis by epigenetically silencing numerous miRNAs, like miR-139-5p, miR-125b, and miR-101, acting as tumor suppressors (Au et al. 2012; Yamaguchi and Hung 2014). This reciprocal interface among miRNAs and epigenetics is involved in abnormal miRNA expression in numerous cancer types.

Long noncoding RNAs (lncRNAs) regulate gene expression by connecting with epigenetic modelers and transcription factors. According to the orientation and genomic location of lncRNAs, they are categorized into sense, antisense, bidirectional, intergenic, and intronic lncRNAs. lncRNAs act as chromatin remodelers and transcriptional activators or repressors and have a role in RNA degradation (El Awady et al., 2006a). Also, they serve as miRNA sponges, thereby influencing the efficacy of translation. Abnormal expression of lncRNAs helps in developing cancer and in its progression. Mechanically, lncRNAs mobilize DNMT3A and DNMT3B to tumorigenic genes and impact the methylation status of their promoters and controlling areas (Yang et al. 2022). lncRNA *ADAMTS9- AS2* reduces the propagation and metastasis of esophageal cancer by recruiting DNMT3s to the promoter of cadherin 3 (CpG islands) (Liu et al. 2020b). Epigenetic mechanisms of miRNAs and lncRNAs are not separated; they frequently collaborate to control gene expression in an organized way. HOTAIR is detected to enhance cervical cancer development through controlling BCL2 via affecting miR-143-3p (Liu et al. 2018b).

In addition, lncRNA *ROR* mobilizes MLL1, a histone remodeler, to the *TIMP3* gene and increases its expression, resulting in proliferation and progression of breast cancer cells (Hu et al. 2021). lncRNAs were reported to recruit chromatin remodeling protein (SWI/SNF complex) to unique genomic areas, thus stimulating the transcription of carcinogenic-related genes. Such as lncRNA *UCA1*, which connects to the BRG1 part of the SWI/SNF compound, ending with suppressing its binding to the promoter of *p21* and stimulating gallbladder tumor (Wang et al. 2014).

lncRNAs directly recruit RNA-modifying compounds, comprising RNA methyltransferases and RNA proofreading compounds, to control RNA alterations and aberrant transforming gene expression, ending up with tumor development. lncRNA *HIF1A-AS2* mobilizes the ADAR enzyme to expedite ADAR1-dependent A-to-I checking, triggering the progression and metastasis of breast cancer (Ma et al. 2019; Picardi et al. 2014). Likewise, lncRNA PCA3 interacts with the ADAR enzyme, enhancing A-to-I correction of *PRUNE2* pre-mRNA and controlling *PRUNE2* gene expression*, acting as a* tumor suppressor, thus augmenting prostate cancer's progression (Salameh et al. 2015).

Circular RNAs (circRNAs) are epigenetic regulators with a closed-loop structure, engaged in different cellular activities, and are plentifully expressed in acute myeloid leukemia (AML) (Issah et al. 2021). In cancer, circRNAs can be uncontrolled via changes in cis-regulatory elements and trans-acting proteins that influence their biogenesis. They act as miRNA sponges, modify the Wnt signaling, and affect the transition of epithelial into mesenchymal cells in cancer progression. For example, circ_0067934 (also identified as circPRKCI) enhances tumor growth and metastasis in hepatocellular carcinoma by working as a sponge for miR-1324, which usually targets the 3′-untranslated region (UTR) of FZD5 mRNA. By segregating miR-1324, circ_0067934 enhances FZD5 expression, thus stimulating the Wnt/β-catenin signaling pathway, accelerating EMT, and augmenting the proliferation and metastasis of HCC cells (Zhu et al. 2018).

Colorectal Cancer and Epigenetics

DNA Methylation Abnormalities in Colorectal Tumorigenesis

Colorectal cancer (CRC) is a life-threatening disease affecting more than 750,000 people per year globally (Brenner et al. 2014). In spite of advances in CRC treatment, including chemotherapy and targeted therapies, the five-year survival rate for metastatic CRC remains below 15%, underscoring the urgent demand for innovative treatment policies (Brenner et al. 2014). One of the characteristics of CRC is aberrant DNA methylations comprising regional hypermethylation and global hypomethylation of gene promoters, as shown in Fig. 4.1a. Remarkably, DNA methylation epigenetic changes often occur more than genetic mutations in CRC, with more than 600 candidate hypermethylated genes recognized (Ng and Yu 2015).

Hypermethylation typically stops tumor suppressor genes and disrupts major signaling pathways, including Wnt, TP53, and TGFβ, hence accelerating cancer onset and its development (Lao and Grady 2011). By contrast, global hypomethylation causes genomic instability through stimulating tandem repeats such as LINE-1 and ALU, which can enhance chromosomal reorganizations and mutations (Gaudet et al. 2003; Rodriguez et al. 2006). Hypomethylation can stimulate numerous other tumorigenic signaling pathways that drive cancer expansion and dissemination in CRC. Remarkably, irregular hypomethylation of proto-oncogenes and other molecules activates the MAPK pathway and the PI3K/Akt pathway. Stimulation of these pathways enhances elevated mitotic activity, survival signaling, and invasive potential. The MAPK pathway transfers external stimuli from growth factor receptors to the nucleus, thus boosting cell cycle progression. The PI3K/Akt pathway is included in supporting cell proliferation and survival through effector signaling cascades (Steelman et al. 2011).

Regional Hypermethylation and CRC

Hypermethylation of CpG islands, mainly in the promoter areas of tumor suppressor genes, hinders gene accessibility and stops its expression. This is leading to loss of crucial regulatory mediators that are responsible for controlling cell proliferation, cell death, and DNA repair, ending up with the initiation of cancer and its development (Tse et al. 2017). This process is catalyzed by DNA methyltransferase enzymes

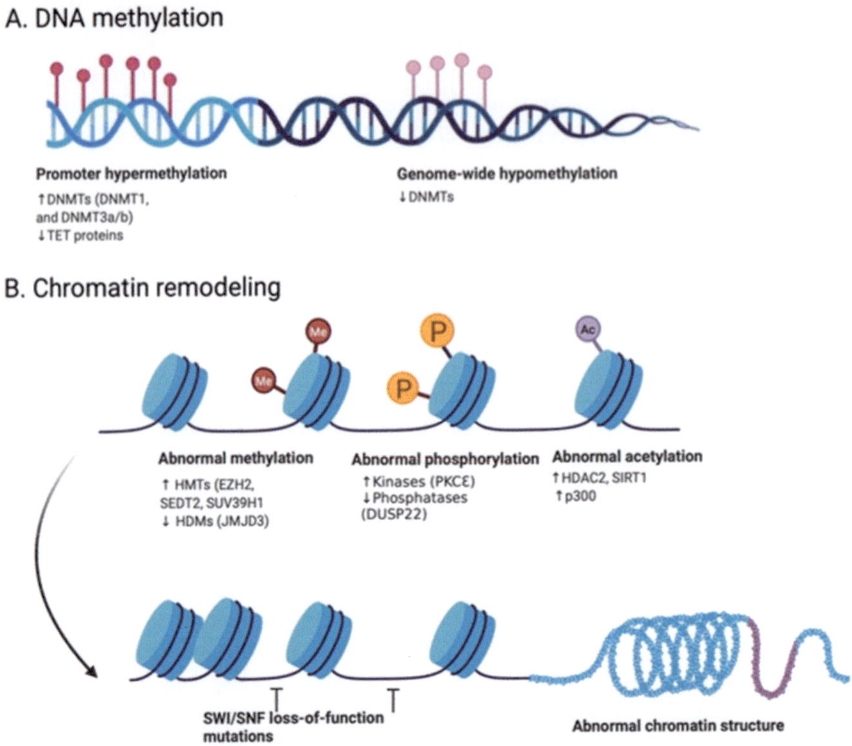

Fig. 4.1 Colorectal cancer epigenetics. (**a**) DNA methylation: Representative of promoter hypermethylation, which consists of high activity of DNA methyltransferases (DNMT1, DNMT3a/b) and low levels of ten-eleven translocation (TET) proteins promoting silence of the gene. Genome-wide hypomethylation is, conversely, found to be associated with low levels of DNMTs and has led to genomic instability. (**b**) Chromatin remodeling: Demonstrates altered histone marks, including elevated methylation (histone methyltransferases [HMTs]: EZH2, SETD2, SUV39H1, and reduced HDMs; JMJD3) and aberrant phosphorylation (elevated kinases: PKCe and reduced phosphatases: DUSP22) and aberrant acetylation (histone acetyltransferases [HDACs]: HDAC2, SIRT1, p300). Loss of the SWI/SNF complex causes irregular chromatin structure, which remains to facilitate distorted gene expression and cancer development

such as DNMT1 and DNMT3a/b, which produce 5-methylcytosine (5mc) by adding a methyl group to the 5-cytosine. Colorectal cancer displays a unique epigenetic phenotype, called CpG island methylator phenotype (CIMP) (Toyota et al. 1999a). It is a broad hypermethylation in numerous CpG-rich regions throughout gene promoters. Toyota et al. first introduced the term "CIMP" in 1999. They recognized a subgroup of CRCs demonstrating recurrent methylation of the Type C site, present only in CRCs, not within normal cells (Toyota et al. 1999b). They also defined a group of methylation-prone loci that describe these tumors, including MLH1, CDKN2A, MINT1, MINT2, and MINT31 (Toyota et al. 1999b). An alternative panel of CIMP markers involving promoter methylation of *CACNA1G, IGF2,*

NEUROG1, RUNX3, and *SOCS1* was later suggested (Weisenberger et al. 2006). However, the classic CIMP panel exhibited powerful prognostic value, mainly in CIMP-positive/microsatellite instability (MSI)-negative CRCs, which were related to the poorest clinical results. At the whole-genome level, CIMP can be additionally grouped into CIMP-high or CIMP-low. CIMP-high cancers are depicted by MSI, *BRAF* mutations, proximal location, serrated histology, and rare *TP53* mutations, whereas CIMP-low cases more frequently harbor *KRAS* mutations (Hinoue et al. 2012). Due to its potent correlations with verified prognostic parameters such as MSI and *BRAF* status, deciding the CIMPs' independent prognostic importance remains challenging (Bae et al. 2013). However, evaluating CIMP status improves molecular subtyping and aids in guiding the prognosis and personalized treatment strategies (Rhee et al. 2017). Consequently, CIMP marker analysis is gradually utilized in clinical practice to recognize cancer subtypes with different biological activities, strengthening its function in precision medicine.

Positively, CIMP-cancers exhibit silence in multiple genes. For example, the MLH1 gene is responsible for DNA repair, while the CDKN2A gene regulates cell cycle mediators. In 10–15% of sporadic CRC cases, MLH1 is methylated and frequently happens independently of deactivating mutations in the MLH1 gene code (Kane et al. 1997). The methylation leads to MLH1 inactivation and induces the hypermutator phenotype, which provokes the risk of tumorigenesis by accumulating genetic alterations that disrupt critical signaling pathways involved in CRC development. For example, hypermutated CRCs with MLH1 silencing demonstrate recurrent BRAF V600E mutations and diminished Wnt signaling. Moreover, these tumors show increased mutations in genes such as RNF43, ARID1A, and PIK3CA, influencing PI3K, RTK-RAS, TGFβ, and DNA break repair pathways (Wang et al. 2020b). Additionally, BRAF V600E activates the transcriptional repressor MAFG, which recruits a corepressor complex including BACH1, CHD8, and DNMT3B to stop the expression of MLH1 and other CIMP genes (Fang et al. 2014).

Moreover, *CDKN2A* is a tumor suppressor gene in different human cancers involving CRC, and its stimulation suppresses carcinogenesis by inducing cell growth arrest and senescence. *CDKN2A* promoter methylation leads to its silencing and dysfunction, which causes irregular cell growth and the initiation of a tumor. It is reported that a hypermethylated *CDKN2A* promoter can predict a shorter survival rate in CRC patients; thus, the prognostic value of *CDKN2A* hypermethylation in CRC has been widely used due to its remarkable relevance to clinical practice (Liang et al. 1999). However, it is recommended that *CDKN2A* hypermethylation could be a predictive parameter of poor prognosis in CRC patients, especially in European and Asian areas, and might expect cancer invasion and metastasis.

Beyond the CpG island methylator phenotype, CRC can also be caused by chromosomal instability (CIN) or microsatellite instability (MSI). CIN is common in sporadic CRC, marked by aneuploidy, allelic imbalances, and frequent mutations in key tumor suppressors such as *APC*, *TP53*, *SMAD2*, and *SMAD4* (Pino and Chung 2010). Impaired chromosome segregation, dysfunctional DNA damage repair, and telomere dysregulation can cause CIN; however, the full spectrum of contributing genes is still incompletely comprehended. Defective mismatch repair, promoting errors in repetitive DNA sequences and subsequent mutations in genes regulating cell proliferation and apoptosis, leads to MSI (Tariq and Ghias 2016).

Global Hypomethylation and CRC

CRC, at the whole-genome level, has 10–40% lower levels of absolute methylation than in normal colonic tissue (Ehrlich 2002; Feinberg and Vogelstein 1983). Ten-eleven translocation methylcytosine dioxygenases (TET enzymes) are responsible for demethylation, oxidizing 5-methylcytosine (5mC) to 5-hydroxymethylcytosine (5hmC), 5-formylcytosine (5fC), and 5-carboxylcytosine (5caC). The intermediates 5fC and 5caC are then eliminated either passively by DNA replication dilution or actively by thymine DNA glycosylase (TDG)-initiated base excision repair (BER). DNA hypomethylation is a common aberrant epigenetic change that mainly influences tandem repeats, especially long interspersed nuclear elements (LINEs), encompassing approximately 17% of the genome in humans, and short interspersed nuclear elements (SINEs) (Debernardi et al. 2021). These elements are abundantly distributed throughout the genome. LINE-1 elements are autonomous retrotransposons responsible for mobilization, though SINEs, such as ALU elements, are non-autonomous and depend on the LINE-1 machine to transpose. With further elaboration, autonomous retrotransposons are genetic elements that can replicate by themselves and insert into new genomic sites. They have complete machinery comprising transcription, translation, reverse transcription, and integration, needed for their life cycle. Conversely, nonautonomous retrotransposons do not have this machinery and rely on proteins from autonomous elements for mobility (Schulman 2012).

These elements in a healthy state are greatly methylated to keep the silence of their transcription, thus preventing genomic disruption. However, in CRC, global hypomethylation happens early in the adenoma-to-carcinoma sequence and is predominantly pronounced in microsatellite stable (MSS) tumors. Therefore, this epigenetic shift reactivates these elements, ending with genomic instability, a characteristic of cancer. Mechanistically, LINE-1 elements encode proteins, involving a reverse transcriptase and an endonuclease, that facilitate their integration into new genomic sites via retrotransposition. This process leads to insertional mutagenesis through disturbing coding sequences or regulatory areas, ending with the inhibition of tumor suppressor genes or stimulation of oncogenes (Elbrashy et al. 2024; Morse et al. 1988). Although ALU elements are shorter and nonautonomous, they can still influence gene expression by affecting splicing patterns, transcription initiation, or the assembly of noncoding RNAs. Furthermore, the reactivation of these elements can produce RNA or protein, which triggers immunogenic or mutagenic responses, further worsening the cancer phenotype by disrupting immune surveillance and enhancing genetic damage (Roulois et al. 2015).

Beyond direct effects of the repetitive element hypomethylation on genetic disruption, reactivated LINE-1 elements can promote colorectal cancer (CRC)-linked signaling pathways such as receptor tyrosine kinase signaling, Wnt/β-catenin, and PI3K/AKT via various mechanisms. For example, low methylation of unique LINE-1 elements in the intronic areas of proto-oncogenes like *MET*, *RAB3IP*, and *CHRM3* causes their unintentional stimulation, enhancing oncogenic signaling in CRC metastasis (Hur et al. 2014). Long noncoding RNAs driven by LINE-1 connect with β-catenin and trigger the Wnt/β-catenin signaling pathway, which activates epithelial-mesenchymal transition (Li et al. 2024). Also, LINE-1 insertions in somatic cells are recognized in main CRC genes involving *APC* and *PIK3CA*, which disrupts their function and causes abnormal activation of Wnt and PI3K/AKT pathways. Furthermore, p53-mutated CRC cases have high expression of LINE-1, which relates inversely with tumor-infiltrating CD3+ T lymphocytes, recommending a role in changing tumor immunogenicity and immune-linked signaling pathways (Nguyen et al. 2020).

Assessment of LINE-1 methylation status is of clinical importance as it serves as a diagnostic and prognostic tool needed for CRC patients' stratification. Patients with hypo-LINE-1-methylation status display remarkably shorter overall survival and higher tumor recurrence rates in comparison with those having higher methylation levels, confirming its significance in predicting clinical outcomes. Moreover, LINE-1 hypomethylation is linked with the localization of colorectal cancer. It indicates a worse prognosis, especially in the proximal region of colon cancers, and connects with higher mortality of colorectal cancer, underscoring the importance of cancer site in explaining LINE-1 methylation status (Mima et al. 2016b). It has been reported that LINE-1 methylation status in a large cohort study is a prognostic biomarker, as the 5-year survival probability of colon cancer progressively declined when LINE-1 methylation levels decreased. Moreover, LINE-1 methylation status may distinguish aggressive carcinomas in colorectal cancers with high microsatellite instability, reinforcing its clinical efficiency. Low LINE-1 methylation in the late stage of CRC is related to worse survival and lower opportunities to stay free from cancer. Also, it could predict if patients respond effectively to chemotherapy such as FOLFOX (Kaneko et al. 2016).

Histone Modifications and Their Role in Colorectal Cancer

Along with DNA methylation, CRC progression is also marked by alterations in histones that can change the structure of chromatin and affect gene expression. There are four core histones (H3, H4, H2A, and H2B) that constitute the basic unit of chromatin, the nucleosome. Histone modifications, as shown in Fig. 4.1b, create separate domains in the genome, one of which is called euchromatin, where DNA is accessible. Heterochromatin means inaccessible chromatin. The pathogenesis of CRC shows prominent changes in the histone acetylation and methylation (Nakazawa et al. 2012).

Histone Acetylation and CRC

Histone acetylation is a reversible process where the lysine residues in the histone are acetylated by histone acetyltransferases (HATs), leading to neutralizing the basic charge of lysine and unfolding the chromatin and activating transcription. HATs include GNAT, MYST, and CBP/p300 families. p300 Acetylates H3K27, and acetylated H3K27 is reported to be remarkably increased in CRC specimens in comparison with controls (Karczmarski et al. 2014). In another study, acetylated H3K12ac and H3K18ac were considerably augmented in mature colonic cancer cells but decreased in less differentiated colonic cancer cells (Ashktorab et al. 2009). Moreover, numerous research demonstrated that there are positive correlations between global histone acetylation and tumor grade, lymph metastasis, survival rate, prognosis, and cancer relapse. For example, acetylated histone H3 (H3Ac) was highly expressed and related to poor overall survival in colorectal cancer tissues (Hashimoto et al. 2013). Additionally, Benard et al. showed that H3K56ac and H4K16ac were significantly expressed in the nucleus and positively correlated with the improved survival rate of CRC, where they have less risk of tumor relapse (Benard et al. 2015), whereas HDACs reverse histone acetylation, causing transcription inhibition. They include three classes of HDACs, with class III being dependent on NAD and belonging to the Sir family. HDAC2 is highly expressed in colon adenocarcinoma in comparison with CRC patients having adenoma, suggesting that HDAC2 expression is strongly associated with the transition of adenoma to adenocarcinoma. Also, they showed that HDAC1 is elevated in 36.4%, HDAC2 in 57.9%, and HDAC3 in 72.9% of tumor samples. These findings indicate that high HDAC levels are linked with poorer survival in CRC patients (Ashktorab et al. 2009). Another study reported that the nuclear expression level of HDAC2 is highly increased in 81.9% of colorectal carcinomas, in comparison with 62.1% in patients with adenomas and 53.1% in normal tissues, underscoring its possible function in cancer progression (Ishihama et al. 2007). Furthermore, SIRT1 (class III HDAC) is highly expressed in 40% of 485 CRC cases, especially those who are CIMP-high (Strahl and Allis 2000).

Histone Methylation and CRC

Histone methylation is another reversible posttranslational alteration that happens on both lysine and arginine residues in the histones' *N*-terminal tail. This process is done by histone methyltransferases (HMTases) and histone demethylases (HDMs), which methylate and demethylate histones, respectively (Islam et al. 2011). Histone methylation has different biological roles in heterochromatin development, DNA breaks, inhibition of the X chromosome, and control of gene transcription. These aberrant cellular roles of histone methylation control the etiopathogenesis of cancers. Abnormal expression of histone methylases and demethylases has been defined in CRC. Remarkably, the expression level of SUV39H1, which is an H3K9 methyltransferase, is increased, leading to CRC development and poor survival in mice (Yokoyama et al. 2013). Furthermore, histone methyltransferase SEDT2 was reported to be reduced in CRC, facilitating its development through influencing alternative splicing (Kornblihtt 2017). While histone methyltransferase EZH2 was elevated in CRC and associated with shorter survival and progressive stage, suggesting that it can be used as a marker of CRC patients' clinical results. Mechanistically, the tumorigenic role of EZH2 is due to its ability to silence anti-oncogenes, comprising E-cadherin, *p27, INK4B-ARF-INK4A*, $p57^{KIP2}$ (Gonzalez et al. 2009), and *BRCA1* (Yu et al. 2007). *MLL2, an* HMTase, is expressed highly in less differentiated and metastatic colon carcinoma cell lines and in CRC tissues in comparison with nearby regular colon epithelial cells (Kang et al. 2007).

Additionally, the histone H3 (H3K9me2) dimethylated lysine 9 level was raised in CRC adenocarcinoma, especially in the nucleus, more than in adenoma (Nakazawa et al. 2012). Additionally, histone 3 is reported to be trimethylated at lysine 9, particularly in invasive areas of CRC tissues, and is linked with lymph node metastasis (Yokoyama et al. 2013). However, the plasma level of H3K9me3 and H4K20me3 was diminished in CRC patients (Gezer et al. 2013). Intriguingly, H3K4me2 is significantly related to CRCs' tumor invasion and late stage as manifested by its high AS (Allred scoring system) score (Hashimoto et al. 2013). Moreover, H3K4me2 is lowly expressed in CRC patients and might act as a prognostic factor (Tamagawa et al. 2012). Also, H4K20me2 is lowly expressed in CRC cells and considered a popular characteristic of CRC (Fraga et al. 2005). In addition, H3K27me3 is shown by immunohistochemical staining (IHC) to be elevated in comparison with controls, and it is thoroughly related to improved survival and delayed disease relapse (Benard et al. 2014b).

Besides, demethylase JMJD3 was lowly expressed in the nucleus of healthy colorectal tissues compared to CRC tissues, thus expecting a poor prognosis in CRC patients (Tokunaga et al. 2016). While lysine-specific demethylase (LSD1) is highly expressed in CRC tissues and strongly associated with progressive TNM grades and metastasis (Ding et al. 2013).

Histone Phosphorylation and CRC

To keep chromatin stability, it is critical to maintain the balance between kinase and phosphatase. Thus, imbalanced phosphorylation could lead to various human diseases, involving colon cancer (Qin et al., 2023). However, there are few studies have investigated the correlation between histone phosphorylation and CRC. For instance, dual-specificity phosphatase 22 (DUSP22) is lowly expressed in CRC samples and displayed poor survival outcome in stage IV CRC patients (Yu et al., 2015). While H2AX histone (p-H2AX) is highly phosphorylated in CRC tissues and serves as an indicator of advanced tumor growth and poor CRC survival (Lee et al., 2015). Additionally, PKCε alters the phosphorylation of MLLP-S303, where its expression level is related to CRCs' progression and prognosis (Chen et al., 2017). It has been found that T-LAK cell-originated protein kinase (TOPK) is highly phosphorylated at Y74 and Y272 residues, which accelerated the tumorigenesis of colon cancer (Xiao et al., 2016).

The Role of Noncoding RNAs in Epigenetic Regulation in CRC

MiRNAs are playing a vital role in the initiation, development, and progression of CRC through modifying oncogenic cellular pathways. Expression of certain miRNA panels is potentially important for CRC diagnosis at the early stages, as they can discriminate CRC from other diseases linked to the colon (Bader El Din et al. 2020). MiRNAs include *onco*miRs (oncogenic miRNAs), which suppress expression of tumor suppressor genes, or *ts*miRs (tumor-suppressing miRNAs), which hinder the expression of oncogenes, as shown in Fig. 4.2a (Bader El Din et al. 2024; Farouk et al. 2020). For example, miR-21 acts as an oncogene by inhibiting PDCD4 and PTEN (tumor suppressors), thus enhancing tumor growth and progression. Additionally, miR-21 enhances tumorigenesis in CRC via promoting the translocation of β-catenin into the nucleus (Shi et al. 2016). However, miR-143 serves as a tumor suppressor miRNA targeting KRAS and ERK1/2 (oncogenes), ending up with suppression of cell proliferation and tumor development in CRC (Chen et al. 2009). Moreover, miRNAs are regulated by DNA methylation, as manifested in a genome-wide analysis of human colon carcinoma cells. Specific miRNAs affect certain CRC-signaling pathways. For instance, miR-135, miR-139, miR-145, and miR-17 affect the β-catenin/Wnt signaling pathway. The let-7 family, miR-18, miR-21, miR-126, miR-143, and miR-200 influence proliferation. MiR-34, miR-133, and miR-195 impact apoptosis. Cell cycle regulation by miR-34, miR-192, miR-203, miR-215, miR-675, and p53 signaling is controlled by miR-34. MiR-141 and miR-200 affect differentiation. Migration and invasion processes are controlled by miR-126, miR-143, miR-196, miR-200, miR-373, and miR-520 (Mehrgou et al. 2021; Farouk et al., 2025).

MiR-21 is a promising biomarker for the noninvasive detection of CRC at early stages. Consistently, miR-21 is dysregulated at the beginning of the transition of adenoma to carcinoma, and its expression is increased in CRC. Additionally, miR-21 is an exosome secreted in plasma or serum by the cancer cells. Plasma miR-21 could distinguish CRC patients from controls, with 90% sensitivity and specificity. Another study confirmed the diagnostic and prognostic potential of serum miR-21 in patients with CRC (Toiyama et al. 2013). In vitro and in vivo studies targeting miR-21 by antisense oligonucleotides (ASOs) cause suppression of the CRC growth and metastasis. Mechanistically, ASOs act by lowering levels of miR-21 and reducing the AKT and ERK pathways. In addition, miR-21 may serve as a therapeutic agent through the stimulation of DUSP8, which is a miR-21 target

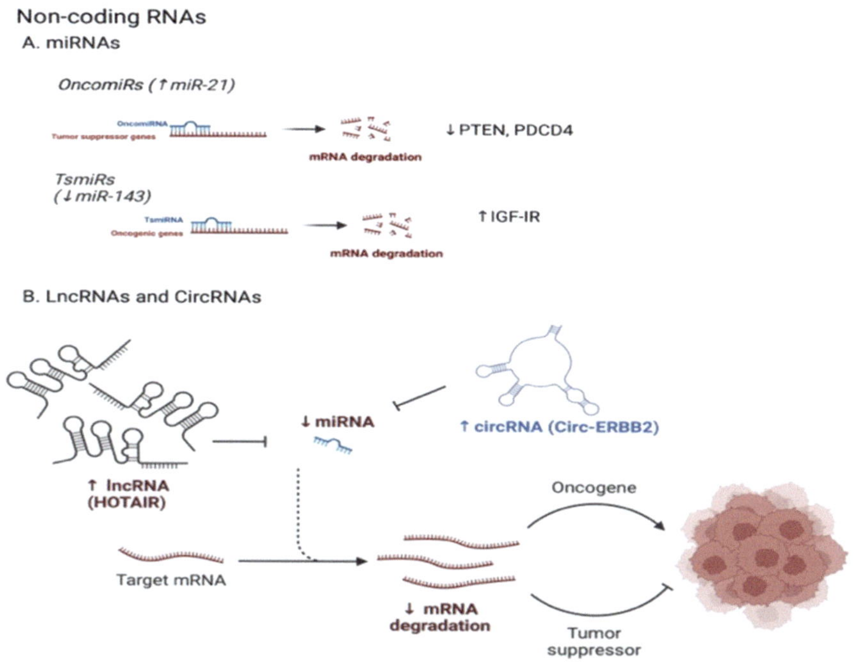

Fig. 4.2 Noncoding RNAs as regulators of colorectal cancer . (**a**) MiRNAs: Malignancy-increasing miRNAs (OncomiRs, e.g., upregulated miR-21) bind tumor suppressor genes and thus can inactivate mRNA and downregulate tumor suppressor proteins such as PTEN and PDCD4. In contrast, reduced tumor-suppressive miRNAs (TsmiRs, e.g., miR-143) cause failure to degrade oncogenic mRNAs, thereby raising the expression of oncogenes such as IGF-IR. (**b**) LncRNAs and CircRNAs: A greater expression of long noncoding RNAs (lncRNAs, e.g., HOTAIR) or circular RNAs (circRNAs, e.g., Circ-ERBB2) can lead to miRNA sequestration and make miRNAs less available. That, in turn, contributes to the reduction of target mRNA degradation by miRNAs, with the consequences of increased oncogene expression and reduced expression of tumor suppressors in order to facilitate tumorigenesis

gene. Locked nucleic acid (LNA), modified oligonucleotides, is another therapeutic way that targets dysregulated miRNAs in CRC (Nedaeinia et al. 2017). These results underscore the importance of miR-21 inhibitors as a novel therapeutic way for CRC therapy. MiR-143 is lowly expressed in CRC tissues and serves as a prognostic marker, with lower levels linked to poorer survival outcomes. Monitoring miR-143 expression may assist in evaluating disease prognosis and stratifying patients based on risk. MiR-143 is downregulated in CRC tissues and blood samples and correlates with the clinical stages and lymph node metastasis. miR-143 could be a therapeutic agent, as overexpression of it suppresses cell proliferation, tumor growth, and angiogenesis, and increases chemosensitivity to oxaliplatin. MiR-143 is a tumor-suppressive agent as it targets the oncogene IGF-IR (Qian et al. 2013). The consistent downregulation of miR-143 in CRC makes it an attractive candidate for diagnosis and targeted therapies through tumor-suppressing microRNA replacement.

Long noncoding RNAs (lncRNAs) have many cellular activities, involving the inactivation of the X chromosome, regulation of gene expression by epigenetic alterations, and control of posttranscriptional modifications (El Awady et al., 2006a; He et al. 2019). The major signaling pathways that are regulated by lncRNAs in CRC progression comprise the Wnt pathway (CCAT-1-L, CCAT-2, and CCAT-S), PI3K/AKT/mTOR (linc01296 and LINC00115), tumor growth (LUNAR1), and the NOTCH pathway (Lnc34a). However, LncRNA-422 and SNHG6 inhibit the CRC progression as they act as tumor suppressors (Doghish et al. 2024).

HOTAIR (Hox transcript antisense intergenic RNA) is an lncRNA with oncogenic function in distinct cancers, involving CRC (Fig. 4.2b). It is highly expressed in CRC tissues and cells, with poor prognosis, and relates to liver metastasis (Borai et al. 2018; Dawood et al. 2018; Kogo et al. 2011; Salum et al. 2018). Mechanistically, *HOTAIR*-associated polycomb repressive complex 2 (PRC2) can trimethylate histone H3 lysine-27 (H3K27me3) of the *HOXD* locus and suppresses gene expression of the tumor suppressor gene *HOXD*, causing chromatin modification and reprogramming, ending with promoting CRC metastasis (Rinn et al. 2007). Additionally, the HOTAIR/miR-1277-5p/ZEB1 axis mediates epithelial-mesenchymal transition in CRC (Weng et al. 2022). Moreover, HOTAIR can function as a competitive endogenous RNA, sponging miR-197 in CRC. Also, HOTAIR controls CRC progression and chemoresistance by changing the miR-203a-3p expression levels, thereby influencing β-catenin/Wnt signaling (Xiao et al. 2018). Furthermore, the diagnostic potential of HOTAIR was reported with optimal performance in distinguishing CRC patients from controls. Additionally, HOTAIR has prognostic value, as its level is linked with the stage of malignancy, lymph node spread, and long-term survival (Nazari et al. 2024). Intriguingly, knocking down HOTAIR suppresses the growth of in vitro CRC cell lines. LncRNAs present in immune cells which are associated with tumors could act as potential biomarkers and therapeutic targets, especially in treatments that modulate immune response by using checkpoint inhibitors (Gao et al. 2023). These results underscore HOTAIR's importance in CRC management and its potential for enhancing patient care.

Circular RNAs (circRNAs) are oncogenes or tumor suppressors through serving as miRNA sponges, binding to RNA-binding proteins, and translating functional peptides. For example, circ-ERBB2 sponges miR-125a-5p and miR-138-5p (Fig. 4.2b), leading to an increase in the expression of eIF4E-binding protein 1 and stimulation of the HIF-1α signaling pathway, ending with the promotion of tumor vascularization, hyperplasia, tissue infiltration, and cellular migration of CRC cells. Additionally, circAGFG1 sponges miR-4262 and miR-185-5p and enhances transcription of CTNNB1 (β-catenin), activating the Wnt/β-catenin pathway, which supports cancer stem cell self-renewal, contributing to tumor maintenance and resistance (Zhang et al. 2020a). While circ-FND3B is downregulated in CRC tissues, acting as a tumor suppressor in normal conditions by reducing miR-937-5p activity and inhibiting tumor growth and metastasis, exogenous delivery of circ-FND3B suppresses CRC progression (Zeng et al. 2020).

Several circRNAs have been identified as promising noninvasive indicators and therapeutic agents in CRC due to their stability and differential expression patterns. For instance, circ_0082182, circ_0000370, and circ_0035445 are diagnostic biomarkers; they distinguish between CRC tissues and normal tissues, and their levels are related to clinicopathological features, recommending their utility in early diagnosis and disease monitoring. Additionally, their stability in bodily fluids like plasma enhances their value as noninvasive biomarkers. Circ_0007031 is elevated in the plasma and tissues of CRC patients; its expression correlates with tumor progression, making it a candidate for liquid biopsy (He et al. 2020a). In vitro and in vivo studies reported that circRNA sponges could suppress cancer cell proliferation, promote apoptosis, and suppress tumor growth by targeting oncogenic miRNAs such as miR-21 and miR-93. Another way of therapy is by targeting oncogenic circRNAs. For instance, silencing the oncogenic circ_0007031 by using siRNAs or antisense oligonucleotides reduces tumor cell proliferation and chemoresistance and restores normal gene expression patterns (Wang et al. 2020c).

Epigenetic Alterations in the Adenoma-Carcinoma Sequence

The progression of CRC encompasses four stages: initiation, promotion, progression, and metastasis. First, irreversible genetic modifications occurred during the chemical carcinogenesis, such as DNA adduct formation. This modification originates from a covalent binding of carcinogens to DNA and is detected in the human colonic mucosa. Of note, the DNA repair system, or apoptosis, can eliminate DNA adducts. Meanwhile, there is a balance between exposure to carcinogens, formation of adducts, and their exclusion. However, continuous formation of DNA adducts results in mutations and can trigger cancerous growths. Such mutations, especially in genes responsible for cell proliferation, cause the protrusion of adenomas (small benign neoplasms) that progress to carcinomas (malignant states), Fig. 4.3 (Keum and Giovannucci 2019). Second, these genetically modified cells proliferate in the promotion stage and form neoplasms. Third, with extra genetic and epigenetic alterations, neoplasms progress and become malignant, invasive, and metastatic. Finally, tumor cells propagate from the original sites to distant areas through blood or lymph (Li et al. 2024). There are two main sources of colorectal cancerous cells: adenomatous and serrated polyps. Around 85–90% of sporadic CRC cases come from

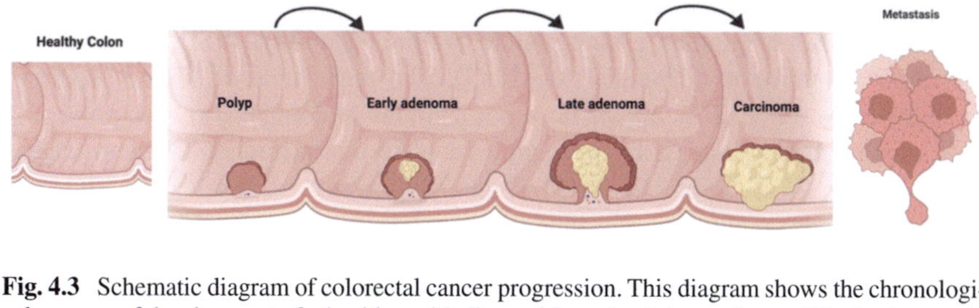

Fig. 4.3 Schematic diagram of colorectal cancer progression. This diagram shows the chronological process of development of a healthy epithelium of the colon into a polyp and further into early adenoma, late adenoma, and carcinoma. The last phase involves metastasis, in which cancer cells are transported outside the initial point of attack. Arrows are used to show the ledgewise change between colorectal tumorigenesis stages

adenomatous polyps. Serrated polyp types include hyperplastic polyp, traditional serrated adenomas, sessile serrated adenomas, and mixed polyps. Serrated polyps represent 10–15% of sporadic CRC patients (Crockett and Nagtegaal 2019).

Hypermethylation of certain genes is an early epigenetic event that creates a favorable environment for initiating tumors (Inno et al. 2014; Lao and Grady 2011). For example, methylation of the MGMT (O6-methylguanine-DNA methyltransferase) promoter is one of the earliest epigenetic changes detected in normal-appearing colonic mucosa and acts as a "field defect" that predisposes to mutagenesis. MGMT is a DNA repair enzyme; its loss causes accumulation of alkylation-induced mutations, where MGMT methylation happens before KRAS mutations in the adenoma-carcinoma sequence (Inno et al. 2014). Also, hypermethylation of adenomatous polyposis coli (APC) appears early, with frequencies significantly higher in colorectal adenomas in comparison with normal tissue (OR 5.76, 95% CI 2.45–13.56). Irregular Wnt/β-catenin signaling activation is caused by methylation of genes like SFRP1, SFRP2, and other Wnt pathway antagonists in adenomatous polyps. In addition, early methylation events include genes such as ITGA4, SLC5A8, and MINT1, which are detected in aberrant crypt cells and early adenomas (Lao and Grady 2011; Patai et al. 2015).

The conversion of early adenomas to advanced ones involves progressive accumulation of methylation events that promote the neuroplasticity of epithelial cells. For instance, promoters of MGMT and CDKN2A/p16 are more hypermethylated in tumorigenic forms of colorectal adenomas than in less tumorigenic forms, with tubular adenomas having lower methylation frequencies. Interestingly, it is reported that over 85% of progressed tumor samples show hypermethylation in ten genes, including *SFRP1*, *SST*, *BNC1*, *MAL*, *SLIT2*, *SFRP2*, *SLIT3*, *ALDH1A3*, *TMEFF2*, and *WIF1* (Patai et al. 2015). Additionally, the CpG Island Methylator Phenotype (CIMP) is identified in advanced tubular adenomas but is not commonly detected in early tubular adenomas, suggesting that extensive coordinated methylation happens later in polyp development. Serrated adenomas, especially sessile serrated adenomas/polyps, are reported to be the precursor lesions for CIMP colorectal cancers, exhibiting high methylation levels of both early and late methylation markers (De Palma et al. 2019; Sakai et al. 2016).

At a late stage, epigenetic alterations become more pronounced during the adenoma-to-carcinoma progression and are linked with invasion and metastasis (Lao and Grady 2011). MLH1 methylation appears exclusively in carcinomas rather than in precursor adenomas, causing microsatellite instability (MSI) and accelerating tumor progression (Zarandi et al. 2017). The prevalence of MLH1 promoter methylation is remarkably higher in tumor samples in comparison with polyp tissues, recommending that MLH1 silencing is a late event in colorectal tumorigenesis (Zarandi et al. 2017). Moreover, TIMP3 methylation is more frequent in advanced colorectal cancers and metastases than in adenomas, facilitating tumor invasion and metastatic behavior by disrupting extracellular matrix regulation (Brueckl et al. 2005). Other late methylation events include silencing of genes like ID4 and IRF8, which show higher frequencies in colon cancers and metastases than in adenomas, providing clonal growth advantages during malignant progression (Kim et al. 2006).

The methylation of CXCL12, a chemokine ligand involved in metastasis, also occurs preferentially in invasive carcinomas (Wendt et al. 2006).

The chronological sequence of epigenetic changes in colorectal cancer development highlights distinct functional roles for early versus late methylation events in tumorigenesis. Early methylation events, especially MGMT and APC, create a mutagenic and proliferative environment that triggers the initial steps of neoplastic conversion. These early changes cause genomic instability and disturb growth control mechanisms, setting the stage for adenoma formation (Liang et al. 2017). In contrast, late methylation events like MLH1 and TIMP3 silencing enhance genomic instability, immune evasion, and the potential for metastasis. The progressive nature of these epigenetic alterations, in combination with genetic mutations, triggers the stepwise histological progression from normal epithelium through adenoma to invasive carcinoma. This coordinated accumulation of epigenetic and genetic changes highlights the crucial function of DNA methylation as both an early initiating event and a late progression factor in CRC development (Nazemalhosseini Mojarad et al. 2013).

Epigenetic Biomarkers for Diagnosis and Prognosis of Colorectal Cancer

There is a growing demand for noninvasive alternatives to traditional screening methods for CRC detection. Blood, stool, and tissue epigenetic biomarkers have become crucial tools for the early detection, prognosis, and treatment prediction of CRC. Detection of circulating tumor DNA (ctDNA) in blood as a DNA methylation alteration is one of the most extensively studied and clinically implemented epigenetic markers. For example, hypermethylated *SEPTIN9*, *SDC2*, and *BCAT1* genes enable sensitive detection with reported sensitivities around 86% and specificities exceeding 97% in large CRC patient cohorts (Chen et al. 2025). These methylation markers detect early tumorigenic changes and have prognostic value by being linked with patient survival outcomes, making them reliable for diagnosis and risk stratification. Methylated SFRP2 is another DNA methylation biomarker that is secreted and acts as a Wnt signaling modulator. It has a sensitivity of 63.8–66.9% and a specificity of 97.3–100% achieving high diagnostic accuracy in CRC (Barták et al. 2017; Tang et al. 2011). In addition to single-gene methylation, multigene panels mixing methylated targets further promote diagnostic accuracy. For instance, panels comprising *mSDC2* with *mSFRP1/2* or *mTFPI2* have exhibited sensitivities of more than 90% for CRC and specificities close to 98% (Lin et al. 2022). Combining *SEPT9*, *SDC2*, and *OSMR* in a methylation marker panel has been shown to enhance detection rates. A panel combining mSEPT9 and mOSMR demonstrated higher sensitivity (around 77%) than conventional serum markers such as carcinoembryonic antigen (CEA), which typically shows sensitivities below 40%. This multimarker technique improves early detection and may better capture tumor heterogeneity. These panels overtake some FDA-approved tests, such as Cologuard®, in identifying early-stage lesions, suggesting a crucial demand for early intervention. Moreover, combining genetic mutations such as *APC* and *KRAS* with epigenetic methylation markers in liquid biopsies increases the extent of tumor detection, acquiring the heterogeneous molecular landscape of CRC (Ahlquist et al. 2000).

Beyond DNA methylation markers, other epigenetic modifications, such as circulating miRNAs, histone modifications, and lncRNAs, are detected in blood, stool, or tissue samples in CRC. For example, miR-21 and miR-92 are stable in plasma and dysregulated in CRC, reflecting posttranscriptional gene regulation, thus acting as diagnostic and prognostic biomarkers (Fathi et al. 2025). In addition, meta-analyses have underscored elevated miR-31 expression levels that have been linked to poorer survival outcomes and reduced responsiveness to anti-EGFR treatment. MiR-20a levels in stool, serum, or tumor tissue have demonstrated diagnostic accuracy in comparison with CEA and CA19-9 (Moody et al. 2019). The combination of several miRNAs into biomarker panels has enhanced the detection of CRC at an early stage, as these panels have high diagnostic precision. A five-miRNA panel involving miR-21, miR-29, miR-92, miR-125, and miR-223 had an AUC of 0.952 (Bader El Din et al. 2021). While a three-miRNA set including miR-29, miR-125, and miR-145 has significantly promoted risk prediction. Additionally, downregulated miRNAs like miR-194 and miR-29b have been related to advanced CRC stages and worse outcomes. Remarkably, a four-miRNA signature (miR-193, miR-210, miR-513, and miR-628) exhibited potential in the detection of CRC at early-onset, with high accuracy within international cohorts (Liu et al. 2018c).

Measurement of circulating nucleosome levels in blood as an indication of histone modifications involving acetylation and methylation marks, offering insight into the chromatin states and tumor biology. Circulating nucleosomes bearing histone H3 lysine 9 acetylation (H3K9ac) and histone H3 lysine 27 trimethylation (H3K27me3) are increased in CRC patients and are associated with tumor presence and its progression. Additionally, many studies have proven that histone modifications are diagnostic biomarkers for CRC. For example, methylated H3K9 and acetylated H3K27, H4K12, and H3K18 are increased in CRC compared to normal colonic mucosa (Ashktorab et al. 2009; Karczmarski et al. 2014). While circulating nucleosomes with methylated H3K9, H3K27, and H4K20 are lower in CRC patients than in controls (Gezer et al. 2013, 2015). Also, histone alterations are crucial in the understanding of CRC progression and survival rate. H3K4 and H3K27 are hypomethylated and linked to worse patient survival rates [77]. While H3K9, H3K20, and H3K27 are highly methylated and related to a good prognosis. The combination of histone modifications offers favorable prognostic factors (Benard et al. 2014a). H3K4, H3K9, and H4K20 are reported to be trimethylated and associated with disease-free survival (DFS) and recurrence-free survival (RFS) in patients with early-stage CRC. This combination offers a better way to stratify patients compared to using individual markers. Another study reported that the expression levels of EZH2, BMI1, SUZ12, and H3K27me3 exhibited prognostic value in CRC (Benard et al. 2014b). H3K56 and H4K16 are among the histone modifications explored as biomarkers in CRC research, with studies examining their association with disease behavior and clinical outcomes (Benard et al. 2015).

Moreover, lncRNA expression levels are related to disease progression. For example, HOTAIR, MALAT1, CCAT1, and MIR31HG expression levels are upregulated in CRC tissues and blood and linked with advanced tumor stages, metastasis, and poorer survival rates. In detail, upregulated HOTAIR levels are related to early-stage CRC and worse prognosis (Kogo et al. 2011). However, CCAT1 and CCAT2 exhibited elevated expression over all CRC stages, acting as potential diagnostic and prognostic markers. Notably, the combination of CCAT1 and HOTAIR achieves high diagnostic accuracy (AUC up to 0.954), especially in early-stage detection (Zhao et al. 2015). Also, LINC01485 and UCA1 show strong discriminatory power between CRC patients and healthy controls, supporting their use as noninvasive biomarkers. In addition, LncRNAs are prognostic indicators, reflecting the tumor aggressiveness and metastatic potential. MALAT1, for instance, is related to shorter disease-free and overall survival in CRC. HOTAIR overexpression relates to venous invasion and distant metastasis (Chen et al. 2020).

Several DNA methylation and miRNA biomarkers in stool have shown potential for noninvasive CRC screening. Methylated vimentin is a well-known diagnostic biomarker having sensitivity of 38.3–81% and specificity of 82–95% (W.-D. Chen et al. 2005). It was commercialized as *ColoSure*™ for early CRC detection. Also, *NDRG4* in stool samples exhibits a high 100% specificity; however, it has moderate sensitivity of 53% (Melotte et al. 2009). The *Cologuard*® test is approved by the FDA and combines *KRAS* mutations, *BMP3* and *NDRG4* methylation, and hemoglobin detection, having 92.3% sensitivity and 86.6% specificity for CRC (Imperiale et al. 2014). Additionally, multi-marker panels, such as those incorporating (SDC2/TFPI2) or (SFRP2/NDRG4/BMP3), further improve diagnostic accuracy, with some achieving sensitivities of more than 90% (Park et al. 2017; Zhang et al. 2021b). Additional methylation markers under investigation include *CDKN2A*, *SEPT9*, and *WIF1*, underscoring their potential for early CRC detection.

In addition, fecal miRNAs are stable and reliable, offering promising noninvasive biomarkers for CRC. A cluster of miR-21, miR-92, and miR-17 was identified to be upregulated in stool samples, while miR-143 and miR-145 were downregulated. These discrepancies correlate with tumor progression, where they can differentiate CRC from benign conditions like ulcerative colitis (UC). Combining two upregulated miR-421 and miR-27, or integrating miRNA data with clinical parameters such as age and hemoglobin levels, has revealed superior performance over fecal immunochemical testing (FIT) alone, with AUCs achieving 0.93 (Duran-Sanchon et al. 2020). Notably, miR-92a exhibits superior sensitivity for distal CRC and advanced adenomas, recommending its use in risk stratification (Wu et al. 2012).

Technologies for CRC biomarker discovery comprise molecular techniques such as liquid biopsy, targeted bisulfite sequencing, methylation-specific PCR, and next-generation sequencing (NGS). Liquid biopsies offer a better identification of the cancer genome in comparison with a single tissue biopsy and include any genetic material from the tumor secreted in body fluids such as blood, feces, or urine. They utilize circulating materials originating from the tumor, like circulating DNA or cells, miRNA, and exosomes, to identify molecular changes that exist in CRC tumors (Nordgård et al. 2018; Rizk et al. 2024). Guardant Health company has developed multimodal assays (e.g., Shield™ test) that combine ctDNA mutation profiling with epigenomic signatures to enhance early CRC detection and monitoring. While RNA sequencing and microarray platforms applied to tissue and liquid biopsy samples show miRNAs and other noncoding RNAs as crucial epigenetic biomarkers. Where high-throughput small RNA sequencing permits comprehensive profiling of miRNA expression patterns related to CRC (Zhang et al. 2025). Similarly, lncRNAs are identified via RNA-seq and confirmed by quantitative PCR. These RNA-based biomarkers are often combined with DNA methylation and mutation data using machine learning pipelines to improve diagnostic and prognostic accuracy (Zhang et al. 2025). Moreover, metabolomics and proteomics technologies complement genomic and epigenomic approaches by profiling metabolic shifts and protein expression alterations related to CRC. Also, mass spectrometry and nuclear magnetic resonance (NMR) spectroscopy detect metabolites in blood or stool samples, demonstrating pathways such as purine metabolism that relate to CRC pathogenesis. Machine learning tools applied to these multi-omic datasets can identify novel metabolism-related biomarker signatures for CRC diagnosis and treatment monitoring. Multi-omics integration, artificial intelligence (AI), and the implementation of uniform validated standards are emerging directions in CRC biomarker discovery (Lişcu et al. 2024). Genomic, epigenomic, transcriptomic, proteomic, and metabolomic data altogether enhance biomarker reliability and clinical utility. AI-driven algorithms improve feature selection and predictive modeling; however, there is an urgent need to have standardized protocols for the validation of biomarkers and its approval (Ruusuvuori et al. 2023).

Epigenetics and Resistance to Chemotherapy in Colorectal Cancer

Epigenetic changes in CRC contribute to resistance against conventional drugs, such as 5-FU and oxaliplatin. These alterations influence the expression of genes in multiple CRC-signaling pathways, including cell proliferation, survival, apoptosis, and DNA repair. For example, the hypermethylated MLH1a tumor suppressor gene causes resistance to 5-FU by prohibiting the detection and repair of DNA damage (Arnold et al. 2003). Additionally, genes responsible for apoptosis, including TGF-beta, DAXX, and p38, and genes involved in pyrimidine metabolism (drug metabolism), comprising cytoP450, EGFR, and P53, are hypermethylated in CRC and downregulated, thus enhancing resistance to 5-FU. 5-aza-2′-deoxycytidine (5-azadC)/ decitabine, a demethylating agent, **restores** CRC cells' sensitivity to 5-FU when combined with 5-FU. This is because 5-azadC induces apoptosis in mismatch repair (MMR)-deficient CRC cells (Arnold et al. 2003). Moreover, hypermethylation of the MEIS1 promoter in CRC causes its downregulation, resulting in increased expression of the DNAQ repair gene FEN1, ending with promoting resistance of CRC cells toward oxaliplatin. Also, the combination of decitabine and oxaliplatin enhances CRCs' sensitivity to the oxaliplatin drug by inhibiting CRC cell proliferation, besides its roles as a DNMT inhibitor for DNA demethylation (Hosokawa et al. 2019; Qin et al. 2009).

Beyond hypermethylation in CRC, the 5-FU drug induces reactive oxygen species (ROS), which promote DNA demethylation through enhancing ten-eleven translocation (TET) enzymes, which oxidize 5-methylcytosine. One of the cellular protection genes against ROS is Nrf2, which, upon stress, translocates to the nucleus and binds to the heme oxygenase 1 (HO-1) promoter. In 5-FU-treated CRC cells, the Nrf2 promoter region is hypomethylated, leading to an increase in its expression and inducing cellular protection against ROS produced by 5-FU treatment, and becoming resistant to 5-FU. To reverse 5-FU resistance, a study proved that the 5-FU-resistant SNUC5 cell line transfected with siRNA or shRNA against Nrf2 or HO-1 exhibited high cytotoxicity and reduced proliferation and restored sensitivity to 5-FU treatment (Kang et al. 2014).

In addition, hypomethylation of the PARPBP gene promoter causes oxaliplatin resistance. Hypomethylation of PARPBP is intermediated by KIF18b, which hinders DNMT3b recruitment to the PARPBP promoter, causing increased PARPBP expression and promoting resistance to oxaliplatin in CRC cells (Hong et al. 2021).

Moreover, genes responsible for histone modifications contribute to the resistance to 5-FU and oxaliplatin (Oxa) in CRC. EZH2 is a histone methyltransferase that trimethylates histone H3 on lysine 27 (H3K27me3). In 5-FU-treated CRC, EZH2 is low expressed, accompanied by high DPYD, and correlated with poor outcomes. DPYD is a gene that encodes dihydropyrimidine dehydrogenase (DPD), which catabolizes 85% of 5-FU into the nontoxic metabolite 5-dihydrofluorouracil (Heggie et al. 1987). CRC cells with high *DPYD levels* and low *EZH2* expression levels are remarkably resistant to 5-FU and oxaliplatin (Li et al. 2022; Wu et al. 2016). LSD1/KDM1A, which is a histone demethylase that removes methyl groups from H3K4me1/2. LSD1 is overexpressed in CRC, causing demethylation of H3K4me2 and increased glutamine synthetase (GS) expression, enhancing 5-FU resistance via metabolic reprogramming and mTOR induction (Zhang et al. 2021a). KDM6A and KDM6B are histone demethylases that remove the trimethyl group from histone H3 lysine 27 (H3K27me3). Oxaliplatin enhances KDM6A/6B expression, which decreases H3K27me3 levels, causing activation of NOTCH2 and NOTCH signaling. This induces stemness-related gene expression, resulting in oxaliplatin resistance. Inhibition of KDM6A/6B by GSK-J4 resensitizes CRC cells to oxaliplatin as it restores H3K27me3 levels (Wang et al. 2020a). Additionally, ELFN1-AS1 (lncRNA) recruits EZH2 and DNMT3a to the MEIS1 promoter, causing histone methylation and DNA methylation that inhibit MEIS1 expression, thus enhancing oxaliplatin resistance (Li et al. 2022).

MiRNAs and lncRNAs dysregulation contribute to oxaliplatin and 5-FU resistance in CRC by controlling gene expression networks that influence apoptosis, drug uptake, DNA repair, and survival signaling pathways. For example, miR-21 is overexpressed in CRC and downregulates mismatch repair proteins *hMSH2* and *hMSH6*, thus impairing DNA damage recognition and reducing G2/M arrest and apoptosis induced by 5-FU, thereby conferring resistance (Valeri et al. 2010). While miR-587 causes resistance by antagonizing apoptosis induced by 5-FU through controlling the expression of PPP2R1B in CRC (Zhang et al. 2015). MiR-26b is downregulated in 5-FU-resistant cells due to promoter hypermethylation; it normally suppresses P-glycoprotein (Pgp), a drug efflux pump. Loss of miR-26b leads to increased Pgp expression and reduced intracellular 5-FU accumulation (Farouk et al. 2022; Wang et al. 2018b). GIHCG (lncRNA) induces CRC cells' migration, invasion, and proliferation and chemoresistance toward both 5-FU and oxaliplatin (Press 2019). While H19 (lncRNA) confers 5-Fu resistance in CRC through SIRT1-intermediated autophagy (Wang et al. 2018a). Moreover, SCARNA2 (lncRNA) is highly expressed in 5-FU and oxaliplatin-resistant CRC cells, increases the expression of target genes of miR-342, including EGFR and BCL2, by sponging miR-342, and afterward enhances the resistance of CRC cells to 5-FU and oxaliplatin (Zhang et al. 2019). In oxaliplatin-resistant CRC cell lines, miR-19a is overexpressed and inhibits the expression of PTEN, which stimulates apoptosis (Zhang et al. 2020b). MiR-29a-3p is increased in oxaliplatin-resistant LoVo cells, which reduces Fem1b levels, a pro-apoptotic protein, causing resistance to oxaliplatin treatment (Su et al. 2023). MiR-153-5p is downregulated in CRC-resistant cells; its overexpression targets *Bcl-2*, which inhibits protective autophagy and restores sensitivity to oxaliplatin (He et al. 2020b). LncRNAs, including *MALAT1*, *NEAT1*, *Linc00152*, *CRNDE*, *GAS5*, *KCNQ1OT1*, *TUG1*, *MEG3*, *PVT1*, and *MEIS1,* affect tumorigenesis and oxaliplatin and 5-FU resistance (Doghish et al. 2024).

Targeting exosomal miRNAs can improve oxaliplatin sensitivity in CRC by disrupting key resistance mechanisms, enhancing drug efficacy, and potentially modulating the tumor microenvironment. Exosomes, which are small vesicles secreted by cells, transport miRNAs that can confer resistance to recipient cells (Borai et al. 2017; Ning et al. 2021). Certain exosomal miRNAs are significantly elevated in CRC patients with oxaliplatin resistance and actively contribute to chemoresistance. For example, oxaliplatin-resistant CRC cells exhibit high expression of exosomal miR-208b, which promotes resistance by targeting PDCD4. This in turn leads to the expansion of regulatory T cells (Tregs) and a reduction in oxaliplatin-based chemosensitivity (Ning et al. 2021). Blocking or downregulating such miRNAs may reverse their pro-resistance effects. Conversely, some exosomal miRNAs are downregulated in resistant CRC patients, and their restoration could resensitize tumor cells to oxaliplatin. A panel of miR-100, miR-92, miR-16, miR-30, miR-144, and let-7 was remarkably dysregulated in oxaliplatin-chemoresistant CRC patients, with miR-100 being upregulated and the others downregulated (Han et al. 2020).

Therapeutic Targeting of the Epigenome in Colorectal Cancer

The Targeted antibody immunotherapy usually binds to antigens on infected or cancerous cells, thereby disrupting their activity and induces their elimination by the immune system (El-Awady et al., 2006b; El-Awady et al., 2009). The current epigenetic therapies include DNMT inhibitors (decitabine, azacitidine) as well as HDAC inhibitors (vorinostat, romidepsin) in CRC. **Decitabine (DAC)** primarily targets CRC by integrating into DNA and then impeding DNA methyltransferases (DNMTs), resulting in global DNA hypomethylation. Demethylation reduces tumor cell proliferation and enhances apoptosis by reactivating silenced tumor suppressor genes as well as immunoregulatory genes. Decitabine can also activate DNA damage response pathways, indirectly modify histone modifications and chromatin structure, and induce the expression of stemness markers and immune checkpoint molecules like PD-L1 and cancer-testis antigens such as NY-ESO-1, thereby enhancing tumor immunogenicity and the response to immunotherapy (Taib et al. 2023). In clinical settings, researchers have assessed decitabine in conjunction with chemotherapy and immunotherapy to surmount resistance and improve CRC treatment efficacy. Preclinical studies have demonstrated that decitabine synergizes with oxaliplatin. This combination promoted reactive oxygen species (ROS) and programmed cell death in treatment-resistant CRC cells (Hosokawa et al. 2019). Low doses of decitabine enhance the PD-1 immune checkpoint blockade efficacy via antigen-processing and presentation gene upregulation. This augments lymphocyte penetration within tumor models, suggesting a promising combinatorial strategy for microsatellite stable CRC patients showing characteristically substandard responsiveness toward monotherapy immunotherapy. For improved patient outcomes, sensitized tumors, and remodeled tumor microenvironments, early-phase clinical trials are investigating decitabine along with immune checkpoint inhibitors and chemotherapeutic agents (e.g., oxaliplatin, 5-FU) (Pawlak et al. 2022; Taib et al. 2021; Yang et al. 2012).

Azacitidine (5-azacytidine) treats CRC primarily by inhibiting DNA methyltransferase (DNMT), leading to hypomethylation and reactivation of tumor suppressor genes. Azacitidine incorporates into DNA and RNA at low doses and inhibits DNMT enzymes, plus CRC commonly shows aberrant DNA methylation patterns, so azacitidine reverses it; thus, it restores normal gene expression that controls cell growth and apoptosis. At higher doses, it exerts direct cytotoxic effects via incorporation into nucleic acids, and this causes cell death in rapidly dividing cancer cells (Pawlak et al. 2022). Studies contrast azacitidine with its analog decitabine to suggest that azacitidine has weaker long-term anticancer effects in CRC cells, and those studies reveal important immediate toxicity yet slight sustained inhibition of proliferation, unlike decitabine, which causes cellular senescence and lasting growth arrest (Pawlak et al. 2022). The epigenetic reprogramming effects of azacitidine may be transient in CRC, indicating that its clinical efficacy might be improved through combination therapies. For better CRC outcomes, several clinical trials combined azacitidine with chemotherapy and immunotherapy. These trials probed the therapy. To sensitize CRC cells by reversing methylation-mediated drug resistance, azacitidine has been combined alongside standard chemotherapeutic agents like 5-FU, along with oxaliplatin. For example, azacitidine can upregulate tumor suppressor genes like CDH1 and SOX17 through demethylating their promoters because this inhibits proliferation as well as metastasis in cancer cells. Such combinations aim to overcome chemoresistance and improve cytotoxic efficacy. Epigenetic modulation by azacitidine can enhance immune responsiveness through raising cancer-testis antigens and promoting antigen presentation, potentially improving responses to immune checkpoint suppressors, including anti-PD-1/PD-L1 antibodies (Li et al. 2004). Early-phase clinical trials are investigating azacitidine combined with immunotherapy in microsatellite stable (MSS) CRC, which typically shows poor response to immunotherapy alone (Li et al. 2004). Additionally, studies also suggest that combining azacitidine with agents like curcumin may overcome resistance seen in decitabine-resistant CRC cells by modulating epigenetic and metabolic pathways (Hosokawa et al. 2021).

Vorinostat treats CRC via inhibition of histone deacetylases (HDACs), which are enzymes that eliminate acetyl groups with chromatin condensation that represses transcription. Vorinostat promotes histone acetylation via HDAC inhibition, leading to a less compact chromatin configuration and reactivation of the tumor suppressor gene. This results in cell growth, differentiation, and apoptosis of cancer cells. Furthermore, vorinostat impacts proteins beyond histones. It governs cell cycles too and causes apoptosis through expression of cyclin-dependent kinase inhibitors like p21 and pro-apoptotic factors. Vorinostat exhibits antiproliferative effects upon CRC cell lines from preclinical studies and synergizes alongside agents like bortezomib, the proteasome inhibitor, to improve tumor cell apoptosis and cell cycle arrest. Vorinostat coupled with capecitabine (a 5-FU prodrug) exhibited cooperative suppression of neoplasm proliferation in CRC xenograft models via alteration of pharmaceutical metabolism enzymes plus increased apoptotic transduction. Vorinostat has undergone clinical evaluation via combination regimens. These regimens increase efficacy and help overcome resistance in CRC. A phase I/II study merged vorinostat with hydroxychloroquine, an autophagy suppressant, in resistant metastatic CRC subjects, and that study depicted safety and possible immune advantages, including diminished regulatory T cells and depleted T cells (Patel et al. 2016). Vorinostat coalesces alongside chemotherapeutic agents such as oxaliplatin to exhibit potential because vorinostat has the ability to sensitize CRC cells to chemotherapy when it impedes pathways such as NRF2 that arbitrate drug resistance (Tanaka et al. 2022). Additionally, they investigate vorinostat coupled with immunotherapies because that improves tumor immunogenicity, although clinical data still transpire.

Also, **romidepsin** is a potent histone deacetylase inhibitor (HDACi) that changes epigenetic regulation by inhibiting class I and II HDAC enzymes. This inhibition also causes increased acetylation of histones and nonhistone proteins, resulting in chromatin relaxation and reactivation of tumor suppressor genes, which induce cell cycle arrest, apoptosis, and differentiation in cancer cells (Marks and Dokmanovic 2005). Additionally, romidepsin downregulates multiple oncogenic signaling pathways that promote cell survival and proliferation in CRC, including PI3K/AKT/mTOR and β-catenin pathways. Romidepsin also induces DNA damage responses, produces reactive oxygen species, and causes mitochondrial dysfunction, further promoting apoptosis. Notably, romidepsin-mediated hypomethylation of the promoter of SFRP1, an inhibitor of β-catenin, contributes to suppression of Wnt signaling, a critical pathway in colorectal tumorigenesis. Preclinical studies demonstrated that romidepsin combined with copanlisib, a PI3K inhibitor, synergistically induces apoptosis in PIK3CA-mutant CRC models through upregulation of GADD45B, a pro-apoptotic gene (Maduscha et al. 2024). Furthermore, romidepsin has been evaluated in combination with chemotherapy agents to overcome resistance mechanisms, given its ability to inhibit survival pathways like PI3K/AKT/mTOR and β-catenin signaling, which are often upregulated in chemoresistant tumors (Mizutani et al. 2010).

Microbiome-Epigenome Interactions in Colorectal Cancer

Gut microbiota exerts epigenetic modifications in human cells by producing low molecular weight substances. Several studies underscore the relationship between microbiota and human diseases. Microbial metabolites modulate the gut epigenome by affecting the methylation of cancer-related genes. Therefore, exploring the mechanisms of gut microbiota-related epigenetic modifications could help in identifying therapeutic targets for CRC. Probiotic bacteria, such as *Bifidobacterium* and *Lactobacillus,* produce folate, where 5-methyltetrahydrofolate is essential for DNA methylation, and 5-formyltetrahydrofolate and 5, 10-methenyltetrahydrofolate are crucial for DNA synthesis (Kopp et al. 2016; Zhou et al. 2013). Moreover, folate is important for *S*-Adenosyl methionine (SAM) production, an intracellular methyl donor that converts cytosine to 5-methylcytosine (5-mC), which is essential for DNA methylation. Indeed, CRC patients with folate deficiency exhibit DNA hypomethylation (Wasson et al. 2006). Individuals administered *Bifidobacterium* exhibit elevated fecal folate levels, confirming this probiotic's capacity to synthesize folate and modulate DNA methylation patterns (Pompei et al. 2007). However, harmful bacteria such as *Helicobacter pylori*, which causes gastric ulcers, can stimulate numerous epigenetic modifications (Cheng et al. 2013). Remarkably, patients with gastritis caused by *Helicobacter pylori* infection show hypermethylation of *CDH1*, *MGMT*, *WIF1*, and *MLH1* promoters *in comparison with* healthy individuals (Kawanaka et al. 2016). Additionally, several studies have potentially recommended that *Fusobacterium nucleatum* could cause epigenetic changes in CRC. For example, CRC patients with *F. nucleatum* infection were related to genomic hypermutation regardless of *CIMP* and *BRAF* mutations (Koi et al. 2018). Another study links the *Fusobacterium* infection and *CIMP* phenotype, wild-type *TP53*, *hMLH1* methylation, genomic hypermutation, and *CHD7/8* mutation (Inamura 2018). *F. nucleatum* DNA is elevated in CRC tissues compared to normal tissues and associated with the high-CpG island methylator phenotype and LINE-1 hypomethylation, thereby contributing to poor CRC survival rate (Mima et al. 2016a). Additionally, *H. hathewayi* induces hypermethylation of tumor suppressor gene promoters by upregulating DNMT expression in CRC patients (Xia et al. 2020).

Short-chain fatty acids (butyrate and acetate) are microbiota-derived metabolites that induce epigenetic modifications in colonic cells. *Firmicutes* metabolize unmetabolized dietary carbohydrates and proteins through fermentation to produce butyrate, which regulates multiple genes responsible for apoptosis and cell cycle control. Butyrate suppresses HDACs and stimulates hyperacetylation of histones, leading to altered expression levels of *CCND3* and *CDKN1A,* which regulate the cell cycle. Additionally, it stimulates ROS in epithelial cells and inhibits the protein complex (NF-*k*B), which controls DNA transcription (Bultman 2017). Butyrate was also found to control the expression of certain miRNAs in CRC, including Let-7, miR-18, miR-25, and miR-17 (Hu et al. 2015). MiR-17 is associated with c-Myc to suppress PTEN and enhance the PI3K-Akt-mTOR pathway, hence increasing the cell survival in the early stage of CRC (adenoma) (Gao et al. 2009).

Beyond DNA methylation, microbiota-derived metabolites like butyrate and propionate directly modify histones to suppress carcinogenic pathways. After infection with *L. monocytogenes*, it induces deacetylation of histone H3K18 in *SMAD1*, *IRF2*, *SMARCA2*, and *CXCL12 genes* in colonic cells (Niller and Minarovits 2016). *L. monocytogenes* facilitates deacetylation by delivering NAD-dependent deacetylase sirtuin 2 into the host nucleus. This epigenetic modification alters the expression of cell cycle-related genes and dampens the host immune response, promoting bacterial invasion. In addition, propionate degrades euchromatic histone lysine N-methyltransferase 2 (EHMT2), reducing H3K9me2 marks on the *Tnfaip1* promoter and inducing apoptosis in CRC cells (Ryu et al. 2022). Microbes could also manipulate host miRNAs to promote tumorigenesis. *F. nucleatum* upregulates oncogenic miR-21 via TLR4/MyD88 signaling while suppressing miR-4802 and miR-18a to activate autophagy and chemoresistance (Yang et al. 2017). Moreover, a study suggested a link between gut microbiota and miRNAs. Seventy-six oncogenic miRNAs (mir-503, mir-182, and mir-17 ~ 92) were expressed at higher levels in CRC tumor tissues than in normal tissues. These tumorigenic miRNAs were associated with the relative abundance of *Proteobacteria, Firmicutes, and Bacteroidetes* (Yuan et al. 2018).

The gut microbiome causes inflammation and modulates various signaling pathways involved in the carcinogenesis of CRC. Gut microbiota are prognostic biomarkers, as they can detect CRC and predict clinical outcomes. Intricate connections evolved during the development of cancer among microorganisms originating from the gut and from the tumor and the host immune cells (Metwally et al. 2025; Scott et al. 2019). Eubiosis refers to the gut microbiome in healthy conditions, where there is variability in the gut bacteria; an equilibrium between pro- and anti-inflammatory mediators; an equilibrium between immune cells and IgA secretion; and a potent mucosal barrier. Imbalance in these parameters causes microbial dysbiosis. The tumor-associated microbiome adversely disrupts gut microbial balance, impairing both local and systemic immune function. This dysbiosis also reduces the effectiveness of systemic therapies, including chemo- and immuno-therapy, as shown in Fig. 4.4.

Studies showed that CRC cases have remarkable alterations in the abundance of certain bacteria in comparison with controls. For instance, CRC cases have an increased abundance of *E. coli*, *F. nucleatum*, and *Providencia,* alongside a reduced presence of *Lactobacillus* and butyrate-producing bacteria such as *Roseburia* and *Fecalibacterium* (Bultman 2014; Burns et al. 2015). Butyrate inhibits NF-κB, a key regulator of immune responses to infection. It was reported that TNF-α activates the NF-κB signaling cascade, inducing IL-6 production which further enhances

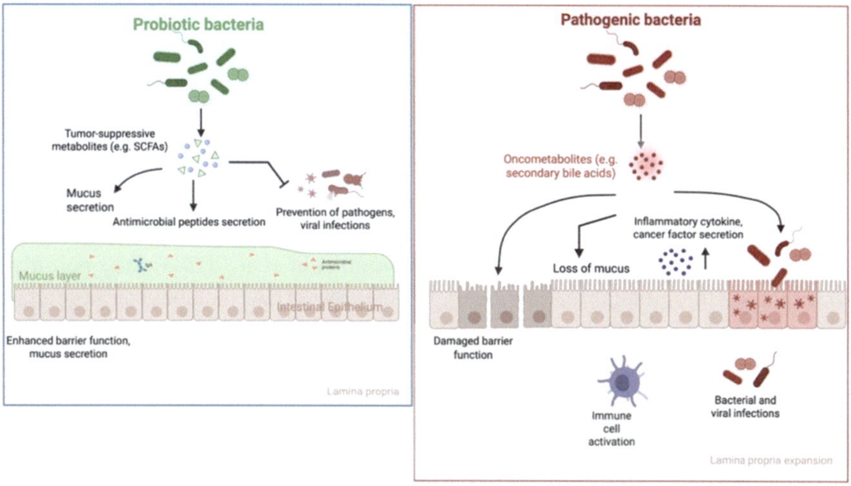

Fig. 4.4 The role of probiotic and pathogenic bacteria in the development of colorectal cancer and inflammation. This diagram explains how intestinal health-maintaining gut bacteria and pro-carcinogenic gut bacteria differ in their functions. In the left panel, probiotic bacteria, beneficial bacteria, are shown to produce tumor-suppressive metabolites, including short-chain fatty acids (SCFAs). These metabolites induce the secretion of mucus and antimicrobial peptides, thus improving the barrier performance and the prevention of cytopathic and viral infections, supporting the health and preservation of the intestinal epithelium and lamina propria. The right-hand panel, pathogenic bacteria, indicates how pathogenic bacteria generate oncometabolites (e.g., secondary bile acids). This results in impaired barrier activity, mucus loss, and extension of secretion of inflammatory cytokines and cancer factors. These processes lead to immune cell activation, bacterial and viral infections, and enlargement of the lamina propria, which altogether favor inflammation and may lead to the development of colorectal cancer

Wnt/β-catenin signaling and upregulates anti-apoptotic genes, contributing to inflammation, pathological changes and tumor growth (El Din et al. 2016). Upon activation, NF-κB translocates to the nucleus, links with target gene sequences, and upregulates the expression of pro-inflammatory factors. Similarly, *Bacteroides thetaiotaomicron* triggers inflammatory signaling by linking with IκB (inhibitor of κB), a regulatory constituent of the NF-κB pathway, hence suppressing NF-κB activation (Metwally et al. 2024; Zou et al. 2018). NF-κB activation represents a key driver of inflammation-related cancer (Anwar et al. 2022), including CRC, with studies detecting constitutive activation in 40% of resected CRC specimens and 67% of human CRC cell lines (Sakamoto et al. 2009). *B. fragilis*, an enterotoxin bacterial strain, is detected in CRC patients (Ahmad et al. 2024). *B. fragilis* toxin stimulates inflammatory cascades, including signal transducer IL-17 and NF-kB signaling in colonic epithelial cells (Wu et al. 2009). In APC Min mice (a model for CRC trials), these inflammatory cascades stimulate CXC chemokine expression, recruiting polymorphonuclear immature myeloid cells (PMN-IMMs) and establishing a pro-inflammatory microenvironment—particularly in the distal colon (Cells et al. 2018). Inflammation serves as a key indicator of pro-carcinogenic microbiota activity in CRC, particularly involving genotoxic polyketide synthase (pks) + *E. coli* (Bonnet et al. 2014), *A. finegoldii* (Moschen et al. 2016), and *E. faecalis* (Wang et al. 2012). In rat models of CRC, *F. nucleatum* accelerates tumorigenesis by activating TLR4 signaling (Wu et al. 2018; Yang et al. 2017). While *Peptostreptococcus anaerobius* promotes carcinogenesis through TLR2/TLR4 pathway activation (Tsoi et al. 2017).

Secondary bile acids influence the host immune system.

They are produced by *B. fragilis*, *Bacteroides vulgatus*, *Clostridium perfringens*, *Eubacterium*, *Lactobacillus*, and *Bifidobacterium* (Liu et al. 2020a). Any perturbation in the gut microbiota constitution impacts bile acid (BA) metabolism. Secondary BAs, particularly deoxycholic acid (DCA), are elevated in the serum of CRC patients and mediate the malignant transition of colorectal adenomas. Additionally, elevated levels of DCA in $Apc^{min/+}$mice result in a leaky intestinal barrier and intestinal inflammation, thereby enhancing intestinal tumorigenesis (Liu et al. 2018a). DCA promotes the development of CRC through the EGFR/ERK1/2/AP-1 axis, which mediates the proliferation and differentiation of colonic cells. Furthermore, DCA enhances the proliferation of CRC cells through activating Wnt signaling and stimulates an inflammatory response, resulting in CRC progression. Besides, DCA activates PKC through plasma membrane perturbation. PKC subsequently triggers p38 MAPK and NF-κB signaling. Activated NF-κB then drives IL-1β gene transcription (Jean-Louis et al. 2006; Metwally and Elbrashy 2024).

Future Perspectives: Epigenetic Editing and Precision Oncology

Epigenetic editing tools such as CRISPR-Cas9 and precision oncology are an emerging avenue for CRC treatment and are summarized in Fig. 4.5. Combining these tools with inhibitors of histone modifications paves the way for an accurate control of certain genomic domains, affecting the survival and proliferation of cells in CRC. Epigenetic targets of CRISPR-Cas9 are named readers, writers, and erasers. DNMTs and histone-modifying enzymes are epigenetic writers; they modify DNA and histones. While bromodomain proteins are epigenetic readers that identify and decipher these modifications. Removing these chemical marks requires epigenetic erasers, including histone deacetylases, which remove these chemical tags. Understanding the relationship between DNA methylation and gene expression in CRC is essential. A study combined human CRC groups with CRC cell lines and used a gRNA library combined with CRISPR-dCas9-based tools to change the DNA methylation pattern of 56 hypermethylated promoters related to epigenetic downregulation in CRC. For example, in this study, the methylated *RSPO2* promoter exhibited a gradual increase along the adenoma-to-carcinoma sequence, recommending that *RSPO2* is a potential tumor suppressor gene in CRC tumorigenesis. Using CRISPR-dCas9 in this study leads to epigenetic reactivation of *RSPO2* in the tumor milieu, which was related to a remarkable loss in cell proliferation in p53

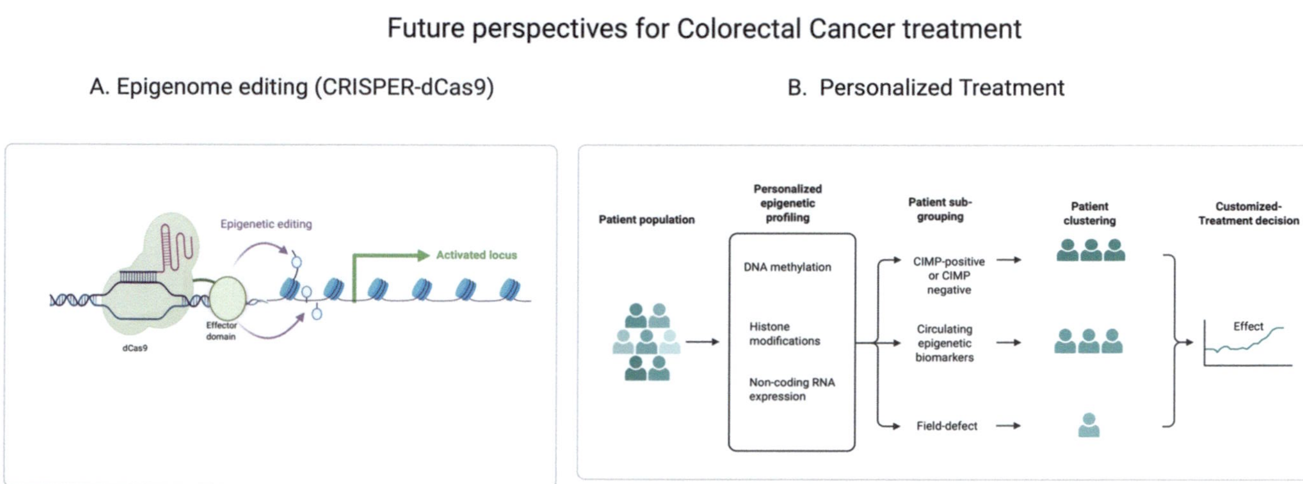

Fig. 4.5 Prospects of colorectal cancer therapy. (**a**) Epigenome editing (CRISPR-dCas9): Graphical representation of targeted epigenetic editing via the introduction of a catalytically inactive Cas9 (dCas9) fused to an effector protein, such that it binds to the desired locus and activates or inhibits gene expression by affecting chromatin structure. (**b**) Personalized treatment: Workflow showing how patient subgrouping (e.g., CIMP status, circulating epigenetic markers, field defects) may be enabled by personalized epigenetic profiling (including DNA methylation, histone modifications, and noncoding RNA expression) and patients then clustered, allowing customized treatment decisions to be developed to optimize the improvement of therapeutic outcomes

knockout cancer cell lines monitored by a high-content cell proliferation screen. This finding is further validated in human samples, where epigenetic silencing of the *RSPO2* gene occurs during intermediate-to-late stages of the adenoma-carcinoma transition (Tejedor et al. 2023). Moreover, epigenetic readers, including bromodomain proteins (BRD2 and BRD4), identify acetylated histone tails and regulate gene expression by recruiting transcription factors. Several bromodomain and extra-terminal domain (BET) inhibitors have been developed, as BET is involved in the cell cycle and cancer progression. Loss of BRD8 leads to cell cycle arrest at the G1 phase and diminished cell proliferation. In addition, its bromodomain binds to histone H4, controlling transcription and keeping its protein stability (Yamaguchi et al. 2023). These studies strongly recommend that these readers be targeted by CRISPR-dCas9.

Beyond DNA methylation, aberrant histone modifications such as H3/H4 hypoacetylation and altered H3 methylation significantly contribute to CRC pathogenesis. HDACs are frequently overexpressed in CRC, which promotes the investigation of HDAC inhibitors (HDACi), which elicit anticancer roles through cell cycle arrest, differentiation, and autophagy. HDACi efficacy in CRC is still limited; however, HDACi are clinically accepted for the treatment of hematologic malignancies. CRISPR/Cas9 functional genomics enables systematic identification of such vulnerabilities. Intriguingly, a CRISPR-Cas9 screen targeting 614 epigenetic mediators identified euchromatic histone methyltransferases EHMT1/2 as synthetic lethal partners of HDACi in CRC. Suppression of EHMT1/2 potentiated HDACi-mediated growth inhibition, with low EHMT2 expression correlating with advanced CRC stages and poor prognosis in clinical samples. Dual pharmacological inhibition of HDAC and EHMT1/2 synergistically reduced proliferation, validating this synthetic lethal interaction. These results highlight CRISPR screening's utility in uncovering targetable epigenetic vulnerabilities, advancing personalized combination therapies for CRC (Bamberg et al. 2022; Brownell et al. 1996).

Epigenetic drugs (epidrugs) target key regulators of CRC progression, but they have limitations, including therapeutic resistance and off-target effects. CRISPR-Cas9 genome editing offers precise genetic modifications but poses significant risks, including unintended mutations from off-target effects, inefficient delivery to solid tumors using current viral/nonviral systems, low homology-directed repair (HDR) efficiency, and potential immunogenicity due to the bacterial Cas9 origin (Hillman 2023). Ethical concerns regarding germline editing further necessitate strict governance. Comprehensive epigenome profiling remains limited, hindering precise programmable gene regulation via epigenome editing. Advancing this field requires rigorous off-target assessments and mechanistic studies. Critical challenges for clinical translation include uncertain persistence of epigenetic modifications, though emerging "hit-and-run" strategies show stable silencing potential, and there remains the need for optimized delivery systems. While adeno-associated virus (AAV) vectors show promise through engineering advances, their immunogenicity and off-target risks require further characterization. Despite limitations, CRISPR/Cas9 holds transformative potential for CRC through improved delivery vectors (lipid/polymeric nanoparticles, exosomes), multi-omics-guided target discovery, and novel tools like base/prime editors. Combining these approaches enables precise epigenetic reprogramming and personalized therapies. However, robust ethical frameworks are essential to ensure safety, equitable access, and responsible clinical translation of these technologies (Wang et al. 2022).

Personalized epigenetics profiling can significantly improve the risk stratification of CRC patients by reflecting the individual epigenomic changes that underlie tumor behavior, prognosis, and response to treatment. Epigenetic alterations, such as DNA methylation, histone modification, and noncoding RNA, are dynamic chromatin modifications that differ from genetic mutations but play essential roles in regulating gene expression programs essential for the development of CRC. Profiles can be combined by clinicians to more accurately stratify patients into risk groups and guide treatment decisions. An important benefit of personalized epigenome profiling is the refinement of tailored genomic risk classifiers that can predict distinct types of cancer with different clinical courses. For instance, the CIMP, characterized by global promoter hypermethylation, identifies a subset of patients with CRC that have distinct molecular phenotypes and poorer survival. It has been demonstrated in studies that CIMP-positive patients have worse survival but respond differently to some types of chemotherapy, such as 5-FU-based chemotherapy, compared to CIMP-negative patients (Hinoue et al. 2012). This gradient supports informed treatment decisions and prognostic counseling. Beyond classification, blood, stool, or tissue samples provide easy and minimally invasive ways to detect and monitor epigenetic changes. As an example, SEPT9 methylation markers are applied in the plasma and NDRG4 markers in the stool to screen early-stage CRC as well as advanced adenomas (Nian et al. 2017; Xiao et al. 2015). The epigenetic markers have the potential of being used to produce personalized epigenetic signatures that can narrow down the risk assessment to individuals with a chance of developing malignant tumors out of benign polyps so that early interventions can be made. Personalized epigenetics profiles additionally capture the phenomena of a so-called field defect; some risk of development of synchronous or metachronous lesions that are indicated by epigenetic aberrations in histologically normal mucosa. The identification of these changes in endoscopic biopsy can help determine the intensity of surveillance and preventive efforts, limiting the number of CRCs and resulting deaths. Noteworthy, epigenetic markers can forecast response and resistance to treatment. To take one example, the methylation status of MLH1 is correlated with microsatellite instability (MSI) and responses to immunotherapy, whereas other epigenetic signatures could reflect chemoresistance (Li et al. 2024). Through a personalized assessment of the epigenetic landscape of each patient, clinicians and researchers will be able to prescribe or choose the epigenetic drugs, including DNA methyltransferase inhibitors or histone deacetylase drugs, that specifically target the reversal of the pathogenic epimutations, leading to better efficacy and reduced toxicity (Jung et al. 2020).

In conclusion, personalized epigenetic profiles are a promising tool, assisting us to know the risk of colorectal cancer, identify the subtypes of tumor early, understand how patients will respond to treatment, and discover new targets of treatment. By adding this to clinical and genetic data, we have a chance to refine the prognosis, better select treatments, and provide better patient outcomes for CRC.

References

Ahlquist, D. A., Skoletsky, J. E., Boynton, K. A., Harrington, J. J., Mahoney, D. W., Pierceall, W. E., Thibodeau, S. N., & Shuber, A. P. (2000). Colorectal cancer screening by detection of altered human DNA in stool: feasibility of a multitarget assay panel. *Gastroenterology, 119*(5), 1219–1227. https://doi.org/10.1053/gast.2000.19580

Ahmad, A., Mahmood, N., Raza, M. A., Mushtaq, Z., Saeed, F., Afzaal, M., Hussain, M., Amjad, H. W., & Al-Awadi, H. M. (2024). Gut microbiota and their derivatives in the progression of colorectal cancer: Mechanisms of action, genome and epigenome contributions. *Heliyon, 10*(8), e29495. https://doi.org/10.1016/j.heliyon.2024.e29495

Anderson, M. W., Reynolds, S. H., You, M., & Maronpot, R. M. (1992). Role of proto-oncogene activation in carcinogenesis. *Environmental Health Perspectives, 98*, 13–24. https://api.semanticscholar.org/CorpusID:9780084

Anwar, M. M., Albanese, C., Hamdy, N. M., & Sultan, A. S. (2022). Rise of the natural red pigment 'prodigiosin' as an immunomodulator in cancer. *Cancer Cell International, 22*(1), 419. https://doi.org/10.1186/s12935-022-02815-4

Arnold, C. N., Goel, A., & Boland, C. R. (2003). Role of hMLH1 promoter hypermethylation in drug resistance to 5-fluorouracil in colorectal cancer cell lines. *International Journal of Cancer, 106*(1), 66–73. https://doi.org/10.1002/ijc.11176

Ashktorab, H., Belgrave, K., Hosseinkhah, F., Brim, H., Nouraie, M., Takkikto, M., Hewitt, S., Lee, E. L., Dashwood, R. H., & Smoot, D. (2009). Global histone H4 acetylation and HDAC2 expression in colon adenoma and carcinoma. *Digestive Diseases and Sciences, 54*(10), 2109–2117. https://doi.org/10.1007/s10620-008-0601-7

Au, S. L. K., Wong, C. C. L., Lee, J. M. F., Fan, D. N. Y., Tsang, F. H., Ng, I. O. L., & Wong, C. M. (2012). Enhancer of zeste homolog 2 epigenetically silences multiple tumor suppressor microRNAs to promote liver cancer metastasis. *Hepatology, 56*(2), 622–631. https://doi.org/10.1002/hep.25679

Bader El Din, N. G., El-Shenawy, R., Moustafa, R. I., Khairy, A., & Farouk, S. (2024). Association between the expression level of miRNA-374a and TGF-β1 in patients with colorectal cancer. *World Academy of Sciences Journal, 6*(6), 1–8. https://doi.org/10.3892/wasj.2024.283

Bader El Din, N. G., Farouk, S., Abdel-Salam, L. O., & Khairy, A. (2021). The potential value of miRNA-223 as a diagnostic biomarker for Egyptian colorectal patients. *European Journal of Gastroenterology and Hepatology, 33*(1), 25–31. https://doi.org/10.1097/MEG.0000000000001961

Bader El Din, N. G., Ibrahim, M. K., El-Shenawy, R., Salum, G. M., Farouk, S., Zayed, N., Khairy, A., & El Awady, M. (2020). MicroRNAs expression profiling in Egyptian colorectal cancer patients. *IUBMB Life, 72*(2), 275–284. https://doi.org/10.1002/iub.2164

Bae, J. M., Kim, J. H., & Kang, G. H. (2013). Epigenetic alterations in colorectal cancer: the CpG Island methylator phenotype. *Histology and Histopathology, 28 5*, 585–595. https://api.semanticscholar.org/CorpusID:29059389

Bamberg, L. V., Heigwer, F., Wandmacher, A. M., Singh, A., Betge, J., Rindtorff, N., Werner, J., Josten, J., Skabkina, O. V., Hinsenkamp, I., Erdmann, G., Röcken, C., Ebert, M. P., Burgermeister, E., Zhan, T., & Boutros, M. (2022). Targeting euchromatic histone lysine methyltransferases sensitizes colorectal cancer to histone deacetylase inhibitors. *International Journal of Cancer, 151*(9), 1586–1601. https://doi.org/10.1002/ijc.34155

Barták, B. K., Kalmár, A., Péterfia, B., Patai, Á. V, Galamb, O., Valcz, G., Spisák, S., Wichmann, B., Nagy, Z. B., Tóth, K., Tulassay, Z., Igaz, P., & Molnár, B. (2017). Colorectal adenoma and cancer detection based on altered methylation pattern of SFRP1, SFRP2, SDC2, and PRIMA1 in plasma samples. *Epigenetics, 12*(9), 751–763. https://doi.org/10.1080/15592294.2017.1356957

Benard, A., Goossens-Beumer, I. J., van Hoesel, A. Q., de Graaf, W., Horati, H., Putter, H., Zeestraten, E. C. M., van de Velde, C. J. H., & Kuppen, P. J. K. (2014a). Histone trimethylation at H3K4, H3K9 and H4K20 correlates with patient survival and tumor recurrence in early-stage colon cancer. *BMC Cancer, 14*, 531. https://doi.org/10.1186/1471-2407-14-531

Benard, A., Goossens-Beumer, I. J., van Hoesel, A. Q., Horati, H., de Graaf, W., Putter, H., Zeestraten, E. C. M., Liefers, G.-J., van de Velde, C. J. H., & Kuppen, P. J. K. (2015). Nuclear expression of histone deacetylases and their histone modifications predicts clinical outcome in colorectal cancer. *Histopathology*, *66*(2), 270–282. https://doi.org/10.1111/his.12534

Benard, A., Goossens-Beumer, I. J., van Hoesel, A. Q., Horati, H., Putter, H., Zeestraten, E. C. M., van de Velde, C. J. H., & Kuppen, P. J. K. (2014b). Prognostic value of polycomb proteins EZH2, BMI1 and SUZ12 and histone modification H3K27me3 in colorectal cancer. *PloS One*, *9*(9), e108265. https://doi.org/10.1371/journal.pone.0108265

Berger, S. L., Kouzarides, T., Shiekhattar, R., & Shilatifard, A. (2009). An operational definition of epigenetics. *Genes & Development*, *23*(7), 781–783. https://doi.org/10.1101/gad.1787609

Bonnet, M., Buc, E., Sauvanet, P., Darcha, C., Dubois, D., Pereira, B., Déchelotte, P., Bonnet, R., Pezet, D., & Darfeuille-Michaud, A. (2014). Colonization of the Human Gut by E. coli and Colorectal Cancer Risk. *Clinical Cancer Research*, *20*(4), 859–867. https://doi.org/10.1158/1078-0432.CCR-13-1343

Borai, I. H., Shaker, Y., Kamal, M. M., Ezzat, W. M., Ashour, E., Afify, M., Gouda, W., & Elbrashy, M. M. (2017). Evaluation of biomarkers in egyptian patients with different grades of nonalcoholic fatty liver disease. *Journal of Clinical and Translational Hepatology*, *5*(2), 109–118. https://doi.org/10.14218/JCTH.2017.00004

Borai, I. H., Shaker, Y. M., Moustafa Kamal, M., Ezzat, W. M., Ashour, E., Afify, M., Gouda, W., & Elbrashy, M. M. (2018). Association of nonalcoholic fatty liver disease grades with the plasma cell antigen-1 (PC-1) gene polymorphism. *Egyptian Journal of Medical Human Genetics*, *19*(3), 241—246. https://doi.org/10.1016/j.ejmhg.2017.12.005

Brenner, H., Kloor, M., & Pox, C. P. (2014). Colorectal cancer. *The Lancet*, *383*(9927), 1490–1502. https://doi.org/10.1016/S0140-6736(13)61649-9

Brownell, J. E., Zhou, J., Ranalli, T., Kobayashi, R., Edmondson, D. G., Roth, S. Y., & Allis, C. D. (1996). Tetrahymena Histone Acetyltransferase A: A Homolog to Yeast Gcn5p Linking Histone Acetylation to Gene Activation. *Cell*, *84*(6), 843–851. https://doi.org/10.1016/S0092-8674(00)81063-6

Brueckl, W. M., Grombach, J., Wein, A., Ruckert, S., Porzner, M., Dietmaier, W., Rümmele, P., Croner, R. S., Boxberger, F., Kirchner, T., Hohenberger, W., Hahn, E. G., & Jung, A. (2005). Alterations in the tissue inhibitor of metalloproteinase-3 (TIMP-3) are found frequently in human colorectal tumours displaying either microsatellite stability (MSS) or instability (MSI). *Cancer Letters*, *223*(1), 137–142. https://doi.org/10.1016/j.canlet.2004.09.037

Bultman, S. J. (2014). Emerging roles of the microbiome in cancer. *Carcinogenesis*, *35*(2), 249–255. https://doi.org/10.1093/carcin/bgt392

Bultman, S. J. (2017). Interplay between diet, gut microbiota, epigenetic events, and colorectal cancer. *Molecular Nutrition \& Food Research*, *61*(1), 1500902. https://doi.org/10.1002/mnfr.201500902

Burns, M. B., Lynch, J., Starr, T. K., Knights, D., & Blekhman, R. (2015). Virulence genes are a signature of the microbiome in the colorectal tumor microenvironment. *Genome Medicine*, *7*(1), 55. https://doi.org/10.1186/s13073-015-0177-8

Cells, C. E., Chung, L., Orberg, E. T., Geis, A. L., Wu, S., Sears, C. L., & Housseau, F. (2018). Bacteroides fragilis Toxin Coordinates a Pro- carcinogenic Inflammatory Cascade via Targeting of Article Bacteroides fragilis Toxin Coordinates a Pro-carcinogenic Inflammatory Cascade via Targeting of Colonic Epithelial Cells. *Cell Host and Microbe*, *23*(2), 203–214.e5. https://doi.org/10.1016/j.chom.2018.01.007

Chen, B., Zhao, H., Hu, H., Shang, H., Wang, H., Yao, Z., Huang, J., Lv, H., Xu, W., Wang, J., Nie, C., Zhao, J., He, Y., Wang, S., & Chen, X. (2025). Circulating DNA methylation-based diagnostic, prognostic, and predictive biomarkers in colorectal cancer. *Scientific Reports*, *15*(1), 13577. https://doi.org/10.1038/s41598-025-95712-5

Chen, S., Zhang, C., & Feng, M. (2020). Prognostic Value of LncRNA HOTAIR in Colorectal Cancer: A Meta-analysis. *Open Medicine (Warsaw, Poland)*, *15*, 76–83. https://doi.org/10.1515/med-2020-0012

Chen, W.-D., Han, Z. J., Skoletsky, J., Olson, J., Sah, J., Myeroff, L., Platzer, P., Lu, S., Dawson, D., Willis, J., Pretlow, T. P., Lutterbaugh, J., Kasturi, L., Willson, J. K. V, Rao, J. S., Shuber, A., & Markowitz, S. D. (2005). Detection in fecal DNA of colon cancer-specific methylation of the nonexpressed vimentin gene. *Journal of the National Cancer Institute*, *97*(15), 1124–1132. https://doi.org/10.1093/jnci/dji204

Chen, X., Guo, X., Zhang, H., Xiang, Y., Chen, J., Yin, Y., Cai, X., Wang, K., Wang, G., Ba, Y., Zhu, L., Wang, J., Yang, R., Zhang, Y., Ren, Z., Zen, K., Zhang, J., & Zhang, C.-Y. (2009). Role of miR-143 targeting KRAS in colorectal tumorigenesis. *Oncogene*, *28*(10), 1385–1392. https://doi.org/10.1038/onc.2008.474

Chen T, Li J, Xu M, Zhao Q, Hou Y, Yao L, Zhong Y, Chou PC, Zhang W, Zhou P, Jiang Y. PKCε phosphorylates MIIP and promotes colorectal cancer metastasis through inhibition of RelA deacetylation. Nat Commun. 2017 Oct 16;8(1):939. https://doi.org/10.1038/s41467-017-01024-2. PMID: 29038521; PMCID: PMC5643311.

Cheng, A. S. L., Li, M. S., Kang, W., Cheng, V. Y., Chou, J.-L., Lau, S. S., Go, M. Y., Lee, C. C., Ling, T. K., Ng, E. K., Yu, J., Huang, T. H., To, K. F., Chan, M. W., Sung, J. J. Y., & Chan, F. K. L. (2013). Helicobacter pylori causes epigenetic dysregulation of FOXD3 to promote gastric carcinogenesis. *Gastroenterology*, *144*(1), 122–133.e9. https://doi.org/10.1053/j.gastro.2012.10.002

Crockett, S. D., & Nagtegaal, I. D. (2019). Terminology, Molecular Features, Epidemiology, and Management of Serrated Colorectal Neoplasia. *Gastroenterology*, *157*(4), 949–966.e4. https://doi.org/10.1053/j.gastro.2019.06.041

Dawood, R. M., El-Meguid, M. A., Ibrahim, M. K., Bader El Din, N. G., Barakat, A., El-Wakeel, K., Alla, M. D. A. A., Wu, G. Y., & El Awady, M. K. (2018). Dysregulation of fibrosis related genes in HCV induced liver disease. *Gene*, *664*, 58–69. https://doi.org/10.1016/j.gene.2018.04.032

Day, C. A., Hinchcliffe, E. H., & Robinson, J. P. (2022). H3K27me3 in Diffuse Midline Glioma and Epithelial Ovarian Cancer: Opposing Epigenetic Changes Leading to the Same Poor Outcomes. *Cells*, *11*(21). https://doi.org/10.3390/cells11213376

De Palma, F. D. E., D'Argenio, V., Pol, J., Kroemer, G., Maiuri, M. C., & Salvatore, F. (2019). The Molecular Hallmarks of the Serrated Pathway in Colorectal Cancer. *Cancers*, *11*(7). https://doi.org/10.3390/cancers11071017

Debernardi, C., Libera, L., Berrino, E., Sahnane, N., Chiaravalli, A. M., Laudi, C., Berselli, M., Sapino, A., Sessa, F., Venesio, T., & Furlan, D. (2021). Evaluation of global and intragenic hypomethylation in colorectal adenomas improves patient stratification and colorectal cancer risk prediction. *Clinical Epigenetics*, *13*(1), 1–11. https://doi.org/10.1186/s13148-021-01135-0

Ding, J., Zhang, Z.-M., Xia, Y., Liao, G.-Q., Pan, Y., Liu, S., Zhang, Y., & Yan, Z.-S. (2013). LSD1-mediated epigenetic modification contributes to proliferation and metastasis of colon cancer. *British Journal of Cancer*, *109*(4), 994–1003. https://doi.org/10.1038/bjc.2013.364

Doghish, A. S., Bakr, M., Gomaa, M., Radwan, A. F., Moussa, R., Abdel-wahab, M. M., Kizilaslan, E. Z., Alhamshry, N. A. A., Ashour, A. E., & Elimam, H. (2024). Pathology - Research and Practice The potential relevance of long non-coding RNAs in colorectal cancer pathogenesis and treatment : A review focus on signaling pathways. *Pathology - Research and Practice*, *253*(November 2023), 155044. https://doi.org/10.1016/j.prp.2023.155044

Duran-Sanchon, S., Moreno, L., Augé, J. M., Serra-Burriel, M., Cuatrecasas, M., Moreira, L., Martín, A., Serradesanferm, A., Pozo, À., Costa, R., Lacy, A., Pellisé, M., Lozano, J. J., Gironella, M., & Castells, A. (2020). Identification and Validation of MicroRNA Profiles in Fecal Samples for Detection of Colorectal Cancer. *Gastroenterology*, *158*(4), 947–957.e4. https://doi.org/10.1053/j.gastro.2019.10.005

Ehrlich, M. (2002). DNA methylation in cancer: too much, but also too little. *Oncogene*, *21*(35), 5400–5413. https://doi.org/10.1038/sj.onc.1205651

El-Awady, M. K., el-Din, N. G. B., el-Garf, W. T., Youssef, S. S., Omran, M. H., el-Abd, J., & Goueli, S. A. (2006a). Antisense oligonucleotide inhibition of hepatitis C virus genotype 4 replication in HepG2 cells. *Cancer Cell International*, *6*(1), 18. https://doi.org/10.1186/1475-2867-6-18

El-Awady, M. K., Tabll, A. A., Atef, K., Yousef, S. S., Omran, M. H., El-Abd, Y., ... & El-Garf, W. T. (2006b). Antibody to E1 peptide of hepatitis C virus genotype 4 inhibits virus binding and entry to HepG2 cells in vitro. World Journal of Gastroenterology: WJG, 12(16), 2530.

El-Awady, M. K., Tabll, A. A., El-Abd, Y. S., Yousif, H., Hegab, M., Reda, M., ... & El Din, N. G. B. (2009). Conserved peptides within the E2 region of Hepatitis C virus induce humoral and cellular responses in goats. Virology Journal, 6(1), 66.

Elbrashy, M. M., Metwally, H., Sakakibara, S., & Kishimoto, T. (2024). Threonine Phosphorylation and the Yin and Yang of STAT1: Phosphorylation-Dependent Spectrum of STAT1 Functionality in Inflammatory Contexts. *Cells*, *13*(18). https://doi.org/10.3390/cells13181531

Eldash, S., Sanad, E. F., Nada, D., & Hamdy, N. M. (2023). *The Intergenic Type LncRNA (LINC RNA) Faces in Cancer with In Silico Scope and a Directed Lens to LINC00511 : A Step toward ncRNA Precision*. *9*(58).

El Din, N. G. B., Farouk, S., El-Shenawy, R., Ibrahim, M. K., Dawood, R. M., Elhady, M. M., ... & El Awady, M. K. (2016). Tumor necrosis factor-α-G308A polymorphism is associated with liver pathological changes in hepatitis C virus patients. *World journal of gastroenterology*, *22*(34), 7767.

Esteller, M. (2005). Aberrant DNA methylation as a cancer-inducing mechanism. *Annual Review of Pharmacology and Toxicology*, *45*, 629–656. https://api.semanticscholar.org/CorpusID:3932142

Fang, M., Ou, J., Hutchinson, L., & Green, M. R. (2014). The BRAF Oncoprotein Functions through the Transcriptional Repressor MAFG to Mediate the CpG Island Methylator Phenotype. *Molecular Cell*, *55*(6), 904–915. https://doi.org/10.1016/j.molcel.2014.08.010

Farouk, S., El-Shenawy, R., Khairy, A. M., & Bader El-Din, N. G. (2022). Overexpression of miRNA 26a and 26b with MMP-9 are valuable diagnostic biomarkers for colorectal cancer patients. *Biomarkers in Medicine*, *17*(3), 159—169. https://doi.org/10.2217/bmm-2022-0861

Farouk, S., Khairy, A., Salem, A. M., Soliman, A. F., & Bader El Din, N. G. (2020). Differential Expression of miR-21, miR-23a, and miR-27a, and Their Diagnostic Significance in Egyptian Colorectal Cancer Patients. *Genetic Testing and Molecular Biomarkers*, *24*(12), 825—834. https://doi.org/10.1089/gtmb.2020.0184

Farouk, S., Elbrashy, M. M., Moustafa, R. I., El-Shenawy, R. M., Helmy, N. M., Khairy, A., & Bader El Din, N. G. (2025). Unique dysregulation of miR-7, miR-15, and miR-203 as novel biomarkers in Egyptian colorectal cancer. Biomarkers in Medicine, 19(20), 999–1008. https://doi.org/10.1080/17520363.2025.2585779

Fathi, S., Aazzane, O., Guendaoui, S., Tawfiq, N., Sahraoui, S., Guessous, F., & Karkouri, M. (2025). A miRNA Signature for Non-Invasive Colorectal Cancer Diagnosis in Morocco: miR-21, miR-29a and miR-92a. *Non-Coding RNA*, *11*(2). https://doi.org/10.3390/ncrna11020026

Feinberg, A. P., & Vogelstein, B. (1983). Hypomethylation distinguishes genes of some human cancers from their normal counterparts. *Nature*, *301*(5895), 89–92. https://doi.org/10.1038/301089a0

Fraga, M. F., Ballestar, E., Villar-Garea, A., Boix-Chornet, M., Espada, J., Schotta, G., Bonaldi, T., Haydon, C., Ropero, S., Petrie, K., Iyer, N. G., Pérez-Rosado, A., Calvo, E., Lopez, J. A., Cano, A., Calasanz, M. J., Colomer, D., Piris, M. A., Ahn, N., … Esteller, M. (2005). Loss of acetylation at Lys16 and trimethylation at Lys20 of histone H4 is a common hallmark of human cancer. *Nature Genetics*, *37*(4), 391–400. https://doi.org/10.1038/ng1531

Gao, P., Tchernyshyov, I., Chang, T.-C., Lee, Y.-S., Kita, K., Ochi, T., Zeller, K. I., De Marzo, A. M., Van Eyk, J. E., Mendell, J. T., & Dang, C. V. (2009). c-Myc suppression of miR-23a/b enhances mitochondrial glutaminase expression and glutamine metabolism. *Nature*, *458*(7239), 762–765. https://doi.org/10.1038/nature07823

Gao, Y., Wang, X., Dong, L., Qu, C., Lu, Q., Wang, P., Xin, M., Zheng, W., Liu, C., & Ning, S. (2023). Identifying immune checkpoint-related lncRNA biomarkers for immunotherapy response and prognosis in cancers. *Scientific Data*, *10*(1), 663. https://doi.org/10.1038/s41597-023-02550-z

Gaudet, F., Hodgson, J. G., Eden, A., Jackson-Grusby, L., Dausman, J., Gray, J. W., Leonhardt, H., & Jaenisch, R. (2003). Induction of Tumors in Mice by Genomic Hypomethylation. *Science, 300*(5618), 489–492. https://doi.org/10.1126/science.1083558

Gezer, U., Ustek, D., Yörüker, E. E., Cakiris, A., Abaci, N., Leszinski, G., Dalay, N., & Holdenrieder, S. (2013). Characterization of H3K9me3- and H4K20me3-associated circulating nucleosomal DNA by high-throughput sequencing in colorectal cancer. *Tumour Biology : The Journal of the International Society for Oncodevelopmental Biology and Medicine*, *34*(1), 329–336. https://doi.org/10.1007/s13277-012-0554-5

Gezer, U., Yörüker, E. E., Keskin, M., Kulle, C. B., Dharuman, Y., & Holdenrieder, S. (2015). Histone Methylation Marks on Circulating Nucleosomes as Novel Blood-Based Biomarker in Colorectal Cancer. *International Journal of Molecular Sciences*, *16*(12), 29654–29662. https://doi.org/10.3390/ijms161226180

Gonzalez, M. E., Li, X., Toy, K., DuPrie, M., Ventura, A. C., Banerjee, M., Ljungman, M., Merajver, S. D., & Kleer, C. G. (2009). Downregulation of EZH2 decreases growth of estrogen receptor-negative invasive breast carcinoma and requires BRCA1. *Oncogene*, *28*(6), 843–853. https://doi.org/10.1038/onc.2008.433

Hamdy, N. M., Basalious, E. B., El-Sisi, M. G., Nasr, M., Kabel, A. M., Nossier, E. S., & Abadi, A. H. (2024a). Advancements in current one-size-fits-all therapies compared to future treatment innovations for better improved chemotherapeutic outcomes: a step-toward personalized medicine. *Current Medical Research and Opinion*, *40*(11), 1943–1961. https://doi.org/10.1080/03007995.2024.2416985

Hamdy, N. M., Zaki, M. B., Abdelmaksoud, N. M., Ismail, R. A., Abd-Elmawla, M. A., Rizk, N. I., Fathi, D., & Abulsoud, A. I. (2025). Insights into the genetic and epigenetic mechanisms governing X-chromosome-linked-miRNAs expression in cancer; a step-toward ncRNA precision. *International Journal of Biological Macromolecules*, *289*, 138773. https://doi.org/10.1016/j.ijbiomac.2024.138773

Hamdy, N. M., Zaki, M. B., Rizk, N. I., Abdelmaksoud, N. M., Abd-Elmawla, M. A., Ismail, R. A., & Abulsoud, A. I. (2024b). Unraveling the ncRNA landscape that governs colorectal cancer: A roadmap to personalized therapeutics. *Life Sciences*, *354*, 122946. https://doi.org/10.1016/j.lfs.2024.122946

Hamid, A. A., Gray, K. P., Shaw, G., MacConaill, L. E., Evan, C., Bernard, B., Loda, M., Corcoran, N. M., Van Allen, E. M., Choudhury, A. D., & Sweeney, C. J. (2019). Compound Genomic Alterations of TP53, PTEN, and RB1 Tumor Suppressors in Localized and Metastatic Prostate Cancer. *European Urology*, *76*(1), 89–97. https://doi.org/10.1016/j.eururo.2018.11.045

Han, J., Sun, W., Liu, R., Zhou, Z., Zhang, H., Chen, X., & Ba, Y. (2020). Plasma Exosomal miRNA Expression Profile as Oxaliplatin-Based Chemoresistant Biomarkers in Colorectal Adenocarcinoma. *Frontiers in Oncology*, *10*, 1495. https://doi.org/10.3389/fonc.2020.01495

Hashimoto, T., Yamakawa, M., Kimura, S., Usuba, O., & Toyono, M. (2013). Expression of acetylated and dimethylated histone H3 in colorectal cancer. *Digestive Surgery*, *30*(3), 249–258. https://doi.org/10.1159/000351444

He, R.-Z., Luo, D.-X., & Mo, Y.-Y. (2019). Emerging roles of lncRNAs in the post-transcriptional regulation in cancer. *Genes & Diseases*, *6*(1), 6–15. https://doi.org/10.1016/j.gendis.2019.01.003

He, X., Ma, J., Zhang, M., Cui, J., & Yang, H. (2020a). Circ_0007031 enhances tumor progression and promotes 5-fluorouracil resistance in colorectal cancer through regulating miR-133b/ABCC5 axis. *Cancer Biomarkers : Section A of Disease Markers*, *29*(4), 531–542. https://doi.org/10.3233/CBM-200023

He, Y., Zhang, L., Tan, F., Wang, L.-F., Liu, D.-H., Wang, R.-J., & Yin, X.-Z. (2020b). MiR-153-5p promotes sensibility of colorectal cancer cells to oxaliplatin via targeting Bcl-2-mediated autophagy pathway. *Bioscience, Biotechnology, and Biochemistry*, *84*(8), 1645–1651. https://doi.org/10.1080/09168451.2020.1760784

Heggie, G. D., Sommadossi, J. P., Cross, D. S., Huster, W. J., & Diasio, R. B. (1987). Clinical pharmacokinetics of 5-fluorouracil and its metabolites in plasma, urine, and bile. *Cancer Research*, *47*(8), 2203–2206.

Hernández-Contreras, K. A., Martínez-Díaz, J. A., Hernández-Aguilar, M. E., Herrera-Covarrubias, D., Rojas-Durán, F., Chi-Castañeda, L. D., García- Hernández, L. I., & Aranda-Abreu, G. E. (2024). Alterations of mRNAs and Non-coding RNAs Associated with Neuroinflammation in Alzheimer's Disease. *Molecular Neurobiology*, *61*(8), 5826–5840. https://doi.org/10.1007/s12035-023-03908-5

Hillman, T. (2023). The use of plant-derived exosome-like nanoparticles as a delivery system of CRISPR/Cas9-based therapeutics for editing long non-coding RNAs in cancer colon cells. *Frontiers in Oncology*, *13*. https://doi.org/10.3389/fonc.2023.1194350

Hinoue, T., Weisenberger, D. J., Lange, C. P. E., Shen, H., Byun, H.-M., Van Den Berg, D., Malik, S., Pan, F., Noushmehr, H., van Dijk, C. M., Tollenaar, R. A. E. M., & Laird, P. W. (2012). Genome-scale analysis of aberrant DNA methylation in colorectal cancer. *Genome Research*, *22*(2), 271–282. https://doi.org/10.1101/gr.117523.110

Hong, B., Lu, R., Lou, W., Bao, Y., Qiao, L., Hu, Y., Liu, K., Chen, J., Bao, D., Ye, M., Fang, Z., Gong, C., & Zhang, X. (2021). KIF18b-dependent hypomethylation of PARPBP gene promoter enhances oxaliplatin resistance in colorectal cancer. *Experimental Cell Research*, *407*(2), 112827. https://doi.org/10.1016/j.yexcr.2021.112827

Hosokawa, M., Seiki, R., Iwakawa, S., & Ogawara, K. (2021). Combination of azacytidine and curcumin is a potential alternative in decitabine-resistant colorectal cancer cells with attenuated deoxycytidine kinase. *Biochemical and Biophysical Research Communications*, *578*, 157–162. https://doi.org/10.1016/j.bbrc.2021.09.041

Hosokawa, M., Tanaka, S., Ueda, K., Iwakawa, S., & Ogawara, K.-I. (2019). Decitabine exerted synergistic effects with oxaliplatin in colorectal cancer cells with intrinsic resistance to decitabine. *Biochemical and Biophysical Research Communications*, *509*(1), 249–254. https://doi.org/10.1016/j.bbrc.2018.12.115

Hu, A., Hong, F., Li, D., Jin, Y., Kon, L., Xu, Z., He, H., & Xie, Q. (2021). Long non-coding RNA ROR recruits histone transmethylase MLL1 to up-regulate TIMP3 expression and promote breast cancer progression. *Journal of Translational Medicine*, *19*(1), 95. https://doi.org/10.1186/s12967-020-02682-5

Hu, S., Liu, L., Chang, E. B., Wang, J.-Y., & Raufman, J.-P. (2015). Butyrate inhibits pro-proliferative miR-92a by diminishing c-Myc-induced miR-17-92a cluster transcription in human colon cancer cells. *Molecular Cancer*, *14*, 180. https://doi.org/10.1186/s12943-015-0450-x

Hur, K., Cejas, P., Feliu, J., Moreno-Rubio, J., Burgos, E., Boland, C. R., & Goel, A. (2014). Hypomethylation of long interspersed nuclear element-1 (LINE-1) leads to activation of proto-oncogenes in human colorectal cancer metastasis. *Gut*, *63*(4), 635–646. https://doi.org/10.1136/gutjnl-2012-304219

Imperiale, T. F., Ransohoff, D. F., Itzkowitz, S. H., Levin, T. R., Lavin, P., Lidgard, G. P., Ahlquist, D. A., & Berger, B. M. (2014). Multitarget stool DNA testing for colorectal-cancer screening. *The New England Journal of Medicine*, *370*(14), 1287–1297. https://doi.org/10.1056/NEJMoa1311194

Inamura, K. (2018). Colorectal Cancers: An Update on Their Molecular Pathology. *Cancers*, *10*(1). https://doi.org/10.3390/cancers10010026

Inno, A., Fanetti, G., Di Bartolomeo, M., Gori, S., Maggi, C., Cirillo, M., Iacovelli, R., Nichetti, F., Martinetti, A., de Braud, F., Bossi, I., & Pietrantonio, F. (2014). Role of MGMT as biomarker in colorectal cancer. *World Journal of Clinical Cases*, *2*(12), 835–839. https://doi.org/10.12998/wjcc.v2.i12.835

Ishihama, K., Yamakawa, M., Semba, S., Takeda, H., Kawata, S., Kimura, S., & Kimura, W. (2007). Expression of HDAC1 and CBP/p300 in human colorectal carcinomas. *Journal of Clinical Pathology*, *60*(11), 1205–1210. https://doi.org/10.1136/jcp.2005.029165

Islam, A. B. M. M. K., Richter, W. F., Jacobs, L. A., Lopez-Bigas, N., & Benevolenskaya, E. V. (2011). Co-regulation of histone-modifying enzymes in cancer. *PLoS ONE*, *6*(8). https://doi.org/10.1371/journal.pone.0024023

Issah, M. A., Wu, D., Zhang, F., Zheng, W., Liu, Y., Fu, H., Zhou, H., Chen, R., & Shen, J. (2021). Epigenetic modifications in acute myeloid leukemia: The emerging role of circular RNAs (Review). *International Journal of Oncology*, *59*(6), 1–15. https://doi.org/10.3892/ijo.2021.5287

Jean-Louis, S., Akare, S., Ali, M. A., Mash, E. A. J., Meuillet, E., & Martinez, J. D. (2006). Deoxycholic acid induces intracellular signaling through membrane perturbations. *The Journal of Biological Chemistry*, *281*(21), 14948–14960. https://doi.org/10.1074/jbc.M506710200

Jung, G., Hernández-Illán, E., Moreira, L., Balaguer, F., & Goel, A. (2020). Epigenetics of colorectal cancer: biomarker and therapeutic potential. *Nature Reviews. Gastroenterology & Hepatology*, *17*(2), 111–130. https://doi.org/10.1038/s41575-019-0230-y

Kane, M. F., Loda, M., Gaida, G. M., Lipman, J., Mishra, R., Goldman, H., Jessup, J. M., & Kolodner, R. (1997). Methylation of the hMLH1 promoter correlates with lack of expression of hMLH1 in sporadic colon tumors and mismatch repair-defective human tumor cell lines. *Cancer Research*, *57*(5), 808–811.

Kaneko, M., Kotake, M., Bando, H., Yamada, T., Takemura, H., & Minamoto, T. (2016). Prognostic and predictive significance of long interspersed nucleotide element-1 methylation in advanced-stage colorectal cancer. *BMC Cancer*, *16*(1), 1–10. https://doi.org/10.1186/s12885-016-2984-8

Kang, K. A., Piao, M. J., Kim, K. C., Kang, H. K., Chang, W. Y., Park, I. C., Keum, Y. S., Surh, Y. J., & Hyun, J. W. (2014). Epigenetic modification of Nrf2 in 5-fluorouracil-resistant colon cancer cells: involvement of TET-dependent DNA demethylation. *Cell Death & Disease*, *5*(4), e1183–e1183. https://doi.org/10.1038/cddis.2014.149

Kang, M. Y., Lee, B. Bin, Kim, Y.-H., Chang, D. K., Kyu Park, S., Chun, H.-K., Song, S. Y., Park, J., & Kim, D.-H. (2007). Association of the SUV39H1 histone methyltransferase with the DNA methyltransferase 1 at mRNA expression level in primary colorectal cancer. *International Journal of Cancer*, *121*(10), 2192–2197. https://doi.org/10.1002/ijc.22953

Karczmarski, J., Rubel, T., Paziewska, A., Mikula, M., Bujko, M., Kober, P., Dadlez, M., & Ostrowski, J. (2014). Histone H3 lysine 27 acetylation is altered in colon cancer. *Clinical Proteomics*, *11*(1), 24. https://doi.org/10.1186/1559-0275-11-24

Kawanaka, M., Watari, J., Kamiya, N., Yamasaki, T., Kondo, T., Toyoshima, F., Ikehara, H., Tomita, T., Oshima, T., Fukui, H., Daimon, T., Das, K. M., & Miwa, H. (2016). Effects of Helicobacter pylori eradication on the development of metachronous gastric cancer after endoscopic treatment: analysis of molecular alterations by a randomised controlled trial. *British Journal of Cancer*, *114*(1), 21–29. https://doi.org/10.1038/bjc.2015.418

Keum, N., & Giovannucci, E. (2019). Global burden of colorectal cancer: emerging trends, risk factors and prevention strategies. *Nature Reviews. Gastroenterology & Hepatology*, *16*(12), 713–732. https://doi.org/10.1038/s41575-019-0189-8

Khan, A. Q., Kuttikrishnan, S., Siveen, K. S., Prabhu, K. S., Shanmugakonar, M., Al-Naemi, H. A., Haris, M., Dermime, S., & Uddin, S. (2019). RAS-mediated oncogenic signaling pathways in human malignancies. *Seminars in Cancer Biology*, *54*, 1–13. https://doi.org/10.1016/j.semcancer.2018.03.001

Kim, Y.-H., Petko, Z., Dzieciatkowski, S., Lin, L., Ghiassi, M., Stain, S., Chapman, W. C., Washington, M. K., Willis, J., Markowitz, S. D., & Grady, W. M. (2006). CpG Island methylation of genes accumulates during the adenoma progression step of the multistep pathogenesis of colorectal cancer. *Genes, Chromosomes & Cancer*, *45*(8), 781–789. https://doi.org/10.1002/gcc.20341

Kogo, R., Shimamura, T., Mimori, K., Kawahara, K., Imoto, S., Sudo, T., Tanaka, F., Shibata, K., Suzuki, A., Komune, S., Miyano, S., & Mori, M. (2011). Long noncoding RNA HOTAIR regulates polycomb-dependent chromatin modification and is associated with poor prognosis in colorectal cancers. *Cancer Research*, *71*(20), 6320–6326. https://doi.org/10.1158/0008-5472.CAN-11-1021

Koi, M., Okita, Y., & Carethers, J. M. (2018). Fusobacterium nucleatum Infection in Colorectal Cancer: Linking Inflammation, DNA Mismatch Repair and Genetic and Epigenetic Alterations. *Journal of the Anus, Rectum and Colon*, *2*(2), 37–46. https://doi.org/10.23922/jarc.2017-055

Kopp, M., Dürr, K., Steigleder, M., Clavel, T., & Rychlik, M. (2016). Development of stable isotope dilution assays for the quantitation of intra- and extracellular folate patterns of Bifidobacterium adolescentis. *Journal of Chromatography. A*, *1469*, 48–59. https://doi.org/10.1016/j.chroma.2016.09.048

Kornblihtt, A. R. (2017). Epigenetics at the base of alternative splicing changes that promote colorectal cancer. *The Journal of Clinical Investigation*, *127*(9), 3281–3283. https://doi.org/10.1172/JCI96497

Lao, V. V., & Grady, W. M. (2011). Epigenetics and colorectal cancer. *Nature Reviews Gastroenterology & Hepatology*, *8*(12), 686–700. https://doi.org/10.1038/nrgastro.2011.173

Lee YC, Yin TC, Chen YT, Chai CY, Wang JY, Liu MC, Lin YC, Kan JY. High expression of phospho-H2AX predicts a poor prognosis in colorectal cancer. Anticancer Res. 2015 Apr;35(4):2447-53. PMID: 25862913.

Li, H., Chiappinelli, K. B., Guzzetta, A. A., Easwaran, H., Azad, N. S., Stearns, V., Pardoll, D. M., & Davidson, N. (2004). *Immune regulation by low doses of the DNA methyltransferase inhibitor 5-azacitidine in common human epithelial cancers ABSTRACT : 5*(3).

Li, Q., Geng, S., Luo, H., Wang, W., Mo, Y.-Q., Luo, Q., Wang, L., Song, G.-B., Sheng, J.-P., Xu, B., & Therapy, T. (2024). Signaling pathways involved in colorectal cancer: pathogenesis and targeted therapy. *Signal Transduction and Targeted Therapy*, *9*(1), 266. https://doi.org/10.1038/s41392-024-01953-7

Li, Y., Gan, Y., Liu, J., Li, J., Zhou, Z., Tian, R., Sun, R., Liu, J., Xiao, Q., Li, Y., Lu, P., Peng, Y., Peng, Y., Shu, G., & Yin, G. (2022). Downregulation of MEIS1 mediated by ELFN1-AS1/EZH2/DNMT3a axis promotes tumorigenesis and oxaliplatin resistance in colorectal cancer. *Signal Transduction and Targeted Therapy*, *7*(1), 87. https://doi.org/10.1038/s41392-022-00902-6

Liang, J.-T., Chang, K.-J., Chen, J.-C., Lee, C.-C., Cheng, Y.-M., Hsu, H.-C., Wu, M.-S., Wang, S.-M., Lin, J.-T., & Cheng, A.-L. (1999). Hypermethylation of the p16 Gene in Sporadic T3N0M0 Stage Colorectal Cancers: Association with DNA Replication Error and Shorter Survival. *Oncology*, *57*(2), 149–156. https://doi.org/10.1159/000012023

Liang, T.-J., Wang, H.-X., Zheng, Y.-Y., Cao, Y.-Q., Wu, X., Zhou, X., & Dong, S.-X. (2017). APC hypermethylation for early diagnosis of colorectal cancer: a meta-analysis and literature review. *Oncotarget*, *8*(28), 46468–46479. https://doi.org/10.18632/oncotarget.17576

Lin, J., Zhang, L., Chen, M., Chen, J., Wu, Y., Wang, T., Lu, Y., Ba, Z., Cheng, X., Xu, R., Tian, T., Sun, A., Zhang, T., & Chen, M. (2022). Evaluation of combined detection of multigene mutation and SDC2/SFRP2 methylation in stool specimens for colorectal cancer early diagnosis. *International Journal of Colorectal Disease*, *37*(6), 1231–1238. https://doi.org/10.1007/s00384-022-04170-2

Ling, C. (2020). Epigenetic regulation of insulin action and secretion - role in the pathogenesis of type 2 diabetes. *Journal of Internal Medicine*, *288*(2), 158–167. https://doi.org/10.1111/joim.13049

Lişcu, H.-D., Verga, N., Atasiei, D.-I., Badiu, D.-C., Dumitru, A. V., Ultimescu, F., Pavel, C., Stefan, R.-E., Manole, D.-C., & Ionescu, A.-I. (2024). Biomarkers in Colorectal Cancer: Actual and Future Perspectives. *International Journal of Molecular Sciences*, *25*(21). https://doi.org/10.3390/ijms252111535

Liu, D., Wu, K., Yang, Y., Zhu, D., Zhang, C., & Zhao, S. (2020b). Long noncoding RNA ADAMTS9-AS2 suppresses the progression of esophageal cancer by mediating CDH3 promoter methylation. *Molecular Carcinogenesis*, *59*(1), 32–44. https://doi.org/10.1002/mc.23126

Liu, H.-N., Liu, T.-T., Wu, H., Chen, Y.-J., Tseng, Y.-J., Yao, C., Weng, S.-Q., Dong, L., & Shen, X.-Z. (2018c). Serum microRNA signatures and metabolomics have high diagnostic value in colorectal cancer using two novel methods. *Cancer Science*, *109*(4), 1185–1194. https://doi.org/10.1111/cas.13514

Liu, L., Dong, W., Wang, S., Zhang, Y., Liu, T., Xie, R., Wang, B., & Cao, H. (2018a). Deoxycholic acid disrupts the intestinal mucosal barrier and promotes intestinal tumorigenesis. *Food & Function*, *9*(11), 5588–5597. https://doi.org/10.1039/c8fo01143e

Liu, M., Jia, J., Wang, X., Liu, Y., Wang, C., & Fan, R. (2018b). Long non-coding RNA HOTAIR promotes cervical cancer progression through regulating BCL2 via targeting miR-143-3p. *Cancer Biology \& Therapy*, *19*(5), 391–399. https://doi.org/10.1080/15384047.2018.1423921

Liu, R., Wu, J., Guo, H., Yao, W., Li, S., Lu, Y., Jia, Y., Liang, X., Tang, J., & Zhang, H. (2023). Post-translational modifications of histones: Mechanisms, biological functions, and therapeutic targets. *MedComm*, *4*(3), e292. https://doi.org/10.1002/mco2.292

Liu, T., Song, X., Khan, S., Li, Y., Guo, Z., Li, C., Wang, S., Dong, W., Liu, W., Wang, B., & Cao, H. (2020a). The gut microbiota at the intersection of bile acids and intestinal carcinogenesis: An old story, yet mesmerizing. *International Journal of Cancer*, *146*(7), 1780–1790. https://doi.org/10.1002/ijc.32563

Liu, Y., Chen, C., Xu, Z., Scuoppo, C., Rillahan, C. D., Gao, J., Spitzer, B., Bosbach, B., Kastenhuber, E. R., Baslan, T., Ackermann, S., Cheng, L., Wang, Q., Niu, T., Schultz, N. D., Levine, R. L., Mills, A. A., & Lowe, S. W. (2016). Deletions linked to TP53 loss drive cancer through p53-independent mechanisms. *Nature*, *531*, 471–475. https://api.semanticscholar.org/CorpusID:4446099

Ma, C., Liu, H., Yi-Feng Chang, I., Wang, W., Chen, Y., Wu, S., Chen, H., Kuo, Y., Shih, C., Li, C., & Tan, B. C. (2019). ADAR1 promotes robust hypoxia signaling via distinct regulation of multiple HIF-1α-inhibiting factors. *EMBO Reports*, *20*(5), 1–16. https://doi.org/10.15252/embr.201847107

Maduscha, K., Schmitz, A., DeStefanis, R., Pasch, C., & Deming, D. (2024). Abstract 2965: GADD45B as a mechanism for enhanced treatment response with the combination of copanlisib and romidepsin in colorectal cancer. *Cancer Research*, *84*(6_Supplement), 2965. https://doi.org/10.1158/1538-7445.AM2024-2965

Maleszewska, M., Wojtas, B., & Kamińska, B. (2018). Deregulation of epigenetic mechanisms in cancer. *Postepy Biochemii*, *64*(2), 148–156.

Marks, P. A., & Dokmanovic, M. (2005). Histone deacetylase inhibitors: discovery and development as anticancer agents. *Expert Opinion on Investigational Drugs*, *14*(12), 1497–1511. https://doi.org/10.1517/13543784.14.12.1497

Mehrgou, A., Ebadollahi, S., Seidi, K., Ayoubi-Joshaghani, M. H., Ahmadieh Yazdi, A., Zare, P., Jaymand, M., & Jahanban-Esfahlan, R. (2021). Roles of miRNAs in Colorectal Cancer: Therapeutic Implications and Clinical Opportunities. *Advanced Pharmaceutical Bulletin*, *11*(2), 233–247. https://doi.org/10.34172/apb.2021.029

Melotte, V., Lentjes, M. H. F. M., van den Bosch, S. M., Hellebrekers, D. M. E. I., de Hoon, J. P. J., Wouters, K. A. D., Daenen, K. L. J., Partouns-Hendriks, I. E. J. M., Stessels, F., Louwagie, J., Smits, K. M., Weijenberg, M. P., Sanduleanu, S., Khalid-de Bakker, C. A. J., Oort, F. A., Meijer, G. A., Jonkers, D. M. A. E., Herman, J. G., de Bruïne, A. P., & van Engeland, M. (2009). N-Myc Downstream-Regulated Gene 4 (NDRG4): A Candidate Tumor Suppressor Gene and Potential Biomarker for Colorectal Cancer. *JNCI: Journal of the National Cancer Institute*, *101*(13), 916–927. https://doi.org/10.1093/jnci/djp131

Metwally, H., & Elbrashy, M. M. (2024). Analysis of Protein Interactions in Patient-Derived Xenografts Using Immunoprecipitation. *Methods in Molecular Biology*, *2806*, 219–227. https://doi.org/10.1007/978-1-0716-3858-3_16

Metwally, H., Elbrashy, M. M., Kayama, H., Okuyama, K., Taniuchi, I., Takeda, K., & Kishimoto, T. (2025). Threonine phosphorylation of STAT1 safeguards gut epithelial integrity and restricts interferon-mediated cytotoxicity. *Proceedings of the National Academy of Sciences*, *122*(30), e2511957122. https://doi.org/10.1073/pnas.2511957122

Metwally, H., Elbrashy, M. M., Ozawa, T., Okuyama, K., White, J. T., Tulyeu, J., Søndergaard, J. N., Wing, J. B., Muratsu, A., Matsumoto, H., Ikawa, M., Kishi, H., Taniuchi, I., & Kishimoto, T. (2024). Threonine phosphorylation of STAT1 restricts interferon signaling and promotes innate inflammatory responses. *Proceedings of the National Academy of Sciences*, *121*(17), e2402226121. https://doi.org/10.1073/pnas.2402226121

Mima, K., Nishihara, R., Qian, Z. R., Cao, Y., Sukawa, Y., Nowak, J. A., Yang, J., Dou, R., Masugi, Y., Song, M., Kostic, A. D., Giannakis, M., Bullman, S., Milner, D. A., Baba, H., Giovannucci, E. L., Garraway, L. A., Freeman, G. J., Dranoff, G., … Ogino, S. (2016a). Fusobacterium nucleatum in colorectal carcinoma tissue and patient prognosis. *Gut*, *65*(12), 1973–1980. https://doi.org/10.1136/gutjnl-2015-310101

Mima, K., Nowak, J. A., Qian, Z. R., Cao, Y., Song, M., Masugi, Y., Shi, Y., da Silva, A., Gu, M., Li, W., Hamada, T., Zhang, X., Wu, K., Meyerhardt, J. A., Baba, H., Giovannucci, E. L., Chan, A. T., Fuchs, C. S., Ogino, S., & Nishihara, R. (2016b). Tumor LINE-1 methylation level and colorectal cancer location in relation to patient survival. *Oncotarget*, *7*(34), 55098–55109. https://doi.org/10.18632/oncotarget.10398

Mizutani, H., Hiraku, Y., Tada-Oikawa, S., Murata, M., Ikemura, K., Iwamoto, T., Kagawa, Y., Okuda, M., & Kawanishi, S. (2010). Romidepsin (FK228), a potent histone deacetylase inhibitor, induces apoptosis through the generation of hydrogen peroxide. *Cancer Science*, *101*(10), 2214–2219. https://doi.org/10.1111/j.1349-7006.2010.01645.x

Moody, L., Dvoretskiy, S., An, R., Mantha, S., & Pan, Y.-X. (2019). The Efficacy of miR-20a as a Diagnostic and Prognostic Biomarker for Colorectal Cancer: A Systematic Review and Meta-Analysis. *Cancers*, *11*(8). https://doi.org/10.3390/cancers11081111

Morse, B., Rotherg, P. G., South, V. J., Spandorfer, J. M., & Astrin, S. M. (1988). Insertional mutagenesis of the myc locus by a LINE-1 sequence in a human breast carcinoma. *Nature*, *333*(6168), 87–90. https://doi.org/10.1038/333087a0

Moschen, A. R., Gerner, R. R., Wang, J., Klepsch, V., Adolph, T. E., Reider, S. J., Hackl, H., Pfister, A., Schilling, J., Moser, P. L., Kempster, S. L., Swidsinski, A., Orth-Höller, D., Weiss, G., Baines, J. F., Kaser, A., & Tilg, H. (2016). Lipocalin 2 Protects from Inflammation and Tumorigenesis Associated with Gut Microbiota Alterations. *Cell Host & Microbe*, *19*(4), 455–469. https://doi.org/10.1016/j.chom.2016.03.007

Nakazawa, T., Kondo, T., Ma, D., Niu, D., Mochizuki, K., Kawasaki, T., Yamane, T., Iino, H., Fujii, H., & Katoh, R. (2012). Global histone modification of histone H3 in colorectal cancer and its precursor lesions. *Human Pathology*, *43*(6), 834–842. https://doi.org/10.1016/j.humpath.2011.07.009

Nazari, M., Babakhanzadeh, E., Mollazadeh, A., Ahmadzade, M., Soleimani, E. M., & Hajimaqsoudi, E. (2024). HOTAIR in cancer: diagnostic, prognostic, and therapeutic perspectives. *Cancer Cell International*, *24*. https://api.semanticscholar.org/CorpusID:274892331

Nazemalhosseini Mojarad, E., Kuppen, P. J., Aghdaei, H. A., & Zali, M. R. (2013). The CpG Island methylator phenotype (CIMP) in colorectal cancer. *Gastroenterology and Hepatology from Bed to Bench*, *6*(3), 120–128.

Nedaeinia, R., Avan, A., Ahmadian, M., Nia, S. N., Ranjbar, M., Sharifi, M., Goli, M., Piroozmand, A., Nourmohammadi, E., Manian, M., Ferns, G. A., Ghayour-Mobarhan, M., & Salehi, R. (2017). Current Status and Perspectives Regarding LNA-Anti-miR Oligonucleotides and microRNA miR-21 Inhibitors as a Potential Therapeutic Option in Treatment of Colorectal Cancer. *Journal of Cellular Biochemistry*, *118*. https://api.semanticscholar.org/CorpusID:28448801

Ng, J. M. K., & Yu, J. (2015). Promoter hypermethylation of tumour suppressor genes as potential biomarkers in colorectal cancer. *International Journal of Molecular Sciences*, *16*(2), 2472–2496. https://doi.org/10.3390/ijms16022472

Nguyen, L. H., Goel, A., & Chung, D. C. (2020). Pathways of Colorectal Carcinogenesis. *Gastroenterology*, *158*(2), 291–302. https://doi.org/10.1053/j.gastro.2019.08.059

Nian, J., Sun, X., Ming, S., Yan, C., Ma, Y., Feng, Y., Yang, L., Yu, M., Zhang, G., & Wang, X. (2017). Diagnostic Accuracy of Methylated SEPT9 for Blood-based Colorectal Cancer Detection: A Systematic Review and Meta-Analysis. *Clinical and Translational Gastroenterology*, *8*(1), e216. https://doi.org/10.1038/ctg.2016.66

Niller, H. H., & Minarovits, J. (2016). Patho-epigenetics of Infectious Diseases Caused by Intracellular Bacteria. *Advances in Experimental Medicine and Biology*, *879*, 107–130. https://doi.org/10.1007/978-3-319-24738-0_6

Ning, T., Li, J., He, Y., Zhang, H., Wang, X., Deng, T., Liu, R., Li, H., Bai, M., Fan, Q., Zhu, K., Ying, G., & Ba, Y. (2021). Exosomal miR-208b related with oxaliplatin resistance promotes Treg expansion in colorectal cancer. *Molecular Therapy : The Journal of the American Society of Gene Therapy*, *29*(9), 2723–2736. https://doi.org/10.1016/j.ymthe.2021.04.028

Noh, J. H., Chang, Y. G., Kim, M. G., Jung, K. H., Kim, J. K., Bae, H. J., Eun, J. W., Shen, Q., Kim, S.-J., Kwon, S. H., Park, W. S., Lee, J. Y., & Nam, S. W. (2013). MiR-145 functions as a tumor suppressor by directly targeting histone deacetylase 2 in liver cancer. *Cancer Letters*, *335*(2), 455–462. https://doi.org/10.1016/j.canlet.2013.03.003

Nordgård, O., Tjensvoll, K., Gilje, B., & Søreide, K. (2018). Circulating tumour cells and DNA as liquid biopsies in gastrointestinal cancer. *The British Journal of Surgery*, *105*(2), e110–e120. https://doi.org/10.1002/bjs.10782

Oike, T., Ogiwara, H., Nakano, T., Yokota, J., & Kohno, T. (2013). Inactivating Mutations in SWI/SNF Chromatin Remodeling Genes in Human Cancer. *Japanese Journal of Clinical Oncology*, *43*(9), 849–855. https://doi.org/10.1093/jjco/hyt101

Park, S.-K., Baek, H. L., Yu, J., Kim, J. Y., Yang, H.-J., Jung, Y. S., Choi, K. Y., Kim, H., Kim, H. O., Jeong, K. U., Chun, H.-K., Kim, K., & Park, D. Il. (2017). Is methylation analysis of SFRP2, TFPI2, NDRG4, and BMP3 promoters suitable for colorectal cancer screening in the Korean population? *Intestinal Research*, *15*(4), 495–501. https://doi.org/10.5217/ir.2017.15.4.495

Patai, Á. V, Valcz, G., Hollósi, P., Kalmár, A., Péterfia, B., Patai, Á., Wichmann, B., Spisák, S., Barták, B. K., Leiszter, K., Tóth, K., Sipos, F., Kovalszky, I., Péter, Z., Miheller, P., Tulassay, Z., & Molnár, B. (2015). Comprehensive DNA Methylation Analysis Reveals a Common Ten-Gene Methylation Signature in Colorectal Adenomas and Carcinomas. *PLOS ONE*, *10*(8), 1–19. https://doi.org/10.1371/journal.pone.0133836

Patel, S., Hurez, V., Nawrocki, S. T., Goros, M., Michalek, J., Sarantopoulos, J., Curiel, T., & Mahalingam, D. (2016). Vorinostat and hydroxychloroquine improve immunity and inhibit autophagy in metastatic colorectal cancer. *Oncotarget*, *7*(37), 59087–59097. https://doi.org/10.18632/oncotarget.10824

Pawlak, A., Chybicka, K., Zioło, E., Strządała, L., & Kałas, W. (2022). The Contrasting Delayed Effects of Transient Exposure of Colorectal Cancer Cells to Decitabine or Azacitidine. *Cancers*, *14*(6). https://doi.org/10.3390/cancers14061530

Picardi, E., D'Erchia, A. M., Gallo, A., Montalvo, A., & Pesole, G. (2014). Uncovering RNA Editing Sites in Long Non-Coding RNAs. *Frontiers in Bioengineering and Biotechnology*, *Volume 2-*. https://doi.org/10.3389/fbioe.2014.00064

Pino, M. S., & Chung, D. C. (2010). The Chromosomal Instability Pathway in Colon Cancer. *Gastroenterology*, *138*(6), 2059–2072. https://doi.org/10.1053/j.gastro.2009.12.065

Pompei, A., Cordisco, L., Amaretti, A., Zanoni, S., Matteuzzi, D., & Rossi, M. (2007). Folate production by bifidobacteria as a potential probiotic property. *Applied and Environmental Microbiology*, *73*(1), 179–185. https://doi.org/10.1128/AEM.01763-06

Press, D. (2019). *Long noncoding RNA GIHCG induces cancer progression and chemoresistance and indicates poor prognosis in colorectal cancer*. 1059–1070.

Qian, X., Yu, J., Yin, Y., He, J., Wang, L., Li, Q., Zhang, L., Qian, X., Yu, J., Yin, Y., He, J., Wang, L., Li, Q., Zhang, L., Li, C., Shi, Z., Xu, Q., & Li, W. (2013). *angiogenesis and sensitizes chemosensitivity to oxaliplatin in colorectal cancers MicroRNA-143 inhibits tumor growth and angiogenesis and sensitizes chemosensitivity to oxaliplatin in colorectal cancers*. *4101*. https://doi.org/10.4161/cc.24477

Qin, T., Jelinek, J., Si, J., Shu, J., & Issa, J.-P. J. (2009). Mechanisms of resistance to 5-aza-2′-deoxycytidine in human cancer cell lines. *Blood*, *113*(3), 659–667. https://doi.org/10.1182/blood-2008-02-140038

Qin S, Kitty I, Hao Y, Zhao F, Kim W. Maintaining Genome Integrity: Protein Kinases and Phosphatases Orchestrate the Balancing Act of DNA Double-Strand Breaks Repair in Cancer. Int J Mol Sci. 2023 Jun 16;24(12):10212. https://doi.org/10.3390/ijms241210212. PMID: 37373360; PMCID: PMC10299397.

Rhee, Y. Y., Kim, K. J., & Kang, G. H. (2017). CpG Island methylator phenotype-high colorectal cancers and their prognostic implications and relationships with the serrated Neoplasia pathway. *Gut and Liver*, *11*(1), 38–46. https://doi.org/10.5009/gnl15535

Rinn, J. L., Kertesz, M., Wang, J. K., Squazzo, S. L., Xu, X., Brugmann, S. A., Goodnough, L. H., Helms, J. A., Farnham, P. J., Segal, E., & Chang, H. Y. (2007). Functional Demarcation of Active and Silent Chromatin Domains in Human <em>HOX</em> Loci by Noncoding RNAs. *Cell*, *129*(7), 1311–1323. https://doi.org/10.1016/j.cell.2007.05.022

Rizk, N. I., Kassem, D. H., Abulsoud, A. I., AbdelHalim, S., Yasser, M. B., Kamal, M. M., & Hamdy, N. M. (2024). Revealing the role of serum exosomal novel long non-coding RNA NAMPT-AS as a promising diagnostic/prognostic biomarker in colorectal cancer patients. *Life Sciences*, *352*, 122850. https://doi.org/10.1016/j.lfs.2024.122850

Rodriguez, J., Frigola, J., Vendrell, E., Risques, R.-A., Fraga, M. F., Morales, C., Moreno, V., Esteller, M., Capellà, G., Ribas, M., & Peinado, M. A. (2006). Chromosomal Instability Correlates with Genome-wide DNA Demethylation in Human Primary Colorectal Cancers. *Cancer Research*, *66*(17), 8462–9468. https://doi.org/10.1158/0008-5472.CAN-06-0293

Roulois, D., Loo Yau, H., Singhania, R., Wang, Y., Danesh, A., Shen, S. Y., Han, H., Liang, G., Jones, P. A., Pugh, T. J., O'Brien, C., & De Carvalho, D. D. (2015). DNA-Demethylating Agents Target Colorectal Cancer Cells by Inducing Viral Mimicry by Endogenous Transcripts. *Cell*, *162*(5), 961–973. https://doi.org/10.1016/j.cell.2015.07.056

Ruusuvuori, P., Valkonen, M., & Latonen, L. (2023). Deep learning transforms colorectal cancer biomarker prediction from histopathology images. *Cancer Cell*, *41*(9), 1543–1545. https://doi.org/10.1016/j.ccell.2023.08.006

Ryu, T. Y., Kim, K., Han, T.-S., Lee, M.-O., Lee, J., Choi, J., Jung, K. B., Jeong, E.-J., An, D. M., Jung, C.-R., Lim, J. H., Jung, J., Park, K., Lee, M.-S., Kim, M.-Y., Oh, S. J., Hur, K., Hamamoto, R., Park, D.-S., … Cho, H.-S. (2022). Human gut-microbiome-derived propionate coordinates proteasomal degradation via HECTD2 upregulation to target EHMT2 in colorectal cancer. *The ISME Journal*, *16*(5), 1205–1221. https://doi.org/10.1038/s41396-021-01119-1

Sakai, E., Fukuyo, M., Ohata, K., Matsusaka, K., Doi, N., Mano, Y., Takane, K., Abe, H., Yagi, K., Matsuhashi, N., Fukushima, J., Fukayama, M., Akagi, K., Aburatani, H., Nakajima, A., & Kaneda, A. (2016). Genetic and epigenetic aberrations occurring in colorectal tumors associated with serrated pathway. *International Journal of Cancer*, *138*(7), 1634–1644. https://doi.org/10.1002/ijc.29903

Sakamoto, K., Maeda, S., Hikiba, Y., Nakagawa, H., Hayakawa, Y., Shibata, W., Yanai, A., Ogura, K., & Omata, M. (2009). Constitutive NF-kappaB activation in colorectal carcinoma plays a key role in angiogenesis, promoting tumor growth. *Clinical Cancer Research : An Official Journal of the American Association for Cancer Research*, *15*(7), 2248–2258. https://doi.org/10.1158/1078-0432.CCR-08-1383

Salameh, A., Lee, A. K., Cardó-Vila, M., Nunes, D. N., Efstathiou, E., Staquicini, F. I., Dobroff, A. S., Marchiò, S., Navone, N. M., Hosoya, H., Lauer, R. C., Wen, S., Salmeron, C. C., Hoang, A., Newsham, I., Lima, L. A., Carraro, D. M., Oliviero, S., Kolonin, M. G., … Arap, W. (2015). PRUNE2 is a human prostate cancer suppressor regulated by the intronic long noncoding RNA *PCA3*. *Proceedings of the National Academy of Sciences*, *112*(27), 8403–8408. https://doi.org/10.1073/pnas.1507882112

Salum, G. M., Bader El Din, N. G., Ibrahim, M. K., Anany, M. A., Dawood, R. M., Khairy, A., & El Awady, M. K. (2017). Vascular Endothelial Growth Factor Expression in Hepatitis C Virus-Induced Liver Fibrosis: A Potential Biomarker. *Journal of Interferon & Cytokine Research : The Official Journal of the International Society for Interferon and Cytokine Research*, *37*(7), 310–316. https://doi.org/10.1089/jir.2016.0127

Salum, G. M., Bader El Din, N. G., Ibrahim, M. K., Dawood, R. M., Farouk, S., & El Awady, M. K. (2018). Correlation between TGF-β1 and c-MET Expression in HCV Genotype 4-Induced Liver Fibrosis. *Journal of Interferon and Cytokine Research*, *38*(12), 552—558. https://doi.org/10.1089/jir.2018.0096

Sandhu, R., Rivenbark, A. G., & Coleman, W. B. (2012). Loss of post-transcriptional regulation of DNMT3b by microRNAs: A possible molecular mechanism for the hypermethylation defect observed in a subset of breast cancer cell lines. *International Journal of Oncology*, *41*(2), 721–732. https://doi.org/10.3892/ijo.2012.1505

Schulman, A. H. (2012). *Hitching a ride : nonautonomous retrotransposons and parasitism as a lifestyle*. https://api.semanticscholar.org/CorpusID:82157785

Scott, A. J., Alexander, J. L., Merrifield, C. A., Cunningham, D., Jobin, C., Brown, R., Alverdy, J., O'Keefe, S. J., Gaskins, H. R., Teare, J., Yu, J., Hughes, D. J., Verstraelen, H., Burton, J., O'Toole, P. W., Rosenberg, D. W., Marchesi, J. R., & Kinross, J. M. (2019). International

Cancer Microbiome Consortium consensus statement on the role of the human microbiome in carcinogenesis. *Gut*, *68*(9), 1624–1632. https://doi.org/10.1136/gutjnl-2019-318556

Shenawy, R. M. E. L., Tohamy, N. B. E., Kassem, L. M. H., & Farouk, S. (2025). *Prognostic value of p53 expression in comparison with CEA , CA15 - 3 and CA125 protein levels in patients with breast cancer*. 1–11. https://doi.org/10.3892/wasj.2025.348

Shi, C., Yang, Y., Xia, Y., Okugawa, Y., Yang, J., Liang, Y., Chen, H., Zhang, P., Wang, F., Han, H., Wu, W., Gao, R., Gasche, C., Qin, H., Ma, Y., & Goel, A. (2016). Novel evidence for an oncogenic role of microRNA-21 in colitis-associated colorectal cancer. *Gut*, *65*(9), 1470–1481. https://doi.org/10.1136/gutjnl-2014-308455

Steelman, L. S., Chappell, W. H., Abrams, S. L., Kempf, R. C., Long, J., Laidler, P., Mijatovic, S., Maksimovic-Ivanic, D., Stivala, F., Mazzarino, M. C., Donia, M., Fagone, P., Malaponte, G., Nicoletti, F., Libra, M., Milella, M., Tafuri, A., Bonati, A., Bäsecke, J., … McCubrey, J. A. (2011). Roles of the Raf/MEK/ERK and PI3K/PTEN/Akt/mTOR pathways in controlling growth and sensitivity to therapy-implications for cancer and aging. *Aging*, *3*(3), 192–222. https://doi.org/10.18632/aging.100296

Strahl, B. D., & Allis, C. D. (2000). The language of covalent histone modifications. *Nature*, *403*(6765), 41–45. https://doi.org/10.1038/47412

Su, Y.-C., Metzen, L. T., Vélez, L. M., Bournique, E., Seldin, M., Buisson, R., Kuo, W.-W., Huang, C.-Y., & Kaiser, P. (2023). Induction of resistance to oxaliplatin in cancer by a microRNA/Fem1B/Gli1 pathway. *American Journal of Cancer Research*, *13*(12), 6011–6025.

Suzuki, H., Toyota, M., Sato, H., Sonoda, T., Sakauchi, F., & Mori, M. (2006). Roles and causes of abnormal DNA methylation in gastrointestinal cancers. *Asian Pacific Journal of Cancer Prevention : APJCP*, *7 2*, 177–185. https://api.semanticscholar.org/CorpusID:22843340

Taib, N., Merhi, M., Inchakalody, V., Mestiri, S., Hydrose, S., Makni-Maalej, K., Raza, A., Sahir, F., Azizi, F., Nizamuddin, P. B., Fernandes, Q., Yoosuf, Z. S. K. M., Almoghrabi, S., Al-Zaidan, L., Shablak, A., Uddin, S., Maccalli, C., Al Homsi, M. U., & Dermime, S. (2023). Treatment with decitabine induces the expression of stemness markers, PD-L1 and NY-ESO-1 in colorectal cancer: potential for combined chemoimmunotherapy. *Journal of Translational Medicine*, *21*(1), 235. https://doi.org/10.1186/s12967-023-04073-y

Taib, N., Merhi, M., Inchakalody, V., Mestiri, S., Raza, A., Uddin, S., Maccalli, C., AlHomsi, M. U., & Dermime, S. (2021). 736 Treatment with decitabine (DAC) induces the expression of stemness markers, PD-L1 and NY-ESO-1 in colorectal cancer. *Journal for ImmunoTherapy of Cancer*, *9*(Suppl 2), A767–A767. https://doi.org/10.1136/jitc-2021-SITC2021.736

Tamagawa, H., Oshima, T., Shiozawa, M., Morinaga, S., Nakamura, Y., Yoshihara, M., Sakuma, Y., Kameda, Y., Akaike, M., Masuda, M., Imada, T., & Miyagi, Y. (2012). The global histone modification pattern correlates with overall survival in metachronous liver metastasis of colorectal cancer. *Oncology Reports*, *27*(3), 637–642. https://doi.org/10.3892/or.2011.1547

Tanaka, S., Hosokawa, M., Tatsumi, A., Asaumi, S., Imai, R., & Ogawara, K. (2022). Improvement of resistance to oxaliplatin by vorinostat in human colorectal cancer cells through inhibition of Nrf2 nuclear translocation. *Biochemical and Biophysical Research Communications*, *607*, 9–14. https://doi.org/10.1016/j.bbrc.2022.03.070

Tang, D., Liu, J., Wang, D., Yu, H., Li, Y., & Zhang, J. (2011). Diagnostic and prognostic value of the methylation status of secreted frizzled-related protein 2 in colorectal cancer. *Clinical and Investigative Medicine. Medecine Clinique et Experimentale*, *34*(2), E88–95. https://doi.org/10.25011/cim.v34i1.15105

Tariq, K., & Ghias, K. (2016). Colorectal cancer carcinogenesis: a review of mechanisms. *Cancer Biology \& Medicine*, *13*, 120–135. https://api.semanticscholar.org/CorpusID:11484406

Tejedor, J. R., Peñarroya, A., Gancedo-Verdejo, J., Santamarina-Ojeda, P., Pérez, R. F., López-Tamargo, S., Díez-Borge, A., Alba-Linares, J. J., González-del-Rey, N., Urdinguio, R. G., Mangas, C., Roberti, A., López, V., Morales-Ruiz, T., Ariza, R. R., Roldán-Arjona, T., Meijón, M., Valledor, L., Cañal, M. J., … Fraga, M. F. (2023). CRISPR/dCAS9-mediated DNA demethylation screen identifies functional epigenetic determinants of colorectal cancer. *Clinical Epigenetics*, *15*(1), 133. https://doi.org/10.1186/s13148-023-01546-1

Toiyama, Y., Takahashi, M., Hur, K., Nagasaka, T., Tanaka, K., Inoue, Y., Kusunoki, M., Boland, C. R., & Goel, A. (2013). Serum miR-21 as a diagnostic and prognostic biomarker in colorectal cancer. *Journal of the National Cancer Institute*, *105*(12), 849–859. https://doi.org/10.1093/jnci/djt101

Tokunaga, R., Sakamoto, Y., Nakagawa, S., Miyake, K., Izumi, D., Kosumi, K., Taki, K., Higashi, T., Imamura, Y., Ishimoto, T., Iwatsuki, M., Baba, Y., Miyamoto, Y., Yoshida, N., Oki, E., Watanabe, M., & Baba, H. (2016). The Prognostic Significance of Histone Lysine Demethylase JMJD3/KDM6B in Colorectal Cancer. *Annals of Surgical Oncology*, *23*(2), 678–685. https://doi.org/10.1245/s10434-015-4879-3

Toyota, M., Ahuja, N., Ohe-Toyota, M., Herman, J. G., Baylin, S. B., & Issa, J.-P. J. (1999a). CpG Island methylator phenotype in colorectal cancer. *Proceedings of the National Academy of Sciences of the United States of America*, *96*(15), 8681—8686. https://doi.org/10.1073/pnas.96.15.8681

Toyota, M., Ho, C., Ahuja, N., Jair, K.-W., Li, Q., Ohe-Toyota, M., Baylin, S. B., & Issa, J.-P. J. (1999b). Identification of Differentially Methylated Sequences in Colorectal Cancer by Methylated CpG Island Amplification1. *Cancer Research*, *59*(10), 2307–2312.

Tse, J. W. T., Jenkins, L. J., Chionh, F., & Mariadason, J. M. (2017). Aberrant DNA Methylation in Colorectal Cancer: What Should We Target? *Trends in Cancer*, *3*(10), 698–712. https://doi.org/10.1016/j.trecan.2017.08.003

Tsoi, H., Chu, E. S. H., Zhang, X., Sheng, J., Nakatsu, G., Ng, S. C., Chan, A. W. H., Chan, F. K. L., Sung, J. J. Y., & Yu, J. (2017). Peptostreptococcus anaerobius Induces Intracellular Cholesterol Biosynthesis in Colon Cells to Induce Proliferation and Causes Dysplasia in Mice. *Gastroenterology*, *152*(6), 1419–1433.e5. https://doi.org/10.1053/j.gastro.2017.01.009

Umapathy, V. R., Natarajan, P. M., & Swamikannu, B. (2024). Molecular and Therapeutic Roles of Non-Coding RNAs in Oral Cancer—A Review. *Molecules*, *29*(10). https://doi.org/10.3390/molecules29102402

Valeri, N., Gasparini, P., Braconi, C., Paone, A., Lovat, F., Fabbri, M., Sumani, K. M., Alder, H., Amadori, D., Patel, T., Nuovo, G. J., Fishel, R., & Croce, C. M. (2010). *MicroRNA-21 induces resistance to 5-fl uorouracil by down-regulating human DNA MutS homolog*. *107*(49). https://doi.org/10.1073/pnas.1015541107

Vidal, E., Sayols, S., Moran, S., Guillaumet-Adkins, A., Schroeder, M. P., Royo, R., Orozco, M., Gut, M., Gut, I., Lopez-Bigas, N., Heyn, H., & Esteller, M. (2017). A DNA methylation map of human cancer at single base-pair resolution. *Oncogene*, *36*(40), 5648–5657. https://doi.org/10.1038/onc.2017.176

Wang, B., Lu, F.-Y., Shi, R.-H., Feng, Y.-D., Zhao, X.-D., Lu, Z.-P., Xiao, L., Zhou, G.-Q., Qiu, J.-M., & Cheng, C.-E. (2018b). MiR-26b regulates 5-FU-resistance in human colorectal cancer via down-regulation of Pgp. *American Journal of Cancer Research*, *8*(12), 2518–2527.

Wang, J., Li, R., He, Y., Yi, Y., Wu, H., & Liang, Z. (2020b). Next-generation sequencing reveals heterogeneous genetic alterations in key signaling pathways of mismatch repair deficient colorectal carcinomas. *Modern Pathology*, 1–11. https://api.semanticscholar.org/CorpusID:220325560

Wang, M., Han, D., Yuan, Z., Hu, H., Zhao, Z., Yang, R., Jin, Y., Zou, C., Chen, Y., Wang, G., Gao, X., & Wang, X. (2018a). Long non-coding RNA H19 confers 5-Fu resistance in colorectal cancer by promoting SIRT1-mediated autophagy. *Cell Death & Disease*, *9*(12), 1149. https://doi.org/10.1038/s41419-018-1187-4

Wang, Q., Chen, X., Jiang, Y., Liu, S., Liu, H., Sun, X., Zhang, H., Liu, Z., Tao, Y., Li, C., Hu, Y., Liu, D., Ye, D., Liu, Y., Wang, M., & Zhang, X. (2020a). Elevating H3K27me3 level sensitizes colorectal cancer to oxaliplatin. *Journal of Molecular Cell Biology*, *12*(2), 125–137. https://doi.org/10.1093/jmcb/mjz032

Wang, S.-W., Gao, C., Zheng, Y.-M., Yi, L., Lu, J.-C., Huang, X.-Y., Cai, J.-B., Zhang, P.-F., Cui, Y.-H., & Ke, A.-W. (2022). Current applications and future perspective of CRISPR/Cas9 gene editing in cancer. *Molecular Cancer*, *21*(1), 57. https://doi.org/10.1186/s12943-022-01518-8

Wang, X., Gong, Y., Jin, B., Wu, C., Yang, J., Wang, L., Zhang, Z., & Mao, Z. (2014). Long noncoding RNA urothelial carcinoma associated 1 induces cell replication by inhibiting BRG1 in 5637 cells. *Oncology Reports*, *32*(3), 1281–1290. https://doi.org/10.3892/or.2014.3309

Wang, X., Yang, Y., Moore, D. R., Nimmo, S. L., Lightfoot, S. A., & Huycke, M. M. (2012). 4-Hydroxy-2-Nonenal Mediates Genotoxicity and Bystander Effects Caused by Enterococcus faecalis–Infected Macrophages. *Gastroenterology*, *142*(3), 543–551.e7. https://doi.org/10.1053/j.gastro.2011.11.020

Wang, Y.-Y., Wang, H., Zhang, J., Chu, Z., Liu, P., Zhang, X., Li, C., & Gu, X. (2020c). Circ_0007031 Serves as a Sponge of miR-760 to Regulate the Growth and Chemoradiotherapy Resistance of Colorectal Cancer via Regulating DCP1A. *Cancer Management and Research*, *12*, 8465–8479. https://api.semanticscholar.org/CorpusID:221880682

Wasson, G. R., McGlynn, A. P., McNulty, H., O'Reilly, S. L., McKelvey-Martin, V. J., McKerr, G., Strain, J. J., Scott, J., & Downes, C. S. (2006). Global DNA and p53 region-specific hypomethylation in human colonic cells is induced by folate depletion and reversed by folate supplementation. *The Journal of Nutrition*, *136*(11), 2748–2753. https://doi.org/10.1093/jn/136.11.2748

Weisenberger, D. J., Siegmund, K. D., Campan, M., Young, J., Long, T. I., Faasse, M. A., Kang, G. H., Widschwendter, M., Weener, D., Buchanan, D., Koh, H., Simms, L., Barker, M., Leggett, B., Levine, J., Kim, M., French, A. J., Thibodeau, S. N., Jass, J., … Laird, P. W. (2006). CpG Island methylator phenotype underlies sporadic microsatellite instability and is tightly associated with BRAF mutation in colorectal cancer. *Nature Genetics*, *38*(7), 787–793. https://doi.org/10.1038/ng1834

Wendt, M. K., Johanesen, P. A., Kang-Decker, N., Binion, D. G., Shah, V., & Dwinell, M. B. (2006). Silencing of epithelial CXCL12 expression by DNA hypermethylation promotes colonic carcinoma metastasis. *Oncogene*, *25*(36), 4986–4997. https://doi.org/10.1038/sj.onc.1209505

Weng, X., Liu, H., Ruan, J., Du, M., Wang, L., Mao, J., Cai, Y., Lu, X., Chen, W., Huang, Y., Zhi, X., & Shan, J. (2022). HOTAIR/miR-1277-5p/ZEB1 axis mediates hypoxia-induced oxaliplatin resistance via regulating epithelial-mesenchymal transition in colorectal cancer. *Cell Death Discovery*, *8*(1), 310. https://doi.org/10.1038/s41420-022-01096-0

Wu, C. W., Ng, S. S. M., Dong, Y. J., Ng, S. C., Leung, W. W., Lee, C. W., Wong, Y. N., Chan, F. K. L., Yu, J., & Sung, J. J. Y. (2012). Detection of miR-92a and miR-21 in stool samples as potential screening biomarkers for colorectal cancer and polyps. *Gut*, *61*(5), 739–745. https://doi.org/10.1136/gut.2011.239236

Wu, R., Nie, Q., Tapper, E. E., Jerde, C. R., Dunlap, G. S., Shrestha, S., Elraiyah, T. A., Offer, S. M., & Diasio, R. B. (2016). Histone H3K27 Trimethylation Modulates 5-Fluorouracil Resistance by Inhibiting PU.1 Binding to the DPYD Promoter. *Cancer Research*, *76*(21), 6362–6373. https://doi.org/10.1158/0008-5472.CAN-16-1306

Wu, S., Rhee, K.-J., Albesiano, E., Rabizadeh, S., Wu, X., Yen, H.-R., Huso, D. L., Brancati, F. L., Wick, E., McAllister, F., Housseau, F., Pardoll, D. M., & Sears, C. L. (2009). A human colonic commensal promotes colon tumorigenesis via activation of T helper type 17 T cell responses. *Nature Medicine*, *15*(9), 1016–1022. https://doi.org/10.1038/nm.2015

Wu, Y., Wu, J., Chen, T., Li, Q., Peng, W., Li, H., Tang, X., & Fu, X. (2018). Fusobacterium nucleatum Potentiates Intestinal Tumorigenesis in Mice via a Toll-Like Receptor 4/p21-Activated Kinase 1 Cascade. *Digestive Diseases and Sciences*, *63*(5), 1210–1218. https://doi.org/10.1007/s10620-018-4999-2

Xia, X., Wu, W. K. K., Wong, S. H., Liu, D., Kwong, T. N. Y., Nakatsu, G., Yan, P. S., Chuang, Y.-M., Chan, M. W.-Y., Coker, O. O., Chen, Z., Yeoh, Y. K., Zhao, L., Wang, X., Cheng, W. Y., Chan, M. T. V., Chan, P. K. S., Sung, J. J. Y., Wang, M. H., & Yu, J. (2020). Bacteria pathogens drive host colonic epithelial cell promoter hypermethylation of tumor suppressor genes in colorectal cancer. *Microbiome*, *8*(1), 108. https://doi.org/10.1186/s40168-020-00847-4

Xiao, W., Zhao, H., Dong, W., Li, Q., Zhu, J., Li, G., Zhang, S., & Ye, M. (2015). Quantitative detection of methylated NDRG4 gene as a candidate biomarker for diagnosis of colorectal cancer. *Oncology Letters*, *9*(3), 1383–1387. https://doi.org/10.3892/ol.2014.2815

Xiao J, Duan Q, Wang Z, Yan W, Sun H, Xue P, Fan X, Zeng X, Chen J, Shao C, Zhu F. Phosphorylation of TOPK at Y74, Y272 by Src increases the stability of TOPK and promotes tumorigenesis of colon. Oncotarget. 2016 Apr 26;7(17):24483-94. https://doi.org/10.18632/oncotarget.8231. PMID: 27016416; PMCID: PMC5029716.

Xiao, Z., Qu, Z., Chen, Z., Fang, Z., Zhou, K., Huang, Z., Guo, X., & Zhang, Y. (2018). LncRNA HOTAIR is a Prognostic Biomarker for the Proliferation and Chemoresistance of Colorectal Cancer via MiR-203a-3p-Mediated Wnt/ß-Catenin Signaling Pathway. *Cellular Physiology and Biochemistry*, *46*(3), 1275–1285. https://doi.org/10.1159/000489110

Xue, W., Kitzing, T., Roessler, S., Zuber, J., Krasnitz, A., Schultz, N., Revill, K., Weissmueller, S., Rappaport, A. R., Simon, J., Zhang, J., Luo, W., Hicks, J., Zender, L., Wang, X. W., Powers, S., Wigler, M., & Lowe, S. W. (2012). A cluster of cooperating tumor-suppressor gene candidates in chromosomal deletions. *Proceedings of the National Academy of Sciences of the United States of America*, *109*(21), 8212–8217. https://doi.org/10.1073/pnas.1206062109

Yamaguchi, H., & Hung, M. C. (2014). Regulation and role of EZH2 in cancer. *Cancer Research and Treatment*, *46*(3), 209–222. https://doi.org/10.4143/crt.2014.46.3.209

Yamaguchi, K., Nakagawa, S., Saku, A., Isobe, Y., Yamaguchi, R., Sheridan, P., Takane, K., Ikenoue, T., Zhu, C., Miura, M., Okawara, Y., Nagatoishi, S., Kozuka-Hata, H., Oyama, M., Aikou, S., Ahiko, Y., Shida, D., Tsumoto, K., Miyano, S., … Furukawa, Y. (2023). Bromodomain protein BRD8 regulates cell cycle progression in colorectal cancer cells through a TIP60-independent regulation of the pre-RC complex. *IScience*, *26*(4), 106563. https://doi.org/10.1016/j.isci.2023.106563

Yang, D., Torres, C. M., Bardhan, K., Zimmerman, M., McGaha, T. L., & Liu, K. (2012). Decitabine and Vorinostat Cooperate To Sensitize Colon Carcinoma Cells to Fas Ligand-Induced Apoptosis In Vitro and Tumor Suppression In Vivo. *The Journal of Immunology*, *188*(9), 4441–4449. https://doi.org/10.4049/jimmunol.1103035

Yang, Y., Weng, W., Peng, J., Hong, L., Yang, L., Toiyama, Y., Gao, R., Liu, M., Yin, M., Pan, C., Li, H., Guo, B., Zhu, Q., Wei, Q., Moyer, M.-P., Wang, P., Cai, S., Goel, A., Qin, H., & Ma, Y. (2017). Fusobacterium nucleatum Increases Proliferation of Colorectal Cancer Cells and Tumor Development in Mice by Activating Toll-Like Receptor 4 Signaling to Nuclear Factor-κB, and Up-regulating Expression of MicroRNA-21. *Gastroenterology*, *152*(4), 851–866.e24. https://doi.org/10.1053/j.gastro.2016.11.018

Yang, Z., Xu, F., Teschendorff, A. E., Zhao, Y., Yao, L., Li, J., & He, Y. (2022). Insights into the role of long non-coding RNAs in DNA methylation mediated transcriptional regulation. *Frontiers in Molecular Biosciences*, *9*, 1067406. https://doi.org/10.3389/fmolb.2022.1067406

Yokoyama, Y., Hieda, M., Nishioka, Y., Matsumoto, A., Higashi, S., Kimura, H., Yamamoto, H., Mori, M., Matsuura, S., & Matsuura, N. (2013). Cancer-associated upregulation of histone H3 lysine 9 trimethylation promotes cell motility in vitro and drives tumor formation in vivo. *Cancer Science*, *104*(7), 889–895. https://doi.org/10.1111/cas.12166

Yu, J., Cao, Q., Mehra, R., Laxman, B., Yu, J., Tomlins, S. A., Creighton, C. J., Dhanasekaran, S. M., Shen, R., Chen, G., Morris, D. S., Marquez, V. E., Shah, R. B., Ghosh, D., Varambally, S., & Chinnaiyan, A. M. (2007). Integrative genomics analysis reveals silencing of beta-adrenergic signaling by polycomb in prostate cancer. *Cancer Cell*, *12*(5), 419–431. https://doi.org/10.1016/j.ccr.2007.10.016

Yu D, Li Z, Gan M, Zhang H, Yin X, Tang S, Wan L, Tian Y, Zhang S, Zhu Y, Lai M, Zhang D. Decreased expression of dual specificity phosphatase 22 in colorectal cancer and its potential prognostic relevance for stage IV CRC patients. Tumour Biol. 2015 Nov;36(11):8531-5. https://doi.org/10.1007/s13277-015-3588-7. Epub 2015 Jun 2. PMID: 26032091.

Yuan, C., Burns, M. B., Subramanian, S., & Blekhman, R. (2018). Interaction between Host MicroRNAs and the Gut Microbiota in Colorectal Cancer. *MSystems*, *3*(3), 10.1128/msystems.00205-17. https://doi.org/10.1128/msystems.00205-17

Zarandi, A., Irani, S., Savabkar, S., Chaleshi, V., Ghavideldarestani, M., Mirfakhraie, R., Khodadoostan, M., Nazemalhosseini-Mojarad, E., & Asadzadeh Aghdaei, H. (2017). Evaluation of promoter methylation status of MLH1 gene in Iranian patients with colorectal tumors and adenoma polyps. *Gastroenterology and Hepatology from Bed to Bench, 10*(Suppl1), S117–S121.

Zeng, W., Liu, Y., Li, W.-T., Li, Y., & Zhu, J.-F. (2020). CircFNDC3B sequestrates miR-937-5p to derepress TIMP3 and inhibit colorectal cancer progression. *Molecular Oncology, 14*(11), 2960–2984. https://doi.org/10.1002/1878-0261.12796

Zhang, C., Lu, J., & Zhang, P. (2016). The Roles of Chromatin Remodeling Proteins in Cancer. *Current Protein \& Peptide Science, 17 5*, 446–454. https://api.semanticscholar.org/CorpusID:21404867

Zhang, J., Zhu, H., Liu, W., & Miao, J. (2025). *Prognostic and predictive molecular biomarkers in colorectal cancer. April*, 1–15. https://doi.org/10.3389/fonc.2025.1532924

Zhang, L., Dong, X., Yan, B., Yu, W., & Shan, L. (2020a). CircAGFG1 drives metastasis and stemness in colorectal cancer by modulating YY1/CTNNB1. *Cell Death & Disease, 11*(7), 542. https://doi.org/10.1038/s41419-020-2707-6

Zhang, P.-F., Wu, J., Wu, Y., Huang, W., Liu, M., Dong, Z.-R., Xu, B.-Y., Jin, Y., Wang, F., & Zhang, X.-M. (2019). The lncRNA SCARNA2 mediates colorectal cancer chemoresistance through a conserved microRNA-342-3p target sequence. *Journal of Cellular Physiology, 234*(7), 10157–10165. https://doi.org/10.1002/jcp.27684

Zhang, W., Yang, C., Wang, S., Xiang, Z., Dou, R., Lin, Z., Zheng, J., & Xiong, B. (2021b). SDC2 and TFPI2 Methylation in Stool Samples as an Integrated Biomarker for Early Detection of Colorectal Cancer. *Cancer Management and Research, 13*, 3601–3617. https://doi.org/10.2147/CMAR.S300861

Zhang, X., Li, Q., Du, A., Li, Y., Shi, Q., Chen, Y., Zhao, Y., Wang, B., & Pan, F. (2021a). Adipocytic Glutamine Synthetase Upregulation via Altered Histone Methylation Promotes 5FU Chemoresistance in Peritoneal Carcinomatosis of Colorectal Cancer. *Frontiers in Oncology, Volume 11*. https://doi.org/10.3389/fonc.2021.748730

Zhang, Y., Liu, X., Zhang, J., Xu, Y., Shao, J., Hu, Y., Shu, P., & Cheng, H. (2020b). Inhibition of miR-19a partially reversed the resistance of colorectal cancer to oxaliplatin via PTEN/PI3K/AKT pathway. *Aging, 12*(7), 5640–5650. https://doi.org/10.18632/aging.102929

Zhang, Y., Talmon, G., & Wang, J. (2015). MicroRNA-587 antagonizes 5-FU-induced apoptosis and confers drug resistance by regulating PPP2R1B expression in colorectal cancer. *Cell Death & Disease, 6*(8), e1845–e1845. https://doi.org/10.1038/cddis.2015.200

Zhao, W., Song, M., Zhang, J., Kuerban, M., & Wang, H. (2015). Combined identification of long non-coding RNA CCAT1 and HOTAIR in serum as an effective screening for colorectal carcinoma. *International Journal of Clinical and Experimental Pathology, 8*(11), 14131–14140.

Zhou, H.-R., Zhang, F.-F., Ma, Z.-Y., Huang, H.-W., Jiang, L., Cai, T., Zhu, J.-K., Zhang, C., & He, X.-J. (2013). Folate polyglutamylation is involved in chromatin silencing by maintaining global DNA methylation and histone H3K9 dimethylation in Arabidopsis. *The Plant Cell, 25*(7), 2545–2559. https://doi.org/10.1105/tpc.113.114678

Zhu, Q., Lu, G., Luo, Z., Gui, F., Wu, J., Zhang, D., & Ni, Y. (2018). CircRNA circ_0067934 promotes tumor growth and metastasis in hepatocellular carcinoma through regulation of miR-1324/FZD5/Wnt/β-catenin axis. *Biochemical and Biophysical Research Communications, 497*(2), 626–632. https://doi.org/10.1016/j.bbrc.2018.02.119

Zou, S., Fang, L., & Lee, M.-H. (2018). Dysbiosis of gut microbiota in promoting the development of colorectal cancer. *Gastroenterology Report, 6*(1), 1–12. https://doi.org/10.1093/gastro/gox031

Chapter 5
Noncoding RNAs: A Novel Frontier in Liver Cancer Research and Therapy: Implications for Precision Oncology

Al-Aliaa M. Sallam, Mahmoud A. Elrebehy, Ibrahim M. Elazab, Mohamed S. Elballal, Ola E. Elazazy, Samah S. Abbas, Manar M. El Tabaa, Shih-Min Hsia, Miguel Angel Chávez-Fumagalli, and Nadia M. Hamdy

Abstract There is an urgent need for reliable noninvasive indicators of the occurrence and course of liver disease. According to research conducted in recent decades, the risk scores for liver-related complications can be determined by utilizing the genetic and epigenetic components involved in the development of liver disease. This might potentially indicate the feasibility of implementing programs for target screening and monitoring of complications.

Precision medicine may be used to treat liver illnesses, thanks to recent advancements in our knowledge of the epigenetics of liver cells. In a multicellular organism, each cell has a distinct phenotype, even if they all have the same genetics. Chromatin

A.-A. M. Sallam (✉)
Biochemistry Department, Faculty of Pharmacy, Badr University in Cairo (BUC), Cairo, Egypt

Biochemistry and Molecular Biology Department, Faculty of Pharmacy, Ain Shams University, Cairo, Egypt
e-mail: Aliaa.sallam@pharma.asu.edu.eg; Aliaa.mohamed@buc.edu.eg

M. A. Elrebehy
Biochemistry Department, Faculty of Pharmacy, Galala University, Suez, Egypt

I. M. Elazab
Biochemistry Department, Faculty of Pharmacy, Tanta University, Tanta, Egypt

M. S. Elballal · O. E. Elazazy
Biochemistry Department, Faculty of Pharmacy, Badr University in Cairo (BUC), Cairo, Egypt

S. S. Abbas
Pharmacology and Toxicology Department, Faculty of Pharmacy, Misr International University (MIU), Cairo, Egypt

M. M. El Tabaa
Pharmacology and Environmental Toxicology Department, Environmental Studies and Research Institute (ESRI), Sadat University, Menoufia, Egypt

N. M. Hamdy et al. (eds.), *From (Epi)genetics-to-Precision (E2P)*, Subcellular Biochemistry 114, https://doi.org/10.1007/978-3-032-08530-6_5

states determined by epigenetic processes are necessary for this heritable yet dynamic cell identity. Genetic, environmental, and metabolic factors that determine DNA accessibility to the transcriptional machinery governing gene expression and cellular states in various liver illnesses can alter the epigenomic landscapes unique to the liver. Noncoding RNAs (ncRNAs) are examples of the epigenetic regulation of chromatin. The coordinated actions of numerous epigenetic factors that modify nucleosome positioning and structure (remodelers), create epigenetic marks in DNA and histones (writers), identify and interpret the marks (readers), and eliminate these marks (erasers) preserve this epigenetic information.

Here, we summarize the literature on how epigenetic changes contribute to the development of liver cancers. Along with talking about the potential of epigenetic therapy approaches, we also address their usefulness as epigenetic biomarkers for the diagnosis and prognosis of hepatocellular carcinoma.

Keywords ncRNA · HCC · Epigenetics · miRNA · circRNA · lncRNA

Abbreviations

AFP	Alpha-fetoprotein
AKAP12	A-kinase anchoring protein 12
Akt	Protein kinase B
ANGPT2	Angiopoietin-2
ANRIL	Antisense noncoding RNA in the inhibitors of CDK4 (INK4) locus
AP-2α	Activating protein 2 alpha
ASOs	Antisense oligonucleotides
CAT1	Cationic amino acid transporter-1

S.-M. Hsia
School of Nutrition and Health Sciences, College of Nutrition, Taipei Medical University, Taipei, Taiwan

Graduate Institute of Metabolism and Obesity Sciences, College of Nutrition, Taipei Medical University, Taipei, Taiwan

School of Food Safety, Taipei Medical University, Taipei, Taiwan

Nutrition Research Center, Taipei Medical University Hospital, Taipei, Taiwan

TMU Research Center for Digestive Medicine, Taipei Medical University, Taipei, Taiwan

M. A. Chávez-Fumagalli
Computational Biology and Chemistry Research Group, Vicerrectorado de Investigación, Universidad Católica de Santa María, Arequipa, Peru

N. M. Hamdy
Biochemistry and Molecular Biology Department, Faculty of Pharmacy, Ain Shams University, Cairo, Egypt
e-mail: nadia_hamdy@pharma.asu.edu.eg

CC	Cholangiocarcinoma
CCAT1	Colon cancer–associated transcript 1
ceRNA	Competitive endogenous RNA
cHCC-CC	Combined hepatocellular carcinoma and cholangiocarcinoma
circRNAs	Circular RNAs
CLDN1	Claudin-1
CRC	Colorectal cancer
CREB	cAMP response element–binding protein
CRISPR/dCas9	Clustered regularly interspaced short palindromic repeats/dead Cas9
CT	Computed tomography
CTNNA1	Catenin alpha 1
DANCR	Differentiation antagonizing nonprotein coding RNA
DFS	Disease-free survival
DNA	Deoxyribonucleic acid
DNMT3A	DNA methyltransferase 3 alpha
DNMTs	DNA methyltransferases
DPYD	Dihydropyrimidine dehydrogenase
DR4	Death receptor 4
E2F1	Transcription factor E2F1
ECM	Extracellular matrix
EGFR	Epidermal growth factor receptor
EMT	Epithelial-to-mesenchymal transition
ETS1	Erythroblastosis virus E26 oncogene homolog 1
EZH2	Enhancer of zeste homolog 2
FBXL2	F-box and leucine-rich repeat protein 2
FGF	Fibroblast growth factor
FLT1	FMS-like tyrosine kinase 1
FRAS1	Fraser extracellular matrix complex subunit 1
FXR	Farnesoid X receptor
GAS5	Growth arrest–specific 5
GNMT	Glycine N-methyltransferase
H3K27me3	Trimethylation of lysine 27 on histone H3
HBV/HCV	Hepatitis B virus/hepatitis C virus
HCC	Hepatocellular carcinoma
HDAC	Histone deacetylase
HEIH	High expression in HCC
HEPN1	Hepatocellular carcinoma downregulated 1
HIF-1α	Hypoxia-inducible factor 1 alpha
HOTAIR	HOX transcript antisense RNA
HOTTIP	HOXA transcript at the distal tip
HOXD10	Homeobox D10
HULC	Highly upregulated in liver cancer
ICC	Intrahepatic cholangiocarcinoma
IFITM3	Interferon-induced transmembrane protein 3

IGFBP4	Insulin-like growth factor–binding protein 4
KLF2	Krüppel-like factor 2
KLF4	Krüppel-like factor 4
LINC01138	Long intergenic nonprotein coding RNA 1138
LNA	Locked nucleic acid
lncRNAs	Long noncoding RNAs
LSD1	Lysine-specific demethylase 1
MAGI2	Membrane-associated guanylate kinase inverted 2
MALAT1	Metastasis-associated lung adenocarcinoma transcript 1
MAP3K8	MAPK kinase kinase 8
MAP K7	Mitogen-activated protein kinase 7
Mcl-1	Myeloid cell leukemia 1
MEG3	Maternally expressed gene 3
miRNAs	MicroRNAs
MMP	Matrix metalloproteinase
MRI	Magnetic resonance imaging
MYC	Myelocytomatosis oncogene
ncRNAs	Noncoding RNAs
NEAT1	Nuclear paraspeckle assembly transcript 1
NEGR1	Neuronal growth regulator 1
NKD2	Naked cuticle homolog 2
OCT4	Octamer-binding transcription factor 4
OS	Overall survival
PD-L1	Programmed death-ligand 1
PDAC	Pancreatic ductal adenocarcinoma
PDCD4	Programmed cell death 4
PGE2	Prostaglandin E2
PI3K	Phosphoinositide 3-kinase
PRC2	Polycomb repressive complex 2
pre-RNA	Premature RNA
PRKACB	Protein kinase cAMP-activated catalytic subunit beta
PRL-3	Phosphatase of regenerating liver-3
PRMT5	Protein arginine methyltransferase 5
PTEN	Phosphatase and tensin homolog
PTPN14	Protein tyrosine phosphatase non-receptor type 14
PVT1	Plasmacytoma variant translocation 1
RHOC	Ras homolog family member C
RNA	Ribonucleic acid
ROCK2	Rho-associated coiled-coil-containing protein kinase 2
RTN4	Reticulon 4
RTNK2	Rhotekin 2
SEMA3A	Semaphorin-3A
siRNA	Small interfering RNA
SNHG6	Small nucleolar RNA host gene 6
SOCS6	Suppressor of cytokine signaling 6

Sox2	SRY-box transcription factor 2
SUZ12	Suppressor of zeste 12
TAMs	Tumor-associated macrophages
TGF-β	Transforming growth factor beta
TGFBR3	Transforming growth factor beta receptor III
TIMP3	Tissue inhibitor of metalloproteinases-3
TRAIL	Tumor necrosis factor (TNF)-related apoptosis-inducing ligand
TSLC1	Tumor suppressor in lung cancer 1
TUG1	Taurine upregulated gene 1
UCA1	Urothelial cancer–associated 1
VEGFA	Vascular endothelial growth factor A
WDR5/MLL	WD repeat domain 5/mixed-lineage leukemia
YAP	Yes-associated protein
ZEB	Zinc finger E-box-binding homeobox
ZEB2	Zinc finger E-box-binding homeobox 2
ZIC1	Zic family member 1
15-PGDH	15-Hydroxyprostaglandin dehydrogenase

Introduction: Overview of Liver Cancers

Liver cancer is a complex disease with multiple types and classifications, primarily defined by cellular origin and morphological characteristics. The classification of liver cancer has been evolving from traditional morphology-based approaches to more functional and molecularly oriented systems, reflecting the complexity of tumorigenesis and the diversity of oncogenic pathways involved.

Liver malignancies are generally divided into primary and secondary categories, each with distinct origins and characteristics. Primary liver cancers arise within the liver itself; in contrast, secondary liver cancers, also known as liver metastases, develop from tumors originating in other organs—such as the colorectum, breast, pancreas, or lungs—that subsequently spread to the liver. Recognizing these differences is critical for accurate diagnosis, treatment planning, and prognosis.

The predominant subtypes of primary liver neoplasms include hepatocellular carcinoma (HCC), cholangiocarcinoma (CC), and the mixed variant known as combined hepatocellular carcinoma and cholangiocarcinoma (cHCC-CC). These entities are further distinguished by their unique clinical and pathological features, along with their specific molecular and genetic profiles.

Hepatocellular Carcinoma (HCC)

HCC is the most common type of primary hepatic malignancy, accounting for approximately 85–90% of cases (D'Avola et al. 2019; Hayat 2005). It arises from hepatocytes, the principal parenchymal cells of the liver, and is strongly associated with chronic liver conditions, such as cirrhosis, hepatitis B and C viral infections, and nonalcoholic fatty liver disease (NAFLD) (D'Avola et al. 2019; Swellam and Hamdy 2012). Histopathologically, HCC exhibits diverse growth patterns, including trabecular, acinar, and solid forms, with the macrotrabecular-massive subtype recognized for its aggressive clinical course (D'Avola et al. 2019).

Cholangiocarcinoma (CC)

Cholangiocarcinoma (CC), which arises from the bile duct epithelium, is the second most common primary liver cancer after HCC (Choi and Choi 2014; Kojiro 2011). It is classified into intrahepatic, perihilar, and distal types, with intrahepatic CC further subdivided into mass-forming, periductal infiltrative, and intraductal papillary variants (Choi and Choi 2014; Kojiro 2011). Among these, the mass-forming subtype is the most prevalent, whereas the periductal infiltrative subtype is associated with a poorer prognosis (Choi and Choi 2014; Kojiro 2011).

Combined Hepatocellular Carcinoma and Cholangiocarcinoma (cHCC-CC)

This rare type of liver cancer exhibits histological and phenotypic features of both HCC and CC, reflecting the potential of mature hepatocytes to transdifferentiate into cholangiocytes (Wang et al. 2020). Importantly, cHCC-CC is not simply a coexistence of HCC and CC but rather a distinct entity with unique genetic alterations and clinical behaviors (Hayat 2005).

The diagnosis of primary liver cancers typically involves imaging modalities such as computed tomography (CT) and magnetic resonance imaging (MRI), along with biomarker assessment, including elevated α-fetoprotein (AFP) levels. Treatment strategies encompass surgical resection, liver transplantation, and minimally invasive approaches, such as radiofrequency ablation and transarterial chemoembolization (TACE) (Addissouky et al. 2024; Wu et al. 2023).

Secondary Liver Cancers

Secondary liver cancers are predominantly metastases from extrahepatic primary tumors, with colorectal cancer being the most frequent source. Other common origins include the lung, breast, pancreas, and stomach (Line 2023; Liu et al. 2017b).

These cancers are of substantial clinical significance due to their association with advanced disease and poor prognosis, with an untreated median survival time of often less than 2 years. Nevertheless, surgical intervention can markedly improve outcomes (Wang et al. 2021b). Prognosis is influenced by factors such as the site of the primary tumor, the extent of hepatic involvement, and the presence of extrahepatic metastases (Wang et al. 2021b).

Diagnostic evaluation relies heavily on CT and MRI for lesion detection and staging (Liu et al. 2017b). More recently, liquid biopsy approaches, such as circulating tumor DNA (ctDNA) analysis, have enhanced early detection, real-time monitoring of disease progression, and treatment response assessment (Wehrle et al. 2024).

The management of secondary liver cancers is multifaceted, integrating surgical and nonsurgical modalities. Surgical resection remains the most effective option for certain metastatic tumors, particularly colorectal and neuroendocrine metastases, achieving 5-year survival rates of up to 42% and 50–75%, respectively (Barreto and Posner 2019; Line 2023). Liver transplantation is considered for carefully selected patients with unresectable neuroendocrine metastases, yielding 5-year survival rates exceeding 80% (Line 2023). Systemic therapies, including chemotherapy and targeted agents, play a crucial role in disseminated disease control, although they are rarely curative. Combination strategies that integrate systemic treatment with surgical approaches can further optimize patient outcomes (Barreto and Posner 2019).

Other Types of Liver Cancer

Hepatoblastoma

Hepatoblastoma is a rare pediatric liver malignancy of embryonal origin, characterized by distinctive histological patterns and predominantly affecting children (D'Avola et al. 2019; Wang et al. 2020).

Rare Tumors

Other uncommon primary liver tumors include angiosarcoma, epithelioid hemangioendothelioma, and biliary cystadenocarcinoma (Choi and Choi 2014; D'Avola et al. 2019). Each of these malignancies presents with unique clinical features and necessitates specialized diagnostic and therapeutic strategies.

The Significance of Noncoding RNAs in the Epigenetic Modulation of Hepatocellular Carcinoma

Noncoding RNAs (ncRNAs) are RNA molecules that do not encode proteins but play crucial roles in regulating gene expression at transcriptional, posttranscriptional, and translational levels. These transcripts arise from the noncoding regions of the genome, which constitute the majority of the human genome. Research on ncRNAs has advanced substantially, revealing their involvement in diverse cellular processes and their implications for both health and disease. The following subsections outline the principal types of ncRNAs and their mechanisms of action in gene regulation.

Long Noncoding RNAs (lncRNAs)

Typically longer than 200 nucleotides, lncRNAs participate in chromatin remodeling, transcriptional regulation, and posttranscriptional control. They function as scaffolds, guides, or decoys, thereby influencing gene expression through interactions with DNA, RNA, and proteins (Younger 2022).

MicroRNAs (miRNAs)

These short, ~22-nucleotide RNAs regulate gene expression posttranscriptionally by binding complementary sequences to target mRNAs, leading to either their degradation or translational repression (Sarropoulou and Fernández 2023; Tripathi and Srivastava 2023).

Piwi-Interacting RNAs (piRNAs)

piRNAs silence transposable elements and maintain genomic stability, particularly in germ cells (Sarropoulou and Fernández 2023; Tripathi and Srivastava 2023).

Small Interfering RNAs (siRNAs)

siRNAs are short double-stranded RNAs that induce RNA interference (RNAi), resulting in the degradation of specific target mRNAs (Wei et al. 2017).

Circular RNAs (circRNAs)

circRNAs form covalently closed-loop structures and can act as miRNA sponges, thereby modulating miRNA availability and downstream gene expression (Plawgo and Raczynska 2022).

Mechanisms of ncRNAs in Gene Regulation

Transcriptional Regulation

ncRNAs influence transcription by modifying the chromatin structure and recruiting transcriptional machinery. For instance, lncRNAs can guide chromatin-modifying complexes to specific genomic loci, thereby regulating histone modifications and DNA methylation (Samoilă et al. 2023; Wei et al. 2017).

Posttranscriptional Regulation

miRNAs and siRNAs primarily act at this level by binding to target mRNAs, resulting in either their degradation or the inhibition of translation. This mechanism enables the fine-tuning of gene expression in response to cellular signals (Plawgo and Raczynska 2022; Sarropoulou and Fernández 2023).

Epigenetic Regulation

ncRNAs contribute to establishing and maintaining epigenetic marks, including DNA methylation and histone modifications, which are essential for developmental gene expression programs and adaptive responses to environmental stimuli (Wei et al. 2017).

Translational Regulation

Certain ncRNAs directly affect protein synthesis by interacting with ribosomes or translation factors, thereby modulating the translation efficiency (Pecoraro et al. 2022).

Implications of Noncoding RNAs in Liver Cancers

Although the regulatory roles of ncRNAs are well established, their full therapeutic potential remains under active investigation. Aberrant ncRNA expression has been linked to cancer and neurodegenerative disorders, underscoring their value as diagnostic biomarkers and therapeutic targets (George et al. 2024; Tripathi and Srivastava

2023). However, the complexity and context-specificity of ncRNA networks present significant challenges for clinical translation. Continued research is needed to elucidate how ncRNAs—including miRNAs, lncRNAs, and circRNAs—play pivotal roles in epigenetic regulation. These molecules modulate cell proliferation, apoptosis, angiogenesis, and metastasis by regulating gene expression and interacting with key oncogenic pathways. Their involvement highlights the promise of ncRNAs as prognostic indicators and therapeutic targets. The following section details specific roles and mechanisms of ncRNAs in the epigenetic regulation of HCC.

Dysregulation of miRNAs in Hepatocellular Carcinoma (HCC)

HCC arises from a complex interplay of genetic and epigenetic modifications. The phosphatidylinositol 3-kinase/protein kinase B (PI3K/Akt) pathway is a major signaling cascade promoting survival and proliferation in HCC. The tumor suppressor PTEN antagonizes this pathway by converting phosphatidylinositol (3,4,5)-trisphosphate to phosphatidylinositol (4,5)-bisphosphate, thereby preventing Akt activation. A loss or downregulation of PTEN leads to persistent PI3K/Akt activation, fueling tumor growth, apoptosis resistance, and disease progression. Wang et al. reported that **miR-137** directly targets and suppresses DNA methyltransferase 3α (DNMT3A) in HCC cells, thereby attenuating proliferation, migration, and invasion through the PTEN/Akt pathway modulation (Ratnasari et al. 2022; Wang et al. 2021a). Similarly, **miR-155-5p** and **miR-203a-3p** promote HCC progression by targeting PTEN, further enhancing PI3K/Akt signaling (Fu et al. 2017; Zhang et al. 2020b). The **miR-29** family (**miR-29a/b/c**), known as tumor suppressors, regulates DNMT3 proteins that shape global DNA methylation. These miRNAs are frequently downregulated in HCC tissues. Notably, **miR-29a** directly targets hypoxia-inducible factor-1α (HIF-1α) and angiopoietin-2 (ANGPT2), both of which drive hypoxia adaptation, angiogenesis, and inflammation (Huang et al. 2022b). Another study identified interferon-induced transmembrane protein 3 (IFITM3) as a direct target of **miR-29a**, with its upregulation linked to enhanced proliferation, migration, and invasion (Liang et al. 2018).

Serum alpha-fetoprotein (AFP), normally expressed only in the fetal liver, is reactivated in ~60% of HCC patients and correlates with poor prognosis. Parpart et al. showed that AFP suppresses **miR-29a/b** expression and upregulates DNMT3A via c-MYC binding, thereby antagonizing the tumor-suppressive function of the **miR-29** family (Parpart et al. 2014). **miR-487a** has emerged as a key oncogenic miRNA in HCC, with elevated expression correlating with poor prognosis. It promotes tumor proliferation and metastasis by targeting tumor suppressors such as membrane-associated guanylate kinase inverted 2 (MAGI2) and activating downstream pathways that enhance malignancy (Chang et al. 2017). Mechanistically, **miR-487a** is regulated by the lncRNA MIR4435-2HG, which is also overexpressed in HCC. MIR4435-2HG enhances **miR-487a** expression, and silencing **miR-487a** mitigates the proliferative effects of MIR4435-2HG, highlighting a functional MIR4435-2HG/**miR-487a** oncogenic axis (Kong et al. 2019). Targeting this pathway presents a potential therapeutic strategy in HCC management.

Recent advances highlight the pivotal role of **miR-873-5p** in regulating liver pathology through its targeting of glycine N-methyltransferase (GNMT), a key enzyme managing methyl-group flux and mitochondrial function in the liver. Fernández-Ramos et al. demonstrated that **miR-873-5p** is markedly upregulated in early fibrotic, cholestatic, and cirrhotic liver models, including bile duct ligation and Mdr2^{-}/$^{-}$ mice, as well as in patient samples, where its levels negatively correlate with hepatic GNMT expression. Therapeutically, the antagonism of **miR-873-5p** restored GNMT-attenuated hepatocyte apoptosis, cholangiocyte proliferation, inflammation, and fibrogenesis while reducing global DNA and histone methylation. Complementing these epigenetic findings, Han et al. reported that **miR-873** is upregulated in HCC tissues and cell lines, where it promotes proliferation and metastasis by directly targeting tumor suppressor in lung cancer 1 (TSLC1) and activating the PI3K/AKT/mTOR pathway. Together, these studies position **miR-873-5p** as a multifaceted oncomiR in the liver, impairing methylation-dependent epigenetic homeostasis and unleashing oncogenic signaling, underscoring its therapeutic promise in fibrotic and malignant liver conditions (Fernández-Ramos et al. 2018; Han et al. 2018a).

miR-302b, a member of the miR-302-367 cluster, plays a significant role in enhancing the chemosensitivity of HCC cells. The overexpression of **miR-302b** in HCC cell lines, such as HepG2 and SMMC-7721, has been shown to increase sensitivity to 5-fluorouracil (5-FU) by targeting and downregulating Mcl-1, an anti-apoptotic protein, and DPYD, an enzyme involved in 5-FU catabolism. This dual targeting leads to increased apoptosis and reduced drug metabolism, thereby enhancing the cytotoxic effects of 5-FU on HCC cells (Cai et al. 2015). Additionally, **miR-302**-mediated reprogramming has been explored as a strategy to improve drug sensitivity in HCC. Transfection of **miR-302** into HCC cells induces the formation of induced pluripotent stem cell (iPSC)–like spheres characterized by high expression of pluripotency markers (NANOG and OCT4), decreased proliferation, increased apoptosis, and reduced expression of epithelial–mesenchymal transition (EMT) markers. This reprogramming is associated with the downregulation of AOF2 (also known as LSD1 or KDM1), leading to increased H3K4 methylation and repression of c-MYC, a key oncogene. These epigenetic modifications contribute to the enhanced drug sensitivity observed in reprogrammed HCC cells (Koga et al. 2014). Collectively, these findings suggest that **miR-302b** not only directly sensitizes HCC cells to chemotherapy by targeting specific genes but also induces epigenetic reprogramming that further enhances drug responsiveness. Dysregulated miRNAs in HCC are collected in Table 5.1.

Table 5.1 Dysregulated miRNAs in HCC

miRNAs	Pattern of regulation	Target	References
miR-103	Upregulation	AKAP12	Han et al. (2018b)
miR-137	Downregulation	DNMT3A	Wang et al. (2021a)
miR-141	Downregulation	ZEB2	Wu et al. (2014)
miR-155	Upregulation	PTEN	Fu et al. (2017) and Ratnasari et al. (2022)
miR-181b	Upregulation	TIMP3	Wang et al. (2010)
miR-18a	Upregulation	KLF4	Liu et al. (2017a)
miR-192-5p	Upregulation	SEMA3A	Yan-Chun et al. (2017)
miR-203a-3p	Upregulation	PTEN	Zhang et al. (2020b)
miR-21	Upregulation	HEPN1	Ratnasari et al. (2022)
miR-29a	Downregulation	HIF-1α, ANGPT2, IFITM3, CLDN1	Huang et al. (2022b), Liang et al. 2018, and Mahati et al. (2017)
miR-29a, miR-29b, miR-29c	Downregulation	DNMT3A	Parpart et al. (2014) and Ratnasari et al. (2022)
miR-346	Upregulation	FBXL2	Yu et al. (2017)
miR-454	Upregulation	SOCS6	Lei et al. (2020)
miR-487a	Upregulation	MAGI2	Chang et al. (2017)
miR-873	Upregulation	TSLC1	Han et al. (2018a)

AKAP12 A-kinase anchor protein 12, *ANGPT2* angiopoietin-2, *CLDN1* claudin-1, *DNMT3A* DNA (cytosine-5)-methyltransferase 3A, *FBXL2* F-box and leucine-rich repeat protein 2, *HEPN1* hepatocellular carcinoma, downregulated 1, *HIF-1α* hypoxia-inducible factor 1 alpha, *IFITM3* interferon-induced transmembrane protein 3, *KLF4* Krüppel-like factor 4, *MAGI2* membrane-associated guanylate kinase, WW, and PDZ domain–containing protein 2, *PTEN* phosphatase and tensin homolog, *SEMA3A* semaphorin-3A, *SOCS6* suppressor of cytokine signaling 6, *TIMP3* metalloproteinase inhibitor 3, *TSLC1* tumor suppressor in lung cancer 1, *ZEB2* zinc finger E-box-binding homeobox 2

Dysregulation of miRNAs in Cholangiocarcinoma (CCA)

While the targets and functions of several miRNAs have been identified in CCA, many dysregulated miRNAs in this malignancy still lack defined biological roles and are active areas of research (Razumilava et al. 2012). Wang and colleagues discovered that some miRNAs (**miR-21**, **miR-31**, and **miR-223**) were significantly dysregulated in CCA patients compared to healthy individuals; however, they lack correlation with the clinicopathological stages of CCA (Karakatsanis et al. 2013). In the following study, the suppression of **miR-21** inhibited CCA cell proliferation and in vivo tumor growth through the induction of cell cycle arrest and apoptosis Wang et al. 2015). Furthermore, PTPN14 and PTEN were recognized as direct and functional targets of **miR-21**, which has been demonstrated to promote CCA cell proliferation and chemotherapy resistance via suppressing PTEN in previous studies (Meng et al. 2006). Another study performed by Selaru et al. confirmed the robust upregulation of **miR-21** in CCA patients compared to healthy individuals, and further investigation of its targets revealed programmed cell death 4 (PDCD4) and TIMP3 to be the potential targets. TIMP3 is a tumor suppressor gene that plays a crucial role in controlling angiogenesis in the CCA microenvironment. These results are consistent with the fact that mRNA levels of TIMP3 were significantly lower in CCA patients (Selaru et al. 2009). Another study investigated the role of the **miR-181b**/TIMP3 axis in angiogenesis and tumor growth in a murine choline-deficient and amino acid–defined diet-induced HCC model after 84 weeks. This study confirmed the TGFβ signaling role in **miR-181b** upregulation. Furthermore, the decrease of miR-181b impeded HCC proliferation and chemotherapy resistance in nude mice (Wang et al. 2010).

miR-370 acts as a tumor suppressor by targeting the oncogene MAPK kinase kinase 8 (MAP3K8). Elevated levels of interleukin-6 (IL-6) can downregulate **miR-370** through promoter hypermethylation, thereby restoring MAP3K8 expression and promoting CCA progression (Nakaoka et al. 2017). Razumilava et al. found increased levels of **miR-25** in CCA patient samples and cell lines. Hedgehog signaling may be responsible for this rise, supported by the decline in **miR-25** expression upon treatment of CCA cell lines with cyclopamine, a specific inhibitor of the Hedgehog signaling pathway. **miR-25** exerts an antiapoptotic effect in CCA by suppressing the expression of TNF-related apoptosis-inducing ligand (TRAIL), a key death receptor involved in apoptosis. Consistently, the inhibition of **miR-25** in cultured CCA cells increased their sensitivity to apoptosis. Computational predictions identified TRAIL death receptor 4 (DR4) as a potential target of **miR-25**, which was validated through luciferase reporter assays and immunoblotting (Razumilava et al. 2012).

Meng et al. managed to determine the differentially expressed miRNAs in Mz-ChA-1 cell xenografts from established CCA in nude mice. Among these miRNAs, they found elevated expressions of **miR-200b**, **miR-141**, **miR-181a**, **miR-181b-1**, **miR-34a**, **miR-224**, and **miR-21**. They determined the targets of three miRNAs as well: **miR-141** (CLOCK, controls cell cycle), **miR-200b** (PTPN12, binds oncogenic products such as c-Abl or Src), and **miR-21** (PTEN) (Meng et al. 2006). Lu et al. observed that **miR-21** promotes CCA growth by targeting and downregulating 15-hydroxyprostaglandin dehydrogenase (15-PGDH), an enzyme responsible for prostaglandin degradation. The suppression of 15-PGDH by **miR-21** leads to the accumulation of prostaglandin E2 (PGE2), which contributes to tumor progression (Lu et al. 2014a). Unlike typical miRNAs that suppress gene expression, **miR-122-5p** enhances the transcription of insulin-like growth factor–binding protein 4 (IGFBP4) by directly binding to its promoter region, thereby attenuating CCA metastasis (Tao et al. 2024).

miR-200b is downregulated in CCA patients and was found to be reversibly correlated with TGF-β levels. Overexpression of **miR-200b** using a mimic significantly reduced the proliferation and migratory capacity of CCA cells by targeting two key genes: TFAP2A, which encodes the transcription factor activating protein 2 alpha (AP-2α), and mitogen-activated protein kinase 7 (MAPK7), a member of the MAP kinase family involved in cell signaling. AP-2α regulates the genes involved in

EMT, a critical process in cancer metastasis whereby epithelial cells acquire mesenchymal traits, enhancing their invasiveness. Similarly, MAPK7 has been implicated in promoting tumor growth and metastatic behavior through its role in cellular responses to growth factors and stress. By downregulating these targets, **miR-200b** impedes EMT-related pathways, thereby suppressing tumor progression. This study addressed the potential of employing gene therapy to limit CCA tumor proliferation (Zhang et al. 2017b). Expression analysis revealed that four members of the **miR-200** family (**miR-200a/b/c** and **miR-429**) were markedly downregulated in CCA. Functional assays demonstrated that restoring **miR-200b** or **miR-200c** expression suppressed the invasive and migratory behavior of CCA cells, both in vitro and in animal models.

Additionally, these miRNAs were shown to affect several tumorigenic features, including tumor initiation and resistance to chemotherapy. Mechanistically, **miR-200b/c** targeted rho-kinase 2 (ROCK2) to inhibit migration and invasion. At the same time, their effects on tumorigenicity were mediated via suppression of the suppressor of zeste 12 (SUZ12), a component of the polycomb repressive complex (Peng et al. 2013). The dysregulated miRNAs in CCA are summarized in Table 5.2.

Table 5.2 Dysregulated miRNAs in CCA

miRNAs	Pattern of regulation	Target	References
Let-7c	Downregulation	EZH2	Xie et al. (2018)
miR-122-5p	Downregulation	IGFBP4	Tao et al. (2024)
miR-141	Upregulation	CLOCK	Meng et al. (2006)
miR-200b	Downregulation	TFAP2A, MAPK7	Zhang et al. (2017b)
miR-200b	Upregulation	PTPN12	Meng et al. (2006)
miR-200b/c	Downregulation	SUZ12/ROCK2	Peng et al. (2013)
miR-21	Upregulation	PTPN14, PTEN, TIMP3, 15-PGDH	Lu et al. (2014b), Meng et al. (2006), Selaru et al. 2009, and Wang et al. (2015)
miR-25	Upregulation	DR4	Razumilava et al. (2012)
miR-26a	Upregulation	GSK-3β	Zhang et al. (2012a)
miR-370	Downregulation	MAP3K8	Nakaoka et al. (2017)
miR-421	Upregulation	FXR	Zhong et al. (2012)
miR-92b	Upregulation		Zhou et al. (2018)

15-PGDH 15-hydroxyprostaglandin dehydrogenase, *CLOCK* circadian locomotor output cycles kaput, *DR4* death receptor 4, *EZH2* enhancer of zeste homolog 2, *FXR* farnesoid X receptor, *GSK-3β* glycogen synthase kinase 3 beta, *IGFBP4* insulin-like growth factor–binding protein 4, *MAP3K8* mitogen-activated protein kinase kinase kinase 8, *MAPK7* mitogen-activated protein kinase 7, *PTEN* phosphatase and tensin homolog, *PTPN12* protein tyrosine phosphatase non-receptor type 12, *PTPN14* protein tyrosine phosphatase non-receptor type 14, *ROCK2* rho-associated protein kinase 2, *SUZ12* polycomb protein SUZ12, *TFAP2A* transcription factor AP-2 alpha, *TIMP3* metalloproteinase inhibitor 3

Table 5.3 Dysregulated miRNAs in metastatic liver cancer

miRNAs	Primary tumor	Dysregulation pattern in liver metastases	Role in liver metastasis	References
miR-122	Colorectal cancer	Up	NEGR1 and CAT1	Dai et al. (2024a) and Iino et al. (2013)
miR-200c		Down	ETS1, FLT1, SOX2, and ZEB1	Hur et al. (2013) and Lu et al. (2014b)
miR-21		Up	PTEN	Zhang et al. (2012b)
miR-371-5p		Down	SOX2	Li et al. (2015)
miR-10b	Breast cancer	Up	HOXD10	Halim et al. (2025)
miR-501-3p	Pancreatic cancer	Up	TGFBR3	Yin et al. (2019)
miR-607	Pancreatic cancer	Up		Jiang et al. (2021)

CAT1 cationic amino acid transporter 1, *ETS1* ETS proto-oncogene 1, *FLT1* Fms-related tyrosine kinase 1, *HOXD10* homeobox D10, *NEGR1* neuronal growth regulator 1, *PTEN* phosphatase and tensin homolog, *SOX2* SRY-box transcription factor 2, *TGFBR3* transforming growth factor beta receptor 3, *ZEB1* zinc finger E-box-binding homeobox 1

Role of miRNAs in Metastatic Liver Cancer from Colorectal, Breast, or Pancreatic Origin (Table 5.3)

Although the role of miRNAs in metastatic liver cancer has not been widely elucidated, there is evidence of the interplay in metastatic stages of colorectal cancer (CRC), breast cancer, and pancreatic ductal adenocarcinoma (PDAC). CRC frequently metastasizes to the liver, occurring in 20–25% of patients at diagnosis and, eventually, in over half, following primary resection. **miR-122**, the liver's dominant miRNA, is notably depleted in primary CRC yet becomes overexpressed in hepatic metastases, where it concurrently represses neuronal growth regulator 1 (NEGR1) and cationic amino acid transporter-1 (CAT1). This dual targeting activates PI3K/AKT signaling, drives M2 macrophage polarization, and enhances tumor cell survival and invasiveness within the hepatic niche (Abd El Fattah et al. 2023; Dai et al. 2024a; Hamdy et al. 2024; Iino et al. 2013). In murine models, hepatocyte-specific delivery of **miR-122** via galactose-modified lipid–calcium phosphate nanoparticles reshapes the liver microenvironment, boosting $CD8^+$ T cell infiltration and reducing immunosuppressive populations to markedly inhibit metastatic colonization (Sendi et al. 2022). Clinically, raised levels of circulating exosomal **miR-122** accurately distinguish CRC patients with hepatic metastases from nonmetastatic and healthy cohorts, highlighting its promise as a noninvasive prognostic biomarker of CRC liver metastasis (Dai et al. 2024a).

In situ hybridization of primary CRC specimens demonstrated markedly reduced **miR-200c** levels at the tumor's invasive front. Functionally, the reexpression of **miR-200c** in CRC cell lines restores E-cadherin, reduces vimentin, and curtails invasion and migration by directly targeting EMT-drivers ZEB1, ETS1, and FMS-like tyrosine kinase (FLT1) (Hur et al. 2013). In orthotopic CRC mouse models, the enforced expression of **miR-200c** in tumor cells completely abrogated hepatic metastases (0% incidence vs. 60% in vector controls), an inhibition fully reversed by the co-expression of the SRY-box transcription factor 2 (Sox2), demonstrating a functional **miR-200c**/Sox2-negative feedback loop that dictates metastatic competency (Lu et al. 2014b).

Another study demonstrated that **miR-371-5p** functions as a potent tumor suppressor in CRC by inhibiting EMT, cancer stem cell traits, and metastatic spread. In CRC cell lines and patient samples, **miR-371-5p** levels were markedly reduced compared to normal mucosa, correlating with higher rates of lymphatic and liver metastases. Mechanistically, the transcription factor SOX17 induces **miR-371-5p** expression, which, in turn, directly targets the 3′-UTR of SOX2 mRNA to suppress its translation. The loss of SOX17 led to increased SOX2, upregulation of pluripotency factors (OCT4 and CD133), enhanced sphere formation, and greater invasive capacity. These effects were fully reversed by restoring the **miR-371-5p** expression (Li et al. 2015). **miR-21** functions as a pivotal oncomiR in CRC liver metastasis, exhibiting significantly higher levels in metastatic lesions than in matched primary tumors. Driven by phosphatase of regenerating liver-3 (PRL-3)-mediated activation of the Src/STAT3 pathway, **miR-21** transcription is upregulated in CRC cells, leading to a direct suppression of the tumor suppressor PTEN and consequent activation of PI3K/AKT signaling. Functionally, enforced **miR-21** expression enhances CRC cell proliferation, migration, and invasion in vitro, whereas **miR-21** inhibition reverses these aggressive phenotypes in PRL-3-overexpressing cells (Zhang et al. 2012a).

miR-10b acts as a master regulator of breast cancer metastasis, including hepatic dissemination. Ma et al. showed that the Twist-driven upregulation of miR-10b in nonmetastatic breast cancer cells initiates invasion and distant metastasis by directly repressing the homeobox gene, homeobox D10 (HOXD10), and thereby derepressing the pro-metastatic GTPase, Ras homolog family member C (RHOC). Clinically, **miR-10b** levels are markedly elevated in the serum and metastatic lesions of patients with liver-involved breast cancer, where high expression correlates with advanced stage and poorer overall survival (Halim et al. 2025). Therapeutically, the systemic administration of MN-anti-**miR-10b** (a nanoparticle-delivered antagomir) effectively inhibits **miR-10b** in vivo, reducing metastatic burden in organs, including the liver, and improving survival in murine models of metastatic breast cancer (Halim et al. 2024).

Yin et al. found that tumor-associated macrophages (TAMs), particularly the M2 subtype, release exosomes enriched with **miR-501-3p**. These exosomes are taken up by the PDAC cells, leading to the downregulation of TGF-β receptor III (TGFBR3). This suppression activates the TGF-β signaling pathway, enhancing the migration, invasion, and metastasis of PDAC cells. In vivo experiments demonstrated that inhibiting **miR-501-3p** in these exosomes reduced tumor growth and

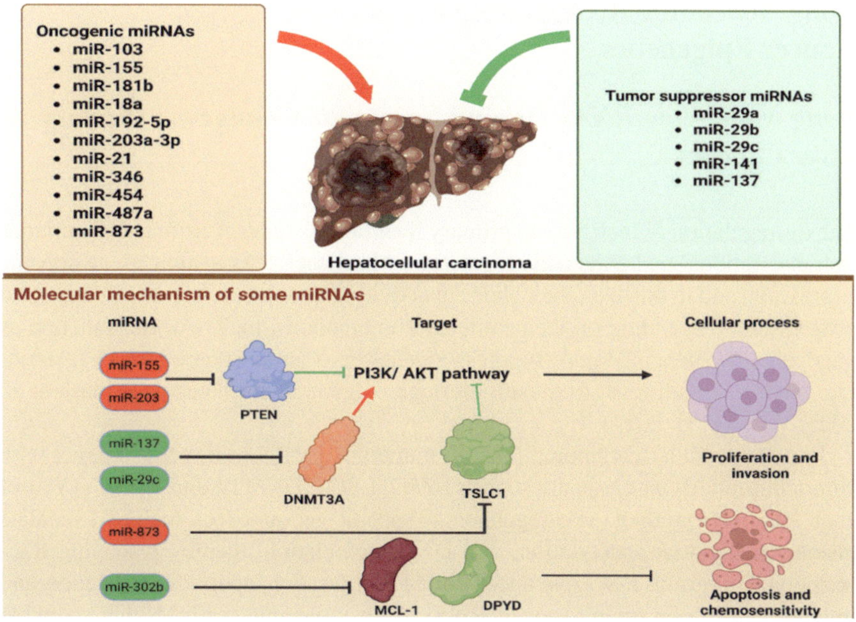

Fig. 5.1 Dysregulated miRNAs and their molecular mechanisms in hepatocellular carcinoma. The top panel shows an overview of dysregulated miRNAs in hepatocellular carcinoma (HCC). Oncogenic miRNAs are upregulated and promote cancer progression by targeting tumor suppressor genes, enhancing cell proliferation, invasion, and survival. Tumor suppressor miRNAs are downregulated in HCC, normally functioning to inhibit oncogene expression and suppress malignant transformation. The bottom panel shows representative molecular mechanisms regulated by selected miRNAs in HCC. Oncogenic miRNAs are shown in red, and tumor suppressor miRNAs are shown in green. *DNMT3A* DNA methyltransferase 3 alpha, *DPYD* dihydropyrimidine dehydrogenase, *Mcl-1* myeloid cell leukemia 1, *PTEN* phosphatase and tensin homolog, *TSLC1* tumor suppressor in lung cancer 1

spread (Yin et al. 2019). Jiang and colleagues demonstrated that **miR-607** is significantly downregulated in both PDAC tissues and patient serum samples. A low **miR-607** expression was associated with more aggressive clinical features, including lymphatic and liver metastasis, perineural invasion, and reduced overall and progression-free survival, identifying it as an independent predictor of poor prognosis. Diagnostic performance analysis revealed that serum **miR-607** had better sensitivity and specificity for distinguishing PDAC than traditional markers like CA19-9 and Carcinoembryonic antigen (CEA), especially when used in combination. Functionally, the overexpression of **miR-607** inhibited PDAC cell migration and invasion by suppressing EMT, suggesting its potential role as both a biomarker and a therapeutic target in PDAC (Jiang et al. 2021). Collectively, dysregulated miRNAs and their molecular mechanisms in HCC are summarized in Fig. 5.1.

Long Noncoding RNAs (lncRNAs) in Liver Cancer Epigenetics

Long Noncoding RNAs Driving Epigenetic Changes in Primary Liver Cancers

The dysregulation of lncRNAs in primary liver tumors reflects profound disruptions in transcriptional and epigenetic regulation. Many lncRNAs exhibit tissue-specific expression, and in the liver, their aberrant activity contributes to tumor initiation and progression. Depending on their molecular interactions, lncRNAs can function as oncogenes by silencing tumor suppressor genes or as tumor suppressors by restraining oncogenic pathways, frequently through epigenetic mechanisms (Eldash et al. 2023; Verma et al. 2023).

In hepatocellular carcinoma (HCC), numerous lncRNAs directly interact with chromatin-modifying enzymes, such as EZH2, LSD1, DNMTs, and HDACs (Yousef et al. 2020). By guiding these regulators to specific genomic loci, lncRNAs mediate histone methylation, acetylation, and DNA methylation, thereby reshaping transcriptional programs that drive uncontrolled proliferation, apoptosis resistance, and immune evasion. In some cases, lncRNAs act as molecular scaffolds that assemble protein complexes into activating or repressive chromatin structures, further fine-tuning the oncogenic gene expression (Wang et al. 2017).

Importantly, lncRNAs display stage-specific expression during HCC progression. Certain lncRNAs are upregulated in early stages, repressing differentiation- and cell cycle–related genes to promote malignant transformation (Romeo et al. 2023; Youness et al. 2024), whereas others become prominent in advanced disease, enhancing angiogenesis, epithelial–mesenchymal transition (EMT), and metastasis through the repression of adhesion molecules or immune regulators (Chatterjee 2024). This temporal modulation indicates that lncRNAs actively shape disease evolution rather than merely reflecting it.

Recent evidence also implicates lncRNAs in metabolic reprogramming via epigenetic regulation, enabling HCC cells to adapt to hypoxic and nutrient-deprived microenvironments. Such adaptations support rapid proliferation and survival under stress, underscoring the multifaceted role of lncRNAs in HCC biology (Li et al. 2024a).

Given their regulatory capacity, many lncRNAs are under investigation as biomarkers for diagnosis, prognosis, and therapeutic response prediction in liver cancer. Therapeutic strategies targeting lncRNAs—using antisense oligonucleotides, RNA interference, or CRISPR/dCas9-mediated epigenetic editing—are emerging as innovative tools in precision oncology (Rahman and Tollefsbol 2021). Future studies will focus on optimizing delivery systems and improving target specificity to fully realize the translational potential of lncRNA-based interventions.

Several representative lncRNAs have been identified as key regulators of HCC epigenetics, each acting through distinct molecular mechanisms.

HOX Transcript Antisense RNA (HOTAIR) HOX transcript antisense RNA (HOTAIR) is one of the most well-investigated oncogenic lncRNAs in hepatocellular carcinoma (Chen et al. 2017a). It advances tumor growth by enlisting the polycomb repressive complex 2 (PRC2), particularly EZH2, to designated gene loci, where it enables H3K27me3-mediated repression of tumor suppressor genes such as p53 and p21 (Park et al. 2021). Its overexpression correlates with poor differentiation, high metastatic potential, and reduced overall survival (Han et al. 2018c).

Highly Upregulated in Liver Cancer (HULC) HULC is a liver-specific lncRNA that promotes hepatocarcinogenesis by sponging miRNAs (e.g., miR-372), thereby relieving repression of oncogenes such as PRKACB. Detectable in serum, HULC serves as a promising noninvasive biomarker for HCC diagnosis and monitoring (Lim et al. 2019; Yasi Pan et al. 2020).

Maternally Expressed Gene 3 (MEG3) MEG3 is a tumor-suppressive lncRNA, frequently downregulated in HCC, primarily due to promoter hypermethylation (Zhuo et al. 2016). MEG3 interacts with DNMTs and transcriptional regulators such as p53 to inhibit proliferation and promote apoptosis. A loss of MEG3 is linked to advanced disease and poor clinical outcomes (L. Zhang et al. 2022).

LINC01138 The recently identified carcinogenic lncRNA, LINC01138, operates by interacting with PRMT5, an arginine methyltransferase, thereby promoting the methylation of histone and nonhistone proteins (Jantrapirom et al. 2021). This relationship facilitates the repression of tumor suppressor genes and enhances proliferation, invasion, and immune evasion in hepatocellular carcinoma (HCC). Preclinical models have demonstrated therapeutic advantages through the inhibition of LINC01138, indicating its viability as a pharmacological target (Yutian Yu et al. 2024).

Differentiation Antagonizing Nonprotein Coding RNA (DANCR) DANCR maintains stemness and blocks hepatocyte differentiation by recruiting EZH2 to the promoters of genes such as CTNNA1 and p21, inducing H3K27me3-mediated silencing. It contributes to tumor initiation and chemoresistance (Zhang et al. 2021c).

Plasmacytoma Variant Translocation 1 (PVT1) Often co-amplified with MYC, PVT1 exerts oncogenic activity by recruiting EZH2 and DNMT1 to tumor suppressor gene promoters, driving DNA hypermethylation and silencing (K. Jin et al. 2019).

Antisense Noncoding RNA in the INK4 Locus (ANRIL) ANRIL interacts with both PRC1 and PRC2, facilitating H3K27me3-mediated repression at the INK4/ARF locus and silencing cell cycle regulators (p15, p16, p14), thereby promoting proliferation (Aguilo et al. 2011).

LINC00662 It functions as an oncogene by establishing a repressive complex with EZH2 and HDAC1 (C. Yuan et al. 2022). It inhibits tumor suppressor genes such as

KLF2 and NKD2, facilitating epithelial–mesenchymal transition (EMT), proliferation, and invasion. Its expression is heightened in the later stages of hepatocellular carcinoma and is regarded as an indicator of aggressiveness (Ogunwobi et al. 2019).

High Expression in HCC (HEIH) HEIH promotes HCC by recruiting EZH2 to silence p16 and p21, crucial mediators of cell cycle arrest and senescence. High HEIH levels are associated with early recurrence after surgical resection (Toh et al. 2019).

Taurine Upregulated Gene 1 (TUG1) TUG1 exerts oncogenic effects, primarily via PRC2-mediated H3K27me3 silencing of tumor suppressors such as KLF2. TUG1 enhances proliferation and angiogenesis, making it a potential therapeutic target (Pathania 2023).

HOXA Transcript at the Distal Tip (HOTTIP) HOTTIP contributes to hepatocarcinogenesis by activating HOXA gene transcription. It recruits the WDR5/MLL complex, which induces H3K4 trimethylation at target promoters, thereby sustaining oncogenic transcriptional programs. Elevated HOTTIP expression is linked to increased proliferation, migration, and therapeutic resistance and is strongly correlated with advanced disease and HCC recurrence (Ghafouri-Fard et al. 2020).

Collectively, these lncRNAs, through their diverse yet specific epigenetic interactions, reshape the transcriptional landscape of hepatocellular carcinoma. Their distinct expression patterns and molecular functions not only advance our understanding of liver cancer biology but also highlight their potential as diagnostic biomarkers and therapeutic targets (Table 5.4).

Function of lncRNAs in Establishing Metastatic Niches in Secondary Liver Cancer

In secondary liver cancers—arising as metastases from colorectal, breast, pancreatic, and other primary tumors—lncRNAs play a pivotal role in shaping a microenvironment conducive to metastatic colonization and survival within hepatic tissue (Tam et al. 2019). Unlike in primary liver cancer, where lncRNAs largely initiate tumorigenic signaling in hepatocytes, in secondary malignancies, they act as mediators of intercellular communication among metastatic tumor cells, hepatic stromal cells, immune cells, and the vasculature (Lin et al. 2018).

The principal function of lncRNAs in this context lies in the epigenetic reprogramming of resident hepatic cells to establish a pre-metastatic niche (Li et al. 2025b). Tumor-derived exosomes enriched with specific lncRNAs are released into the circulation and taken up by Kupffer cells, hepatic stellate cells, and endothelial cells (Zhou et al. 2022). Once internalized, these lncRNAs alter gene expression through interactions with chromatin modifiers or miRNAs, leading to changes in cytokine secretion, extracellular matrix (ECM) remodeling, and immune cell recruitment (Chen et al. 2017b). For example, colorectal cancer–derived lncRNAs

Table 5.4 Key oncogenic and tumor-suppressive lncRNAs driving epigenetic modifications in hepatocellular carcinoma

lncRNA	Function	Epigenetic interactors	Main targets	Effect in HCC	References
HOTAIR	Oncogenic	PRC2 (EZH2)	p53, p21	Promotes metastasis	Chen et al. (2017a), Han et al. (2018c), and Park et al. 2021)
HULC	Oncogenic	miR-372, CREB pathway	PRKACB	Proliferation, antiapoptosis	Lim et al. (2019) and Pan et al. (2020)
MEG3	Tumor suppressor	DNMT1, p53	p53, Rb	Inhibits proliferation	Zhang et al. (2022) and Zhuo et al. (2016)
LINC01138	Oncogenic	PRMT5	AR-dependent genes	Immune evasion, growth	Jantrapirom et al. (2021) and Yu et al. (2024)
DANCR	Oncogenic	EZH2	p21, CTNNA1	Stemness, dedifferentiation	Zhang et al. (2021c)
PVT1	Oncogenic	EZH2, DNMT1	p15, p16	Proliferation	Jin et al. (2019)
ANRIL	Oncogenic	PRC1/2	INK4 locus	Cell cycle escape	Aguilo et al. (2011)
LINC00662	Oncogenic	EZH2, HDAC1	KLF2, NKD2	Invasion, EMT	Ogunwobi et al. (2019) and Yuan et al. (2022)
HEIH	Oncogenic	EZH2	p16, p21	Enhanced growth	Toh et al. (2019)
TUG1	Oncogenic	PRC2	KLF2	Angiogenesis	Pathania (2023)
HOTTIP	Oncogenic	MLL/ WDR5	HOXA genes	Proliferation, migration	Ghafouri-Fard et al. (2020)

ANRIL antisense noncoding RNA in the INK4 locus, *AR* androgen receptor–dependent genes, *CTNNA1* catenin alpha 1, *CREB* cAMP response element–binding protein pathway, *DANCR* differentiation antagonizing nonprotein coding RNA, *DNMT1* DNA methyltransferase 1, *EZH2* enhancer of zeste homolog 2, *HDAC1* histone deacetylase 1, *HEIH* high expression in hepatocellular carcinoma, *HOTAIR* HOX transcript antisense intergenic RNA, *HOTTIP* HOXA transcript at the distal tip, *HULC* highly upregulated in liver cancer, *INK4* inhibitors of CDK4 locus, *KLF2* Krüppel-like factor 2, *LINC00662* long intergenic nonprotein coding RNA 662, *LINC01138* long intergenic nonprotein coding RNA 1138, *MEG3* maternally expressed gene 3, *MLL* mixed-lineage leukemia protein, *NKD2* naked cuticle homolog 2, *p15* cyclin-dependent kinase inhibitor 2B, *p53* tumor protein p53, *PRC1* polycomb repressive complex 1, *PRC2* polycomb repressive complex 2, *PRKACB* protein kinase cAMP-activated catalytic subunit beta, *PRMT5* protein arginine methyltransferase 5, *PVT1* plasmacytoma variant translocation 1, *Rb* retinoblastoma protein, *TUG1* taurine upregulated gene 1, *WDR5* WD repeat–containing protein 5

have been shown to induce fibrotic and immunosuppressive changes in the liver, thereby promoting metastatic cell survival and outgrowth (Tam et al. 2019).

lncRNAs also regulate angiogenesis within metastatic sites. Acting as competing endogenous RNAs (ceRNAs) or scaffolds for histone modification complexes, they enhance the expression of pro-angiogenic factors such as VEGFA and angiopoietins in liver endothelial cells, driving the neovascularization essential for metastatic expansion (Yao et al. 2023).

Another critical role of lncRNAs is the suppression of local immune surveillance. Tumor-derived lncRNAs can downregulate chemokines that recruit cytotoxic T lymphocytes or polarize macrophages toward an M2 phenotype, thereby facilitating tissue remodeling and tumor progression (Mohapatra et al. 2021). Such immunomodulatory effects enable metastatic cells to evade immune eradication and establish a long-term presence in the liver (Li et al. 2021b).

Furthermore, some lncRNAs modulate metabolic adaptability in metastatic cells (Z. Li and Sun 2020). The liver offers a unique metabolic environment, requiring tumor cells to adapt for successful metastasis (Bergers and Fendt 2021). lncRNAs can modulate pathways, including fatty acid oxidation, glucose absorption, and oxidative stress responses, ensuring metastatic cells fulfill their energy requirements and sustain redox equilibrium throughout colonization (Sellitto et al. 2021).

Collectively, lncRNAs orchestrate a complex network of epigenetic, immunological, and metabolic mechanisms that facilitate the establishment, adaptation, and progression of secondary liver malignancies (Li et al. 2021b). Understanding these functions may provide novel avenues for therapeutic intervention—targeting not only metastatic tumor cells directly but also the lncRNA-mediated remodeling of the hepatic microenvironment (Shi et al. 2024).

Numerous lncRNAs have been identified as essential regulators of liver metastasis, originating from colorectal, pancreatic, gastric, and breast cancers. They act either locally within metastatic cells or distally via extracellular vesicles to reprogram the hepatic niche. Key examples include (Table 5.5):

H19 Initially recognized for its function in embryonic development, H19 is reactivated in numerous tumors and is significantly correlated with liver metastasis, especially in colorectal and gastric malignancies. H19 is frequently encapsulated in exosomes and internalized by liver stromal cells. Upon internalization, it stimulates fibroblast activation and angiogenesis by modulating the VEGF and TGF-β pathways (Liu et al. 2019). H19 functions as a sponge for the miR-200 family, thereby indirectly activating EMT-related transcription factors (e.g., ZEB1 and ZEB2) that facilitate invasion and metastasis (Lin et al. 2016).

Metastasis-Associated Lung Adenocarcinoma Transcript 1 (MALAT1) **MALAT1 is** one of the most extensively studied lncRNAs in metastasis. MALAT1 regulates alternative splicing, nuclear architecture, and transcription (Arun et al. 2020). In liver metastases, it facilitates tumor–endothelial adhesion, angiogenesis, and immune evasion. It also recruits EZH2 to repress tumor suppressor genes, sustaining proliferation and invasiveness (Tang et al. 2025).

Small Nucleolar RNA Host Gene 6 (SNHG6) In CRC, SNHG6 is significantly upregulated in liver-metastatic tissues and facilitates the activation of the TGF-β/Smad signaling pathway (Wang et al. 2019). It functions as a ceRNA to sequester

Table 5.5 Key lncRNAs involved in liver metastasis

lncRNA	Primary cancer origin	Mechanism of action	Role in liver metastasis	References
H19	Colorectal, gastric	miRNA sponging (miR-200), exosomal signaling	ECM remodeling, EMT, angiogenesis	Lin et al. (2016) and Liu et al. (2019)
MALAT1	Multiple	EZH2 interaction, immune evasion	Endothelial adhesion, angiogenesis	Arun et al. (2020) and Tang et al. (2025)
SNHG6	Colorectal	ceRNA (miR-101-3p), TGF-β/Smad activation	EMT, invasion, liver colonization	Khan et al. (2023) and Wang et al. (2019)
UCA1	Colorectal, bladder	EZH2 binding, immune suppression	Chemoresistance, niche modulation	Hu et al. (2016) and Ramli et al. (2021)
LINC01296	Colorectal, pancreatic	MMP induction, PD-L1 upregulation	Invasion, immune evasion	Jun et al. (2023) and Wu et al. (2017)
CCAT1	Colorectal	MYC regulation, chromatin remodeling	Tumor stemness, growth in the liver microenvironment	Correnti et al. (2022), Xiang et al. (2014), and Zhao et al. (2022)
GAS5	Colorectal, gastric	Downregulation leads to β-catenin and YAP activation	Promotes hepatic niche permissiveness; impairs suppression	Li et al. (2024b) and Yu and Hann (2019)
NEAT1	Pancreatic, gastric	Induces M2 macrophage polarization; cytokine release	Shapes the immunosuppressive microenvironment	Jing et al. (2022) and Yuan et al. (2025)
TUC338	HCC, colorectal	Activates EGFR/FGF pathways; ECM modulation	Enhances tumor–host cell interaction	Drew and Machesky (2021) and Jin et al. (2017)
PVT1	Colorectal, pancreatic	Upregulates HIF-1α and MYC; supports hypoxia response	Promotes angiogenesis and metastatic colonization	Jin et al. 2019 and Zhang et al. (2021b)

CCAT1 colon cancer–associated transcript 1, *ECM* extracellular matrix, *EGFR/FGF* epidermal growth factor receptor/fibroblast growth factor, *EMT* epithelial–mesenchymal transition, *EZH2* enhancer of zeste homolog 2, *GAS5* growth arrest–specific transcript 5, *H19* H19-imprinted maternally expressed transcript, *HCC* hepatocellular carcinoma, *HIF-1α* hypoxia-inducible factor 1-alpha, *LINC01296* long intergenic nonprotein coding RNA 1296, *MALAT1* metastasis-associated lung adenocarcinoma transcript 1, *MMP* matrix metalloproteinase, *MYC* myelocytomatosis oncogene, *NEAT1* nuclear-enriched abundant transcript 1, *PD-L1* programmed death-ligand 1, *PVT1* plasmacytoma variant translocation 1, *SNHG6* small nucleolar RNA host gene 6, *TGF-β/Smad* transforming growth factor beta/SMAD family, *TUC338* transcript upregulated in cancer 338, *UCA1* urothelial cancer–associated 1, *YAP* yes-associated protein

tumor-suppressive miRNAs such as miR-101-3p, thus upregulating ZEB1, a crucial regulator of epithelial–mesenchymal transition and metastasis. This augments cellular motility and facilitates liver colonization (Khan et al. 2023).

Urothelial Cancer–Associated 1 (UCA1) UCA1 facilitates metastatic characteristics through many pathways. In liver metastases, it augments chemoresistance, glucose metabolism, and cellular survival (Ramli et al. 2021). UCA1 can bind to EZH2 and epigenetically decrease p27 expression, thereby facilitating continued cell cycle. It has also been associated with pre-metastatic niche conditioning through the modulation of hepatic stellate cells (Hu et al. 2016).

LINC01296 LINC01296 is significantly elevated in metastatic colorectal, bladder, and pancreatic cancers, promoting metastasis via increasing matrix metalloproteinase (MMP) production, thereby enabling extracellular matrix (ECM) breakdown and invasion (Eissa et al. 2003; El-Mesallamy et al. 2013; Wu et al. 2017). It is also associated with immune suppression via the modulation of PD-L1 expression on neoplastic cells (Jun et al. 2023).

Colon Cancer–Associated Transcript 1 (CCAT1) Colon cancer–associated transcript 1 (CCAT1) is overexpressed in CRC and is significantly associated with liver metastases. It functions by enhancing MYC oncogene expression and engaging with chromatin regulators to activate metastatic genes (Franz et al. 2023). CCAT1 additionally promotes a stem-like phenotype that enhances tumor adaptability in the liver, is overexpressed in colorectal cancer, and strongly correlates with liver metastasis (Zhao et al. 2022). It acts by upregulating MYC oncogene expression and interacting with chromatin regulators to activate metastatic genes (Xiang et al. 2014). CCAT1 also supports a stem-like phenotype that aids tumor adaptability in the liver (Correnti et al. 2022).

Growth Arrest–Specific 5 (GAS5) In contrast to numerous carcinogenic lncRNAs, GAS5 operates as a tumor suppressor. Its expression is often diminished in metastatic colorectal and gastric malignancies (Yu and Hann 2019). The absence of GAS5 results in the stimulation of proliferative and metastatic pathways, including YAP and β-catenin signaling. This downregulation promotes cancer cell survival, invasion, and establishment within the hepatic milieu (Li et al. 2024b).

Nuclear Paraspeckle Assembly Transcript 1 (NEAT1) Nuclear paraspeckle assembly transcript 1 (NEAT1) is pivotal in regulating the tumor immune microenvironment (Jing et al. 2022). In pancreatic and gastric malignancies with liver metastases, NEAT1 expression is associated with enhanced M2 macrophage polarization and the secretion of pro-inflammatory cytokines, thus creating an immunosuppressive and tumor-promoting microenvironment in the liver (Yuan et al. 2025).

TUC338 Initially recognized in primary hepatocellular carcinoma, TUC338 has also been associated with liver metastases originating from colorectal cancer (Jin et al. 2017). It governs tumor–host interactions by stimulating the FGF and EGFR pathways and alters the extracellular matrix (ECM), rendering the hepatic niche more amenable to metastatic colonization (Drew and Machesky 2021).

Plasmacytoma Variant Translocation 1 (PVT1) PVT1 is overexpressed in multiple metastatic malignancies, such as colorectal and pancreatic cancers. It increases the expression of HIF-1α, facilitating angiogenesis and aiding cancer cell adaptability to hypoxic environments in the liver (Zhang et al. 2021b). PVT1 is linked to MYC amplification, which contributes to aggressive tumor characteristics and liver metastasis (Jin et al. 2019).

Differential Expression and Mechanisms of lncRNAs between Primary and Secondary Liver Cancers

lncRNAs are crucial modulators of cancer advancement, with their functions varying markedly between primary liver malignancies, including HCC and ICC (K. Jin et al. 2019), and secondary (metastatic) liver cancers that arise from tumors in the colon, pancreas, breast, or other organs (Tam et al. 2019). The variation in their expression patterns and molecular pathways indicates both the tissue of origin and the tumor's development stage (Nielsen et al. 2002) (Fig. 5.2 and Table 5.6).

Distinct lncRNA Expression Profiles

In primary liver malignancies, numerous lncRNAs are unique to hepatocytes or cholangiocytes, and their dysregulation frequently arises from intrinsic hepatic injuries, such as chronic inflammation, HBV/HCV infection, or lipotoxicity. Long noncoding RNAs such as HULC, HOTAIR, MALAT1, and HEIH are persistently increased in hepatocellular carcinoma and function as epigenetic regulators of genes that govern proliferation, apoptosis, and immune evasion (Unfried and Fortes 2020).

Conversely, secondary liver tumors obtain their lncRNA profiles from the initial tumor source but alter them when entering the hepatic microenvironment. CCAT1, PVT1, and GAS5 are lncRNAs expressed in colorectal, gastric, and pancreatic malignancies, respectively, which are reprogrammed to facilitate metastasis after liver colonization. These lncRNAs frequently facilitate immunological regulation, angiogenesis, and metabolic adaptability, which are separate from the proliferation and survival mechanisms activated in primary tumors (Fu et al. 2016; Kladi-Skandali et al. 2015).

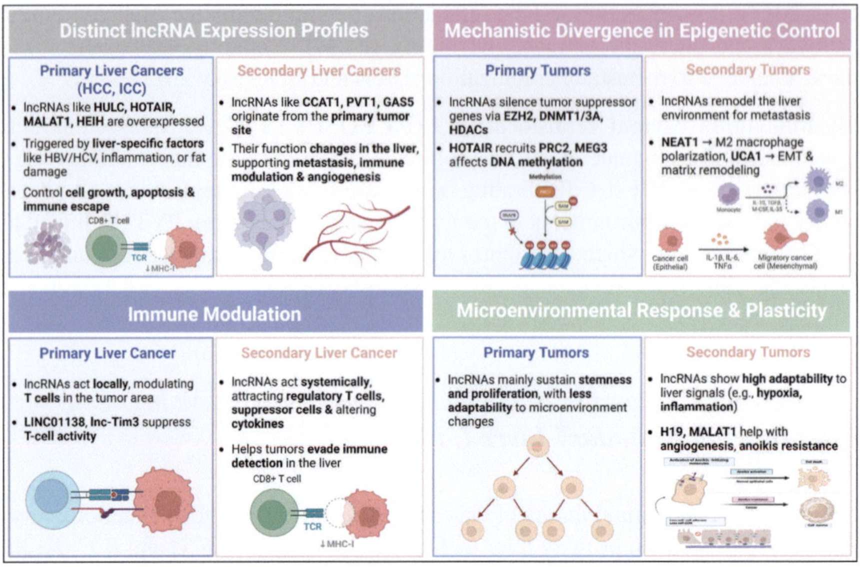

Fig. 5.2 Comparative roles of lncRNAs in primary vs. secondary liver cancers: expression, mechanisms, and tumor microenvironment interactions. The figure emphasizes four fundamental aspects of lncRNA function: (1) expression profiles: liver-specific lncRNAs in primary cancers compared to metastasis-related lncRNAs in secondary cancers, (2) epigenetic control: gene silencing in primary tumors vs. microenvironmental remodeling in secondary tumors, (3) immune modulation: local T cell suppression in primary cancers vs. systemic immune evasion in secondary cancers, and (4) microenvironmental plasticity: sustained proliferation in primary tumors vs. enhanced adaptability and colonization in secondary tumors. *CCAT1* colon cancer–associated transcript 1, *CTLA-4* cytotoxic T lymphocyte–associated protein 4, *DNMT1* DNA methyltransferase 1, *DNMT3A* DNA methyltransferase 3 alpha, *EZH2* enhancer of zeste homolog 2, *GAS5* growth arrest–specific transcript 5, *HBV* hepatitis B virus, *HCV* hepatitis C virus, *HDACs* histone deacetylases, *HEIH* high expression in hepatocellular carcinoma, *HOTAIR* HOX transcript antisense intergenic RNA, *HULC* highly upregulated in liver cancer, *H19* H19-imprinted maternally expressed transcript, *ICC* intrahepatic cholangiocarcinoma, *IL-6* interleukin-6, *IL-1β* interleukin-1 beta, *LINC01138* long intergenic nonprotein coding RNA 1138, *MALAT1* metastasis-associated lung adenocarcinoma transcript 1, *MEG3* maternally expressed gene 3, *NEAT1* nuclear-enriched abundant transcript 1, *PRC2* polycomb repressive complex 2, *PVT1* plasmacytoma variant translocation 1, *TCR* T cell receptor, *TGF-β* transforming growth factor beta, *TNFα* tumor necrosis factor alpha, *UCA1* urothelial cancer–associated 1

Mechanistic Divergence in Epigenetic Control

lncRNAs in primary liver tumors predominantly facilitate the epigenetic silencing of tumor suppressor genes. They engage with epigenetic modifiers such as EZH2, DNMT1/3A, and HDACs, resulting in chromatin modification that facilitates unregulated cellular proliferation (Mann 2014). For instance, HOTAIR attracts PRC2 to inhibit tumor suppressor loci, while MEG3 regulates DNA methylation patterns in hepatocytes (Bian et al. 2017).

Table 5.6 Comparative analysis of lncRNA expression patterns and epigenetic mechanisms in primary vs. secondary liver cancers

Feature	Primary liver cancer	Secondary liver cancer (metastasis)	References
Origin of lncRNAs	Hepatocyte/ cholangiocyte-specific	Derived from primary tumor cells	Unfried and Fortes (2020)
Key lncRNAs	HULC, HOTAIR, HEIH, MEG3	CCAT1, PVT1, GAS5, NEAT1, UCA1	Fu et al. (2016 and Kladi-Skandali et al. (2015)
Main functions	Epigenetic silencing, stemness, local immune evasion	Niche formation, immune suppression, angiogenesis	Bian et al. (2017) and Rahma and Hodi (2019)
Mechanistic focus	Histone methylation, DNA methylation	ECM remodeling, immune modulation, hypoxia response	Mann (2014)
Immune interaction	Local checkpoint regulation	Systemic immunosuppression and macrophage polarization	Luo et al. (2020)
Plasticity and adaptation	Limited to hepatic oncogenesis	High adaptability to the liver microenvironment	Cao et al. (2016), Lv et al. (2021a), and Zhao et al. (2019)

CCAT1 colon cancer–associated transcript 1, *ECM* extracellular matrix, *GAS5* growth arrest–specific transcript 5, *HEIH* high expression in hepatocellular carcinoma, *HOTAIR* HOX transcript antisense RNA, *HULC* highly upregulated in liver cancer, *MEG3* maternally expressed gene 3, *NEAT1* nuclear-enriched abundant transcript 1, *PVT1* plasmacytoma variant translocation 1, *UCA1* urothelial carcinoma–associated 1

In secondary liver tumors, lncRNAs have a significant role in modifying the metastatic microenvironment. This encompasses altering the extracellular matrix, augmenting vascular permeability, and modulating interactions with immune and stromal cells (Zeng et al. 2022). lncRNAs, such as NEAT1, facilitate M2 macrophage polarization, whereas UCA1 enhances epithelial–mesenchymal transition (EMT) and matrix remodeling, which are activities that receive less emphasis in primary liver cancers (Mohapatra et al. 2021; Shu et al. 2022).

Immune Modulation: Local vs. Systemic

lncRNAs in primary HCC regulate immunological checkpoints and T cell functionality locally within the tumor microenvironment. Instances encompass lnc-Tim3 and LINC01138, which directly inhibit cytotoxic T cell function (Lin et al. 2023).

In metastatic liver tumors, lncRNAs frequently facilitate systemic immunosuppression by modifying cytokine signaling or attracting immunosuppressive cells, including regulatory T cells and myeloid-derived suppressor cells. This systemic regulation is crucial for metastatic tumor cells to escape immune surveillance in an alien hepatic environment (Luo et al. 2020).

Microenvironmental Response and Plasticity

lncRNAs in metastatic tumors have significant plasticity, reacting to stimuli from the hepatic stroma, hypoxia, and inflammation (Cao et al. 2016). For instance, H19 and MALAT1 are variably produced in reaction to hypoxic stress in secondary tumors, facilitating angiogenesis and conferring resistance to anoikis (Zhao et al. 2019). In contrast, lncRNAs in primary HCC frequently play a role in sustaining stemness and facilitating cell cycle progression, lacking an equivalent microenvironmental flexibility (Lv et al. 2021a).

Clinical Implications and Therapeutic Targeting of lncRNAs in Liver Cancer

lncRNAs are emerging as powerful clinical biomarkers and therapeutic targets in liver cancer, particularly HCC. Due to their tissue-specific expression and regulatory roles in epigenetics, proliferation, metastasis, and immune evasion, lncRNAs offer several potential clinical applications (Shah and Sarkar 2024; Unfried et al. 2021) (Table 5.7).

Diagnostic Biomarkers

Some lncRNAs are consistently upregulated or downregulated in liver cancer tissue and can be detected in patient serum or exosomes, making them suitable for noninvasive diagnosis. Their expression can distinguish early-stage tumors from cirrhosis or chronic liver disease (Lee et al. 2019; Sun et al. 2018).

Prognostic Biomarkers

Levels of certain lncRNAs correlate with tumor grade, metastatic potential, and overall survival. Thus, they help stratify patients into risk categories and guide clinical decisions (Zhang et al. 2017a).

Predictive Markers of Therapy Response

Some lncRNAs are involved in pathways that mediate chemoresistance or immunotherapy resistance, making them useful in predicting which patients would benefit from certain treatments (Chen et al. 2022).

Table 5.7 Clinically relevant lncRNAs and their targeting

lncRNA	Function	Clinical implication	Therapeutic strategy	Current status	References
HOTAIR	Oncogenic; recruits PRC2 to silence tumor suppressor genes	Poor prognosis; metastasis predictor	ASOs or siRNA to silence HOTAIR	Preclinical studies in HCC	Comet et al. (2016) and Lennox and Behlke (2016)
MALAT1	Promotes proliferation, angiogenesis, and EMT	Correlates with recurrence and metastasis	ASOs or locked nucleic acid (LNA) inhibitors	Preclinical liver and lung cancer models	Zhang et al. (2019)
MEG3	Tumor suppressor; regulates p53 and DNA methylation	Downregulated in HCC; low levels linked to poor survival	Gene therapy to restore MEG3	Preclinical studies	Dong et al. (2019) and Zhang et al. (2022)
GAS5	Inhibits cell growth; sponge for miR-21	Reduced in HCC; predicts doxorubicin response	Restoration by vector delivery or demethylation agents	Experimental stage	Grossi et al. (2023) and Li et al. (2019)
LINC01138	Oncogenic; activates PRMT5	Associated with immune suppression	siRNA to inhibit LINC01138 expression	Preclinical	Guo et al. (2021) and Zhang et al. (2024)
ZFAS1	Promotes CRC metastasis to the liver	Biomarker for metastatic colorectal cancer	Blocking exosomal packaging	In vitro and animal models	Mehrab Mohseni et al. (2024) and Mohseni et al. (2024)
H19	Facilitates EMT and immune evasion	Diagnostic marker for early HCC and cirrhosis	ASOs and miRNA decoys	Early clinical testing in other cancers	Ariel et al. (1998) and Rojas et al. (2022)
UC001kfo	Inhibits proliferation and metastasis	Predicts survival; downregulated in HCC	Restoration therapy	Preclinical HCC models	Pan et al. (2025)

ASOs antisense oligonucleotides, *CRC* colorectal cancer, *EMT* epithelial–mesenchymal transition, *GAS5* growth arrest–specific transcript 5, *H19* H19-imprinted maternally expressed transcript, *HCC* hepatocellular carcinoma, *HOTAIR* HOX transcript antisense RNA, *LINC01138* long intergenic nonprotein coding RNA 1138, *MALAT1* metastasis-associated lung adenocarcinoma transcript 1, *MEG3* maternally expressed gene 3, *P53* tumor protein p53, *PRC2* polycomb repressive complex 2, *siRNA* small interfering RNA, *UC001kfo* uncharacterized noncoding RNA UC001kfo, *ZFAS1* zinc finger antisense 1

Direct Therapeutic Targets

Targeting oncogenic lncRNAs (e.g., HOTAIR and MALAT1) or restoring tumor-suppressive lncRNAs (e.g., MEG3 and GAS5) can reverse malignant phenotypes (Momtazmanesh and Rezaei 2021).

Circular RNAs and Their Epigenetic Roles in Liver Cancer

Biogenesis of Circular RNAs

A plant pathogen called viroid was the first evidence for the presence of RNA molecules with a circular conformation. Viroids are covalently linked, single-stranded circular RNA molecules with high stability against degradation (Sanger et al. 1976). Their circular conformation was biochemically verified by the resistance to 5′ phosphorylation of the RNA molecule, accompanied by the electron micrograph of rod-shaped RNA molecules (Sanger et al. 1976). Circular RNAs (circRNAs) are mainly derived from two main regions: from the intronic regions of premature RNA (pre-RNA) (mitochondrial RNA, transfer RNA, or ribosomal RNA), which were seen in yeast and archaea, or from the exons of pre-mRNA generated by spliceosomes (eukaryotic circRNAs biogenesis) (Salgia et al. 2003; Tabak et al. 1987).

Regarding eukaryotes, circRNAs are mainly generated from the exons of protein-coding genes. Pre-mRNA consists of introns (noncoding sequence) and exons (coding sequence). The generation of mature and translatable mRNA molecules requires the removal of these introns that interspace the exons via a complex ribonucleoprotein machinery called the spliceosome in a process termed canonical splicing (Wahl et al. 2009). On the other hand, when two or more exons covalently bind together to form a closed circle of RNA, the process is called back-splicing. Although competition exists between both the splicing events, circRNAs are typically generated, together with their corresponding linear mRNA, but are often found in low concentrations. circRNA levels tend to accumulate in slowly-dividing, long-lived cells (Xiao et al. 2020). circRNA biogenesis is highly regulated and controlled by various factors. For instance, the introns flanking the circRNA-producing exons were longer than the introns flanking the non-circRNA-producing exons. Additionally, the presence of certain elements in the introns enhances the biogenesis of circRNAs from the exons they are flanking: *Alu* elements in humans and *SINE* elements in mice (Yang et al. 2022).

Epigenetic Regulatory Roles of circRNAs in Liver Cancer

It was initially thought that circRNAs had no exact biological function; however, they were then found to contribute to the progression of numerous diseases as well as many physiological events. CircRNAs serve multiple functions, mainly acting as sponges for microRNAs (miRNAs), regulating the expression of their precursor genes by modifying RNA polymerase activity, binding to proteins, and being translated into regulatory proteins (Ruiz Esparza and Velázquez Flores 2023). Numerous circRNAs were differentially expressed in cancers and normal tissues, such as in esophageal cancer and breast cancer, indicating their possible role in the pathogenesis of these malignancies (P. Shi et al. 2018). It was shown that circRNAs were involved in various cancers, such as lung cancer, gastric cancer, and thyroid cancer (Li et al. 2025a; Wu et al. 2025; Zhu et al. 2025).

Regarding liver cancer, circRNAs were deregulated and involved in several hallmarks of liver cancer (Louis et al. 2022). The most observed role of circRNA in cancers is in sponging miRNAs, thereby regulating gene expression at the posttranscriptional level. A study has shown that hsa_circ_0001073 was significantly upregulated in liver cancer cell lines (HepG2, Huh-7, and Hep) compared to normal human liver cell lines (LO2 and MIHA). The study showed that hsa_circ_0001073 promoted proliferation, invasion, and migration and inhibited apoptosis of liver cancer cell lines. It was shown that hsa_circ_0001073 could enhance the activity of the phosphoinositide 3-kinase (PI3K) pathway by sponging miR-511-5p and subsequently elevating the expression of PI3K pathway components, such as PI3K, p-PI3K, and p-AKT (Fei et al. 2024).

Another study showed that hsa_circ_0032683 acted as a tumor suppressor in liver cancer by sponging miR-338-5p. The gene expression of hsa_circ_0032683 was significantly downregulated in liver cancer tissues and cell lines (HCCLM3, MHCC97-L, Hep3B, SMMC-7721, Huh7, Bel-7402, and MHCC97-H). Overexpressing hsa_circ_0032683 by the lentivirus hsa_circ_0032683 vector decreased the proliferation and induced apoptosis in liver cancer cell lines. Luciferase reporter assays confirmed that hsa_circ_0032683 can sponge miR-338-5p. The study also showed that miR-338-5p was inhibiting the translation of reticulon 4 (RTN4) by binding to its mRNA. RTN4 was involved in promoting apoptosis and decreasing proliferation in liver cancer. Overexpressing hsa_circ_0032683 resulted in decreased miR-338-5p expression and elevated RTN4 protein level, thus decreasing cellular proliferation and inducing apoptosis (Shen et al. 2022).

A study has shown that hsa_circ_0007874 was shown to be downregulated in liver cancer compared to normal control. It was found that hsa_circ_0007874 could sponge miR-541-5p, and their direct binding was verified by luciferase reporter assays. Downregulated hsa_circ_0007874 resulted in an increased gene expression of miR-541-5p; this resulted in a decrease in the expression of miR-541-5p's putative target, zinc family member 1 (ZIC1). ZIC1 could be acting as a tumor suppressor in liver cancer by inhibiting β-catenin and inhibiting epithelial–mesenchymal transition (Li et al. 2021a). Another circRNA that was ameliorating liver cancer by targeting and inhibiting the Wnt/β-catenin pathway was hsa_circ_0004018. The expression of hsa_circ_0004018 in human liver cancer tumor samples was higher than in normal tissues. The same expression pattern of hsa_circ_0004018 was also found in liver cancer cell lines (Huh7, Bel7402, SNU182, Hep3B, and SNU449) compared to the normal liver cell line (HL7702). Bioinformatic analyses showed that hsa_circ_0004018 could sponge and downregulate the oncomiR miR-626. The interaction was then confirmed by luciferase reporter assays. It was shown that hsa_circ_0004018 ameliorated liver cancer by downregulating miR-626 and, subsequently, upregulating its target mRNA, Dickkopf 3 (DKK3). Upregulated DKK3 is an inhibitor of the β-catenin pathway, thus inhibiting cellular proliferation (Zhu et al. 2020).

hsa_circRNA_104348 was found to be upregulated in liver cancer tissues compared to normal liver tissues. The elevated expression of hsa_circRNA_104348 was also seen in liver cancer cell lines (Hep3B, SK-HEP-1, MHCC97L, MHCC97H, and Huh7) compared to the normal liver cell line (HEK293). It was shown that hsa_circRNA_104348 promoted cell proliferation, invasion, and migration while inhibiting the apoptosis of liver cancer cells. The level of hsa_circRNA_104348 was inversely proportional to miR-187-3p, suggesting a binding relationship. Luciferase reporter assays confirmed a binding interaction between hsa_circRNA_104348 and miR-187-3p. Rhotekin 2 (RTNK2) was shown to be a direct target of miR-187-3p, and it was related to the overexpression of cell cycle–related proteins and overproliferation. Therefore, upregulated hsa_circRNA_104348 sponged and decreased the cellular level of miR-187-3p (tumor suppressor), which, in turn, increased the level of RTNK2 mRNA and induced cellular proliferation (Huang et al. 2020).

Metastatic Colonization of Liver Mediated by circRNAs (Secondary Liver Cancer)

circRNAs were found to be included in either mediating or suppressing metastatic colonization of the liver from other cancers, resulting in secondary liver tumors. circRNAs could be differentially expressed between primary colon cancer cells and colon cancer cells that metastasize to the liver. Two antimetastatic circRNAs were identified: hsa_circRNA_100873 and hsa_circRNA_100872, which originate from the same gene, *PPFIA1*; they were shown to be significantly downregulated in liver-metastatic colon cancer cells (KM12L4) compared to the primary colon cancer cell line (KM12C), suggesting their role in preventing metastasis. Knocking down hsa_circRNA_100873 and hsa_circRNA_100872 was shown to increase the in vitro migration and invasion ability of colon cancer cells (KM12C) and in vivo too. However, the overexpression of hsa_circRNA_100873 and hsa_circRNA_100872 resulted in decreased invasion and proliferation of the colon cancer cells. It was shown that hsa_circRNA_100873 and hsa_circRNA_100872 exerted their antimetastatic effect by sponging miR-155-5p, which was an oncomiR and was involved in the progression of colon cancer migration and invasion into the liver by inhibiting caudal-type homeobox 1 (CDX1). CDX1 could be a tumor suppressor that was shown to inhibit metastasis by downregulating the level of β-catenin (Ji et al. 2022).

Regarding breast cancer metastasizing to the liver, hsa_circ_0007255 was shown to be overexpressed in the primary breast cancer tissues and the liver metastasis tissues compared to the normal non-tumor tissues. Silencing of hsa_circ_0007255 resulted in the inhibition of glucose metabolism in breast cancer cells, as well as decreased invasion and migration of breast cancer cells (MCF-7, BT474, and HCC1806). The metastatic inducing effect of hsa_circ_0007255 could be due to its sponging effect on miR-335. This was verified by the fact that miR-335 was downregulated in primary and liver-metastatic breast cancer. miR-355 targets the mRNA of two important metabolic enzymes that mediate glucose uptake and lactate production: aldolase and octamer-binding transcription factor 4 (OCT4). Therefore, overexpressed hsa_circ_0007255 sponged the tumor suppressor miR-335 and, in turn, increased the mRNA levels of two metabolic proteins that are involved in enhancing cancer progression and migration: aldolase and OCT4 (Huang et al. 2022a).

Another study focusing on the role of circRNAs in metastasis showed that hsa_circ_0043278 was upregulated in metastatic colorectal cancer compared to primary colorectal cancer tissues. The overexpression of hsa_circ_0043278 with expression vectors could significantly increase the adhesion, migration, and invasive capacities of primary CRC cells. The metastasis-promoting effect of hsa_circ_0043278 could be due to the upregulation of the expression of Fraser extracellular matrix complex subunit 1 (FRAS1). hsa_circ_0043278 was found to be sponging miR-761 and miR-192 3p, both in vivo and in vitro, but without direct interaction with FRAS1. Since circRNAs can also bind to ribonucleoproteins to some extent, hsa_circ_102049 was found to bind to DGCR8 microprocessor complex subunit (DGCR8) and alter its cellular compartmentalization. Cells overexpressing hsa_circ_0043278 showed higher cytoplasmic levels of DGCR8 and lower nuclear levels compared to control cells, suggesting that the decrease in the levels of mature miR-761 and miR-192 3p is not just by the sponging effect of hsa_circ_0043278 but also by the alteration of their biogenesis in the nucleus (Zhi et al. 2021).

An additional circRNA involved in mediating metastasis was hsa_circ_0060467, which was found to be highly expressed in primary breast cancer tissues and breast cancer liver metastasis tissues compared to normal tissues. Knocking down hsa_circ_0060467 showed a significant decrease in breast cancer cell proliferation and migration. Injecting mice with MDA-MB-231 overexpressing hsa_circ_0060467 by the lentivirus vector led to increased metastatic liver nodules, suggesting that hsa_circ_0060467 could enhance liver metastasis and result in secondary liver cancer originating from breast cancer. It was shown that hsa_circ_0060467 exerted its effect by sponging miR-1205 and thus increasing its putative target, E2F transcription factor 1 (E2F1) (Zeng et al. 2023). To sum up, the roles of circRNAs in primary

Table 5.8 Role of circRNAs in primary and secondary liver cancers

Tumor type	circRNA	Expression	Target miRNA sponged	miRNA target	Type of study	References
Primary	hsa_circ_0001073	↑	miR-511-5p	PI3K	In vitro and in vivo	Fei et al. (2024)
	hsa_circ_0032683	↓	miR-338-5p	RTN4	In vitro and in vivo	Shen et al. (2022)
	hsa_circ_0007874	↓	miR-541-5p	ZIC1	In vitro and in vivo	Li et al. (2021a)
	hsa_circ_0004018	↓	miR-626	DKK3	In vitro and in vivo	Zhu et al. (2020)
	hsa_circRNA_104348	↑	miR-187-3p	RTNK2	In vitro and in vivo	Huang et al. (2020)
Secondary (metastatic)	hsa_circRNA_100872 hsa_circRNA_100873	↓	miR-155-5p	CDX1	In vitro and in vivo	Ji et al. (2022)
	hsa_circ_0007255	↑	miR-355	Aldolase and OCT4	In vitro and in vivo	Huang et al. (2022a)
	hsa_circ_0043278	↑	miR-761 and miR-192-3p	FRAS1	In vitro and in vivo	Zhi et al. (2021)
	hsa_circ_0060467	↑	miR-1205	E2F1	In vitro and in vivo	Zeng et al. (2023)

CDX1 caudal-type homeobox 1, *DKK3* dickkopf-related protein 3, *E2F1* E2F transcription factor 1, *FRAS1* Fraser extracellular matrix complex subunit 1, *OCT4* octamer-binding transcription factor 4, *PI3K* phosphoinositide 3-kinase, *RTN4* reticulon 4, *RTNK2* reticulon kinase 2, *ZIC1* Zic family member 1

and secondary liver cancers are presented in Table 5.8. Biogenesis, functions, and roles of circRNAs in HCC are illustrated in Fig. 5.3.

Shown here are the circRNAs involved in primary and secondary liver cancers, including their expression levels, sponged miRNAs, downstream mRNA targets, and the type of study (in vivo or in vitro) supporting their functional roles.

CDX1: caudal-type homeobox 1, DKK3: dickkopf 3, E2F1: E2F transcription factor 1, FRAS1: Fraser extracellular matrix complex subunit 1, OCT4: octamer-binding transcription factor 4, PI3K: phosphoinositide 3-kinase, RTN4: reticulon 4, RTNK2: rhotekin 2, and ZIC1: Zic family member 1.

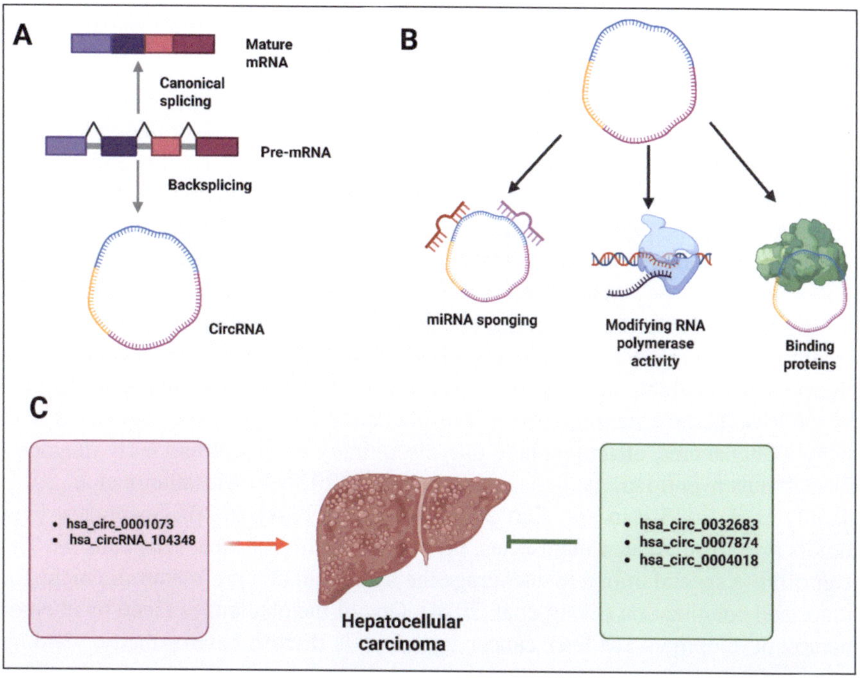

Fig. 5.3 Biogenesis, functions, and roles of circular RNAs in hepatocellular carcinoma. (**a**) Circular RNA biogenesis. Canonical pre-mRNA splicing generates linear mRNAs, while back-splicing of exons results in the formation of circRNAs. These are covalently closed RNA loops lacking 5′ caps and 3′ poly(A) tails. (**b**) Functions of circRNAs include acting as miRNA sponges, interacting with RNA-binding proteins, and regulating transcription. (**c**) Roles of circRNAs in hepatocellular carcinoma. Some circRNAs function as oncogenes, promoting proliferation, migration, and invasion, while others act as tumor suppressors

ncRNA Biomarkers in Precision Medicine

Precision medicine is a new approach for the treatment and prevention of diseases that relies on analyzing the individual variability (genomic, transcriptomic, proteomic, and metabolomic) to identify patient subgroups that will benefit from certain treatments. It involves treating each patient according to their unique genetic and immunological variations, which usually affect the disease course and the response to medications. Patients should ideally undergo an initial evaluation to profile their condition at the time of diagnosis, along with biomarker testing to determine the primary important pathophysiological pathways, disease course prediction, and risk stratification for specific organ damage. This procedure ought to assist in directing the selection of suitable therapy, monitoring, and preventing and predicting flare-ups (Fasano et al. 2023).

Diagnostic Potential of Circulating ncRNAs in Distinguishing Primary and Secondary Liver Cancers

The liver ranks as the sixth most prevalent organ of primary cancer in humans, usually due to cirrhosis and inflammation. The liver's anatomical position, architecture, and metabolic and immunosuppressive milieu make it a frequent site of metastasis from other organs, particularly the colon (Li et al. 2021b). Hepatocellular carcinoma (HCC), which makes up 80–90% of primary liver malignancies, is the fourth most common cause of cancer-related deaths globally. Primary liver malignancies are expected to rise in the coming decades. HCC's predisposing factors include hepatitis B virus (HBV) infection, hepatitis C virus (HCV) infection, alcoholic liver disease, and nonalcoholic fatty liver disease with steatosis. Few people are eligible for curative therapy, including liver transplantation and surgical excision. As with many malignancies, efforts include interdisciplinary treatment and early diagnosis through unconventional methods (Campbell et al. 2021; El-Mesallamy et al. 2011; Hamdy et al. 2025; Lin and Kao 2023; Mejia and Pasko 2020). Secondary liver tumors arise from distant metastases, often from gastrointestinal (GI) cancers. The liver offers a special milieu to encourage the formation of a pre-metastatic niche for tumor cell colonization (Dong et al. 2024). One of the challenges faced by chemotherapy development for liver cancer treatment is disease heterogeneity, yielding only minimal extensions of survival rates and slight enhancements in quality of life (Li et al. 2021b).

The discovery of effective biomarkers for the monitoring and early detection of HCC remains inadequate, as existing circulating biomarkers exhibit low sensitivity and variable specificity across different cutoff points, especially when evaluated longitudinally or in conjunction with other serum biomarkers. On the other hand, prognostic markers might soon have a bigger clinical role (Hammad et al. 2022; Piñero et al. 2020). The current HCC stratification criteria include tumor burden, cancer-related features, liver function, and physical state. Tumor nodules are detected by ultrasonography, CT, MRI, and the modified Child–Pugh categorization. Yet, their accuracy is being questioned; α-fetoprotein (AFP), the commonly used serum biomarker for HCC, also has significant sensitivity/specificity limitations in terms of HCC diagnosis and staging (Chan et al. 2024; Kumada et al. 2020). Moreover, the current therapeutic outcomes for advanced HCC are poor. The primary reason is that HCC patients in the early stages frequently exhibit no evident symptoms, and most HCC patients are detected after their disease has progressed to a late stage. As a result, it is critical to identify suitable noninvasive biomarkers for the early detection of HCC (C. Yang et al. 2023). ncRNAs in the bloodstream can have diagnostic value. Given their function in genomic regulation, measuring circulating miRNAs has potential as a noninvasive diagnosis for human cancers (Chan et al. 2024).

A single miRNA may not always be able to differentiate between progressing liver disorders and HCC. An investigation uncovered interesting miRNA expression patterns in HCV patients. miR-484 reduced advanced fibrosis compared to moderate fibrosis and HCC but increased HCC compared to cirrhosis. Furthermore, miR-524-5p increased in both cirrhosis and HCC, but miR-615-5p increased in cirrhosis only. Additionally, miR-524 has shown substantial potential for discriminating between cirrhosis and fibrosis. The study of these miRNAs may be useful in the staging and diagnosis of HCV hepatic progression, offering a considerable understanding of the illness (El-Maraghy et al. 2020).

Regarding lncRNAs, they have received great attention as diagnostic biomarkers in HCC because of their potential for early diagnosis and better patient outcomes. Studies show that certain lncRNAs have altered expression patterns in HCC, making them eligible for diagnostic purposes. Integrating lncRNA expression values with conventional biomarkers using machine learning approaches resulted in a diagnostic model with 100% sensitivity and 97% specificity, pointing out the potential for improved diagnostic capabilities (Samir et al. 2024). In a small cohort, the higher ratio of LINC00152 to GAS5 was correlated with a higher risk of mortality (Samir et al. 2024).

HULC, the human hepatocellular carcinoma–upregulated lncRNA, was reported to be elevated in HCC patients, with a correlation to the tumor stage. Notably, its diagnostic specificity was matched with AFP (Lumkul et al. 2024). As new diagnostic lncRNAs for hepatitis B virus–associated HCC, ADAMTSL4-AS1, SOCS2-AS1, and AC067931 are found to be significantly upregulated in HCC in comparison to chronic hepatitis B and liver cirrhosis, especially in peripheral blood mononuclear cells (Guan et al. 2024). Moreover, two possible biomarkers for HCC are serum mRNA PTTG1 and serum lncRNA PTTG3P, whose combination demonstrated high diagnostic sensitivity and specificity for HCC (Cao et al. 2024).

Besides, exosome-derived lncRNAs could act as biomarkers for HCC. For instance, LINC00853 shows excellent diagnostic capacity for all-stage HCC, with greater sensitivity/specificity than AFP, particularly in early-stage and AFP-negative HCC, indicating its potential as an innovative diagnostic biomarker (Kim et al. 2020). Moreover, a panel consisting of DLEU2, SNHG1, MALAT1, and HOTTIP, derived from extracellular vesicles, showed promising potential for first-stage HCC diagnosis (Kim et al. 2021). The use of lncRNAs as HCC biomarkers is relatively new; however, significant results were demonstrated by some studies. Yet, the lack of clinical validation hinders the use of ncRNAs as HCC biomarkers (Chan et al. 2024).

As for circRNAs, their association with tumor development has been established. circMTO1 was reported to be downregulated significantly, with implications against patient survival, as it suppresses HCC progression, and it may serve as a biomarker for poor prognosis (D. Han et al. 2017). Similarly, circSETD3 is implicated in tumor suppression via the MAPK pathway and may signify prognosis (Chan et al. 2024) (D. Han et al. 2017).

Prognostic Value of ncRNAs for Assessing Liver Metastasis Risk

Even while cancer treatment modalities like surgery, targeted treatments, chemotherapy, and radiotherapy have advanced significantly, metastasis and recurrence still present major clinical difficulties. Hence, it is crucial to identify clinical markers linked to prognosis in order to make precise diagnoses and create customized treatments for individuals with liver cancer (Doghish et al. 2023; Wang and Pu 2023). Assessing liver cancer response after therapy and modifying the treatment approach are critical to improving survival. Clinical monitoring of liver cancer following treatment relies on blood markers and imaging. Morphological evaluation cannot measure small tumors and has low reproducibility, making it inadequate for cancer evaluation after immunotherapy or targeted treatment (Liu et al. 2023). Significant work has been conducted over the last few decades to develop indicators for the prognostic forecasting of HCC. But as of yet, no biomarkers have been proven to accurately predict the survival of HCC patients, mainly because of the disease's heterogeneous causes, which include immunological complications, alcohol use, and viral infections (Ou et al. 2018; Woo and Kim 2018).

lncRNAs, among other ncRNAs, play a crucial role in regulating various pathologies, including cancer. Furthermore, a few studies have shown their potential value as biomarkers for the prognosis of HCC. Cancer cells exhibited a substantial increase in the expression of DYNLL1-AS1, while a substantial decrease was found with RP11-116D2.1. Moreover, longer survival times were correlated with higher levels of RP11-116D2.1 and lower expressions of DYNLL1-AS1. This 2-lncRNA signature was implicated in the wound healing of HCC, where a fibrin clot forms at the site of the wound and neutrophils infiltrate early in the wound healing response; these immune cells can produce a multitude of chemokines and cytokines (Liu et al. 2022).

It was shown that six lncRNAs mediate transcription regulation and chromatin remodeling, which, in turn, regulate several pathways relevant to the cell cycle and stress response. These lncRNAs are MIR22HG, CTD-2139B15.2, CTC-297 N7.9, RP11-343 N15.5, RP11-589 N15.2, and RP11-479G22.8, and they were considered independent predictors of patients' overall survival in HCC (Wu et al. 2018).

The presence of liver metastases in early-stage colorectal cancer patients was substantially correlated with the high expression levels of GAS5 and YIYA lncRNAs, signifying their likely prognostic biomarkers role (Kong et al. 2016). Moreover, the lncRNAs CRNDE and SNHG7 were confirmed as significant prognostic biomarkers for evaluating the risk of liver metastasis in concurrent colorectal–liver metastasis patients since they correlated with various tumor parameters, such as differentiation, lymph node metastases, invasion depth, and also higher recurrence and shorter overall survival (Zhang et al. 2020a). Again, HULC was overexpressed in several malignancies, correlated with overall poor survival and elevated risk for metastasis (Ding et al. 2019).

According to reports, the lncRNA translation regulation long noncoding RNA 1 (TRERNA1) was elevated in a number of malignancies, and it speeds up tumor metastasis by inducing the epithelial–mesenchymal transition (EMT). Together with HIF-1α and E-cadherin, TRERNA1 can classify HCC patients into distinct risk categories for survival prediction despite being a poor prognosis biomarker (Qian et al. 2021).

Regarding tRNA-derived small RNAs (tsRNAs), two models were developed: a signature of seven tsRNAs for prognosis (ts-N64, ts-N44, ts-N37, ts-N36, ts-N22, ts-N21, and ts-N20) and another of five tsRNAs (ts-N102, ts-N94, ts-N84, ts-N37, and ts-N7) for diagnosis. Both models demonstrated high accuracy in the differentiation of liver cancer patients and their outcome prediction (Zuo et al. 2022).

Some miRNAs could be described as metastasis-specific; such miRNAs include decreased let-7i, increased miR-885-5p, and miR-10b. They are associated with a higher likelihood of liver metastases and a worse outcome in colorectal cancer (Hur et al. 2015). Additionally, miR-21 is one of the potential prognostic biomarkers in HCC. Its higher levels are linked to a worse response to treatment and more rapid systemic progression (Nie et al. 2022; Stechele et al. 2023). Together, the use of ncRNA as diagnostic and prognostic markers in HCC is described in Fig. 5.4.

Therapeutic Targeting of ncRNAs in Primary and Secondary Liver Cancers

In accordance with the swift progress in cancer research and the continuous development of molecular tools, ncRNAs have become a leading focus in the field of cancer research (Hamdy et al. 2025). The regulatory role and molecular mechanism of ncRNAs in HCC have been widely investigated. A multitude of ncRNAs have been recognized as either promoters or suppressors of HCC growth. For example, miRNA-15a-3p is downregulated in HCC tissues, and its diminished levels were positively associated with distant metastases and poor prognosis. The lncRNA MCM3AP-AS1 exhibits superior expression in HCC and correlates with larger, advanced tumor stage and reduced survival of patients. The expression of circ-TRIM33-12 is diminished in HCC tissues and cell lines compared to normal controls, and its level may serve as an independent prognostic factor for long-term patient survival (Xue et al. 2022).

Therapeutic targeting of ncRNAs is an emerging and promising strategy for treating primary and secondary liver cancers, especially HCC. In order to increase therapeutic efficacy and longevity, research is investigating the inhibition of multiple molecular targets, which could expand the appropriate patient population (Bo Wang et al. 2024).

ncRNAs can regulate various cancer hallmarks, where they can influence cell proliferation, apoptosis, invasion, metastasis, and drug resistance in HCC and other liver cancers. They could also contribute to chemotherapeutic resistance and shape the tumor microenvironment. One example of a tumor suppressor is miR-638, which targets EZH2 and induces apoptosis and autophagy in liver cancer cells. This results in the suppression of tumor growth (Jesenko et al. 2023; Slaby et al. 2017; Zhang et al. 2021a).

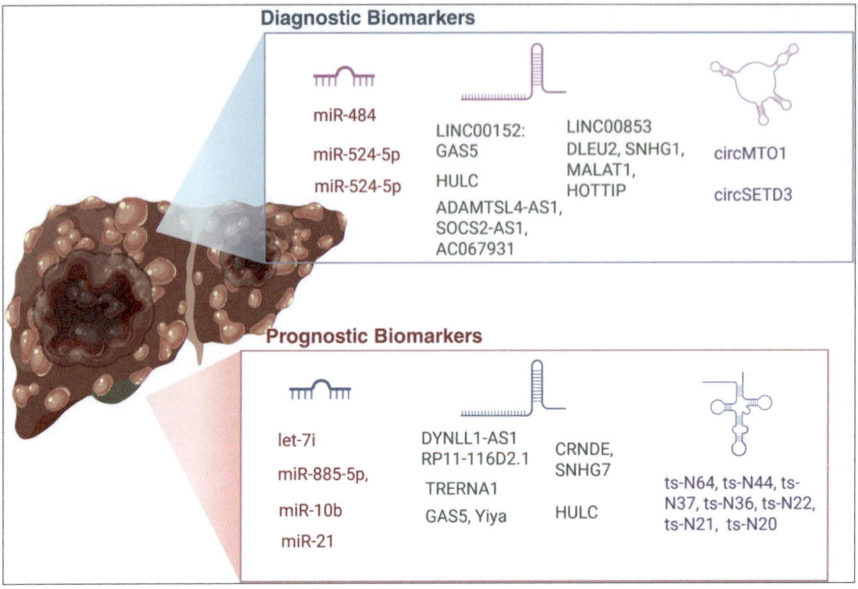

Fig. 5.4 The use of ncRNAs as diagnostic and prognostic markers in HCC. The figure highlights the significance of ncRNA markers in precision medicine, as they facilitate disease identification and predict patient outcomes. It illustrates the key molecular biomarkers used in the diagnosis and prognosis of hepatocellular carcinoma. Diagnostic biomarkers dysregulated in HCC include miRNAs, such as miR-484, miR-524-5p, and miR-615-5p; lncRNAs, such as LINC00152, GAS5, HULC, ADAMTSL4-AS1, SOCS2-AS1, AC067931, LINC00853, DLEU2, SNHG1, MALAT1, and HOTTIP; and circRNAs, such as circMTO1 and circSETD3, while ncRNAs that are considered prognostic biomarkers feature miRNAs, such as let-7i, miR-885-5p, miR-10b, and miR-21; lncRNAs, such as DYNL1-AS1, RP11–116D2.1, TRERNA1, GAS5, Yiya, CRNDE, SNHG7, and HULC; and several tsRNAs, such as ts-N64, ts-N44, ts-N37, ts-N36, ts-N22, ts-N21, and ts-N20. *CCAT1* colon cancer–associated transcript 1, *ECM* extracellular matrix, *GAS5* growth arrest–specific transcript 5, *HEIH* high expression in hepatocellular carcinoma, *HOTAIR* HOX transcript antisense RNA, *HULC* highly upregulated in liver cancer, *MEG3* maternally expressed gene 3, *NEAT1* nuclear-enriched abundant transcript 1, *PVT1* plasmacytoma variant translocation 1, *UCA1* urothelial carcinoma–associated 1

Therapeutic approaches using ncRNAs include the use of specific miRNA mimics (such as miR-34 and miR-16) and inhibitors (anti-miRs); such methods can be utilized to either restore or hinder the functions of particular miRNAs. For instance, miR-34 is known to be a tumor suppressor for its repressing effects on glucose metabolism and lower expression levels in hepatic tumors; it also prevents metastasis and induces apoptosis (Wei et al. 2018; Zhang et al. 2018). Thus, the administration of miR-34 mimic would probably propose beneficial responses in liver cancers (Daige et al. 2014). Interestingly, this mimic reached phase I clinical trials as MRX34, and the study concluded that MRX34 treatment with dexamethasone premedication showed anticancer efficacy and was associated with acceptable safety (Beg et al. 2017). miR-16 is also a valuable therapeutic target since it can hinder

HCC progression by modulating NF-κB signaling and inducing apoptosis. Moreover, it could target pathways involved in tumor suppression as heme oxygenase-1 and genes as cyclin-B1 (Eltayeb et al. 2024; Ghafouri-Fard et al. 2022). Another approach is using ncRNAs in combination with conventional therapies as tyrosine kinase inhibitors to provide an added advantage of reducing chemoresistance (Vianello et al. 2024). Exosomal ncRNAs could also be used as delivery vehicles as well as modulators of the tumor microenvironment (Vianello et al. 2024). Regarding other ncRNAs, such as circRNAs and lncRNAs, both offer interesting therapeutic targets, yet such areas require further research to fully elucidate possible modalities for their clinical usage.

ncRNA-Epigenetic Patient Stratification in Liver Cancer Precision Based on ncRNA Profiles

Epigenetic variation, which includes DNA methylation, histone modifications, and microRNAs, plays a significant role in influencing gene expression and cell function without altering the genomic sequence. Immune profiling, which uses methods, including flow cytometry, mass cytometry, transcriptomics, microarray analysis, and single-cell RNA sequencing, can assist in determining a person's unique response to a treatment and may even be able to forecast results. Additionally, the discovery of new serum and urine biomarkers would make it possible to stratify patients based on assessments of possible medication response and long-term prognosis projections (Fasano et al. 2023).

In the field of liver cancer, ncRNAs and epigenetic alterations are being increasingly acknowledged as valuable tools for patient stratification and precision treatment. Recent studies have shed light on the potential of noncoding RNA expression profiles and epigenetic fingerprints to enhance diagnosis, prognosis, and risk classification, thus enabling more individualized therapy approaches.

Comprehensive genomic, transcriptomic, and epigenomic profiling in hepatoblastoma, a form of pediatric liver cancer, revealed unique methylation patterns and ncRNA overexpression. This led to the development of the first molecular risk stratification system, which outperforms conventional clinical risk models and could lead to the development of novel therapies (Carrillo-Reixach et al. 2020). Moreover, the mapping of millions of expression quantitative trait loci (eQTLs) that alter ncRNA expression has been accomplished by genome-wide analysis. A significant number of these eQTLs are associated with patient survival and coincide with urinary.

h established disease risk loci, which lends support to their utilization in precision stratification, combining serum lncRNAs levels of HULC, HOTAIR, and UCA1 significantly enhanced the diagnostic sensitivity and specificity in discriminating liver cancer from other liver diseases, promoting early detection for better tailored treatment (Lumkul et al. 2024). Also, the use of signature ncRNAs represented as panels, like multi-lncRNA panels, improves risk stratification and early detection. A study demonstrated that seven lncRNAs combined were able to efficiently predict survival in HCC patients, stratifying them into high-/low-risk groups and independent of conventional factors (Yan et al. 2019).

Tools to Study ncRNA-Epigenetic Interactions in Liver Tumors

Scientists are using new molecular biology techniques like cross-linking and immunoprecipitation sequencing (CLIP-seq) and chromatin isolation by RNA purification sequencing (ChIRP-seq) to study how noncoding RNA (ncRNA) affects epigenetics in liver tumors. Both tools can work nicely together for figuring out the complex ways in which ncRNAs affect the epigenetic landscapes of liver cancers. CLIP-seq reveals how RNA and proteins work together, whereas ChIRP-seq illustrates how RNA, DNA, and chromatin interact. Accordingly, each tool has its own strengths (Louis et al. 2022; Tian and Hu 2020).

CLIP-seq It is a method that uses UV cross-linking and immunoprecipitation to find protein–RNA interactions across the whole transcriptome. This is especially useful for learning how ncRNAs interact with RNA-binding proteins that change epigenetic states, such as those that change chromatin or add methyl groups. It lets researchers accurately map these interactions in living things (Ule et al. 2018).

ChIRP-seq It makes it possible to separate chromatin-bound ncRNAs and the DNA and protein complexes that go with them. It works very well for researching lncRNAs like HOTAIR and MALAT1, which are involved in epigenetic regulation in liver tumors (Amicone et al. 2015). ChIRP-seq shows where some ncRNAs attach to the genome, which helps us understand how they work in turning genes on or off by changing chromatin (Bell et al. 2018).

Single-Cell Resolution Researchers are looking at new ways to use CLIP-seq and ChIRP-seq to get single-cell resolution so that they can learn more about the differences in liver tumor microenvironments and find unusual but important ncRNA interactions (Rotem et al. 2015).

CRISPR-Based Techniques for Epigenetic Editing of ncRNAs in Liver Cancer

CRISPR-based technologies have been very useful for altering epigenetics, especially for controlling ncRNAs that are linked to liver cancer. By combining transcriptional modulators with catalytically dead Cas9 (dCas9), it is possible to turn ncRNA expression on or off without changing the DNA sequence (Dominguez et al.

2016). This is especially helpful for targeting lncRNAs and miRNAs to affect how tumors grow and spread or resist drugs (Esteller 2011).

Target Specificity and Off-Target Effects

Off-target binding is a big problem with CRISPR-based systems, especially when dCas9 is combined with epigenetic modifiers. These mechanisms do not break DNA, but they can cause nontarget genes to become dysregulated by accidentally bringing in transcriptional regulators to homologous sequences. Researchers are working on better guided RNA (gRNA) design tools, off-target prediction algorithms, and high-fidelity Cas9 variations to make the process more targeted (Guo et al. 2023).

Delivery Challenges

Getting CRISPR-dCas9 parts into liver cells safely and effectively is still a big problem. Viral vectors, including adeno-associated viruses (AAVs), are good at delivering things, but they have problems with packaging and may cause an immune response. Nonviral approaches, including lipid nanoparticles, are safer but do not always work as efficiently. Researchers are still working on developing liver-targeted delivery systems that can carry massive dCas9-effector fusions while keeping immune reactions to a minimum (Liu et al. 2021).

Temporal and Spatial Control

It is important to fine-tune the activity of CRISPR-based epigenetic editors to avoid long-term impacts that are not wanted. Inducible systems with drug- or light-controlled activators, like doxycycline-inducible promoters, let you control when genes are turned on and off. Adding tissue-specific promoters can also help with spatial control, which is important when trying to target genes that do different things in different types of cells or tissues (Khajanchi and Saha 2022).

Heterogeneity of ncRNA Functions

The role of ncRNAs depends a lot on the situation, and it can change between different forms of cancer and even across different types of HCC. For example, one lncRNA may stop tumors from growing in one situation but help them grow in another. To find functionally relevant targets for CRISPR modulation, it is important to fully map ncRNA regulatory networks. Combining CRISPR screening with single-cell sequencing might help make this easier to understand (Lv et al. 2021b).

Synergistic Approaches and Combination Therapies

In the future, CRISPR-based epigenetic editing may also be used alongside other treatments, like chemotherapy, immunotherapy, or RNA-based medicines. For instance, using CRISPRa to restore the expression of ncRNA that suppresses tumors may make liver tumors more sensitive to immunotherapy or stop them from becoming resistant. You could also use CRISPR to change the tumor microenvironment by changing ncRNAs that affect how immune cells get into the tumor or how fibrous it is (Palaz et al. 2024).

Clinical Translation and Biomarker Development

Before going onto clinical trials, strong preclinical models are needed to test CRISPR-edited targets. Patient-derived xenografts (PDX) and organoids are potential ways to assess the safety and effectiveness of ncRNA-targeted epigenetic treatments. Creating reliable indicators to keep an eye on ncRNA expression and CRISPR activity can also help us figure out how well treatments are working (Gu et al. 2025).

Translating ncRNA-Based Epigenetic Findings into Clinical Practice

One of the most exciting and difficult things to undertake in precision medicine is to use what we learn about ncRNA in regular medical practice. miRNAs, lncRNAs, and circRNAs are types of ncRNAs that influence the remodeling of chromatin, the splicing of RNA, and the regulation of genes after transcription. These processes change how diseases like cancer, neurodegeneration, metabolic disorders, and heart problems progress. But only a few ncRNA tests or drugs have reached patients, which indicates how large the gap is between research and practice (Hamdy 2011; Yin et al. 2022).

Analytical Validity and Assay Harmonization

Different labs use different extraction chemicals, library preparation, and normalization for gold-standard qPCR and next-generation sequencing. Without established operating methods, lab coefficients of variation can exceed 30%, making results difficult to duplicate and regulatory approval slower. International projects like microRNA quality control (miRQC) are standardizing reference materials, spike-in controls, and bioinformatic pipelines. Best practice suggestions must be sponsored and used by smaller clinical labs to become ISO-accredited diagnoses (Byron et al. 2016).

From Single-Analyte to Multi-Omic Signatures

In disease biology, one ncRNA is rarely the key contributor. Multi-omic panels that incorporate ncRNAs, DNA methylation, histone marks, proteomics, and clinical data are better at early cancer detection and therapy monitoring than single-marker tests. Next, these panels will be used in health care systems to improve predictive algorithms with new patient data. This requires safe data lakes, federated analytics to protect patient data, and unambiguous AI interpretability so that clinicians can understand model recommendations (Salido-Guadarrama et al. 2023).

Delivery Platforms for ncRNA Therapeutics

Safe, cell-type-specific administration is the biggest pharmacological challenge. Lipid nanoparticles (used in mRNA vaccines) have a high payload capacity but accumulate in the liver and spleen, viral vectors have long-term expression but are immunogenic, and chemical conjugates (e.g., GalNAc) are restricted to hepatocytes. Ultrasound-responsive microbubbles for spatiotemporally regulated release, biodegradable polymeric micelles that shed "stealth" polyethylene glycol (PEG) layers inside tumors, and exosome-mimetic vesicles with tailored surface ligands are promising developments. They will compete in preclinical trials over the next 5 years, and regulators will require early toxicology packages (Tenchov et al. 2021).

Reducing Off-Target Effects and Immune Activation

Even correctly built antisense oligonucleotides can connect to incorrect transcripts or activate toll-like receptors. Before animal studies, deep mutational scanning and superior in silico target prediction (CRISPR-Cleave and RNAfold) provide full off-target maps. Two new chemical modifications, 2′-O-methyl and locked nucleic acid backbones, decrease innate immune system activation and boost nuclease resilience (Wai et al. 2024).

Adaptive Clinical Trial Designs and Real-World Evidence

Regular phase I–III pipelines do not work for rapid oligonucleotide chemistry. Basket, umbrella, and platform trials look at many ncRNA agents or tumor subtypes at the same time and use Bayesian statistics to put patients in the most promising arms. Post-marketing registries gather long-term safety and effectiveness data, especially for unusual side effects like pseudo-allergy caused by complement activation. Regulators are increasingly accepting hybrid evidence for quick approval. This is in line with a trend toward evaluating new therapies over their whole life cycle (Dai et al. 2024b).

Overcoming Challenges in Developing ncRNA-Targeted Precision Medicine

Developing precision medicine that targets noncoding RNA (ncRNA) is hard for a number of reasons, such as how exact the targeting is, how it gets delivered, and what consequences it might have on other targets.

Specificity and Off-Target Effects

Because base pairing is not ideal, ncRNAs often have more than one target. High-throughput sequencing and bioinformatics have come a long way, making it possible to find ncRNA expression patterns that are unique to certain diseases. This makes it possible to create treatments that affect only the ncRNAs linked to the disease. Scientists are using chemical changes, such as locked nucleic acids and phosphorothioate backbones, to make binding more selective and cut down on interactions with things that are not the target (Wang et al. 2022).

Efficient Delivery Systems

Indeed, ncRNAs are easy to break down and need to be delivered quickly. Researchers are working on lipid nanoparticles (LNPs), viral vectors, and conjugates (like GalNAc) to improve how cells take up drugs and target specific tissues (Winkle et al. 2021). Nanoparticles, liposomes, and exosome-based delivery technologies have made it much easier to get ncRNA therapies to targeted tissues while reducing degradation and immunological responses (Yang et al. 2024).

Chemical Modifications to Enhance Tolerability and Immune Response

Immune responses can be triggered by modified nucleic acids. Changing the chemistry of antisense oligonucleotides and small interfering RNAs can make them more stable and less likely to have unwanted effects, which can lead to better therapeutic outcomes; for example, chemical changes like 2′-O-methylation assist the body in avoiding detection by the immune system and make the drug easier to use (Anwar et al. 2023).

Understanding Functional Mechanisms

Functional annotation of lncRNAs is still not finished. To figure out what ncRNAs do, we need advanced sequencing and bioinformatic techniques. This is important for targeting therapies (Gong et al. 2021).

By combining these scientific and ethical strategies, researchers can enhance both the effectiveness and integrity of their work in ncRNA-based precision medicine.

References

Abd El Fattah, Y. K., Abulsoud, A. I., AbdelHamid, S. G., AbdelHalim, S., & Hamdy, N. M. (2023). CCDC144NL-AS1/hsa-miR-143-3p/HMGA2 interaction: In-silico and clinically implicated in CRC progression, correlated to tumor stage and size in case-controlled study; step toward ncRNA precision. *International Journal of Biological Macromolecules, 253*, 126739.

Addissouky, T. A., Sayed, I. E. T. E., Ali, M. M. A., Wang, Y., Baz, A. E., Khalil, A. A., & Elarabany, N. (2024). Latest advances in hepatocellular carcinoma management and prevention through advanced technologies. *Egyptian Liver Journal, 14*(1), 2. doi:https://doi.org/10.1186/s43066-023-00306-3

Aguilo, F., Zhou, M.-M., & Walsh, M. J. (2011). Long noncoding RNA, polycomb, and the ghosts haunting INK4b-ARF-INK4a expression. *Cancer Research, 71*(16), 5365–5369.

Amicone, L., Citarella, F., & Cicchini, C. (2015). Epigenetic regulation in hepatocellular carcinoma requires long noncoding RNAs. *BioMed research international, 2015*(1), 473942.

Anwar, S., Mir, F., & Yokota, T. (2023). Enhancing the effectiveness of oligonucleotide therapeutics using cell-penetrating peptide conjugation, chemical modification, and carrier-based delivery strategies. *Pharmaceutics, 15*(4), 1130.

Ariel, I., Miao, H., Ji, X., Schneider, T., Roll, D., De Groot, N.,… Ayesh, S. (1998). Imprinted H19 oncofetal RNA is a candidate tumour marker for hepatocellular carcinoma. *Molecular pathology, 51*(1), 21.

Arun, G., Aggarwal, D., & Spector, D. L. (2020). MALAT1 long non-coding RNA: functional implications. *Non-coding RNA, 6*(2), 22.

Barreto, J. C., & Posner, M. C. (2019). Chapter 134 - Management of Secondary Hepatic Neoplasms. In C. J. Yeo (Ed.), *Shackelford's Surgery of the Alimentary Tract, 2 Volume Set (Eighth Edition)* (pp. 1565–1574). Philadelphia: Elsevier.

Beg, M. S., Brenner, A. J., Sachdev, J., Borad, M., Kang, Y.-K., Stoudemire, J.,… Hong, D. S. (2017). Phase I study of MRX34, a liposomal miR-34a mimic, administered twice weekly in patients with advanced solid tumors. *Investigational new drugs, 35*, 180–188.

Bell, J. C., Jukam, D., Teran, N. A., Risca, V. I., Smith, O. K., Johnson, W. L.,… Straight, A. F. (2018). Chromatin-associated RNA sequencing (ChAR-seq) maps genome-wide RNA-to-DNA contacts. *Elife, 7*, e27024.

Bergers, G., & Fendt, S.-M. (2021). The metabolism of cancer cells during metastasis. *Nature Reviews Cancer, 21*(3), 162–180.

Bian, E.-B., Wang, Y.-Y., Yang, Y., Wu, B.-M., Xu, T., Meng, X.-M.,… Xiong, Z.-G. (2017). Hotair facilitates hepatic stellate cells activation and fibrogenesis in the liver. *Biochimica et Biophysica Acta (BBA)-Molecular Basis of Disease, 1863*(3), 674–686.

Byron, S. A., Keuren-Jensen, V., Kendall, R., Engelthaler, D. M., Carpten, J. D., & Craig, D. W. (2016). Translating RNA sequencing into clinical diagnostics: opportunities and challenges. *Nature reviews genetics, 17*(5), 257–271.

Cai, D., He, K., Chang, S.E Tong, D., & Huang, C. (2015). MicroRNA-302b enhances the sensitivity of hepatocellular carcinoma cell lines to 5-FU via targeting Mcl-1 and DPYD. *International Journal of Molecular Sciences, 16*(10), 23668–23682.

Campbell, C., Wang, T., McNaughton, A. L., Barnes, E., & Matthews, P. C. (2021). Risk factors for the development of hepatocellular carcinoma (HCC) in chronic hepatitis B virus (HBV) infection: a systematic review and meta-analysis. *Journal of viral hepatitis, 28*(3), 493–507.

Cao, M.-X., Jiang, Y.-P., Tang, Y.-L., & Liang, X.-H. (2016). The crosstalk between lncRNA and microRNA in cancer metastasis: orchestrating the epithelial-mesenchymal plasticity. *Oncotarget, 8*(7), 12472.

Cao, S., Zhong, F., Chen, X., Ke, S., Zhong, X., Li, T.,… Wang, H. (2024). The combination of serum lncRNA PTTG3P and mRNA PTTG1 serves as a diagnostic and prognostic marker for hepatocellular carcinoma. *Molecular Medicine Reports, 31*(2), 44.

Carrillo-Reixach, J., Torrens, L., Simon-Coma, M., Royo, L., Domingo-Sàbat, M., Abril-Fornaguera, J.,… Arnal, M. (2020). Epigenetic footprint enables molecular risk stratification of hepatoblastoma with clinical implications. *Journal of Hepatology, 73*(2), 328–341.

Chan, Y.-T., Zhang, C., Wu, J., Lu, P., Xu, L., Yuan, H.,… Wang, N. (2024). Biomarkers for diagnosis and therapeutic options in hepatocellular carcinoma. *Molecular cancer, 23*(1), 189.

Chang, R.-M., Xiao, S., Lei, X., Yang, H., Fang, F., & Yang, L.-Y. (2017). miRNA-487a promotes proliferation and metastasis in hepatocellular carcinoma. *Clinical Cancer Research, 23*(10), 2593–2604.

Chatterjee, S. (2024). Interplay Between Epithelial-Mesenchymal Transition and Cancer Stem Cells: Unveiling Potential Therapeutic Avenues.

Chen, B., Dragomir, M. P., Yang, C., Li, Q., Horst, D., & Calin, G. A. (2022). Targeting non-coding RNAs to overcome cancer therapy resistance. *Signal transduction and targeted therapy, 7*(1), 121.

Chen, W.-M., Chen, W.-D., Jiang, X.-M., Jia, X.-F., Wang, H.-M., Zhang, Q.-J.,… Zhao, H.-B. (2017a). HOX transcript antisense intergenic RNA represses E-cadherin expression by binding to EZH2 in gastric cancer. *World journal of gastroenterology, 23*(33), 6100.

Chen, Y. G., Satpathy, A. T., & Chang, H. Y. (2017b). Gene regulation in the immune system by long noncoding RNAs. *Nature immunology, 18*(9), 962–972.

Choi, J., & Choi, B. I. (2014). Other Malignant Tumors of the Liver. In B. I. Choi (Ed.), *Radiology Illustrated: Hepatobiliary and Pancreatic Radiology* (pp. 169–231). Berlin, Heidelberg: Springer Berlin Heidelberg.

Comet, I., Riising, E. M., Leblanc, B., & Helin, K. (2016). Maintaining cell identity: PRC2-mediated regulation of transcription and cancer. *Nature Reviews Cancer, 16*(12), 803–810.

Correnti, M., Binatti, E., Gammella, E., Invernizzi, P., & Recalcati, S. (2022). The emerging role of tumor microenvironmental stimuli in regulating metabolic rewiring of liver cancer stem cells. *Cancers, 15*(1), 5.

D'Avola, D., Bilbao, J. I., & Sangro, B. (2019). Hepatocellular Carcinoma: Essentials Interventional Radiologists Need to Know. *Cardiovasc Intervent Radiol, 42*(9), 1262–1270. doi:https://doi.org/10.1007/s00270-019-02221-w

Dai, S., Zhuang, H., Li, Z., Chen, Z., Chai, Y., & Zhou, Q. (2024a). miR-122/NEGR1 axis contributes colorectal cancer liver metastasis by PI3K/AKT pathway and macrophage modulation. *Journal of Translational Medicine, 22*(1), 1060.

Dai, W., Qiao, X., Fang, Y., Guo, R., Bai, P., Liu, S.,… Na, Z. (2024b). Epigenetics-targeted drugs: current paradigms and future challenges. *Signal Transduction and Targeted Therapy, 9*(1), 332.

Daige, C. L., Wiggins, J. F., Priddy, L., Nelligan-Davis, T., Zhao, J., & Brown, D. (2014). Systemic delivery of a miR34a mimic as a potential therapeutic for liver cancer. *Molecular cancer therapeutics, 13*(10), 2352–2360.

Ding, Y., Sun, C., Li, J., Hu, L., Li, M., Liu, J.,… Xiong, S. (2019). The significance of long non-coding RNA HULC in predicting prognosis and metastasis of cancers: a meta-analysis. *Pathology & Oncology Research, 25*, 311–318.

Doghish, A. S., Elballal, M. S., Elazazy, O., Elesawy, A. E., Elrebehy, M. A., Shahin, R. K.,… Sallam, A.-A. M. (2023). The role of miRNAs in liver diseases: Potential therapeutic and clinical applications. *Pathology-Research and Practice*, 154375.

Dominguez, A. A., Lim, W. A., & Qi, L. S. (2016). Beyond editing: repurposing CRISPR–Cas9 for precision genome regulation and interrogation. *Nature reviews Molecular cell biology, 17*(1), 5–15.

Dong, D., Yu, X., Xu, J., Yu, N., Liu, Z., & Sun, Y. (2024). Cellular and molecular mechanisms of gastrointestinal cancer liver metastases and drug resistance. *Drug Resistance Updates*, 101125.

Dong, H., Zhang, Y., Xu, Y., Ma, R., Liu, L., Luo, C., & Jiang, W. (2019). Downregulation of long non-coding RNA MEG3 promotes proliferation, migration, and invasion of human hepatocellular carcinoma cells by upregulating TGF-β1. *Acta Biochimica et Biophysica Sinica, 51*(6), 645–652.

Drew, J., & Machesky, L. M. (2021). The liver metastatic niche: modelling the extracellular matrix in metastasis. *Disease Models & Mechanisms, 14*(4), dmm048801.

Eissa, S., Swellam, M., El-Mosallamy, H., Mourad, M. S., Hamdy, N., Kamel, K., … El-Ahmady, O. (2003). Diagnostic value of urinary molecular. *Anticancer Res, 23*, 4347–4356.

El-Maraghy, S. A., Adel, O., Zayed, N., Yosry, A., El-Nahaas, S. M., & Gibriel, A. A. (2020). Circulatory miRNA-484, 524, 615 and 628 expression profiling in HCV mediated HCC among Egyptian patients; implications for diagnosis and staging of hepatic cirrhosis and fibrosis. *Journal of Advanced Research, 22*, 57–66.

El-Mesallamy, H. O., Hamdy, N. M., Rizk, H. H., & El-Zayadi, A.-R. (2011). Apelin serum level in Egyptian patients with chronic hepatitis C. *Mediators of inflammation, 2011*(1), 703031.

El-Mesallamy, H. O., Hamdy, N. M., Zaghloul, A. S., & Sallam, A. M. (2013). Clinical value of circulating lipocalins and insulin-like growth factor axis in pancreatic cancer diagnosis. *Pancreas, 42*(1), 149–154.

Eldash, S., Sanad, E. F., Nada, D., & Hamdy, N. M. (2023). The intergenic type LncRNA (LINC RNA) faces in cancer with in silico scope and a directed lens to LINC00511: a step toward ncRNA precision. *Non-coding RNA, 9*(5), 58.

Esteller, M. (2011). Non-coding RNAs in human disease. *Nature reviews genetics, 12*(12), 861–874.

MF Eltayeb, M. M Amer, Hassan, A., Elmetwalli, D. A., Nasr-Eldin, M., & Elmetwalli, A. (2024). Modulation of miR-16 and HO-1 in Egyptian Liver Cancer Patients. *Benha Journal of Applied Sciences, 9*(3), 155–159.

Fasano, S., Milone, A., Nicoletti, G. F., Isenberg, D. A., & Ciccia, F. (2023). Precision medicine in systemic lupus erythematosus. *Nature Reviews Rheumatology, 19*(6), 331–342.

Fei, D., Wang, F., Wang, Y., Chen, J., Chen, S., Fan, L.,… Jiang, X. (2024). Circular RNA ACVR2A promotes the progression of hepatocellular carcinoma through mir-511-5p targeting PI3K-Akt signaling pathway. *Molecular Cancer, 23*(1), 159. doi:https://doi.org/10.1186/s12943-024-02074-z

Fernández-Ramos, D., Fernández-Tussy, P., Lopitz-Otsoa, F., Gutiérrez-de-Juan, V., Navasa, N., Barbier-Torres, L.,… Arbelaiz, A. (2018). MiR-873-5p acts as an epigenetic regulator in early stages of liver fibrosis and cirrhosis. *Cell death & disease, 9*(10), 958.

Franz, C., Wuehrl, M., Hartmann, S., Klupp, F., Schmidt, T., & Schneider, M. (2023). Long non-coding RNAs CCAT1 and CCAT2 in colorectal liver metastases are tumor-suppressive via MYC interaction and might predict patient outcomes. *Plos one, 18*(6), e0286486.

Fu, M., Zou, C., Pan, L., Liang, W., Qian, H., Xu, W.,… Zhang, X. (2016). Long noncoding RNAs in digestive system cancers: Functional roles, molecular mechanisms, and clinical implications. *Oncology reports, 36*(3), 1207–1218.

Fu, X., Wen, H., Jing, L., Yang, Y., Wang, W., Liang, X.,… Tian, T. (2017). Micro RNA-155-5p promotes hepatocellular carcinoma progression by suppressing PTEN through the PI 3K/Akt pathway. *Cancer science, 108*(4), 620–631.

George, T. P., Subramanian, S., & Supriya, M. H. (2024). A brief review of noncoding RNA. *Egyptian Journal of Medical Human Genetics, 25*(1), 98. doi:https://doi.org/10.1186/s43042-024-00553-y

Ghafouri-Fard, S., Dashti, S., & Taheri, M. (2020). The HOTTIP (HOXA transcript at the distal tip) lncRNA: Review of oncogenic roles in human. *Biomedicine & Pharmacotherapy, 127*, 110158.

Ghafouri-Fard, S., Khoshbakht, T., Hussen, B. M., Abdullah, S. T., Taheri, M., & Samadian, M. (2022). A review on the role of mir-16-5p in the carcinogenesis. *Cancer Cell International, 22*(1), 342.

Gong, Y., Zhu, W., Sun, M., & Shi, L. (2021). Bioinformatics analysis of long non-coding RNA and related diseases: an overview. *Frontiers in genetics, 12*, 813873.

Grossi, I., Marchina, E., De Petro, G., & Salvi, A. (2023). The biological role and translational implications of the long non-coding RNA GAS5 in breast cancer. *Cancers, 15*(13), 3318.

Gu, A., Li, J., Li, M. Y., & Liu, Y. (2025). Patient-derived xenograft model in cancer: establishment and applications. *MedComm, 6*(2), e70059.

Guan, W., Zhang, C., Miao, T., Dong, C., Li, L., Yuan, X.,… Sun, M. (2024). The Potential of the lncRNAs ADAMTSL4-AS1, AC067931 and SOCS2-AS1 in Peripheral Blood Mononuclear Cells as Novel Diagnostic Biomarkers for Hepatitis B Virus-Associated Hepatocellular Carcinoma. *Journal of Hepatocellular Carcinoma*, 1221–1233.

Guo, C., Ma, X., Gao, F., & Guo, Y. (2023). Off-target effects in CRISPR/Cas9 gene editing. *Frontiers in Bioengineering and Biotechnology, 11*, 1143157.

Guo, J.-H., Yin, S.-S., Liu, H., Liu, F., & Gao, F.-H. (2021). Tumor microenvironment immune-related lncRNA signature for patients with melanoma. *Annals of Translational Medicine, 9*(10), 857.

Halim, A., Al-Qadi, N., Kenyon, E., Conner, K. N., Mondal, S. K., Medarova, Z., & Moore, A. (2024). Inhibition of miR-10b treats metastatic breast cancer by targeting stem cell-like properties. *Oncotarget, 15*, 591.

Halim, A., Kim, B., Kenyon, E., & Moore, A. (2025). miR-10b as a Clinical Marker and a Therapeutic Target for Metastatic Breast Cancer. *Technology in Cancer Research & Treatment, 24*, 15330338251339256.

Hamdy, N. M. (2011). Relationship between pro-anti-inflammatory cytokines, T-cell activation and CA 125 in obese patients with heart failure. *Medical science monitor: international medical journal of experimental and clinical research, 17*(3), CR173.

Hamdy, N. M., Sallam, A.-A. M., Elazazy, O., Kabel, A. M., Salama, R. M., Gouhar, S. A.,… Darwish, S. F. (2025). LincRNA-miR interactions in hepatocellular carcinoma: comprehensive review and in silico analysis: a step toward ncRNA precision. *Naunyn-Schmiedeberg's Archives of Pharmacology*, 1–28.

Hamdy, N. M., Zaki, M. B., Rizk, N. I., Abdelmaksoud, N. M., Abd-Elmawla, M. A., Ismail, R. A., & Abulsoud, A. I. (2024). Unraveling the ncRNA landscape that governs colorectal cancer: a roadmap to personalized therapeutics. *Life sciences, 354*, 122946.

Hammad, R., Aglan, R. B., Mohammed, S. A., Awad, E. A.-e., Elsaid, M. A., Bedair, H. M.,… Rushdi, A. (2022). Cytotoxic T cell expression of leukocyte-associated immunoglobulin-like receptor-1 (LAIR-1) in viral hepatitis C-mediated hepatocellular carcinoma. *International journal of molecular sciences, 23*(20), 12541.

Han, D., Li, J., Wang, H., Su, X., Hou, J., Gu, Y.,… Huang, M. (2017). Circular RNA circMTO1 acts as the sponge of microRNA-9 to suppress hepatocellular carcinoma progression. *Hepatology, 66*(4), 1151–1164.

Han, G., Zhang, L., Ni, X., Chen, Z., Pan, X., Zhu, Q.,… Wang, X. (2018a). MicroRNA-873 promotes cell proliferation, migration, and invasion by directly targeting TSLC1 in hepatocellular carcinoma. *Cellular Physiology and Biochemistry, 46*(6), 2261–2270.

Han, L.-L., Yin, X.-R., & Zhang, S.-Q. (2018b). miR-103 promotes the metastasis and EMT of hepatocellular carcinoma by directly inhibiting LATS2. *International Journal of Oncology, 53*(6), 2433–2444.

Han, T.-S., Ban, H. S., Hur, K., & Cho, H.-S. (2018c). The epigenetic regulation of HCC metastasis. *International journal of molecular sciences, 19*(12), 3978.

Hayat, M. A. (2005). 1 - Liver Carcinoma. In M. A. Hayat (Ed.), *Handbook of Immunohistochemistry and in Situ Hybridization of Human Carcinomas* (Vol. 3, pp. 131–151): Academic Press.

Hu, J.-J., Song, W., Zhang, S.-D., Shen, X.-H., Qiu, X.-M., Wu, H.-Z.,… He, M.-L. (2016). HBx-upregulated lncRNA UCA1 promotes cell growth and tumorigenesis by recruiting EZH2 and repressing p27Kip1/CDK2 signaling. *Scientific Reports, 6*(1), 23521.

Huang, G., Liang, M., Liu, H., Huang, J., Li, P., Wang, C.,… Jiang, X. (2020). CircRNA hsa_circRNA_104348 promotes hepatocellular carcinoma progression through modulating miR-187-3p/RTKN2 axis and activating Wnt/β-catenin pathway. *Cell Death Dis, 11*(12), 1065. doi:https://doi.org/10.1038/s41419-020-03276-1

Huang, J., Deng, X., Chen, X., Chang, Z., Lu, Q., Tang, A., & Liu, P. (2022a). Circular RNA KIF4A Promotes Liver Metastasis of Breast Cancer by Reprogramming Glucose Metabolism. *J Oncol, 2022*, 8035083. doi:https://doi.org/10.1155/2022/8035083

Huang, Y.-H., Lian, W.-S., Wang, F.-S., Wang, P.-W., Lin, H.-Y., Tsai, M.-C., & Yang, Y.-L. (2022b). MiR-29a curbs hepatocellular carcinoma incidence via targeting of HIF-1α and ANGPT2. *International Journal of Molecular Sciences, 23*(3), 1636.

Hur, K., Toiyama, Y., Schetter, A. J., Okugawa, Y., Harris, C. C., Boland, C. R., & Goel, A. (2015). Identification of a metastasis-specific MicroRNA signature in human colorectal cancer. *Journal of the National Cancer Institute, 107*(3), dju492.

Hur, K., Toiyama, Y., Takahashi, M., Balaguer, F., Nagasaka, T., Koike, J.,… Goel, A. (2013). MicroRNA-200c modulates epithelial-to-mesenchymal transition (EMT) in human colorectal cancer metastasis. *Gut, 62*(9), 1315–1326.

Iino, I., Kikuchi, H., Miyazaki, S., Hiramatsu, Y., Ohta, M., Kamiya, K.,… Konno, H. (2013). Effect of mi R-122 and its target gene cationic amino acid transporter 1 on colorectal liver metastasis. *Cancer science, 104*(5), 624–630.

Jantrapirom, S., Koonrungsesomboon, N., Yoshida, H., M. Candeias, Pruksakorn, D., & Lo Piccolo, L. (2021). Long noncoding RNA-dependent methylation of nonhistone proteins. *Wiley Interdisciplinary Reviews: RNA, 12*(6), e1661.

Jesenko, T., Brezar, S. K., Cemazar, M., Biasin, A., Tierno, D., Scaggiante, B.,… Truong, N. H. (2023). Targeting non-coding RNAs for the development of novel hepatocellular carcinoma therapeutic approaches. *Pharmaceutics, 15*(4), 1249.

Ji, H., Kim, T. W., Lee, W. J., Jeong, S. D., Cho, Y. B., & Kim, H. H. (2022). Two circPPFIA1s negatively regulate liver metastasis of colon cancer via miR-155-5p/CDX1 and HuR/RAB36. *Molecular Cancer, 21*(1), 197. doi:https://doi.org/10.1186/s12943-022-01667-w

Jiang, D., Yuan, X., Ni, J., Shen, L., Cai, M., & Xu, L. (2021). Low Serum miR-607 Level as a Potential Diagnostic and Prognostic Biomarker in Patients of Pancreatic Ductal Adenocarcinoma: A Preliminary Study. *Canadian Journal of Gastroenterology and Hepatology, 2021*(1), 8882129.

Jin, K., Wang, S., Zhang, Y., Xia, M., Mo, Y., Li, X.,… He, Y. (2019). Long non-coding RNA PVT1 interacts with MYC and its downstream molecules to synergistically promote tumorigenesis. *Cellular and Molecular Life Sciences, 76*, 4275–4289.

Jin, W., Chen, L., Cai, X., Zhang, Y., Zhang, J., Ma, D.,… Yu, F. (2017). Long non-coding RNA TUC338 is functionally involved in sorafenib-sensitized hepatocarcinoma cells by targeting RASAL1. *Oncology reports, 37*(1), 273–280.

Jing, D., Zhu, F., Xu, Z., Zhang, G., & Zhou, G. (2022). The role of long noncoding RNA (lncRNA) nuclear-enriched abundant transcript 1 (NEAT1) in immune diseases. *Transplant Immunology, 75*, 101716.

Jun, L., Fang, Y., & Han, F. (2023). LncRNA LINC01296 regulates PD-L1 to affect the immune escape of nasopharyngeal carcinoma cells. *Immunological Journal, 39*(10), 847–856.
Karakatsanis, A., Papaconstantinou, I., Gazouli, M., Lyberopoulou, A., Polymeneas, G., & Voros, D. (2013). Expression of microRNAs, miR-21, miR-31, miR-122, miR-145, miR-146a, miR-200c, miR-221, miR-222, and miR-223 in patients with hepatocellular carcinoma or intrahepatic cholangiocarcinoma and its prognostic significance. *Molecular carcinogenesis, 52*(4), 297–303.
Khajanchi, N., & Saha, K. (2022). Controlling CRISPR with small molecule regulation for somatic cell genome editing. *Molecular Therapy, 30*(1), 17–31.
Khan, K., Irfan, M., Sattar, A. A., Faiz, M. B., Rahman, A.U Athar, H., … Cho, W. C. (2023). LncRNA SNHG6 role in clinicopathological parameters in cancers. *European Journal of Medical Research, 28*(1), 363.
Kim, S. S., Baek, G. O., Ahn, H. R., Sung, S., Seo, C. W., Cho, H. J.,… Eun, J. W. (2020). Serum small extracellular vesicle-derived LINC00853 as a novel diagnostic marker for early hepatocellular carcinoma. *Molecular oncology, 14*(10), 2646–2659.
Kim, S. S., Baek, G. O., Son, J. A., Ahn, H. R., Yoon, M. K., Cho, H. J.,… Eun, J. W. (2021). Early detection of hepatocellular carcinoma via liquid biopsy: Panel of small extracellular vesicle-derived long noncoding RNAs identified as markers. *Molecular oncology, 15*(10), 2715–2731.
Kladi-Skandali, A., Michaelidou, K., Scorilas, A., & Mavridis, K. (2015). Long noncoding RNAs in digestive system malignancies: a novel class of cancer biomarkers and therapeutic targets? *Gastroenterology research and practice, 2015*(1), 319861.
Koga, C., Kobayashi, S., Nagano, H., Tomimaru, Y., Hama, N., Wada, H.,… Ishii, H. (2014). Reprogramming using microRNA-302 improves drug sensitivity in hepatocellular carcinoma cells. *Annals of Surgical Oncology, 21*, 591–600.
Kojiro, M. (2011). Pathology of Hepatocellular Carcinoma. In X. W. Wang, J. W. Grisham, & S. S. Thorgeirsson (Eds.), *Molecular Genetics of Liver Neoplasia* (pp. 37–48). New York, NY: Springer New York.
Kong, H., Wu, Y., Zhu, M., Zhai, C., Qian, J., Gao, X.,… Zhu, H. (2016). Long non-coding RNAs: novel prognostic biomarkers for liver metastases in patients with early stage colorectal cancer. *Oncotarget, 7*(31), 50428.
Kong, Q., Liang, C., Jin, Y., Pan, Y., Tong, D., Kong, Q., & Zhou, J. (2019). The lncRNA MIR4435-2HG is upregulated in hepatocellular carcinoma and promotes cancer cell proliferation by upregulating miRNA-487a. *Cellular & Molecular Biology Letters, 24*, 1–7.
Kumada, T., Toyoda, H., Tada, T., Yasuda, S., & Tanaka, J. (2020). Changes in background liver function in patients with hepatocellular carcinoma over 30 years: comparison of child-pugh classification and albumin bilirubin grade. *Liver cancer, 9*(5), 518–528.
Lee, Y. R., Kim, G., Tak, W. Y., Jang, S. Y., Kweon, Y. O., Park, J. G.,… Park, S. Y. (2019). Circulating exosomal noncoding RNAs as prognostic biomarkers in human hepatocellular carcinoma. *International journal of cancer, 144*(6), 1444–1452.
Lei, Z., Tang, X., Si, A., Yang, P., Wang, L., Luo, T.,… Cheng, Z. (2020). microRNA-454 promotes liver tumor-initiating cell expansion by regulating SOCS6. *Experimental Cell Research, 390*(1), 111955.
Lennox, K. A., & Behlke, M. A. (2016). Cellular localization of long non-coding RNAs affects silencing by RNAi more than by antisense oligonucleotides. *Nucleic acids research, 44*(2), 863–877.
Li, A., Wang, R., Zhao, Y., Zhao, P., & Yang, J. (2024a). Crosstalk between epigenetics and metabolic reprogramming in metabolic dysfunction-associated steatotic liver disease-induced hepatocellular carcinoma: a new sight. *Metabolites, 14*(6), 325.
Li, B., Hu, X., Yang, Y., Zhu, M., Zhang, J., Wang, Y.,… Wu, J. (2019). GAS5/miR-21 axis as a potential target to rescue ZCL-082-induced autophagy of female germline stem cells in vitro. *Molecular Therapy-Nucleic Acids, 17*, 436–447.
Li, D., Zhang, J., Yang, J., Wang, J., Zhang, R., Li, J., & Zhang, R. (2021a). CircMTO1 suppresses hepatocellular carcinoma progression via the miR-541-5p/ZIC1 axis by regulating Wnt/β--

catenin signaling pathway and epithelial-to-mesenchymal transition. *Cell Death Dis, 13*(1), 12. doi:https://doi.org/10.1038/s41419-021-04464-3

Li, P., Ma, X., & Huang, D. (2024b). Role of the lncRNA/Wnt signaling pathway in digestive system cancer: a literature review. *European Journal of Medical Research, 29*(1), 447.

Li, X., Liu, Y., Zheng, Q., Liu, W., Li, H., Rao, S.,… Jiang, Y. (2025a). Circular RNA circ_0004470 accelerates the occurrence of lung cancer by promoting DNA damage and cell cycle arrest. *Journal of Biological Chemistry, 301*(5), 108456. https://doi.org/10.1016/j.jbc.2025.108456

Li, X., Ramadori, P., Pfister, D., Seehawer, M., Zender, L., & Heikenwalder, M. (2021b). The immunological and metabolic landscape in primary and metastatic liver cancer. *Nature Reviews Cancer, 21*(9), 541–557.

Li, Y., Lv, Z., He, G., Wang, J., Zhang, X., Lu, G.,… Ding, Y. (2015). The SOX17/miR-371-5p/SOX2 axis inhibits EMT, stem cell properties and metastasis in colorectal cancer. *Oncotarget, 6*(11), 9099.

Li, Y., Wang, H., Mao, D., Che, X., Chen, Y., & Liu, Y. (2025b). Understanding pre-metastatic niche formation: implications for colorectal cancer liver metastasis. *Journal of Translational Medicine, 23*(1), 340.

Li, Z., & Sun, X. (2020). Non-coding RNAs operate in the crosstalk between cancer metabolic reprogramming and metastasis. *Frontiers in Oncology, 10*, 810.

Liang, Y., Li, E., Min, J., Gong, C., Gao, J., Ai, J.,… Wu, L. (2018). miR-29a suppresses the growth and metastasis of hepatocellular carcinoma through IFITM3. *Oncology Reports, 40*(6), 3261-3272.

Lim, L. J., Wong, S. Y., Huang, F., Lim, S., Chong, S. S., Ooi, L. L.,… Lee, C. G. (2019). Roles and regulation of long noncoding RNAs in hepatocellular carcinoma. *Cancer Research, 79*(20), 5131–5139.

Lin, C.-L., & Kao, J.-H. (2023). Development of hepatocellular carcinoma in treated and untreated patients with chronic hepatitis B virus infection. *Clinical and Molecular Hepatology, 29*(3), 605.

Lin, C.-W., Lin, P.-Y., & Yang, P.-C. (2016). Noncoding RNAs in Tumor Epithelial-to-Mesenchymal Transition. *Stem cells international, 2016*(1), 2732705.

Lin, Q., Zhu, J., & Hu, Y. (2023). Involvement of lncRNAs in the tumor microenvironment: a new property of tumor immunity. *Cancer biology & medicine, 20*(8), 553.

Lin, Y.-H., Wu, M.-H., Yeh, C.-T., & Lin, K.-H. (2018). Long non-coding RNAs as mediators of tumor microenvironment and liver cancer cell communication. *International journal of molecular sciences, 19*(12), 3742.

Line, P.-D. (2023). 217C20Secondary neoplasms of the liver. In Liver, Gall Bladder, and Bile Ducts (pp. 0): Oxford University Press. Retrieved from https://doi.org/10.1093/med/9780192862457.003.0020. doi:10.1093/med/9780192862457.003.0020

Liu, H., Zhu, J., Guo, L., Zhang, H., Liu, S., & Kou, X. (2022). Identification of a Two-lncRNA Signature with Prognostic and Diagnostic Value for Hepatocellular Carcinoma. *J Oncol, 2022*, 2687455. doi:https://doi.org/10.1155/2022/2687455

Liu, J., Qu, J., Xu, L., Qiao, C., Shao, G., Liu, X.,… Zhang, J. (2023). Prediction of liver cancer prognosis based on immune cell marker genes. *Frontiers in Immunology, 14*, 1147797.

Liu, L., Cai, X., Liu, E., Tian, X., & Tian, C. (2017a). MicroRNA-18a promotes proliferation and metastasis in hepatocellular carcinoma via targeting KLF4. *Oncotarget, 8*(40), 68263.

Liu, R., Li, X., Zhu, W., Wang, Y., Zhao, D., Wang, X.,… Lai, G. (2019). Cholangiocyte-derived exosomal long noncoding RNA H19 promotes hepatic stellate cell activation and cholestatic liver fibrosis. *Hepatology, 70*(4), 1317–1335.

Liu, W., Li, L., Jiang, J., Wu, M., & Lin, P. (2021). Applications and challenges of CRISPR-Cas gene-editing to disease treatment in clinics. *Precision clinical medicine, 4*(3), 179–191.

Liu, Y., Sun, C., & Chen, B. (2017b). Screening and Identification of Molecular Marker for Metastatic Liver Cancer. In X. Qin, J. Xu, & Y. Zhong (Eds.), *Multidisciplinary Management of Liver Metastases in Colorectal Cancer: Early Diagnosis and Treatment* (pp. 59–91). Dordrecht: Springer Netherlands.

Louis, C., Leclerc, D., & Coulouarn, C. (2022). Emerging roles of circular RNAs in liver cancer. *JHEP Reports, 4*(2), 100413.

Lu, L., Byrnes, K., Han, C., Wang, Y., & Wu, T. (2014a). miR-21 targets 15-PGDH and promotes cholangiocarcinoma growth. *Molecular Cancer Research, 12*(6), 890–900.

Lu, Y.-X., Yuan, L., Xue, X.-L., Zhou, M., Liu, Y., Zhang, C.,… Li, X.-N. (2014b). Regulation of colorectal carcinoma stemness, growth, and metastasis by an miR-200c-Sox2–negative feedback loop mechanism. *Clinical Cancer Research, 20*(10), 2631–2642.

Lumkul, L., Jantaree, P., Jaisamak, K., Wongkummool, W., Lapisatepun, W., Orrapin, S.,… Chaiyawat, P. (2024). Combinatorial gene expression profiling of serum HULC, HOTAIR, and UCA1 lncRNAs to differentiate hepatocellular carcinoma from liver diseases: a systematic review and meta-analysis. *International Journal of Molecular Sciences, 25*(2), 1258.

Luo, Y., Yang, J., Yu, J., Liu, X., Yu, C., Hu, J.,… Ma, X. (2020). Long non-coding RNAs: emerging roles in the immunosuppressive tumor microenvironment. *Frontiers in oncology, 10*, 48.

Lv, D., Chen, L., Du, L., Zhou, L., & Tang, H. (2021a). Emerging regulatory mechanisms involved in liver cancer stem cell properties in hepatocellular carcinoma. *Frontiers in cell and developmental biology, 9*, 691410.

Lv, E., Sheng, J., Yu, C., Rao, D., & Huang, W. (2021b). LncRNA influence sequential steps of hepatocellular carcinoma metastasis. *Biomedicine & Pharmacotherapy, 136*, 111224.

Mahati, S., Xiao, L., Yang, Y., Mao, R., & Bao, Y. (2017). miR-29a suppresses growth and migration of hepatocellular carcinoma by regulating CLDN1. *Biochemical and biophysical research communications, 486*(3), 732–737.

Mann, D. A. (2014). Epigenetics in liver disease. *Hepatology, 60*(4), 1418–1425.

Mehrab Mohseni, M., Zamani, H., Momeni, M., & Shirvani-Farsani, Z. (2024). An update on the molecular mechanisms of ZFAS1 as a prognostic, diagnostic, or therapeutic biomarker in cancers. *Discover Oncology, 15*(1), 219.

Mejia, J. C., & Pasko, J. (2020). Primary liver cancers: intrahepatic cholangiocarcinoma and hepatocellular carcinoma. *Surgical Clinics, 100*(3), 535–549.

Meng, F., Henson, R., Lang, M., Wehbe, H., Maheshwari, S., Mendell, J. T.,… Patel, T. (2006). Involvement of human micro-RNA in growth and response to chemotherapy in human cholangiocarcinoma cell lines. *Gastroenterology, 130*(7), 2113–2129.

Mohapatra, S., Pioppini, C., Ozpolat, B., & Calin, G. A. (2021). Non-coding RNAs regulation of macrophage polarization in cancer. *Molecular cancer, 20*, 1–15.

Mohseni, M. M., Zamani, H., Momeni, M., & Shirvani-Farsani, Z. (2024). An update on the molecular mechanisms of ZFAS1 as a prognostic, diagnostic, or therapeutic biomarker in cancers. *Discover Oncology, 15*, 219.

Momtazmanesh, S., & Rezaei, N. (2021). Long non-coding RNAs in diagnosis, treatment, prognosis, and progression of glioma: a state-of-the-art review. *Frontiers in Oncology, 11*, 712786.

Nakaoka, T., Saito, Y., & Saito, H. (2017). Aberrant DNA methylation as a biomarker and a therapeutic target of cholangiocarcinoma. *International Journal of Molecular Sciences, 18*(6), 1111.

Nie, G., Lian, N., Peng, D., Lu, J., & Li, B. (2022). Prognostic value of exosomal noncoding RNA in hepatocellular carcinoma: a meta-analysis. *Carcinogenesis, 43*(8), 754–765.

Nielsen, T. O., West, R. B., Linn, S. C., Alter, O., Knowling, M. A., O'Connell, J. X.,… Pollack, J. R. (2002). Molecular characterisation of soft tissue tumours: a gene expression study. *The Lancet, 359*(9314), 1301–1307.

Ogunwobi, O. O., Harricharran, T., Huaman, J., Galuza, A., Odumuwagun, O., Tan, Y.,… Nguyen, M. T. (2019). Mechanisms of hepatocellular carcinoma progression. *World journal of gastroenterology, 25*(19), 2279.

Ou, H., Huang, Y., Xiang, L., Chen, Z., Fang, Y., Lin, Y.,… Yang, D. (2018). Circulating Tumor Cell Phenotype Indicates Poor Survival and Recurrence After Surgery for Hepatocellular Carcinoma. *Dig Dis Sci, 63*(9), 2373–2380. doi:https://doi.org/10.1007/s10620-018-5124-2

Palaz, F., Ozsoz, M., Zarrinpar, A., & Sahin, I. (2024). CRISPR in targeted therapy and adoptive T cell immunotherapy for hepatocellular carcinoma. *Journal of Hepatocellular Carcinoma*, 975–995.

Pan, Y., Chen, H., & Yu, J. (2020). Biomarkers in hepatocellular carcinoma: current status and future perspectives. *Biomedicines, 8*(12), 576.

Pan, Y., Liu, Q., Li, Q., & Ren, Z. (2025). Mechanistic insights into Uc001kfo-induced hepatocellular carcinoma metastasis. *Discover Oncology, 16*(1), 260.

Park, S. H., Fong, K.-W., Mong, E., Martin, M. C., Schiltz, G. E., & Yu, J. (2021). Going beyond Polycomb: EZH2 functions in prostate cancer. *Oncogene, 40*(39), 5788–5798.

Parpart, S., Roessler, S., Dong, F., Rao, V., Takai, A., Ji, J.,… Tang, Z. Y. (2014). Modulation of miR-29 expression by alpha-fetoprotein is linked to the hepatocellular carcinoma epigenome. *Hepatology, 60*(3), 872–883.

Pathania, A. S. (2023). Crosstalk between Noncoding RNAs and the Epigenetics Machinery in Pediatric Tumors and Their Microenvironment. *Cancers, 15*(10), 2833.

Pecoraro, V., Rosina, A., & Polacek, N. (2022). Ribosome-associated ncRNAs (rancRNAs) adjust translation and shape proteomes. *Non-Coding RNA, 8*(2), 22.

Peng, F., Jiang, J., Yu, Y., Tian, R., Guo, X., Li, X.,… Shi, C. (2013). Direct targeting of SUZ12/ROCK2 by miR-200b/c inhibits cholangiocarcinoma tumourigenesis and metastasis. *British journal of cancer, 109*(12), 3092–3104.

Piñero, F., Dirchwolf, M., & Pessôa, M. G. (2020). Biomarkers in hepatocellular carcinoma: diagnosis, prognosis and treatment response assessment. *Cells, 9*(6), 1370.

Plawgo, K., & Raczynska, K. D. (2022). Context-Dependent Regulation of Gene Expression by Non-Canonical Small RNAs. *Non-Coding RNA, 8*(3). Retrieved from doi:https://doi.org/10.3390/ncrna8030029

Qian, Y., Li, Y., Ge, Y., Song, W., & Fan, H. (2021). Elevated LncRNA TRERNA1 correlated with activation of HIF-1α predicts poor prognosis in hepatocellular carcinoma. *Pathology-Research and Practice, 227*, 153612.

Rahma, O. E., & Hodi, F. S. (2019). The intersection between tumor angiogenesis and immune suppression. *Clinical cancer research, 25*(18), 5449–5457.

Rahman, M. M., & Tollefsbol, T. O. (2021). Targeting cancer epigenetics with CRISPR-dCAS9: Principles and prospects. *Methods, 187*, 77–91.

Ramli, S., Sim, M. S., Guad, R. M., Gopinath, S. C. B., Subramaniyan, V., Fuloria, S.,… Wu, Y. S. (2021). Long noncoding RNA UCA1 in gastrointestinal cancers: molecular regulatory roles and patterns, mechanisms, and interactions. *Journal of oncology, 2021*(1), 5519720.

Ratnasari, N., Lestari, P., Renovaldi, D., Raditya Ningsih, J., Qoriansas, N., Wardana, T.,… Triwikatmani, C. (2022). Potential plasma biomarkers: miRNA-29c, miRNA-21, and miRNA-155 in clinical progression of Hepatocellular Carcinoma patients. *PLoS One, 17*(2), e0263298.

Razumilava, N., Bronk, S. F., Smoot, R. L., Fingas, C. D., Werneburg, N. W., Roberts, L. R., & Mott, J. L. (2012). miR-25 targets TNF-related apoptosis inducing ligand (TRAIL) death receptor-4 and promotes apoptosis resistance in cholangiocarcinoma. *Hepatology, 55*(2), 465–475.

Rojas, Á., Gil-Gómez, A., de la Cruz-Ojeda, P., Muñoz-Hernández, R., Sánchez-Torrijos, Y., Gallego-Durán, R.,… Gato-Zambrano, S. (2022). Long non-coding RNA H19 as a biomarker for hepatocellular carcinoma. *Liver International, 42*(6), 1410–1422.

Romeo, M., Dallio, M., Scognamiglio, F., Ventriglia, L., Cipullo, M., Coppola, A.,… Federico, A. (2023). Role of non-coding RNAs in hepatocellular carcinoma progression: from classic to novel clinicopathogenetic implications. *Cancers, 15*(21), 5178.

Rotem, A., Ram, O., Shoresh, N., Sperling, R. A., Goren, A., Weitz, D. A., & Bernstein, B. E. (2015). Single-cell ChIP-seq reveals cell subpopulations defined by chromatin state. *Nature biotechnology, 33*(11), 1165–1172.

Ruiz Esparza Garrido, R. & Velázquez Flores, M. (2023). Circular RNAs: the next level of gene regulation. *Am J Transl Res, 15*(10), 6122–6135.

Salgia, S. R., Singh, S. K., Gurha, P., & Gupta, R. (2003). Two reactions of Haloferax volcanii RNA splicing enzymes: joining of exons and circularization of introns. *Rna, 9*(3), 319–330. doi:https://doi.org/10.1261/rna.2118203

Salido-Guadarrama, I., Romero-Cordoba, S. L., & Rueda-Zarazua, B. (2023). Multi-omics mining of lncRNAs with biological and clinical relevance in cancer. *International Journal of Molecular Sciences, 24*(23), 16600.

Samir, A., Abdeldaim, A., Mohammed, A., Ali, A., Alorabi, M., Hussein, M. M.,… Abdelhafiz, A. S. (2024). Analysis of four long non-coding RNAs for hepatocellular carcinoma screening and prognosis by the aid of machine learning techniques. *Scientific reports, 14*(1), 29582.

Samoilă, I., Şelaru, A., Nazarie, S., Costache, M., & Dinescu, S. (2023). Chapter 2 - Regulatory non-coding RNAs-biogenesis, mechanisms of action and role in gene expression regulation. In J. Sztuba-Solinska (Ed.), *Navigating Non-Coding RNA* (pp. 47–88): Academic Press.

Sanger, H. L., Klotz, G., Riesner, D., Gross, H. J., & Kleinschmidt, A. K. (1976). Viroids are single-stranded covalently closed circular RNA molecules existing as highly base-paired rod-like structures. *Proc Natl Acad Sci U S A, 73*(11), 3852–3856. doi:https://doi.org/10.1073/pnas.73.11.3852

Sarropoulou, E., & Fernández, I. (2023). Epigenetic Regulation of Gene Expression by Noncoding RNAs. In *Epigenetics in Aquaculture* (pp. 65–93).

Selaru, F. M., Olaru, A. V., Kan, T., David, S., Cheng, Y., Mori, Y.,… Agarwal, R. (2009). MicroRNA-21 is overexpressed in human cholangiocarcinoma and regulates programmed cell death 4 and tissue inhibitor of metalloproteinase 3. *Hepatology, 49*(5), 1595–1601.

Sellitto, A., Pecoraro, G., Giurato, G., Nassa, G., Rizzo, F., Saggese, P.,… Tarallo, R. (2021). Regulation of metabolic reprogramming by long non-coding RNAs in cancer. *Cancers, 13*(14), 3485.

Sendi, H., Yazdimamaghani, M., Hu, M., Sultanpuram, N., Wang, J., Moody, A. S.,… Li, L. (2022). Nanoparticle delivery of miR-122 inhibits colorectal cancer liver metastasis. *Cancer Research, 82*(1), 105–113.

Shah, M., & Sarkar, D. (2024). HCC-Related lncRNAs: roles and mechanisms. *International journal of molecular sciences, 25*(1), 597.

Shen, H., Li, H., & Zhou, J. (2022). Circular RNA hsa_circ_0032683 inhibits the progression of hepatocellular carcinoma by sponging microRNA-338-5p. *Bioengineered, 13*(2), 2321–2335. doi:https://doi.org/10.1080/21655979.2021.2024961

Shi, P., Sun, J., He, B., Song, H., Li, Z., Kong, W.,… Xue, H. (2018). Profiles of differentially expressed circRNAs in esophageal and breast cancer. *Cancer Manag Res, 10*, 2207–2221. doi:https://doi.org/10.2147/cmar.S167863

Shi, X., Wang, X., Yao, W., Shi, D., Shao, X., Lu, Z.,… Wang, X. (2024). Mechanism insights and therapeutic intervention of tumor metastasis: latest developments and perspectives. *Signal transduction and targeted therapy, 9*(1), 192.

Shu, B., Zhang, R.-Z., Zhou, Y.-X., He, C., & Yang, X. (2022). METTL3-mediated macrophage exosomal NEAT1 contributes to hepatic fibrosis progression through Sp1/TGF-β1/Smad signaling pathway. *Cell death discovery, 8*(1), 266.

Slaby, O., Laga, R., & Sedlacek, O. (2017). Therapeutic targeting of non-coding RNAs in cancer. *Biochemical journal, 474*(24), 4219–4251.

Stechele, M., Link, H., Hirner-Eppeneder, H., Alunni-Fabbroni, M., Wildgruber, M., Salvermoser, L.,… Rössler, D. (2023). Circulating miR-21 as a prognostic biomarker in HCC treated by CT-guided high-dose rate brachytherapy. *Radiation Oncology, 18*(1), 125.

Sun, L., Su, Y., Liu, X., Xu, M., Chen, X., Zhu, Y.,… Wei, C. (2018). Serum and exosome long non coding RNAs as potential biomarkers for hepatocellular carcinoma. *Journal of Cancer, 9*(15), 2631.

Swellam, M., & Hamdy, N. M. (2012). Association of nonalcoholic fatty liver disease with a single nucleotide polymorphism on the gene encoding leptin receptor. *IUBMB life, 64*(2), 180–186.

Tabak, H. F., Van der Horst, G., Kamps, A. M. J. E., & Arnberg, A. C. (1987). Interlocked RNA circle formation by a self-splicing yeast mitochondrial group I intron. *Cell, 48*(1), 101–110. https://doi.org/10.1016/0092-8674(87)90360-6

Tam, C., Wong, J. H., Tsui, S. K. W., Zuo, T., Chan, T. F., & Ng, T. B. (2019). LncRNAs with miRNAs in regulation of gastric, liver, and colorectal cancers: updates in recent years. *Applied microbiology and biotechnology, 103*, 4649–4677.

Tang, L., Peng, S. C., Zhuang, X. W., He, Y., Song, Y. X., Nie, H.,… He, M. L. (2025). Tumor Metastasis: Mechanistic Insights and Therapeutic Intervention. *MedComm–Oncology, 4*(1), e70012.

Tao, L., Wang, Y., Shen, Z., Cai, J., Zheng, J., Xia, S.,… Jin, R. (2024). Activation of IGFBP4 via unconventional mechanism of miRNA attenuates metastasis of intrahepatic cholangiocarcinoma. *Hepatology International, 18*(1), 91–107.

Tenchov, R., Bird, R., Curtze, A. E., & Zhou, Q. (2021). Lipid nanoparticles— from liposomes to mRNA vaccine delivery, a landscape of research diversity and advancement. *Acs Nano, 15*(11), 16982–17015.

Tian, C., & Hu, G. (2020). Chromatin isolation by RNA purification (ChIRP) and its applications. In *Epigenetics methods* (pp. 507–521): Elsevier.

Toh, T. B., Lim, J. J., & Chow, E. K.-H. (2019). Epigenetics of hepatocellular carcinoma. *Clinical and translational medicine, 8*(1), 13.

Tripathi, P., & Srivastava, S. P. (2023). Chapter 9 - Noncoding RNAs in human health and diseases. In M. Garg, G. Sethi, & A. K. Pandey (Eds.), *Transcription and Translation in Health and Disease* (pp. 139–152): Academic Press.

Ule, J., Hwang, H.-W., & Darnell, R. B. (2018). The future of cross-linking and immunoprecipitation (CLIP). *Cold Spring Harbor perspectives in biology, 10*(8), a032243.

Unfried, J. P., & Fortes, P. (2020). LncRNAs in HCV infection and HCV-related liver disease. *International Journal of Molecular Sciences, 21*(6), 2255.

Unfried, J. P., Sangro, P., Prats-Mari, L., Sangro, B., & Fortes, P. (2021). The landscape of lncRNAs in hepatocellular carcinoma: a translational perspective. *Cancers, 13*(11), 2651.

Verma, S., Sahu, B. D., & Mugale, M. N. (2023). Role of lncRNAs in hepatocellular carcinoma. *Life sciences, 325*, 121751.

Vianello, C., Monti, E., Leoni, I., Galvani, G., Giovannini, C., Piscaglia, F.,… Fornari, F. (2024). Noncoding RNAs in Hepatocellular Carcinoma: potential applications in combined therapeutic strategies and promising candidates of treatment response. *Cancers, 16*(4), 766.

Wahl, M. C., Will, C. L., & Lührmann, R. (2009). The spliceosome: design principles of a dynamic RNP machine. *Cell, 136*(4), 701–718. doi:https://doi.org/10.1016/j.cell.2009.02.009

Wai, H. A., Svobodova, E., Herrera, N. R., Douglas, A. G., Holloway, J. W., Baralle, F. E.,… Baralle, D. (2024). Tailored antisense oligonucleotides designed to correct aberrant splicing reveal actionable groups of mutations for rare genetic disorders. *Experimental & molecular medicine, 56*(8), 1816–1825.

Wang, B., Hsu, S.-H., Majumder, S., Kutay, H., Huang, W., Jacob, S. T., & Ghoshal, K. (2010). TGFβ-mediated upregulation of hepatic miR-181b promotes hepatocarcinogenesis by targeting TIMP3. *Oncogene, 29*(12), 1787–1797.

Wang, B., & Pu, R. (2023). Association between glycolysis markers and prognosis of liver cancer: a systematic review and meta-analysis. *World Journal of Surgical Oncology, 21*(1), 390.

Wang, B., Qi, F.-Z., Chen, P., Qian, L.-M., Su, H.-S., Wang, Y.,… Li, D. (2024). Hypoxia-activated selectivity-improved anti-PKM2 antibody combined with prodrug TH-302 for potentiated targeting therapy in hepatocellular carcinoma. *International journal of biological sciences, 20*(5), 1634.

Wang, C., Wang, L., Ding, Y., Lu, X., Zhang, G., Yang, J.,… Xu, L. (2017). LncRNA structural characteristics in epigenetic regulation. *International journal of molecular sciences, 18*(12), 2659.

Wang, G., Wang, Q., Liang, N., Xue, H., Yang, T., Chen, X.,… He, X. (2020). Oncogenic driver genes and tumor microenvironment determine the type of liver cancer. *Cell Death Dis, 11*(5), 313. doi:https://doi.org/10.1038/s41419-020-2509-x

Wang, J., Tian, T., Li, X., & Zhang, Y. (2022). Noncoding RNAs emerging as drugs or drug targets: their chemical modification, bio-conjugation and intracellular regulation. *Molecules, 27*(19), 6717.

Wang, J., Wang, Z., Yuan, J., Wang, Q., & Shen, X. (2021a). Upregulation of miR-137 expression suppresses tumor growth and progression via interacting with DNMT3a through inhibiting the PTEN/Akt signaling in HCC. *OncoTargets and therapy*, 165–176.

Wang, L.-J., He, C.-C., Sui, X., Cai, M.-J., Zhou, C.-Y., Ma, J.-L.,… Zhu, Q. (2015). MiR-21 promotes intrahepatic cholangiocarcinoma proliferation and growth in vitro and in vivo by targeting PTPN14 and PTEN. *Oncotarget, 6*(8), 5932.

Wang, X., Lai, Q., He, J., Li, Q., Ding, J., Lan, Z.,… Zhao, X. (2019). LncRNA SNHG6 promotes proliferation, invasion and migration in colorectal cancer cells by activating TGF-β/Smad signaling pathway via targeting UPF1 and inducing EMT via regulation of ZEB1. *International journal of medical sciences, 16*(1), 51.

Wang, Z. G., He, Z. Y., Chen, Y. Y., Gao, H., & Du, X. L. (2021b). Incidence and survival outcomes of secondary liver cancer: a Surveillance Epidemiology and End Results database analysis. *Transl Cancer Res, 10*(3), 1273–1283. doi:https://doi.org/10.21037/tcr-20-3319

Wehrle, C. J., Tocci, N. X., Sun, K., Jiao, C., Hong, H., Gross, A.,… Aucejo, F. (2024). Utility of circulating tumor DNA in secondary liver malignancies: What we know and what is to come. *J Surg Oncol*. doi:https://doi.org/10.1002/jso.27838

Wei, J. W., Huang, K., Yang, C., & Kang, C. S. (2017). Non-coding RNAs as regulators in epigenetics (Review). *Oncol Rep, 37*(1), 3–9. doi:https://doi.org/10.3892/or.2016.5236

Wei, W., Tang, H., & Tang, L. (2018). MicroRNA-34a inhibits metastasis in liver cancer cells. *Oncology letters, 16*(6), 6960–6965.

Winkle, M., El-Daly, S. M., Fabbri, M., & Calin, G. A. (2021). Noncoding RNA therapeutics—challenges and potential solutions. *Nature reviews Drug discovery, 20*(8), 629–651.

Woo, H. G., & Kim, Y. J. (2018). Multiplatform Genomic Roadmap of Hepatocellular Carcinoma: A Matter of Molecular Heterogeneity. *Hepatology, 68*(5), 2029–2032. doi:https://doi.org/10.1002/hep.29925

Wu, G., Ding, C., Shi, C.-L., Zhang, X.-C., Lei, G.-R., Wang, S.,… Shi, T.-F. (2025). Circular RNA 14580 promotes papillary thyroid cancer progression by interacting with CTSW/fibrosis singling pathway. *International Journal of Biological Macromolecules, 286*, 138538. https://doi.org/10.1016/j.ijbiomac.2024.138538

Wu, J., Cheng, G., Zhang, C., Zheng, Y., Xu, H., Yang, H., & Hua, L. (2017). Long noncoding RNA LINC01296 is associated with poor prognosis in prostate cancer and promotes cancer-cell proliferation and metastasis. *OncoTargets and therapy*, 1843–1852.

Wu, S.-M., Ai, H.-W., Zhang, D.-Y., Han, X.-Q., Pan, Q., Luo, F.-L., & Zhang, X.-L. (2014). MiR-141 targets ZEB2 to suppress HCC progression. *Tumor Biology, 35*, 9993–9997.

Wu, X., Lokken, R. P., & Mehta, N. (2023). Optimal treatment for small HCC (<3 cm): Resection, liver transplantation, or locoregional therapy? *JHEP Rep, 5*(8), 100781. doi:https://doi.org/10.1016/j.jhepr.2023.100781

Wu, Y., Wang, P. S., Wang, B. G., Xu, L., Fang, W. X., Che, X. F.,… Li, Z. (2018). Genomewide identification of a novel six-LncRNA signature to improve prognosis prediction in resectable hepatocellular carcinoma. *Cancer Med, 7*(12), 6219–6233. doi:https://doi.org/10.1002/cam4.1854

Xiang, J.-F., Yin, Q.-F., Chen, T., Zhang, Y., Zhang, X.-O., Wu, Z.,… Lu, X. (2014). Human colorectal cancer-specific CCAT1-L lncRNA regulates long-range chromatin interactions at the MYC locus. *Cell research, 24*(5), 513–531.

Xiao, M. S., Ai, Y., & Wilusz, J. E. (2020). Biogenesis and Functions of Circular RNAs Come into Focus. *Trends Cell Biol, 30*(3), 226–240. doi:https://doi.org/10.1016/j.tcb.2019.12.004

Xie, Y., Zhang, H., Guo, X.-J., Feng, Y.-C., He, R.-Z., Li, X.,… Zhu, F. (2018). Let-7c inhibits cholangiocarcinoma growth but promotes tumor cell invasion and growth at extrahepatic sites. *Cell death & disease, 9*(2), 249.

Xue, C., Gu, X., Bao, Z., Su, Y., Lu, J., & Li, L. (2022). The mechanism underlying the ncRNA dysregulation pattern in hepatocellular carcinoma and its tumor microenvironment. *Frontiers in Immunology, 13*, 847728.

Yan-Chun, L., Hong-Mei, Y., Zhi-Hong, C., Qing, H., Yan-Hong, Z., & Ji-Fang, W. (2017). MicroRNA-192-5p promote the proliferation and metastasis of hepatocellular carcinoma cell by targeting SEMA3A. *Applied Immunohistochemistry & Molecular Morphology, 25*(4), 251–260.

Yan, J., Zhou, C., Guo, K., Li, Q., & Wang, Z. (2019). A novel seven-lncRNA signature for prognosis prediction in hepatocellular carcinoma. *Journal of cellular biochemistry, 120*(1), 213–223.

Yang, C., Zhang, H., Zhang, L., Zhu, A. X., Bernards, R., Qin, W., & Wang, C. (2023). Evolving therapeutic landscape of advanced hepatocellular carcinoma. *Nature reviews Gastroenterology & hepatology, 20*(4), 203–222.

Yang, L., Wilusz, J. E., & Chen, L. L. (2022). Biogenesis and Regulatory Roles of Circular RNAs. *Annu Rev Cell Dev Biol, 38*, 263–289. doi:https://doi.org/10.1146/annurev-cellbio-120420-125117

Yang, Q., Li, S., Ou, H., Zhang, Y., Zhu, G., Li, S., & Lei, L. (2024). Exosome-based delivery strategies for tumor therapy: an update on modification, loading, and clinical application. *Journal of nanobiotechnology, 22*(1), 41.

Yao, C., Wu, S., Kong, J., Sun, Y., Bai, Y., Zhu, R.,… Zheng, L. (2023). Angiogenesis in hepatocellular carcinoma: mechanisms and anti-angiogenic therapies. *Cancer biology & medicine, 20*(1), 25–43.

Yin, X., Lin, H., Lin, L., Miao, L., He, J., & Zhuo, Z. (2022). LncRNAs and CircRNAs in cancer. *MedComm, 3*(2), e141.

Yin, Z., Ma, T., Huang, B., Lin, L., Zhou, Y., Yan, J.,… Chen, S. (2019). Macrophage-derived exosomal microRNA-501-3p promotes progression of pancreatic ductal adenocarcinoma through the TGFBR3-mediated TGF-β signaling pathway. *Journal of Experimental & Clinical Cancer Research, 38*, 1–20.

Youness, R. A., Hassan, H. A., Abaza, T., Hady, A. A., El Magdoub, H. M., Ali, M.,… Hamdy, N. M. (2024). A comprehensive insight and in silico analysis of CircRNAs in hepatocellular carcinoma: a step toward ncRNA-based precision medicine. *Cells, 13*(15), 1245.

Younger, S. T. (2022). Noncoding RNA†. In G. M. Blackburn, M. Egli, M. J. Gait, & J. K. Watts (Eds.), *Nucleic Acids in Chemistry and Biology* (pp. 0). The Royal Society of Chemistry.

Yousef, M. H., El-Fawal, H. A., & Abdelnaser, A. (2020). Hepigenetics: a review of epigenetic modulators and potential therapies in hepatocellular carcinoma. *BioMed Research International, 2020*(1), 9593254.

Yu, Q., Yang, X., Duan, W., Li, C., Luo, Y., & Lu, S. (2017). miRNA-346 promotes proliferation, migration and invasion in liver cancer. *Oncology letters, 14*(3), 3255–3260.

Yu, Y., & Hann, S. S. (2019). Novel tumor suppressor lncRNA growth arrest-specific 5 (GAS5) in human cancer. *OncoTargets and therapy*, 8421–8436.

Yu, Y., Wang, J., Guo, Q., & Luo, H. (2024). LINC01134: a pivotal oncogene with promising predictive maker and therapeutic target in hepatocellular carcinoma. *Frontiers in Oncology, 14*, 1265762.

Yuan, C., Ding, Y., Zhuang, Y., Zhang, C., Han, L., Li, W.,… Zhang, E. (2022). Copy number amplification-activated long non-coding RNA LINC00662 epigenetically inhibits BIK by interacting with EZH2 to regulate tumorigenesis in non-small cell lung cancer. *Journal of Cancer, 13*(5), 1640.

Yuan, Q., Jia, L., Yang, J., & Li, W. (2025). The role of macrophages in liver metastasis: mechanisms and therapeutic prospects. *Frontiers in immunology, 16*, 1542197.

Zeng, H., Hou, Y., Zhou, X., Lang, L., Luo, H., Sun, Y.,… Liu, Y. (2022). Cancer-associated fibroblasts facilitate premetastatic niche formation through lncRNA SNHG5-mediated angiogenesis and vascular permeability in breast cancer. *Theranostics, 12*(17), 7351.

Zeng, Y., Du, W., Huang, Z., Wu, S., Ou, X., Zhang, J.,… Tang, H. (2023). Hsa_circ_0060467 promotes breast cancer liver metastasis by complexing with eIF4A3 and sponging miR-1205. *Cell Death Discov, 9*(1), 153. doi:https://doi.org/10.1038/s41420-023-01448-4

Zhang, C., Xu, A., Liu, R., Liu, M., Zhao, W., Yao, A.,… Zhao, K. (2024). LINC01138 expresses two novel isoforms and functions as a repressive factor in glioma cells. *Heliyon, 10*(12).

Zhang, C., Yin, D., Sun, S., & Han, L. (2017a). The use of lncRNA analysis for stratification management of prognostic risk in patients with NSCLC. *Eur Rev Med Pharmacol Sci, 21*(1), 115–119.

Zhang, D., Li, H., Jiang, X., Cao, L., Wen, Z., Yang, X., & Xue, P. (2017b). Role of AP-2α and MAPK7 in the regulation of autocrine TGF-β/miR-200b signals to maintain epithelial-mesenchymal transition in cholangiocarcinoma. *Journal of hematology & oncology, 10*, 1–11.

Zhang, H., Liang, H., Wu, S., Zhang, Y., & Yu, Z. (2021a). MicroRNA-638 induces apoptosis and autophagy in human liver cancer cells by targeting enhancer of zeste homolog 2 (EZH2). *Environmental Toxicology and Pharmacology, 82*, 103559.

Zhang, H. F., Wang, Y. C., & Han, Y. D. (2018). MicroRNA-34a inhibits liver cancer cell growth by reprogramming glucose metabolism. *Molecular Medicine Reports, 17*(3), 4483–4489.

Zhang, J., Han, C., & Wu, T. (2012a). MicroRNA-26a promotes cholangiocarcinoma growth by activating β-catenin. *Gastroenterology, 143*(1), 246–256. e248.

Zhang, J., Xiao, Z., Lai, D., Sun, J., He, C., Chu, Z.,… Wang, J. (2012b). miR-21, miR-17 and miR-19a induced by phosphatase of regenerating liver-3 promote the proliferation and metastasis of colon cancer. *British journal of cancer, 107*(2), 352–359.

Zhang, L., Zhao, F., Li, W., Song, G., Kasim, V., & Wu, S. (2022). The biological roles and molecular mechanisms of long non-coding RNA MEG3 in the hallmarks of cancer. *Cancers, 14*(24), 6032.

Zhang, M.-X., Zhang, L.-Z., Fu, L.-M., Yao, H.-H., Tan, L., Feng, Z.-H.,… Shu, G.-N. (2021b). Positive feedback regulation of lncRNA PVT1 and HIF2α contributes to clear cell renal cell carcinoma tumorigenesis and metastasis. *Oncogene, 40*(37), 5639–5650.

Zhang, P., Shi, L., Song, L., Long, Y., Yuan, K., Ding, W., & Deng, L. (2020a). LncRNA CRNDE and lncRNA SNHG7 are promising biomarkers for prognosis in synchronous colorectal liver metastasis following hepatectomy. *Cancer Management and Research*, 1681–1692.

Zhang, S., Zhang, S., Zhou, P., Wei, X., Mao, X., Lin, S., & Liu, C. (2019). LncRNA MALAT1 affects high glucose-induced endothelial cell proliferation, apoptosis, migration and angiogenesis by regulating the PI3K/Akt signaling pathway. *European Review for Medical & Pharmacological Sciences, 23*(19).

Zhang, Y., Zheng, Y., & Zhu, G. (2020b). MiR-203a-3p targets PTEN to promote hepatocyte proliferation by regulating PI3K/Akt pathway in BRL-3A cells. *Bioscience, Biotechnology, and Biochemistry, 84*(4), 725–733.

Zhang, Z., Shu, L., Hu, M., Zhou, X., Yang, F., & Zhou, X.-H. (2021c). Emerging role of lncRNA DANCR in progenitor cells: beyond cancer. *European Review for Medical & Pharmacological Sciences, 25*(3).

Zhao, J., Li, L., Han, Z.-Y., Wang, Z.-X., & Qin, L.-X. (2019). Long noncoding RNAs, emerging and versatile regulators of tumor-induced angiogenesis. *American journal of cancer research, 9*(7), 1367.

Zhao, W., Dai, S., Yue, L., Xu, F., Gu, J., Dai, X., & Qian, X. (2022). Emerging mechanisms progress of colorectal cancer liver metastasis. *Frontiers in Endocrinology, 13*, 1081585.

Zhi, Q., Wan, D., Ren, R., Xu, Z., Guo, X., Han, Y.,… Wang, Y. (2021). Circular RNA profiling identifies circ102049 as a key regulator of colorectal liver metastasis. *Mol Oncol, 15*(2), 623–641. doi:https://doi.org/10.1002/1878-0261.12840

Zhong, X.-Y., Yu, J.-H., Zhang, W.-G., Wang, Z.-D., Dong, Q., Tai, S.,… Li, H. (2012). MicroRNA-421 functions as an oncogenic miRNA in biliary tract cancer through down-regulating farnesoid X receptor expression. *Gene, 493*(1), 44–51.

Zhou, M.-H., Zhou, H.-W., Liu, M., & Sun, J.-Z. (2018). The role of miR-92b in cholangiocarcinoma patients. *The International Journal of Biological Markers, 33*(3), 293–300.

Zhou, Q., Luo, H., Wang, X., Li, P., Kong, H., & He, B. (2022). Downregulation of miR-210 Promoted Apoptosis of Hippocampal Neurons by Negatively Regulating the TLR4/NF-κB1 Signaling Pathway in a Rat Model of Status Epilepticus. *Neuropsychiatric Disease and Treatment*, 1763–1770.

Zhu, P., Liang, H., Huang, X., Zeng, Q., Liu, Y., Lv, J., & Ming, L. (2020). Circular RNA Hsa_circ_0004018 Inhibits Wnt/β-Catenin Signaling Pathway by Targeting microRNA-626/DKK3 in Hepatocellular Carcinoma. *Onco Targets Ther, 13*, 9351–9364. doi:https://doi.org/10.2147/ott.S254997

Zhu, X., Zhang, X., Qin, Y., Chen, Y., Feng, X., Deng, S.,… Peng, Y. (2025). Circular RNA circATM binds PARP1 to suppress Wnt/β-catenin signaling and induce cell cycle arrest in gastric cancer cells. *Journal of Advanced Research*. https://doi.org/10.1016/j.jare.2025.04.033

Zhuo, H., Tang, J., Lin, Z., Jiang, R., Zhang, X., Ji, J.,… Sun, B. (2016). The aberrant expression of MEG3 regulated by UHRF1 predicts the prognosis of hepatocellular carcinoma. *Molecular carcinogenesis, 55*(2), 209–219.

Zuo, Y., Chen, S., Yan, L., Hu, L., Bowler, S., Zitello, E.,… Deng, Y. (2022). Development of a tRNA-derived small RNA diagnostic and prognostic signature in liver cancer. *Genes & diseases, 9*(2), 393–400.

Chapter 6
The Epigenetic Angle in the Precision Medicine Era for Blood Disorder Advancements

Nadia M. Hamdy, Ahmed I. Abulsoud, Nehal I. Rizk, Tohada M. AL-Noshokaty, Mohamed Bakr Zaki, Ahmed S. Sultan, and Ahmed S. Doghish

Abstract This work examined the integration of epigenetics and precision medicine in the management of various blood disorders, including anemias, antiphospholipid syndrome, hemochromatosis, hemophilia, leukemia, lymphoma, multiple myeloma, porphyria, thalassemia, thrombocytopenia, thrombocytosis, polycythe-

---

The original version of the chapter has been revised. A correction to this chapter can be found at https://doi.org/10.1007/978-3-032-08530-6_9

N. M. Hamdy
Biochemistry and Molecular Biology Department, Faculty of Pharmacy,
Ain Shams University, Cairo, Egypt
e-mail: nadia_hamdy@pharma.asu.edu.eg

A. I. Abulsoud
Faculty of Pharmacy, Biochemistry Department, Heliopolis University, Cairo, Egypt

Biochemistry and Molecular Biology Department, Faculty of Pharmacy (Boys),
Al-Azhar University, Cairo, Egypt

N. I. Rizk
Department of Biochemistry, Faculty of Pharmacy and Drug Technology,
Egyptian Chinese University, Cairo, Egypt

T. M. AL-Noshokaty
Biochemistry and Molecular Biology Department, Faculty of Pharmacy (Boys),
Al-Azhar University, Cairo, Egypt

M. Bakr Zaki
Department of Biochemistry, Faculty of Pharmacy, University of Sadat City, Menoufia, Egypt

Department of Biochemistry, Faculty of Pharmacy, Menoufia National University, Menofia, Egypt

A. S. Sultan
Department of Oncology and Diagnostic Sciences, College of Dental Surgery,
University of Maryland Baltimore, Baltimore, MD, USA

Biochemistry Department, Faculty of Science, Alexandria University, Alexandria, Egypt

N. M. Hamdy et al. (eds.), *From (Epi)genetics-to-Precision (E2P)*, Subcellular Biochemistry 114, https://doi.org/10.1007/978-3-032-08530-6_6

mia, von Willebrand disease, and coagulopathy. It begins with an overview of key concepts and the significance of precision medicine in treating blood diseases, supported by current statistics. The role of noncoding RNAs (ncRNAs) is highlighted, detailing their mechanisms of action and clinical implications as potential biomarkers and therapeutic targets. Additionally, the chapter explores natural products used in personalized medicine, examining their sources, mechanisms, and successful case studies in blood disorders. A comprehensive review of recent clinical trials provides insights into the impact of innovative therapies and FDA approvals on treatment protocols, emphasizing the importance of combination therapies. Future directions address emerging research technologies such as clustered regularly interspaced short palindromic repeats (CRISPR) and ethical considerations surrounding genetic testing and patient consent. The synthesis of findings underscores the contributions of epigenetics and precision medicine to blood disease treatment, advocating for interdisciplinary research and ongoing education to enhance patient care and outcomes.

Keywords Epigenetics · PM · Blood disorders · Blood cancer · MicroRNAs · lncRNAs · Biomarkers · Leukemia · Multiple myeloma · Therapeutic targets

Abbreviations

AAV	Adeno-associated virus
AHA	Acquired hemophilia A
AML	Acute myeloid leukemia
APS	Antiphospholipid syndrome
AUC	Area under the curve
CLL	Chronic lymphocytic leukemia
CTCL	Cutaneous T cell lymphoma
CXC	C-X-C motif chemokine ligand
DLBCL	Large B cell lymphoma
DNMT	DNA methyltransferase
DNMTis	DNA methyltransferase inhibitors
DVT	Deep vein thrombosis
EKLF	Erythroid Krüppel-like factor
FcγRIIA	Fc gamma receptor Iia
FLT3	Fms-like tyrosine kinase 3
HbF	Fetal hemoglobin
HDAC	Histone deacetylases

A. S. Doghish (✉)
Biochemistry and Molecular Biology Department, Faculty of Pharmacy (Boys), Al-Azhar University, Cairo, Egypt

Department of Biochemistry, Faculty of Pharmacy, Badr University in Cairo (BUC), Cairo, Egypt
e-mail: ahmed_doghish@azhar.edu.eg

HDACis	Histone deacetylases inhibitors
HH	Hereditary hemochromatosis
HMAs	Hypomethylating agents
HMGA2	High mobility group AT-hook 2
HOTAIRM1	Hox antisense intergenic RNA myeloid 1
LIN28B	Lin-28 homolog B
lncRNA	Long noncoding RNA
MDS	Myelodysplastic syndromes
MEG3	Maternally expressed gene 3
miRNA	MicroRNA
MM	Multiple myeloma
mRNA	Messenger RNA
MYCBP	MYC-binding protein
ncRNAs	Noncoding RNAs
piRNA	Piwi-interacting RNA
PRC2	Polycomb repressive complex 2
PTCL	Peripheral T cell lymphoma
SCD	Sickle cell disease
TLR4	Toll-like receptor 4
TrxG/MLL complex	Trithorax-group/mixed-lineage leukemia complex

Introduction

A crucial area in comprehending the intricate processes underpinning numerous diseases is epigenetics, which is the study of heritable alterations in gene expression that do not comprise alterations to the underlying DNA sequence. We cannot emphasize enough its importance in precision medicine, especially in the field of hematology. The goal of precision medicine is to eliminate the cookie-cutter approach to healthcare by customizing medical interventions for each patient based on their unique set of symptoms, medical history, and personal preferences. A potential way forward for better blood problem diagnosis, treatment, and management is the use of epigenetic knowledge in precision medicine (Jones and Baylin 2007; Sashida and Iwama 2012; Johnson et al. 2021).

Many different diseases and ailments can impact the body's ability to make and use blood cells, and these disorders can cause serious health problems or even death. Among these disorders are anemias, antiphospholipid syndrome, hemochromatosis, hemophilia, leukemia, lymphoma, multiple myeloma, porphyria, thalassemia, thrombocytopenia, thrombocytosis, polycythemia, von Willebrand disease, and others related to blood clotting. Different approaches to treatment are necessary for each of these conditions because of the specific difficulties they cause. Environmental influences, genetic predispositions, and epigenetic changes, such as DNA methylation, histone modification, and noncoding RNA expression, all interact in a complicated way to shape the epigenetic landscape of these disorders (Pujadas and Feinberg 2012; Sashida and Iwama 2012; Goodell and Godley 2013).

The function of epigenetic changes in the etiology of blood diseases has recently been elucidated by epigenetic research (Eldash et al. 2023). Specific epigenetic alterations have been associated with disease development and therapeutic efficacy in certain cancers, such as leukemia and lymphoma. One factor that contributes to the malignant transformation is abnormal DNA methylation patterns, which can either activate oncogenes or mute tumor suppressor genes. The same holds true for disorders such as von Willebrand disease and hemophilia; new opportunities for gene therapy have emerged as a result of research into the epigenetic regulation of clotting factor genes (Goodell and Godley 2013; Swystun and James 2017; Sermer et al. 2019; Liu et al. 2020).

Epigenetic biomarkers also hold great promise for use in diagnosis and outcome prediction. Targeted medicines that target epigenetic changes are becoming more feasible as researchers work to find epigenetic patterns linked to particular blood illnesses. One example is the investigation of inhibitors of DNA methylation and histone deacetylation as potential treatments for a range of hematologic malignancies. Initial findings indicate that these agents may improve the effectiveness of current therapies (Walker et al. 2011; García-Giménez et al. 2017; Cruz-Rodriguez et al. 2018; Bewersdorf et al. 2019; Gambacorta et al. 2019; Evangelidis et al. 2024).

Epigenetics has the potential to revolutionize precision medicine for several blood illnesses, and this chapter will go into those implications. Anemias, antiphospholipid syndrome, hemochromatosis, hemophilia, leukemia, lymphoma, multiple myeloma, porphyria, thalassemia, thrombocytopenia, thrombocytosis, polycythemia, von Willebrand disease, and coagulopathy will be covered, along with the current knowledge of epigenetic mechanisms involved in their pathogenesis. We will also discuss new medicinal approaches that use epigenetic information to better serve patients. Our goal is to give a thorough overview of how epigenetics can change the game for precision medicine in blood disorders by connecting the dots between research on the subject and its practical use in the clinic.

Overview of (Epi)genetics and Precision in Blood Diseases

Genomic mechanisms that affect DNA templates rather than DNA sequence changes are called "epigenetics." Epigenetic modification controls protein complexes that write, read, or erase. This comprises DNA methylation and covalent histone modification. Epigenetic writers add unique chemical modifications to DNA or histones called markers. Readers understand methyl-CpG-binding domain proteins (MBPs) in modified proteins. Epigenetic markers are erased by chromatin enzymes. Epigenetic dysregulation is widespread in rare cancers. This involves faulty DNA and RNA methylation and posttranslational modification (Biswas and Rao 2018; Singh et al. 2022; Yu et al. 2024b).

Geneticist-tailored prevention and therapeutic methods account for individual heterogeneity based on biological reasons and other aspects in precision medicine. The ideal "precision medicine" would use clinicopathological markers and molecular profiling to create personalized diagnosis, prognosis, and treatment recommendations. Pharmacogenetics defines "precision medicine" as using molecular and clinical data to develop vaccines and drugs. Precision medicine seeks particular cures, not causes. The Human Genome Project data has improved tailored therapy. Doctors can choose better the patient's medications, when they have the patient's genetic profile (Hamburg and Collins 2010; Katsnelson 2013; Collins and Varmus 2015).

Epigenetic processes may enhance the transcriptional expression of oncogenes and activate oncogenic signaling pathways when risk factors for hematologic malignancies, such as metabolic syndrome and chronic inflammation, are present. Epigenetic regulators are crucial to normal hematopoiesis, according to recent experimental results. One example is the hematopoietic stem cell (HSC) development–promoting PRC1 member BMI1. This protein is essential for HSC maintenance efficiency and provides HSCs with resistance to oxidative stress by negatively regulating ROS signaling. Preventing HSC senescence also requires the activation of the PRC2 component EZH2. It was found that HSC formation, maintenance, proliferation, and differentiation are dependent on the HAT activity of the multidomain protein monocytic leukemia zinc finger (MOZ). Also, the hematopoietic transcription factor GATA-1 cannot do its job without the HAT CREB-binding protein (CBP). A new GATA-1, ZBP-89, physically binds to HAT p300 and plays an important role in erythroid differentiation, according to more recent research. This suggests that epigenetic processes work together for HSC maintenance and likely for the maintenance of LSCs derived from them as well, as PRC1 and 2 can recruit DNMT1 and PRC2 can recruit HDAC1 and 2 (Woo et al. 2011; Nakamura et al. 2012; Yang 2015; Nwabo Kamdje et al. 2024).

On the flip side, certain epigenetic regulators' (namely HATs) significance and functions in typical hematopoiesis remain debatable. The main source of the conflicting data came from loss-of-function studies, like the one by Kimbrel and colleagues. Their findings show that hematopoiesis is largely unaffected by inactivating mutations that affect HAT activity. What is more surprising is that even in mutants with active HAT enzymatic activity, significant defects in hematopoiesis were observed when CBP and p300 were nullizygous, indicating that these enzymes are unnecessary for this process. Therefore, we recognize the significance of investigating the function of epigenetics in the pathophysiology of blood diseases (Kimbrel et al. 2009; Memari et al. 2018).

Role of ncRNAs in Blood Diseases

Introduction to Noncoding RNAs

In 2001, the draft sequence of the human genome revealed fewer protein-coding genes than expected, indicating that only a small proportion of the genome encodes proteins. Despite exons comprising only 2% of the genome, around 75–80% is transcriptionally active (Nurk et al. 2022). Post-genomic studies have identified various noncoding RNA (ncRNAs) classes and their roles in cellular processes and diseases, including cancer and blood disorders. Noncoding RNAs are involved in mammalian development and hematopoietic cells. They contribute to inherited and acquired disorders of the hematopoietic system and offer potential therapeutic applications in blood diseases (Sherman and Salzberg 2020).

Types of Noncoding RNAs

MicroRNAs

MicroRNAs (miRNAs) are the best-studied ncRNAs. These single-stranded RNAs have a length of about 20 to 24 nucleotides and are important posttranscriptional regulators of gene expression. These small RNAs do not code for proteins but instead bind to their target mRNA through the 3'-untranslated region (Li et al. 2023). Then, the degradation of the target mRNA or the suppression of its translation is reached. The complex process of disease pathogenesis is often associated with the aberrant expression of a large number of oncogenic miRNAs, tumor suppressor miRNAs, and other tumor-specific microRNAs (Soni and Biswas 2021). Nearly 1500 miRNAs have been described so far, and during the last two decades, their potential prognostic or diagnostic role as well as therapeutic applications have been studied in a large number of diseases. As a result, miRNA-based agents are used in phase I and II clinical trials for the treatment of hematological malignancies (Roy et al. 2023). Due to the critical role of miRNAs in red and white blood cell differentiation, a promising consideration for miRNA-based agents arises. miRNAs regulate the expression of specific proteins by binding to mRNA in a precise manner (Yu et al. 2024a). Basically, it is necessary for the tumor-suppressing or oncogenic effects of miRNAs that they participate in the modulation of genes involved in critical biological processes, like cell cycle control, differentiation, and apoptosis. In addition, several studies highlight the role of miRNAs in the tumor microenvironment and during the metabolic reprogramming occurring in the tumor (Doghish et al. 2022).

miRNAs are small, evolutionarily well-conserved ncRNA genes. These genes, up to 22 nucleotides, regulate the transition from mRNA to protein. miRNAs account for about 3% of the genome's total estimated transcriptome and may target more than 60% of protein-encoding genes (Fromm et al. 2022). It is expected that, as potential regulators of mRNA metabolism, miRNAs will also play a significant role in the erythropoiesis and megakaryopoiesis processes. It is currently believed that miRNAs can play a role in the differentiation of hematopoietic stem cells into erythrocytes and megakaryocytes (Iaquinta et al. 2021). miRNAs often play the role of central regulators of multicellular hematopoiesis. Using some in vivo mouse models, the effect of miR-144 RNA on erythropoiesis was demonstrated (DeVeale et al. 2021). It was found that these mice lack endogenous miR-144/451 in their erythrocytes, resulting in anemia and erythropoiesis failure (Xu et al. 2020).

Long Noncoding RNAs

Long noncoding RNAs (lncRNAs) are transcripts longer than 200 nucleotides that do not encode proteins. These transcripts can be considered a new and historical class of lncRNA due to the unexpected discovery that most of the human genome is transcribed into RNA molecules that do not encode proteins (Palazzo and Koonin 2020). For a long time now, it was assumed that the vast majority of genetic information presented to mammalian cells was not present, even though the human and other genomes are transcribed into vast numbers of noncoding transcripts, which range from a handful of small RNAs, like ribosomal RNAs and tRNAs, by diverged classes of small RNAs such as miRNAs, to many species of medium long-range and cis-acting long noncoding RNAs. lncRNAs are an important class of coding molecules in eukaryotic genomes, but today, the understanding of their roles is provocative compared to the state of miRNAs (Paloviita and Vuoristo 2022). lncRNAs have been shown to participate in numerous genomic regulatory modes through their secondary and tertiary structures and their interacting proteins (Graf and Kretz 2020). These various architectural mechanisms of lncRNAs can be used to explain the growing number of papers on lncRNAs and the mechanism of action of molecular networks. Since they directly include phenotypes to receive outside information, the intergenic and intronic lncRNAs express important features in the specificity of cell types.

Mechanisms of Action of ncRNAs and Their Roles in Gene Regulation

Noncoding RNAs have been shown to be involved in many diseases. In general, ncRNAs can regulate target gene expression through different mechanisms. lncRNAs, which are important for the regulation of protein-coding gene expression, function at the epigenetic, transcriptional, and posttranscriptional levels. Short ncRNAs, including miRNA, have been used for the treatment of diseases by interfering with target gene expression. miRNA can be used to target genes and even regulate large gene networks (Pisignano and Ladomery 2021).

In this way, strong gene expression regulation activity is obtained. Such gene regulation is accomplished through different mechanisms. The most important mechanism to be mentioned is the RNA interference pathway, which includes the binding of target genes by complementary pairing of ncRNAs. Because ncRNAs can be modified by recombination, ncRNAs themselves can be mature derivatives of ribonucleoprotein (Braga et al. 2020; Deogharia and Gurha 2022).

miRNAs and lncRNAs are pivotal noncoding RNAs involved in gene regulation, particularly the suppression of gene expression at the posttranscriptional level. Both regulatory, small ncRNAs and lncRNAs transcribed in a context- (or cell-) specific manner orchestrate gene expression patterns essential for the maintenance of cellular identities, cell proliferation, and differentiation (Statello et al. 2021). Cellular functions are controlled by complex networks of macromolecules that manage a wide variety of processes, including cellular adhesion, division, transcription, migration, apoptosis, and cytoskeletal remodeling. miRNAs regulate gene expression at the posttranscriptional level through interaction with their target mRNAs, resulting in the suppression of gene expression and/or translational inhibition (Smutny et al. 2021).

Piwi-interacting RNAs (piRNAs) are a class of short, single-stranded ncRNAs, 26–31 nucleotides in length, originally discovered in association with PIWI proteins during spermatogenesis. piRNAs function predominantly by silencing transposons and other targets at the transcriptional level and play a key role in the biogenesis of mature male gametes. Recent studies, however, have identified piRNAs in various somatic tissues, highlighting their broader posttranscriptional regulatory roles in cellular processes. These findings suggest that piRNAs, like miRNAs and lncRNAs, contribute to gene regulatory networks beyond germline cells, particularly through their interaction with PIWI proteins (Abramowicz and Story 2020; Burgos et al. 2021).

Noncoding RNAs are an ancient RNA species, key to gene regulation and genetic complexity. They are transcribed by different RNA polymerases: RNA polymerase II produces microRNAs, enhancer ncRNAs, and initiating ncRNAs, while RNA polymerase III synthesizes structurally related ncRNAs, including 5S rRNA, tRNA, RNase P, Alu, 7SK, SINE ncRNAs, and Y RNA. Pol III genes are often found within ancient retroelement families linked to genetic syndromes (Bhatti et al. 2021). Additionally, ultraconserved nucleotides in transcribed LINE ncRNAs highlight evolutionary conservation across distant species, with polymorphisms in these regions contributing to genetic diseases (Momozawa and Mizukami 2021).

The human genome contains a rich diversity of ncRNA genes, including those transcribed from chromosomal regions associated with conserved sequences, such as exons, tRNAs, or ultraconserved regions. RNA polymerase I and III work alongside Pol II to transcribe ribosomal RNA (rRNA) genes, which are processed into mature rRNA molecules (Bhatti et al. 2021). Posttranscriptional regulation by miRNAs occurs after their maturation from pre-miRNAs, influencing processes like cell senescence (Zhang et al. 2021a). RNA and DNA transcriptional interplay is critical for diverse cellular functions, including antiviral responses, as seen in ncRNA-mediated protection against the West Nile virus (Cai et al. 2022).

ncRNAs in Hematopoiesis and Blood Disorders (Table 6.1)

miRNAs in Hematopoiesis and Blood Disorders

Multiple miRNAs are expressed in hematopoietic cells and are involved in the differentiation of specific cell types. They act as crucial fine-tuners of gene expression. Recent findings suggest that miRNA genes are involved in a number of

Table 6.1 Roles and mechanisms of noncoding RNAs in hematopoiesis and blood disorders

Type of ncRNA	Example (Ref.)	Role in hematopoiesis/blood disorders	Mechanism of action
miRNA	miR-144 (Ling et al. 2024)	Regulates erythropoiesis. Mice lacking miR-144/451 in erythrocytes exhibit anemia and erythropoiesis failure	Binds to target mRNAs, leading to degradation or translational suppression
	miR-21 (Gaál 2022)	Acts as a fine-tuner of gene expression in both normal and malignant hematopoietic cells	Negatively correlates with several mRNAs, potentially involved in blood tumor progression
	miR-125a (Ghafouri-Fard et al. 2021)	Highly expressed in AML; its knockout delays leukemogenesis	Modulates gene expression related to hematological malignancies
	miR-17-92 cluster (Bauer et al. 2020)	Critical for hematopoiesis and maintenance of hematopoietic stem cells; loss or overexpression leads to blood malignancies	Disrupts hematopoiesis by regulating multiple target genes involved in cell differentiation
	miR-142-3p (Huang et al. 2023)	Expressed in myelogenous cells; involved in hematopoietic regulation	Part of a cluster with miR-142-5p, influencing gene expression patterns in hematopoiesis
lncRNA	EKLF-AS (Xue et al. 2024)	Positively regulates EKLF mRNA abundance during erythropoiesis; associated with fetal liver erythropoiesis	Suppresses nonsense-mediated decay of EKLF mRNA, enhancing its stability and expression
	HOTAIRM1 (Chen et al. 2017a, b)	Important for granulocytic/macrophage lineage specification; absence leads to impaired myeloid differentiation and increased apoptosis	Interacts with transcriptional co-regulators (TrxG/MLL complex and PRC2) to regulate target gene transcription, particularly Hoxa1
	MEG3 (Sommerkamp, Renders et al. 2017)	Acts as a tumor suppressor and is involved in erythroid differentiation; knockdown affects hemoglobin synthesis	Binds to oncogenic miRNAs, suppresses Wnt/β-catenin signaling, and regulates splicing of differentiation-associated genes
	GATA1-regulated lncRNA erythroid pro-survival (Alvarez-Dominguez et al. 2014)	Involved in erythropoiesis with specific isoforms; regulates the survival of erythroid cells	Directly regulated by GATA1, influencing transcriptional activity in erythroid cells
piRNA	Various piRNAs (Abramowicz and Story 2020)	Play roles in silencing transposons and regulating gene expression; emerging evidence suggests involvement in hematopoietic regulation	Interact with PIWI proteins to silence transposons at the transcriptional level and regulate posttranscriptional processes

AML acute myeloid leukemia, *EKLF* erythroid Krüppel-like factor, *HOTAIRM1* Hox antisense intergenic RNA myeloid 1, *TrxG/MLL complex* trithorax-group/mixed-lineage leukemia complex, *MEG3* maternally expressed gene 3, *PRC2* polycomb repressive complex 2, *mRNA* messenger RNA, *miRNA* microRNA, *lncRNA* long noncoding RNA, *piRNA* Piwi-interacting RNA blood-related pathologies, such as leukemia, lymphoma, and other blood diseases. The discovery of the miRNA function contributes to understanding the molecular mechanisms involved and generating new methods for treating these diseases (Roy et al. 2022). The prostate-tissue-enriched miR-125b has recently been found in cells comprising the hematopoietic system (Peng et al. 2021).

The expression of miR-21 in normal and malignant hematopoietic cells shows that miR-21 is a fine-tuner of gene expression in both normal and malignant mouse and human hematopoietic cells. Its expression negatively correlates with the expression of several mRNAs that are targets of oncogenic miR-21 in humans (Gaál 2022). Further research is needed to determine whether the overexpression of miR-21 plays a role in the initiation, maintenance, or progression of a subset of blood tumors, as found in solid tumors. The hematopoietic miR-142-3p has also been found in myelogenous cells. It maps to chromosomal region 17q22, and a cluster of miR-142 genes exists with miR-142-5p, another miRNA that is expressed in the hematopoietic system, with the clustering of miR-142-3p (Galka-Marciniak et al. 2022; Huang et al. 2023).

miRNAs are key regulators of hematopoiesis and leukemogenesis. Among them, the miR-17-92 cluster likely represents one of the best-studied hematopoietic miRNA clusters that are critically underlying the physiology and homeostasis of hematopoiesis, maintenance of hematopoietic stem cells (HSCs), differentiation of common lymphoid progenitors, and B- and T lymphocyte lineage development. The loss of the miR-17-92 cluster or the ectopic overexpression of its principal oncogenic targets accelerates the development of different blood cell malignancies through the disruption of hematopoiesis. Besides the miR-17-92 cluster, a few other miRNAs have also been shown to have roles in hematological disorders, including miR-125a and miR-126. For example, miR-125a is highly expressed in acute myeloid leukemia (AML) and postpones leukemogenesis when knocked out in the AML-initiating cell population (Bauer et al. 2020; Ghafouri-Fard et al. 2021; Katsaraki et al. 2021).

lncRNAs in Hematopoiesis and Blood Disorders

Increasing evidence has revealed the regulatory roles of lncRNAs in hematopoiesis and their association with blood disorders (Wang et al. 2020). Furthermore, the functions of lncRNAs, including survival, proliferation, apoptosis, differentiation, migration, adhesion, and self-renewal programs, allow cells to adapt readily to their location and perform stable homeostasis (Sebastian-delaCruz et al. 2021). Nevertheless, only a small proportion of these have been characterized or investigated. Changes in the expression of various lncRNAs have been linked to different series of acquisitions of a diverse array of characteristics of hematological malignancies (Lodde et al. 2020). Most lncRNAs have been detected in many types of hematological malignancies, and some have been demonstrated to regulate biological behaviors, including the promotion or suppression of leukemia pathogenesis, that play pivotal roles in normal hematopoiesis (El Mesallamy et al. 2014; Arman and Möröy 2020). However, much still remains to be discovered. The essential roles of lncRNAs and their regulatory mechanisms in blood disorders are highlighted. This paves the way for future research on the roles of novel lncRNAs and their functions. The exploration and investigation of the lncRNA realm has witnessed an extraordinary expansion due to remarkable advancements facilitated by various state-of-the-art next-generation RNA-sequencing techniques (Alsaedy et al. 2022; Prabhakaran et al. 2024).

In reality, it is unequivocally documented that lncRNAs are intricately implicated in numerous cancer types, notably blood disorders. Furthermore, alterations in the expression levels of various lncRNAs have been inexorably linked to the multifold acquisition of characteristics emblematic of hematological malignancies (Al-Hawary et al. 2024). The ubiquitous presence of lncRNAs within the gamut of hematological malignancies has been widely observed, with some demonstrating a remarkable capacity to regulate essential biological behaviors. This includes the precise orchestration of leukemia pathogenesis, intricately woven into the fabric of normal hematopoiesis (Carrasco-Leon et al. 2021). Nevertheless, there are vast domains that remain untapped, waiting to be discovered, unraveling the enigma that shrouds lncRNAs. Traversing the scientific landscape, the emphasis is on the indispensable roles fulfilled by lncRNAs and the captivating regulatory mechanisms within blood disorders. The efforts resonate with a visionary tone, paving the path for future research exploring the roles played by novel lncRNAs and their manifold functions.

In the past decades, genetic analyses of genome-wide association studies and exome sequencing have been used to identify many candidate ncRNAs associated with diverse human complex traits or diseases through the identification of disease-associated selective sweeps in the intergenic and intronic regions of the human genome. Blood-related traits have been no exception, in which a large number of lncRNAs mediate a variety of biological processes and participate in a diverse array of diseases. An increasing number of studies have also revealed that noncoding regulatory mechanisms of ncRNAs are important in different biological contexts, particularly in the development of immune cells and blood lineage, by controlling gene expression at both transcriptional and posttranscriptional levels (Guan et al. 2022; Abubakar et al. 2024).

As with protein-coding genes and miRNAs, many lncRNAs have also been found to be important in hematopoiesis and the development of blood cells. Below are the examples of specific lncRNAs involved in various aspects of normal and pathological hematopoiesis and the mechanisms by which they control those processes (Ghafouri-Fard et al. 2021). One of the highly differentially expressed lncRNAs during erythropoiesis, GATA1-regulated lncRNA erythroid pro-survival, has two erythroid-specific isoforms. The long-terminal repeat of murine erythroleukemia virus, which contains a crippling point mutation, was first placed into intron 8 of EKLF in fibroblasts, and then the EKLF disruptor exon was targeted into the reconstituted viral hemizygous mice. In the resulting EKLF mutants, fetal liver erythropoiesis was blocked at the transition from the basophilic to the chromatin-globin stages (Alvarez-Dominguez et al. 2014).

A significant collateral phenotype of the EKLF megainternon mutant mice was the upregulation of multiple erythroid lncRNAs, most notably the EKLF antisense transcript. This lncRNA is antisense to EKLF and positively regulates EKLF mRNA abundance by suppressing nonsense-mediated decay. GATA1 directly binds to the promoter region and the upstream exonic enhancer of the EKLF-AS lncRNA, indicating its erythroid-specific upregulation. The RNA with both nuclear and cytoplasmic fractions bind GATA1 through its N-terminal ZnF region, and this interaction is downregulated during differentiation. Accumulation of Klf1 mRNA is integral to chromatin binding and erythropoiesis-inhibitory erythroid cytotoxic T lymphocyte lines. Moreover, high GATA1 expression and low miR-144 levels are required for the efficient upregulation of EKLF-AS during erythropoiesis (Drissen et al. 2005; Xue et al. 2024).

HOTAIRM1, which stands for Hox antisense intergenic RNA myeloid 1, is one of the first lncRNAs that have been functionally characterized. It is transcribed from an intergenic region between the Hoxa1 and Hoxa2 genes in mice, running antisense with respect to the Hoxa1 gene (Chen et al. 2017a, b). The locus between Hoxa1 and Hoxa2 shows a condensed chromatin structure with repressive histone marks under steady-state conditions. The upregulation of HOTAIRM1 accompanies changes in this chromatin, described as intergenic transcription of co-repressed genes (Xu et al. 2014). HOTAIRM1 interacts with transcriptional co-regulators called the TrxG/MLL complex and the PRC2 repressor complex, suggesting that the lncRNA may fine-tune target gene transcription through the recruitment and release of chromatin modifiers acting on those genes. One of those genes is Hoxa1, which later in the process becomes derepressed, gains activating histone marks, and becomes transcriptionally activated when myeloid cells differentiate (Wang and Dostie 2017).

It has been shown that HOTAIRM1 is important for granulocytic/macrophage lineage specification of hematopoietic stem/progenitor cells during the early stages of differentiation. When HOTAIRM1 is absent, the cells have less propensity to become myeloid or mature into granulocytic or macrophage lineages. They experience increased apoptosis and G1 cell cycle arrest. In mice, hematopoiesis was significantly impaired when HOTAIRM1 activity was suppressed during bone marrow transplantation. This finding suggests that HOTAIRM1 promotes myeloid cells not only under steady-state conditions but also when the system goes through the regeneration process. Finally, the expression of HOTAIRM1 was found to be upregulated during bone marrow regeneration in mice, showing a similar dynamic pattern as observed in human myeloid differentiation (Rea et al. 2020).

Maternally expressed gene 3 (MEG3) is an lncRNA that is highly conserved between humans and mice. It acts as a tumor suppressor gene in several kinds of cancer, and its loss of function frequently occurs due to epigenetic silencing in tumors (Zheng et al. 2018). MEG3 binds to and inactivates oncogenic microRNA, suppresses Wnt/β-catenin signaling, and hence inhibits multiple myeloma tumor growth and enhances sensitivity. So far, there have only been a few studies about the roles of MEG3 in hematopoiesis. MEG3 is upregulated during erythroid differentiation, and its knockdown attenuates erythroid differentiation and hemoglobin synthesis. MEG3 has been shown to be physically associated with multiple splicing regulatory proteins, including a well-known splicing factor that mediates the inclusion of exon 16 into the mature mRNA in differentiating cells, and the knockdown of MEG3 increased the exon 16 exclusion. This data suggests that MEG3 might function to regulate the splicing of multiple differentiation-associated genes, leading to proper erythroid differentiation (Sommerkamp et al. 2017; Sun et al. 2020b; Zhao et al. 2023).

The Role of lncRNAs and miRNAs as Prognostic and Diagnostic Markers in Blood Diseases

ncRNAs have drawn a lot of attention in the field of disease prognosis because of their potential as prognostic signs and diagnostic biomarkers. They are effective tools to monitor the progress, severity, and effect of therapies for different types of diseases, as they can control the regulation of several genes (Trzybulska et al. 2018; Terrinoni et al. 2019). ncRNAs are stable in bodily fluids like blood and urine; thus, they are great options as noninvasive prognostic and diagnostic indicators. Researchers may be able to anticipate the progress of a disease, its metastases, and its responses to treatment by assessing the gene expression levels of particular ncRNAs (Trzybulska et al. 2018). ncRNAs are very helpful in predicting treatment response and relapse. Their expression profiles can reveal which subjects are more prone to experience disease progression or who might respond to a specific therapy. Herein, we reviewed the current knowledge on the utility of ncRNAs as clinical markers, as shown in Table 6.2.

lncRNAs as Prognostic and Diagnostic Markers in Blood Diseases

In AML, the level of lncRNA KCNQ1OT1 was elevated in AML patients with low survival time, and it was markedly linked with severity (Jia et al. 2018). A group of seven lncRNAs, named NFE4, LINC-00989, LINC-02062, AC-006460.2, AL-353796.1, PSMB8-AS1, and AC-000120.1, was disclosed as prognostic biomarkers for AML. These lncRNAs had shown remarkably accurate prediction of 1-, 3-, and 5-year survival (AUC = 0.84, 0.80, and 0.89, respectively) (Zhu et al. 2023). Besides, the oncogenic lncRNA PANDAR was addressed in diagnosing and prognosing AML. The upregulated levels of PANDAR were linked with high bone marrow blasts and short overall survival (Yang et al. 2018). Wang and his colleagues developed a risk score based on the expression level of the lncRNA LRRC75A-AS1. The former score was described as a powerful predictor for prognosing AML (Wang et al. 2018b). Also, the diagnostic and prognostic performance of the lncRNA

Table 6.2 The role of lncRNAs and miRNAs as prognostic and diagnostic markers in blood diseases

ncRNA	Location	Disease	Regulation	Role	References
KCNQ1OT1	Chr: 11p15.5	AML	Up	Prognostic marker Low survival time, and it was markedly linked with risk grade	Jia et al. (2018)
PANDAR	Chr: 6p21.2		Up	Diagnostic and prognostic marker Linked with high bone marrow blasts and short overall survival	Yang et al. (2018)
CCDC26	Chr: 8q24.21		Up	Prognostic marker Linked with old age, anemic state, severity, and short survival	Chen et al. (2019)
LINC00265	Chr: 7p14.1		Down	Diagnostic marker AUC= 0.74, sensitivity = 85%, specificity = 64% Correlated with age	Xiao et al. (2022)
LINC00467	Chr: 1q32.3			Diagnostic marker AUC = 0.72, sensitivity = 100%, specificity = 50% Correlated with age and leucocytes	
UCA1	Chr: 19p13.12			Diagnostic marker AUC = 0.66, sensitivity = 90%, specificity = 50% Correlated with leucocytes	
SNHG1	Chr: 11q12.3		Up	Diagnostic marker AUC = 0.66, sensitivity = 95%, specificity = 49%	
ANRIL	Chr: 9p21.3		Up	Diagnostic and prognostic marker AUC = 0.886 Associated with a low survival rate and a bad prognosis	Tan et al. (2021)
SNHG7 SNHG12	Chr: 9q34.3 Chr: 1p35.3		Up	Prognostic marker Associated with survival time, disease-free survival time, and leucocytes	Shi et al. (2020)
HOTAIR	Chr: 12q13.13		Up	Prognostic marker Associated with a high number of leucocytes, low hemoglobin, and short survival time	Wu et al. (2015)

LINC00899	Chr: 22q13.31		Up	Diagnostic and prognostic marker AUC = 0.807 Associated with adverse outcomes and remission	Wang et al. (2018a)
HOXB-AS3	Chr: 17q21.32		Up	Prognostic marker Associated with a bad prognosis and low survival time	Huang et al. (2019)
PRAL	Chr: 17p13.1	MM	Down	Diagnostic and prognostic marker Associated with stage scoring and overall and disease-free survival	Xiao et al. (2018)
TUG1	Chr: 22q12.2		Up	Diagnostic and prognostic marker AUC = 0.79 Correlated with clinicopathological markers such as staging of the disease, β_2-microglobulin, total protein, albumin, globulin, and bone injury	Yin et al. (2019)
NEAT1	Chr: 11q13.1		Up	Diagnostic and prognostic marker Correlated with remission, overall survival, and disease-free time	Yu et al. (2020)

(continued)

ncRNA	Location	Disease	Regulation	Role	References
XIST	Chr: Xq13.2	APS	Up	Diagnostic marker AUC = 0.7 Association with other clinicopathological features	Abd-Elmawla et al. (2023)
MEG3	Chr: 14q32.2		Up	Prognostic marker Association with autoantibody titers, complications, and other inflammatory markers	Rashad et al. (2022)
lncDC	Chr: 17q23.1	Thrombocytopenia	Down	Diagnostic marker Specificity = 78% Inversely proportional to platelet levels	Jiang et al. (2023) and Hamdy et al. (2023)
THRIL	Chr: 12q24.31		Up	Diagnostic marker Specificity = 97%	Jiang et al. (2023)
NEAT1	Chr: 11q13.1		Up	Diagnostic marker Inversely proportional to platelet levels	Hamdy et al. (2023)
MALAT1	Chr: 11q13.1		Up	Diagnostic marker 100% specificity, 63% sensitivity	Ayoub et al. (2020)
MEG3	Chr: 14q32.2		Up	Prognostic marker Identifying bleeding (AUC = 0.8) and determining steroidal drug responses (AUC = 0.85)	Naguib et al. (2023)
GAS5	Chr: 1q25.1		Down	Diagnostic marker Sensitivity = 88.8% Associated with fast recovery	Hodeib et al. (2023)
RUNX1	Chr: 21q22.12		Up	Diagnostic marker Sensitivity = 77.7% Associated with fast recovery	
PVT1	Chr: 8q24.21		Down	Diagnostic marker	Yu et al. (2021)
IFNG-AS1	Chr: 12q15		Up	Diagnostic and prognostic marker Inversely associated with platelets and linked with treatment responsiveness, reflecting its prognostic value	Ali et al. (2022)
GAS5	Chr: 1q25.1				

miR-181a-3p	Chr: 1q32.1	AML	Up	Diagnostic and prognostic marker Associated with recovery and inversely proportional to NF-κB	Qiang et al. (2020)
miR-9-5p miR-155-5p	Chr: 1q22 Chr: 21q21.3		Up	Prognostic marker Associated with an adverse prognosis	Chuang et al. (2015)
miR-203	Chr: 14q32.33		Down	Prognostic marker Associated with a good prognostic outcome	
miR-335	Chr:		Up	Prognostic marker Associated with low survival time and adverse outcomes	Lin et al. (2015)
miR-100	Chr: 11q24.1		Up	Prognostic marker Associated with low survival time and a bad prognosis	Bai et al. (2012)
let-7b miR-18a	Chr: 22q13.31 Chr: 13q31.3	MM	Up	Prognostic marker Linked with a bad prognosis	Manier et al. (2017)
miR-125b-5p	Chr: 11q24.1		Up	Diagnostic and prognostic marker AUC = 0.95	Jiang et al. (2018)
miR-490-3p	Chr: 7q33			Diagnostic marker AUC = 0.86	
miR-203	Chr: Chr: 14q32.33		Up	Diagnostic marker	Gupta et al. (2019)
miR-4449	Chr: 4q12		Up	Diagnostic marker AUC = 0.88	Shen et al. (2017)
miR-30d	Chr: 8q24.22		Down	Diagnostic and prognostic marker	Zhu et al. (2018)
miR-483-5p	Chr: 11p15.5		Up	Diagnostic and prognostic marker Sensitivity = 58%, specificity = 90 % Correlated with disease severity and prognosis	Qu et al. (2014)
miR-20a	Chr: 13q31.3		Down	Diagnostic marker Sensitivity = 63%, specificity = 85%	
miR-19a miR-4254	Chr: 13q31.3 Chr: 1p35.2		Down	Diagnostic marker Sensitivity = 91.7 %, specificity = 90.5 %	Hao et al. (2015)

(continued)

Table 6.2 (continued)

ncRNA	Location	Disease	Regulation	Role	References
miR-21	Chr: 17q23.1	DLBCL	Up	Prognostic marker Associated with short overall survival	Zheng et al. (2016)
miR-224	Chr: Xq28		down		
miR-199a miR-146b-5p miR-224	Chr: 19p13.2 Chr: 10q24.32 Chr: Xq28		Down	Prognostic marker Disease-free survival	
miR-222	Chr: Xp11.3		Up		
miR-21 miR-127 miR-34a	Chr: 17q23.1 Chr: 14q32.2 Chr: 1p36.22		Down	Prognostic marker	Roehle et al. (2008)
miR-195 miR-let7g	Chr: 17p13.1 Chr: 3p21.2		Up	Prognostic marker	
miR-200c-3p	Chr: 12p13.31	Polycythemia	Down	Diagnostic marker	Wang et al. (2023)
miR-125a-5p	Chr: 19q13.41	Thrombocytopenia	Up	Prognostic marker Associated with responding to steroidal treatments	Naguib et al. (2023)
miR-106b-5p miR-200c-3p	Chr: 7q22.1 Chr: 12p13.31		Up	Diagnostic and prognostic marker Tracking the stages of the disease	Qian et al. (2018)
let-7a-5p	Chr: 9q22.32		Up	Diagnostic marker Specificity=96.8%, sensitivity=100%	Eltarouty et al. (2024)

AML acute myeloid leukemia, *APS* antiphospholipid syndrome, *AUC* area under the curve, *DLBCL* large B cell lymphoma, *MM* multiple myeloma

CCDC26 was evaluated by a preceding study. The level of CCDC26 in newly diagnosed and relapsed AML was elevated relative to the healthy group and linked with old age, anemic state, severity, and short survival (Chen et al. 2019).

Another signature of differently expressed lncRNAs was also suggested as a good indicator for identifying AML and tracking its progression, where low levels of CCDC26 and high levels of SNHG1 were observed in the exosomes of AML patients. Furthermore, the combination of these lncRNAs showed more powerful diagnostic performance (Xiao et al. 2022). The oncogenic lncRNA SMIM20-1 was correlated with a bad prognosis in AML (Liu et al. 2022a). Even more, the widely studied lncRNA ANRIL was studied in 178 AML patients who were newly diagnosed. ANRIL, which was markedly upregulated, could differentiate between AML patients and normal subjects (AUC=0.886). In addition, these high levels were markedly linked with low survival rates and a bad prognosis (Tan et al. 2021). The lncRNAs SNHG7 and SNHG12 were linked with overall survival time and disease-free survival time (Shi et al. 2020). Furthermore, Wu et al. revealed that high levels of HOTAIR were accompanied by a high number of leucocytes, low hemoglobin, and short survival time (Wu et al. 2015). Alongside, the upregulated LINC00899 was able to differentiate between AML and normal control (AUC=0.807) and was also associated with adverse outcomes. LINC00899 was markedly decreased in AML patients who were completely in remission (Wang et al. 2018a). The lncRNA HOXB-AS3 was also associated with a bad prognosis and low survival time due to its effects on modulating cell cycle progress (Huang et al. 2019). In another study, 66 lncRNAs were investigated in AML. The lncRNAs LINC01003, CTD-2234N14, RP1-137K24, and RP11-834C111 were addressed as potential prognostic markers, which may be attributed to their modulatory effect on mTOR, apoptotic mechanisms, and other immunological pathways (Zeng et al. 2020). The downregulation of MEG3 could also play a role in identifying AML (Yao et al. 2017).

Regarding MM, the lncRNA NR046683 has a close connection with its prognosis. NR046683 was expressed in different MM cell lines, as well as drug-resistant ones. Patients with varying MM subgroups and stages showed markedly varied levels of NR046683. Furthermore, NR046683 upregulation was tightly associated with β2-microglobulin and was also accompanied by some chromosomal abnormalities (Dong et al. 2019). Another study elucidated that the lncRNA PRAL could be valuable in diagnosing and predicting the clinical outcome, where it was markedly suppressed in MM cell lines, associated with stage scoring and overall and disease-free survival (Xiao et al. 2018). In addition, the lncRNA TUG1 was suggested as an innovative diagnostic marker for MM (AUC=0.79). It was correlated with clinicopathological markers as a staging of the disease, β_2-microglobulin, total protein, albumin, globulin, and bone injury, thus reflecting its prognostic value too (Yin et al. 2019). Consistently, the lncRNA NEAT1 was elevated in MM and demonstrated outstanding value in distinguishing normal donors from MM. NEAT1 was also correlated with remission, overall survival, and disease-free time (Yu et al. 2020).

In APS, the lncRNA XIST was able to distinguish between patients and normal subjects. Furthermore, it was able to differentiate between primary and secondary APS. These findings were based on its association with other clinicopathological features (Abd-Elmawla et al. 2023). In parallel, upregulated MEG3 was also a crucial prognostic biomarker in APS, based on its association with autoantibody titers, complications, and other biochemical inflammatory markers (Rashad et al. 2022). It is noteworthy that MEG3 has crucial impacts on other immunological diseases (Mehana et al. 2022).

In thrombocytopenia, diverse lncRNAs prevailed and were suggested as crucial biomarkers for diagnosis and prognosis. The low-expressed lncDC possessed an increased specificity of 78%, whereas the highly expressed THRIL possessed a specificity of 97%. In addition, their combination showed better performance than each lncRNA separately, where the AUC reached 0.86 (Jiang et al. 2023). The upregulated lncRNAs NEAT1 and lncDC were also highlighted in diagnosing thrombocytopenia, especially since their expression was inversely proportional to the platelet levels (Hamdy et al. 2023). In the same direction, MALAT1 and THRIL, which were elevated in thrombocytopenic children, exhibited 100% specificity, as well as 63% and 73% sensitivity, respectively. Moreover, their combination showed better performance than each lncRNA alone (Ayoub et al. 2020). According to the results of ROC analysis, the upregulated MEG3 could be able to differentiate thrombocytopenia, and MEG3 showed good performance in identifying bleeding, with AUC=0.80, and determining steroidal drug responses, with AUC=0.85 (Naguib et al. 2023). Two other lncRNAs, GAS5 and RUNX1, which were downregulated and upregulated, respectively, showed great performance in the prognosis of the disease. GAS5 and RUNX1 could recognize patients who accomplish disease recovery with a sensitivity of 88.8% and 77.7%, respectively (Hodeib et al. 2023). The lncRNA PVT1 was also addressed in diagnosing thrombocytopenia, where its levels were markedly depressed (Yu et al. 2021). In addition, the highly expressed IFNG-AS1 and GAS5 were suggested as indicators for thrombocytopenia. Moreover, their levels were inversely associated with platelets and linked with treatment responsiveness, reflecting their prognostic value (Ali et al. 2022).

In β-thalassemia, upregulated MALAT1 was detected at the earlier stages of diagnosis and was associated with splenomegaly. Both MALAT1 and MIAT were also elevated and linked with blood transfusion demand at young ages (Li et al. 2015b).

miRNAs as Prognostic and Diagnostic Markers in Blood Diseases

Due to its high degree of heterogeneity, AML requires precise classification of risks to provide patients with the best prospective predictions. MicroRNAs play important roles in the pathophysiology and prognosis of AML and have a wide range of biological functions; consequently, they could be valuable for prognostic purposes. Bousquet et al. revealed the potential of miR-181a-3p to diagnose AML patients. Furthermore, miR-181a-3p was associated with AML prognosis, where patients who completed the chemotherapeutic course had low levels of miR-181a-3p, and it was inversely proportional to NF-κB (Qiang et al. 2020). The highly expressed miR-9-5p and miR-155-5p in AML are linked with an adverse prognosis, while high levels of miR-203 exhibit good prognostic outcomes (Chuang et al. 2015). Another study on AML revealed that the oncomiR-335 was correlated with low survival time and adverse outcomes (Lin et al. 2015). Similarly, highly expressed miR-100 was accompanied by low survival time and a bad prognosis (Bai et al. 2012).

In MM, exosomes can be extracted from Peripheral Blood Mononuclear Cells (PBMC) and have been demonstrated as effective indicators of disease diagnosis and the prediction of prognosis. Accordingly, Manier and his colleagues demonstrated that the miRNAs let-7b and miR-18a that were extracted from exosomes were markedly linked with disease-free survival and overall survival time (Manier et al. 2017). In another report, the combined usage of the two miRNAs, miR-34a and let-7e, is capable of distinguishing MM patients, with a sensitivity and specificity of 80% and 86%, respectively. The report added that the downregulation of miR-744 and let-7e was accompanied by a bad prognosis and high mortality rate in MM (Kubiczkova et al. 2014). Jiang and his coworkers disclosed that miR-125b-5p and miR-490-3p could be innovative diagnostic biomarkers in MM, where they exhibited AUC= 0.95 and 0.86, respectively. Moreover, elevated miR-125b-5p levels were linked with an adverse prognosis (Jiang et al. 2018). The oncogenic miR-203 was also addressed as a noninvasive biomarker for diagnosing MM (Gupta et al. 2019).

Upregulated levels of miR-4449 were observed in MM patients, and the ROC analysis revealed its diagnostic performance, where the AUC was 0.885 (Shen et al. 2017). Zhu et al. revealed that miR-30d could be crucial for MM patients' diagnosis and therapy tracking, based on its modulatory effect on PI3K/AKT (Zhu et al. 2018). miR-483-5p and miR-20a possessed a high degree of diagnostic precision of MM, with a sensitivity of 58% and 63%, respectively, as well as a specificity of 90% and 85%, respectively. Furthermore, miR-483-5p was correlated with disease severity and prognosis (Qu et al. 2014). The ROC analysis disclosed that miR-19a and miR-4254 together were capable of differentiating MM (sensitivity= 91.7 % and specificity=90.5 %). Particularly, miR-19a is suggested as a prognostic indicator based on its correlation with overall survival and disease-free survival (Hao et al. 2015).

In large B cell lymphoma (DLBCL), short overall survival was associated with upregulated miR-21 and downregulated miR-224. Disease-free survival was lower in the case of downregulated miR-199a/b, miR-146b-5p, miR-224, and upregulated miR-222 (Zheng et al. 2016). In another study, a battery of miRNAs was associated with prognosis in DLBCL; some of them possess tumor-suppressing roles, such as miR-21, miR-127, and miR-34a, while others are oncogenic, such as miR-195 and miR-let7g (Roehle et al. 2008). In polycythemia, Wang et al. addressed the diagnostic utility of the highly expressed miR-122-5p, miR-423-5p, miR-4433b-3p, miR-1291, and miR-106b-5p as well as the low-expressed miR-200c-3p (Wang et al. 2023). In thrombocytopenia patients, the upregulated levels of miR-125a-5p are associated with responding to steroidal treatments (Naguib et al. 2023). Another study revealed that downregulated miR-130a is observed in thrombocytopenia (Cui et al. 2016). Qian et al. revealed that miR-106b-5p and miR-200c-3p are innovative biomarkers for tracking the stages of the disease (Qian et al. 2018). Eltarouty et al. found that the miRNA let-7a-5p, which was upregulated in thrombocytopenia, could be a promising diagnostic marker. let-7a-5p achieved a specificity of 96.8% and a sensitivity of 100% at values less than 18.5 (Eltarouty et al. 2024).

Therapeutic Interventions of ncRNAs in Blood Disorders

Understanding the various functions of ncRNAs has made it possible to investigate DNA, RNA, and proteins in novel approaches. For the treatment of illnesses, ncRNAs represent an intriguing new class of medicines, particularly when used against undruggable targets. Despite the promising anticipated effects of using ncRNAs against diverse diseases, several obstacles still need to be overcome, such as their immunogenic nature (Anwar et al. 2022), low penetration into cells, unwanted effects, and quick degradation and elimination. Significant changes to overcome these obstacles are presented by developing delivery carriers, introducing chemical changes, and using a rational design (Grillone et al. 2024; Nemeth et al. 2024).

Many features of ncRNA physiology make them very appealing as therapeutic goals, including accessibility to several proteins that were undruggable. RNA-based therapeutics utilize the diverse consequences of oligonucleotides, which bind to RNAs in a base-pairing way, which, in turn, could lead to specific RNA-degrading, alternate-splicing-redirecting, protein-producing, or impaired RNA repair (Mostafa et al. 2015). While most of these approaches have been established for coding mRNAs, almost all RNA kinds could be similarly targeted (Roberts et al. 2020) (Kole et al. 2012). The splicing oligonucleotide could be utilized to excise exons that are encoding crucial lncRNA functional domain, thus depicting oncogenic lncRNA disabled to perform their roles. Regarding the steric-blocking oligonucleotide approach, they prevent the interaction of the lncRNAs with their targets (Schmitt and Chang 2016). In spite of these approaches not being well-established, they markedly broaden the promising scope of using ncRNAs in the therapeutic management of diseases.

lncRNA-based treatments are now being developed primarily for two conditions: rare genetic diseases and cancer. There are primarily two strategies in use: knocking down the lncRNA from the NAT subtype and disrupting the lncRNA/PRC2 interaction (Khorkova et al. 2015). Also, small interfering RNAs (siRNAs), which range from 20 to 28 nucleotides, can cleave the mRNA precisely via a cytoplasmic pathway known as RNA interference (RNAi). Due to its special advantages, such as unique specificity, unlimited range of targets, and high efficiency, siRNA has emerged as a powerful tool for cancer therapeutic gene silencing. siRNA has become a potent tool for silencing the target genes because of its distinct specificity, limitless variety of targets, and great effectiveness. In addition, anti-miRNA antisense, which is also called antagomir, and miRNA-replacing therapy with synthetic miRNAs or miRNA mimics are innovative therapeutic tools in treating diverse diseases (Liu et al. 2008; Naidu et al. 2015).

Suppressing Overexpressed miRNAs

It is a therapeutic approach to regulate gene expression and treat diseases, particularly cancers and other disorders, by hindering miRNAs that are involved in the pathogenesis (Hamdy et al. 2024b). For instance, miR-126 was disclosed as one of the genes implicated in AML. Upregulated miR-126 was considered a regulator of stemness-associated genes in leukemia and was also linked with adverse outcomes. Earlier studies reported that depleting miR-126 by introducing anti-miR-126 in nanoparticulates disturbs AML-associated genes, suggesting its role as a new treatment plan (Dorrance et al. 2015).

In the same direction, Chen et al. reported that the implementation of anti-miR-9 inhibits the proliferation and viability of cancer cells and triggers programmed cell death, while stimulating the expression of miR-9 showed opposite effects (Chen et al. 2013). Another study focused on the oncogenic miR-22 in AML development and its clinical utility. Within genetically modified mice with high levels of miR-22, a marked disruption in the hematopoietic cell differentiation was revealed, which subsequently progressed to malignancy. On the contrary, suppressing miR-22 showed marked attenuation in the proliferation of these malignant cells. It is noteworthy that miR-22 suppresses the TET2 gene, which plays a crucial function in the DNA methylation process, and is thus crucial in controlling gene regulation and sustaining regular biological processes (Song et al. 2013). Consistently, Jiang et al. found that delivering anti-miR-22 in nanoparticles suppresses AML development by modulating genes such as Fms-like tyrosine kinase 3 (FLT3) and MYC-binding protein (MYCBP) (Jiang et al. 2016). Aberrations of MYCBP lead to the dysregulation of cell cycle progress and apoptosis (Li et al. 2019), whereas disturbed FLT3 affects the differentiation of the myeloid cells, as shown in Table 6.3 (Gilliland and Griffin 2002).

The oncogenic miR-125 was also raised as a potential factor in leukemia, and the sponging of miR-125 improves hemopoiesis. The suppression of miR-125 stimulates the expression of the lin28A gene, which is involved in stem cells' normal regulation (Chaudhuri et al. 2012). Knocking down miR-126 significantly increases leukemia cells' sensitivity to chemotherapeutic treatments (Li et al. 2015a). In addition, upregulated miR-126 was linked to lower survival rates, an increased likelihood of relapse, and genes associated with leukemic cell proliferation. Knocking down miR-126 using lentivirus decreased its clonogenic capability and abolished leukemic cells but, this time, without exhibiting comparable inhibitory effects on healthy bone marrow cells (De Leeuw et al. 2014). Zhang et al. found that miR-21-5p could be promising in treating AML, where its suppression inhibits cell cycle progression at different phases (Zhang et al. 2021b). Liu et al. demonstrated the utility

Table 6.3 Role of miRNAs in the therapeutic intervention of blood diseases

miRNAs	Disease	Regulation	Role in disease pathogenesis	Therapeutic intervention	References
miR-126	AML	Up	Regulator of stemness-associated genes Linked with adverse outcomes	Anti-miR-126 nanoparticles alter AML-associated genes	Dorrance et al. (2015)
miR-9			Promotes proliferation	Anti-miR-9 inhibits proliferation and viability; triggers apoptosis	Chen et al. (2013)
miR-22			Promotes disrupted hematopoietic cell differentiation progressed into leukemia Suppresses the TET2 gene Modulates FLT3and MYCBP	Anti-miR-22 nanoparticles suppress AML; attenuate proliferation	Song et al. (2013) and Jiang et al. (2016)
miR-125			Modulates hematopoiesis by regulating the Lin28A gene, which is involved in stem cells' normal regulation	Suppressed miR-125 stimulates Lin28A gene	Chaudhuri et al. (2012)
miR-126			Promotes genes associated with leukemia development, associated with relapse and low survival rates	Lenti-miR-126 decreased clonogenic capability and abolished leukemic cells Elevates sensitivity to treatments	Li et al. (2015a) and De Leeuw et al. (2014)
mir-196b			Blocks the differentiation of bone marrow progenitors Enhances proliferation	Anti-mir-196 promotes normal differentiation of bone marrow progenitors	Popovic et al. (2009)
miR-155	CLL		Promotes CLL and is linked with severe stages	Anti-miR-155 counteracts CLL progression	Cui et al. (2014)
miR-222-3p			Promotes proliferation Inhibits apoptosis Modulates Axin2-Wnt/β-catenin	Suppressing miR-222-3p decreases proliferation and triggers apoptosis	Liu et al. (2022b)
miR-148a	Thrombosis		Promotes thrombosis after platelet stimulation	Anti-miR-148a modulates platelet FcγRIIA signaling and decreases thrombosis	Zhou et al. (2015)
miR-374b-5p	Thrombosis		Promotes venous thrombosis	Anti-miR-374b-5p leads to elevation of interleukin-10 and promotes venous thrombi development	Zhang et al. (2020b)

miR-29a/b	AML	Down	Promotes irregular division of bone marrow blasts, hindering apoptosis	Synthetic miR-29a/b/c improves the differentiation of blast cells and the myeloid precursors and promotes the apoptotic genes	Gong et al. (2014) and Garzon et al. (2009)
miR-193a			Promotes proliferation Inhibits apoptosis Modulates the PTEN/PI3K axis	Synthetic miR-193a exerts antiproliferative actions and promotes apoptosis and granulocyte differentiation	Gao et al. (2011)
miR-125b	MM		Promotes MM pathogenesis	miR-125b mimic stimulates the tumor suppressor miR-34a, promotes apoptosis	Misso et al. (2019)
miR-9	AML		Promotes oncogenes LIN28B and HMGA2 – suppress apoptosis	Activated miR-9 by hypomethylation triggers apoptotic effects, alleviates proliferation, suppresses the oncogenic LIN28B and HMGA2	Mittal et al. (2019) and Emmrich et al. (2014)
miR-199b			Promotes myeloproliferation	Activating miR-199b promotes apoptosis	Favreau et al. (2015)
miR-143			Modulates autophagy genes ATG-7 and ATG-2B	Ectopic miR-143 improves the efficiency of cytarabine	Zhang et al. (2020a)
miR-497-5p	CLL			Coadministration of miR-497-5p and venetoclax revealed high effectiveness in preventing the progression of AML with little toxicity and chemoresistance	Pan et al. (2024)
miR-125-5p	Thrombocytopenia		Promotes CXCL13, which is implicated in immune cell trafficking and chronic inflammation	miR-125-5p mimics decreased CXCL-13	
miR-30c	Thrombosis		Promotes thrombosis	Lenti-miR-30c inhibits thrombosis by suppressing PAI-1	Luo et al. (2016)
miR-103a-3p				miR-103a-3p mimics suppress DVT through the repression of CXCL-12	
miR-495				miR-495 mimics mitigate thrombosis by inhibiting TLR-4 and IL-1 receptor	Tang et al. (2018)

AML acute myeloid leukemia, *CLL* chronic lymphocytic leukemia, *CXC* C-X-C motif chemokine ligand, *DVT* deep vein thrombosis, *FcγRIIA* Fc gamma receptor IIa, *FLT3* Fms-like tyrosine kinase 3, *HMGA2* high mobility group AT-hook 2, *LIN28B* Lin-28 homolog B, *MYCBP* MYC-binding protein, *TLR4* toll-like receptor 4 of the highly expressed miR-222-3p as a therapeutic target in AML by modulating Axin2 as well as Wnt/β-catenin (Liu et al. 2022b).

miR-155 is an oncogene that has been proposed as a possible therapeutic target and is elevated in a variety of tumor types, including leukemia (Schneider et al. 2016). Marcucci et al. reported that elevated miR-155 correlated with a low rate of recovery and short time of survival. Prospective treatment targeting of miR-155 is made possible by the abundance of novel drugs with an inhibitory action on miRNAs in the clinical setting (Marcucci et al. 2013). In bone marrow progenitors, the activation of mir-196b results in a partial blockade in differentiation as well as a rise in the proliferating ability and viability. However, transfection with anti-mir-196 exerts opposite effects (Popovic et al. 2009).

In chronic lymphocytic leukemia (CLL), miR-155 was elevated and linked with severe stages, whereas injection with anti-miR-155 counteracts CLL progression (Cui et al. 2014). In thrombosis, miR-148a-3p was disclosed as a thrombosis stimulator by modulating platelet T cell ubiquitin ligand-2 gene expression; thus, anti-miR-148a modulates platelet FcγRIIA signaling and decreases thrombosis (Zhou et al. 2015). In another study, the introduction of anti-miR-374b-5p leads to the elevation of interleukin (IL)-10 and promotes venous thrombi development (Zhang et al. 2020b).

Restoring Under-Expressed miRNAs

Another promising therapeutic approach for the management of diseases associated with downregulated or missed miRNAs is to restore these miRNAs. This could be achieved by using miRNA mimics, synthetic pri-miRNAs, nanoparticles, vectoring in lentivirus, and others (Kotowska-Zimmer et al. 2021).

Improper division, apoptotic suppression, and the differentiating abnormalities of blood cells are identified as crucial triggers of AML (Kamal et al. 2016). According to earlier research, miR-29a/b may have antitumor effects in leukemia pathogenesis. Gong et al. addressed the absence of miR-29a/b/c in PBMCs of AML subjects, leading to irregular division of bone marrow blasts, hindering apoptosis and alterations in the cell-differentiating mechanisms. On the flip side, the restoration of the miR-29 family by injecting synthetic miR-29a/b/c revealed a marked improvement in differentiating blast cells and the myeloid precursors, as well as promoting the apoptotic genes. The implication of the miR-29 family in AML could be attributed to its effect as upstream regulators of Akt2 and CCND2 (Gong et al. 2014). In the same direction, Garzon and his coworkers addressed that the reintroduction of *miR-29b mimics* exerts promoting effects on apoptosis in the AML experimental model (Garzon et al. 2009).

One of the miRNAs that possess antitumor activities is miR-193a. Injecting synthetic miR-193a exerts antiproliferative actions and promotes apoptosis and granulocyte differentiation. These findings demonstrate the critical function that methylated-suppressed miR-193a plays in myeloid leukemogenesis, as well as the curative value of elevating the miR-193a level in AML patients via c-kit-positive (Gao et al. 2011). Li et al. added that the silencing of miR-193a promotes AML by targeting the PTEN/PI3K axis (Li et al. 2013). In AML, the tumor-suppressing miR-9 was recognized to have essential roles in hematopoiesis and hematological diseases. Mittal and his coworkers suggested that the restimulation of miR-9 by hypomethylation and its forced expression triggered apoptotic effects and alleviated proliferative effects (Mittal et al. 2019). In addition, Emmrich et al. disclosed that the antitumor effects of miR-9 are attributed to the suppression of oncogenes LIN28B and HMGA2 (Emmrich et al. 2014).

Drugs that restore the tumor-suppressive miR-199b triggered apoptosis in AML, whereas miR-199b deletion can result in myeloproliferation (Favreau et al. 2015). Autophagy targeting as a cancer therapeutic approach has drawn a lot of interest (Atta et al. 2023). In AML, the ectopic expression of miR-143 improves the efficiency of cytarabine by targeting the autophagic genes ATG7 and ATG2B (Zhang et al. 2020a). Regarding the tumor-suppressing miR-497-5p in AML, its coadministration via liposomes with venetoclax revealed high effectiveness in preventing the progression of AML with little toxicity and less exposure to chemoresistance (Pan et al. 2024).

The most prevalent genetic abnormality in CLL is the deletion of miR-15 and miR-16, which encourages BCL2 and ROR1 expressions (Rassenti et al. 2017). Since miR-125b is widely expressed and often disrupted in a variety of cancerous growths, it has drawn particular attention in the field of cancer science. Depending on the type of tumor, it may exhibit oncosuppressive or oncogenic activity. Its antitumor function in MM has been previously established. The administration of miR-125b mimics stimulates the tumor suppressor miR-34a that subsequently modulates diverse signaling in MM (Misso et al. 2019).

A chemokine called CXCL13 is essential for immune cell function, especially for driving B cells to areas where immunological responses take place. CXCL13 is being researched as a possible target for therapeutic intervention in autoimmune illnesses and malignancies because of its crucial function in immune cell trafficking and chronic inflammation. CXCL13 inhibition may be able to lessen inflammation and stop immune cells from accumulating improperly. In immune thrombocytopenia, the transfection with miR-125-5p mimics decreased the CXCL13 level, which is implicated in the disease pathogenesis (Li et al. 2015b).

In thrombosis, the elevation of miR-30c using lentivirus suppresses PAI-1, thus inhibiting thrombosis (Luo et al. 2016). Moreover, introducing miR-338-5p mimics leads to suppressing interleukin-6 and hindering thrombosis incidence (Zhang et al. 2020c). Elevated levels of miR-103a-3p suppress DVT through the repression of CXCL12 (Sun et al. 2020a). The administration of miR-495 mimics encourages proliferation, along with suppressing the apoptosis of femoral vein endothelium, thus mitigating thrombosis by inhibiting TLR-4 and interleukin-1 receptor (Tang et al. 2018).

Suppressing Overexpressed lncRNAs

Ahmadi et al. found that the highly expressed MALAT1 could be a therapeutic target in treating CLL (Ahmadi et al. 2018). Hofmans et al. stated that the knockdown of highly expressed lncRNAs THADA-4, ACOT9-1, and NRIR could be innovative in treatment plans for juvenile myelomonocytic leukemia (Hofmans et al. 2021). In another report, BLACAT1, KCNQ1OT1, and HOTAIR were suggested as valuable therapeutic targets in AML, where it was elevated in AML patients and associated with different hallmarks of the disease (Wu et al. 2020) (Jia et al. 2018) (Wu et al. 2015). In addition, the upregulated levels of LINC01257 in AML rationalize its usage as a treatment target. Interestingly, the delivery of antiLINC01257 in lipid-nanoparticles was effective in reducing the level of LINC01257 within 48 hours of therapy, and the total number of cells decreased considerably by 55 %. On the other hand, the administration of anti-LINC01257 did not show any adverse effects in healthy controls, reflecting that the suppression of oncogenic lncRNAs is a safe avenue for treating AML (Connerty et al. 2021). Ma et al. revealed that knocking down CRNDE showed promising effects in hindering AML progression and elongated the survival time in the experimental model. The effects could be referred to the modulatory effect of CRNDE on miR-181 and the subsequent modulation of NOTCH-2 (Ma et al. 2020). Another research found that knocking down the lncRNA ZEB2-AS1 hinders the progression of AML via its modulatory effects on ZEB2/E--cadherin (Wang et al. 2018c).

The locked nucleic acid (LNA)-gapmeR antisense oligonucleotide technique was established in counteracting MALAT1 and NEAT1 as a therapeutic intervention in MM. Amodio et al. revealed that the introduction of anti-MALAT1 yields promising effects in counteracting MM. Repression of MALAT1 results in a marked suppression of two antioxidant genes: Nrf11 and Nrf2, which subsequently elevate the transcription of KEAP1. All of these events lead to the suppression of the antioxidant genes and the promotion of oxidative insults (Amodio et al. 2018). The targeting of NEAT1 also suppresses the proliferative ability of MM cells while promoting the apoptotic mechanisms. Furthermore, anti-NEAT1 augments the effect of diverse chemotherapeutics (Taiana et al. 2020).

Restoring Under-Expressed lncRNAs

The newly discovered lncRNA MAGI2-AS3 was reported in diverse diseases. In AML, it was identified as one of the oncosuppressive genes, and it was downregulated. However, lentivirus-forced expression of the former lncRNA impaired the progression of the leukemic cells, thus acting as an innovative alternative to enhance the management of AML (Chen et al. 2020). It is the same situation in NEAT1, which was markedly downregulated in AML, while its overexpression hinders the proliferation of leukemic cells and promotes its programmed death. These effects were suggested to be attributed to the regulatory role of NEAT1 on DNMT3A (Duan et al. 2020). The lncRNA HBS1L-MYB was highlighted as an essential factor in the therapeutic management of fetal hemoglobin aberrations and thus could be valuable in managing sickle cell anemia and β-thalassemia (Morrison et al. 2018).

Nevertheless, RNA-based therapies' progress must overcome many challenges to be implemented into clinical applications. One of the major concerns faced by oligonucleotide treatments is the efficient delivery of these molecules to specific cells. In an attempt to solve this concern, these oligonucleotides are chemically modified to enhance their performance. Additionally, they can be coupled to a range of carrier molecules, including nanoparticles, antibodies, conjugation with lipids, or others, that can improve transportation to the cells of interest, facilitate intracellular absorption, and decrease renal elimination from the bloodstream (Ling et al. 2013; Roberts et al. 2020).

The establishment of an RNA-based pharmacological framework will accelerate the development of the drugs. Thus, oligonucleotide-based therapies can be delivered against the disease-promoting ncRNAs, thereby broadening the range of druggable goals and paving the road for customized treatments and personalized therapeutics in treatment strategies.

Natural Products in Personalized Medicine

Natural products have been utilized for centuries in various cultures to support health and manage conditions, including those related to blood health. Natural products, derived from various biological sources, including plants, fungi, and marine organisms, have been utilized for centuries in traditional medicine systems across the globe. Their therapeutic potential is increasingly recognized in modern medicine, especially as adjuncts or alternatives to conventional pharmaceutical drugs. This chapter explores the sources, types, and applications of natural products in the treatment of blood diseases, emphasizing their role in personalized medicine. While the efficacy of these natural remedies varies, many have been studied for their potential benefits in supporting blood-related functions.

Sources of Natural Products

Natural products are primarily sourced from three main categories: plants, microorganisms, and animals.

Plants

Historically, plants have been the most significant source of natural products. They synthesize a wide array of secondary metabolites, such as alkaloids, flavonoids, terpenoids, and phenolics, which exhibit various biological activities. For example, paclitaxel, derived from the Pacific yew tree, is a well-known anticancer drug used to treat various cancers, including breast and ovarian cancers. Additionally, compounds like vincristine and vinblastine, extracted from the Madagascar periwinkle plant, are essential in treating leukemia and lymphoma (Chunarkar-Patil et al. 2024).

Microorganisms

The discovery of penicillin from the *Penicillium* fungus marked a revolutionary moment in medicine. Microorganisms continue to be a rich source of bioactive compounds. For instance, tetracycline and doxorubicin are antibiotics derived from soil bacteria that have played crucial roles in treating infections and cancer (Dzobo 2022; Jabeena et al. 2014).

Animals

Natural products can also be sourced from animals. For example, toxins from certain frogs have led to the development of potent analgesics that are significantly more effective than traditional opioids (Jabeena et al. 2014). These diverse sources contribute to a vast repository of bioactive compounds that can be harnessed for therapeutic purposes.

Types of Natural Products Used in Blood Diseases

Natural products encompass a variety of compounds that can be classified based on their pharmacological effects. In the context of blood diseases, several key types are noteworthy:

Flavonoids

These compounds are known for their anti-inflammatory and antioxidant properties. They can modulate immune responses and have been studied for their potential in managing conditions like anemia and clotting disorders. For instance, quercetin, found in many fruits and vegetables, has shown efficacy in reducing inflammation associated with blood disorders (Pillai and Antony 2024).

Terpenoids

Compounds like artemisinin, derived from the sweet wormwood plant, are crucial in treating malaria—a blood disease caused by parasites. Artemisinin's mechanism involves targeting the parasite's ability to metabolize hemoglobin, thus providing an effective treatment option (Jabeena et al. 2014).

Phenolic Compounds

These include substances like curcumin found in turmeric. Curcumin has been studied for its anti-inflammatory effects and potential role in managing conditions such as sickle cell disease (SCD) due to its ability to modulate hemoglobin levels (Chunarkar-Patil et al. 2024; Pillai and Antony 2024). The diversity of these natural products underscores their potential application across a range of blood-related conditions.

Mechanisms of Action

The therapeutic effects of natural products can be attributed to their multifaceted mechanisms of action. Many natural products exert their effects through immunomodulation—altering immune responses to improve outcomes in diseases such as leukemia or lymphoma. For example, curcumin has been shown to inhibit inflammatory cytokines like TNFα, which play a role in various blood disorders (Bulaj et al. 2016). This property not only helps reduce inflammation but also enhances the efficacy of conventional therapies when used concomitantly. Resveratrol, another potent natural product found in grapes and berries, has demonstrated anti-inflammatory effects that can benefit patients with blood disorders by improving endothelial function and reducing oxidative stress (Pillai and Antony 2024).

These mechanisms highlight how natural products can complement traditional therapies by enhancing patient outcomes through improved immune regulation, as shown in Table 6.4.

Examples

Turmeric (*Curcuma longa*)

Source and Composition: Turmeric is a flowering plant belonging to the ginger family, and its rhizome is widely used as a spice and medicinal herb. The primary active compound in turmeric is curcumin, which is responsible for its vibrant yellow color and numerous health benefits (Govindarajan and Stahl 1980).

Table 6.4 Different natural compounds' potential use in blood disorders

Natural compound	Source	Blood disorder	References
Curcumin	*Curcuma longa*	Leukemia management/ anticoagulant	Shanbhag (2017))
Gingerol	*Zingiber officinale*	Inhibits platelet aggregation	Omoregie et al. (2013)
Ginsenosides	*Panax ginseng*	Modulates immune responses in leukemia patients	Qi et al. (2020)
Allicin	*Allium sativum*	Inhibits platelet aggregation	Turner (1990), Jabeena et al. (2014), and Bulaj et al. (2016)
Omega-3 fatty acids	Natural oils	Decreases platelet aggregation	Superko et al. (2013)
Nitrates	*Beta vulgaris*	Improves blood flow	Babarykin et al. (2019) and Mirmiran et al. (2020)
Ginkgo biloba	Ginkgo tree	Anticoagulant	Erdinyler et al. (1996) and Savović et al. (2005)

Therapeutic Uses: Curcumin has been extensively studied for its anti-inflammatory, antioxidant, and anticancer properties. In the context of blood diseases, curcumin has shown promise in managing leukemia by potentially enhancing the effectiveness of conventional treatments. Research suggests that curcumin may help slow disease progression and reduce the side effects associated with chemotherapy (Hamdy et al. 2024a). Additionally, its anticoagulant properties can help prevent blood clot formation, making it beneficial for patients at risk of thrombosis (Shanbhag 2017). Curcumin has been shown to inhibit the AKT signaling pathway, which plays a crucial role in cell survival and proliferation. In AML cells, curcumin treatment leads to a dose-dependent suppression of AKT phosphorylation, thereby disrupting its activity. This inhibition results in decreased activation of downstream targets involved in cell growth and survival, such as PRAS40, 4E-BP1, and p70S6K. The inactivation of AKT is critical because its activation is associated with increased tumor progression and resistance to therapy (Dzobo 2022; Zhou et al. 2021a; Entezari et al. 2024). Curcumin also promotes apoptosis (programmed cell death) in leukemia cells through various mechanisms. It regulates the expression of apoptosis-related proteins, including Bcl-2 (an antiapoptotic protein) and cleaved-caspase-3 (a marker of apoptosis). By decreasing Bcl-2 levels and increasing cleaved-caspase-3 levels, curcumin effectively triggers the apoptotic cascade in AML cells (Zhou et al. 2021a). Curcumin has been found to induce cell cycle arrest at the G2/M phase, preventing leukemia cells from proliferating. This effect is mediated by alterations in cyclin-dependent kinases (CDKs) and cyclins, specifically through the upregulation of p21 (a cyclin-dependent kinase inhibitor) and the downregulation of cyclin D1 (Zhou et al. 2021a).

Ginger (*Zingiber officinale*)

Source and Composition: Ginger is a flowering plant whose rhizome is commonly used as a spice and herbal remedy. It contains bioactive compounds such as gingerol, which contribute to its medicinal properties (White 2007).

Therapeutic Uses: Ginger has been recognized for its anti-inflammatory and blood-thinning effects. It may help reduce the risk of clotting by inhibiting platelet aggregation, thus serving as a natural anticoagulant (Omoregie et al. 2013). Ginger contains bioactive compounds such as gingerol and shogaol, which have been shown to inhibit the activity of cyclooxygenase-1 (COX-1). COX-1 is an enzyme involved in the conversion of arachidonic acid (AA) into thromboxane A2 (TXA2), a potent promoter of platelet aggregation. By inhibiting COX-1, ginger reduces the production of TXA2, thereby diminishing platelet activation and aggregation. Alongside inhibiting COX-1, ginger reduces the formation of thromboxane B2 (TXB2), a stable metabolite of TXA2. This reduction further contributes to decreased platelet aggregation and activation. The inhibition of TXB2 formation is crucial because it directly correlates with reduced platelet responsiveness to aggregating agents (Nurtjahja-Tjendraputra et al. 2003; Liao et al. 2012). Ginger may also exert antiplatelet effects by inhibiting phospholipase A2, an enzyme that releases arachidonic acid from membrane phospholipids. By blocking this initial step in the AA cascade, ginger can prevent subsequent pathways that lead to platelet activation and aggregation. This action not only contributes to its antiplatelet properties but also suggests potential anti-inflammatory benefits (Marx et al. 2015). Various studies have shown that ginger extracts can directly inhibit platelet aggregation induced by different agonists, such as adenosine diphosphate (ADP), epinephrine, and collagen. For instance, clinical trials have reported that ginger significantly reduces platelet aggregation in response to these agonists, indicating its broad-spectrum antiplatelet activity (AlAskar et al. 2020).

Ginseng (*Panax ginseng*)

Source and Composition: Ginseng is a well-known herbal remedy derived from the roots of the *Panax* species of plants. It contains various active compounds called ginsenosides, which are believed to contribute to its health benefits (Kiefer and Pantuso 2003).

Therapeutic Uses: Ginseng has been traditionally used to boost energy levels and enhance immune function. Recent studies have explored its potential role in supporting patients with leukemia by modulating immune responses and possibly improving treatment outcomes when used alongside conventional therapies. However, more research is needed to establish definitive clinical benefits (Qi et al. 2020). Ginseng has been shown to enhance the proliferation of immune cells, including natural killer (NK) cells and monocytes. A study highlighted that fractions of North American ginseng could activate these immune cells in the bone marrow and spleen, potentially improving the body's ability to combat hematological malignancies, including leukemia. Moreover, ginsenosides, the active compounds in ginseng, have demonstrated the ability to induce apoptosis in various leukemia cell lines. For instance, studies have shown that ginsenosides can promote apoptosis in acute nonlymphocytic leukemia cells through mechanisms involving increased intracellular cyclic adenosine monophosphate (cAMP) levels and interferon induction (Azike 2014; (Radwan et al. 2016).

Garlic (*Allium sativum*)

Source and Composition: Garlic is a bulbous plant known for its culinary uses and medicinal properties. It contains several sulfur-containing compounds, including allicin, which are responsible for its health benefits (Milner 1996).

Therapeutic Uses: Garlic is recognized for its ability to reduce cholesterol levels and improve cardiovascular health. Its anticoagulant properties can inhibit platelet aggregation, making it useful for preventing blood clots. However, individuals with bleeding disorders or those undergoing surgery should exercise caution when using garlic supplements due to their potential to increase bleeding risks. Garlic inhibits platelet aggregation through the modulation of thromboxane A2 synthesis, reducing the risk of thrombosis. It also enhances fibrinolytic activity, promoting the breakdown of clots (Turner 1990; Jabeena et al. 2014; Bulaj et al. 2016). Garlic has been shown to inhibit the activity of cyclooxygenase (COX) enzymes, particularly COX-1. This inhibition reduces the formation of thromboxane A2 (TXA2), a potent promoter of platelet aggregation. By decreasing TXA2 levels, garlic effectively diminishes platelet activation and aggregation, which is crucial for preventing thrombotic events (Rahman 2007). Garlic also enhances the levels of cyclic adenosine monophosphate (cAMP) and cyclic guanosine monophosphate (cGMP) within platelets. Elevated cAMP and cGMP levels lead to decreased platelet activation and aggregation by promoting vasodilation and inhibiting platelet activation pathways. This effect is partly due to garlic's ability to activate nitric oxide synthase (NOS), leading to an increased production of nitric oxide (NO), which further enhances cGMP levels (Irfan et al. 2018).

Omega-3 Fatty Acids

Source and Composition: Omega-3 fatty acids are polyunsaturated fats found in high concentrations in fish oil, flaxseeds, chia seeds, and walnuts. The primary types include eicosapentaenoic acid (EPA) and docosahexaenoic acid (DHA) (Covington 2004).

Therapeutic Uses: Omega-3 fatty acids are known for their anti-inflammatory properties and cardiovascular benefits. They can help improve blood circulation and reduce the risk of clot formation by decreasing platelet aggregation. Additionally, omega-3s have been studied for their potential role in managing conditions such as anemia by improving overall blood health (Superko et al. 2013). Omega-3 fatty acids inhibit several signaling pathways involved in platelet activation. For example, they can reduce the activation of cyclooxygenase enzymes, which are responsible for converting AA into TXA2. Additionally, omega-3s may attenuate the activation of integrin αIIbβ3 (GPIIb/IIIa), a receptor critical for platelet aggregation and thrombus formation. This inhibition leads to decreased platelet adhesion and aggregation. Omega-3 fatty acids have been shown to influence various coagulation factors, including fibrinogen and plasminogen activator inhibitor-1 (PAI-1). A decrease in fibrinogen levels has been observed with omega-3 supplementation, which can contribute to reduced thrombus formation. Moreover, lowering PAI-1 levels can enhance fibrinolysis, further supporting the antithrombotic effects of omega-3s (Adili et al. 2018).

Beetroot (*Beta vulgaris*)

Source and Composition: Beetroot is a root vegetable rich in essential nutrients, including iron, folate, vitamin C, and nitrates (Ceclu and Nistor 2020).

Therapeutic Uses: Beetroots have been linked to improved blood flow due to their high nitrate content, which can enhance nitric oxide production in the body, leading to vasodilation and improved blood flow. This mechanism enhances exercise performance and lowers blood pressure. Moreover, this effect can aid in oxygen delivery to tissues and improve overall cardiovascular health. The iron content also supports hemoglobin production, making beetroots beneficial for individuals with anemia (Babarykin et al. 2019; Mirmiran et al. 2020).

Ginkgo biloba

Source and Composition: *Ginkgo biloba* is derived from the leaves of the ginkgo tree, one of the oldest living tree species. It contains flavonoids and terpenoids that contribute to its medicinal properties (Sierpina et al. 2003).

Therapeutic Uses: *Ginkgo biloba* has been used traditionally to enhance cognitive function and improve circulation. Some studies suggest that it may possess anticoagulant properties by inhibiting thrombin activity—an enzyme crucial for blood clotting—thereby reducing the risk of thrombosis. However, further research is needed to fully understand its efficacy in clinical settings (Erdinyler et al. 1996; Savović et al. 2005).

Personalized Medicine Approach

Personalized medicine aims to tailor medical treatments based on individual patient characteristics, including genetic makeup, lifestyle factors, and specific disease profiles. The incorporation of natural products into this paradigm is particularly promising.

By understanding individual responses to specific natural compounds—such as how different patients metabolize curcumin or resveratrol—clinicians can optimize treatment plans that enhance efficacy while minimizing side effects. Furthermore, lifestyle factors such as diet can influence the effectiveness of natural products. For instance, patients adhering to diets rich in flavonoids may experience better outcomes when treated with flavonoid-based therapies due to synergistic effects on health. The integration of natural products into personalized medicine not only enhances therapeutic options but also empowers patients by involving them in their treatment decisions (Alzeer 2023).

Challenges and Future Directions

Despite the promising potential of natural products in personalized medicine for blood diseases, several challenges remain (Dzobo 2022; Pillai and Antony 2024):

Standardization

The variability in the concentration and efficacy of active compounds within natural products poses significant challenges for clinical application. Establishing standardized extracts with consistent potency is crucial for ensuring reliable therapeutic outcomes.

Regulatory Approval

Navigating the regulatory landscape for natural products can be complex. Many natural compounds lack the extensive clinical trial data required for approval as pharmaceuticals. Increased collaboration between researchers and regulatory bodies is necessary to facilitate this process.

Sustainability

The sustainability of sourcing natural products is another concern. Overharvesting certain plants can lead to ecological imbalances. Therefore, sustainable practices must be adopted to ensure long-term availability without compromising biodiversity.

Future research should focus on overcoming these challenges by exploring innovative extraction methods, conducting rigorous clinical trials, and promoting sustainable harvesting practices.

Epigenetic Drugs and Their Impact on Treatment Protocol

A number of epigenetic medications have been approved by the FDA, expressly to treat blood illnesses, including myelodysplastic syndromes (MDS) and various hematologic malignancies, such as azacitidine, decitabine, romidepsin, vorinostat, Casgevy, Lyfgenia, and imetelstat.

Azacitidine, a hypomethylating drug that targets DNA methyltransferase (DNMT) enzymes, was approved in May 2004. It has demonstrated effectiveness in enhancing hematologic parameters, such as red blood cell counts and transfusion independence, and is used to treat MDS. According to clinical trials, about 51% of patients see an improvement in their red blood cell lineage; total response rates show that patients with high-risk MDS benefit greatly (Dai et al. 2024). For patients with acute myeloid leukemia (AML) who are not suitable for intense therapy, oral azacitidine, which was approved in September 2020, provides a practical substitute. Clinical studies have demonstrated that azacitidine enhances patients' quality of life by dramatically increasing overall survival and relapse-free survival while reducing severe adverse events (Dai et al. 2024).

Decitabine is another hypomethylating medication that was approved in June 2006 and is recommended for all MDS subtypes. About 17% of treated individuals have been demonstrated to experience responses, with a significant percentage of them obtaining full or partial responses. Usually given intravenously, the medication can cause myelosuppression, but its side effects are usually tolerable (Sasaki et al. 2022). INQOVI®, the oral combination of decitabine with the cytidine deaminase inhibitor, cedazuridine, is intended to increase the oral bioavailability of decitabine. Despite decitabine's effectiveness, when taken orally, it is quickly broken down by cytidine deaminase before entering the bloodstream. By addressing the drawbacks of intravenous administration, this formulation seeks to improve therapeutic efficacy and adherence (Dhillon 2020).

Romidepsin and vorinostat are also approved for use in treating some hematologic cancers, including cutaneous T cell lymphoma (CTCL). They work by inhibiting histone deacetylase (HDAC), which causes histones to become more acetylated and tumor suppressor genes to become active again (Moskowitz and Horwitz 2017).

Casgevy and Lyfgenia, two innovative gene treatments, were recently approved by the FDA to treat sickle cell disease (SCD) in patients 12 years of age and older. These treatments, which use cutting-edge genome-editing technologies—specifically, Clustered Regularly Interspaced Short Palindromic Repeats (CRISPR)/Cas9 for Casgevy and a lentiviral vector for Lyfgenia—represent the first cell-based gene therapies for sickle cell disease. By altering the patient's own blood stem cells to produce a hemoglobin type that lessens red blood cell sickling, both treatments seek to relieve symptoms and avoid disease-related consequences (Leonard and Tisdale 2024).

Imetelstat, indicated for people with transfusion-dependent anemia and low- to intermediate-1-risk MDS, has also received FDA approval. Being the first of its kind, this medication is noteworthy because it offers a novel mode of action by blocking telomerase, which may enhance the generation of red blood cells and lessen the need for transfusions. In order to improve the quality of life for patients who do not respond to conventional treatments, this approval is very crucial (Platzbecker et al. 2024).

Regarding future FDA approvals for epigenetic medications in the treatment of blood malignancies, there are a number of encouraging developments. At the MD Anderson Cancer Center, a new DNA hypomethylating drug, guadecitabine (SGI-110), created especially for patients with AML or high-risk MDS, is being studied in a clinical trial (NCT01261312) headed by Dr. Hagop Kantarjian and funded by Astex Pharmaceuticals (Garcia-Manero et al. 2019). If the medication shows notable efficacy and safety in these populations, the trial's findings may open the door for future FDA approvals.

Epigenetic alteration has a substantial impact on blood disease treatment regimens, opening up novel therapeutic options that may improve patient outcomes. The following are the main points about how epigenetic changes affect therapeutic approaches:

Reversibility and Targeting

Histone modifications and DNA methylation are examples of reversible epigenetic alterations. Without changing the underlying DNA sequence, this feature enables targeted therapeutic interventions that can change gene expression. Opportunities to reverse pathological states in blood illnesses, especially hematological malignancies like AML and MDS, are made possible by this reversibility (Wouters and Delwel 2016). Promising treatment approaches have resulted from the discovery of certain drugs that target epigenetic modifiers. Histone deacetylase inhibitors (HDACis) and DNA methyltransferase inhibitors (DNMTis), for example, have demonstrated clinical effectiveness in treating specific blood malignancies, improving overall survival and remission rates (Kelly et al. 2010).

Histone deacetylases, enzymes that remove acetyl groups from histones and non-histone proteins, are inhibited by HDACis. Increased acetylation from this inhibition results in a more flexible chromatin shape and improved gene expression. This process is essential for controlling genes that are frequently dysregulated in cancer cells and are involved in cell cycle progression, apoptosis, and differentiation (Tang et al. 2013). In addition, by encouraging the expression of proapoptotic molecules and suppressing antiapoptotic proteins, HDACis can cause cancer cells to undergo apoptosis and cell death. This action is especially helpful in treating blood malignancies, which are characterized by unchecked cell proliferation (Eckschlager et al. 2017). Patients with specific forms of anemia may benefit from HDACis because they can increase the expression of genes required for erythropoiesis (Kim et al. 2020).

DNA methylation inhibitors, particularly hypomethylating agents (HMAs) like azacitidine and decitabine, have shown significant effectiveness in treating MDS. These compounds block DNMT, which causes DNA to get demethylated and tumor suppressor genes that had been silenced to become active again. This procedure is essential in MDS, as aberrant hypermethylation plays a role in the development of the illness and hematologic abnormalities. HMAs can promote normal hematopoiesis in MDS patients by reversing DNA methylation and restoring the expression of genes that control cell proliferation and differentiation (Stomper et al. 2021). In high-risk MDS patients, azacitidine has been demonstrated to increase overall survival when compared to supportive treatment or traditional chemotherapy. Significant clinical responses have resulted from the treatment, including partial and total remissions in a subgroup of individuals (Kalinkova et al. 2022). DNMTis can undo abnormal methylation patterns that inhibit vital genes related to iron metabolism or erythropoiesis and, therefore, have a novel therapeutic target for anemias (Yang et al. 2024).

In hemophilia, epigenetic changes have an impact on the expression of genes linked to coagulation factors, which is essential for properly controlling hemophilia. DNA methylation has the ability to suppress gene expression, which impacts the synthesis of coagulation components, while histone protein alterations can modify chromatin structure, which affects the accessibility and expression levels of genes. Noncoding RNAs (ncRNAs) may also be involved in the pathophysiology of diseases such as acquired hemophilia A (AHA) and have involvement in the posttranscriptional regulation of gene expression (Coriu et al. 2022).

Hereditary hemochromatosis (HH), which is mainly brought on by mutations in the HFE gene that result in an excessive buildup of iron in the body, can be treated using epigenetic changes (Hamdy et al. 2025). Having a better understanding of these changes might assist in customizing treatment plans and enhancing patient results. Also, the expression of genes related to iron metabolism can be impacted by epigenetic modifications, including DNA methylation and histone modifications. For example, iron absorption and storage may be impacted by epigenetic changes that affect the genes that control hepcidin, a crucial hormone in iron homeostasis (Rametta et al. 2020).

Combination Therapies

Synergistic Approaches

Treatment results can be enhanced by the synergistic effects of combining epigenetic medications with traditional therapy. For instance, research has shown that patients with MDS and AML may experience higher clinical response rates when DNMTis are used in conjunction with cytotoxic treatments (Bennett and Licht 2018).

The cornerstone of traditional hemophilia gene therapy is the delivery of functional copies of the faulty genes that produce clotting factors (VIII for hemophilia A and IX for hemophilia B) into the liver cells using adeno-associated virus (AAV) vectors. By achieving these factors' long-term expression, this approach may lessen or perhaps completely eliminate the requirement for frequent factor replacement therapy (Miesbach et al. 2022). Unfortunately, the efficacy of AAV-mediated gene therapy may be limited by preexisting immunity to AAV capsids. By fostering tolerance to these vectors, epigenetic techniques may improve the therapeutic genes' long-term expression (Sternberg et al. 2024).

Epigenetic medications, such as HDACis and DNMTis (azacitidine and decitabine), can improve the efficacy of conventional anemic treatments. To increase the overall effectiveness of traditional medicines, such as erythropoiesis-stimulating drugs, for example, these agents can encourage the reexpression of genes that may be silenced in disease conditions (Kim et al. 2020).

Overcoming Resistance

Treatment resistance in blood malignancies is frequently caused by epigenetic changes. Clinicians may be able to improve patient responses by addressing these changes and regaining sensitivity to treatments that were previously ineffective (Wajapeyee and Gupta 2021). To increase the effectiveness of treatment, HDACis are frequently used in conjunction with other anticancer medications. For instance, they can boost results by making cancer cells more sensitive to radiation or chemotherapy. HDACis and conventional chemotherapeutic drugs can work in concert to increase overall response rates in hematological malignancies, according to clinical research (Zhou et al. 2021b). By focusing on several pathways implicated in the development of the disease, epigenetic medications like azacitidine and romidepsin can improve the efficacy of conventional therapies. In patients with refractory diseases, for example, research suggests that combining these drugs with chemotherapy can assist in overcoming resistance and increase overall response rates (Amengual 2020).

Personalized and Targeted Medicine

More individualized treatment plans tailored to the specific needs of each patient are made possible by our growing understanding of epigenetic mechanisms. Better-targeted treatments that take into account each patient's disease's distinct epigenetic landscape may result from this precision medicine approach (Rasool et al. 2015). Epigenetic therapies may be more appropriate for elderly individuals or those with particular risk factors for AML, according to research comparing hypomethylating drugs such as azacitidine and decitabine with conventional chemotherapy. These substances have been investigated as substitutes for harsh chemotherapy, which may be too harmful for weaker patients. Although research is ongoing, these medicines offer a promising alternative to existing medications in cases where they may not be effective in improving results for anemia patients (Dombret and Itzykson 2017).

New epigenetic targets that may be used therapeutically have been discovered recently. For example, modifying gamma-globin expression via epigenetic techniques has been suggested as a treatment approach for thalassemia and sickle cell disease (Venkatesan et al. 2020). The FDA has approved two HDACis, vorinostat and romidepsin, for the treatment of peripheral T cell lymphoma (PTCL) and CTCL. Patients with these diseases have demonstrated that these medications are effective in causing remission (Lu et al. 2023).

Epigenetic medications have demonstrated promise in raising fetal hemoglobin (HbF) levels, especially those that target DNMT1 and HDAC. This is important because elevated HbF can help patients with SCD. Although clinical trials have shown that inhibitors of these enzymes can result in significant increases in HbF and total hemoglobin levels, the potential adverse effects of HDACis' nonspecific activities make clinical application still challenging (Lavelle et al. 2018).

By increasing targeting, bioavailability, and controlled release, developments in nanomedicine are boosting the administration of epigenetic medications. Drug delivery methods based on nanocarriers are being developed to maximize therapeutic efficacy and reduce adverse effects. These methods have the potential to increase the efficacy of epigenetic medications by efficiently delivering them straight to the target cells (Li et al. 2024).

Without changing the DNA sequence itself, CRISPR/Cas9 and epigenetic editing enable targeted changes that can upregulate or downregulate particular genes. CRISPR/dCas9 systems, for instance, can alter epigenetic markers to boost the expression of clotting factors like factor VIII or IX (Bartova 2024).

Summary and Conclusion

The review discussed various studies highlighting the significance of microRNAs and lncRNAs as biomarkers and therapeutic targets in hematological malignancies, particularly in leukemia, multiple myeloma, and lymphomas, highlighting their significance as diagnostic and prognostic markers. Specific miRNAs, such as miR-125b, miR-497-5p, miR-30d, and miR-19a, are noted for their contributions to pathogenesis and therapeutic effectiveness, aiding in treatment decision-making. The function of these miRNAs emphasizes their potential in improving outcomes and understanding the molecular mechanisms underlying these conditions.

Various lncRNA signatures used in assessing patient outcomes and guiding therapeutic decisions were also mentioned in detail. Additionally, the exploration of epigenetic modifications as targets for treatment was a point of view, illustrating a shift toward tailored therapies based on molecular profiles.

References

Abd-Elmawla, M. A., Y. A. Elsabagh and N. M. Aborehab (2023). "Association of XIST/miRNA155/Gab2/TAK1 cascade with the pathogenesis of anti-phospholipid syndrome and its effect on cell adhesion molecules and inflammatory mediators." Scientific Reports **13**(1): 18790.

Abramowicz, A. and M. D. Story (2020). "The long and short of it: the emerging roles of non-coding RNA in small extracellular vesicles." Cancers **12**(6): 1445.

Abubakar, M., M. Hajjaj, Z. e. Z. Naqvi, H. Shanawaz, A. Naeem, S. S. N. Padakanti, C. Bellitieri, R. Ramar, F. Gandhi and A. Saleem (2024). "Non-coding RNA-mediated gene regulation in cardiovascular disorders: current insights and future directions." Journal of Cardiovascular Translational Research **17**(4): 739–767.

Adili, R., M. Hawley and M. Holinstat (2018). "Regulation of platelet function and thrombosis by omega-3 and omega-6 polyunsaturated fatty acids." Prostaglandins Other Lipid Mediat **139**: 10–18.

Ahmadi, A., S. Kaviani, M. Yaghmaie, H. Pashaiefar, M. Ahmadvand, M. Jalili, K. Alimoghaddam, M. Eslamijouybari and A. Ghavamzadeh (2018). "Altered expression of MALAT1 lncRNA in chronic lymphocytic leukemia patients, correlation with cytogenetic findings." Blood research **53**(4): 320.

Al-Hawary, S. I. S., R. O. Saleh, S. G. Taher, S. M. Ahmed, A. Hjazi, A. Yumashev, P. Ghildiyal, M. T. Qasim, A. Alawadi and A. Ihsan (2024). "Tumor-derived lncRNAs; Behind-the-scenes mediators that modulate the immune system and play a role in cancer pathogenesis." Pathology-Research and Practice: 155123.

AlAskar, A., N. A. Shaheen, A. H. Khan, N. AlGhasham, M. A. Mendoza, D. B. Matar, G. Gmati, M. AlJeraisy and A. AlSuhaibani (2020). "Effect of daily ginger consumption on platelet aggregation." Journal of Herbal Medicine **20**: 100316.

Ali, M. A., S. K. Hussein, A. A. Khalifa, A. M. El Amin Ali, M. S. Farhan, A. A. Ibrahim Amin and E. A. Mohamed (2022). "The Ifng antisense RNA 1 (IFNG-AS1) and growth arrest-specific transcript 5 (GAS5) are novel diagnostic and prognostic markers involved in childhood ITP." Frontiers in Molecular Biosciences **9**: 1007347.

Alsaedy, H. K., A. R. Mirzaei and R. A. H. Alhashimi (2022). "Investigating the structure and function of Long Non-Coding RNA (LncRNA) and its role in cancer." Cellular, Molecular Biomedical Reports **2**(4): 245–253.

Alvarez-Dominguez, J. R., W. Hu, B. Yuan, J. Shi, S. S. Park, A. A. Gromatzky, A. v. Oudenaarden and H. F. Lodish (2014). "Global discovery of erythroid long noncoding RNAs reveals novel regulators of red cell maturation." Blood, The Journal of the American Society of Hematology **123**(4): 570–581.

Alzeer, J. (2023). "Integrating medicine with lifestyle for personalized and holistic healthcare." Journal of Public Health and Emergency **7**.

Amengual, J. E. (2020). "Can we use epigenetics to prime chemoresistant lymphomas?" Hematology Am Soc Hematol Educ Program **2020**(1): 85–94.

Amodio, N., M. A. Stamato, G. Juli, E. Morelli, M. Fulciniti, M. Manzoni, E. Taiana, L. Agnelli, M. E. G. Cantafio and E. Romeo (2018). "Drugging the lncRNA MALAT1 via LNA gapmeR ASO inhibits gene expression of proteasome subunits and triggers anti-multiple myeloma activity." Leukemia **32**(9): 1948–1957.

Anwar, M. M., C. Albanese, N. M. Hamdy and A. S. Sultan (2022). "Rise of the natural red pigment 'prodigiosin' as an immunomodulator in cancer." Cancer Cell International **22**(1): 419.

Arman, K. and T. Möröy (2020). "Crosstalk between MYC and lncRNAs in hematological malignancies." Frontiers in Oncology **10**: 579940.

Atta, H., N. Alzahaby, N. M. Hamdy, S. H. Emam, A. Sonousi and L. Ziko (2023). "New trends in synthetic drugs and natural products targeting 20S proteasomes in cancers." Bioorganic Chemistry **133**: 106427.

Ayoub, S. E., E. M. Hefzy, R. G. Abd El-Hmid, N. A. Ahmed, A. A. Khalefa, D. Y. Ali and M. A. Ali (2020). "Analysis of the expression profile of long non-coding RNAs MALAT1 and THRIL in children with immune thrombocytopenia." IUBMB life **72**(9): 1941–1950.

Azike, C. G. (2014). American ginseng modulation of immune function and phytochemical analysis, The University of Western Ontario (Canada).

Babarykin, D., G. Smirnova, I. Pundinsh, S. Vasiljeva, G. Krumina and V. Agejchenko (2019). "Red beet (Beta vulgaris) impact on human health." Journal of biosciences and medicines **7**(3): 61–79.

Bai, J., A. Guo, Z. Hong and W. Kuai (2012). "Upregulation of microRNA-100 predicts poor prognosis in patients with pediatric acute myeloid leukemia." OncoTargets and therapy: 213-219.

Bartova, E. (2024). "Epigenetic and gene therapy in human and veterinary medicine." Environ Epigenet **10**(1): dvae006.

Bauer, M., C. Vaxevanis, N. Heimer, H. K. Al-Ali, N. Jaekel, M. Bachmann, C. Wickenhauser and B. Seliger (2020). "Expression, regulation and function of microRNA as important players in the transition of MDS to secondary AML and their cross talk to RNA-binding proteins." International Journal of Molecular Sciences **21**(19): 7140.

Bennett, R. L. and J. D. Licht (2018). "Targeting Epigenetics in Cancer." Annu Rev Pharmacol Toxicol **58**: 187–207.

Bewersdorf, J. P., R. Shallis, M. Stahl and A. M. Zeidan (2019). "Epigenetic therapy combinations in acute myeloid leukemia: what are the options?" Ther Adv Hematol **10**: 2040620718816698.

Bhatti, G. K., N. Khullar, I. S. Sidhu, U. S. Navik, A. P. Reddy, P. H. Reddy and J. S. Bhatti (2021). "Emerging role of non-coding RNA in health and disease." Metabolic brain disease **36**: 1119–1134.

Biswas, S. and C. M. Rao (2018). "Epigenetic tools (The Writers, The Readers and The Erasers) and their implications in cancer therapy." European Journal of Pharmacology **837**: 8–24.

Braga, E. A., M. V. Fridman, A. A. Moscovtsev, E. A. Filippova, A. A. Dmitriev and N. E. Kushlinskii (2020). "LncRNAs in ovarian cancer progression, metastasis, and main pathways: ceRNA and alternative mechanisms." International journal of molecular sciences **21**(22): 8855.

Bulaj, G., M. M. Ahern, A. Kuhn, Z. S. Judkins, R. C. Bowen and Y. Chen (2016). "Incorporating Natural Products, Pharmaceutical Drugs, Self-Care and Digital/Mobile Health Technologies into Molecular-Behavioral Combination Therapies for Chronic Diseases." Curr Clin Pharmacol **11**(2): 128–145.

Burgos, M., A. Hurtado, R. Jiménez and F. J. Barrionuevo (2021). "Non-coding RNAs: lncRNAs, miRNAs, and piRNAs in sexual development." Sexual Development **15**(5–6): 335–350.

Cai, L., J. Xuan, Q. Lin, J. Wang, S. Liu, F. Xie, L. Zheng, B. Li, L. Qu and J. Yang (2022). "Pol3Base: a resource for decoding the interactome, expression, evolution, epitranscriptome and disease variations of Pol III-transcribed ncRNAs." Nucleic acids research **50**(D1): D279-D286.

Carrasco-Leon, A., A. Amundarain, N. Gomez-Echarte, F. Prosper and X. Agirre (2021). "The role of lncRNAs in the pathobiology and clinical behavior of multiple myeloma." Cancers **13**(8): 1976.

Ceclu, L. and O.-V. Nistor (2020). "Red beetroot: Composition and health effects—A review." J. Nutr. Med. Diet Care **6**(1): 1-9.

Chaudhuri, A. A., A. Y.-L. So, A. Mehta, A. Minisandram, N. Sinha, V. D. Jonsson, D. S. Rao, R. M. O'Connell and D. Baltimore (2012). "Oncomir miR-125b regulates hematopoiesis by targeting the gene Lin28A." Proceedings of the National Academy of Sciences **109**(11): 4233–4238.

Chen, C., P. Wang, W. Mo, Y. Zhang, W. Zhou, T. Deng, M. Zhou, X. Chen, S. Wang and C. Wang (2019). "lncRNA-CCDC26, as a novel biomarker, predicts prognosis in acute myeloid leukemia." Oncology letters **18**(3): 2203–2211.

Chen, L., X. Fan, J. Zhu, X. Chen, Y. Liu and H. Zhou (2020). "LncRNA MAGI2-AS3 inhibits the self-renewal of leukaemic stem cells by promoting TET2-dependent DNA demethylation of the LRIG1 promoter in acute myeloid leukaemia." RNA biology **17**(6): 784–793.

Chen, P., C. Price, Z. Li, Y. Li, D. Cao, A. Wiley, C. He, S. Gurbuxani, R. B. Kunjamma and H. Huang (2013). "miR-9 is an essential oncogenic microRNA specifically overexpressed in mixed lineage leukemia–rearranged leukemia." Proceedings of the National Academy of Sciences **110**(28): 11511–11516.

Chen, Z.-H., W.-T. Wang, W. Huang, K. Fang, Y.-M. Sun, S.-R. Liu, X.-Q. Luo and Y.-Q. Chen (2017a). "The lncRNA HOTAIRM1 regulates the degradation of PML-RARA oncoprotein and myeloid cell differentiation by enhancing the autophagy pathway." Cell Death & Differentiation **24**(2): 212–224.

Chen, Z.-H., W.-T. Wang, W. Huang, K. Fang, Y.-M. Sun, S.-R. Liu, X.-Q. Luo and Y.-Q. Chen (2017b). "The lncRNA HOTAIRM1 regulates the degradation of PML-RARA oncoprotein and myeloid cell differentiation by enhancing the autophagy pathway." Cell Death and Differentiation **24**(2): 212–224.

Chuang, M., Y. Chiu, W. Chou, H. Hou, E. Chuang and H. Tien (2015). "A 3-microRNA scoring system for prognostication in de novo acute myeloid leukemia patients." Leukemia **29**(5): 1051–1059.

Chunarkar-Patil, P., M. Kaleem, R. Mishra, S. Ray, A. Ahmad, D. Verma, S. Bhayye, R. Dubey, H. N. Singh and S. Kumar (2024). "Anticancer Drug Discovery Based on Natural Products: From Computational Approaches to Clinical Studies." Biomedicines **12**(1): 201.

Collins, F. S. and H. Varmus (2015). "A new initiative on precision medicine." N Engl J Med **372**(9): 793–795.

Connerty, P., E. Moles, C. E. de Bock, N. Jayatilleke, J. L. Smith, S. Meshinchi, C. Mayoh, M. Kavallaris and R. B. Lock (2021). "Development of siRNA-loaded lipid nanoparticles targeting long non-coding RNA LINC01257 as a novel and safe therapeutic approach for t (8; 21) pediatric acute myeloid leukemia." Pharmaceutics **13**(10): 1681.

Coriu, D., N. Dirzu, A. A. Zimta, B. C. Hambley, X. Liu, L. Nemes, G. Ghiaur, J. Mahlangu and C. Tomuleasa (2022). "Editorial: Modern Approaches to Hemophilia Management: Gene Therapy and Beyond." Front Med (Lausanne) **9**: 859710.

Covington, M. B. (2004). "Omega-3 fatty acids." American family physician **70**(1): 133–140.

Cruz-Rodriguez, N., A. L. Combita and J. Zabaleta (2018). "Epigenetics in Hematological Malignancies." Methods Mol Biol **1856**: 87–101.

Cui, B., L. Chen, S. Zhang, M. Mraz, J.-F. Fecteau, J. Yu, E. M. Ghia, L. Zhang, L. Bao and L. Z. Rassenti (2014). "MicroRNA-155 influences B-cell receptor signaling and associates with aggressive disease in chronic lymphocytic leukemia." Blood, The Journal of the American Society of Hematology **124**(4): 546–554.

Cui, Y.-L., B. Wang, H.-M. Gao, Y.-H. Xing, J. Li, H.-J. Li, Z. Lin and Y.-Q. Wang (2016). "Interleukin-18 and miR-130a in severe sepsis patients with thrombocytopenia." Patient preference and adherence: 313–319.

Dai, W., X. Qiao, Y. Fang, R. Guo, P. Bai, S. Liu, T. Li, Y. Jiang, S. Wei, Z. Na, X. Xiao and D. Li (2024). "Epigenetics-targeted drugs: current paradigms and future challenges." Signal Transduct Target Ther **9**(1): 332.

De Leeuw, D. C., F. Denkers, M. C. Olthof, A. P. Rutten, W. Pouwels, G. Jan Schuurhuis, G. J. Ossenkoppele and L. Smit (2014). "Attenuation of microRNA-126 expression that drives CD34+ 38− stem/progenitor cells in acute myeloid leukemia leads to tumor eradication." Cancer research **74**(7): 2094–2105.

Deogharia, M. and P. Gurha (2022). "The "guiding" principles of noncoding RNA function." Wiley Interdisciplinary Reviews: RNA **13**(4): e1704.

DeVeale, B., J. Swindlehurst-Chan and R. Blelloch (2021). "The roles of microRNAs in mouse development." Nature Reviews Genetics **22**(5): 307–323.

Dhillon, S. (2020). "Decitabine/Cedazuridine: First Approval." Drugs **80**(13): 1373–1378.

Doghish, A. S., A. Ismail, H. A. El-Mahdy, M. A. Elkady, M. A. Elrebehy and A.-A. M. Sallam (2022). "A review of the biological role of miRNAs in prostate cancer suppression and progression." International journal of biological macromolecules **197**: 141–156.

Dombret, H. and R. Itzykson (2017). "How and when to decide between epigenetic therapy and chemotherapy in patients with AML." Hematology Am Soc Hematol Educ Program **2017**(1): 45–53.

Dong, H., S. Jiang, Y. Fu, Y. Luo, R. Gui and J. Liu (2019). "Upregulation of lncRNA NR_046683 serves as a prognostic biomarker and potential drug target for multiple myeloma." Frontiers in Pharmacology **10**: 45.

Dorrance, A. M., P. Neviani, G. J. Ferenchak, X. Huang, D. Nicolet, K. S. Maharry, H. G. Ozer, P. Hoellarbauer, J. Khalife and E. B. Hill (2015). "Targeting leukemia stem cells in vivo with antagomiR-126 nanoparticles in acute myeloid leukemia." Leukemia **29**(11): 2143–2153.

Drissen, R., M. von Lindern, A. Kolbus, S. Driegen, P. Steinlein, H. Beug, F. Grosveld, S. J. M. Philipsen and c. biology (2005). "The erythroid phenotype of EKLF-null mice: defects in hemoglobin metabolism and membrane stability." Molecular and cellular biology **25**(12): 5205–5214.

Duan, M.-Y., M. Li, H. Tian, G. Tang, Y.-C. Yang and N.-C. Peng (2020). "Down-regulation of lncRNA NEAT1 regulated by miR-194-5p/DNMT3A facilitates acute myeloid leukemia." Blood Cells, Molecules, and Diseases **82**: 102417.

Dzobo, K. The Role of Natural Products as Sources of Therapeutic Agents for Innovative Drug Discovery, Comprehensive Pharmacology. 2022:408–22. doi: https://doi.org/10.1016/B978-0-12-820472-6.00041-4. Epub 2022 Jun 9.

Eckschlager, T., J. Plch, M. Stiborova and J. Hrabeta (2017). "Histone Deacetylase Inhibitors as Anticancer Drugs." Int J Mol Sci **18**(7).

El Mesallamy, H. O., W. M. Rashed, N. M. Hamdy and N. Hamdy (2014). "High-dose methotrexate in Egyptian pediatric acute lymphoblastic leukemia: the impact of ABCG2 C421A genetic polymorphism on plasma levels, what is next?" Journal of Cancer Research and Clinical Oncology **140**(8): 1359–1365.

Eldash, S., E. F. Sanad, D. Nada and N. M. Hamdy (2023). "The Intergenic Type LncRNA (LINC RNA) Faces in Cancer with In Silico Scope and a Directed Lens to LINC00511: A Step toward ncRNA Precision." Non-Coding RNA **9**(5): 58.

Eltarouty, M., M. A. Eldesoky, F. B. Rashidi, S. Matbouly and N. El-Khazargy (2024). "The Expression Profile of MicroRNA LET-7a-5p in Children with Idiopathic Thrombocytopenic Purpura." Egyptian Journal of Chemistry **67**(9): 65–75.

Emmrich, S., J. Katsman-Kuipers, K. Henke, M. Khatib, R. Jammal, F. Engeland, F. Dasci, C. Zwaan, M. Den Boer and L. Verboon (2014). "miR-9 is a tumor suppressor in pediatric AML with t (8; 21)." Leukemia **28**(5): 1022–1032.

Entezari, M., A. Tayari, M. D. A. Paskeh, S. K. Kheirabad, S. Naeemi, A. Taheriazam, H. Dehghani, S. Salimimoghadam, M. Hashemi, S. Mirzaei and S. Samarghandian (2024). "Curcumin in treatment of hematological cancers: Promises and challenges." J Tradit Complement Med **14**(2): 121–134.

Erdinyler, D., Y. Karakoy, S. Toplan, S. Onen, A. Sukyasyan, T. Beger and Ç. Demiroğlu (1996). "The effect of Ginkgo biloba glycoside on the blood viscosity and erythrocyte deformability." Clinical Hemorheology and Microcirculation **16**(3): 271–276.

Evangelidis, N., N. Kotsiou, P. Evangelidis, V. I. Alevizopoulos, I. Dermitzakis, S. Chissan, S. Vakalopoulou and E. Gavriilaki (2024). "Genetics and Epigenetics in Acquired Hemophilia A: From Bench to Bedside." Curr Issues Mol Biol **46**(6): 5147–5160.

Favreau, A. J., R. E. McGlauflin, C. W. Duarte and P. Sathyanarayana (2015). "miR-199b, a novel tumor suppressor miRNA in acute myeloid leukemia with prognostic implications." Experimental hematology & oncology **5**: 1–11.

Fromm, B., E. Høye, D. Domanska, X. Zhong, E. Aparicio-Puerta, V. Ovchinnikov, S. U. Umu, P. J. Chabot, W. Kang and M. Aslanzadeh (2022). "MirGeneDB 2.1: toward a complete sampling of all major animal phyla." Nucleic acids research **50**(D1): D204-D210.

Gaál, Z. (2022). "Implication of microRNAs in carcinogenesis with emphasis on hematological malignancies and clinical translation." International Journal of Molecular Sciences **23**(10): 5838.

Galka-Marciniak, P., Z. Kanduła, A. Tire, W. Wegorek, K. Gwozdz-Bak, L. Handschuh, M. Giefing, K. Lewandowski and P. Kozlowski (2022). "Mutations in the miR-142 gene are not common in myeloproliferative neoplasms." Scientific Reports **12**(1): 10924.

Gambacorta, V., D. Gnani, L. Vago and R. Di Micco (2019). "Epigenetic Therapies for Acute Myeloid Leukemia and Their Immune-Related Effects." Front Cell Dev Biol **7**: 207.

Gao, X., J. Lin, Y. Li, L. Gao, X. Wang, W. Wang, H. Kang, G. Yan, L. Wang and L. Yu (2011). "MicroRNA-193a represses c-kit expression and functions as a methylation-silenced tumor suppressor in acute myeloid leukemia." Oncogene **30**(31): 3416–3428.

García-Giménez, J. L., M. Seco-Cervera, T. O. Tollefsbol, C. Romá-Mateo, L. Peiró-Chova, P. Lapunzina and F. V. Pallardó (2017). "Epigenetic biomarkers: Current strategies and future challenges for their use in the clinical laboratory." Crit Rev Clin Lab Sci **54**(7–8): 529–550.

Garcia-Manero, G., G. Roboz, K. Walsh, H. Kantarjian, E. Ritchie, P. Kropf, C. O'Connell, R. Tibes, S. Lunin, T. Rosenblat, K. Yee, W. Stock, E. Griffiths, J. Mace, N. Podoltsev, J. Berdeja, E. Jabbour, J. J. Issa, Y. Hao, H. N. Keer, M. Azab and M. R. Savona (2019). "Guadecitabine (SGI-110) in patients with intermediate or high-risk myelodysplastic syndromes: phase 2 results from a multicentre, open-label, randomised, phase 1/2 trial." Lancet Haematol **6**(6): e317-e327.

Garzon, R., C. E. Heaphy, V. Havelange, M. Fabbri, S. Volinia, T. Tsao, N. Zanesi, S. M. Kornblau, G. Marcucci and G. A. Calin (2009). "MicroRNA 29b functions in acute myeloid leukemia." Blood, The Journal of the American Society of Hematology **114**(26): 5331–5341.

Ghafouri-Fard, S., V. Niazi and M. Taheri (2021). "Role of miRNAs and lncRNAs in hematopoietic stem cell differentiation." Non-coding RNA research **6**(1): 8–14.

Gilliland, D. G. and J. D. Griffin (2002). "The roles of FLT3 in hematopoiesis and leukemia." Blood, The Journal of the American Society of Hematology **100**(5): 1532–1542.

Gong, J., J. Yu, H. Lin, X. Zhang, X. Yin, Z. Xiao, F. Wang, X. Wang, R. Su and C. Shen (2014). "The role, mechanism and potentially therapeutic application of microRNA-29 family in acute myeloid leukemia." Cell Death & Differentiation **21**(1): 100–112.

Goodell, M. A. and L. A. J. B. Godley, The Journal of the American Society of Hematology (2013). "Perspectives and future directions for epigenetics in hematology." **121**(26): 5131–5137.

Govindarajan, V. and W. H. Stahl (1980). "Turmeric—chemistry, technology, and quality." Critical Reviews in Food Science & Nutrition **12**(3): 199–301.

Graf, J. and M. Kretz (2020). "From structure to function: Route to understanding lncRNA mechanism." BioEssays **42**(12): 2000027.

Grillone, K., G. Caridà, F. Luciano, A. Cordua, M. T. Di Martino, P. Tagliaferri and P. Tassone (2024). "A systematic review of non-coding RNA therapeutics in early clinical trials: a new perspective against cancer." Journal of Translational Medicine **22**(1): 731.

Guan, X., Y. Sun and C. Zhang (2022). "LncRNAs in blood cells: Roles in cell development and potential pathogenesis in hematological malignancies." Critical Reviews in Oncology/Hematology **180**: 103849.

Gupta, N., R. Kumar, T. Seth, B. Garg, H. C. Sati and A. Sharma (2019). "Clinical significance of circulatory microRNA-203 in serum as novel potential diagnostic marker for multiple myeloma." Journal of Cancer Research and Clinical Oncology **145**: 1601–1611.

Hamburg, M. A. and F. S. Collins (2010). "The path to personalized medicine." N Engl J Med **363**(4): 301–304.

Hamdy, N. M., E. B. Basalious, M. G. El-Sisi, M. Nasr, A. M. Kabel, E. S. Nossier and A. H. Abadi (2024a). "Advancements in current one-size-fits-all therapies compared to future treatment innovations for better improved chemotherapeutic outcomes: a step-toward personalized medicine." Current Medical Research and Opinion **40**(11): 1943–1961.

Hamdy, N. M., M. B. Zaki, N. M. Abdelmaksoud, R. A. Ismail, M. A. Abd-Elmawla, N. I. Rizk, D. Fathi and A. I. Abulsoud (2025). "Insights into the genetic and epigenetic mechanisms governing X-chromosome-linked-miRNAs expression in cancer; a step-toward ncRNA precision." International Journal of Biological Macromolecules **289**: 138773.

Hamdy, N. M., M. B. Zaki, N. I. Rizk, N. M. Abdelmaksoud, M. A. Abd-Elmawla, R. A. Ismail and A. I. Abulsoud (2024b). "Unraveling the ncRNA landscape that governs colorectal cancer: A roadmap to personalized therapeutics." Life Sciences **354**: 122946.
Hamdy, S. M., M. S. Ali, R. G. Abd El-Hmid, N. K. Abdelghaffar and O. O. Abdelaleem (2023). "Role of long non coding RNAs, NEAT1 and Lnc-DC expression in pediatric immune thrombocytopenic purpura." Reports of Biochemistry & Molecular Biology **11**(4): 635.
Hao, M., M. Zang, E. Wendlandt, Y. Xu, G. An, D. Gong, F. Li, F. Qi, Y. Zhang and Y. Yang (2015). "Low serum mi R-19a expression as a novel poor prognostic indicator in multiple myeloma." International journal of cancer **136**(8): 1835–1844.
Hodeib, H., D. El Amrousy, E. Elaskary, N. Hablas, A. Youssef and D. Abdelhai (2023). "lncRNA GAS5 and RUNX1 Genes in Children With Primary Immune Thrombocytopenia." Journal of Pediatric Hematology/Oncology **45**(3): e395-e400.
Hofmans, M., T. Lammens, B. Depreter, Y. Wu, M. Erlacher, A. Caye, H. Cavé, C. Flotho, V. de Haas and C. M. Niemeyer (2021). "Long non-coding RNAs as novel therapeutic targets in juvenile myelomonocytic leukemia." Scientific reports **11**(1): 2801.
Huang, H.-H., F.-Y. Chen, W.-C. Chou, H.-A. Hou, B.-S. Ko, C.-T. Lin, J.-L. Tang, C.-C. Li, M. Yao and W. Tsay (2019). "Long non-coding RNA HOXB-AS3 promotes myeloid cell proliferation and its higher expression is an adverse prognostic marker in patients with acute myeloid leukemia and myelodysplastic syndrome." BMC cancer **19**: 1–14.
Huang, W., D. Paul, G. A. Calin and R. Bayraktar (2023). "miR-142: A Master Regulator in Hematological Malignancies and Therapeutic Opportunities." Cells **13**(1): 84.
Iaquinta, M. R., C. Lanzillotti, C. Mazziotta, I. Bononi, F. Frontini, E. Mazzoni, L. Oton-Gonzalez, J. C. Rotondo, E. Torreggiani and M. Tognon (2021). "The role of microRNAs in the osteogenic and chondrogenic differentiation of mesenchymal stem cells and bone pathologies." Theranostics **11**(13): 6573.
Irfan, M., M. Kim, H.-W. Kwon, M. H. Rhee and H.-K. Kim (2018). "Antiplatelet effects of garlic and chitosan: a comparative study between fermented and non-fermented preparations." Biomedical Science Letters **24**(3): 280–284.
Jabeena, S., M. A. Hanifa, M. M. Khanb and R. W. K. Qadric (2014). "Natural products sources and their active compounds on disease prevention: A." health **6**(7).
Jia, Z.-W., Y. Li, G.-R. Cui, H.-B. Zhao, P.-Y. Li and J.-M. Luo (2018). "Expression and clinical significance of LncRNA KCNQ1OT1 in patients with acute myeloid leukemia." Zhongguo Shi Yan Xue Ye Xue Za Zhi **26**(3): 653–657.
Jiang, N., Y. Wang, Q. Feng, S. Wang, S. Leng, X. Zhang, Q. Liu, J. Peng and X. Li (2023). "Plasma long noncoding RNAs lncDC and THRIL as potential diagnostic markers of adult primary immune thrombocytopenia." International Journal of Laboratory Hematology **45**(4): 481–488.
Jiang, X., C. Hu, S. Arnovitz, J. Bugno, M. Yu, Z. Zuo, P. Chen, H. Huang, B. Ulrich and S. Gurbuxani (2016). "miR-22 has a potent anti-tumour role with therapeutic potential in acute myeloid leukaemia." Nature communications **7**(1): 11452.
Jiang, Y., Y. Luan, H. Chang and G. Chen (2018). "The diagnostic and prognostic value of plasma microRNA-125b-5p in patients with multiple myeloma." Oncology letters **16**(3): 4001–4007.
Johnson, K. B., W. Q. Wei, D. Weeraratne, M. E. Frisse, K. Misulis, K. Rhee, J. Zhao and J. L. Snowdon (2021). "Precision Medicine, AI, and the Future of Personalized Health Care." Clin Transl Sci **14**(1): 86–93.
Jones, P. A. and S. B. Baylin (2007). "The epigenomics of cancer." Cell **128**(4): 683–692.
Kalinkova, L., A. Sevcikova, V. Stevurkova, I. Fridrichova and S. Ciernikova (2022). "Targeting DNA Methylation in Leukemia, Myelodysplastic Syndrome, and Lymphoma: A Potential Diagnostic, Prognostic, and Therapeutic Tool." Int J Mol Sci **24**(1).
Kamal, A. M., N. M. Hamdy, H. M. Hegab and H. O. El-Mesallamy (2016). "Expression of thioredoxin-1 (TXN) and its relation with oxidative DNA damage and treatment outcome in adult AML and ALL: A comparative study." Hematology **21**(10): 567–575.

Katsaraki, K., P. Karousi, P. I. Artemaki, A. Scorilas, V. Pappa, C. K. Kontos and S. G. Papageorgiou (2021). "MicroRNAs: Tiny regulators of gene expression with pivotal roles in normal B-Cell development and B-Cell chronic lymphocytic leukemia." Cancers **13**(4): 593.

Katsnelson, A. (2013). "Momentum grows to make 'personalized' medicine more 'precise'." Nat Med **19**(3): 249.

Kelly, T. K., D. D. De Carvalho and P. A. Jones (2010). "Epigenetic modifications as therapeutic targets." Nat Biotechnol **28**(10): 1069–1078.

Khorkova, O., J. Hsiao and C. Wahlestedt (2015). "Basic biology and therapeutic implications of lncRNA." Advanced drug delivery reviews **87**: 15–24.

Kiefer, D. and T. Pantuso (2003). "Panax ginseng." American family physician **68**(8): 1539–1542.

Kim, M. Y., B. Yan, S. Huang and Y. Qiu (2020). "Regulating the Regulators: The Role of Histone Deacetylase 1 (HDAC1) in Erythropoiesis." Int J Mol Sci **21**(22).

Kimbrel, E. A., M. E. Lemieux, X. Xia, T. N. Davis, V. I. Rebel and A. L. Kung (2009). "Systematic in vivo structure-function analysis of p300 in hematopoiesis." Blood **114**(23): 4804–4812.

Kole, R., A. R. Krainer and S. Altman (2012). "RNA therapeutics: beyond RNA interference and antisense oligonucleotides." Nature reviews Drug discovery **11**(2): 125–140.

Kotowska-Zimmer, A., M. Pewinska and M. Olejniczak (2021). "Artificial miRNAs as therapeutic tools: Challenges and opportunities." Wiley Interdisciplinary Reviews: RNA **12**(4): e1640.

Kubiczkova, L., F. Kryukov, O. Slaby, E. Dementyeva, J. Jarkovsky, J. Nekvindova, L. Radova, H. Greslikova, P. Kuglik and E. Vetesnikova (2014). "Circulating serum microRNAs as novel diagnostic and prognostic biomarkers for multiple myeloma and monoclonal gammopathy of undetermined significance." Haematologica **99**(3): 511.

Lavelle, D., J. D. Engel and Y. Saunthararajah (2018). "Fetal Hemoglobin Induction by Epigenetic Drugs." Semin Hematol **55**(2): 60–67.

Leonard, A. and J. F. Tisdale (2024). "A new frontier: FDA approvals for gene therapy in sickle cell disease." Mol Ther **32**(2): 264–267.

Li, C., F. Tan, Q. Pei, Z. Zhou, Y. Zhou, L. Zhang, D. Wang and H. Pei (2019). "Non-coding RNA MFI2-AS1 promotes colorectal cancer cell proliferation, migration and invasion through miR-574-5p/MYCBP axis." Cell proliferation **52**(4): e12632.

Li, J.-Q., S.-Y. Hu, Z.-Y. Wang, J. Lin, S. Jian, Y.-C. Dong, X.-F. Wu, D. Lan and L.-J. Cao (2015b). "MicroRNA-125-5p targeted CXCL13: a potential biomarker associated with immune thrombocytopenia." American Journal of Translational Research **7**(4): 772.

Li, T., Y. Chen and S. Li (2024). "The Advances in the Development of Epigenetic Modifications Therapeutic Drugs Delivery Systems." Int J Nanomedicine **19**: 10623–10637.

Li, Y., L. Gao, X. Luo, L. Wang, X. Gao, W. Wang, J. Sun, L. Dou, J. Li and C. Xu (2013). "Epigenetic silencing of microRNA-193a contributes to leukemogenesis in t (8; 21) acute myeloid leukemia by activating the PTEN/PI3K signal pathway." Blood, The Journal of the American Society of Hematology **121**(3): 499–509.

Li, Y., H. Zhang, B. Hu, P. Wang, W. Wang and J. Liu (2023). "Post-transcriptional regulation of erythropoiesis." Blood Science **5**(3): 150–159.

Li, Z., P. Chen, R. Su, Y. Li, C. Hu, Y. Wang, S. Arnovitz, M. He, S. Gurbuxani and Z. Zuo (2015a). "Overexpression and knockout of miR-126 both promote leukemogenesis." Blood, The Journal of the American Society of Hematology **126**(17): 2005–2015.

Liao, Y.-R., Y.-L. Leu, Y.-Y. Chan, P.-C. Kuo and T.-S. Wu (2012). "Anti-Platelet Aggregation and Vasorelaxing Effects of the Constituents of the Rhizomes of Zingiber officinale." Molecules **17**(8): 8928–8937.

Lin, X., Z. Wang, R. Zhang and W. Feng (2015). "High serum microRNA-335 level predicts aggressive tumor progression and unfavorable prognosis in pediatric acute myeloid leukemia." Clinical and Translational Oncology **17**: 358–364.

Ling, H., M. Fabbri and G. A. Calin (2013). "MicroRNAs and other non-coding RNAs as targets for anticancer drug development." Nature reviews Drug discovery **12**(11): 847–865.

Ling, L., F. Wang, Y. Li, S. He, F. Wu, L. Yang, L. Xu, T. Wang, S. Zhou and F. Yang (2024). "Depletion of miR-144/451 alleviates anemia in β-thalassemic mice." Blood Advances **8**(10): 2565–2570.

Liu, S., Z. Sun, M. Zhu, M. Liu, M. Wei, X. Pan and S. Huang (2022a). "Prognostic value and potential mechanism of long non-coding RNA Lnc-SMIM20-1 in acute myeloid leukemia." Expert Review of Anticancer Therapy **22**(8): 875–885.

Liu, X. L., H. Q. Liu, J. Li, C. Y. Mao, J. T. He and X. Zhao (2020). "Role of epigenetic in leukemia: From mechanism to therapy." Chem Biol Interact **317**: 108963.

Liu, Z., A. Sall and D. Yang (2008). "MicroRNA: an emerging therapeutic target and intervention tool." International journal of molecular sciences **9**(6): 978–999.

Liu, Z., L. Zhong, W. Dan, X. Chu, C. Liu, X. Luo, Z. Zhang, Y. Lu, P. Wan and X. Wang (2022b). "miRNA-222-3p enhances the proliferation and suppresses the apoptosis of acute myeloid leukemia cells by targeting Axin2 and modulating the Wnt/β-catenin pathway." Biochemical and Biophysical Research Communications **620**: 83–91.

Lodde, V., G. Murgia, E. R. Simula, M. Steri, M. Floris and M. L. Idda (2020). "Long noncoding RNAs and circular RNAs in autoimmune diseases." Biomolecules & therapeutics **10**(7): 1044.

Lu, G., S. Jin, S. Lin, Y. Gong, L. Zhang, J. Yang, W. Mou and J. Du (2023). "Update on histone deacetylase inhibitors in peripheral T-cell lymphoma (PTCL)." Clin Epigenetics **15**(1): 124.

Luo, M., R. Li, M. Ren, N. Chen, X. Deng, X. Tan, Y. Li, M. Zeng, Y. Yang and Q. Wan (2016). "Hyperglycaemia-induced reciprocal changes in miR-30c and PAI-1 expression in platelets." Scientific reports **6**(1): 36687.

Ma, X., W. Zhang, M. Zhao, S. Li, W. Jin and K. Wang (2020). "Oncogenic role of lncRNA CRNDE in acute promyelocytic leukemia and NPM1-mutant acute myeloid leukemia." Cell Death Discovery **6**(1): 121.

Manier, S., C.-J. Liu, H. Avet-Loiseau, J. Park, J. Shi, F. Campigotto, K. Z. Salem, D. Huynh, S. V. Glavey and B. Rivotto (2017). "Prognostic role of circulating exosomal miRNAs in multiple myeloma." Blood, The Journal of the American Society of Hematology **129**(17): 2429–2436.

Marcucci, G., K. S. Maharry, K. H. Metzeler, S. Volinia, Y.-Z. Wu, K. Mrózek, D. Nicolet, J. Kohlschmidt, S. P. Whitman and J. H. Mendler (2013). "Clinical role of microRNAs in cytogenetically normal acute myeloid leukemia: miR-155 upregulation independently identifies high-risk patients." Journal of clinical oncology **31**(17): 2086–2093.

Marx, W., D. McKavanagh, A. L. McCarthy, R. Bird, K. Ried, A. Chan and L. Isenring (2015). "The Effect of Ginger (Zingiber officinale) on Platelet Aggregation: A Systematic Literature Review." PLoS One **10**(10): e0141119.

Mehana, N. A., H. R. Ghaiad, M. Hassan, Y. A. Elsabagh, S. Labib and M. A. Abd-Elmawla (2022). "LncRNA MEG3 regulates the interplay between Th17 and Treg cells in Behçet's disease and systemic lupus erythematosus." Life Sciences **309**: 120965.

Memari, F., Z. Joneidi, B. Taheri, S. F. Aval, A. Roointan and N. Zarghami (2018). "Epigenetics and Epi-miRNAs: Potential markers/therapeutics in leukemia." Biomed Pharmacother **106**: 1668–1677.

Miesbach, W., R. Klamroth, J. Oldenburg and A. Tiede (2022). "Gene Therapy for Hemophilia-Opportunities and Risks." Dtsch Arztebl Int **119**(51–52): 887–894.

Milner, J. A. (1996). "Garlic: its anticarcinogenic and antiumorigenic properties." Nutrition reviews **54**(11): S82.

Mirmiran, P., Z. Houshialsadat, Z. Gaeini, Z. Bahadoran and F. Azizi (2020). "Functional properties of beetroot (Beta vulgaris) in management of cardio-metabolic diseases." Nutrition & metabolism **17**: 1–15.

Misso, G., M. R. Zarone, A. Lombardi, A. Grimaldi, A. M. Cossu, C. Ferri, M. Russo, D. C. Vuoso, A. Luce and H. Kawasaki (2019). "miR-125b upregulates miR-34a and sequentially activates stress adaption and cell death mechanisms in multiple myeloma." Molecular Therapy-Nucleic Acids **16**: 391–406.

Mittal, N., L. Li, Y. Sheng, C. Hu, F. Li, T. Zhu, X. Qiao and Z. Qian (2019). "A critical role of epigenetic inactivation of miR-9 in EVI1 high pediatric AML." Molecular cancer **18**: 1–6.

Momozawa, Y. and K. Mizukami (2021). "Unique roles of rare variants in the genetics of complex diseases in humans." Journal of human genetics **66**(1): 11–23.

Morrison, T. A., I. Wilcox, H.-Y. Luo, J. J. Farrell, R. Kurita, Y. Nakamura, G. J. Murphy, S. Cui, M. H. Steinberg and D. H. Chui (2018). "A long noncoding RNA from the HBS1L-MYB intergenic region on chr6q23 regulates human fetal hemoglobin expression." Blood Cells, Molecules, and Diseases **69**: 1–9.

Moskowitz, A. J. and S. M. Horwitz (2017). "Targeting histone deacetylases in T-cell lymphoma." Leuk Lymphoma **58**(6): 1306–1319.

Mostafa, A. M., N. M. Hamdy, H. O. El-Mesallamy and S. Z. Abdel-Rahman (2015). "Glucagon-like peptide 1 (GLP-1)-based therapy upregulates LXR-ABCA1/ABCG1 cascade in adipocytes." Biochemical and Biophysical Research Communications **468**(4): 900–905.

Naguib, M., S. El Sawy, L. Rashed, M. AlHelf and M. Abdelgwad (2023). "Long non-coding RNA maternally expressed gene 3, miR-125a-5p, CXCL13, and NF-kB in patients with immune thrombocytopenia." Genes & Immunity **24**(2): 108–115.

Naidu, S., P. Magee and M. Garofalo (2015). "MiRNA-based therapeutic intervention of cancer." Journal of hematology & oncology **8**: 1–8.

Nakamura, S., M. Oshima, J. Yuan, A. Saraya, S. Miyagi, T. Konuma, S. Yamazaki, M. Osawa, H. Nakauchi, H. Koseki and A. Iwama (2012). "Bmi1 confers resistance to oxidative stress on hematopoietic stem cells." PLoS One **7**(5): e36209.

Nemeth, K., R. Bayraktar, M. Ferracin and G. A. Calin (2024). "Non-coding RNAs in disease: from mechanisms to therapeutics." Nature Reviews Genetics **25**(3): 211–232.

Nurk, S., S. Koren, A. Rhie, M. Rautiainen, A. V. Bzikadze, A. Mikheenko, M. R. Vollger, N. Altemose, L. Uralsky and A. Gershman "The complete sequence of a human genome., 2022, Science **376**(6588): 44–53.

Nurtjahja-Tjendraputra, E., A. J. Ammit, B. D. Roufogalis, V. H. Tran and C. C. Duke (2003). "Effective anti-platelet and COX-1 enzyme inhibitors from pungent constituents of ginger." Thromb Res **111**(4–5): 259–265.

Nwabo Kamdje, A. H., H. P. Dongmo Fogang and P. N. Mimche (2024). "Role of epigenetic in cancer biology, in hematologic malignancies and in anticancer therapy." Front Mol Med **4**: 1426454.

Omoregie, S. N., F. O. Omoruyi, V. F. Wright, L. Jones and P. V. Zimba (2013). "Antiproliferative activities of lesser galangal (Alpinia officinarum Hance Jam1), turmeric (Curcuma longa L.), and ginger (Zingiber officinale Rosc.) against acute monocytic leukemia." Journal of medicinal food **16**(7): 647–655.

Palazzo, A. F. and E. V. Koonin (2020). "Functional long non-coding RNAs evolve from junk transcripts." Cell **183**(5): 1151–1161.

Paloviita, P. and S. Vuoristo (2022). "The non-coding genome in early human development – Recent advancements." Seminars in Cell & Developmental Biology **131**: 4–13.

Pan, Q., X. Xin, S. Mahto, Y. Dong, V. Kumar, R. K. Hyde, N. Gupta, V. R. Bhatt and R. I. Mahato (2024). "Anti-CLL1 liposome loaded with miR-497-5p and venetoclax as a novel therapeutic strategy in acute myeloid leukemia." Molecular Therapy **32**(11): 4058–4074.

Peng, B., P. Y. Theng and M. T. Le (2021). "Essential functions of miR-125b in cancer." Cell proliferation **54**(2): e12913.

Pillai, M. G. and H. Antony (2024). "Harnessing the immunomodulatory potential of natural products in precision medicine—a comprehensive review." Exploration of Drug Science **2**(3): 339–360.

Pisignano, G. and M. Ladomery (2021). "Post-transcriptional regulation through long non-coding rnas (lncrnas)." Non-coding RNA **7**(2): 29.

Platzbecker, U., V. Santini, P. Fenaux, M. A. Sekeres, M. R. Savona, Y. F. Madanat, M. Diez-Campelo, D. Valcarcel, T. Illmer, A. Jonasova, P. Belohlavkova, L. J. Sherman, T. Berry, S. Dougherty, S. Shah, Q. Xia, L. Sun, Y. Wan, F. Huang, A. Ikin, S. Navada, F. Feller,

R. S. Komrokji and A. M. Zeidan (2024). "Imetelstat in patients with lower-risk myelodysplastic syndromes who have relapsed or are refractory to erythropoiesis-stimulating agents (IMerge): a multinational, randomised, double-blind, placebo-controlled, phase 3 trial." Lancet **403**(10423): 249–260.

Popovic, R., L. E. Riesbeck, C. S. Velu, A. Chaubey, J. Zhang, N. J. Achille, F. E. Erfurth, K. Eaton, J. Lu and H. L. Grimes (2009). "Regulation of mir-196b by MLL and its overexpression by MLL fusions contributes to immortalization." Blood, The Journal of the American Society of Hematology **113**(14): 3314–3322.

Prabhakaran, R., R. Thamarai, S. Sivasamy, S. Dhandayuthapani, J. Batra, C. Kamaraj, K. Karthik, M. A. Shah and S. Mallik (2024). "Epigenetic frontiers: miRNAs, long non-coding RNAs and nanomaterials are pioneering to cancer therapy." Epigenetics & Chromatin **17**(1): 31.

Pujadas, E. and A. P. Feinberg (2012). "Regulated noise in the epigenetic landscape of development and disease." Cell **148**(6): 1123–1131.

Qi, W., X. Yan, X. Xu, B. Song, L. Sun, D. Zhao and L. Sun (2020). "The effects of cytarabine combined with ginsenoside compound K synergistically induce DNA damage in acute myeloid leukemia cells." Biomedicine & Pharmacotherapy **132**: 110812.

Qian, C., W. Yan, T. Li, Q. Cui, P. Liu, M. Gu, J. Guo, W. Zhang, C. Ren and T. Wu (2018). "Differential expression of MiR-106b-5p and MiR-200c-3p in newly diagnosed versus chronic primary immune thrombocytopenia patients based on systematic analysis." Cellular Physiology and Biochemistry **45**(1): 301–318.

Qiang, P., Q. Pan, C. Fang, C. Fozza, K. Song, Y. Dai, W. Chang, W. Chen, W. Yao and W. Zhu (2020). "MicroRNA-181a-3p as a diagnostic and prognostic biomarker for acute myeloid leukemia." Mediterranean journal of hematology and infectious diseases **12**(1).

Qu, X., M. Zhao, S. Wu, W. Yu, J. Xu, J. Xu, J. Li and L. Chen (2014). "Circulating microRNA 483-5p as a novel biomarker for diagnosis survival prediction in multiple myeloma." Medical oncology **31**: 1–8.

Radwan, S. M., N. M. Hamdy, H. M. Hegab and H. O. El-Mesallamy (2016). "Beclin-1 and hypoxia-inducible factor-1α genes expression: Potential biomarkers in acute leukemia patients." Cancer Biomarkers **16**(4): 619–626.

Rahman, K. (2007). "Effects of garlic on platelet biochemistry and physiology." Mol Nutr Food Res **51**(11): 1335–1344.

Rametta, R., M. Meroni and P. Dongiovanni (2020). "From Environment to Genome and Back: A Lesson from HFE Mutations." Int J Mol Sci **21**(10).

Rashad, N., U. A. Khalil, A. M. El-Helaly, M. H. Soliman and M. M. Sherif (2022). "LncRNA MEG3 in Promoting Antiphospholipid Syndrome Nephropathy in Patients with Systemic Lupus Erythematosus." The Egyptian Journal of Hospital Medicine **87**(1): 1591–1598.

Rasool, M., A. Malik, M. I. Naseer, A. Manan, S. Ansari, I. Begum, M. H. Qazi, P. Pushparaj, A. M. Abuzenadah, M. H. Al-Qahtani, M. A. Kamal and S. Gan (2015). "The role of epigenetics in personalized medicine: challenges and opportunities." BMC Med Genomics **8** (Suppl 1): S5.

Rassenti, L. Z., V. Balatti, E. M. Ghia, A. Palamarchuk, L. Tomasello, P. Fadda, Y. Pekarsky, G. F. Widhopf, T. J. Kipps and C. M. Croce (2017). "MicroRNA dysregulation to identify therapeutic target combinations for chronic lymphocytic leukemia." Proceedings of the National Academy of Sciences **114**(40): 10731–10736.

Rea, J., V. Menci, P. Tollis, T. Santini, A. Armaos, M. G. Garone, F. Iberite, A. Cipriano, G. G. Tartaglia and A. Rosa (2020). "HOTAIRM1 regulates neuronal differentiation by modulating NEUROGENIN 2 and the downstream neurogenic cascade." Cell death & disease **11**(7): 527.

Roberts, T. C., R. Langer and M. J. Wood (2020). "Advances in oligonucleotide drug delivery." Nature reviews Drug discovery **19**(10): 673–694.

Roehle, A., K. P. Hoefig, D. Repsilber, C. Thorns, M. Ziepert, K. O. Wesche, M. Thiere, M. Loeffler, W. Klapper and M. Pfreundschuh (2008). "MicroRNA signatures characterize diffuse large B-cell lymphomas and follicular lymphomas." British journal of haematology **142**(5): 732–744.

Roy, B., S. Dutta and C. Mukherjee (2023). "Role of non-coding RNAs in development and metastasis of solid tumours." Frontiers in Cell and Developmental Biology **11**: 1281200.

Roy, B., E. Lee, T. Li and M. Rampersaud (2022). "Role of miRNAs in neurodegeneration: From disease cause to tools of biomarker discovery and therapeutics." Genes **13**(3): 425.

Sasaki, K., E. Jabbour, G. Montalban-Bravo, F. Darbaniyan, K. A. Do, C. Class, N. J. Short, R. Kanagal-Shamana, T. Kadia, G. Borthakur, N. Pemmaraju, J. Cortes, F. Ravandi, Y. Alvarado, K. Chien, R. Komrokji, M. A. Sekeres, D. P. Steensma, A. DeZern, G. Roboz, K. Soltysiak, H. Yang, H. M. Kantarjian and G. Garcia-Manero (2022). "Low-Dose Decitabine versus Low-Dose Azacitidine in Lower-Risk MDS." NEJM Evid **1**(10): EVIDoa2200034.

Sashida, G. and A. Iwama (2012). "Epigenetic regulation of hematopoiesis." Int J Hematol **96**(4): 405–412.

Savović, J., B. Wider and E. Ernst (2005). "Effects of Ginkgo biloba on blood coagulation parameters: a systematic review of randomised clinical trials." Evidence-Based Integrative Medicine **2**: 167–176.

Schmitt, A. M. and H. Y. Chang (2016). "Long noncoding RNAs in cancer pathways." Cancer cell **29**(4): 452–463.

Schneider, E., A. Staffas, L. Röhner, K. Krowiorz, M. Heuser, K. Döhner, L. Bullinger, H. Döhner, L. Fogelstrand and A. Rouhi (2016). "MicroRNA-155 is upregulated in MLL-rearranged AML but its absence does not affect leukemia development." Experimental Hematology **44**(12): 1166–1171.

Sebastian-delaCruz, M., I. Gonzalez-Moro, A. Olazagoitia-Garmendia, A. Castellanos-Rubio and I. Santin (2021). "The role of lncRNAs in gene expression regulation through mRNA stabilization." Non-coding RNA **7**(1): 3.

Sermer, D., L. Pasqualucci, H. G. Wendel, A. Melnick and A. Younes (2019). "Emerging epigenetic-modulating therapies in lymphoma." Nat Rev Clin Oncol **16**(8): 494–507.

Shanbhag, V. K. L. (2017). "Curcumin in chronic lymphocytic leukemia – A review." Asian Pacific Journal of Tropical Biomedicine **7**(6): 505–512.

Shen, X., Y. Ye, J. Qi, W. Shi, X. Wu, H. Ni, H. Cong and S. Ju (2017). "Identification of a novel microRNA, miR-4449, as a potential blood based marker in multiple myeloma." Clinical Chemistry and Laboratory Medicine (CCLM) **55**(5): 748–754.

Sherman, R. M. and S. L. Salzberg (2020). "Pan-genomics in the human genome era." Nature Reviews Genetics **21**(4): 243–254.

Shi, J., W. Ding and H. Lu (2020). "Identification of long non-coding RNA SNHG family as promising prognostic biomarkers in acute Myeloid Leukemia." OncoTargets and therapy: 8441–8450.

Sierpina, V. S., B. Wollschlaeger and M. Blumenthal (2003). "Ginkgo biloba." American family physician **68**(5): 923–926.

Singh, M., V. Kumar, N. Sehrawat, M. Yadav, M. Chaudhary, S. K. Upadhyay, S. Kumar, V. Sharma, S. Kumar, N. Dilbaghi and A. K. Sharma (2022). "Current paradigms in epigenetic anticancer therapeutics and future challenges." Semin Cancer Biol **83**: 422–440.

Smutny, T., L. Hyrsova, A. Braeuning, M. Ingelman-Sundberg and P. Pavek (2021). "Transcriptional and post-transcriptional regulation of the pregnane X receptor: a rationale for interindividual variability in drug metabolism." Archives of Toxicology **95**: 11–25.

Sommerkamp, P., S. Renders, L. Ladel, K. Schönberger, P. Zeisberger, A. Przybylla, N. Cabezas-Wallscheid and A. Trumpp (2017). "The role of the lncRNA Meg3 in the adult hematopoietic compartment." Experimental Hematology **53**: S65.

Song, S. J., K. Ito, U. Ala, L. Kats, K. Webster, S. M. Sun, M. Jongen-Lavrencic, K. Manova-Todorova, J. Teruya-Feldstein and D. E. Avigan (2013). "The oncogenic microRNA miR-22 targets the TET2 tumor suppressor to promote hematopoietic stem cell self-renewal and transformation." Cell stem cell **13**(1): 87–101.

Soni, D. K. and R. Biswas (2021). "Role of non-coding RNAs in post-transcriptional regulation of lung diseases." Frontiers in Genetics **12**: 767348.

Statello, L., C.-J. Guo, L.-L. Chen and M. Huarte (2021). "Gene regulation by long non-coding RNAs and its biological functions." Nature reviews Molecular cell biology **22**(2): 96–118.
Sternberg, A. R., C. Martos-Rus, R. J. Davidson, X. Liu and L. A. George (2024). "Pre-clinical evaluation of an enhanced-function factor VIII variant for durable hemophilia A gene therapy in male mice." Nat Commun **15**(1): 7193.
Stomper, J., J. C. Rotondo, G. Greve and M. Lubbert (2021). "Hypomethylating agents (HMA) for the treatment of acute myeloid leukemia and myelodysplastic syndromes: mechanisms of resistance and novel HMA-based therapies." Leukemia **35**(7): 1873–1889.
Sun, H., G. Peng, H. Wu, M. Liu, G. Mao, X. Ning, H. Yang and J. Deng (2020b). "Long non-coding RNA MEG3 is involved in osteogenic differentiation and bone diseases." Biomedical reports **13**(1): 15–21.
Sun, S., S. Chai, F. Zhang and L. Lu (2020a). "Overexpressed microRNA-103a-3p inhibits acute lower-extremity deep venous thrombosis via inhibition of CXCL12." IUBMB life **72**(3): 492–504.
Superko, H. R., S. M. Superko, K. Nasir, A. Agatston and B. C. Garrett (2013). "Omega-3 fatty acid blood levels: clinical significance and controversy." Circulation **128**(19): 2154–2161.
Swystun, L. L. and P. D. James (2017). "Genetic diagnosis in hemophilia and von Willebrand disease." Blood Rev **31**(1): 47–56.
Taiana, E., V. Favasuli, D. Ronchetti, K. Todoerti, F. Pelizzoni, M. Manzoni, M. Barbieri, S. Fabris, I. Silvestris and M. E. Gallo Cantafio (2020). "Long non-coding RNA NEAT1 targeting impairs the DNA repair machinery and triggers anti-tumor activity in multiple myeloma." Leukemia **34**(1): 234–244.
Tan, Z., K. Zhu, Y. Yin and Z. Luo (2021). "Long non-coding RNA ANRIL is a potential indicator of disease progression and poor prognosis in acute myeloid leukemia." Molecular Medicine Reports **23**(2): 1–1.
Tang, J., H. Yan and S. Zhuang (2013). "Histone deacetylases as targets for treatment of multiple diseases." Clin Sci (Lond) **124**(11): 651–662.
Tang, K.-C., Z.-P. Yang, Q. Zeng, J. Wang, F. Guo and Y. Zhao (2018). "Effect of miR-495 on lower extremity deep vein thrombosis through the TLR4 signaling pathway by regulation of IL1R1." Bioscience reports **38**(6): BSR20180598.
Terrinoni, A., C. Calabrese, D. Basso, A. Aita, S. Caporali, M. Plebani and S. Bernardini (2019). "The circulating miRNAs as diagnostic and prognostic markers." Clinical Chemistry and Laboratory Medicine (CCLM) **57**(7): 932–953.
Trzybulska, D., E. Vergadi and C. Tsatsanis (2018). "miRNA and other non-coding RNAs as promising diagnostic markers." Ejifcc **29**(3): 221.
Turner, M. (1990). "Garlic and circulatory disorders." Journal of the Royal Society of Health **110**(3): 90–93.
Venkatesan, V., S. Srinivasan, P. Babu and S. Thangavel (2020). "Manipulation of Developmental Gamma-Globin Gene Expression: an Approach for Healing Hemoglobinopathies." Mol Cell Biol **41**(1).
Wajapeyee, N. and R. Gupta (2021). "Epigenetic Alterations and Mechanisms That Drive Resistance to Targeted Cancer Therapies." Cancer Res **81**(22): 5589–5595.
Walker, A. L., S. Steward, T. A. Howard, N. Mortier, M. Smeltzer, Y. D. Wang and R. E. Ware (2011). "Epigenetic and molecular profiles of erythroid cells after hydroxyurea treatment in sickle cell anemia." Blood **118**(20): 5664–5670.
Wang, F.-Y., Z.-Y. Gu and C.-J. Gao (2020). "Emerging role of long non-coding RNAs in normal and malignant hematopoiesis." Chinese Medical Journal **133**(04): 462–473.
Wang, F., X. Tian, J. Zhou, G. Wang, W. Yu, Z. Li, Z. Fan, W. Zhang and A. Liang (2018b). "A three-lncRNA signature for prognosis prediction of acute myeloid leukemia in patients." Molecular medicine reports **18**(2): 1473–1484.
Wang, S., J. Ma, H. Qiu, S. Liu, S. Zhang, H. Liu, P. Zhang, R.-l. Ge, G. Li and S. Cui (2023). "Plasma exosomal microRNA expression profiles in patients with high-altitude polycythemia." Blood Cells, Molecules, and Diseases **98**: 102707.

Wang, X. Q. and J. Dostie (2017). "Reciprocal regulation of chromatin state and architecture by HOTAIRM1 contributes to temporal collinear HOXA gene activation." Nucleic acids research **45**(3): 1091–1104.

Wang, Y., Y. Li, H. Song and G. Sun (2018a). "Long non-coding RNA LINC00899 as a novel serum biomarker for diagnosis and prognosis prediction of acute myeloid leukemia." Eur Rev Med Pharmacol Sci **22**(21): 7364–7370.

Wang, Y., Q. Zhou and J.-J. Ma (2018c). "High expression of lnc-CRNDE presents as a biomarker for acute myeloid leukemia and promotes the malignant progression in acute myeloid leukemia cell line U937." European Review for Medical & Pharmacological Sciences **22**(3).

White, B. (2007). "Ginger: an overview." American family physician **75**(11): 1689–1691.

Woo, A. J., J. Kim, J. Xu, H. Huang and A. B. Cantor (2011). "Role of ZBP-89 in human globin gene regulation and erythroid differentiation." Blood **118**(13): 3684–3693.

Wouters, B. J. and R. Delwel (2016). "Epigenetics and approaches to targeted epigenetic therapy in acute myeloid leukemia." Blood **127**(1): 42–52.

Wu, S., C. Zheng, S. Chen, X. Cai, Y. Shi, B. Lin and Y. Chen (2015). "Overexpression of long non-coding RNA HOTAIR predicts a poor prognosis in patients with acute myeloid leukemia." Oncology letters **10**(4): 2410–2414.

Wu, X., J. Wang, D. Li and Z. Xu (2020). "Expression and Diagnostic Value of lncRNA BLACAT1 in Peripheral Blood of Patients with Acute Myeloid Leukemia." Clinical Laboratory **86**(4).

Xiao, G., Y. Li, Y. Wang, B. Zhao, Z. Zou, S. Hou, X. Jia, X. Liu, Y. Yao and J. Wan (2018). "LncRNA PRAL is closely related to clinical prognosis of multiple myeloma and the bortezomib sensitivity." Experimental cell research **370**(2): 254–263.

Xiao, Q., C. Lin, M. Peng, J. Ren, Y. Jing, L. Lei, Y. Tao, J. Huang, J. Yang and M. Sun (2022). "Circulating plasma exosomal long non-coding RNAs LINC00265, LINC00467, UCA1, and SNHG1 as biomarkers for diagnosis and treatment monitoring of acute myeloid leukemia." Frontiers in Oncology **12**: 1033143.

Xu, L., F. Wu, L. Yang, F. Wang, T. Zhang, X. Deng, X. Zhang, X. Yuan, Y. Yan and Y. Li (2020). "miR-144/451 inhibits c-Myc to promote erythroid differentiation." The FASEB Journal **34**(10): 13194–13210.

Xu, M., G.-N. Zhao, X. Lv, G. Liu, L. Y. Wang, D.-L. Hao, J. Wang, D.-P. Liu and C.-C. Liang (2014). "CTCF controls HOXA cluster silencing and mediates PRC2-repressive higher-order chromatin structure in NT2/D1 cells." Molecular and cellular biology.

Xue, L., K. Mukherjee, K. A. Kelley and J. J. Bieker (2024). "Generation, characterization, and use of EKLF (Klf1)/CRE knock-in mice for cell-restricted analyses." Frontiers in Hematology **2**: 1292589.

Yang, L., J.-D. Zhou, T.-J. Zhang, J.-C. Ma, G.-F. Xiao, Q. Chen, Z.-Q. Deng, J. Lin, J. Qian and D.-M. Yao (2018). "Overexpression of lncRNA PANDAR predicts adverse prognosis in acute myeloid leukemia." Cancer management and research: 4999–5007.

Yang, Q., L. Chen, H. Zhang, M. Li, L. Sun, X. Wu, H. Zhao, X. Qu, X. An and T. Wang (2024). "DNMT1 regulates human erythropoiesis by modulating cell cycle and endoplasmic reticulum stress in a stage-specific manner." Cell Death Differ **31**(8): 999–1012.

Yang, X. J. (2015). "MOZ and MORF acetyltransferases: Molecular interaction, animal development and human disease." Biochim Biophys Acta **1853**(8): 1818–1826.

Yao, H., P. Sun, M. Duan, L. Lin, Y. Pan, C. Wu, X. Fu, H. Wang, L. Guo and T. Jin (2017). "microRNA-22 can regulate expression of the long non-coding RNA MEG3 in acute myeloid leukemia." Oncotarget **8**(39): 65211.

Yin, Q., X. Shen, X. Cui and S. Ju (2019). "Elevated serum lncRNA TUG1 levels are a potential diagnostic biomarker of multiple myeloma." Experimental hematology **79**: 47–55. e42.

Yu, H., S. Peng, X. Chen, S. Han and J. Luo (2020). "Long non-coding RNA NEAT1 serves as a novel biomarker for treatment response and survival profiles via microRNA-125a in multiple myeloma." Journal of clinical laboratory analysis **34**(9): e23399.

Yu, L., L. Zhang, Z. Jiang and B. Yu (2021). "Decreasing lncRNA PVT1 causes Treg/Th17 imbalance via NOTCH signaling in immune thrombocytopenia." Hematology **26**(1): 734–740.

Yu, X., Y. Zhang, F. Luo, Q. Zhou and L. Zhu (2024a). "The role of microRNAs in the gastric cancer tumor microenvironment." Molecular Cancer **23**(1): 170.

Yu, X., H. Zhao, R. Wang, Y. Chen, X. Ouyang, W. Li, Y. Sun and A. Peng (2024b). "Cancer epigenetics: from laboratory studies and clinical trials to precision medicine." Cell Death Discovery **10**(1): 28.

Zeng, H., H. Wu, M. Yan, L. Tang, X. Guo and X. Zhao (2020). "Characterization of a 4 lncRNAs-based prognostic risk scoring system in adults with acute myeloid leukemia." Leukemia Research **88**: 106261.

Zhang, H., J. Kang, L. Liu, L. Chen, S. Ren and Y. Tao (2020a). "MicroRNA-143 sensitizes acute myeloid leukemia cells to cytarabine via targeting ATG7-and ATG2B-dependent autophagy." Aging (Albany NY) **12**(20): 20111.

Zhang, L., L. Yu, Y. Liu, S. Wang, Z. Hou and J. Zhou (2021b). "miR-21-5p promotes cell proliferation by targeting BCL11B in Thp-1 cells." Oncology letters **21**(2): 1–1.

Zhang, Q., W. Dou, C. N. T. Taning, G. Smagghe and J.-J. Wang (2021a). "Regulatory roles of microRNAs in insect pests: prospective targets for insect pest control." Current Opinion in Biotechnology **70**: 158–166.

Zhang, Y., X. Miao, Z. Zhang, R. Wei, S. Sun, G. Liang, H. Li, C. Chu, L. Zhao and X. Zhu (2020b). "miR-374b-5p is increased in deep vein thrombosis and negatively targets IL-10." Journal of Molecular and Cellular Cardiology **144**: 97–108.

Zhang, Y., Z. Zhang, R. Wei, X. Miao, S. Sun, G. Liang, C. Chu, L. Zhao, X. Zhu and Q. Guo (2020c). "IL (Interleukin)-6 contributes to deep vein thrombosis and is negatively regulated by miR-338-5p." Arteriosclerosis, thrombosis, and vascular biology **40**(2): 323–334.

Zhao, A., H. Zhou, J. Yang, M. Li and T. Niu (2023). "Epigenetic regulation in hematopoiesis and its implications in the targeted therapy of hematologic malignancies." Signal Transduction and Targeted Therapy **8**(1): 71.

Zheng, Q., Z. Lin, J. Xu, Y. Lu, Q. Meng, C. Wang, Y. Yang, X. Xin, X. Li and H. Pu (2018). "Long noncoding RNA MEG3 suppresses liver cancer cells growth through inhibiting β-catenin by activating PKM2 and inactivating PTEN." Cell death & disease **9**(3): 253.

Zheng, Z., X. Li, Y. Zhu, W. Gu, X. Xie and J. Jiang (2016). "Prognostic significance of MiRNA in patients with diffuse large B-Cell lymphoma: a meta-analysis." Cellular Physiology and Biochemistry **39**(5): 1891–1904.

Zhou, H., Y. Ning, G. Zeng, C. Zhou and X. Ding (2021a). "Curcumin promotes cell cycle arrest and apoptosis of acute myeloid leukemia cells by inactivating AKT." Oncol Rep **45**(4): 11.

Zhou, M., M. Yuan, M. Zhang, C. Lei, O. Aras, X. Zhang and F. An (2021b). "Combining histone deacetylase inhibitors (HDACis) with other therapies for cancer therapy." Eur J Med Chem **226**: 113825.

Zhou, Y., S. Abraham, P. Andre, L. C. Edelstein, C. A. Shaw, C. A. Dangelmaier, A. Y. Tsygankov, S. P. Kunapuli, P. F. Bray and S. E. McKenzie (2015). "Anti–miR-148a regulates platelet FcγRIIA signaling and decreases thrombosis in vivo in mice." Blood, The Journal of the American Society of Hematology **126**(26): 2871–2881.

Zhu, B., H. Chen, X. Zhang, Y. Pan, R. Jing, L. Shen, X. Wang, S. Ju, C. Jin and H. Cong (2018). "Serum miR-30d as a novel biomarker for multiple myeloma and its antitumor role in U266 cells through the targeting of the MTDH/PI3K/Akt signaling pathway." International journal of oncology **53**(5): 2131–2144.

Zhu, Y., J. He, Z. Li and W. Yang (2023). "Cuproptosis-related lncRNA signature for prognostic prediction in patients with acute myeloid leukemia." BMC bioinformatics **24**(1): 37.

Chapter 7
A Comprehensive Guide on Epigenetics' Precision Role in Uncommon "Rare" Cancers' Biology and Treatment Frontiers

Nadia M. Hamdy, Ahmed S. Doghish, Nehal I. Rizk, Nourhan M. Abdelmaksoud, Tohada M. AL-Noshokaty, Mohamed Bakr Zaki, John Frederick Kennedy, and Ahmed I. Abulsoud

Abstract The significance of epigenetics and precision medicine in the context of uncommon cancers, referred to as "rare cancers," is investigated in this chapter.

A comprehensive review of the current literature on noncoding RNAs (ncRNAs), natural products, and clinical trials specific to uncommon cancers was conducted. The role of ncRNAs in tumor biology is elaborated, revealing their dual influence on tumor suppression and oncogenesis, alongside current therapeutic strategies targeting these molecules. Recent clinical trial outcomes were analyzed to assess therapeutic poten-

---

The original version of the chapter has been revised. A correction to this chapter can be found at https://doi.org/10.1007/978-3-032-08530-6_9

N. M. Hamdy
Biochemistry and Molecular Biology Department, Faculty of Pharmacy, Ain Shams University, Cairo, Egypt
e-mail: nadia_hamdy@pharma.asu.edu.eg

A. S. Doghish
Department of Biochemistry, Faculty of Pharmacy, Badr University in Cairo (BUC), Cairo, Egypt

Biochemistry and Molecular Biology Department, Faculty of Pharmacy (Boys), Al-Azhar University, Cairo, Egypt

N. I. Rizk
Faculty of Pharmacy, Biochemistry Department, Egyptian Chinese University, Cairo, Egypt

N. M. Abdelmaksoud · T. M. AL-Noshokaty
Faculty of Pharmacy, Biochemistry Department, Heliopolis University, Cairo, Egypt

M. Bakr Zaki
Department of Biochemistry, Faculty of Pharmacy, University of Sadat City, Sadat City, Menoufia, Egypt

Department of Biochemistry, Faculty of Pharmacy, Menoufia National University, Menofia, Egypt

J. F. Kennedy
Chembiotech Laboratories, Kyrewood House, Tenbury Wells, Worcestershire, UK

N. M. Hamdy et al. (eds.), *From (Epi)genetics-to-Precision (E2P)*, Subcellular Biochemistry 114, https://doi.org/10.1007/978-3-032-08530-6_7

tials and innovations. The role of epigenetic mechanisms in rare cancer development and treatment, particularly focusing on how natural products can modify epigenetic mechanisms, such as DNA methylation, histone modifications, and the regulation of ncRNAs, was discussed in detail. The chapter emphasizes the potential of these natural products as therapeutic agents, or "epi-drugs," in cancer prevention and treatment.

Keywords Epigenetics · PM · Uncommon cancers · micrornas · lncRNAs · Biomarkers · Therapeutic targets

Abbreviations

ACC	Adrenocortical carcinoma
AGAP2-AS1	AGAP2 antisense RNA 1
AML	Acute myeloid leukemia
ANGPTL4	Angiopoietin-like 4
ANP32A	Acidic nuclear phosphoprotein 32 family member A
ANRIL	Antisense noncoding RNA in the inhibitor of cyclin-dependent kinase 4 (INK4) locus
ARL2	ADP ribosylation factor-like GTPase 2
ATG12	Autophagy-related gene 12
ATM	Ataxia telangiectasia mutated
ATRA	All-trans retinoic acid
BANCR	BRAF-regulated lncRNA 1
BCL-2	B cell lymphoma 2
BDNF	Brain-derived neurotrophic factor
BMP	Bone morphogenetic protein
CASC9	Cancer susceptibility candidate 9
CCA	Cholangiocarcinoma
CCND1	Cyclin D1
CDK6	Cyclin-dependent kinase 6
CDKN1A	Cyclin-dependent kinase inhibitor 1A
circ-CCAC1	CCA-associated circRNA 1
CK-1α	Casein kinase-1α
CLL	Chronic lymphocytic leukemia
CRNDE	Colorectal neoplasia differentially expressed
CSF1	Colony-stimulating factor 1
CTGF	Connective tissue growth factor
CX3CL1	Chemokine (C-X3-C motif) ligand 1

A. I. Abulsoud (✉)
Biochemistry and Molecular Biology Department, Faculty of Pharmacy (Boys), Al-Azhar University, Cairo, Egypt

Faculty of Pharmacy, Biochemistry Department, Heliopolis University, Cairo, Egypt
e-mail: ahmed.abulsoud@hu.edu.eg

CXCR4	C-X-C chemokine receptor type 4
DANCR	Differentiation antagonizing non-protein coding RNA
DDLPS	Dedifferentiated liposarcoma (LPS)
DEPDC1B	DEP domain containing protein 1B
DLEU1	Deleted in leukemia 1
DLX6-AS1	DLX6 antisense RNA 1
DNMT3A	DNA methyltransferase 3 alpha
E2F1	E2F transcription factor 1
EGR2	Early growth response protein 2
EMT	Epithelial-mesenchymal transition
EphA1	Ephrin receptor A1
EPIC1	Epigenetically induced lncRNA1
ERRFI1	ERBB receptor feedback inhibitor 1
EZH2	Zeste homolog 2
FBP1	Fructose-1,6-bisphosphatase
FENDER	FOXF1 adjacent noncoding developmental regulatory RNA
GAS5	Growth arrest-specific 5
GBM	Glioblastoma multiforme
GREM1	Gremlin 1
HCC	Hepatocellular carcinoma
HDAC-1	Histone deacetylase 1
HIF-1α	Hypoxia-inducible factor 1-alpha
HMGB1	High-mobility group box 1
HNPRK	Heterogenous nuclear ribonucleoprotein K
HOTAIR	HOX transcript antisense RNA
HOTAIRM1	HOXA transcript antisense RNA, myeloid-specific 1
HOTTIP	HOXA transcript at the distal tip lncRNA
HOXA13	Homeobox A13
HULC	Highly upregulated in liver cancer lncRNA
IKAROS	IKZF1 gene
IL-6	Interleukin 6
ITGA4	Integrin subunit alpha 4
JNK	C-Jun N-terminal kinase
KCNQ1	Potassium voltage-gated channel subfamily Q member 1
KLF2	Krüppel-like factor 2
LASP1	LIM and SH3 protein 1
LATS2	Large tumor suppressor kinase 2
LEF1	Lymphoid enhancer-binding factor 1
lincRNA-p21	Long intergenic noncoding RNA p21
lncNB1	Neuroblastoma 1
LPS	Liposarcoma
MALAT1	Metastasis-associated lung adenocarcinoma transcript 1
MAP3K8	Mitogen-activated protein kinase kinase 8
MAPK	Mitogen-activated protein kinase
MAZ	MYC-associated zinc finger protein
MCL1	Myeloid cell leukemia 1

MEG3	Maternally expressed 3
MET	MET proto-oncogene, receptor tyrosine kinase
MFAP4	Microfibril-associated protein 4
MIAT	Myocardial infarction-associated transcript
MIR17HG	miR-17-92a-1 cluster host gene
MKNK1	MAPK-interacting serine/threonine kinase 1
MPM	Malignant pleural mesothelioma
NB	Neuroblastoma
NBAT-1	Neuroblastoma-associated transcript-1
NEAT1	Nuclear-enriched abundant transcript 1
NF-κB	Nuclear factor-kappa B
NLK	Nemo-like kinase
NPM1	Nucleophosmin 1
NRP-1	Neuropilin 1
Octt4	Octamer-binding transcription factor 4
OS	Overall survival
OSMR	Oncostatin M receptor
PCAT6	Prostate cancer-associated transcript 6
PDCD4	Programmed cell death 4
PDGFR	Platelet-derived growth factor receptor
PDPN	Podoplanin
PEG3	Paternally expressed gene 3
PFS	Progression-free survival
PTEN	Phosphatase and tensin homolog
PI3K	Phosphoinositide 3-kinase
PIWIL4	Piwi-like RNA-mediated gene silencing 4
PLAC2	Placenta-specific protein 2
PLK1	Polo-like kinase 1
PRC1	Protein regulator of cytokinesis 1
PRINS	Psoriasis-associated nonprotein coding RNA induced by stress
PTEN	Phosphatase and tensin homolog
PTTG1	Pituitary tumor-transforming gene-1
PUMA	P53-upregulated modulator of apoptosis
PVT1	Plasmacytoma variant translocation 1
RAS	Rat sarcoma
RASSF1	Ras association domain family member 1
RB	Retinoblastoma
RCBTB1	Regulator of chromosome condensation 1 (RCC1) and broad complex, tramtrack and bric-a-brac (BTB) domain containing protein 1
RP	Ribosomal protein
RPL36	Ribosomal protein L36 gene
Sam68	Src-associated in mitosis with a molecular weight of 68 kDa
SH2B3	SH2B adapter protein 3
SHIP1	Src homology 2 (SH2) domain containing inositol polyphosphate 5-phosphatase 1

SIRT1	Sirtuin 1
SMAD4	SMAD family member 4
SMARCA4	SWItch/Sucrose Non-Fermentable (SWI/SNF)-related, matrix-associated, actin-dependent regulator Of chromatin, subfamily A, member 4
SNHG1	Small nucleolar RNA host gene 1
SOX9	SRY-box transcription factor 9
SP1	Specificity protein 1
SPRY2	Sprouty RTK signaling antagonist 2
SPRY4-IT1	Sprouty RTK signaling antagonist 4-intronic transcript 1
STAT3	Signal transducer and activator of transcription 3
STS	Soft-tissue sarcomas
TAF1D	TATA-box-binding protein-associated factor, RNA polymerase I subunit D
Tap63	Tumor suppressor homolog of p53
TDRG1	Testis developmental related gene 1
TGCTs	Testicular germ cell tumors
TGF-β	Transforming growth factor-β
TIMP3	TIMP metallopeptidase inhibitor 3
TP53INP1	Tumor protein 53-induced nuclear protein 1
TRIM8	Tripartite motif 8
TSEN15	TRNA-splicing endonuclease subunit Sen15
TSKU	Tsukushi, small leucine-rich proteoglycan
TTN-AS1	TTN antisense RNA 1
UCA1	Urothelial cancer-associated 1
VCAN	Versican
VEGF	Vascular endothelial growth factor
XIST	X-inactive specific transcript
YAP1	Yes-associated protein 1
ZBTB4	Zinc finger and BTB domain-containing 4

Introduction

Background

Carcinogenesis is a complex, multistage process in which normal cells experience a sequence of genetic and epigenetic changes, eventually resulting in cancer development and progression. This transformation unfolds through various essential stages and mechanisms that establish the defining features of cancer cells (Hanahan and Weinberg 2011). Traditional approaches, such as radiation, surgery, and chemotherapy, are being supplemented by the exploration of natural products that demonstrate potential in modulating cancer through epigenetic mechanisms (Costa et al. 2023).

Uncommon cancers present significant sociomedical challenges due to their rarity and unique biological characteristics. These cancers often exhibit distinct clinical features that warrant focused study in cancer biology, revealing insights that could enhance our understanding of both rare and common tumors. Rare cancers are typically defined as those affecting fewer than 40,000 new patients annually in the United States or fewer than 6 in 100,000 people in Europe (Vivekanandhan et al. 2023). Collectively, however, they account for a substantial portion of cancer diagnoses, with nearly one-third of all cancer patients being diagnosed with a rare tumor type. This prevalence underscores the importance of studying these cancers, as they can illuminate mechanisms relevant to more common types (Gatta et al. 2011).

Uncommon cancers often arise from specific anatomical locations and exhibit unique phenotypes. For instance, certain tumors may result from lineage-dependent mutational events that are specific to particular cell types or tissues (Gatta et al. 2011). This specificity can lead to distinct pathways of tumorigenesis that differ from those seen in more prevalent cancers (Gatta et al. 2011; Vivekanandhan et al. 2023). Additionally, some rare tumors are characterized by metaplastic changes, particularly involving secretory cells. In these cases, the cancer may reflect adaptations at the stem-cell level and involve coevolutionary processes not observed in typical common cancers. This complexity necessitates a deeper investigation into the molecular alterations driving their transformation and growth (Vivekanandhan et al. 2023).

The study of rare tumors is crucial for identifying the molecular changes that underlie their development. Many rare tumors display a more uniform genetic profile compared to common cancers, which often exhibit significant heterogeneity among cases (Tognon et al. 2022). This uniformity can facilitate the identification of key molecular pathways involved in tumor growth and treatment resistance. For example, mutations in metabolic enzymes like isocitrate dehydrogenases (IDHs) have been linked to rare brain tumors, such as glioblastoma, but have also been found in other cancer types, suggesting broader implications for cancer biology (Solomou et al. 2023). Understanding these connections can lead to novel therapeutic strategies applicable to various cancer types (Solomou et al. 2023).

Epigenetics encompasses dynamic changes that influence gene transcription, including mechanisms like chromatin remodeling, histone modifications, and DNA methylation. These modifications play critical roles in regulating cellular processes, such as growth, differentiation, and tissue-specific gene expression (Pathak et al. 2023). Specifically, noncoding RNAs (ncRNAs), which include microRNAs (miRNAs) and long noncoding RNAs (lncRNAs), are significant players in epigenetic regulation and have been linked to various cancers due to their roles in posttranscriptional gene regulation (Fathi Dizaji 2020).

For instance, miRNAs can regulate gene expression by binding to messenger RNAs (mRNAs), leading to their degradation or translational repression. Dysregulation of miRNAs is common in cancers and contributes to processes such as proliferation and metastasis (Hamdy et al. 2024b). On the other hand, lncRNAs can modulate gene expression by interacting with DNA, RNA, or proteins, functioning as decoys, scaffolds, or guides for transcription regulators. Aberrant expression of lncRNAs has been associated with multiple cancer types and is increasingly recognized as a target for therapeutic intervention (Arun et al. 2018).

The potential of natural products in cancer treatment lies in their ability to influence epigenetic modifications. Many natural compounds have been identified that can restore normal ncRNA expression patterns or modulate their activity, thereby enhancing the efficacy of existing therapies (Huang et al. 2023a). Additionally, compounds derived from plants have shown promise in altering miRNA profiles associated with cancer progression, potentially reversing resistance to conventional therapies (Ki et al. 2024).

Aim

This chapter aims to explore the role of epigenetic mechanisms in rare cancer development and treatment, particularly focusing on how natural products can modify epigenetic mechanisms, such as DNA methylation, histone modifications, and the regulation of ncRNAs. The chapter emphasizes the potential of these natural products as therapeutic agents, or "epi-drugs," in cancer prevention and treatment. The chapter discusses how noncoding RNAs, including miRNAs and lncRNAs, are involved in various cellular processes, such as cell proliferation, apoptosis, and DNA damage response. These ncRNAs can act as epigenomic modifiers themselves and are often deregulated in cancer, affecting the expression of tumor suppressor genes and contributing to tumorigenesis. Targeting ncRNAs with natural compounds is presented as a promising approach for cancer therapy, potentially opening new avenues for drug development.

Overview of (Epi)genetics and Precision Medicine in Cancer

Tailored Cancer Care: Harnessing the Epigenome

Any tumor that affects just a tiny fraction of the population, regardless of the age group, is considered rare. A rare disease, often known as an orphan disease, is defined by the National Institutes of Health (NIH) as a disease that affects fewer than 200,000 people annually in the United States. When considered independently, rare malignancies place an enormous strain on the human species, regardless of where they live (Mano et al. 2010; Mallone et al. 2013).

The term "epigenetics" refers to a set of biological mechanisms that regulate DNA templates through chromatin, rather than through alterations to the DNA sequence itself. Epigenetic modification controls protein complexes that can be categorized as writers, readers, or erasers. This includes DNA methylation and covalent histone modification. Epigenetic markers are unique epigenetic chemical alterations that epigenetic writers introduce to DNA or histones. The specialized domains of modified proteins are recognized and understood by readers, which are methyl-CpG-binding domain proteins (MBPs). Enzymes that alter chromatin act as erasers by erasing epigenetic marks. There is a high frequency of epigenetic dysregulation across uncommon cancer types. This includes DNA and RNA methylation abnormalities, as well as aberrant posttranslational modification activities (Biswas and Rao 2018; Singh et al. 2022; Yu et al. 2024).

Precision medicine is characterized by geneticist-defined, tailored preventative and treatment approaches that account for individual variability based on underlying biological causes and other factors. The ideal "precision medicine" would combine traditional clinicopathological indicators with cutting-edge molecular profiling to develop individualized plans for diagnosis, prognosis, and treatment (Hamdy et al. 2024a). The field of pharmacogenetics defines "precision medicine" as the process of finding and creating vaccines and medications by combining molecular and clinical data. The focus of precision medicine is on finding specific treatments rather than root causes. The field of customized medicine has been greatly enhanced by the data collected from the Human Genome Project. When doctors get access to a patient's genetic profile, they are better able to choose the right treatment for their condition and ensure that their patients take it as prescribed (Hamburg and Collins 2010; Katsnelson 2013; Collins and Varmus 2015).

The therapeutic care of cancer patients has already been influenced by targeting therapy based on aberrant genetic alterations. As an example, epidermal growth factor receptor (EGFR) kinase inhibitors have been effective in treating tumors with EGFR gene mutations. In medulloblastoma and glioblastoma, the cancer genomic landscape identified 138 "mutation (Mut)-driver" genes (Sokolov et al. 2024). The majority of cancers have mutations, but research suggests that on average, just 2 to 8 "driver gene" alterations—participants in 12 signaling pathways that control cell survival, genome maintenance, and cell fate—are present in the typical tumor. When it comes to human tumors, aberrant epigenetic alterations are more common than gene mutations. As an example, compared to mutational inactivation, the epigenetic silence of the well-known driver genes, cyclin-dependent kinase inhibitor 2A (CDK2NA) and DNA mismatch repair protein (MLH1), is far more common (Sharma et al. 2007; Beggs et al. 2013; Vogelstein et al. 2013).

One of the most important factors in the development and advancement of cancer is the accumulation of genetic and epigenetic changes (Guo et al. 2008). Epigenetic changes have not yet been thoroughly investigated for their potential use in customized treatment despite the obvious importance of epigenetic "machinery" disturbance in cancer progression. Human cancer patients may benefit from epigenome-based personalized treatment if cancer epigenome landscapes can be identified. Clinical trials for new therapeutic medicines that target epigenetic machinery have been initiated as a result of our growing understanding of the epigenetic alterations that cause cancer (Costa et al. 2023; Sahafnejad et al. 2023).

Epigenetic Signatures: Predicting Chemotherapy Outcomes

As part of personalized medicine, biomarkers that indicate how a patient will react to chemotherapy can also be found. Analyzing methylation patterns can help evaluate the efficacy of chemotherapeutic drugs or the results of clinical trials. Tumor-specific treatment response indicators in various malignancies have been discovered by DNA methylation profiling. By removing carcinogenic alkyl groups from guanine's O6 position, methylated-DNA-protein-cysteine methyltransferase (MGMT) prevents G-A transitions and the creation of harmful cross-links in DNA following DNA damage. Glioma patients often experience MGMT silencing through methylation. Research has linked MGMT methylation to an improved response to chemotherapy, based on alkylators, longer overall survival, and delayed disease progression. One important factor in maintaining chromosomal integrity is the checkpoint with forkhead and ring finger domains (CHFR) gene, which acts as an early mitotic checkpoint. At the G2/M checkpoint, CHFR regulates cell cycle progression. A mitotic arrest occurs when the expression of CHFR is elevated. In humans, the methylation of CHFR is a sensitive indicator for taxanes in oral squamous cell carcinoma, endometrial cancer, and gastric, cervical, lung, and esophageal malignancies. The overall survival rate was higher in the CHFR methylation group compared to the CHFR unmethylated group in patients with gastric cancer who were treated with docetaxel. A gene that falls under the Fanconi anemia family is FANCF. Oxaliplatin resistance is indicated by MLH1 methylation. Overall survival was longer in the MLH1 unmethylated group compared to the MLH1 methylated group in gastric cancer patients treated with oxaliplatin (Esteller et al. 2000; Heyn and Esteller 2012; Li et al. 2015b).

The three DNA methyltransferases (DNMTs), DNMT1, DNMT3A, and DNMT3B, are responsible for establishing and maintaining DNA methylation patterns. De novo methylation is the job of DNMT3A and DNMT3B, whereas DNMT1 is in charge of maintaining preexisting methylation patterns. Two DNA demethylating drugs that inhibit DNMT1 and DNMT3B are 5-azacytidine and 5-aza-2′deoxycytidine, also known as decitabine. The original intent of these demethylating medicines was as chemotherapeutic agents; however, their lethal levels caused considerable toxicity when employed in that capacity (Kim et al. 2010; Ahuja et al. 2014).

Among the many modifications that can be applied to histone tails are acetylation, methylation, and ubiquitination. These alterations control gene expression by interacting with proteins associated with chromatin, which identify areas of euchromatin that are actively involved in transcription and those of heterochromatin that are not. Potentially, the DNA sequence dictates chromatin layouts and functions. Promoter, enhancer, and gene body histone alterations can activate or repress transcription, depending on their dynamics. Hypomethylation at the DNA level is common in promoters that include histone marks, such as acetylated histone H3 and, more specifically, di- or tri-methylated histone H3 lysine 4 (H3K4me2 and H3K4me3). Therefore, transcriptional initiation occurs in these promoter areas, which are also enriched for RNA polymerase II (RNAPII) (Zhou et al. 2011; Laille et al. 2014).

Innovative Cancer Therapies: The Role of Nonprotein Coding RNAs (ncRNAs)

ncRNAs are RNA molecules that can function but do not include protein-coding genes. Depending on their size, they are categorized into small interfering RNAs (siRNAs), micro-ribonucleic acids (miRNAs), piwi-interacting RNAs (piRNAs), and long noncoding RNAs (lncRNAs).

Tiny ncRNAs, about 18 to 25 nucleotides in length, target the 3′-untranslated regions (UTRs) of messenger RNAs (mRNAs) and either degrade them or prevent their translation. Close to 40%–60% of human genes are supposedly regulated by microRNAs. The number of mRNAs that each miRNA targets and the number of miRNAs that target each mRNA are both open questions. Based on their function, which is mostly dictated by the impact of their target mRNAs on cancer progression, these miRNAs can be categorized into either tumor-suppressing miRNAs or oncogenic miRNAs. On its own or in conjunction with existing targeted medicines, anticancer medications based on microRNAs are now under development. So far, numerous studies have shown that miRNAs can increase sensitivity to immunotherapy, radiation, and chemotherapy (Garofalo et al. 2014; Guo and Yan 2015; Yan and Guo 2015).

Restoring the activity of an miRNA that has been downregulated or deleted is possible with the help of miRNA mimics. Possible treatments for oncogenic miRNAs include small-molecule inhibitors, miRNA masking, anti-miRNA oligonucleotides, and miRNA sponges. An effective strategy for preventing miRNAs from binding to their targets is the use of anti-miRNA oligonucleotides (AMOs). This suggests that microRNAs (miRNAs) may be a promising new target for AMOs in cancer treatment (Weiler et al. 2006; Bader et al. 2011). The delivery of miRNA-targeting drugs is still an obstacle to overcome before moving on to clinical applications, despite the promising early results and promising future. MiRNAs are formed from genome-encoded hairpin-shaped precursors and only require a partial sequence match to inhibit target gene expression, in contrast to siRNAs, which are often derived from synthetic or foreign double-stranded RNAs (dsRNAs) and require perfect matching to cleave a specific target mRNA. Therefore, there are inherent parallels and distinctions between miRNA- and siRNA-based therapeutic approaches (Li et al. 2009).

Researchers have just scratched the surface of the lncRNA's potential as a medicinal agent. It is possible to create single-stranded oligonucleotides with the specificity to either degrade the antisense transcript and therefore induce transcriptional derepression of the gene or prevent the interaction between the antisense transcript and the sense gene mRNA. Natural derepression and upregulation of antisense transcripts can also be achieved with double-stranded siRNAs but only if the antisense transcript is targeted in a location that does not overlap with the sense transcript. Hybridization of oligonucleotides can occur either on or off the target, leading to unexpected and undesirable outcomes. All oligonucleotide antisense methods showed the majority of the harmful effects (Bennett and Swayze 2010; Modarresi et al. 2012; Qi and Du 2013).

Overview of ncRNAs Relevant to Uncommon Cancers

In human cancers, ncRNAs are essential because they can act as suppressors or oncogenes to control the development and spread of cancer. The roles of different noncoding RNAs in different types of uncommon cancer are summarized in Table 7.1.

Chronic Lymphocytic Leukemia (CLL)

Because of its delayed clinical course, CLL has often been characterized as a disease in which malignant B cells accumulate rather than proliferate due to a dysregulated apoptosis (El Mesallamy et al. 2014). A crucial element of CLL characterization is the proper equilibrium between the two biological processes. Although poor apoptosis is the source of the buildup of malignant B cells in the blood, bone marrow, and lymphoid organs, a smaller and actively reproducing population still exists, primarily in the bone marrow and lymph nodes (LNs) (Messmer, Messmer et al. 2005, Herndon et al. 2017). This intricate process is regulated at multiple levels by miRNAs and lncRNAs. The miR-17/92 polycistronic miRNA cluster, which includes **miR-17**, **miR-18a**, **miR-19a**, **miR-19b**, **miR-20a**, and **miR-92a**, is one of the most well-studied clusters and is overexpressed in several lymphoid malignancies, including CLL (Bomben et al. 2012). Through a variety of targets, such as cyclin-dependent kinase inhibitor A1 (CDKN1A), connective tissue growth factor (CTGF), early growth response protein 2 (EGR2), E2Fs, DNA-binding protein IKAROS, phosphatase and tensin homolog (PTEN), signal transducer and activator of transcription (STATs), tumor protein p53-inducible nuclear protein 1 (TP53INP1), tripartite motif–containing 8 (TRIM8), and zinc finger and BTB domain-containing 4 (ZBTB4), the miR-17/92 cluster controls cell cycle progression and proliferation (Mogilyansky and Rigoutsos 2013). E2F5 is engaged in the G1 arrest (Majumder et al. 2016), whereas ZBTB4 and TP53INP1 are implicated in the regulation of apoptosis via CDKN1A and TP53 (Weber et al. 2008).

Myocardial infarction-associated transcript (**MIAT**) is an lncRNA that has a significant functional role in the growth of CLL. Malignant mature B cells, such as original leukemic cells isolated from CLL patients, established CLL cell lines, and cells of non-Hodgkin lymphoma origin overexpress MIAT. MIAT expression is

Table 7.1 The role of different noncoding RNAs in the pathogenesis of uncommon cancers

Cancer type	ncRNA	Target	Mechanism	References
CLL	miR-17/92 cluster	CDKN1A, CTGF, EGR2, E2Fs, IKAROS, PTEN, STATs, TP53INP1, TRIM8, and ZBTB4	Controls cell cycle progression and proliferation Promotes G1 arrest and regulates apoptosis	Weber et al. (2008), Bomben et al. (2012), Mogilyansky and Rigoutsos (2013), and Majumder et al. (2016)
	MIAT	OCT4	Prevents apoptosis and promotes the growth of monoclonal malignant B cells	Sattari et al. (2016)
	DLEU1 and DLEU2	–	Positively regulate NF-κB activity	Garding et al. (2013)
	lincRNA-p21	–	Induces apoptosis and cell cycle arrest	Blume et al. (2015) and Adriaens et al. (2016)
	NEAT1	–	Induces apoptosis and cell cycle arrest	Blume et al. (2015) and Adriaens et al. (2016)
	HULC	CCND1 and BCL-2	Reduces the expression of HULC, causing increased apoptosis	Peng et al. (2016)
ACC	miR-100	CXCR7 and PLK1	Triggers the MAPK signaling pathway and accelerates the growth of the tumor	Shi et al. (2010) and Chifu et al. (2020)
	miR-181b	Bcl-2 and ATM	Inhibits the development of ACC cells and stimulates DNA damage response to double-strand breaks	De Martino et al. (2013) and Crona and Beuschlein (2019))
	miR-195	CCND1 and CDK6	Progresses the G1 phase and the G1/S transition of the cell cycle	Hadjadj et al. (2017)
	miR-483-5p	MKNK1 and KCNQ1	Suppresses tumor cell proliferation, migration, and invasion and triggers apoptosis	Özata et al. (2011), Liu et al. (2019b), and Chen et al. (2020)

(continued)

Table 7.1 (continued)

Cancer type	ncRNA	Target	Mechanism	References
	PRINS	G1P3	Downregulates the Wnt/β-catenin pathway	Szegedi et al. (2010) and Glover et al. (2015)
	MALAT1	–	Induces ACC carcinogenesis	Shiyanbola et al. (2020)
	H19	–	Induces ACC carcinogenesis	Di Fazio et al. (2022)
	GAS5	–	Induces ACC carcinogenesis	Decmann et al. (2020)
	CRNDE	–	Controls the Notch1 and PI3K/Akt signaling pathways	De Martino et al. (2013) and Ronchi et al. (2015)
	circ-CCAC1	–	Promotes ACC cell invasion, migration, and proliferation	Li et al. (2020b)
RB	miR-491-3p	Vimentin, N-cadherin, and E-cadherin	Promotes apoptosis by suppressing RB cell invasion, migration, and proliferation	Hu et al. (2021)
	miR-340	–	Promotes RB cell invasion, migration, and proliferation	Fuchs et al. (2001)
	miR-613	E2F5	Causes cell cycle arrest and inhibits RB cell invasion, migration, and proliferation	Zhang et al. (2017c)
	miR-936	HDAC9	Suppresses RB cell proliferation, colony formation, migration, and invasion in vitro and decreases tumor growth in vivo by regulating the PI3K/Akt pathway	Xu et al. (2020)
	miR-101-3p	EZH2 and HDAC9	Inhibits RB cell growth, causing cell cycle arrest	Zheng et al. (2020)
	MIAT	miR-665/LASP1 axis	Induces RB cell invasion, migration, and proliferation	Grossi et al. (2015)

(continued)

Table 7.1 (continued)

Cancer type	ncRNA	Target	Mechanism	References
	XIST	miR-191-5p/ BDNF axis	Suppresses apoptosis and encourages invasion, migration, and proliferation	Xu et al. (2021)
	MIR17HG	miR-155-5p/ HIF-1α axis	Increases RB cell invasion, migration, and proliferation	Yan et al. (2022)
	HOTAIR	miR-20b-5p/ RRM2 axis	Activates the PI3K/ Akt pathway, promoting RB cell proliferation and inhibiting apoptosis	Fu et al. (2022)
	LINC00205	miR-665/ HMGB1 axis	Enhances RB progression	Zhang et al. (2020a)
	CASC9	E2F3 miR-145-5p/ E2F3 axis	Promotes invasion, apoptosis, EMT, and proliferation	Zhang et al. (2020b)
MPM	miR-16	CCND1 and Bcl2	Has a tumor-suppressing effect	Reid et al. (2013)
	miR-31	PPP6C/PP6C	Suppresses colony formation, migration, invasion, and proliferation	Ivanov et al. (2010)
	miR-29c	Adiponectin	Suppresses colony formation, migration, invasion, and proliferation	Pass et al. (2010)
	let-7	c-MYC, EphA2, and RAS	Inhibits MPM formation	Ramírez-Salazar et al. (2014) and Riquelme et al. (2014)
	miR-145	OCT4	Decreases the clonogenicity, viability, and migration potential of mesothelioma cells	Cioce et al. (2014)
	miR-193a-3p	–	Inhibits MPM growth, both in vitro and in vivo	Williams et al. (2015)
	PVT1	c-MYC	Decreases apoptosis by promoting cellular proliferation and reduces sensitivity to cisplatin	Riquelme et al. (2014)

(continued)

Table 7.1 (continued)

Cancer type	ncRNA	Target	Mechanism	References
	GAS5	PDPN	Promotes cell proliferation	Renganathan et al. (2014)
NB	miR-204	N-MYC	Prevents NB carcinogenesis	Ooi et al. (2018)
	miR-2110	TSKU	Enhances patient survival	Zhao et al. (2018b)
	miR-184	N-MYC	Induces apoptosis	Chen and Stallings (2007))
	miR-221	N-MYC and NLK	Regulates cell cycle and NB cell proliferation	He et al. (2017)
	miR-203	Sam68	Prevents NB cell invasion, migration, and proliferation	Zhao et al. (2015a)
	miR-337-3p and miR-584-5p	MMP14	Suppress NB cell proliferation, invasion, metastasis, and angiogenesis	Xiang et al. (2015a, b)
	miR-449a	MFAP4, PKP4, TSEN15, CDK6, and LEF1	Inhibits cell growth and survival by causing cell differentiation and cell cycle arrest	Zhao et al. (2015b)
	miR-506-3p and miR-124-3p	CDK4 and STAT3	Suppress tumor proliferation	Silber et al. (2008) and Ko et al. (2015)
	miR-193b	MCL1, CCND1, and N-MYC	Promotes G1 cell cycle arrest and NB cell death	Roth et al. (2018)
	NBAT-1	SOX9, OSMR, VCAN, and EZH2	Suppresses tumor proliferation	Pandey et al. (2014) and Pandey and Kanduri (2015)
	DLX6-AS1	–	Induces lung adenocarcinoma invasion, migration, and proliferation	Zhang et al. (2019a) and Li et al. (2020a)
	lncNB1	DEPDC1B and E2F1	Regulates the cell cycle and decreases apoptosis	Liu et al. (2019c)
	SNHG1	miR338-3p/PLK4 axis and TAF1D	Encourages invasion, migration, and proliferation	Sahu et al. (2016) and Zhang et al. (2019b)
Cholangiocarcinoma	miR-21	TIMP3 and PDCD4	Induces carcinogenesis	Selaru et al. (2009)

(continued)

Table 7.1 (continued)

Cancer type	ncRNA	Target	Mechanism	References
	miR-26a	β-catenin	Increases CCA cell proliferation	Zhang et al. (2012a)
	miR-494	CDK6, CDK4, CCND1, cyclin E2, HDAC-1, RB transcriptional corepressor 1, PLK1, PTTG1, cyclin B1, CDK2, and DNA topoisomerase IIα	Prevents cell cycle progression	Mott et al. (2007) and Olaru et al. (2011)
	miR-370	MAP3K8 and IL-6	Promotes CCA growth	Meng et al. (2008)
	miR-148a and miR-152	CDKN2A and RASSF1	Suppress tumor proliferation	Salati and Braconi (2019))
	miR-21	E-cadherin, vimentin, and N-cadherin	Regulates tumor growth	Liu et al. (2015, 2017a)
	ANRIL	EZH2 and ERRFI1	Accelerates CCA carcinogenesis	Yu et al. (2020)
	DANCR	EZH2 and FBP1	Stimulates CCA cell proliferation and migration	Wang et al. (2019)
	SNHG1	CDKN1A	Promotes CCA cell proliferation	Yu et al. (2018a, b)
	AGAP2-AS1	CDKN1A	Promotes CCA cell proliferation	Ji et al. (2021)
	PVT1	ANGPTL4	Promotes CCA cell proliferation	Yu et al. (2018a, b)
	NEAT1	E-cadherin	Promotes CCA cell proliferation	Zhang et al. (2018a)
	FENDER	Survivin	Regulates CCA cell proliferation, migration, and invasion	Qin et al. (2019)
	SPRY4-IT1	KLF2 and LATS2	Promotes CCA cell proliferation	Xu et al. (2018)
	DLEU1	miR-149-5p/ YAP1 axis	Has a critical regulatory role in cell signaling	Liu et al. (2019a) and Sugihara et al. (2019)

(continued)

Table 7.1 (continued)

Cancer type	ncRNA	Target	Mechanism	References
	TTN-AS1	miR-320a in an Ago2-dependent manner	Stimulates angiogenesis via NRP-1, the EMT, and cell cycle progression	Zhu et al. (2020)
	H19	let-7a/let-7b/IL-6 axis	Controls CCA cell motility and invasion	Wang et al. (2016)
	HULC	miR-372/ miR-373/CXCR4 axis	Controls CCA cell motility and invasion	Wang et al. (2016)
	PCAT6	miR-326/RohA signaling	Influences macrophages, including ROS generation, mitochondrial stress response, and M2 polarization, which, in turn, contribute to the immunological response of CCA	Tu et al. (2021)
TGCT	miR-302/367	Rb family and CDK1NA	Accelerates the G1 to S transition of the cell cycle and increases proliferation	Wang et al. (2013)
	miR-371-373	LATS2	Encourages tumor growth and cell division	McPherson et al. (2004)
	miR-223-3p	FBXW7	Suppresses apoptosis and encourages cell growth	Liu et al. (2017a)
	miR-449a-b	E2F1	Promotes apoptosis and inhibits cell proliferation by upregulating p53	Lizé et al. (2010)
	miR-514a-3p	PEG3	Promotes apoptosis by promoting the NF-κB pathway	Özata et al. (2017)
	miR-125b	CSF1 and CX3CL1	Suppresses tumor growth	Batool et al. (2018)
	TDRG1	–	Enhances tumor growth, progression, and cisplatin resistance	Lobo et al. (2019) and Peng et al. (2019)

(continued)

Table 7.1 (continued)

Cancer type	ncRNA	Target	Mechanism	References
	H19	miR-106b-5p, which typically negatively regulates TDRG1	Promotes cell survival under cisplatin-based chemotherapeutic conditions	Wei et al. (2018)
	SPRY4-IT1	–	Increases cell growth, migration, and invasion	Das et al. (2018)
	HOTTIP	miR-128-3p/HOXA13	Promotes cell proliferation	Su et al. (2019)
GBM	miR-21	PDCD4, ANP32A, SMARCA4, PTEN, SPRY2, Tap63, and HNPRK	Has an antiapoptotic effect and increases tumor cell proliferation and tumor growth	Chan et al. (2005), Ciafrè et al. (2005), Silber et al. (2008), Zhou et al. (2010), Gaur et al. (2011), Li et al. (2011), and Schramedei et al. (2011)
	miR-221/222	PUMA	Prevents apoptosis and increases cell viability	Chen et al. (2008) and Zhou et al. (2010)
	miR-34a	Receptor tyrosine kinases, MET, NOTCH1 and 2, CDK6, CCND1, and SIRT1	Controls cell invasion, self-renewal, proliferation, and survival in GBM	Gillies and Lorimer (2007), Zhang et al. (2010), and Chen et al. (2012)
	miR-296	VEGF	Promotes endothelial cell tube formation and improves tumor vascularization	Yao et al. (2015) and Cheng et al. (2017)
	miR-125b	MAZ	MAZ is overexpressed when miR-125b is repressed, which encourages tumor vascularization	Würdinger et al. (2008)
	miR-218	HIF-2α	Inhibits GBM tumor neovascularization and cell survival	Smits et al. (2012)
	miR-93	Integrin-β8	Promotes blood vessel development, sphere formation, and cell survival	Mathew et al. (2014)

(continued)

Table 7.1 (continued)

Cancer type	ncRNA	Target	Mechanism	References
	CRNDE	miR-384 / PIWIL4/STAT3 axis	Promotes glioma growth in vivo and stimulates cell proliferation and migration in vitro	Brown et al. (1991), Guttman et al. (2009), and Wang et al. (2018)
	MEG3	–	Inhibits glioma cell proliferation and encourages death and autophagy	Zhao et al. (2018a)
	PLAC2	RPL36/STAT3	Suppresses glioma growth in vivo in a xenograft model, decreases cell proliferation, and promotes G1/S arrest	Qin et al. (2017) and Zhang et al. (2017a)
	lncRNA-ATB	p38/MAPK pathway	Improves TGF-β-mediated invasion of glioma cells	Hu et al. (2018)
	EPIC1	Cdc20	Suppresses apoptosis, increases cell viability, makes cells less sensitive to TMZ, and increases cell invasion	Wang et al. (2020)
	PVT1	miR-128-3p/ GREM1/BMP2 and BMP4	Promotes the growth of tumors and gliomas by controlling the BMP signaling system	Tang et al. (2019)
AML	miR-155	SHIP1 and PU.1	Promotes AML carcinogenesis	Vigorito et al. (2007) and Xue et al. (2014)
	miR-183-5p	Erbin	Increases AML cell proliferation	Zheng et al. (2019)
	miR-29b	DNMT3A, DNMT3B, and DNMT1	Causes DNA hypomethylation and tumor suppressor gene reactivation	Garzon et al. (2009)
	miR-7977	Poly(rC)-binding protein	Interferes with normal hematopoiesis and causes mesenchymal stem cells to produce hematopoietic growth factors	Horiguchi et al. (2016)
	hsa_ circ_0004277	SH2B3	Suppresses AML carcinogenesis	Li et al. (2017)

(continued)

Table 7.1 (continued)

Cancer type	ncRNA	Target	Mechanism	References
	circPVT1	MYC	Promotes AML carcinogenesis	Abbate et al. (2018)
	has_ circ_0075001	NPM1	Promotes AML carcinogenesis	Hirsch et al. (2017)
	HOTAIRM1	ITGA4 (CD49d) and ITGAX (CD11c)	Regulates myeloid differentiation and maturation in early AML	Zhang et al. (2014)
	PVT1	MYC	Causes necrosis and death in AML cells	Salehi and Sharifi (2018) and Salehi et al. (2018)
	UCA1	p27kip1	Promotes tumor cells' proliferation	Hughes et al. (2015)
Sarcoma	miR-155	CK-1α	Promotes cell cycle progression, cell proliferation, and colony formation	Zhang et al. (2012b), Vincenzi et al. (2015), and Kapodistrias et al. (2017)
	miR-26a-2	RCBTB1 and HOXA5	Increases tumor cells' proliferation, survival, and migration	Lee et al. (2013, 2015)
	miR-143/ miR-145	BCL2, type II topoisomerase, PLK1, and PRC1	Prevents proliferation and causes apoptosis	Ugras et al. (2011)
	miR-145 and miR-451	–	Induce apoptosis, slow down cell cycle progression, and reduce cellular proliferation	Gits et al. (2014)
	miR-193b	PDGFR, TGF-β, SMAD4, and YAP1	Promotes LPS cell proliferation, differentiation, and carcinogenesis	Mazzu et al. (2019)
	H19	YAP1	Promotes osteosarcoma cell viability	Chan et al. (2014)
	MALAT1	PI3K/Akt signaling pathway	Induces tumor cells' proliferation	Dong et al. (2015)
	BANCR	JNK and Wnt/β-catenin signaling pathways	Decreases tumor cells' proliferation	He and Zhang (2015))

(continued)

Table 7.1 (continued)

Cancer type	ncRNA	Target	Mechanism	References
	HOTAIR	MMP2, MMP9, ATG12, and STAT3	Promotes tumor cells' invasion and metastasis	Wang et al. (2015a) and Bao et al. (2017)
	ANRIL	HIF-1α	Boosts cancer cell migration and reduces apoptosis	Wei et al. (2016)

ACC adrenocortical carcinoma, *AGAP2-AS1* AGAP2 antisense RNA 1, *AML* acute myeloid leukemia, *ANGPTL4* angiopoietin-like 4, *ANP32A* acidic nuclear phosphoprotein 32 family member A, *ANRIL* antisense noncoding RNA in the INK4 locus, *ATG12* autophagy-related gene 12, *ATM* ataxia-telangiectasia mutated, *BANCR* BRAF-regulated lncRNA 1, *BCL-2* B cell lymphoma 2, *BDNF* brain-derived neurotrophic factor, *BMP2* bone morphogenetic protein-2, *CASC9* cancer susceptibility candidate 9, *CCA* cholangiocarcinoma, *CCND1* cyclin D1, *CDK6* cyclin-dependent kinase 6, *CDKN1A* cyclin-dependent kinase inhibitor 1A, *circ-CCAC1* CCA-associated circRNA 1, *CK-1α* casein kinase-1α, *CRNDE* colorectal neoplasia differentially expressed, *CSF1* colony-stimulating factor 1, *CTGF* connective tissue growth factor, *CX3CL1* chemokine (C-X3-C motif) ligand 1; *CXCR4* C-X-C chemokine receptor type 4, *DANCR* differentiation antagonizing nonprotein coding RNA, *DEPDC1B* DEP domain containing protein 1B, *DLEU1* deleted in leukemia 1, *DLX6-AS1* DLX6 antisense RNA 1, *DNMT3A* DNA methyltransferase 3 alpha, *E2F1* E2F transcription factor 1, *EGR2* early growth response protein 2, *EMT* epithelial-mesenchymal transition, *EphA1* ephrin receptor A1, *EPIC1* epigenetically induced lncRNA1, *ERRFI1* ERBB receptor feedback inhibitor 1, *EZH2* zeste homolog 2, *FBP1* fructose-1,6-bisphosphatase, *FENDER* FOXF1 adjacent noncoding developmental regulatory RNA, *GAS5* growth arrest-specific 5, *GBM* glioblastoma multiforme, *GREM1* gremlin 1, *HDAC-1* histone deacetylase 1, *HIF-1α* hypoxia-inducible factor 1-alpha, *HMGB1* high-mobility group box 1, *HNPRK* heterogenous nuclear ribonucleoprotein K, *HOTAIR* HOX transcript antisense RNA, *HOTAIRM1* HOXA transcript antisense RNA, myeloid-specific 1, *HOTTIP* HOXA transcript at the distal tip lncRNA, *HOXA13* homeobox A13, *HULC* highly upregulated in liver cancer lncRNA, *IKAROS* IKZF1 gene, *IL-6* interleukin 6, *ITGA4* integrin subunit alpha 4, *JNK* C-Jun N-terminal kinase, *KCNQ1* potassium voltage-gated channel subfamily Q member 1, *KLF2* Krüppel-like factor 2, *LASP1* LIM and SH3 protein 1, *LATS2* large tumor suppressor kinase 2, *LEF1* lymphoid enhancer-binding factor 1, *lncNB1* neuroblastoma 1, *LPS* liposarcoma, *MALAT1* metastasis-associated lung adenocarcinoma transcript 1, *MAP3K8* mitogen-activated protein kinase kinase 8, *MAPK* mitogen-activated protein kinase, *MAZ* MYC-associated zinc finger protein, *MCL1* myeloid cell leukemia 1, *MEG3* maternally expressed 3, *MET* MET proto-oncogene, receptor tyrosine kinase, *MFAP4* microfibril-associated protein 4, *MIAT* myocardial infarction-associated transcript, *MIR17HG* MiR-17-92a-1 cluster host gene, *MKNK1* MAPK-interacting serine/threonine kinase 1, *MPM* malignant pleural mesothelioma, *NB* neuroblastoma, *NBAT-1* neuroblastoma-associated transcript-1, *NEAT1* nuclear-enriched abundant transcript 1, *NF-κB* nuclear factor-kappa B, *NLK* Nemo-like kinase, *NPM1* nucleophosmin 1, *NRP-1* neuropilin 1, *Oct4* octamer-binding transcription factor 4, *OSMR* oncostatin M receptor, *PCAT6* prostate cancer-associated transcript 6, *PDCD4* programmed cell death 4, *PDGFR* platelet-derived growth factor receptor, *PDPN* podoplanin, *PEG3* paternally expressed gene 3, *PI3K* phosphoinositide 3-kinase, *PIWIL4* piwi-like RNA-mediated gene silencing 4, *PKP4* plakophilin 4, *PLAC2* placenta-specific protein 2, *PLK1* polo-like kinase 1, *PPP6C* protein phosphatase 6 catalytic subunit, *PRC1* protein regulator of cytokinesis 1, *PRINS* psoriasis-associated nonprotein coding RNA induced by stress, *PTEN* phosphatase and tensin homolog, *PTTG1* pituitary tumor-transforming gene-1, *PUMA* P53-upregulated modulator of apoptosis, *PVT1* plasmacytoma variant translocation 1, *RAS* rat sarcoma, *RASSF1* Ras association domain family member 1, *RB* retinoblastoma, *RCBTB1* RCC1 and BTB domain containing protein 1, *RPL36* ribosomal protein

(continued)

Table 7.1 (continued)

L36 gene, *Sam68* Src-associated in mitosis with a molecular weight of 68 kDa, *SH2B3* SH2B adapter protein 3, *SHIP1* Src homology 2 (SH2) domain containing inositol polyphosphate 5-phosphatase 1, *SIRT1* sirtuin 1, *SMAD4* SMAD family member 4, *SMARCA4* SWI/SNF-related, matrix-associated, actin-dependent regulator of chromatin, subfamily A, member 4, *SNHG1* small nucleolar RNA host gene 1, *SOX9* SRY-box transcription factor 9, *SPRY4-IT1* sprouty RTK signaling antagonist 4-intronic transcript 1, *STAT3* signal transducer and activator of transcription 3, *TAF1D* TATA-box-binding protein-associated factor, RNA polymerase I subunit D, *TDRG1* testis developmental related gene 1, *TGF-β* transforming growth factor-β, *TIMP3* TIMP metallopeptidase inhibitor 3, *TP53INP1* tumor protein 53-induced nuclear protein 1, *TRIM8* tripartite motif 8, *TSEN15* TRNA-splicing endonuclease subunit Sen15, *TSKU* Tsukushi, small leucine-rich proteoglycan, *TTN-AS1* TTN antisense RNA 1, *UCA1* urothelial cancer-associated 1, *VCAN* versican, *VEGF* vascular endothelial growth factor, *XIST* X-inactive specific transcript, *YAP1* yes-associated protein 1, *ZBTB4* zinc finger and BTB domain-containing 4

directly regulated by octamer-binding transcription factor 4 (OCT4) and is correlated with the aggressiveness of the disease. By preventing apoptosis, MIAT overexpression promotes the growth of monoclonal malignant B cells with a naïve B cell phenotype, which advances CLL (Sattari et al. 2016). The multiple drug resistance (MDR) of chromosome 13 contains two lncRNAs that are transcribed head-to-head: deleted in leukemia 1 (**DLEU1**) and **DLEU2** (Bullrich et al. 2001). Their homozygous deletion, which is only partially caused by the loss of miR-15a and miR-16-1 found in DLEU2's intron 4, has a significant impact on the homeostasis and regulation of the normal cluster of differentiation (CD)5+ B cells. Indeed, recent research revealed that DLEU1 and DLEU2 regulate the transcription of nearby potential tumor suppressor genes, which could function as nuclear factor-kappa B (NF-κB) activity-positive regulators (Garding et al. 2013). Long intergenic noncoding RNA p21 (**lincRNA-p21**), located 17 kb upstream of p21 and nuclear-enriched abundant transcript 1 (**NEAT1**) are two p53-regulated lncRNAs induced in response to DNA damage identified in primary CLL cells only when a functional form of p53 is present (Blume et al. 2015). Both lncRNAs act in p53-mediated processes, specifically inducing apoptosis and cell cycle arrest (Adriaens et al. 2016). Additionally, it is crucial for the structural integrity of paraspeckles, which are protein-rich nuclear organelles based on a particular lncRNA scaffold that sequesters component proteins and RNAs to affect gene regulation (Eldash et al. 2023). Increased cellular proliferation, enhanced clonogenic survival, and decreased apoptosis are the results of hypoxic activation of NEAT1, which also promotes the development of paraspeckles (Choudhry et al. 2015). In fact, through the weakened ataxia telangiectasia mutated and Rad3-related (ATR) pathway, elevated NEAT1 levels are linked to improved chemosensitivity and death in irradiated CLL cells (Kwok et al. 2015). In the short arm of chromosome 6, the lncRNA highly upregulated in liver cancer (**HULC**) functions as an endogenous sponge, downregulating a number of miRNAs, such as miR-372 and miR-200a-3p (Wang et al. 2010). Its expression levels are directly associated with lower overall survival (OS), progression-free survival (PFS), and clinical stages, particularly in Diffuse large B-cell lymphoma (DLBCL) patients. Reduced expression of lncRNA HULC causes increased apoptosis through the downregulation of cyclin D1 (CCND1) and B cell lymphoma 2 (BCL-2) proteins (Peng et al. 2016).

Adrenocortical Carcinoma (ACC)

By controlling the CXC chemokine receptor type 7 (CXCR7), **miR-100** exhibits anticancer action (Ge et al. 2021). By triggering the MAPK signaling pathway, the chemokine receptor CXCR7 can accelerate the growth of tumors (Chifu et al. 2020). A crucial mitotic regulator at the G2/M transition, polo-like kinase 1 (PLK1) mRNA is one of the miR-100 targets (Shi et al. 2010). One of the targets of **miR-181b** linked to carcinogenesis is Bcl-2, a crucial apoptosis regulator. In vitro, Bcl-2 has been shown to inhibit the development of ACC SW-13 cells (Wu et al. 2015). The gene known as ataxia telangiectasia mutated (ATM), a crucial stimulator of the DNA damage response to double-strand breaks, is another target of miR-181b. In adrenocortical carcinoma (ACC), 4% of cases show an ATM's lack of functionality (De Martino et al. 2013; Crona and Beuschlein 2019). Multiple cancer types are linked to **miR-195**. By focusing on CCND1 and cyclin-dependent kinase 6 (CDK6), miR-195 contributes to the pathophysiology of hepatocellular carcinoma (HCC) (Furuta et al. 2013). CDK6 is essential for the progression of the G1 phase and the G1/S transition of the cell cycle and is controlled by cyclins, more especially by CCND proteins (Youness et al. 2024). High levels of CDK6 have been linked to a shorter time to relapse and a lower patient survival rate in ACC (Hadjadj et al. 2017). However, it is still unclear whether miR-195 acts through ADP ribosylation factor-like GTPase 2 (ARL2) or CDK6 in ACC. Meanwhile, in vitro experiments using ACC human cells have shown the unique involvement of miR-195 in relation to DICER and TARBP2 (Caramuta et al. 2013). Many types of cancer are influenced by **miR-483-5p**. The MAPK-interacting serine/threonine kinase 1 (MKNK1), which suppresses Wilm's tumor cell proliferation and triggers apoptosis, is one of its recognized targets (Liu et al. 2019b). Additionally, it has been demonstrated that miR-483-5p suppresses potassium voltage-gated channel subfamily Q member 1 (KCNQ1), which promotes the growth, migration, and invasion of esophageal cancer (Chen et al. 2020). The gene KCNQ1 produces a potassium channel protein that is found in a variety of tissues. Since there is evidence that miR-483-5p expression and cell proliferation are correlated in vitro, it is possible that miR-483-5p plays a similar role in ACC (Özata et al. 2011).

H19 (Di Fazio et al. 2022), growth arrest-specific 5 (**GAS5**) (Decmann et al. 2020), metastasis-associated lung adenocarcinoma transcript 1 (**MALAT1**), and psoriasis-associated nonprotein coding RNA induced by stress (**PRINS**) are among the identified lncRNAs that have a known carcinogenic effect. In particular, metastatic ACC is linked to PRINS (Glover et al. 2015). The overexpression of PRINS in psoriatic patients' epidermal cells was initially reported (Sonkoly et al. 2005). PRINS controls the apoptotic gene G1P3 in epidermal cells (Szegedi, Sonkoly et al. 2010). The Wnt/β-catenin pathway, which is essential for the pathogenicity of ACC, has been demonstrated to be upregulated when PRINS is silenced (Glover et al. 2015). Likewise, another study revealed that MALAT1 was overexpressed in ACC compared to ACA, indicating a pathophysiological role for MALAT1 in the formation of ACC (Shiyanbola et al. 2020). Colorectal neoplasia differentially expressed (**CRNDE**) is one of the upregulated lncRNAs that is overexpressed in a variety of human malignancies (Zhang et al. 2018b). It has been demonstrated to be connected with the invasion, migration, and proliferation of cells by controlling the neurogenic locus notch homolog protein 1 (Notch1) and PI3K/Akt signaling pathway (Du et al. 2017; Li et al. 2018). The pathophysiology of adrenocortical tumors is thought to involve both signaling pathways (De Martino et al. 2010; Ronchi et al. 2015). Nevertheless, in contrast to other tumor types (Zhang et al. 2018b), CRNDE does not appear to be associated with the prognosis in individuals with ACC (Buishand et al. 2020).

A recent work on circular ncRNAs revealed that CCA-associated circRNA 1 **(circ-CCAC1)** regulates miR-514a-5p to promote ACC cell invasion, migration, and proliferation. According to this study, circ-CCAC1, which has been shown to function as an oncogene, was overexpressed in ACC cell lines and tissue samples and was linked to a poor prognosis (Li et al. 2020b).

Retinoblastoma (RB)

The most common intraocular malignant tumor is RB, which makes up 3% of all pediatric cancers (Rao and Honavar 2017; Dimaras and Corson 2019). It has been discovered that RB samples exhibit aberrant expression of a variety of miRNAs. Numerous cancer-related processes, such as the epithelial-mesenchymal transition (EMT), cell survival and proliferation, apoptosis, metastasis, and angiogenesis, can be impacted by these changed miRNAs (Delsin et al. 2019). Artificial overexpression of **miR-491-3p** boosted apoptosis while significantly suppressing RB cell invasion, migration, and proliferation. miR-491-3p changed the morphology of RB cell lines, producing more adherent cells. It also markedly decreased vimentin and N-cadherin expression while boosting E-cadherin, indicating that it may suppress EMT (Hu et al. 2021). The most widely expressed miRNA, **miR-340**, was linked to tumor growth, optic nerve invasion, and the International Classification of Retinoblastoma (ICRB) stages. They suggested that miR-340 might encourage the invasion, migration, and proliferation of RB cells. After modulating WIF1, miR-340 functions as an oncogenic miRNA to promote RB cell invasion, migration, and proliferation (Fuchs et al. 2001). According to in vitro research, the overexpression of **miR-613** caused cell cycle arrest and inhibited RB cell invasion, migration, and proliferation. Additionally, the overexpression of miR-613 stopped RB cells from growing into tumors in vivo. They also demonstrated that miR-613 directly targets E2F5. Their results demonstrated that miR-613 downregulates E2F5 to function as a tumor suppressor in RB (Zhang et al. 2017c). In both RB tissues and cells, low levels of **miR-936** were associated with tumor lymph node metastasis (TNM) stage, differentiation, and lymph node metastasis because the overexpression of miR-936 increased apoptosis, and exogenous miR-936 suppressed RB cell proliferation, colony formation, migration, and invasion in vitro and decreased tumor growth in vivo. It was discovered that miR-936 directly targeted the histone deacetylase 9 (HDAC9) mRNA. Through the suppression of HDAC9 expression in RB cells, the exogenous production of miR-936 reduced the activity of the PI3K/Akt pathway, both in vitro and in vivo. miR-936 downregulation was linked to a poor outcome in RB patients, but the overexpression of the gene reduced RB aggression by lowering HDAC9 mRNA and, ultimately, deactivating the PI3K/Akt pathway (Xu et al. 2020). The exogenous overexpression of **miR-101-3p** dramatically reduced viability and resulted in cell cycle arrest in WERI-Rb-1 and Y79 cells, as demonstrated by 3-(4,5-dimethylthiazol-2-yl)-2,5-diphenyltetrazolium bromide (MTT) and flow cytometry (FC) experiments. The antitumor efficacy of miR-101-3p in RB was validated by in vivo mouse studies. miR-101-3p inhibited RB cell growth by directly targeting zeste homolog 2 (EZH2) and HDAC9. Restoring EZH2 or HDAC9 expression in WERI-RB-1 and Y79 cells reversed the antiproliferative effects of miR-101-3p (Zheng et al. 2020).

Cell lines and RB tissues showed high levels of **MIAT** expression, but RB tissues showed downregulated levels of miR-665. Furthermore, decreased RB cell invasion, migration, and proliferation were caused by either MIAT downregulation or miR-665 overexpression. Additionally, LIM and SH3 protein 1 (LASP1) was proposed as an miR-665 target gene. The inhibitory effects of MIAT knockdown on Y79 cell invasion, migration, and proliferation were reversed by both the overexpression of LASP1 and the downregulation of miR-665. Furthermore, by controlling the miR-665/LASP1 axis, MIAT silencing inhibited RB proliferation (Grossi et al. 2015). The expression of the X-inactive specific transcript gene (**XIST**) was markedly elevated in RB cells and tissues. In RB cells, the high expression of XIST prevented apoptosis and encouraged invasion, migration, and proliferation. By functioning as a competing endogenous RNA (ceRNA) and preventing the expression of miR-191-5p, XIST indirectly raised the production of BDNF. They came to the conclusion that RB tissues had higher levels of XIST expression and that XIST might regulate the miR-191-5p/brain-derived neurotrophic factor (BDNF) axis to influence invasion, migration, apoptosis, and proliferation (Xu et al. 2021). In RB, hypoxia-inducible factor 1α (HIF-1α) also functions as an oncoprotein (Radwan et al. 2016). When it was knocked down, proapoptotic proteins like Bcl-2-associated protein x (Bax), caspase-9, and caspase-3 were expressed more, which caused cancer cells to undergo apoptosis (Gao et al. 2014). Additionally, by promoting matrix metalloproteinase-9 (MMP-9) production, HIF-1α enhanced RB cell invasion (Li and Zheng 2017). miR-17-92a-1 cluster host gene (**MIR17HG**) boosted RB cell invasion, migration, and proliferation. One MIR17HG target, miR-155-5p, had the ability to prevent RB cell invasion, migration, and proliferation. Additionally, MIR17HG increased HIF-1α gene and protein expression in RB cells. When combined, these findings indicated that MIR17HG's carcinogenic effects in RB were mediated by the miR-155-5p/HIF-1α axis (Yan et al. 2022). Fu et al. (2022) demonstrated that HOX transcript antisense RNA (**HOTAIR**) was elevated in RB cells relative to normal retinal cells in order to examine the underlying mechanism of the lncRNA HOTAIR in RB development. In addition to acting as an miR-20b-5p ceRNA, HOTAIR may also boost RRM2 expression. In order to promote RB cell proliferation and inhibit apoptosis, HOTAIR competitively attached to miR-20b-5p, upregulating RRM2 and activating the PI3K/Akt pathway (Fu et al. 2022). Shorter overall life periods and worse clinicopathological characteristics were linked to the increased expression of **LINC00205** in RB cells and tissues. Mechanistically, LINC00205 functioned as a ceRNA to sponge miR-665 in RB cells, increasing the production of the miR-665 target mRNA high-mobility group box 1 (HMGB1). The activation of the miR-665/HMGB1 pathway lessened the impact of LINC00205 knockdown in RB cells. Accordingly, RB prognosis, diagnosis, and possibly, treatment may benefit by targeting the LINC00205/miR-665/HMGB1 axis (Zhang et al. 2020a). Zhang et al. looked at how the lncRNA cancer susceptibility candidate 9 (**CASC9**) affected the malignant characteristics of RB cells, including invasion, apoptosis, EMT, and proliferation. The overexpression of CASC9 notably enhanced the RB cells' malignant phenotype. CSC9 knockdown, on the other hand, enhanced apoptosis and reduced the malignant potential of RB cells. Dual luciferase reporter assays and RNA immunoprecipitation showed that CASC9 could control E2F3 by sponging miR-145-5p (Zhang et al. 2020b).

Malignant Pleural Mesothelioma (MPM)

The mesothelial cells of the pleura lining the lung are the sites of 80% of MPM, an uncommon but aggressive kind of cancer. The peritoneal and pericardial mesothelial cells can also develop into mesotheliomas (Remon et al. 2015). Although the levels of **miR-16** in the plasma of MPM patients and controls were comparable, Kirschner et al. observed that it was markedly downregulated in all main tumor specimens (Kirschner et al. 2012). Increased expression of its related target mRNAs, such as cyclin D1 (CCND1) and B cell leukemia/lymphoma 2 protein (Bcl2), was linked to the decreased expression of this miRNA (Reid et al. 2013). After transfecting two mesothelioma cell lines with **miR-31**, Ivanov et al. discovered that the cells' capacity for colony formation, migration, invasion, and proliferation was significantly diminished when compared to controls (Ivanov et al. 2010). PPP6C/PP6C, a protein phosphatase 6C that is overexpressed in the great majority of MPM, is one of its targets. Its 3'UTR contains three possible binding sites for hsa-miR-31 (Ivanov et al. 2010). When Pass and colleagues looked into the function of **miR-29c** in MPM, they discovered that it was more expressed in the epithelioid subtype and was linked to a favorable prognosis (Pass et al. 2010). The overexpression of hsa-miR-29c resulted in a reduction in colony formation, migration, and proliferation. In patients who received cytoreductive surgery, hsa-miR-29c was also discovered to be an independent predictor of time to progression. This was consistent with the clinical data showing that MPMs with lower levels of hsa-miR-29c expression were more aggressive. The fact that this miRNA seems to be essential for preserving a "balance" between methylation and demethylation in MPM was one of the more intriguing discoveries. When comparing MPM to matched normal peritoneum, they discovered that DNA methyltransferase 3 (DNMT3) was overexpressed (Pass et al. 2010). One gene that was discovered to be increased in response to hsa-miR-29c overexpression was adiponectin. By blocking the transcription factor specificity protein 1 (SP1) and its downstream target DNMT1, adiponectin has been shown to be able to control methyltransferases by reducing the methylation of CpG islands around the Wnt inhibitory factor 1 (WIF1) promoter (Pass et al. 2010).

The chromosomal clusters of lethal-7 (**let-7**) are frequently reported to be deleted in numerous malignancies, and let-7 expression levels are typically low in cancer (Calin et al. 2004). c-MYC is a crucial regulator of let-7 (Chang et al. 2008), and MPM pathogenesis is linked to c-MYC amplification (Riquelme et al. 2014). It is interesting to note that the miRNA study in MPM has revealed both the overexpression and the downregulation of different let-7 family members. For example, it has been discovered that hsa-let-7a, hsa-let-7b*, and hsa-let-7 g are increased and are possible biomarkers in MPM (Guled et al. 2009; Ramírez-Salazar et al. 2014; Ak et al. 2015). It has been shown that hsa-let-7a plays a functional role in the inhibition of MPM formation. Ephrin receptor A1 (EphA1) signaling via the EphA2 raises let-7 (let-7a1, let-7a3, let-7f1, and let-7f2), which, in turn, causes let-7-mediated regulation of RAS (H-RAS, K-RAS, and N-RAS) (Khodayari et al. 2011). It has been demonstrated that mesothelioma dramatically downregulates **miR-145**, as was covered in a previous study. Blandino and colleagues used a reexpression technique to show that raising the amounts of this miRNA decreased the clonogenicity, viability, and migration potential of mesothelioma cell lines by a mechanism that triggered replicative senescence (Cioce et al. 2014). The transcription factor OCT4 was one of the key targets that this miRNA was influencing, and the investigators confirmed this by using data mining and bioinformatics (Cioce et al. 2014). Two contradictory reports of **miR-193a-3p** in mesothelioma have been made. Benjamin et al. discovered that hsa-miR-193-3p was overexpressed in MPM in their initial findings (Benjamin et al. 2010). On the other hand, a recent work by Reid et al. showed that this miRNA is indeed markedly downregulated in tumors (Williams et al. 2015). Regarding this, Williams et al. have also demonstrated that hsa-miR-193-3p is a tumor suppressor and that MPM growth may be inhibited both in vitro and in vivo by functionally restoring the levels of this miRNA (Williams et al. 2015).

Initially discovered in Burkitt's lymphoma (Graham and Adams 1986), plasmacytoma variant translocation 1 (**PVT1**) is an lncRNA that has now been demonstrated to be necessary for elevated MYC protein levels in cancer cells (Tseng et al. 2014). PVT1 and c-MYC (8q24) share a chromosomal region, and MPM has commonly been linked to gains in this chromosome (Taniguchi et al. 2007; Gray et al. 2009; Takeda et al. 2012; Riquelme et al. 2014). Further research by Wistuba and associates has demonstrated that in 8q24-amplified MPM cell lines, the reduction of PVT1 but not c-MYC increases apoptosis, decreases cellular proliferation, and increases sensitivity to cisplatin (Riquelme et al. 2014). Compared to normal mesothelial cells, MPM cell lines showed a much lower expression of **GAS5** lncRNA, which would also be consistent with GAS5's potential function as a tumor suppressor. GAS5's function in cell proliferation was highlighted by the shorter cell cycle observed when it was inhibited in MPM cells (Renganathan et al. 2014). Podoplanin (PDPN) expression was positively correlated with the GAS5 expression level when compared to established mesothelioma markers (Renganathan et al. 2014). Podoplanin is a type-1 transmembrane glycoprotein that resembles sialomucin and is thought to be involved in metastasis and cell motility (Dang et al. 2014).

Neuroblastoma (NB)

Of all pediatric cancer deaths, 15% are caused by NB, the most prevalent extracranial malignant solid tumor in children. It has a wide range of pathological, biological, genetic, and clinical features (Gurney et al. 1975). **miR-204** was identified as a tumor suppressor in a recent investigation of miR-mRNA interactions for network modeling in NB carcinogenesis. N-MYC proto-oncogene protein mRNA was directly bound by miR-204 expression, which suppressed its expression and prevented NB carcinogenesis (Ooi et al. 2018). Research has shown that **miR-2110** and its target TSKU have a strong inverse relationship. It has also been shown that both low miR-2110 and high TSKU mRNA are significantly associated with poor patient survival (Zhao et al. 2018b). Similarly, Chen and Stallings (2007) demonstrated that N-MYC-amplified NB cells exhibited an under-expression of a particular collection of miRs and that N-MYC targeting restored these essential miRs with functional roles (e.g., **miR-184** in inducing apoptosis). According to these results, N-MYC may trigger NB pathogenesis by controlling miRs that encourage apoptosis, NB cell differentiation, and other processes. Particularly in the subgroup of NB patients at a high risk of progression, Banelli and colleagues found that **miR-34b-3p**, **miR-34b-5p**, **miR-34c-5p**, and **miR-124-2-3p** were significantly hypermethylated and subsequently downregulated (Parodi et al. 2016). On the other hand, **miR-221** has been demonstrated to enhance N-MYC expression through direct targeting of Nemo-like kinase (NLK), cell cycle regulation, and NB cell proliferation (He et al. 2017). By specifically targeting Sam68 translation, **miR-203** prevented NB cell invasion, migration, and proliferation (Zhao et al. 2015a). Similarly, by inhibiting MMP14, **miR-337-3p** and **miR-584-5p** have been shown to suppress NB cell proliferation, invasion, metastasis, and angiogenesis (Xiang et al. 2015a, b). **miR-449a** is a member of this family of tumor suppressor miRs, and it has been demonstrated that miR-449a inhibits cell growth and survival by causing cell differentiation and cell cycle arrest, two convergent mechanisms (Zhao et al. 2015b). miR-449a primarily targets microfibril-associated protein 4 MFAP4, plakophilin-4 (PKP4), and tRNA-splicing endonuclease subunit 15 (TSEN15) to control NB cell differentiation, but it also downregulates CDK6 and lymphoid enhancer-binding factor 1 (LEF1) to influence cell cycle arrest (Zhao et al. 2015b). Unmatched differentiation-inducing power is possessed by **miR-506-3p** and **miR-124-3p**, which also decrease tumors by downregulating their targets, CDK4, and signal transducer and activator of transcription 3 (STAT3) (Silber et al. 2008; Ko et al. 2015). In NB, **miR-193b** was substantially less expressed, which was correlated with higher cell survival and proliferation (Roth et al. 2018). Functionally, miR-193b preferentially inhibits the production of MCL1, CCND1, and N-MYC, causing G1 cell cycle arrest and NB cell death.

The lncRNA neuroblastoma-associated transcript-1 (**NBAT-1)** has been found to be a separate predictor of clinical prognosis in NB patients (Pandey et al. 2014; Mondal et al. 2018). The NBAT-1 promoter was discovered to be hypermethylated in high-risk patients, indicating that epigenetic regulation plays a role in controlling the production of this lncRNA in NB cells as well (Pandey et al. 2014). It is noteworthy that NBAT-1 suppresses tumors by regulating a number of genes, such as SRY-related HMG box gene 9 (SOX9), oncostatin M receptor (OSMR), and versican, that are known to have a role in the emergence of a variety of malignancies. Specifically, NBAT-1 inhibits gene expression through epigenetic processes by interacting with EZH2 (Pandey et al. 2014; Pandey and Kanduri 2015). The neuron-restrictive silencer factor/the repressor element-1-silencing transcription factor (NRSF/REST) pathway is upregulated in NB cells with low NBAT-1 expression, which lowers the expression of important neuron-specific genes (Mitra et al. 2021). Both NB tissues and cell lines have been shown to exhibit the upregulation of DLX6 antisense RNA 1 (**DLX6-AS1)** (Zhang et al. 2019a). This finding was further supported by the finding that poor differentiation, advanced NB stage, and consequently, a bad outcome were positively connected with an increased expression of DLX6-AS1 (Zhang et al. 2019a, Li et al. 2020a). Furthermore, lung adenocarcinoma invasion, migration, and proliferation have all been linked to DLX6-AS1 overexpression (Li et al. 2015a). DLX6-AS1 knockdown delays invasion in vitro, reduces tumor growth in vivo, and causes neuronal differentiation and cell apoptosis (Zhang et al. 2019a). These findings suggest that this lncRNA may contribute to the development and spread of NB in an oncogenic manner.

Only a percentage of NB tumors exhibit increased expression of lncRNA, which is substantially expressed in neuroblastoma 1 (**lncNB1**). An examination of the effects of lncNB1 on the control of gene expression revealed that this transcript upregulates the expression of the DEPDC1B gene and the E2F1 protein (Liu et al. 2019c). DEPDC1B increases the stability of the N-MYC protein by inducing ERK phosphorylation. Specifically, by interacting with the ribosomal protein RPL35, lncNB1 increases the production of the E2F1 protein, which, in turn, triggers the transcription of the DEPDC1B gene. Because inhibiting one of these factors results in a lower proportion of NB cells in the S phase of the cell cycle and a higher number of apoptotic cells, this regulation is essential for NB cell survival and proliferation (Liu et al. 2019c). Zhang et al. investigated the mechanism small nucleolar RNA host gene 1 (**SNHG1)** regulates in NB and showed that it interacts with miR-338-3p, which has been shown to target proto-oncogenes in several malignancies (Zhang et al. 2019b). PLK4 is regulated by miR338-3p in NB cells, most likely by blocking its expression. Conversely, SNHG1 expression causes miR338-3p to be downregulated, which results in PLK4 overexpression, which encourages invasion, migration, and proliferation (Zhang et al. 2019b). SNHG1 is upregulated in

MYCN-amplified NB, according to Sahu et al. (2016). In both cell-line and high-risk NB patients, SNHG1 was shown to have a strong positive correlation with MYCN-amplified and the coding gene TAF1D. Last, they suggested using SNHG1 expression levels as a predictive biomarker to estimate NB patients' clinical outcomes (Sahu et al. 2016).

Cholangiocarcinoma (CCA)

A malignant tumor called CCA develops from the hepatic progenitor cells or biliary epithelial cells (Ilyas and Gores 2013). In separate cohorts of human primary CCA patients, **miR-21** overexpression was verified, and its carcinogenic function was connected, at least in part, to the suppression of the expression of tissue inhibitors of metalloproteinases (TIMP) metallopeptidase inhibitor 3 and programmed cell death 4 (PDCD4) (Selaru et al. 2009). It has also been demonstrated that **miR-26a** is elevated in human CCA, which activates β-catenin-dependent genes and increases CCA cell proliferation (Zhang et al. 2012a). The characteristics of tissue-dependent miRNA expression are highlighted by the intriguing discovery that miR-26 was an onco-suppressor in HCC, the other kind of primary liver cancer (Wang et al. 2012b). In CCA samples, the expression of **miR-494** is downregulated, and it can prevent cell cycle progression (Olaru et al. 2011). CDK6, CDK4, CCND1, cyclin E2, HDAC-1, RB transcriptional corepressor 1, PLK1, pituitary tumor-transforming 1 (PTTG1), cyclin B1, CDK2, and DNA topoisomerase IIα are among the genes that are globally deregulated when miR-494 is expressed (Mott et al. 2007). Furthermore, through the methylation-dependent control of **miR-370** and related targets, including the mRNA of the oncogene mitogen-activated protein kinase kinase 8 (MAP3K8), IL-6 can promote CCA growth (Meng et al. 2008). In human CCA cells, forced IL-6 expression can negatively affect **miR-148a** and **miR-152**, which raises DNA methyltransferase 1 levels and silences important tumor suppressor genes like cyclin-dependent kinase inhibitor 2α (CDKN2A, also called p16INK4a) and Ras association domain family member 1 (RASSF1) (Salati and Braconi 2019). Oishi et al.'s study, which used integrative pathway analysis, explained why a subtype of intrahepatic CCA with stem-like gene expression patterns is clinically aggressive by connecting **miR-200c** signaling to the EMT (Oishi et al. 2012). A low amount of E-cadherin and high levels of vimentin and N-cadherin were also linked to the overexpression of **miR-21**, indicating that a single miRNA can maintain tumor growth by influencing multiple pathways (Liu et al. 2015, 2017a).

By binding to EZH2; recruiting to the promoter region of ERBB receptor feedback inhibitor 1 (ERRFI1), the tumor suppressor gene of CCA; and then directly decreasing ERRFI1 expression through the mediation of H3K27me3 demethylation changes, lncRNA antisense noncoding RNA in the inhibitor of cyclin-dependent kinase 4 (INK4) locus **(ANRIL)** has been shown to accelerate CCA carcinogenesis (Yu et al. 2020). By attaching to EZH2 and catalyzing the fructose-1,6-bisphosphatase 1 (FBP1) promoter region H3K27me3, which can partially restrict FBP1 expression, the lncRNA differentiation antagonizing nonprotein coding RNA (**DANCR**) stimulates CCA cell proliferation and migration (Wang et al. 2019). Additionally, in CCA, EZH2 can be recruited at the promoter regions of CDKN1A, ANGPTL4, and E-cadherin, respectively, by **SNHG1** (Yu et al. 2018a, b) **AGAP2-AS1** (Ji et al. 2021), **PVT1** (Yu et al. 2018a, b), and **NEAT1** (Zhang et al. 2018a), which raises H3K27me3 levels at these loci. In order to suppress survivin expression and prevent CCA cells from proliferating, migrating, and invading, the lncRNA FOXF1 adjacent noncoding developmental regulatory RNA **(FENDER)** may enlist SETDB1 to attach to the survivin gene promoter and cause H3K9 methylation (Qin et al. 2019). The lncRNA sprouty RTK signaling antagonist 4-intronic transcript 1 **(SPRY4-IT1)** may serve as a scaffold to attract EZH2, LSD1, and DNMT1 to the tumor suppressor KLF2 and LATS2 promoter regions for epigenetic silencing according to Xu et al. (2018). An increased expression of the lncRNA **DLEU1** has been identified in CCA tissues and cells, and the transcription factor YY1 transcriptionally enhances its expression. Yes-associated protein 1 (YAP1) is a transcriptional coactivator that has a critical regulatory role in cell signaling and operates as a pro-oncogene in many malignancies, particularly in CCA (Liu et al. 2019a; Sugihara et al. 2019). DLEU1 functions as a sponge to competitively bind miR-149-5p, thus promoting the production of the oncogene YAP1.

It has been revealed that the lncRNA **TTN-AS1** downregulates the expression of miR-320a in CCA cells and sponges miR-320a in an Ago2-dependent manner. TTN-AS1 stimulates angiogenesis via NRP-1, the EMT, and cell cycle progression by downregulating miR-320a. However, miR-320a may partially offset the impact of TTN-AS1 overexpression, following the transfection of miR-320a mimics into CCA cells (Zhu et al. 2020). Wang et al. (2016) verified that oxidative stress raised lncRNAs H19 and HULC, which, in turn, controlled CCA cell motility and invasion by increasing IL-6 and CXCR4 levels by sponging let-7a/let-7b and miR-372/miR-373, respectively. While prostate cancer-associated transcript 6 (**PCAT6**) was highly expressed in macrophages, PCAT6 levels were significantly higher in patients with CCA. By sponging miR-326 and activating RohA signaling, PCAT6 influenced macrophages, including ROS generation, mitochondrial stress response, and M2 polarization, which, in turn, contributed to the immunological response of CCA (Tu et al. 2021).

Testicular Germ Cell Carcinoma/Tumor (TGCT)

Solid tumors that affect men between the ages of 15 and 35 are known as TGCTs (Houldsworth et al. 2006). The elevated expression of two clusters, **miR-302/367** and **miR-371-373**, suggests that they may function as oncogenes and that they are continuously dysregulated in TGCTs. By targeting the Rb family and CDK1NA, miRNA-302/367 has been shown to accelerate the G1 to S transition of the cell cycle and increase proliferation at the molecular level (Wang et al. 2013). It has been demonstrated that **miR-372** and **miR-373** are elevated in TGCT and function as oncogenes that encourage tumor growth and cell division. The repression of the tumor suppressor large tumor suppressor homolog 2 (LATS2) is the molecular mechanism at play. In flies (Xu et al. 1995) and mice (McPherson et al. 2004), the deletion of the serine-threonine kinase LATS2 speeds up cellular proliferation and tumor growth. By suppressing FBXW7, a substrate-recognition part of the SCF-ubiquitin-ligase complex, a tumor suppressor that targets several oncoproteins and oncogenic transcription factors for ubiquitination-mediated proteolysis, **miR-223-3p** plays an oncogenic function in TGCT (Shimizu et al. 2018). Moreover, the overexpression of miR-223-3p suppresses apoptosis and encourages cell growth. Interestingly, Liu et al. (2017b) found that F-box and WD repeat domain-containing 7 (FBXW7) mRNA expression in TGCT was inversely correlated with miR-223-3p.

Normal testes, lungs, and trachea have also been discovered to have high amounts of **miR-449a-b**, whereas testicular cancer and other cancer cells exhibit substantial downregulation of this gene. miR-449a is thought to have a tumor-suppressive function by upregulating p53 activity and aggressively encouraging apoptosis, which inhibits cell proliferation. Cell proliferation requires the transcription factor E2F1, which significantly upregulates miR-449a and miR-449b. miR-449a can reverse the course of the cell cycle by lowering the expression of the cell cycle protein CDK6 (Lizé et al. 2010). The loss of **miR-514a-3p** in TGCT raises the expression of its target gene, paternally expressed gene 3 (PEG3), according to in vitro functional investigations by Özata et al. (2017). It has been proposed that PEG3 may play a role in the development of TGCT since it promotes the NF-κB pathway, which prevents cells from dying. Tumor-derived chemokines CSF1 and CX3CL1, which regulate the recruitment of macrophages to the neoplastic sites and hence promote tumor growth, are produced in greater quantities when **miR-125b** is downregulated, as described in TGCT (Batool et al. 2018).

TGCTs were found to express the transcript of the XIST gene, establishing the first connection between lncRNA and TGCTs (Looijenga et al. 1997). Recent proposals for tissue biomarkers for TGCTs that could distinguish seminomas from non-seminomatous tumors include **XIST** expression and the XIST-promoter demethylation status (Lobo et al. 2019). The transcript of testis developmental-related gene 1 (**TDRG1**) is among the most well-characterized lncRNAs linked to TGCTs. Seminoma has been discovered to have elevated TDRG1, which enhances tumor growth, progression, and cisplatin resistance (Lobo et al. 2019; Peng et al. 2019). It has been demonstrated that the expression of lncRNA **H19** is connected to the expression of lncRNA TDRG1 in TGCTs. The paternally imprinted gene that encodes this lncRNA is typically only expressed by the maternal allele in males. Since TGCTs arise during the early stages of embryogenesis, when biallelic expression of H19 ordinarily occurs, it is possible that the upregulation of H19 in TGCTs is a result of the loss of imprinting (Mishina et al. 1996; Kawakami et al. 2006). By functioning as a molecular sponge and sequestering miR-106b-5p, which typically negatively regulates TDRG1, it has been suggested that H19 increases the expression of TDRG1 in the well-established cisplatin-resistant TCam-2 cell line, thereby promoting cell survival under cisplatin-based chemotherapeutic conditions (Wei et al. 2018).

Several carcinogenic signaling pathways in melanoma are positively regulated by the lncRNA **SPRY4-IT1**, which functions as an miRNA sponge (Mazar et al. 2014). According to a recent study, human TGCTs had high levels of SPRY4-IT1 expression. The transient knockdown of SPRY4-IT1 in two TGCT cell lines led to a significant drop in Akt phosphorylation along with a decrease in cell growth, migration, and invasion (Das et al. 2018). In the testicular embryonal cancer cell line NT2, the HOXA transcript at the distal tip lncRNA (**HOTTIP**) was found to be highly expressed. While its overexpression encouraged cell proliferation, its knockdown hindered it. According to the same study, HOTTIP may function as a sponge for the antiproliferative miRNA miR-128-3p, thus enhancing the production of the oncogenic transcription factor HOXA13 (Su et al. 2019).

Glioblastoma Multiforme (GBM)

The most common and deadly primary brain tumor in adults is glioma, sometimes referred to as GBM. Of all the main brain cancers, gliomas have the greatest fatality rate and are extremely aggressive tumors (Bagherian et al. 2020). **miR-21**, which is significantly increased in GBM and has expression levels that correspond with the WHO tumor grade, was one of the first miRNAs to be associated with glioma malignancy (Zhang et al. 2017b). Tumor suppressor genes, including PDCD4, acidic nuclear phosphoprotein 32 family member A (ANP32A), SWItch/Sucrose Non-Fermentable (SWI/SNF)-related, matrix-associated, actin-dependent regulator of chromatin, subfamily A, member 4 (SMARCA4), phosphatase and tensin homolog (PTEN), and sprouty RTK signaling antagonist 2 (SPRY2), have been demonstrated to be targeted by miR-21. In immunodeficient mice, the inhibition of miR-21 results in a marked decrease in tumor cell proliferation and tumor growth (Chan et al. 2005; Ciafrè et al. 2005; Silber et al. 2008; Li et al. 2011). Furthermore, miR-21 inhibits the tumor suppressor homolog of p53 (Tap63) and heterogenous nuclear ribonucleoprotein K (HNPRK) to produce antiapoptotic effects (Zhou et al. 2010; Gaur et al. 2011; Schramedei et al. 2011). Targeting the tumor suppressor p27 and p53-elevated modulator of apoptosis (PUMA), which can cause cell death by binding to Bcl-2 and Bcl-xL, **miR-221/222** is also upregulated in GBM. By suppressing PUMA expression, the overexpression of miR-221/222 can prevent apoptosis and increase cell viability (Chen et al. 2008; Zhou et al. 2010). It has been demonstrated that **miR-34a**, a significantly downregulated miRNA in GBM, directly inhibits the expression of receptor tyrosine kinases and important cell fate regulators, including MET, neurogenic locus notch homolog protein 1 (NOTCH1) and 2, CDK6, CCND1, and SIRT1, thereby controlling cell invasion, self-renewal, and proliferation and survival in GBM (Gillies and Lorimer 2007; Zhang et al. 2010; Chen et al. 2012). It has been observed that a number of other miRNAs, such as **miR-15b, miR-26a, miR-125b, miR-148a, miR-153, miR-181, miR-184, miR-1218, miR-363, and miR-582-5p**, control cell survival and proliferation and stimulate the formation of tumors (Sun et al. 2008; Bueno and Malumbres 2011).

When endothelial cells are activated with pro-angiogenic growth factors, including vascular endothelial growth factor (VEGF), or when glioma cells are present, **miR-296** is significantly increased. Higher endothelial cell tube formation and improved tumor vascularization have been linked to higher miR-296 expression. Tumor angiogenesis is inhibited by miR-296 knockdown (Yao et al. 2015; Cheng et al. 2017). The angiomiR **miR-125b** targets MYC-associated zinc finger protein (MAZ), a transcription factor that controls VEGF, and is markedly downregulated in GBM-associated endothelial cells. MAZ is overexpressed when miR-125b is repressed, which encourages tumor vascularization (Würdinger et al. 2008). However, **miR-218** has been shown to inhibit GBM tumor neovascularization and cell survival by targeting HIF-2α (Smits et al. 2012). By inhibiting integrin-β8, **miR-93** promotes blood vessel development, sphere formation, and cell survival. According to a study by Fang et al. (Mathew et al. 2014), the overexpression of miR-93 in GBM cells promoted vasculogenesis in mouse xenografts and improved endothelial cell proliferation and tube formation in vitro.

Originally discovered to be overexpressed in colorectal cancer, **CRNDE** was later discovered to be increased in numerous other malignancies as well (Hamdy et al. 2024c). In gliomas, particularly GBM, it is significantly elevated (Zhang et al. 2017b). It is possible that CRNDE works as an oncogene because its knockdown inhibits glioma growth in vivo and attenuates cell proliferation and migration in vitro (Brown et al. 1991; Guttman et al. 2009). By serving as a ceRNA sponge for miR-384 and preventing the activation of miR-384 target piwi-like RNA-mediated gene silencing 4 (PIWIL4) and its downstream protein STAT3, CRNDE accelerates the malignant growth of gliomas (Wang et al. 2015b). Furthermore, it has been demonstrated that CRNDE controls GSC invasion, migration, and proliferation by sponging miR-186 (Zheng et al. 2016). An lncRNA called maternally expressed 3 (**MEG3**) is markedly downregulated in gliomas and is linked to poor overall survival, recurrence, advanced tumor grade, and IDH wild-type status (Zheng et al. 2015). In vitro, MEG3 overexpression inhibits glioma cell proliferation and encourages death and autophagy (Zhao et al. 2018a). Several miRNAs have been found to have direct binding sites on MEG3, indicating that it may operate as a ceRNA, even though the precise mechanism of action of MEG3 in GBM is still unknown (Wang et al. 2012a; Zhao et al. 2018a). According to a different study by Hu et al., the lncRNA placenta-specific protein 2 (**PLAC2**) targets ribosomal protein (RP) L36 and interacts with STAT1 to cause cell cycle arrest in gliomas. According to the authors, PLAC2 overexpression suppresses glioma growth in vivo in a xenograft model, decreases cell proliferation, and promotes G1/S arrest through the modulation of RPL36 expression (Qin et al. 2017; Zhang et al. 2017a).

It has been demonstrated more recently that transforming growth factor-β (TGF-β)-induced invasion of glioma cells is facilitated by **lncRNA-ATB**, which is activated by TGF-β. The activation of the NF-κB pathway and the translocation of p65 into the nucleus caused by the overexpression of lncRNA-ATB made it easier for TGF-β to invade glioma cells. Furthermore, by activating the p38/MAPK pathway, lncRNA-ATB improved TGF-β-mediated invasion of glioma cells (Hu et al. 2018). Another lncRNA that has been shown to impact a number of significant cellular processes is epigenetically induced lncRNA1 (**EPIC1**), which is substantially elevated in gliomas. By targeting Cdc20, its inhibition causes apoptosis, reduces cell viability, makes cells more sensitive to Temozolomide (TMZ), and dramatically reduces cell invasion (Wang et al. 2020). Similarly, it has been demonstrated that the lncRNA PVT1 promotes the growth of tumors and gliomas by controlling the bone morphogenetic protein (BMP) signaling system. By controlling gremlin 1 (GREM1), PVT1 affected the expression of BMP2 and BMP4 by acting as a sponge of miR-128-3p. In a mouse model, PVT1 knockdown resulted in smaller tumors and reduced glioma cell invasion, migration, and proliferation in vitro (Tang et al. 2019).

Acute Myeloid Leukemia (AML)

AML is a cancerous growth of the bone marrow's immature myeloid hematopoietic cells (Kamal et al. 2016). With increasing morbidity and mortality in elderly individuals, it is a very diverse disease (Döhner et al. 2015). By targeting SHIP1 and suppressing the expression of the transcription factor PU.1, **miR-155**, an oncogenic miRNA, may contribute to the pathophysiology of AML (Vigorito et al. 2007; Xue et al. 2014). The neural precursor cell-expressed, developmentally downregulated protein 8 (NEDD8)-dependent ubiquitin ligases partially regulated the activity of NF-κB, which, in turn, regulated this miRNA (Gatto et al. 2008; Milhollen et al. 2010). By activating the MAPK/ERK and PI3K/Akt FoxO3a pathways, the study discovered that Erbb2 interacting protein (Erbin) was the target of **miR-183-5p**, which negatively regulated Erbin expression and increased AML cell proliferation (Zheng et al. 2019). One significant epigenetic alteration in the pathophysiology of AML is aberrant DNA methylation. DNMT1 and DNMT3 are the two primary types of DNA methyltransferases. While the latter carries out de novo methylation, the former preserves methylation (Jones and Baylin 2002). According to Garzon et al., **miR-29b** caused DNA hypomethylation and tumor suppressor gene reactivation by directly targeting DNMT3A and DNMT3B and indirectly targeting DNMT1 (Garzon et al. 2009). A zinc finger-like structural transcription factor called SP1 was directly linked to the DNMT1 promoter region to initiate transcription, thus mediating the indirect repression of DNMT1 (Liu et al. 2008). By suppressing SP1 expression, **miR-29b** interferes with the transcription of DNMT1 that is dependent on SP1 (Chen et al. 2013). Hypomethylating tumor suppressor P115INK4b is another instance of DNMT inhibition that may lessen a mouse model's vulnerability to myeloid leukemia (Wolff et al. 2003). However, SP1 and KIT may downregulate miR-29b alike. Receptor tyrosine kinase (KIT) overexpression encourages the growth of malignant cells and has been shown in a number of tumors, including AML (Lemmon and Schlessinger 2010). According to Liu et al., increased SP1 expression and decreased MYC-dependent miR-29b expression were the outcomes of aberrant KIT activation. SP1 subsequently interacted with the NF-κB/HDAC complex to further suppress miR-29b expression and transactivate KIT (Liu et al. 2010). AML leukemogenesis may be aided by some exosomal miRNAs. According to a recent study, AML exosomes have higher levels of **miR-7977** than exosomes from healthy CD34+ cells. By suppressing the poly(rC)-binding protein, it may play a crucial role in interfering with normal hematopoiesis. Additionally, it caused mesenchymal stem cells to produce hematopoietic growth factors abnormally, creating an environment that was unfriendly to normal stem cells (Horiguchi et al. 2016).

The expression level of **hsa_circ_0004277**, one of the most drastically downregulated circRNAs in AML, was recovered in patients who experienced complete remission and was equal to that of healthy controls after remission. However, if the patient developed relapse-refractory symptoms, the expression level significantly decreased (Li et al. 2017). The overexpression of **circPVT1**, another circRNA, in AML with oncogene MYC amplification may indicate that circPVT1 has an effect on AML patients' survival (Abbate et al. 2018). Circular RNAs of NPM1 were found by Hirsch et al. In contrast to AML cell lines, they discovered that the circular NPM1 transcript, or **has_circ_0075001**, was less expressed in healthy volunteers. Its expression was positively connected with overall NPM1 expression but not with the presence of an NPM1 mutation (Hirsch et al. 2017).

By influencing the expression levels of integrin genes such as ITGA4 (CD49d) and ITGAX (CD11c), **HOTAIRM1** lncRNA was shown to be a regulator of myeloid differentiation and maturation in the early investigations of lncRNAs in AML. All-trans retinoic acid (ATRA)-induced granulocyte differentiation would be prevented by knocking down HOTAIRM1 (Zhang et al. 2014). Additional lncRNAs are involved in controlling the cell cycle, apoptosis, and proliferation of AML cells. lncRNA **PVT1** is a common example (Zeng et al. 2015). On the chromosome, PVT1's coding sequence lies next to MYC. About 10% of AML patients had functional acquisition of MYC and PVT1 as a result of 8q24.21 amplification (Tseng et al. 2014). The overexpression of PVT1 may cause necrosis and death in AML cell lines, most likely by suppressing the expression of c-MYC (Salehi and Sharifi 2018; Salehi et al. 2018). Another lncRNA that may be able to control AML cell proliferation is urothelial cancer-associated 1 (**UCA1**), which short hairpin RNA can silence to cause G1 cell cycle arrest and noticeably reduced cell growth. By suppressing the expression of the cell cycle regulator p27kip1, UCA1 may encourage proliferation (Hughes et al. 2015). Similarly, Wang et al.'s experiment with the U937 cell line showed that CRNDE may coordinate the proliferation and differentiation of AML cells (Wang et al. 2018).

Sarcoma

Soft-tissue sarcomas (STSs), an uncommon and diverse class of tumors originating from connective tissue, can develop anywhere in the body. Liposarcoma (LPS) is one of the most prevalent of the more than 50 distinct forms of STSs. The unique variations in tumor biology that underlie the subtypes of STSs lead to varying oncologic outcomes and responses to pharmacologic therapy (Dancsok et al. 2017). By controlling casein kinase-1α (CK-1α), which, in turn, increases β-catenin signaling and CCND1 expression, **miR-155**, an oncomiR in dedifferentiated LPS (DDLPS), plays a critical role in cell cycle progression, cell proliferation, and colony formation (Zhang et al. 2012b; Vincenzi et al. 2015; Kapodistrias et al. 2017). In three distinct LPS cell lines (SW872, LPS141, and LP6), the overexpression of **miR-26a-2** increased the cells' proliferation, survival, and migration while posttranscriptionally controlling RCBTB1 to prevent adipocyte differentiation and death (Lee et al. 2013). Additionally, employing antagomiRs to suppress miR-26a-2 in LPS cells produced the opposite phenotype, indicating the function of this miRNA in controlling such processes (Lee et al. 2013). By suppressing the expression of HOXA5, miR-26a also indirectly influences p53-independent and caspase-dependent apoptosis, according to another study (Lee et al. 2015). Two separate and co-transcribed miRNAs with a reputation for tumor suppressor properties make up the **miR-143/miR-145** cluster (Almeida and Calin 2016). This cluster is widely distributed in adipose tissues, and among them, miR-143 also contributes to LPS carcinogenesis and adipocyte differentiation (Esau et al. 2004). This cluster is hardly expressed in LPS, and by controlling the expression levels of BCL2, type II topoisomerase, PLK1, and PRC1, miR-143 reexpression prevented proliferation and caused apoptosis (Ugras et al. 2011).

According to reports, LPS even exhibits low expression of **miR-145 and miR-451** (Gits et al. 2014). The authors demonstrated that in LPS cell lines, overexpressing miR-145 and miR-451 resulted in apoptosis, slowed cell cycle progression, and reduced cellular proliferation (Gits et al. 2014). Additionally, **miR-193b** is an example of a tumor suppressor miRNA in LPS. By targeting genes associated with the FAK-SRC-CRKL signaling, this miRNA appears to influence key characteristics, including adipogenesis in human adipose-derived stem cells, and it is under-expressed in DDLPS in comparison to normal fat. It has also been demonstrated that miR-193b targets several genes, including platelet-derived growth factor (PDGFR), TGF-β, mothers against decapentaplegic homolog 4 (SMAD4), and YAP1, to promote LPS cell proliferation, differentiation, and carcinogenesis (Mazzu et al. 2019).

Mechanistically, the oncogene YAP1 and the hedgehog signaling pathway have a favorable effect on **H19** (Chan et al. 2014). After the hedgehog signaling pathway is inhibited or the YAP1 gene is knocked down, human osteosarcoma cell viability is reduced, and H19 is downregulated (Chan et al. 2014). Human osteosarcoma tissue has been shown to exhibit significant expression of **MALAT1** (Dong et al. 2015). Human osteosarcoma cell lines were prevented from proliferating by MALAT1 knockdown through the PI3K/Akt signaling pathway (53). However, osteosarcoma MG-63 cells have a very low level of BRAF-regulated lncRNA 1 (**BANCR**) (He and Zhang 2015). Through the downregulation of the Wnt/β-catenin signaling pathway and the increase of phosphorylated c-Jun N-terminal kinases (JNKs), BANCR overexpression decreased cell proliferation (He and Zhang 2015). **HOTAIR** was often overexpressed in osteosarcoma samples. The link between MMP2, MMP9, and HOTAIR may be the mechanism behind the considerable prevention of invasion and metastasis shown by HOTAIR inhibition (Wang et al. 2015a). In chondrosarcoma cell lines and tissues, HOTAIR expression was elevated.

A poor prognosis and tumor stage were confirmed to be associated with high HOTAIR expression (Bao et al. 2017). Human chondrosarcoma cells showed growth suppression, both in vitro and in vivo, when HOTAIR was knocked down. By causing G0/G1 arrest and death, downregulating HOTAIR inhibited the development of human chondrosarcoma cells. In the meantime, autophagy was suppressed by HOTAIR suppression (Bao et al. 2017). By attracting the enhancer of EZH2 and DNMT1 to the miR-454-3p promoter regions, HOTAIR mechanistically causes DNA methylation of miR-454-3p, significantly suppressing miR-454-3p expression. Additionally, miR-454-3p targets autophagy-related gene 12 (ATG12) and STAT3, which inhibit autophagy and trigger apoptosis produced by HOTAIR deficiency (Bao et al. 2017). According to a different study, human osteosarcoma tissue had higher levels of ANRIL expression than the surrounding normal tissue (Wei et al. 2016). Through a direct connection between HIF-1α and the putative hypoxia response element upstream of the ANRIL gene, hypoxia can upregulate ANRIL expression in vitro. ANRIL then boosted cancer cell migration and reduced apoptosis. Furthermore, ANRIL was downregulated by HIF-1α reduction in hypoxic environments (Wei et al. 2016).

Clinical Trials (CTs): Innovations and Outcomes

CTs targeting uncommon cancers, often referred to as rare cancers, present unique challenges and opportunities within the landscape of oncology. Rare cancers are typically defined by their low incidence rates—generally fewer than 6 cases per 100,000 people per year—which encompass a diverse array of tumor types. Collectively, these cancers account for approximately 25% of all cancer diagnoses, yet they receive disproportionately less research funding compared to more common cancers. This disparity is primarily due to the limited patient populations available for clinical studies, which complicates recruitment and the statistical power necessary for robust trial outcomes (Panageas 2015).

The design and execution of CTs for rare cancers necessitate innovative approaches that diverge from traditional methodologies. Conventional trial designs often rely on large sample sizes to achieve statistically significant results, a strategy that is impractical when dealing with small populations. As a response, researchers have begun to implement alternative strategies that can enhance the feasibility and efficacy of clinical trials in this field. For instance, the development of rare cancer registries allows for better tracking of patient outcomes and facilitates access to ongoing trials. Furthermore, adaptive trial designs have gained traction; these allow modifications to the trial protocol based on interim results, thereby optimizing patient enrollment and treatment pathways (Adashek and Kurzrock 2023). One notable initiative is the National Cancer Institute Molecular Analysis for Therapy Choice (NCI-MATCH) trial, which aims to match patients with rare cancers to targeted therapies based on genomic profiling of their tumors. This trial has been pivotal in demonstrating the potential for molecularly targeted treatments in rare cancer populations. By analyzing over 4000 genetic variants across 143 genes, NCI-MATCH seeks to enroll at least 25% of its participants from rare cancer demographics, showcasing a model that could be replicated in future studies (Christyani et al. 2024).

In addition to genomic profiling, biomarker-driven trials have emerged as a promising avenue for rare cancer research. These trials focus on identifying specific genetic mutations or alterations within tumors that can be targeted by existing therapies (Hamdy et al. 2025). The TRACK (remote clinical trial for rare cancers) clinical trial exemplifies this approach by enrolling patients with rare cancers and providing them with comprehensive biomarker testing, which informs personalized treatment recommendations based on their unique tumor characteristics. Despite these advancements, significant barriers remain in the realm of rare cancer clinical trials. The paucity of expertise in treating these cancers can lead to insufficiently powered studies and limited access to novel therapies. Moreover, many clinical centers are hesitant to initiate trials for rare cancers due to concerns about recruitment feasibility and the potential for inadequate data collection. This reluctance can perpetuate a cycle where patients with rare cancers have fewer opportunities for participation in clinical research (Maki 2024).

Recent studies have also highlighted the importance of real-world data in enhancing CT designs for rare tumors. By utilizing historical data from previous studies or observational cohorts as external control groups, researchers can improve the robustness of their findings without relying solely on placebo-controlled designs. This strategy not only increases the ethical viability of trials but also provides a more accurate reflection of treatment efficacy in diverse patient populations. Moreover, ongoing efforts by organizations like the European Organization for Research and Treatment of Cancer (EORTC) are crucial in advancing research into rare cancers through large-scale trials that focus on specific tumor types, such as gastrointestinal stromal tumors (GIST) and sarcomas. These studies aim to refine treatment protocols and enhance patient outcomes through rigorous evaluation of new therapeutic strategies (Maki 2024).

Unique Challenges in Epigenetic Drug Trials

One of the primary challenges in conducting CTs for rare tumors is the small patient population. As highlighted in various studies, the limited number of available patients can lead to insufficient sample sizes, making it difficult to achieve statistically significant results (Panageas 2015). This situation is exacerbated by the lengthy time frames required for patient accrual, which can extend over a decade without yielding conclusive data (Bogaerts 2025). Consequently, many trials may fail to demonstrate clinical benefit, simply due to inadequate participant numbers (Atta et al. 2023). Additionally, the heterogeneity of rare cancers complicates trial design. Each rare cancer can exhibit a wide range of biological behaviors and responses to treatments, which necessitates a more tailored approach to clinical research (Maki 2024). This variability can impact the choice of endpoints and statistical methods used in trials, further complicating the design process.

Innovative Trial Designs

To address these challenges, researchers are exploring alternative trial designs that are better suited for rare tumors. Some promising approaches include:

Adaptive Trial Design

These allow modifications to be made to the trial protocol based on interim results. For example, if early data indicate that a treatment is ineffective, the trial can be adjusted or halted (Gaddipati et al. 2012). This flexibility can help optimize patient enrollment and resource allocation.

Use of External Control Arms

Instead of using placebo groups, which can be ethically problematic, given the limited treatment options for patients with rare cancers, researchers are advocating for external control arms derived from previous studies or real-world data (Christyani et al. 2024). This approach not only addresses ethical concerns but also reduces the number of participants needed for effective comparisons.

Bayesian Approaches

These statistical methods allow for the incorporation of prior knowledge into the analysis, which can be particularly useful when dealing with small sample sizes typical in rare cancer trials. Bayesian designs can enhance the interpretability of results by providing a more nuanced understanding of treatment effects (Panageas 2015).

Registry-Based Trials

The establishment of rare cancer registries facilitates multicenter studies that can pool data from various institutions. This strategy enhances patient recruitment and allows for larger datasets while still focusing on specific rare tumor types (Gaddipati et al. 2012).

Regulatory Considerations

Regulatory agencies have begun to recognize the unique challenges posed by rare cancer trials and are adapting their guidelines accordingly. The National Cancer Institute's NCI-MATCH trial exemplifies this shift by utilizing genomic profiling to match patients with targeted therapies based on molecular abnormalities rather than relying solely on traditional trial designs (Panageas 2015). Such initiatives highlight a growing acceptance of innovative methodologies that prioritize patient outcomes over conventional statistical rigor.

FDA-Approved Epigenetic Drugs (Table 7.2)

To date, several epigenetic drugs (epi-drugs) have been approved by the US Food and Drug Administration (FDA) for the treatment of various cancers. These drugs primarily target DNA methylation and histone deacetylation. DNA methyltransferase inhibitors, such as 5-azacytidine and decitabine, work by inhibiting the enzymes responsible for adding methyl groups to DNA, leading to the reactivation of silenced genes. Histone deacetylase inhibitors, such as vorinostat and romidepsin, target enzymes that remove acetyl groups from histones, resulting in increased gene expression. While these drugs have shown promise in treating certain cancers, their efficacy in rare cancers may vary (Costa et al. 2023).

DNA Methyltransferase Inhibitors (DNMTis)

DNMTis inhibit the addition of methyl groups to the DNA wrapped around histone proteins by DNMT enzymes (Silverman et al. 2002; Kantarjian et al. 2006).The first epigenetic drug approved by the FDA was azacitidine (Vidaza®) in 2004 for the treatment of all subtypes of myelodysplastic syndromes (MDS) and chronic myelomonocytic leukemia (CML) (Kaminskas et al. 2005), followed by decitabine (Dacogen®), approved in 2006 to treat MDS (Steensma 2009). Both azacitidine and decitabine are two analogs of the cytidine nucleoside in which the carbon atom in position 5, in the pyrimidine ring, has been replaced by a nitrogen (Derissen et al. 2013). These compounds were planned as cytotoxic agents, but it was found that low-dose exposure could cause DNA demethylation by inhibiting the DNMT1 enzyme responsible for maintaining DNA methylation (Silverman et al. 2002; Kantarjian et al. 2006; Stresemann and Lyko 2008). These so-called hypomethylating agents have been used in myeloid malignancies, even though 50% of patients do not respond initially or during repeated cycles of treatment (Zhao et al. 2021; Šimoničová et al. 2022).

Guadecitabine is a second-generation DNA methylation inhibitor being developed for AML and MDS treatment. It consists of a dinucleotide of decitabine and

Table 7.2 Epigenetic drugs approved by the FDA. (Costa et al. 2023)

Epigenetic target	Compound	Clinical name	Disease	Side effects	Company	Approved by (year)
DNMT1	Azacitidine	Vidaza®	MDS	Neutropenia, thrombocytopenia, nausea, emesis, diarrhea, and constipation	Pharmion Corporation	US FDA (2004)
	Decitabine	Dacogen®	MDS	Prolonged myelosuppression (neutropenia and thrombocytopenia)	Janssen Pharmaceuticals	US FDA (2006)
HDACs class I and HDAC6	Suberoylanilide hydroxamic acid (SAHA)	Vorinostat®	CTCL	Fatigue, nausea, diarrhea, and thrombocytopenia	Merck	US FDA (2006)
HDAC6	Romidepsin	Istodax®	CTCL	Fatigue, nausea, leukopenia, granulocytopenia, and thrombocytopenia	Celgene	US FDA (2009)
Pan-HDACi	Belinostat	Beleodaq®	T cell lymphoma	Nausea, vomiting, fatigue, pyrexia, and anemia	Topo Target	US FDA (2014)
	Panobinostat	Farydak®	Multiple myeloma	Diarrhea, peripheral neuropathy, asthenia, fatigue, neutropenia, thrombocytopenia, and lymphocytopenia	Novartis	US FDA (2015)
HDAC1, HDAC2, HDAC3, and HDAC10	Tucidinostat	Epidaza®	PTCL	Fatigue, anorexia, thrombocytopenia, leukopenia, and neutropenia	Chipscreen Biosciences	China FDA (2014)

DNMTi DNA methyltransferase inhibitors, *MDS* myelodysplastic syndromes, *CML* chronic myelomonocytic leukemia, *AML* acute myeloid leukemia, *HDACi* histone deacetylase inhibitor, *PTCL* peripheral T cell lymphoma deoxyguanosine, which is resistant to cytidine deaminase, the enzyme that is responsible for decitabine inactivation (Stomper et al. 2021). In comparison to azacitidine and decitabine, guadecitabine has higher stability, better safety profile, and higher ease of administration (subcutaneous) (Daher-Reyes et al. 2019).

Histone Deacetylase (HDAC) Inhibitors (HDACis)

The first HDACi was suberoylanilide hydroxamic acid (SAHA, vorinostat®) approved in 2006 by the FDA for the treatment of manifestations of cutaneous T cell lymphoma (CTCL) (Duvic and Vu 2007) (Duvic et al. 2007; Olsen et al. 2007). After SAHA approval, a depsipeptide natural product from the bacterium *Chromobacterium violaceum*, named romidepsin, was approved in 2006 for CTCL and peripheral T cell lymphoma (PTCL) treatment (McClure et al. 2018). Belinostat became the third FDA-approved HDACi for T cell lymphoma (Campbell and Thomas 2017). In 2015, panobinostat arose as the first HDACi approved for a non-lymphoma cancer by the FDA and also the European Medicines Agency (EMA). Panobinostat is an oral pan-HDACi recommended for relapsed or refractory multiple myeloma (MM) management (Berdeja et al. 2021), the first time that an HDACi was proposed and accepted for non-lymphoma cancer treatment, refreshing the possibility of designing inhibitors for all HDACs, bearable enough to benefit humans (McClure et al. 2018). These drugs appear to inhibit oncogenesis by promoting apoptosis and cell cycle arrest and altering the DNA damage pathway (Chun 2015).

Tucidinostat (chidamide) is an oral subtype of selective HDACi. This drug inhibits class I HDACs (HDAC1, HDAC2, and HDAC3) and a class IIb HDAC (HDAC10). It was approved in 2014 as a second-line therapy for peripheral relapsed or refractory T cell lymphoma by the China-FDA. In Japan, tucidinostat was approved in 2021 for relapsed or refractory adult T cell lymphoma treatment under the name Hiyasta (Sun et al. 2022). Valproic acid is an FDA-approved antiepileptic drug that also presents inhibitory HDAC class I and II activity (Lunke et al. 2021). Currently, valproic acid is in phase III clinical trial as a potential drug to treat cervical and ovarian malignancies and has been proposed in combination regimens with chemotherapy and radiotherapy (Krauze et al. 2015; Suraweera et al. 2018; Tsai et al. 2021).

Other Epigenetic Drugs (Table 7.2)

The combined use of epigenetic drugs with conventional therapies is gaining prominence due to their potential in increasing tumor cells' sensitivity to classical chemotherapy, improving the therapeutic effect. A phase Ib/II clinical trial showed that the small-molecule eprenetapopt, in combination with azacytidine, improved clinical response rates and molecular remissions in patients with TP53 mutant MDS and oligoblastic AML (Sallman et al. 2021). A selective BCL-2 inhibitor (venetoclax) has recently been approved for use, in combination with hypomethylating agents (azacitidine or decitabine), showing promising results for the treatment of AML in patients who are ineligible to receive intensive chemotherapy (Manda et al. 2021). The use of decitabine associated with carboplatin has shown greater efficacy in the treatment of ovarian cancer when compared with conventional therapies by increasing the sensitivity of the tumor cells (Fan et al. 2014; Fu et al. 2015).

In the same context, a phase II study showed that patients with AML or MDS who underwent idarubicin therapy with a high-dose continuous infusion of Ara-C (cytarabine) in combination with vorinostat® showed an overall response rate of 85% and a 76% complete response to combination therapy (Garcia-Manero et al. 2012). A multicenter phase II trial showed that the combined use of vorinostat® with bortezomib and dexamethasone showed an overall response of 81.3% in patients with relapsed MM, although more studies are necessary to further optimize HDACi-based combinations, in order to improve the tolerability and increase the efficacy of combination therapy (Brown et al. 2021). A multicenter phase II trial showed that the triple use of belinostat, carboplatin, and paclitaxel was well-tolerated and demonstrated clinical benefits in patients with recurrent epithelial ovarian cancer (Finkler et al. 2008). Combination therapy of drugs with the ability to modulate the epigenome and conventional therapies demonstrates increased efficacy and tolerability, requiring lower dosages of each agent and reducing the side effects caused by conventional chemotherapies (Oing et al. 2019).

Natural Products and Their Efficacy

Natural products have emerged as a significant focus in cancer research, particularly for their potential in treating rare cancers through epigenetic modulation. Epigenetics refers to the heritable changes in gene expression that do not involve alterations to the underlying DNA sequence, and it plays a crucial role in cancer development and progression. Natural compounds can influence epigenetic mechanisms, such as DNA methylation and histone modification, thereby restoring normal gene function and inhibiting cancer cell growth.

Mechanisms of Action

Natural products exert their anticancer effects through various mechanisms, primarily by modulating epigenetic regulators. For instance, compounds like curcumin and resveratrol have been shown to inhibit the activity of DNMTs and HDACs. DNMTs are responsible for adding methyl groups to DNA, which can silence tumor suppressor genes, while HDACs remove acetyl groups from histones, leading to a more compact and inactive chromatin structure. By inhibiting these enzymes, natural products can reactivate silenced genes that are critical for controlling cell proliferation and apoptosis (Carlos-Reyes et al. 2019; Huang et al. 2023b).

Examples of Natural Products in Rare Cancer Treatments (Table 7.3)

Curcumin

Curcumin, derived from turmeric, has been extensively studied for its anticancer properties. It has demonstrated the ability to inhibit DNMTs and HDACs, leading to the reactivation of tumor suppressor genes. In clinical settings, curcumin has been used in combination with standard chemotherapy drugs to enhance the efficacy against various cancers, including rare forms like cholangiocarcinoma (bile duct cancer). Clinical trials have explored curcumin's role in sensitizing tumor cells to chemotherapy agents, such as gemcitabine (Naeem et al. 2022).

Paclitaxel (Taxol)

Paclitaxel, originally derived from the bark of the Pacific yew tree (*Taxus brevifolia*), is widely used in treating various cancers, including rare types like Kaposi's sarcoma and Ewing sarcoma. It functions by stabilizing microtubules and

Table 7.3 Natural products modulating epigenetics in rare cancers

Natural product	Source	Mechanism of action	Rare cancer type
Curcumin	Turmeric (*Curcuma longa*)	Inhibits DNMTs and HDACs and promotes apoptosis	Cholangiocarcinoma, Ewing sarcoma
Paclitaxel (Taxol)	Pacific yew tree (*Taxus brevifolia*)	Disrupts microtubule function and inhibits cell division	Kaposi's sarcoma, Ewing sarcoma
Vincristine	Madagascar periwinkle (*Catharanthus roseus*)	Inhibits microtubule formation	Acute lymphoblastic leukemia, desmoplastic small round-cell tumors
Camptothecin	Chinese tree (*Camptotheca acuminata*)	Inhibits topoisomerase I, preventing DNA replication	Rare ovarian cancers, neuroblastoma
Resveratrol	Grapes, berries	Antioxidant effects; modulates cell cycle regulation	Multiple myeloma, rare breast cancers
Sulforaphane	Broccoli and other cruciferous vegetables	Induces phase II detoxifying enzymes and inhibits HDACs	Neuroblastoma, rare soft-tissue tumors
Genistein	Soybeans	Inhibits tyrosine kinases and affects estrogen receptors	Rare breast cancer types
Quercetin	Onions and apples	Antioxidant properties; inhibits cell proliferation	Rare lung cancers

preventing their depolymerization, thereby inhibiting cell division. Clinical trials have established its efficacy in treating soft-tissue sarcomas and other rare malignancies. Ongoing studies continue to explore combinations of paclitaxel with other agents to enhance its therapeutic effects against resistant cancer types (Gill et al. 1999).

Trabectedin (Yondelis)

Trabectedin is another natural product derived from the sea squirt *Ecteinascidia turbinata*. It is primarily used to treat soft-tissue sarcomas and ovarian cancer but has shown activity against rare tumors like desmoplastic small round-cell tumor (DSRCT). Trabectedin interacts with DNA and modulates the expression of genes involved in cell cycle regulation and apoptosis (Sanfilippo et al. 2015; Palmerini et al. 2022).

Resveratrol

Found in grapes and berries, resveratrol exhibits anticancer properties by modulating epigenetic factors, such as histone acetylation. Clinical trials have investigated its use in various cancer types, including rare conditions like multiple myeloma. Resveratrol's ability to enhance the effectiveness of conventional therapies while reducing side effects makes it a candidate for further research in rare cancers. This compound has demonstrated the ability to modulate histone acetylation and influence noncoding RNA (miRNA) expression, which plays a crucial role in regulating gene expression. Resveratrol's epigenetic effects contribute to its anticancer properties by promoting apoptosis and inhibiting cell proliferation. Clinical studies are exploring its use in multiple myeloma and rare breast cancers, where it may enhance the effectiveness of standard treatments (Selmin et al. 2020; Jöhrer and Çiçek 2021).

Sulforaphane

Sulforaphane, a compound found in cruciferous vegetables like broccoli, has been studied for its potential against prostate cancer and other malignancies. It acts as an HDAC inhibitor and has shown promise in preclinical models of rare cancers, such as neuroblastoma. Ongoing clinical trials are assessing its efficacy as an adjunct therapy alongside traditional treatments (Dorna et al. 2023).

Genistein

Found in soybeans, genistein acts as an inhibitor of DNMTs, leading to the hypomethylation of tumor suppressor genes. This compound also interacts with estrogen receptors, making it relevant for hormone-sensitive cancers. Its potential application in rare breast cancer types is being investigated through various clinical trials focusing on its ability to restore normal epigenetic regulation (Shukla et al. 2019).

Quercetin

Quercetin has been shown to regulate the activity of DNMTs and HDACs, thereby influencing gene expression related to cancer progression. Its antioxidant properties also contribute to its anticancer effects. Research into quercetin's application in rare lung cancers is ongoing, with studies indicating its potential to enhance the efficacy of conventional therapies through epigenetic modulation (Dorna et al. 2023).

Future Prospects and Ethical Considerations

Challenges in Research and Treatment

These diseases may occur in any age group, gender, or organ. Etiologic data on rare cancers are limited, and the pathological and molecular patterns are still obscure. The dilemma with these rare tumors is that their entire burden on the population has not been effectively assessed. Even when defined more predictably by considering the peculiarities of pathogenesis and patient outcome, rare tumors constitute about 22 percent of all reported cases of cancers, comprising all tumors occurring in childhood and adolescence (Schaefer 2012; Trama et al. 2012).

In comparison to more common tumors, rare cancers typically receive less funding and scientific attention. Because rare cancers are so uncommon, patients with these diseases face unique obstacles, such as a lack of readily available cancer registries and tissue banks, a lack of trust in clinical decision-making, a delay in diagnosis or inaccurate diagnosis, and an inadequate supply of clinical expertise and appropriate treatments. Additionally, researchers encounter challenges when conducting clinical investigations due to the limited number of participants in the sample. Patients diagnosed with rare tumors have an average prognosis that is impacted by all these factors. Research on rare tumors is often limited to case reports, anecdotal data, case series from a single institution, or, at most, tiny multicenter series. Since the features of the underlying population of all identical tumors are not revealed by such selected studies, the results from these studies can be misleading (Hassan et al. 2006; Miller 2010; Gatta et al. 2011).

It becomes difficult to perform clinical planning when there is minimal scientific support for patients and healthcare providers, even though many rare tumors can be incurable. Clinical trials with sufficient statistical power to address important issues in a reasonable amount of time require international cooperation. Advances in diagnosis, therapy, and prognosis, as well as important insights into the pathophysiology of uncommon malignancies, are all within reach with more robust studies into these diseases (Podnos et al. 2006, Institute of Medicine Committee on Accelerating Rare Diseases and Orphan Product 2010).

Histopathological analysis is the gold standard for determining the diagnosis of rare malignancies. Tissue diagnosis is essential for cancer treatment, but there is still a significant chance of interpretational errors, anywhere from 25% to 40%, in everyday practice. Subjectivity and inexperience on the part of specialists, variability among and between observers, and the submission of insufficient tumor samples are important factors impacting diagnostic accuracy. There has been no decrease in mortality rates for uncommon tumors despite the proposal of new treatments (Cowie 2006; Sidiropoulos et al. 2012).

One challenge in managing uncommon malignancies in adults is the disease's rarity. Few published studies have examined the best methods for detecting and treating uncommon tumors. The therapy options for uncommon cancers are getting better, but doctors still have a long way to go (Raghavan 2013; Raghavan et al. 2017).

Innovations in Personalized Cancer Treatment Plans

As immunochemistry experts converged on the oncogene paradigm in the 1980s, cancer was recognized as a hereditary illness. Disrupting biochemical processes within, between, and on cell surfaces may treat cancer. Researchers focused on somatic genetic alterations, not germline abnormalities, that cause excessive cell proliferation and division that causes tumors. Candidate mutations and protein expression in patients were used to identify genetic categories. The use of monoclonal antibodies to block gene activity became the focus. These block the genetic signaling system that produces tumor-promoting proteins. It was once a single cancer type but now has many subgroups. Integrating biomarker testing and therapy in "theragnostics" has changed the regulatory approval procedure to enable testing technologies in treatment decisions and provisioning (Kerr et al. 2021; Christyani et al. 2024).

In order to make tumor cells more immunogenic, the Vigil personalized cancer vaccine uses tumor tissue that the patient has given consent to be transfected with granulocyte-macrophage colony-stimulating factor and bidirectional short hairpin RNA that targets the furin gene DNA (Anwar et al. 2022). Both the actual survival rate (73% vs. 23%) and the median overall survival rate (731 days vs. 207 days) were significantly higher in the Vigil treatment group in the Ewing sarcoma phase I clinical trial compared to the control group. Despite having satisfactory treatment safety, another cancer vaccine, axalimogene filolisbac (ADXS11-001), had disappointing results when administered to patients with squamous cell carcinoma of the anorectal canal. The overall response rate (ORR) was only 3.4%, and the 6-month progression-free survival rate was 15.5%. Patients with anal cancer may require a combination of treatments, including ADXS11-001, in order to see a substantial improvement in their condition (Ghisoli et al. 2016; Eng et al. 2020).

Summary and Conclusions

This chapter delved into the intersection of epigenetics and precision medicine, particularly in the realm of rare cancers, which are often overlooked in mainstream research. By conducting a thorough review of the existing literature, the chapter highlighted the significance of noncoding RNAs (ncRNAs) as well as other epigenetic mechanisms in tumor biology, illustrating their complex roles in tumor suppression vs. oncogenesis. This duality underscores the potential of epigenetics as a therapeutic target in the precision era, with various strategies currently being explored to modulate their activity.

The chapter also examined recent clinical trials focusing on rare cancers, evaluating the therapeutic potentials of novel interventions and the innovations emerging from these studies. A key focus is placed on the epigenetic mechanisms that contribute to the development and progression of these uncommon malignancies that could be targets/hits for drug design/treatment. Moreover, the discussion included how natural products can influence epigenetic modifications, such as DNA methylation and histone modifications, as well as their role in regulating ncRNAs.

Per, the concept of "epi-drugs" is now introduced; therefore, emphasizing the promise of natural products as therapeutic agents that can modify epigenetic landscapes is also challenging. This approach not only holds potential for cancer treatment but also for prevention and better prognosis with lower side effects, suggesting a proactive strategy in managing rare cancers.

Overall, the chapter advocates for a deeper exploration of epigenetic and ncRNA-targeted therapies in the context of precision medicine, aiming to enhance treatment outcomes for patients with rare cancers.

References

Abbate AL, D. Tolomeo, I. Cifola, M. Severgnini, A. Turchiano, B. Augello, G. Squeo, D. A. P, D. Traversa, G. Daniele, A. Lonoce, M. Pafundi, M. Carella, O. Palumbo, A. Dolnik, D. Muehlematter, J. Schoumans, N. Van Roy, G. De Bellis, G. Martinelli, G. Merla, L. Bullinger, C. Haferlach and C. T. Storlazzi (2018). "MYC-containing amplicons in acute myeloid leukemia: genomic structures, evolution, and transcriptional consequences." Leukemia **32**(10): 2152–2166.

Adashek, J. J. and R. Kurzrock (2023). "Home-run trials for rare cancers: giving the right drug(s) to the right patients at the right time and in the right place." npj Precision Oncology **7**(1): 129.

Adriaens, C., L. Standaert, J. Barra, M. Latil, A. Verfaillie, P. Kalev, B. Boeckx, P. W. G. Wijnhoven, E. Radaelli, W. Vermi, E. Leucci, G. Lapouge, B. Beck, J. van den Oord, S. Nakagawa, T. Hirose, A. A. Sablina, D. Lambrechts, S. Aerts, C. Blanpain and J.-C. Marine (2016). "p53 induces formation of NEAT1 lncRNA-containing paraspeckles that modulate replication stress response and chemosensitivity." Nature Medicine **22**(8): 861–868.

Ahuja, N., H. Easwaran and S. B. Baylin (2014). "Harnessing the potential of epigenetic therapy to target solid tumors." J Clin Invest **124**(1): 56–63.

Ak, G., S. C. Tomaszek, F. Kosari, M. Metintas, J. R. Jett, S. Metintas, H. Yildirim, E. Dundar, J. Dong, M. C. Aubry, D. A. Wigle and C. F. Thomas Jr (2015). "MicroRNA and mRNA

Features of Malignant Pleural Mesothelioma and Benign Asbestos-Related Pleural Effusion." BioMed Research International **2015**(1): 635748.

Almeida, M. I. and G. A. Calin (2016). "The miR-143/miR-145 cluster and the tumor microenvironment: unexpected roles." Genome Med **8**(1): 29.

Anwar, M. M., C. Albanese, N. M. Hamdy and A. S. Sultan (2022). "Rise of the natural red pigment 'prodigiosin' as an immunomodulator in cancer." Cancer Cell International **22**(1): 419.

Arun, G., S. D. Diermeier and D. L. Spector (2018). "Therapeutic targeting of long non-coding RNAs in cancer." Trends in molecular medicine **24**(3): 257–277.

Atta, H., N. Alzahaby, N. M. Hamdy, S. H. Emam, A. Sonousi and L. Ziko (2023). "New trends in synthetic drugs and natural products targeting 20S proteasomes in cancers." Bioorganic Chemistry **133**: 106427.

Bader, A. G., D. Brown, J. Stoudemire and P. Lammers (2011). "Developing therapeutic microRNAs for cancer." Gene Ther **18**(12): 1121–1126.

Bagherian, A., R. Mardani, B. Roudi, M. Taghizadeh, H. R. Banfshe, A. Ghaderi, A. Davoodvandi, S. Shamollaghamsari, M. R. Hamblin and H. Mirzaei (2020). "Combination Therapy with Nanomicellar-Curcumin and Temozolomide for In Vitro Therapy of Glioblastoma Multiforme via Wnt Signaling Pathways." J Mol Neurosci **70**(10): 1471–1483.

Bao, X., T. Ren, Y. Huang, K. Sun, S. Wang, K. Liu, B. Zheng and W. Guo (2017). "Knockdown of long non-coding RNA HOTAIR increases miR-454-3p by targeting Stat3 and Atg12 to inhibit chondrosarcoma growth." Cell Death Dis **8**(2): e2605.

Batool, A., Y. Q. Wang, X. X. Hao, S. R. Chen and Y. X. Liu (2018). "A miR-125b/CSF1-CX3CL1/tumor-associated macrophage recruitment axis controls testicular germ cell tumor growth." Cell Death Dis **9**(10): 962.

Beggs, A. D., A. Jones, M. El-Bahrawy, M. Abulafi, S. V. Hodgson and I. P. Tomlinson (2013). "Whole-genome methylation analysis of benign and malignant colorectal tumours." J Pathol **229**(5): 697-704.

Benjamin, H., D. Lebanony, S. Rosenwald, L. Cohen, H. Gibori, N. Barabash, K. Ashkenazi, E. Goren, E. Meiri, S. Morgenstern, M. Perelman, I. Barshack, Y. Goren, T. B. Edmonston, A. Chajut, R. Aharonov, Z. Bentwich, N. Rosenfeld and D. Cohen (2010). "A Diagnostic Assay Based on MicroRNA Expression Accurately Identifies Malignant Pleural Mesothelioma." The Journal of Molecular Diagnostics **12**(6): 771–779.

Bennett, C. F. and E. E. Swayze (2010). "RNA targeting therapeutics: molecular mechanisms of antisense oligonucleotides as a therapeutic platform." Annu Rev Pharmacol Toxicol **50**: 259–293.

Berdeja, J. G., J. P. Laubach, J. Richter, S. Stricker, A. Spencer, P. G. Richardson and A. Chari (2021). "Panobinostat from bench to bedside: rethinking the treatment paradigm for multiple myeloma." Clinical Lymphoma Myeloma and Leukemia **21**(11): 752–765.

Biswas, S. and C. M. Rao (2018). "Epigenetic tools (The Writers, The Readers and The Erasers) and their implications in cancer therapy." European Journal of Pharmacology **837**: 8–24.

Blume, C. J., A. Hotz-Wagenblatt, J. Hüllein, L. Sellner, A. Jethwa, T. Stolz, M. Slabicki, K. Lee, A. Sharathchandra, A. Benner, S. Dietrich, C. C. Oakes, P. Dreger, D. te Raa, A. P. Kater, A. Jauch, O. Merkel, M. Oren, T. Hielscher and T. Zenz (2015). "p53-dependent non-coding RNA networks in chronic lymphocytic leukemia." Leukemia **29**(10): 2015–2023.

Bogaerts, J. (2025) Methodological challenges Designs of clinical studies, European Organisation for Research and Treatment of Cancer.

Bomben, R., S. Gobessi, M. Dal Bo, S. Volinia, D. Marconi, E. Tissino, D. Benedetti, A. Zucchetto, D. Rossi, G. Gaidano, G. Del Poeta, L. Laurenti, D. G. Efremov and V. Gattei (2012). "The miR-17~92 family regulates the response to Toll-like receptor 9 triggering of CLL cells with unmutated IGHV genes." Leukemia **26**(7): 1584–1593.

Brown, C. J., A. Ballabio, J. L. Rupert, R. G. Lafreniere, M. Grompe, R. Tonlorenzi and H. F. Willard (1991). "A gene from the region of the human X inactivation centre is expressed exclusively from the inactive X chromosome." Nature **349**(6304): 38–44.

Brown, S., C. Pawlyn, A.-L. Tillotson, D. Sherratt, L. Flanagan, E. Low, G. J. Morgan, C. Williams, M. Kaiser and F. E. Davies (2021). "Bortezomib, vorinostat, and dexamethasone combination therapy in relapsed myeloma: results of the phase 2 MUK four trial." Clinical Lymphoma Myeloma and Leukemia **21**(3): 154–161. e153.

Bueno, M. J. and M. Malumbres (2011). "MicroRNAs and the cell cycle." Biochim Biophys Acta **1812**(5): 592–601.

Buishand, F. O., Y. Liu-Chittenden, Y. Fan, A. Tirosh, S. K. Gara, D. Patel, D. Meerzaman and E. Kebebew (2020). "Adrenocortical tumors have a distinct, long, non-coding RNA expression profile and LINC00271 is downregulated in malignancy." Surgery **167**(1): 224–232.

Bullrich, F., H. Fujii, G. Calin, H. Mabuchi, M. Negrini, Y. Pekarsky, L. Rassenti, H. Alder, J. C. Reed, M. J. Keating, T. J. Kipps and C. M. Croce (2001). "Characterization of the 13q14 Tumor Suppressor Locus in CLL: Identification of ALT1, an Alternative Splice Variant of the LEU2 Gene1." Cancer Research **61**(18): 6640–6648.

Calin, G. A., C. Sevignani, C. D. Dumitru, T. Hyslop, E. Noch, S. Yendamuri, M. Shimizu, S. Rattan, F. Bullrich, M. Negrini and C. M. Croce (2004). "Human microRNA genes are frequently located at fragile sites and genomic regions involved in cancers." Proceedings of the National Academy of Sciences **101**(9): 2999–3004.

Campbell, P. and C. M. Thomas (2017). "Belinostat for the treatment of relapsed or refractory peripheral T-cell lymphoma." Journal of Oncology Pharmacy Practice **23**(2): 143–147.

Caramuta, S., L. Lee, D. M. Ozata, P. Akçakaya, H. Xie, A. Höög, J. Zedenius, M. Bäckdahl, C. Larsson and W. O. Lui (2013). "Clinical and functional impact of TARBP2 over-expression in adrenocortical carcinoma." Endocr Relat Cancer **20**(4): 551–564.

Carlos-Reyes, Á., J. S. López-González, M. Meneses-Flores, D. Gallardo-Rincón, E. Ruíz-García, L. A. Marchat, H. Astudillo-de la Vega, O. N. Hernández de la Cruz and C. López-Camarillo (2019). "Dietary Compounds as Epigenetic Modulating Agents in Cancer." Frontiers in Genetics **10**.

Chan, J. A., A. M. Krichevsky and K. S. Kosik (2005). "MicroRNA-21 Is an Antiapoptotic Factor in Human Glioblastoma Cells." Cancer Research **65**(14): 6029–6033.

Chan, L. H., W. Wang, W. Yeung, Y. Deng, P. Yuan and K. K. Mak (2014). "Hedgehog signaling induces osteosarcoma development through Yap1 and H19 overexpression." Oncogene **33**(40): 4857–4866.

Chang, T.-C., D. Yu, Y.-S. Lee, E. A. Wentzel, D. E. Arking, K. M. West, C. V. Dang, A. Thomas-Tikhonenko and J. T. Mendell (2008). "Widespread microRNA repression by Myc contributes to tumorigenesis." Nature Genetics **40**(1): 43–50.

Chen, L., J. Zhang, L. Han, A. Zhang, C. Zhang, Y. Zheng, T. Jiang, P. Pu, C. Jiang and C. Kang (2012). "Downregulation of miR-221/222 sensitizes glioma cells to temozolomide by regulating apoptosis independently of p53 status." Oncol Rep **27**(3): 854–860.

Chen, P., C. Price, Z. Li, Y. Li, D. Cao, A. Wiley, C. He, S. Gurbuxani, R. B. Kunjamma, H. Huang, X. Jiang, S. Arnovitz, M. Xu, G. M. Hong, A. G. Elkahloun, M. B. Neilly, M. Wunderlich, R. A. Larson, M. M. Le Beau, J. C. Mulloy, P. P. Liu, J. D. Rowley and J. Chen (2013). "miR-9 is an essential oncogenic microRNA specifically overexpressed in mixed lineage leukemia-rearranged leukemia." Proc Natl Acad Sci U S A **110**(28): 11511–11516.

Chen, Y., W. Liu, T. Chao, Y. Zhang, X. Yan, Y. Gong, B. Qiang, J. Yuan, M. Sun and X. Peng (2008). "MicroRNA-21 down-regulates the expression of tumor suppressor PDCD4 in human glioblastoma cell T98G." Cancer Letters **272**(2): 197–205.

Chen, Y. and R. L. Stallings (2007). "Differential patterns of microRNA expression in neuroblastoma are correlated with prognosis, differentiation, and apoptosis." Cancer Res **67**(3): 976–983.

Chen, Y., H. Wang, S. Zhu and X. Lan (2020). "miR-483–5p promotes esophageal cancer progression by targeting KCNQ1." Biochemical and Biophysical Research Communications **531**(4): 615–621.

Cheng, Z., Z. Li, K. Ma, X. Li, N. Tian, J. Duan, X. Xiao and Y. Wang (2017). "Long Non-coding RNA XIST Promotes Glioma Tumorigenicity and Angiogenesis by Acting as a Molecular Sponge of miR-429." Journal of Cancer **8**(19): 4106–4116.

Chifu, I., B. Heinze, C. T. Fuss, K. Lang, M. Kroiss, S. Kircher, C. L. Ronchi, B. Altieri, A. Schirbel, M. Fassnacht and S. Hahner (2020). "Impact of the Chemokine Receptors CXCR4 and CXCR7 on Clinical Outcome in Adrenocortical Carcinoma." Frontiers in Endocrinology **11**.

Choudhry, H., A. Albukhari, M. Morotti, S. Haider, D. Moralli, J. Smythies, J. Schödel, C. M. Green, C. Camps, F. Buffa, P. Ratcliffe, J. Ragoussis, A. L. Harris and D. R. Mole (2015). "Erratum: Tumor hypoxia induces nuclear paraspeckle formation through HIF-2α dependent transcriptional activation of NEAT1 leading to cancer cell survival." Oncogene **34**(34): 4546–4546.

Christyani, G., M. Carswell, S. Qin and W. Kim (2024). "An Overview of Advances in Rare Cancer Diagnosis and Treatment." **25**(2).

Chun, P. (2015). "Histone deacetylase inhibitors in hematological malignancies and solid tumors." Archives of pharmacal research **38**: 933–949.

Ciafrè, S. A., S. Galardi, A. Mangiola, M. Ferracin, C. G. Liu, G. Sabatino, M. Negrini, G. Maira, C. M. Croce and M. G. Farace (2005). "Extensive modulation of a set of microRNAs in primary glioblastoma." Biochemical and Biophysical Research Communications **334**(4): 1351–1358.

Cioce, M., F. Ganci, V. Canu, A. Sacconi, F. Mori, C. Canino, E. Korita, B. Casini, G. Alessandrini, A. Cambria, M. A. Carosi, R. Blandino, V. Panebianco, F. Facciolo, P. Visca, S. Volinia, P. Muti, S. Strano, C. M. Croce, H. I. Pass and G. Blandino (2014). "Protumorigenic effects of mir-145 loss in malignant pleural mesothelioma." Oncogene **33**(46): 5319–5331.

Collins, F. S. and H. Varmus (2015). "A new initiative on precision medicine." N Engl J Med **372**(9): 793–795.

Costa, P., S. L. A. Sales, D. P. Pinheiro, L. Q. Pontes, S. S. Maranhão, C. Pessoa, G. P. Furtado and C. L. M. Furtado (2023). "Epigenetic reprogramming in cancer: From diagnosis to treatment." Front Cell Dev Biol **11**: 1116805.

Cowie, F. (2006). "Treatment of rare cancers: gastrointestinal stromal tumours." Br J Hosp Med (Lond) **67**(7): 361–364.

Crona, J. and F. Beuschlein (2019). "Adrenocortical carcinoma — towards genomics guided clinical care." Nature Reviews Endocrinology **15**(9): 548–560.

Daher-Reyes, G. S., B. M. Merchan and K. W. Yee (2019). "Guadecitabine (SGI-110): an investigational drug for the treatment of myelodysplastic syndrome and acute myeloid leukemia." Expert Opinion on Investigational Drugs **28**(10): 835–849.

Dancsok, A. R., K. Asleh-Aburaya and T. O. Nielsen (2017). "Advances in sarcoma diagnostics and treatment." Oncotarget **8**(4): 7068–7093.

Dang, Q., J. Liu, J. Li and Y. Sun (2014). "Podoplanin: a novel regulator of tumor invasion and metastasis." Medical Oncology **31**(9): 24.

Das, M. K., K. Furu, H. F. Evensen, P. Haugen Ø and T. B. Haugen (2018). "Knockdown of SPRY4 and SPRY4-IT1 inhibits cell growth and phosphorylation of Akt in human testicular germ cell tumours." Sci Rep **8**(1): 2462.

De Martino, M. C., A. Al Ghuzlan, S. Aubert, G. Assié, J.-Y. Scoazec, S. Leboulleux, C. Do Cao, R. Libè, C. Nozières, M. Lombès, F. Pattou, F. Borson-Chazot, S. Hescot, C. Mazoyer, J. Young, I. Borget, A. Colao, R. Pivonello, J.-C. Soria, J. Bertherat, M. Schlumberger, L. Lacroix and E. Baudin (2013). "Molecular Screening for a Personalized Treatment Approach in Advanced Adrenocortical Cancer." The Journal of Clinical Endocrinology & Metabolism **98**(10): 4080–4088.

De Martino, M. C., P. M. van Koetsveld, R. Pivonello and L. J. Hofland (2010). "Role of the mTOR Pathway in Normal and Tumoral Adrenal Cells." Neuroendocrinology **92**(Suppl. 1): 28–34.

Decmann, A., P. Perge, P. I. Turai, A. Patócs and P. Igaz (2020) "Non-Coding RNAs in Adrenocortical Cancer: From Pathogenesis to Diagnosis." Cancers **12** DOI: https://doi.org/10.3390/cancers12020461.

Delsin, L. E. A., K. B. Salomao, J. A. Pezuk and M. S. Brassesco (2019). "Expression profiles and prognostic value of miRNAs in retinoblastoma." J Cancer Res Clin Oncol **145**(1): 1–10.

Derissen, E. J., J. H. Beijnen and J. H. Schellens (2013). "Concise drug review: azacitidine and decitabine." The oncologist **18**(5): 619–624.

Di Fazio, P., F. D. Rusche, S. Roth, A. Pehl, S. Wächter, I. Mintziras, D. K. Bartsch and K. Holzer (2022). "Long Non-Coding RNA H19 Expression Correlates with Autophagy Process in Adrenocortical Carcinoma." Cancer Investigation **40**(3): 254–267.

Dimaras, H. and T. W. Corson (2019). "Retinoblastoma, the visible CNS tumor: A review." J Neurosci Res **97**(1): 29–44.

Döhner, H., D. J. Weisdorf and C. D. Bloomfield (2015). "Acute Myeloid Leukemia." N Engl J Med **373**(12): 1136–1152.

Dong, Y., G. Liang, B. Yuan, C. Yang, R. Gao and X. Zhou (2015). "MALAT1 promotes the proliferation and metastasis of osteosarcoma cells by activating the PI3K/Akt pathway." Tumour Biol **36**(3): 1477–1486.

Dorna, D., A. Grabowska and J. Paluszczak (2023). "Natural products modulating epigenetic mechanisms by affecting histone methylation/demethylation: Targeting cancer cells." British Journal of Pharmacology.

Du, D. X., D. B. Lian, B. H. Amin and W. Yan (2017). "Long non-coding RNA CRNDE is a novel tumor promoter by modulating PI3K/AKT signal pathways in human gastric cancer." Eur Rev Med Pharmacol Sci **21**(23): 5392–5398.

Duvic, M., R. Talpur, X. Ni, C. Zhang, P. Hazarika, C. Kelly, J. H. Chiao, J. F. Reilly, J. L. Ricker and V. M. Richon (2007). "Phase 2 trial of oral vorinostat (suberoylanilide hydroxamic acid, SAHA) for refractory cutaneous T-cell lymphoma (CTCL)." Blood **109**(1): 31–39.

Duvic, M. and J. Vu (2007). "Vorinostat: a new oral histone deacetylase inhibitor approved for cutaneous T-cell lymphoma." Expert opinion on investigational drugs **16**(7): 1111–1120.

El Mesallamy, H. O., W. M. Rashed, N. M. Hamdy and N. Hamdy (2014). "High-dose methotrexate in Egyptian pediatric acute lymphoblastic leukemia: the impact of ABCG2 C421A genetic polymorphism on plasma levels, what is next?" Journal of Cancer Research and Clinical Oncology **140**(8): 1359–1365.

Eldash, S., E. F. Sanad, D. Nada and N. M. Hamdy (2023). "The Intergenic Type LncRNA (LINC RNA) Faces in Cancer with In Silico Scope and a Directed Lens to LINC00511: A Step toward ncRNA Precision." Non-Coding RNA **9**(5): 58.

Eng, C., M. Fakih, M. Amin, V. Morris, H. S. Hochster, P. M. Boland and H. J. O. Uronis (2020). "A phase II study of axalimogene filolisbac for patients with previously treated, unresectable, persistent/recurrent loco-regional or metastatic anal cancer." **11**(15): 1334.

Esau, C., X. Kang, E. Peralta, E. Hanson, E. G. Marcusson, L. V. Ravichandran, Y. Sun, S. Koo, R. J. Perera, R. Jain, N. M. Dean, S. M. Freier, C. F. Bennett, B. Lollo and R. Griffey (2004). "MicroRNA-143 regulates adipocyte differentiation." J Biol Chem **279**(50): 52361–52365.

Esteller, M., J. Garcia-Foncillas, E. Andion, S. N. Goodman, O. F. Hidalgo, V. Vanaclocha, S. B. Baylin and J. G. Herman (2000). "Inactivation of the DNA-repair gene MGMT and the clinical response of gliomas to alkylating agents." N Engl J Med **343**(19): 1350–1354.

Fan, H., X. Lu, X. Wang, Y. Liu, B. Guo, Y. Zhang, W. Zhang, J. Nie, K. Feng and M. Chen (2014). "Low-dose decitabine-based chemoimmunotherapy for patients with refractory advanced solid tumors: a phase I/II report." Journal of immunology research **2014**(1): 371087.

Fathi Dizaji, B. (2020). "Strategies to target long non-coding RNAs in cancer treatment: progress and challenges." Egyptian Journal of Medical Human Genetics **21**: 1–15.

Finkler, N., D. Dizon, P. Braly, J. Micha, U. Lassen, P. Celano, R. Glasspool, E. Crowley, P. Buhl-Jensen and R. Penson (2008). "Phase II multicenter trial of the histone deactylase inhibitor (HDACi) belinostat, carboplatin and paclitaxel (BelCaP) in patients (pts) with relapsed epithelial ovarian cancer (EOC)." Journal of Clinical Oncology **26**(15_suppl): 5519–5519.

Fu, K., K. Zhang and X. Zhang (2022). "LncRNA HOTAIR facilitates proliferation and represses apoptosis of retinoblastoma cells through the miR-20b-5p/RRM2/PI3K/AKT axis." Orphanet J Rare Dis **17**(1): 119.

Fu, X., Y. Zhang, X. Wang, M. Chen, Y. Wang, J. Nie, Y. Meng and W. Han (2015). "Low dose decitabine combined with taxol and platinum chemotherapy to treat refractory/recurrent ovarian cancer: an open-label, single-arm, phase I/II study." Current Protein and Peptide Science **16**(4): 329–336.

Fuchs, B., K. Zhang, A. Schabel, M. E. Bolander and G. Sarkar (2001). "Identification of twenty-two candidate markers for human osteogenic sarcoma." Gene **278**(1–2): 245–252.
Furuta, M., K.-i. Kozaki, K. Tanimoto, S. Tanaka, S. Arii, T. Shimamura, A. Niida, S. Miyano and J. Inazawa (2013). "The Tumor-Suppressive miR-497-195 Cluster Targets Multiple Cell-Cycle Regulators in Hepatocellular Carcinoma." PLOS ONE **8**(3): e60155.
Gaddipati, H., K. Liu, A. Pariser and R. Pazdur (2012). "Rare Cancer Trial Design: Lessons from FDA Approvals." Clinical Cancer Research **18**(19): 5172–5178.
Gao, Y., M. Jing, R. Ge, Z. Zhou and Y. Sun (2014). "Inhibition of hypoxia inducible factor 1α by siRNA-induced apoptosis in human retinoblastoma cells." J Biochem Mol Toxicol **28**(9): 394–399.
Garcia-Manero, G., F. P. Tambaro, N. B. Bekele, H. Yang, F. Ravandi, E. Jabbour, G. Borthakur, T. M. Kadia, M. Y. Konopleva and S. Faderl (2012). "Phase II trial of vorinostat with idarubicin and cytarabine for patients with newly diagnosed acute myelogenous leukemia or myelodysplastic syndrome." Journal of clinical oncology **30**(18): 2204–2210.
Garding, A., N. Bhattacharya, R. Claus, M. Ruppel, C. Tschuch, K. Filarsky, I. Idler, M. Zucknick, M. Caudron-Herger, C. Oakes, V. Fleig, I. Keklikoglou, D. Allegra, L. Serra, S. Thakurela, V. Tiwari, D. Weichenhan, A. Benner, B. Radlwimmer, H. Zentgraf, S. Wiemann, K. Rippe, C. Plass, H. Döhner, P. Lichter, S. Stilgenbauer and D. Mertens (2013). "Epigenetic Upregulation of lncRNAs at 13q14.3 in Leukemia Is Linked to the In Cis Downregulation of a Gene Cluster That Targets NF-kB." PLOS Genetics **9**(4): e1003373.
Garofalo, M., G. D. Leva and C. M. Croce (2014). "MicroRNAs as anti-cancer therapy." Curr Pharm Des **20**(33): 5328–5335.
Garzon, R., S. Liu, M. Fabbri, Z. Liu, C. E. Heaphy, E. Callegari, S. Schwind, J. Pang, J. Yu, N. Muthusamy, V. Havelange, S. Volinia, W. Blum, L. J. Rush, D. Perrotti, M. Andreeff, C. D. Bloomfield, J. C. Byrd, K. Chan, L. C. Wu, C. M. Croce and G. Marcucci (2009). "MicroRNA-29b induces global DNA hypomethylation and tumor suppressor gene reexpression in acute myeloid leukemia by targeting directly DNMT3A and 3B and indirectly DNMT1." Blood **113**(25): 6411–6418.
Gatta, G., J. M. Van Der Zwan, P. G. Casali, S. Siesling, A. P. Dei Tos, I. Kunkler, R. Otter, L. Licitra, S. Mallone and A. Tavilla (2011). "Rare cancers are not so rare: the rare cancer burden in Europe." European journal of cancer **47**(17): 2493–2511.
Gatto, G., A. Rossi, D. Rossi, S. Kroening, S. Bonatti and M. Mallardo (2008). "Epstein-Barr virus latent membrane protein 1 trans-activates miR-155 transcription through the NF-kappaB pathway." Nucleic Acids Res **36**(20): 6608–6619.
Gaur, A. B., S. L. Holbeck, N. H. Colburn and M. A. Israel (2011). "Downregulation of Pdcd4 by mir-21 facilitates glioblastoma proliferation in vivo." Neuro Oncol **13**(6): 580–590.
Ge, Y., J. Shu, G. Shi, F. Yan, Y. Li and H. Ding (2021). "miR-100 Suppresses the Proliferation, Invasion, and Migration of Hepatocellular Carcinoma Cells via Targeting CXCR7." Journal of Immunology Research **2021**(1): 9920786.
Ghisoli, M., M. Barve, R. Mennel, C. Lenarsky, S. Horvath, G. Wallraven, B. O. Pappen, S. Whiting, D. Rao, N. Senzer and J. Nemunaitis (2016). "Three-year Follow up of GMCSF/bi-shRNA(furin) DNA-transfected Autologous Tumor Immunotherapy (Vigil) in Metastatic Advanced Ewing's Sarcoma." Mol Ther **24**(8): 1478–1483.
Gill, P. S., A. Tulpule, B. M. Espina, S. Cabriales, J. Bresnahan, M. Ilaw, S. Louie, N. F. Gustafson, M. A. Brown and C. Orcutt (1999). "Paclitaxel is safe and effective in the treatment of advanced AIDS-related Kaposi's sarcoma." Journal of clinical oncology **17**(6): 1876–1876.
Gillies, J. K. and I. A. Lorimer (2007). "Regulation of p27Kip1 by miRNA 221/222 in glioblastoma." Cell Cycle **6**(16): 2005–2009.
Gits, C. M., P. F. van Kuijk, M. B. Jonkers, A. W. Boersma, M. Smid, W. F. van Ijcken, J. M. Coindre, F. Chibon, C. Verhoef, R. H. Mathijssen, M. A. den Bakker, J. Verweij, S. Sleijfer and E. A. Wiemer (2014). "MicroRNA expression profiles distinguish liposarcoma subtypes and implicate miR-145 and miR-451 as tumor suppressors." Int J Cancer **135**(2): 348–361.

Glover, A. R., J. T. Zhao, J. C. Ip, J. C. Lee, B. G. Robinson, A. J. Gill, P. S. H. Soon and S. B. Sidhu (2015). "Long noncoding RNA profiles of adrenocortical cancer can be used to predict recurrence." Endocrine-Related Cancer **22**(1): 99–109.
Graham, M. and J. M. Adams (1986). Chromosome 8 breakpoint far 3′ of the c-myc oncogene in a Burkitt's lymphoma 2;8 variant translocation is equivalent to the murine pvt-1 locusThe EMBO Journal **5**(11): 2845–2851.
Gray, S. G., D. A. Fennell, L. Mutti and K. J. O'Byrne (2009). "In Arrayed Ranks: Array Technology in the Study of Mesothelioma." Journal of Thoracic Oncology **4**(3): 411–425.
Grossi, F., M. G. Dal Bello, S. Salvi, R. Puzone, U. Pfeffer, V. Fontana, A. Alama, E. Rijavec, G. Barletta, C. Genova, C. Sini, G. B. Ratto, M. Taviani, M. Truini and D. F. Merlo (2015). "Expression of Ribonucleotide Reductase Subunit-2 and Thymidylate Synthase Correlates with Poor Prognosis in Patients with Resected Stages I-III Non-Small Cell Lung Cancer." Dis Markers **2015**: 302649.
Guled, M., L. Lahti, P. M. Lindholm, K. Salmenkivi, I. Bagwan, A. G. Nicholson and S. Knuutila (2009). "CDKN2A, NF2, and JUN are dysregulated among other genes by miRNAs in malignant mesothelioma—A miRNA microarray analysis." Genes, Chromosomes and Cancer **48**(7): 615–623.
Guo, M., J. Ren, M. V. Brock, J. G. Herman and H. E. Carraway (2008). "Promoter methylation of HIN-1 in the progression to esophageal squamous cancer." Epigenetics **3**(6): 336–341.
Guo, M. and W. Yan (2015). "Epigenetics of gastric cancer." Methods Mol Biol **1238**: 783–799.
Gurney, J. G., M. A. Smith and J. A. Ross (1975). "Cancer among infants." Cancer incidence survival among children adolescents: United States SEER Program **1995**: 149–156.
Guttman, M., I. Amit, M. Garber, C. French, M. F. Lin, D. Feldser, M. Huarte, O. Zuk, B. W. Carey, J. P. Cassady, M. N. Cabili, R. Jaenisch, T. S. Mikkelsen, T. Jacks, N. Hacohen, B. E. Bernstein, M. Kellis, A. Regev, J. L. Rinn and E. S. Lander (2009). "Chromatin signature reveals over a thousand highly conserved large non-coding RNAs in mammals." Nature **458**(7235): 223–227.
Hadjadj, D., S. J. Kim, T. Denecker, L. Ben Driss, J. C. Cadoret, C. Maric, G. Baldacci and F. Fauchereau (2017). "A hypothesis-driven approach identifies CDK4 and CDK6 inhibitors as candidate drugs for treatments of adrenocortical carcinomas." Aging (Albany NY) **9**(12): 2695–2716.
Hamburg, M. A. and F. S. Collins (2010). "The path to personalized medicine." N Engl J Med **363**(4): 301–304.
Hamdy, N. M., E. B. Basalious, M. G. El-Sisi, M. Nasr, A. M. Kabel, E. S. Nossier and A. H. Abadi (2024a). "Advancements in current one-size-fits-all therapies compared to future treatment innovations for better improved chemotherapeutic outcomes: a step-toward personalized medicine." Current Medical Research and Opinion **40**(11): 1943–1961.
Hamdy, N. M., A. S. Doghish, A. A. El-Husseiny, D. Fathi, N. I. Rezk, N. M. Abdelmaksoud and A. I. Abulsoud (2024b). MicroRNA: Exploring Their Multifaceted Role in Cancer, Springer.
Hamdy, N. M., M. B. Zaki, N. M. Abdelmaksoud, R. A. Ismail, M. A. Abd-Elmawla, N. I. Rizk, D. Fathi and A. I. Abulsoud (2025). "Insights into the genetic and epigenetic mechanisms governing X-chromosome-linked-miRNAs expression in cancer; a step-toward ncRNA precision." International Journal of Biological Macromolecules **289**: 138773.
Hamdy, N. M., M. B. Zaki, N. I. Rizk, N. M. Abdelmaksoud, M. A. Abd-Elmawla, R. A. Ismail and A. I. Abulsoud (2024c). "Unraveling the ncRNA landscape that governs colorectal cancer: A roadmap to personalized therapeutics." Life Sciences **354**: 122946.
Hanahan, D. and R. A. Weinberg (2011). "Hallmarks of cancer: the next generation." cell **144**(5): 646–674.
Hassan, R., R. Alexander, K. Antman, P. Boffetta, A. Churg, D. Coit, P. Hausner, R. Kennedy, H. Kindler, M. Metintas, L. Mutti, M. Onda, H. Pass, A. Premkumar, V. Roggli, D. Sterman, P. Sugarbaker, R. Taub and C. Verschraegen (2006). "Current treatment options and biology of peritoneal mesothelioma: meeting summary of the first NIH peritoneal mesothelioma conference." Ann Oncol **17**(11): 1615–1619.

He, N. and Z. Zhang (2015). "Baicalein suppresses the viability of MG-63 osteosarcoma cells through inhibiting c-MYC expression via Wnt signaling pathway." Mol Cell Biochem **405**(1–2): 187–196.

He, X. Y., Z. L. Tan, Q. Mou, F. J. Liu, S. Liu, C. W. Yu, J. Zhu, L. Y. Lv, J. Zhang, S. Wang, L. M. Bao, B. Peng, H. Zhao and L. Zou (2017). "microRNA-221 Enhances MYCN via Targeting Nemo-like Kinase and Functions as an Oncogene Related to Poor Prognosis in Neuroblastoma." Clin Cancer Res **23**(11): 2905–2918.

Herndon, T. M., S. S. Chen, N. S. Saba, J. Valdez, C. Emson, M. Gatmaitan, X. Tian, T. E. Hughes, C. Sun, D. C. Arthur, M. Stetler-Stevenson, C. M. Yuan, C. U. Niemann, G. E. Marti, G. Aue, S. Soto, M. Z. H. Farooqui, S. E. M. Herman, N. Chiorazzi and A. Wiestner (2017). "Direct in vivo evidence for increased proliferation of CLL cells in lymph nodes compared to bone marrow and peripheral blood." Leukemia **31**(6): 1340–1347.

Heyn, H. and M. Esteller (2012). "DNA methylation profiling in the clinic: applications and challenges." Nat Rev Genet **13**(10): 679–692.

Hirsch, S., T. J. Blätte, S. Grasedieck, S. Cocciardi, A. Rouhi, M. Jongen-Lavrencic, P. Paschka, J. Krönke, V. I. Gaidzik, H. Döhner, R. F. Schlenk, F. Kuchenbauer, K. Döhner, A. Dolnik and L. Bullinger (2017). "Circular RNAs of the nucleophosmin (NPM1) gene in acute myeloid leukemia." Haematologica **102**(12): 2039–2047.

Horiguchi, H., M. Kobune, S. Kikuchi, M. Yoshida, M. Murata, K. Murase, S. Iyama, K. Takada, T. Sato, K. Ono, A. Hashimoto, A. Tatekoshi, Y. Kamihara, Y. Kawano, K. Miyanishi, N. Sawada and J. Kato (2016). "Extracellular vesicle miR-7977 is involved in hematopoietic dysfunction of mesenchymal stromal cells via poly(rC) binding protein 1 reduction in myeloid neoplasms." Haematologica **101**(4): 437–447.

Houldsworth, J., J. E. Korkola, G. J. Bosl and R. S. K. Chaganti 2006 "Biology and Genetics of Adult Male Germ Cell Tumors." Journal of Clinical Oncology **24**(35): 5512–5518.

Hu, Y.-W., C.-M. Kang, J.-J. Zhao, Y. Nie, L. Zheng, H.-X. Li, X. Li, Q. Wang and Y.-R. Qiu (2018). "LncRNA PLAC2 down-regulates RPL36 expression and blocks cell cycle progression in glioma through a mechanism involving STAT1." Journal of Cellular and Molecular Medicine **22**(1): 497–510.

Hu, Y., M. Zhao, L. Li, J. Ding, Y. M. Gui and T. W. Wei (2021). "miR-491-3p is Downregulated in Retinoblastoma and Inhibit Tumor Cells Growth and Metastasis by Targeting SNN." Biochem Genet **59**(2): 453–474.

Huang, Y., Y. Hou, P. Qu and Y. Dai (2023a). Combating cancer with natural products: Non-coding RNA and RNA modification, Frontiers Media SA. **14:** 1149777.

Huang, Y., Y. Hou, P. Qu and Y. Dai (2023b). "Editorial: Combating cancer with natural products: Non-coding RNA and RNA modification." Frontiers in Pharmacology **14**.

Hughes, J. M., I. Legnini, B. Salvatori, S. Masciarelli, M. Marchioni, F. Fazi, M. Morlando, I. Bozzoni and A. Fatica (2015). "C/EBPα-p30 protein induces expression of the oncogenic long non-coding RNA UCA1 in acute myeloid leukemia." Oncotarget **6**(21): 18534–18544.

Ilyas, S. I. and G. J. Gores (2013). "Pathogenesis, Diagnosis, and Management of Cholangiocarcinoma." Gastroenterology **145**(6): 1215–1229.

Institute of Medicine Committee on Accelerating Rare Diseases, R. and D. Orphan Product (2010). The National Academies Collection: Reports funded by National Institutes of Health. Rare Diseases and Orphan Products: Accelerating Research and Development. M. J. Field and T. F. Boat. Washington (DC), National Academies Press (US) Copyright © 2010, National Academy of Sciences.

Ivanov, S. V., C. M. V. Goparaju, P. Lopez, J. Zavadil, G. Toren-Haritan, S. Rosenwald, M. Hoshen, A. Chajut, D. Cohen and H. I. Pass (2010). "Pro-tumorigenic Effects of miR-31 Loss in Mesothelioma *." Journal of Biological Chemistry **285**(30): 22809–22817.

Ji, H., J. Wang, B. Lu, J. Li, J. Zhou, L. Wang, S. Xu, P. Peng, X. Hu and K. Wang (2021). "SP1 induced long non-coding RNA AGAP2-AS1 promotes cholangiocarcinoma proliferation via silencing of CDKN1A." Molecular Medicine **27**(1): 10.

Jöhrer, K. and S. S. Çiçek (2021). "Multiple myeloma inhibitory activity of plant natural products." Cancers **13**(11): 2678.

Jones, P. A. and S. B. Baylin (2002). "The fundamental role of epigenetic events in cancer." Nat Rev Genet **3**(6): 415–428.

Kamal, A. M., N. M. Hamdy, H. M. Hegab and H. O. El-Mesallamy (2016). "Expression of thioredoxin-1 (TXN) and its relation with oxidative DNA damage and treatment outcome in adult AML and ALL: A comparative study." Hematology **21**(10): 567–575.

Kaminskas, E., A. Farrell, S. Abraham, A. Baird, L.-S. Hsieh, S.-L. Lee, J. K. Leighton, H. Patel, A. Rahman and R. Sridhara (2005). "Approval summary: azacitidine for treatment of myelodysplastic syndrome subtypes." Clinical cancer research **11**(10): 3604–3608.

Kantarjian, H., J. P. J. Issa, C. S. Rosenfeld, J. M. Bennett, M. Albitar, J. DiPersio, V. Klimek, J. Slack, C. De Castro and F. Ravandi (2006). "Decitabine improves patient outcomes in myelodysplastic syndromes: results of a phase III randomized study." Cancer: Interdisciplinary International Journal of the American Cancer Society **106**(8): 1794–1803.

Kapodistrias, N., K. Mavridis, A. Batistatou, P. Gogou, V. Karavasilis, I. Sainis, E. Briasoulis and A. Scorilas (2017). "Assessing the clinical value of microRNAs in formalin-fixed paraffin-embedded liposarcoma tissues: Overexpressed miR-155 is an indicator of poor prognosis." Oncotarget **8**(4): 6896–6913.

Katsnelson, A. (2013). "Momentum grows to make 'personalized' medicine more 'precise'." Nat Med **19**(3): 249.

Kawakami, T., C. Zhang, Y. Okada and K. Okamoto (2006). "Erasure of methylation imprint at the promoter and CTCF-binding site upstream of H19 in human testicular germ cell tumors of adolescents indicate their fetal germ cell origin." Oncogene **25**(23): 3225–3236.

Kerr, A., C. Key Chekar, E. Ross, J. Swallow and S. Cunningham-Burley (2021). Personalised cancer medicine: Future crafting in the genomic era, Manchester University Press.

Khodayari, N., K. A. Mohammed, E. P. Goldberg and N. Nasreen (2011). "EphrinA1 inhibits malignant mesothelioma tumor growth via let-7 microRNA-mediated repression of the RAS oncogene." Cancer Gene Therapy **18**(11): 806–816.

Ki, M.-R., S. Youn, D. H. Kim and S. P. Pack (2024). "Natural Compounds for Preventing Age-Related Diseases and Cancers." International Journal of Molecular Sciences **25**(14): 7530.

Kim, H., J. Park, Y. Jung, S. H. Song, S. W. Han, D. Y. Oh, S. A. Im, Y. J. Bang and T. Y. Kim (2010). "DNA methyltransferase 3-like affects promoter methylation of thymine DNA glycosylase independently of DNMT1 and DNMT3B in cancer cells." Int J Oncol **36**(6): 1563–1572.

Kirschner, M. B., Y. Y. Cheng, B. Badrian, S. C. Kao, J. Creaney, J. J. B. Edelman, N. J. Armstrong, M. P. Vallely, A. W. Musk, B. W. S. Robinson, B. C. McCaughan, S. Klebe, S. E. Mutsaers, N. van Zandwijk and G. Reid (2012). "Increased Circulating miR-625-3p: A Potential Biomarker for Patients With Malignant Pleural Mesothelioma." Journal of Thoracic Oncology **7**(7): 1184–1191.

Ko, H. Y., J. Lee, Y. S. Lee, Y. Choi, B. A. Ali, A. A. Al-Khedhairy and S. Kim (2015). "Bioimaging of transcriptional activity of microRNA124a during neurogenesis." Biotechnol Lett **37**(11): 2333–2340.

Krauze, A. V., S. D. Myrehaug, M. G. Chang, D. J. Holdford, S. Smith, J. Shih, P. J. Tofilon, H. A. Fine and K. Camphausen (2015). "A phase 2 study of concurrent radiation therapy, temozolomide, and the histone deacetylase inhibitor valproic acid for patients with glioblastoma." International Journal of Radiation Oncology* Biology* Physics **92**(5): 986–992.

Kwok, M., N. Davies, A. Agathanggelou, E. Smith, E. Petermann, E. Yates, J. Brown, A. Lau and T. Stankovic (2015). "Synthetic lethality in chronic lymphocytic leukaemia with DNA damage response defects by targeting the ATR pathway." The Lancet **385**: S58.

Laille, E., M. R. Savona, B. L. Scott, T. E. Boyd, Q. Dong and B. Skikne (2014). "Pharmacokinetics of different formulations of oral azacitidine (CC-486) and the effect of food and modified gastric pH on pharmacokinetics in subjects with hematologic malignancies." J Clin Pharmacol **54**(6): 630–639.

Lee, D. H., S. Amanat, C. Goff, L. M. Weiss, J. W. Said, N. B. Doan, A. Sato-Otsubo, S. Ogawa, C. Forscher and H. P. Koeffler (2013). "Overexpression of miR-26a-2 in human liposarcoma is correlated with poor patient survival." Oncogenesis **2**(5): e47.

Lee, D. H., C. Forscher, D. Di Vizio and H. P. Koeffler (2015). "Induction of p53-independent apoptosis by ectopic expression of HOXA5 in human liposarcomas." Sci Rep **5**: 12580.

Lemmon, M. A. and J. Schlessinger (2010). "Cell signaling by receptor tyrosine kinases." Cell **141**(7): 1117–1134.

Li, C., Y. Feng, G. Coukos and L. Zhang (2009). "Therapeutic microRNA strategies in human cancer." Aaps j **11**(4): 747–757.

Li, C., S. Wang and C. Yang (2020a). "RETRACTED: Long non-coding RNA DLX6-AS1 regulates neuroblastoma progression by targeting YAP1 via miR-497-5p." Life Sciences **252**: 117657.

Li, D., P. Chen, X.-Y. Li, L.-Y. Zhang, W. Xiong, M. Zhou, L. Xiao, F. Zeng, X.-L. Li, M.-H. Wu and G.-Y. Li (2011). "Grade-Specific Expression Profiles of miRNAs/mRNAs and Docking Study in Human Grade I–III Astrocytomas." OMICS: A Journal of Integrative Biology **15**(10): 673–682.

Li, J., P. Li, W. Zhao, R. Yang, S. Chen, Y. Bai, S. Dun, X. Chen, Y. Du, Y. Wang, W. Zang, G. Zhao and G. Zhang (2015a). "Expression of long non-coding RNA DLX6-AS1 in lung adenocarcinoma." Cancer Cell International **15**(1): 48.

Li, W., R. Liu, D. Wei, W. Zhang, H. Zhang, W. Huang and L. Hao (2020b). "Circular RNA circ-CCAC1 Facilitates Adrenocortical Carcinoma Cell Proliferation, Migration, and Invasion through Regulating the miR-514a-5p/C22orf46 Axis." BioMed Research International **2020**(1): 3501451.

Li, W., C. Zhong, J. Jiao, P. Li, B. Cui, C. Ji and D. Ma (2017) "Characterization of hsa_circ_0004277 as a New Biomarker for Acute Myeloid Leukemia via Circular RNA Profile and Bioinformatics Analysis." International Journal of Molecular Sciences **18(3):597.**

Li, Y., Y. Yang, Y. Lu, J. G. Herman, M. V. Brock, P. Zhao and M. Guo (2015b). "Predictive value of CHFR and MLH1 methylation in human gastric cancer." Gastric Cancer **18**(2): 280–287.

Li, Y. Y. and Y. L. Zheng (2017). "Hypoxia promotes invasion of retinoblastoma cells in vitro by upregulating HIF-1α/MMP9 signaling pathway." Eur Rev Med Pharmacol Sci **21**(23): 5361–5369.

Li, Z., Y. Tang, W. Xing, W. Dong and Z. Wang (2018). "LncRNA, CRNDE promotes osteosarcoma cell proliferation, invasion and migration by regulating Notch1 signaling and epithelial-mesenchymal transition." Experimental and Molecular Pathology **104**(1): 19–25.

Liu, C. H., Q. Huang, Z. Y. Jin, C. L. Zhu, Z. Liu and C. Wang (2017a). "miR-21 and KLF4 jointly augment epithelial-mesenchymal transition via the Akt/ERK1/2 pathway." Int J Oncol **50**(4): 1109–1115.

Liu, J., H. Shi, X. Li, G. Chen, C. Larsson and W. O. Lui (2017b). "miR-223-3p regulates cell growth and apoptosis via FBXW7 suggesting an oncogenic role in human testicular germ cell tumors." Int J Oncol **50**(2): 356–364.

Liu, J., X. Zhao, K. Wang, X. Zhang, Y. Yu, Y. Lv, S. Zhang, L. Zhang, Y. Guo, Y. Li, A. Yang, R. Zhang and J. Li (2019a). "A novel YAP1/SLC35B4 regulatory axis contributes to proliferation and progression of gastric carcinoma." Cell Death & Disease **10**(6): 452.

Liu, K., B. He, J. Xu, Y. Li, C. Guo, Q. Cai and S. Wang (2019b). "miR-483-5p Targets MKNK1 to Suppress Wilms' Tumor Cell Proliferation and Apoptosis In Vitro and In Vivo." Med Sci Monit **25**: 1459–1468.

Liu, P. Y., A. E. Tee, G. Milazzo, K. M. Hannan, J. Maag, S. Mondal, B. Atmadibrata, N. Bartonicek, H. Peng, N. Ho, C. Mayoh, R. Ciaccio, Y. Sun, M. J. Henderson, J. Gao, C. Everaert, A. J. Hulme, M. Wong, Q. Lan, B. B. Cheung, L. Shi, J. Y. Wang, T. Simon, M. Fischer, X. D. Zhang, G. M. Marshall, M. D. Norris, M. Haber, J. Vandesompele, J. Li, P. Mestdagh, R. D. Hannan, M. E. Dinger, G. Perini and T. Liu (2019c). "The long noncoding RNA lncNB1 promotes tumorigenesis by interacting with ribosomal protein RPL35." Nature Communications **10**(1): 5026.

Liu, S., Z. Liu, Z. Xie, J. Pang, J. Yu, E. Lehmann, L. Huynh, T. Vukosavljevic, M. Takeki, R. B. Klisovic, R. A. Baiocchi, W. Blum, P. Porcu, R. Garzon, J. C. Byrd, D. Perrotti, M. A. Caligiuri, K. K. Chan, L. C. Wu and G. Marcucci (2008). "Bortezomib induces DNA hypomethylation and silenced gene transcription by interfering with Sp1/NF-kappaB-dependent DNA methyltransferase activity in acute myeloid leukemia." Blood **111**(4): 2364–2373.

Liu, S., L. C. Wu, J. Pang, R. Santhanam, S. Schwind, Y. Z. Wu, C. J. Hickey, J. Yu, H. Becker, K. Maharry, M. D. Radmacher, C. Li, S. P. Whitman, A. Mishra, N. Stauffer, A. M. Eiring, R. Briesewitz, R. A. Baiocchi, K. K. Chan, P. Paschka, M. A. Caligiuri, J. C. Byrd, C. M. Croce, C. D. Bloomfield, D. Perrotti, R. Garzon and G. Marcucci (2010). "Sp1/NFkappaB/HDAC/miR-29b regulatory network in KIT-driven myeloid leukemia." Cancer Cell **17**(4): 333–347.

Liu, Z., Z. Y. Jin, C. H. Liu, F. Xie, X. S. Lin and Q. Huang (2015). "MicroRNA-21 regulates biological behavior by inducing EMT in human cholangiocarcinoma." Int J Clin Exp Pathol **8**(5): 4684–4694.

Lizé, M., S. Pilarski and M. Dobbelstein (2010). "E2F1-inducible microRNA 449a/b suppresses cell proliferation and promotes apoptosis." Cell Death Differ **17**(3): 452–458.

Lobo, J., S. P. Nunes, A. J. M. Gillis, D. Barros-Silva, V. Miranda-Gonçalves, A. v. d. Berg, M. Cantante, R. Guimarães, R. Henrique, C. Jerónimo and L. H. J. Looijenga (2019) "XIST-Promoter Demethylation as Tissue Biomarker for Testicular Germ Cell Tumors and Spermatogenesis Quality." Cancers **11** DOI: https://doi.org/10.3390/cancers11091385.

Looijenga, L. H., A. J. Gillis, R. J. van Gurp, A. J. Verkerk and J. W. Oosterhuis (1997). "X inactivation in human testicular tumors. XIST expression and androgen receptor methylation status." Am J Pathol **151**(2): 581–590.

Lunke, S., S. Maxwell, I. Khurana, H. KN, J. Okabe, K. Al-Hasani and A. El-Osta (2021). "Epigenetic evidence of an Ac/Dc axis by VPA and SAHA." Clinical Epigenetics **13**: 1–14.

Majumder, S., A. Bhowal, S. Basu, P. Mukherjee, U. Chatterji and S. Sengupta (2016). "Deregulated E2F5/p38/SMAD3 Circuitry Reinforces the Pro-Tumorigenic Switch of TGFβ Signaling in Prostate Cancer." Journal of Cellular Physiology **231**(11): 2482–2492.

Maki, R. G. (2024). "Trials and Tribulations in Rare Cancer Clinical Research." Journal of Clinical Oncology **42**(8): 865–867.

Mallone, S., R. De Angelis, J. M. van der Zwan, A. Trama, S. Siesling, G. Gatta and R. Capocaccia (2013). "Methodological aspects of estimating rare cancer prevalence in Europe: the experience of the RARECARE project." Cancer Epidemiol **37**(6): 850–856.

Manda, S., B. Anz III, C. Benton, E. R. Broun, H. Yimer, J. Renshaw, G. F. Geils Jr, J. Cruz, S. Fanning and J. Melear (2021). "Treatment initiation of venetoclax in combination with azacitidine or decitabine in an outpatient setting in patients with untreated acute myeloid leukemia." Blood **138**: 1265.

Mano, M. S., R. J. Arai and P. M. Hoff (2010). "Rare tumors research in emerging countries." Rare Tumors **2**(3): e49.

Mathew, L. K., N. Skuli, V. Mucaj, S. S. Lee, P. O. Zinn, P. Sathyan, H. Z. Imtiyaz, Z. Zhang, R. V. Davuluri, S. Rao, S. Venneti, P. Lal, J. D. Lathia, J. N. Rich, B. Keith, A. J. Minn and M. C. Simon (2014). "miR-218 opposes a critical RTK-HIF pathway in mesenchymal glioblastoma." Proc Natl Acad Sci U S A **111**(1): 291–296.

Mazar, J., W. Zhao, A. M. Khalil, B. Lee, J. Shelley, S. S. Govindarajan, F. Yamamoto, M. Ratnam, M. N. Aftab, S. Collins, B. N. Finck, X. Han, J. S. Mattick, M. E. Dinger and R. J. Perera (2014). "The functional characterization of long noncoding RNA SPRY4-IT1 in human melanoma cells." Oncotarget; Vol 5, No 19.

Mazzu, Y. Z., Y. Hu, Y. Shen, T. Tuschl and S. Singer (2019). "miR-193b regulates tumorigenesis in liposarcoma cells via PDGFR, TGFβ, and Wnt signaling." Sci Rep **9**(1): 3197.

McClure, J. J., X. Li and C. J. Chou (2018). "Advances and challenges of HDAC inhibitors in cancer therapeutics." Advances in cancer research **138**: 183–211.

McPherson, J. P., L. Tamblyn, A. Elia, E. Migon, A. Shehabeldin, E. Matysiak-Zablocki, B. Lemmers, L. Salmena, A. Hakem, J. Fish, F. Kassam, J. Squire, B. G. Bruneau, M. P. Hande and R. Hakem (2004). "Lats2/Kpm is required for embryonic development, proliferation control and genomic integrity." Embo j **23**(18): 3677–3688.

Meng, F., H. Wehbe-Janek, R. Henson, H. Smith and T. Patel (2008). "Epigenetic regulation of microRNA-370 by interleukin-6 in malignant human cholangiocytes." Oncogene **27**(3): 378–386.

Messmer, B. T., D. Messmer, S. L. Allen, J. E. Kolitz, P. Kudalkar, D. Cesar, E. J. Murphy, P. Koduru, M. Ferrarini, S. Zupo, G. Cutrona, R. N. Damle, T. Wasil, K. R. Rai, M. K. Hellerstein and N. Chiorazzi (2005). "In vivo measurements document the dynamic cellular kinetics of chronic lymphocytic leukemia B cells." The Journal of Clinical Investigation **115**(3): 755–764.

Milhollen, M. A., T. Traore, J. Adams-Duffy, M. P. Thomas, A. J. Berger, L. Dang, L. R. Dick, J. J. Garnsey, E. Koenig, S. P. Langston, M. Manfredi, U. Narayanan, M. Rolfe, L. M. Staudt, T. A. Soucy, J. Yu, J. Zhang, J. B. Bolen and P. G. Smith (2010). "MLN4924, a NEDD8-activating enzyme inhibitor, is active in diffuse large B-cell lymphoma models: rationale for treatment of NF-{kappa}B-dependent lymphoma." Blood **116**(9): 1515–1523.

Miller, R. C. J. R.T (2010). Problems in rare tumor study: a call for papers, SAGE Publications Sage UK: London, England. **2:** 46–47.

Mishina, M., O. Ogawa, H. Kinoshita, H. Oka, K. Okumura, K. Mitsumori, Y. Kakehi, A. E. Reeve and O. Yoshida (1996). "Equivalent Parental Distribution of Frequently Lost Alleles and Biallelic Expression of the H19 Gene in Human Testicular Germ Cell Tumors." Japanese Journal of Cancer Research **87**(8): 816–823.

Mitra, S., S. V. Muralidharan, M. Di Marco, P. K. Juvvuna, S. T. Kosalai, S. Reischl, D. Jachimowicz, S. Subhash, I. Raimondi, L. Kurian, M. Huarte, P. Kogner, M. Fischer, J. I. Johnsen, T. Mondal and C. Kanduri (2021). "Subcellular Distribution of p53 by the p53-Responsive lncRNA NBAT1 Determines Chemotherapeutic Response in Neuroblastoma." Cancer Res **81**(6): 1457–1471.

Modarresi, F., M. A. Faghihi, M. A. Lopez-Toledano, R. P. Fatemi, M. Magistri, S. P. Brothers, M. P. van der Brug and C. Wahlestedt (2012). "Inhibition of natural antisense transcripts in vivo results in gene-specific transcriptional upregulation." Nat Biotechnol **30**(5): 453–459.

Mogilyansky, E. and I. Rigoutsos (2013). "The miR-17/92 cluster: a comprehensive update on its genomics, genetics, functions and increasingly important and numerous roles in health and disease." Cell Death & Differentiation **20**(12): 1603–1614.

Mondal, T., P. K. Juvvuna, A. Kirkeby, S. Mitra, S. T. Kosalai, L. Traxler, F. Hertwig, S. Wernig-Zorc, C. Miranda, L. Deland, R. Volland, C. Bartenhagen, D. Bartsch, S. Bandaru, A. Engesser, S. Subhash, T. Martinsson, H. Carén, L. M. Akyürek, L. Kurian, M. Kanduri, M. Huarte, P. Kogner, M. Fischer and C. Kanduri (2018). "Sense-Antisense lncRNA Pair Encoded by Locus 6p22.3 Determines Neuroblastoma Susceptibility via the USP36-CHD7-SOX9 Regulatory Axis." Cancer Cell **33**(3): 417–434.e417.

Mott, J. L., S. Kobayashi, S. F. Bronk and G. J. Gores (2007). "mir-29 regulates Mcl-1 protein expression and apoptosis." Oncogene **26**(42): 6133–6140.

Naeem, A., P. Hu, M. Yang, J. Zhang, Y. Liu, W. Zhu and Q. Zheng (2022). "Natural Products as Anticancer Agents: Current Status and Future Perspectives." Molecules **27**(23): 8367.

Oing, C., M. Skowron, C. Bokemeyer and D. Nettersheim (2019). "Epigenetic treatment combinations to effectively target cisplatin-resistant germ cell tumors: past, present, and future considerations." Andrology **7**(4): 487–497.

Oishi, N., M. R. Kumar, S. Roessler, J. Ji, M. Forgues, A. Budhu, X. Zhao, J. B. Andersen, Q. H. Ye, H. L. Jia, L. X. Qin, T. Yamashita, H. G. Woo, Y. J. Kim, S. Kaneko, Z. Y. Tang, S. S. Thorgeirsson and X. W. Wang (2012). "Transcriptomic profiling reveals hepatic stem-like gene signatures and interplay of miR-200c and epithelial-mesenchymal transition in intrahepatic cholangiocarcinoma." Hepatology **56**(5): 1792–1803.

Olaru, A. V., G. Ghiaur, S. Yamanaka, D. Luvsanjav, F. An, I. Popescu, S. Alexandrescu, S. Allen, T. M. Pawlik, M. Torbenson, C. Georgiades, L. R. Roberts, G. J. Gores, A. Ferguson-Smith, M. I. Almeida, G. A. Calin, E. Mezey and F. M. Selaru (2011). "MicroRNA down-regulated in human cholangiocarcinoma control cell cycle through multiple targets involved in the G1/S checkpoint." Hepatology **54**(6): 2089–2098.

Olsen, E. A., Y. H. Kim, T. M. Kuzel, T. R. Pacheco, F. M. Foss, S. Parker, S. R. Frankel, C. Chen, J. L. Ricker and J. M. Arduino (2007). "Phase IIb multicenter trial of vorinostat in patients with persistent, progressive, or treatment refractory cutaneous T-cell lymphoma." Journal of clinical oncology **25**(21): 3109–3115.

Ooi, C. Y., D. R. Carter, B. Liu, C. Mayoh, A. Beckers, A. Lalwani, Z. Nagy, S. De Brouwer, B. Decaesteker, T. T. Hung, M. D. Norris, M. Haber, T. Liu, K. De Preter, F. Speleman, B. B. Cheung and G. M. Marshall (2018). "Network Modeling of microRNA-mRNA Interactions in Neuroblastoma Tumorigenesis Identifies miR-204 as a Direct Inhibitor of MYCN." Cancer Res **78**(12): 3122–3134.

Özata, D. M., S. Caramuta, D. Velázquez-Fernández, P. Akçakaya, H. Xie, A. Höög, J. Zedenius, M. Bäckdahl, C. Larsson and W. O. Lui (2011). "The role of microRNA deregulation in the pathogenesis of adrenocortical carcinoma." Endocr Relat Cancer **18**(6): 643–655.

Özata, D. M., X. Li, L. Lee, J. Liu, D. Warsito, P. Hajeri, I. Hultman, O. Fotouhi, S. Marklund, L. Ährlund-Richter, C. C. Juhlin, C. Larsson and W. O. Lui (2017). "Loss of miR-514a-3p regulation of PEG3 activates the NF-kappa B pathway in human testicular germ cell tumors." Cell Death Dis **8**(5): e2759.

Palmerini, E., R. Sanfilippo, G. Grignani, A. Buonadonna, A. Romanini, G. Badalamenti, V. Ferraresi, B. Vincenzi, A. Comandone and A. Pizzolorusso (2022). "Transcription regulators and ultra-rare and other rare translocation-related sarcomas treated with trabectedin: a proof of principle from a post-hoc analysis." Frontiers in Oncology **12**: 1042479.

Panageas, K. S. (2015). "Clinical trial design for rare cancers: why a less conventional route may be required." Expert Rev Clin Pharmacol **8**(6): 661–663.

Pandey, G. K. and C. Kanduri (2015). "Fighting Neuroblastomas with NBAT1." Oncoscience **2**(2): 79–80.

Pandey, Gaurav K., S. Mitra, S. Subhash, F. Hertwig, M. Kanduri, K. Mishra, S. Fransson, A. Ganeshram, T. Mondal, S. Bandaru, M. Östensson, Levent M. Akyürek, J. Abrahamsson, S. Pfeifer, E. Larsson, L. Shi, Z. Peng, M. Fischer, T. Martinsson, F. Hedborg, P. Kogner and C. Kanduri (2014). "The Risk-Associated Long Noncoding RNA <em>NBAT-1</em> Controls Neuroblastoma Progression by Regulating Cell Proliferation and Neuronal Differentiation." Cancer Cell **26**(5): 722–737.

Parodi, F., R. Carosio, M. Ragusa, C. Di Pietro, M. Maugeri, D. Barbagallo, F. Sallustio, G. Allemanni, M. P. Pistillo, I. Casciano, A. Forlani, F. P. Schena, M. Purrello, M. Romani and B. Banelli (2016). "Epigenetic dysregulation in neuroblastoma: A tale of miRNAs and DNA methylation." Biochim Biophys Acta **1859**(12): 1502–1514.

Pass, H. I., C. Goparaju, S. Ivanov, J. Donington, M. Carbone, M. Hoshen, D. Cohen, A. Chajut, S. Rosenwald, H. Dan, S. Benjamin and R. Aharonov (2010). "hsa-miR-29c* Is Linked to the Prognosis of Malignant Pleural Mesothelioma." Cancer Research **70**(5): 1916–1924.

Pathak, A., S. Tomar and S. Pathak (2023). "Epigenetics and Cancer: A Comprehensive Review." Asian Pacific Journal of Cancer Biology **8**(1): 75–89.

Peng, D., J. Wei, Y. Gan, J. Yang, X. Jiang, R. Kitazawa, Y. Xiang, Y. Dai and Y. Tang (2019). "Testis developmental related gene 1 regulates the chemosensitivity of seminoma TCam-2 cells to cisplatin via autophagy." J Cell Mol Med **23**(11): 7773–7784.

Peng, W., J. Wu and J. Feng (2016). "Long noncoding RNA HULC predicts poor clinical outcome and represents pro-oncogenic activity in diffuse large B-cell lymphoma." Biomed Pharmacother **79**: 188–193.

Podnos, Y. D., N. C. Tsai, D. Smith and J. D. Ellenhorn (2006). "Factors affecting survival in patients with anal melanoma." Am Surg **72**(10): 917–920.

Qi, P. and X. Du (2013). "The long non-coding RNAs, a new cancer diagnostic and therapeutic gold mine." Mod Pathol **26**(2): 155–165.

Qin, N., G.-F. Tong, L.-W. Sun and X.-L. Xu (2017). "Long Noncoding RNA MEG3 Suppresses Glioma Cell Proliferation, Migration, and Invasion by Acting as a Competing Endogenous RNA of miR-19a." Oncology Research **25**(9).

Qin, X., M. Lu, Y. Zhou, G. Li and Z. Liu (2019). "LncRNA FENDRR represses proliferation, migration and invasion through suppression of survivin in cholangiocarcinoma cells." Cell Cycle **18**(8): 889–897.

Radwan, S. M., N. M. Hamdy, H. M. Hegab and H. O. El-Mesallamy (2016). "Beclin-1 and hypoxia-inducible factor-1α genes expression: Potential biomarkers in acute leukemia patients." Cancer Biomarkers **16**(4): 619–626.

Raghavan, D. (2013). "A structured approach to uncommon cancers: what should a clinician do?" Ann Oncol **24**(12): 2932–2934.

Raghavan, D., M. S. Ahluwalia, C. D. Blanke, J. Brown, E. S. Kim, G. H. Reaman and M. A. Sekeres (2017). Textbook of uncommon cancer, John Wiley & Sons.

Ramírez-Salazar, E. G., L. C. Salinas-Silva, M. E. Vázquez-Manríquez, L. V. Gayosso-Gómez, M. C. Negrete-Garcia, S. L. Ramírez-Rodriguez, R. Chávez, E. Zenteno, P. Santillán, J. Kelly-García and B. Ortiz-Quintero (2014). "Analysis of microRNA expression signatures in malignant pleural mesothelioma, pleural inflammation, and atypical mesothelial hyperplasia reveals common predictive tumorigenesis-related targets." Experimental and Molecular Pathology **97**(3): 375–385.

Rao, R. and S. G. Honavar (2017). "Retinoblastoma." Indian J Pediatr **84**(12): 937–944.

Reid, G., M. E. Pel, M. B. Kirschner, Y. Y. Cheng, N. Mugridge, J. Weiss, M. Williams, C. Wright, J. J. B. Edelman, M. P. Vallely, B. C. McCaughan, S. Klebe, H. Brahmbhatt, J. A. MacDiarmid and N. van Zandwijk (2013). "Restoring expression of miR-16: a novel approach to therapy for malignant pleural mesothelioma." Annals of Oncology **24**(12): 3128–3135.

Remon, J., N. Reguart, J. Corral and P. Lianes (2015). "Malignant pleural mesothelioma: New hope in the horizon with novel therapeutic strategies." Cancer Treatment Reviews **41**(1): 27–34.

Renganathan, A., J. Kresoja-Rakic, N. Echeverry, G. Ziltener, B. Vrugt, I. Opitz, R. A. Stahel and E. Felley-Bosco (2014). "GAS5 long non-coding RNA in malignant pleural mesothelioma." Molecular Cancer **13**(1): 119.

Riquelme, E., M. B. Suraokar, J. Rodriguez, B. Mino, H. Y. Lin, D. C. Rice, A. Tsao and I. I. Wistuba (2014). "Frequent Coamplification and Cooperation between C-MYC and PVT1 Oncogenes Promote Malignant Pleural Mesothelioma." Journal of Thoracic Oncology **9**(7): 998–1007.

Ronchi, C. L., S. Sbiera, B. Altieri, S. Steinhauer, V. Wild, M. Bekteshi, M. Kroiss, M. Fassnacht and B. Allolio (2015). "Notch1 pathway in adrenocortical carcinomas: correlations with clinical outcome." Endocrine-Related Cancer **22**(4): 531–543.

Roth, S. A., H. Hald Ø, S. Fuchs, C. Løkke, I. Mikkola, T. Flægstad, J. Schulte and C. Einvik (2018). "MicroRNA-193b-3p represses neuroblastoma cell growth via downregulation of Cyclin D1, MCL-1 and MYCN." Oncotarget **9**(26): 18160–18179.

Sahafnejad, Z., S. Ramazi and A. Allahverdi (2023). "An Update of Epigenetic Drugs for the Treatment of Cancers and Brain Diseases: A Comprehensive Review." Genes (Basel) **14**(4).

Sahu, D., C. L. Hsu, C. C. Lin, T. W. Yang, W. M. Hsu, S. Y. Ho, H. F. Juan and H. C. Huang (2016). "Co-expression analysis identifies long noncoding RNA SNHG1 as a novel predictor for event-free survival in neuroblastoma." Oncotarget **7**(36): 58022–58037.

Salati, M. and C. Braconi (2019). "Noncoding RNA in Cholangiocarcinoma." Semin Liver Dis **39**(1): 13–25.

Salehi, M. and M. Sharifi (2018). "Induction of apoptosis and necrosis in human acute erythroleukemia cells by inhibition of long non-coding RNA PVT1." Mol Biol Res Commun **7**(2): 89–96.

Salehi, M., M. Sharifi and M. Bagheri (2018). "Knockdown of Long Noncoding RNA Plasmacytoma Variant Translocation 1 with Antisense Locked Nucleic Acid GapmeRs Exerts Tumor-Suppressive Functions in Human Acute Erythroleukemia Cells Through Downregulation of C-MYC Expression." Cancer Biotherapy and Radiopharmaceuticals **34**(6): 371–379.

Sallman, D. A., A. E. DeZern, G. Garcia-Manero, D. P. Steensma, G. J. Roboz, M. A. Sekeres, T. Cluzeau, K. L. Sweet, A. McLemore and K. L. McGraw (2021). "Eprenetapopt (APR-246) and azacitidine in TP53-mutant myelodysplastic syndromes." Journal of Clinical Oncology **39**(14): 1584–1594.

Sanfilippo, R., P. Dileo, J.-Y. Blay, A. Constantinidou, A. Le Cesne, C. Benson, L. Vizzini, M. Contu, G. G. Baldi and A. P. Dei Tos (2015). "Trabectedin in advanced synovial sarcomas: a multicenter retrospective study from four European institutions and the Italian Rare Cancer Network." Anti-Cancer Drugs **26**(6): 678–681.

Sattari, A., H. Siddiqui, F. Moshiri, A. Ngankeu, T. Nakamura, T. J. Kipps and C. M. Croce (2016). "Upregulation of long noncoding RNA MIAT in aggressive form of chronic lymphocytic leukemias." Oncotarget **7**(34): 54174–54182.

Schaefer, R. (2012). "Rare Cancers Europe: joining forces to tackle a common problem." Rare Tumors **4**(2): e24.

Schramedei, K., N. Mörbt, G. Pfeifer, J. Läuter, M. Rosolowski, J. M. Tomm, M. von Bergen, F. Horn and K. Brocke-Heidrich (2011). "MicroRNA-21 targets tumor suppressor genes ANP32A and SMARCA4." Oncogene **30**(26): 2975–2985.

Selaru, F. M., A. V. Olaru, T. Kan, S. David, Y. Cheng, Y. Mori, J. Yang, B. Paun, Z. Jin, R. Agarwal, J. P. Hamilton, J. Abraham, C. Georgiades, H. Alvarez, P. Vivekanandan, W. Yu, A. Maitra, M. Torbenson, P. J. Thuluvath, G. J. Gores, N. F. LaRusso, R. Hruban and S. J. Meltzer (2009). "MicroRNA-21 is overexpressed in human cholangiocarcinoma and regulates programmed cell death 4 and tissue inhibitor of metalloproteinase 3." Hepatology **49**(5): 1595–1601.

Selmin, O. I., M. G. Donovan, B. J. Stillwater, L. Neumayer and D. F. Romagnolo (2020). "Epigenetic regulation and dietary control of triple negative breast cancer." Frontiers in nutrition **7**: 159.

Sharma, S. V., D. W. Bell, J. Settleman and D. A. Haber (2007). "Epidermal growth factor receptor mutations in lung cancer." Nat Rev Cancer **7**(3): 169–181.

Shi, W., N. M. Alajez, C. Bastianutto, A. B. Y. Hui, J. D. Mocanu, E. Ito, P. Busson, K.-W. Lo, R. Ng, J. Waldron, B. O'Sullivan and F.-F. Liu (2010). "Significance of Plk1 regulation by miR-100 in human nasopharyngeal cancer." International Journal of Cancer **126**(9): 2036–2048.

Shimizu, K., N. T. Nihira, H. Inuzuka and W. Wei (2018). "Physiological functions of FBW7 in cancer and metabolism." Cellular Signalling **46**: 15–22.

Shiyanbola, O., H. Hardin, R. Hu, J. C. Eickhoff and R. V. Lloyd (2020). "Long Noncoding RNA Expression in Adrenal Cortical Neoplasms." Endocrine Pathology **31**(4): 385–391.

Shukla, S., D. Penta, P. Mondal and S. M. Meeran (2019). "Epigenetics of breast cancer: clinical status of epi-drugs and phytochemicals." Breast cancer metastasis and drug resistance: challenges and progress: 293–310.

Sidiropoulos, K., D. Glotsos, S. Kostopoulos, P. Ravazoula, I. Kalatzis, D. Cavouras and J. Stonham (2012). "Real time decision support system for diagnosis of rare cancers, trained in parallel, on a graphics processing unit." Comput Biol Med **42**(4): 376–386.

Silber, J., D. A. Lim, C. Petritsch, A. I. Persson, A. K. Maunakea, M. Yu, S. R. Vandenberg, D. G. Ginzinger, C. D. James, J. F. Costello, G. Bergers, W. A. Weiss, A. Alvarez-Buylla and J. G. Hodgson (2008). "miR-124 and miR-137 inhibit proliferation of glioblastoma multiforme cells and induce differentiation of brain tumor stem cells." BMC Med **6**: 14.

Silverman, L. R., E. P. Demakos, B. L. Peterson, A. B. Kornblith, J. C. Holland, R. Odchimar-Reissig, R. M. Stone, D. Nelson, B. L. Powell and C. M. DeCastro (2002). "Randomized controlled trial of azacitidine in patients with the myelodysplastic syndrome: a study of the cancer and leukemia group B." Journal of Clinical oncology **20**(10): 2429–2440.

Šimoničová, K., Ľ. Janotka, H. Kavcová, Z. Sulová, A. Breier and L. Messingerova (2022). "Different mechanisms of drug resistance to hypomethylating agents in the treatment of myelodysplastic syndromes and acute myeloid leukemia." Drug Resistance Updates **61**: 100805.

Singh, M., V. Kumar, N. Sehrawat, M. Yadav, M. Chaudhary, S. K. Upadhyay, S. Kumar, V. Sharma, S. Kumar, N. Dilbaghi and A. K. Sharma (2022). "Current paradigms in epigenetic anticancer therapeutics and future challenges." Semin Cancer Biol **83**: 422–440.

Smits, M., T. Wurdinger, B. van het Hof, J. A. R. Drexhage, D. Geerts, P. Wesseling, D. P. Noske, W. P. Vandertop, H. E. de Vries and A. Reijerkerk (2012). "Myc-associated zinc finger protein

(MAZ) is regulated by miR-125b and mediates VEGF-induced angiogenesis in glioblastoma." The FASEB Journal **26**(6): 2639–2647.

Sokolov, D., N. Sharda, A. Banerjee, K. Denisenko, E. B. Basalious, H. Shukla, J. Waddell, N. M. Hamdy and A. Banerjee (2024). "Differential signaling pathways in medulloblastoma: nano-biomedicine targeting non-coding epigenetics to improve current and future therapeutics." Current pharmaceutical design **30**(1): 31–47.

Solomou, G., A. Finch, A. Asghar and C. Bardella (2023). "Mutant IDH in gliomas: Role in cancer and treatment options." Cancers **15**(11): 2883.

Sonkoly, E., Z. Bata-Csorgo, A. Pivarcsi, H. Polyanka, A. Kenderessy-Szabo, G. Molnar, K. Szentpali, L. Bari, K. Megyeri, Y. Mandi, A. Dobozy, L. Kemeny and M. Szell (2005). "Identification and Characterization of a Novel, Psoriasis Susceptibility-related Noncoding RNA gene, PRINS *." Journal of Biological Chemistry **280**(25): 24159–24167.

Steensma, D. P. (2009). "Decitabine treatment of patients with higher-risk myelodysplastic syndromes." Leukemia research **33**: S12-S17.

Stomper, J., J. C. Rotondo, G. Greve and M. Lübbert (2021). "Hypomethylating agents (HMA) for the treatment of acute myeloid leukemia and myelodysplastic syndromes: mechanisms of resistance and novel HMA-based therapies." Leukemia **35**(7): 1873–1889.

Stresemann, C. and F. Lyko (2008). "Modes of action of the DNA methyltransferase inhibitors azacytidine and decitabine." International journal of cancer **123**(1): 8–13.

Su, Y., L.-L. Zhou, Y.-Q. Zhang and L.-Y. Ni (2019). "Long noncoding RNA HOTTIP is associated with male infertility and promotes testicular embryonal carcinoma cell proliferation." Molecular Genetics & Genomic Medicine **7**(9): e870.

Sugihara, T., H. Isomoto, G. Gores and R. Smoot (2019). "YAP and the Hippo pathway in cholangiocarcinoma." J Gastroenterol **54**(6): 485–491.

Sun, F., H. Fu, Q. Liu, Y. Tie, J. Zhu, R. Xing, Z. Sun and X. Zheng (2008). "Downregulation of CCND1 and CDK6 by miR-34a induces cell cycle arrest." FEBS Letters **582**(10): 1564–1568.

Sun, Y., J. H. Hong, Z. Ning, D. Pan, X. Fu, X. Lu and J. Tan (2022). "Therapeutic potential of tucidinostat, a subtype-selective HDAC inhibitor, in cancer treatment." Frontiers in Pharmacology **13**: 932914.

Suraweera, A., K. J. O'Byrne and D. J. Richard (2018). "Combination therapy with histone deacetylase inhibitors (HDACi) for the treatment of cancer: achieving the full therapeutic potential of HDACi." Frontiers in oncology **8**: 92.

Szegedi, K., E. Sonkoly, N. Nagy, I. B. Németh, Z. Bata-Csörgő, L. Kemény, A. Dobozy and M. Széll (2010). "The anti-apoptotic protein G1P3 is overexpressed in psoriasis and regulated by the non-coding RNA, PRINS." Experimental Dermatology **19**(3): 269–278.

Takeda, M., T. Kasai, Y. Enomoto, M. Takano, K. Morita, E. Kadota, N. Iizuka, H. Maruyama and A. Nonomura (2012). "Genomic gains and losses in malignant mesothelioma demonstrated by FISH analysis of paraffin-embedded tissues." Journal of Clinical Pathology **65**(1): 77.

Tang, F., H. Wang, E. Chen, E. Bian, Y. Xu, X. Ji, Z. Yang, X. Hua, Y. Zhang and B. Zhao (2019). "LncRNA-ATB promotes TGF-β-induced glioma cells invasion through NF-κB and P38/MAPK pathway." Journal of Cellular Physiology **234**(12): 23302–23314.

Taniguchi, T., S. Karnan, T. Fukui, T. Yokoyama, H. Tagawa, K. Yokoi, Y. Ueda, T. Mitsudomi, Y. Horio, T. Hida, Y. Yatabe, M. Seto and Y. Sekido (2007). "Genomic profiling of malignant pleural mesothelioma with array-based comparative genomic hybridization shows frequent non-random chromosomal alteration regions including JUN amplification on 1p32." Cancer Science **98**(3): 438–446.

Tognon, M., G. A. Calin and L. Del Valle (2022). Epigenetics and molecular genetics of rare tumors, Frontiers Media SA. **12:** 1049140.

Trama, A., S. Mallone, S. Ferretti, F. Meduri, R. Capocaccia and G. Gatta (2012). "The burden of rare cancers in Italy: the surveillance of rare cancers in Italy (RITA) project." Tumori **98**(5): 550–558.

Tsai, H.-C., K.-C. Wei, P.-Y. Chen, C.-Y. Huang, K.-T. Chen, Y.-J. Lin, H.-W. Cheng, Y.-R. Chen and H.-T. Wang (2021). "Valproic acid enhanced temozolomide-induced anticancer activity in human glioma through the p53–PUMA apoptosis pathway." Frontiers in Oncology **11**: 722754.

Tseng, Y.-Y., B. S. Moriarity, W. Gong, R. Akiyama, A. Tiwari, H. Kawakami, P. Ronning, B. Reuland, K. Guenther, T. C. Beadnell, J. Essig, G. M. Otto, M. G. O'Sullivan, D. A. Largaespada, K. L. Schwertfeger, Y. Marahrens, Y. Kawakami and A. Bagchi (2014). "PVT1 dependence in cancer with MYC copy-number increase." Nature **512**(7512): 82–86.

Tu, J., F. Wu, L. Chen, L. Zheng, Y. Yang, X. Ying, J. Song, C. Chen, X. Hu, Z. Zhao and J. Ji (2021). "Long Non-Coding RNA PCAT6 Induces M2 Polarization of Macrophages in Cholangiocarcinoma via Modulating miR-326 and RhoA-ROCK Signaling Pathway." Frontiers in Oncology **10**.

Ugras, S., E. Brill, A. Jacobsen, M. Hafner, N. D. Socci, P. L. Decarolis, R. Khanin, R. O'Connor, A. Mihailovic, B. S. Taylor, R. Sheridan, J. M. Gimble, A. Viale, A. Crago, C. R. Antonescu, C. Sander, T. Tuschl and S. Singer (2011). "Small RNA sequencing and functional characterization reveals MicroRNA-143 tumor suppressor activity in liposarcoma." Cancer Res **71**(17): 5659–5669.

Vigorito, E., K. L. Perks, C. Abreu-Goodger, S. Bunting, Z. Xiang, S. Kohlhaas, P. P. Das, E. A. Miska, A. Rodriguez, A. Bradley, K. G. Smith, C. Rada, A. J. Enright, K. M. Toellner, I. C. Maclennan and M. Turner (2007). "microRNA-155 regulates the generation of immunoglobulin class-switched plasma cells." Immunity **27**(6): 847–859.

Vincenzi, B., M. Iuliani, A. Zoccoli, F. Pantano, M. Fioramonti, D. De Lisi, A. M. Frezza, C. Rabitti, G. Perrone, A. Onetti Muda, A. Russo, A. Giordano, D. Santini, A. P. Dei Tos and G. Tonini (2015). "Deregulation of dicer and mir-155 expression in liposarcoma." Oncotarget **6**(12): 10586–10591.

Vivekanandhan, S., D. Bahr, A. Kothari, M. A. Ashary, M. Baksh and E. Gabriel (2023). "Immunotherapies in rare cancers." Molecular Cancer **22**(1): 23.

Vogelstein, B., N. Papadopoulos, V. E. Velculescu, S. Zhou, L. A. Diaz, Jr. and K. W. Kinzler (2013). "Cancer genome landscapes." Science **339**(6127): 1546–1558.

Wang, B., Y. Su, Q. Yang, D. Lv, W. Zhang, K. Tang, H. Wang, R. Zhang and Y. Liu (2015a). "Overexpression of Long Non-Coding RNA HOTAIR Promotes Tumor Growth and Metastasis in Human Osteosarcoma." Mol Cells **38**(5): 432–440.

Wang, J., X. Liu, H. Wu, P. Ni, Z. Gu, Y. Qiao, N. Chen, F. Sun and Q. Fan (2010). "CREB up-regulates long non-coding RNA, HULC expression through interaction with microRNA-372 in liver cancer." Nucleic Acids Res **38**(16): 5366–5383.

Wang, J., S. Yang, Q. Ji, Q. Li, F. Zhou, Y. Li, F. Yuan, J. Liu, Y. Tian, Y. Zhao and Y. Zheng (2020). "Long Non-coding RNA EPIC1 Promotes Cell Proliferation and Motility and Drug Resistance in Glioma." Mol Ther Oncolytics **17**: 130–137.

Wang, N., C. Zhang, W. Wang, J. Liu, Y. Yu, Y. Li, M. Zhang, X. Ge, Q. Li and L. Miao (2019). "Long noncoding RNA DANCR regulates proliferation and migration by epigenetically silencing FBP1 in tumorigenesis of cholangiocarcinoma." Cell Death & Disease **10**(8): 585.

Wang, P., Z. Ren and P. Sun (2012a). "Overexpression of the long non-coding RNA MEG3 impairs in vitro glioma cell proliferation." Journal of cellular biochemistry **113**(6): 1868–1874.

Wang, W.-T., H. Ye, P.-P. Wei, B.-W. Han, B. He, Z.-H. Chen and Y.-Q. Chen (2016). "LncRNAs H19 and HULC, activated by oxidative stress, promote cell migration and invasion in cholangiocarcinoma through a ceRNA manner." Journal of Hematology & Oncology **9**(1): 117.

Wang, X. W., N. H. Heegaard and H. Orum (2012b). "MicroRNAs in liver disease." Gastroenterology **142**(7): 1431–1443.

Wang, Y., C. Melton, Y.-P. Li, A. Shenoy, X.-X. Zhang, D. Subramanyam and R. Blelloch (2013). "miR-294/miR-302 Promotes Proliferation, Suppresses G1-S Restriction Point, and Inhibits ESC Differentiation through Separable Mechanisms." Cell Reports **4**(1): 99–109.

Wang, Y., Y. Wang, J. Li, Y. Zhang, H. Yin and B. Han (2015b). "CRNDE, a long-noncoding RNA, promotes glioma cell growth and invasion through mTOR signaling." Cancer Letters **367**(2): 122–128.

Wang, Y., Q. Zhou and J. J. Ma (2018). "High expression of lnc-CRNDE presents as a biomarker for acute myeloid leukemia and promotes the malignant progression in acute myeloid leukemia cell line U937." Eur Rev Med Pharmacol Sci **22**(3): 763–770.

Weber, A., J. Marquardt, D. Elzi, N. Forster, S. Starke, A. Glaum, D. Yamada, P. A. Defossez, J. Delrow, R. N. Eisenman, H. Christiansen and M. Eilers (2008). "Zbtb4 represses transcription of P21CIP1 and controls the cellular response to p53 activation." The EMBO Journal **27**(11): 1563–1574.

Wei, J., Y. Gan, D. Peng, X. Jiang, R. Kitazawa, Y. Xiang, Y. Dai, Y. Tang and J. Yang (2018). "Long non-coding RNA H19 promotes TDRG1 expression and cisplatin resistance by sequestering miRNA-106b-5p in seminoma." Cancer Med **7**(12): 6247–6257.

Wei, X., C. Wang, C. Ma, W. Sun, H. Li and Z. Cai (2016). "Long noncoding RNA ANRIL is activated by hypoxia-inducible factor-1α and promotes osteosarcoma cell invasion and suppresses cell apoptosis upon hypoxia." Cancer Cell Int **16**: 73.

Weiler, J., J. Hunziker and J. Hall (2006). "Anti-miRNA oligonucleotides (AMOs): ammunition to target miRNAs implicated in human disease?" Gene Ther **13**(6): 496–502.

Williams, M., M. B. Kirschner, Y. Y. Cheng, J. Hanh, J. Weiss, N. Mugridge, C. M. Wright, A. Linton, S. C. Kao, J. J. B. Edelman, M. P. Vallely, B. C. McCaughan, W. Cooper, S. Klebe, R. C. Y. Lin, H. Brahmbhatt, J. MacDiarmid, N. v. Zandwijk and G. Reid (2015). "miR-193a-3p is a potential tumor suppressor in malignant pleural mesothelioma." Oncotarget **6**(27).

Wolff, L., M. T. Garin, R. Koller, J. Bies, W. Liao, M. Malumbres, L. Tessarollo, D. Powell and C. Perella (2003). "Hypermethylation of the Ink4b locus in murine myeloid leukemia and increased susceptibility to leukemia in p15(Ink4b)-deficient mice." Oncogene **22**(58): 9265–9274.

Wu, Y., W. Wang, W. Hu, W. Xu, G. Xiao, Q. Nie, K. Ouyang and S. Chen (2015). "MicroRNA-205 suppresses the growth of adrenocortical carcinoma SW-13 cells via targeting Bcl-2." Oncol Rep **34**(6): 3104–3110.

Würdinger, T., B. A. Tannous, O. Saydam, J. Skog, S. Grau, J. Soutschek, R. Weissleder, X. O. Breakefield and A. M. Krichevsky (2008). "miR-296 regulates growth factor receptor overexpression in angiogenic endothelial cells." Cancer Cell **14**(5): 382–393.

Xiang, X., H. Mei, H. Qu, X. Zhao, D. Li, H. Song, W. Jiao, J. Pu, K. Huang, L. Zheng and Q. Tong (2015a). "miRNA-584-5p exerts tumor suppressive functions in human neuroblastoma through repressing transcription of matrix metalloproteinase 14." Biochim Biophys Acta **1852**(9): 1743–1754.

Xiang, X., H. Mei, X. Zhao, J. Pu, D. Li, H. Qu, W. Jiao, J. Zhao, K. Huang, L. Zheng and Q. Tong (2015b). "miRNA-337-3p suppresses neuroblastoma progression by repressing the transcription of matrix metalloproteinase 14." Oncotarget **6**(26): 22452–22466.

Xu, L., W. Li, Q. Shi, M. Wang, H. Li, X. Yang and J. Zhang (2020). "MicroRNA-936 inhibits the malignant phenotype of retinoblastoma by directly targeting HDAC9 and deactivating the PI3K/AKT pathway." Oncol Rep **43**(2): 635–645.

Xu, T., W. Wang, S. Zhang, R. A. Stewart and W. Yu (1995). "Identifying tumor suppressors in genetic mosaics: the Drosophila lats gene encodes a putative protein kinase." Development **121**(4): 1053–1063.

Xu, Y., Z. Fu, X. Gao, R. Wang and Q. Li (2021). "Long non-coding RNA XIST promotes retinoblastoma cell proliferation, migration, and invasion by modulating microRNA-191-5p/brain derived neurotrophic factor." Bioengineered **12**(1): 1587–1598.

Xu, Y., Y. Yao, X. Jiang, X. Zhong, Z. Wang, C. Li, P. Kang, K. Leng, D. Ji, Z. Li, L. Huang, W. Qin and Y. Cui (2018). "SP1-induced upregulation of lncRNA SPRY4-IT1 exerts oncogenic properties by scaffolding EZH2/LSD1/DNMT1 and sponging miR-101-3p in cholangiocarcinoma." Journal of Experimental & Clinical Cancer Research **37**(1): 81.

Xue, H., L. M. Hua, M. Guo and J. M. Luo (2014). "SHIP1 is targeted by miR-155 in acute myeloid leukemia." Oncol Rep **32**(5): 2253–2259.

Yan, J., Y. X. Deng, Y. L. Cai and W. D. Cong (2022). "LncRNA MIR17HG promotes the proliferation, migration, and invasion of retinoblastoma cells by up-regulating HIF-1α expression via sponging miR-155-5p." Kaohsiung J Med Sci **38**(6): 554–564.

Yan, W. and M. Guo (2015). "Epigenetics of colorectal cancer." Methods Mol Biol **1238**: 405–424.

Yao, Y., J. Ma, Y. Xue, P. Wang, Z. Li, J. Liu, L. Chen, Z. Xi, H. Teng, Z. Wang, Z. Li and Y. Liu (2015). "Knockdown of long non-coding RNA XIST exerts tumor-suppressive functions in human glioblastoma stem cells by up-regulating miR-152." Cancer Letters **359**(1): 75–86.

Youness, R. A., H. A. Hassan, T. Abaza, A. A. Hady, H. M. El Magdoub, M. Ali, J. Vogel, M. Thiersch, M. Gassmann, N. M. Hamdy and M. A. Abooufy (2024). "A Comprehensive Insight and In Silico Analysis of CircRNAs in Hepatocellular Carcinoma: A Step toward ncRNA-Based Precision Medicine." Cells **13**(15): 1245.

Yu, X., H. Zhao, R. Wang, Y. Chen, X. Ouyang, W. Li, Y. Sun and A. Peng (2024). "Cancer epigenetics: from laboratory studies and clinical trials to precision medicine." Cell Death Discovery **10**(1): 28.

Yu, Y., Q. Chen, X. Zhang, J. Yang, K. Lin, C. Ji, A. Xu, L. Yang and L. Miao (2020). "Long non-coding RNA ANRIL promotes the malignant progression of cholangiocarcinoma by epigenetically repressing ERRFI1 expression." Cancer Science **111**(7): 2297–2309.

Yu, Y., M. Zhang, J. Liu, B. Xu, J. Yang, N. Wang, S. Yan, F. Wang, X. He, G. Ji, Q. Li and L. Miao (2018a). "Long Non-coding RNA <em>PVT1</em> Promotes Cell Proliferation and Migration by Silencing <em>ANGPTL4</em> Expression in Cholangiocarcinoma." Molecular Therapy Nucleic Acids **13**: 503–513.

Yu, Y., M. Zhang, N. Wang, Q. Li, J. Yang, S. Yan, X. He, G. Ji and L. Miao (2018b). "Epigenetic silencing of tumor suppressor gene CDKN1A by oncogenic long non-coding RNA SNHG1 in cholangiocarcinoma." Cell Death & Disease **9**(7): 746.

Zeng, C., X. Yu, J. Lai, L. Yang, S. Chen and Y. Li (2015). "Overexpression of the long non-coding RNA PVT1 is correlated with leukemic cell proliferation in acute promyelocytic leukemia." Journal of Hematology & Oncology **8**(1): 126.

Zhang, C., J.-Y. Li, F.-Z. Tian, G. Zhao, H. Hu, Y.-F. Ma and Y.-L. Yang (2018a). "Long Noncoding RNA NEAT1 Promotes Growth and Metastasis of Cholangiocarcinoma Cells." Oncology Research **26**(6).

Zhang, C. Z., J. X. Zhang, A. L. Zhang, Z. D. Shi, L. Han, Z. F. Jia, W. D. Yang, G. X. Wang, T. Jiang, Y. P. You, P. Y. Pu, J. Q. Cheng and C. S. Kang (2010). "MiR-221 and miR-222 target PUMA to induce cell survival in glioblastoma." Mol Cancer **9**: 229.

Zhang, H. Y., M. Q. Xing, J. Guo, J. C. Zhao, X. Chen, Z. Jiang, H. Zhang and Q. Dong (2019a). "Long noncoding RNA DLX6-AS1 promotes neuroblastoma progression by regulating miR-107/BDNF pathway." Cancer Cell Int **19**: 313.

Zhang, J., C. Han and T. Wu (2012a). "MicroRNA-26a promotes cholangiocarcinoma growth by activating β-catenin." Gastroenterology **143**(1): 246–256.e248.

Zhang, J., M. Yin, G. Peng and Y. Zhao (2018b). "CRNDE: An important oncogenic long non-coding RNA in human cancers." Cell Proliferation **51**(3): e12440.

Zhang, L., X. Liang and Y. Li (2017a). "Long non-coding RNA MEG3 inhibits cell growth of gliomas by targeting miR-93 and inactivating PI3K/AKT pathway Retraction in /10.3892/or.2020.7828." Oncol Rep **38**(4): 2408–2416.

Zhang, N., F. L. Liu, T. S. Ma and Z. Z. J. Zhang (2019b). "LncRNA SNHG1 contributes to tumorigenesis and mechanism by targeting miR-338-3p to regulate PLK4 in human neuroblastoma." Eur Rev Med Pharmacol Sci **23**(20): 8971–8983.

Zhang, P., K. Bill, J. Liu, E. Young, T. Peng, S. Bolshakov, A. Hoffman, Y. Song, E. G. Demicco, D. L. Terrada, C. J. Creighton, M. L. Anderson, A. J. Lazar, G. G. Calin, R. E. Pollock and D. Lev (2012b). "MiR-155 is a liposarcoma oncogene that targets casein kinase-1α and enhances β-catenin signaling." Cancer Res **72**(7): 1751–1762.

Zhang, S., J. Long and Y. Hu (2020a). "Long noncoding RNA LINC00205 enhances the malignant characteristics of retinoblastoma by acting as a molecular sponge of microRNA-665 and consequently increasing HMGB1 expression." Biochem Biophys Res Commun **526**(2): 396–403.

Zhang, T., J. Yang, F. Gong, L. Li and A. Li (2020b). "Long non-coding RNA CASC9 promotes the progression of retinoblastoma via interacting with miR-145-5p." Cell Cycle **19**(18): 2270–2280.

Zhang, X., S. M. Weissman and P. E. Newburger (2014). "Long intergenic non-coding RNA HOTAIRM1 regulates cell cycle progression during myeloid maturation in NB4 human promyelocytic leukemia cells." RNA Biol **11**(6): 777–787.

Zhang, Y., N. Cruickshanks, M. Pahuski, F. Yuan, A. Dutta, D. Schiff, B. Purow and R. Abounader (2017b). "Noncoding RNAs in glioblastoma." Exon Publications: 95–130.

Zhang, Y., X. Zhu, X. Zhu, Y. Wu, Y. Liu, B. Yao and Z. Huang (2017c). "MiR-613 suppresses retinoblastoma cell proliferation, invasion, and tumor formation by targeting E2F5." Tumour Biol **39**(3): 1010428317691674.

Zhao, D., Y. Tian, P. Li, L. Wang, A. Xiao, M. Zhang and T. Shi (2015a). "MicroRNA-203 inhibits the malignant progression of neuroblastoma by targeting Sam68." Mol Med Rep **12**(4): 5554–5560.

Zhao, G., Q. Wang, S. Li and X. Wang (2021). "Resistance to hypomethylating agents in myelodysplastic syndrome and acute myeloid leukemia from clinical data and molecular mechanism." Frontiers in Oncology **11**: 706030.

Zhao, H., X. Wang, X. Feng, X. Li, L. Pan, J. Liu, F. Wang, Z. Yuan, L. Yang, J. Yu, R. Su, Y. Zhang and L. Ma (2018a). "Long non-coding RNA MEG3 regulates proliferation, apoptosis, and autophagy and is associated with prognosis in glioma." Journal of Neuro-Oncology **140**(2): 281–288.

Zhao, Z., X. Ma, D. Sung, M. Li, A. Kosti, G. Lin, Y. Chen, A. Pertsemlidis, T. H. Hsiao and L. Du (2015b). "microRNA-449a functions as a tumor suppressor in neuroblastoma through inducing cell differentiation and cell cycle arrest." RNA Biol **12**(5): 538–554.

Zhao, Z., V. Partridge, M. Sousares, S. D. Shelton, C. L. Holland, A. Pertsemlidis and L. Du (2018b). "microRNA-2110 functions as an onco-suppressor in neuroblastoma by directly targeting Tsukushi." PLoS One **13**(12): e0208777.

Zheng, J., X. D. Li, P. Wang, X. B. Liu, Y. X. Xue, Y. Hu, Z. Li, Z. Q. Li, Z. H. Wang and Y. H. Liu (2015). "CRNDE affects the malignant biological characteristics of human glioma stem cells by negatively regulating miR-186." Oncotarget **6**(28): 25339–25355.

Zheng, J., X. Liu, P. Wang, Y. Xue, J. Ma, C. Qu and Y. Liu (2016). "CRNDE Promotes Malignant Progression of Glioma by Attenuating miR-384/PIWIL4/STAT3 Axis." Mol Ther **24**(7): 1199–1215.

Zheng, Q., Q. Zhu, C. Li, S. Hao, J. Li, X. Yu, D. Qi and Y. Pan (2020). "microRNA-144 functions as a diagnostic and prognostic marker for retinoblastoma." Clinics (Sao Paulo) **75**: e1804.

Zheng, Z., X. Zheng, Y. Zhu, X. Gu, W. Gu, X. Xie, W. Hu and J. Jiang (2019). "miR-183-5p Inhibits Occurrence and Progression of Acute Myeloid Leukemia via Targeting Erbin." Mol Ther **27**(3): 542–558.

Zhou, V. W., A. Goren and B. E. Bernstein (2011). "Charting histone modifications and the functional organization of mammalian genomes." Nat Rev Genet **12**(1): 7–18.

Zhou, X., Y. Ren, L. Moore, M. Mei, Y. You, P. Xu, B. Wang, G. Wang, Z. Jia, P. Pu, W. Zhang and C. Kang (2010). "Downregulation of miR-21 inhibits EGFR pathway and suppresses the growth of human glioblastoma cells independent of PTEN status." Lab Invest **90**(2): 144–155.

Zhu, H., B. Zhai, C. He, Z. Li, H. Gao, Z. Niu, X. Jiang, J. Lu and X. Sun (2020). "LncRNA TTN-AS1 promotes the progression of cholangiocarcinoma via the miR-320a/neuropilin-1 axis." Cell Death & Disease **11**(8): 637.

Chapter 8
An Overview of Targeting Some Cancer Hallmarks with Plant Polyphenols: A Step Toward Precision

Nadia M. Hamdy, Amr Amin, Gamal Eldein Fathy Abd-ellatef, Youssef Abdalla, Ali Abdalla, Dana Ahmad Saqr, Yi Lu, Wei Wu, Mariam I. Gamal El-Din, and Mohamed El-Shazly

Abstract Polyphenols (PPs) are secondary metabolites that are present in more than 80% of plants. They possess a plethora of medicinal properties by modulating key pathological pathways, including epigenetics and those involved in oncogenesis or tumorigenesis. PPs have been shown to inhibit tumor cell proliferation, metastasis, and cancer cell resistance to chemotherapeutic medications. Their coadministration, either as plant-rich extracts or in advanced pharmaceutical formulations like nano-formulations, significantly impacts the tolerability, efficiency, and cytotoxic-

---

N. M. Hamdy
Biochemistry and Molecular Biology Department, Faculty of Pharmacy, Ain Shams University, Cairo, Egypt
e-mail: nadia_hamdy@pharma.asu.edu.eg

A. Amin
College of Medicine, University of Sharjah, Sharjah, United Arab Emirates

Faculty of Science, Cairo University, Cairo, Egypt

The University of Chicago, Chicago, IL, USA

G. E. F. Abd-ellatef
Therapeutic Chemistry Department, Pharmaceutical and Drug Industries Research Division, National Research Centre, Giza, Egypt

Y. Abdalla
University of Nicosia Medical School, Nicosia, Cyprus

A. Abdalla
Department of Health Sciences, University of Michigan, Ann Arbor, MI, USA

D. A. Saqr
Department of Basic Medical Sciences, College of Medicine, University of Sharjah, Sharjah, United Arab Emirates

Y. Lu · W. Wu
School of Pharmaceutical Sciences, Key Laboratory of Smart Drug Delivery of MOE and National Key Laboratory of Advanced Drug Formulations for Overcoming Delivery Barriers, Fudan University, Shanghai, China

N. M. Hamdy et al. (eds.), *From (Epi)genetics-to-Precision (E2P)*, Subcellular Biochemistry 114, https://doi.org/10.1007/978-3-032-08530-6_8

ity of traditional chemotherapeutic drugs. Both clinical and preclinical studies, including in vitro and in vivo models, demonstrated the potent anticancer activity of PPs. Combining PPs with conventional chemotherapeutic agents has led to a significant improvement in the efficiency and safety index of these agents. Advances in PP nano-formulations have enhanced their bioavailability and therapeutic efficacy. The current chapter highlights the potential of PPs and recent advances in their formulation for targeting cancer.

Keywords PP · Flavonoids · Cancer · Antioxidants · Nanotechnology · Chemotherapy · MDR · Drug delivery · Epigenetics

Abbreviations

ADME	Absorption, distribution, metabolism, excretion
BBB	Blood-brain barrier
BC	Breast cancer
Bcl-2	B-cell lymphoma 2
BDPNP	Biodegradable polymeric nanoparticles (NPs)
BSA	Bovine serum albumin
COX-2	Cyclooxygenase 2
EGFR	Epidermal growth factor receptor
GLP-1	Glucagon-like peptide 1
GNT	Green nanotechnology
HAHGs	Hyaluronic acid-based hydrogels
HSA	Human serum albumin
HDAC4	Histone deacetylase 4
HPNPs	Hybrid polymer nanoparticles
IGF-1	Insulin-like growth factor 1
JAK	Janus kinase
LBNC	Lipid-based nanocarriers
LC3-II	Mmicrotubule-associated protein 1A/1B light chain 3
MAPKs	Mitogen-activated protein kinases
MDR	Multidrug resistance

M. I. Gamal El-Din
Pharmacognosy Department, Faculty of Pharmacy, Ain Shams University, Cairo, Egypt

Quadram Institute Bioscience, Norwich Research Park, Norwich, United Kingdom

M. El-Shazly (✉)
Pharmacognosy Department, Faculty of Pharmacy, Ain Shams University, Cairo, Egypt

Graduate Institute of Natural Products, College of Pharmacy, Kaohsiung Medical University, Kaohsiung, Taiwan
e-mail: mohamed.elshazly@pharma.asu.edu.eg

MMP-2, MMP-9	Matrix metalloproteinases 2 and 9
MOFs	Metal-organic frameworks.
MPNs	Metal-polyphenol networks
mTOR	Mammalian target of rapamycin
NA	Not applicable
NF-κB	Nuclear factor kappa B
NPs	Nanoparticles
P-gp	P-glycoprotein
PARP-1	Poly(ADP-ribose) polymerase 1
PEG	Polyethylene glycol
PI3K	Phosphoinositide 3-kinase
PLA	Polylactic acid
PLGA	Poly(lactic-co-glycolic acid)
PPs	Polyphenols
Rb	Retinoblastoma
ROS	Reactive oxygen species
SDGs	Sustainable Development Goals
VEGF	Vascular endothelial growth factor

Introduction

Background

Cancer remains one of the most pressing global health challenges, with high prevalence and mortality rates affecting millions worldwide. In 2022 alone, an estimated 20 million new cancer cases were reported, resulting in approximately 9.7 million deaths. This immense burden is expected to increase dramatically, with projections suggesting that by 2050, cancer cases could reach 35 million globally, primarily driven by aging populations and lifestyle factors such as diet, physical inactivity, and smoking.

Cancer encompasses a broad spectrum of diseases characterized by uncontrolled cell growth, invasion of surrounding tissues, and the potential to spread to distant organs (metastasis). The most common types of cancer, affecting specific organs or tissues, include lung, breast, colorectal, prostate, and liver cancer (Desai et al. 2024). These malignancies originate when normal cells undergo genetic mutations that disrupt key regulatory processes governing cell division, growth, and death. As the mutations accumulate, cells lose their ability to respond to growth-inhibiting signals, leading to unchecked proliferation. Physiologically, cancer cells exhibit altered energy metabolism, enabling rapid growth and survival even in unfavorable conditions. This is coupled with their ability to evade apoptosis (programmed cell death), resist immune detection, and promote angiogenesis—the formation of new blood vessels that support tumor growth (Brown and Giaccia 1998; Gao and Ge 2023).

One of the key mechanisms linking cancer development to cellular damage is oxidative stress. Oxidative stress occurs when there is an imbalance between the production of reactive oxygen species (ROS) and the body's ability to neutralize these free radicals with antioxidants. ROS are chemically reactive molecules that can damage proteins, lipids, and DNA within cells, leading to mutations and genomic instability—hallmarks of cancer (Singh et al. 2024). Chronic oxidative stress has been implicated in the initiation, promotion, and progression of various cancer types. Oxidative damage to DNA can result in mutations that activate oncogenes or inactivate tumor suppressor genes, triggering the uncontrolled cell division seen in cancer. Moreover, oxidative stress creates an inflammatory environment that further promotes tumor growth. Cancer cells often exploit oxidative stress pathways to their advantage, adapting to high ROS levels by enhancing their antioxidant defenses. This enables them to survive and thrive in conditions that would typically be harmful to normal cells. For instance, ROS can activate signaling pathways such as nuclear factor kappa B (NF-κB) and mitogen-activated protein kinases (MAPKs), which play critical roles in cell proliferation, survival, and inflammation—all processes that cancer cells use to grow and invade surrounding tissues (Iqbal et al. 2024).

Cancer treatment and prevention strategies have evolved significantly over the years, integrating both traditional and modern approaches. The primary treatment modalities for cancer include surgery, chemotherapy, radiation therapy, and immunotherapy. Surgery remains a cornerstone for localized tumors, especially when complete removal is possible, while chemotherapy and radiation target rapidly dividing cancer cells systemically and locally, respectively. Immunotherapy, a more recent advancement, leverages the body's immune system to recognize and destroy cancer cells, offering hope for more personalized and less toxic treatments (Ando et al. 2021). Prevention strategies, on the other hand, focus on reducing the risk of developing cancer through lifestyle modifications and the use of chemopreventive agents. Primary prevention emphasizes avoiding risk factors like smoking, excessive alcohol consumption, and poor diet, while secondary prevention focuses on early detection through screening programs for cancers such as breast, cervical, and colorectal. Meanwhile, cancer chemopreventive strategies include biochemical processes using various natural, synthetic, or biological agents to minimize progression or postpone the incidence of malignancy (Tabrez et al. 2020). Despite advances in conventional strategies and treatments like chemotherapy, radiation, and surgery, these methods often come with significant side effects and limitations, including the development of resistance in some cancer types. Hence, there is a growing interest in exploring complementary and alternative therapies that can enhance treatment efficacy while reducing adverse effects.

The study of phytonutrients and their biological activities has attracted significant attention in the last few decades due to their potential role in promoting health and preventing disease (Fayez et al. 2022). These naturally occurring compounds, particularly polyphenols (PPs), are marketed as nutraceuticals, with some formulated into different dosage forms as therapeutic drugs (Cirillo et al. 2016). PPs, as a diverse group of naturally occurring compounds derived from plants, have gained attention for their multiple health benefits concerning safety, tolerability, and therapeutic properties for the treatment of several noncommunicable diseases such as cancer (Abbas et al. 2017). They are among the most widely recognized chemopreventive agents.

PPs in Clinical and Preclinical Studies

PPs have garnered significant attention in cancer research due to their potential as natural, bioactive compounds that can influence multiple pathways involved in cancer development and progression. Extensive studies have investigated their effects on key processes such as cell proliferation, apoptosis, angiogenesis, and immune modulation. These mechanisms highlight the potential of PPs as therapeutic agents in both cancer prevention and treatment. The following tables (Tables 8.1 and 8.2) summarize the key findings from literature reviews and clinical trials, respectively.

PPs from a Biochemical Point of View

PPs are a diverse group of naturally occurring compounds produced by plants in response to stress as part of their defense system (Singh et al. 2021). These compounds are synthesized through secondary metabolism, primarily via the phenylpropanoid pathway, originating from the amino acid L-phenylalanine (Fig. 8.1A, B). Structurally, PPs differ based on the number of aromatic rings and the number and position of phenolic groups (Vittorio et al. 2017; Singh et al. 2021). Typically, they comprise at least two phenolic groups, arranged in various complex configurations, often leading to high molecular weights. PPs can be categorized into several groups—such as flavonoids, phenolic acids, and lignans—based on their chemical structures and associated health benefits (Asensi et al. 2011).

PPs include flavonoids and phenolic acids (Kaczmarek 2020), as well as nonflavonoids such as stilbenes and lignans (Rambaran 2020). Phenolic acids can be classified into two major subclasses: hydroxybenzoic acid derivatives and hydroxycinnamic acid derivatives (Rashmi and Negi 2020). Flavonoids can also be classified into several subclasses differing in their chemical structure, degree of unsaturation, and the oxidation of their carbon rings (Ku et al. 2020).

Flavonoids typically possess two aromatic rings connected by a three-carbon bridge and often form a heterocyclic ring (Ku et al. 2020). They are classified based on the oxidation state of one ring and the position of the other ring into several categories, including flavonols (such as quercetin and kaempferol), flavones (such as apigenin and luteolin), flavanones (such as naringenin and hesperetin), isoflavones (such as genistein and daidzein), anthocyanins (such as cyanidin and delphinidin), and flavanols (such as epicatechin and epigallocatechin gallate [EGCG]). Anthocyanins are water-soluble pigments responsible for the red, purple, and blue colors in many fruits, vegetables, and flowers such as cyanidin in red berries, blueberries, blackberries, raspberries, red cabbage, and red onions; delphinidin in blueberries, Concord grapes, and cranberries; and malvidin in grapes, blueberries, and red wine. Other examples of anthocyanins include pelargonidin in strawberries, raspberries, and red or orange flowers; peonidin in cranberries, cherries, and black

Table 8.1 Overview of systematic and qualitative reviews on PPs for cancer treatment

Study objective(s)	Study design	Study methodology	Finding(s)	References
Potential benefits of natural PPs for cancer treatment	Qualitative review analysis	Analyzed the anticancer effects of PPs, highlighting their bioavailability, toxicity, and metabolic fate when administered in vivo	Inhibition of cancer cell multiplication, inflammation, and resistance to radiation. Promising benefits in treating melanoma and lung cancer	Asensi et al. (2011)
Benefits of PPs obtained by natural means for cancer treatment	Systematic review analysis	Conducted using NCBI's PubMed database to search clinical trials on humans to understand the application of PPs for various cancers in vitro and in vivo Analyzing NPs' benefits as interesting dosage forms for PPs	Beneficial effect when PPs were used as an add-on therapy to standard cancer treatment. Biological effects improved when PPs were formulated as NPs, reaching targeted tumor cells	Khushnud and Mousa (2013)
Anticancer ability of PPs for treating skin cancer	Systematic review analysis	8254 review and research papers were selected and screened to assess the anticancer potential of PPs for treating skin cancer	Quercetin and punicalagin showed beneficial effects in cutaneous cancer and melanoma	Sajadimajd et al. (2020)
Efficiency of PPs in cancer care	Qualitative review analysis	Discussed the pros and cons of various strategies used to present the efficacy of PPs in cancer care	PPs cannot be used individually in cancer treatment but can be used in combination therapy with anticancer medications	Vittorio et al. (2016)
Protective and safety effects of nano-encapsulated PPs in colorectal cancer	Qualitative review analysis	Conducted using encapsulated PPs to assess their chemopreventive properties	PPs in combination with standard treatment showed beneficial effects	Santos et al. (2013)
Benefits of epigallocatechin in nano-formulation for cancer prevention	Qualitative review analysis	Reviewed the use of different epigallocatechin gallate components in cancer care	Epigallocatechin gallate isolated from green tea showed anticarcinogenic and antidiabetic properties, giving it therapeutic potential in cancer care	Yang et al. (2019)

Table 8.2 Clinical trials (CTs) using PPs as an intervention for cancer treatment

NCT number/ study status	Study title	Conditions/phases	Interventions	Study design
NCT02728349 Completed	Tolerance and pharmacokinetic study of chlorogenic acid to advanced glioblastoma	Glioblastoma phase 1	Drug: chlorogenic acid	Allocation: NA Intervention model: single_group Masking: none
NCT06214598 Recruiting	Anti-inflammatory diet in BC patients on aromatase inhibitors	Breast cancer	Dietary supplement: supplement, placebo	Allocation: randomized Intervention model: parallel Masking: triple (participant, investigator, outcomes_ assessor)
NCT03482401 Completed	Disposition of dietary PP and methylxanthines in mammary tissues from breast cancer patients	Breast cancer	Dietary supplement: polyphenol	Allocation: randomized Intervention model: parallel Masking: none
NCT03994055 Active, not recruiting	Effect of an anti-inflammatory diet on patients with cervical cancer	Uterine cervical neoplasm, pelvic inflammatory disease	Other: omega-3 fatty acids, antioxidants, soluble fiber, probiotics, and lactose restriction	Allocation: randomized Intervention model: parallel Masking: none
NCT02245204 Completed	Phase I study of chlorogenic acid for injection for tolerance and pharmacokinetic of advanced cancers	Advanced cancer phase 1	Drug: chlorogenic acid	allocation: NA Intervention model: single_group Masking: none Primary purpose: treatment
NCT01916239 Completed	Pomegranate extract supplementation in colorectal cancer patients	Colorectal cancer phase 2	Dietary supplement: standard pomegranate extract formulation-1, -2	Allocation: randomized Intervention model: parallel Masking: single (investigator)

(continued)

Table 8.2 (continued)

NCT number/ study status	Study title	Conditions/phases	Interventions	Study design
NCT01360320 Completed	Minimizing the risk of metachronous adenomas of the colorectum with green tea extract—a miracle	Colorectal adenomas phase 2	Dietary supplement: green tea extract of *Camellia sinensis*, followed by a placebo	Allocation: randomized Intervention model: parallel Masking: quadruple (participant, care provider, investigator, outcomes assessor)
NCT01912820 Completed	Effect of quercetin on green tea PP uptake in prostate tissue from patients with prostate cancer undergoing surgery	Adenocarcinoma of the prostate, stage I, IIA, IIB, III, and IV prostate cancer phase 1	Dietary supplement: green tea extract Drug: quercetin Other: placebo	Allocation: randomized Intervention model: parallel Masking: double (participant, investigator) Primary purpose: prevention
NCT01823562 Active, not recruiting	Absorption and metabolism of lyophilized black raspberry food products in men with prostate cancer undergoing surgery	Stage I, IIA, IIB, and III prostate cancer phase 1	Drug: lyophilized black raspberry confection Dietary supplement: dietary intervention	Allocation: non-randomized Intervention model: parallel Masking: none
NCT03824652 Recruiting	Walnuts for power: PP, omega-3 fatty acids, weight loss, and energy	Prostate cancer phase 2	Other: walnuts	Allocation: randomized Intervention model: parallel Masking: none
NCT00676780 Completed	Green tea extract and prostate cancer	Prostate cancer phase 2	Drug: polyphenone (EGCG)	Allocation: NA Intervention model: single_group Masking: none

Data were retrieved from https://clinicaltrials.gov/search?cond=Cancer&intr=PP&aggFilters=phase:3%202%201%200%20NA,status:rec%20act%20com Accessed October 1, 2024
NA not applicable, *PP* polyphenol, *BC* breast cancer

rice; petunidin in berries like black currants and grapes; and stilbenes (such as resveratrol) and lignans (such as secoisolariciresinol and matairesinol).

PPs are widely distributed in plant-based foods, making them accessible through a variety of dietary sources. These bioactive compounds are particularly abundant in

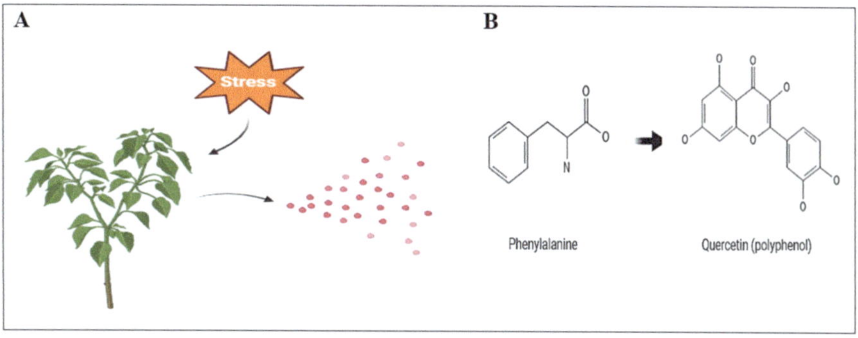

Fig. 8.1 PP release from plants in response to stress (**A**); quercetin, as a polyphenol example, is derived from the precursor amino acid phenylalanine (**B**)

nuts, fruits, vegetables, tea, and other plant-derived products. Each source provides unique polyphenolic profiles that contribute to distinct health benefits.

Nuts are highly regarded for their health benefits, attributed to their content of phenolics, fatty acids, vegetable proteins, phytosterols, and vitamins. Nuts are a rich source of PPs such as flavonoids and tannins, which contribute to their therapeutic properties (Tomás-Barberán et al. 2016). Pecans, pistachios, walnuts, and baru almonds possess the highest content of phenolic compounds in comparison to other nuts. Pecans are rich in both flavonoids and tannins, walnuts are high in flavonoids, and baru almonds are rich in tannins. The flavonoids and PPs participate in the anti-oxidant and anti-inflammatory effects of these nuts (Carrera-Quintanar et al. 2018). These effects can help reduce the risk of metabolic disorders such as diabetes, as well as cardiovascular and inflammatory diseases (Rambaran 2020). Moreover, the antioxidant and anti-inflammatory effects of flavonoids and PPs protect the body against tumors and cancer by suppressing mutations caused by aging and environmental factors (de Lima Cherubim et al. 2020). PPs can activate the generation of oxygen-consuming agents, including glutathione, at the cellular level (León-González et al. 2015).

Fruits, especially *Citrus* fruits such as lemon, orange, bitter lemon, mandarin, and grapefruit, are rich in vitamin C and flavonoids such as flavanones (Srinivasan et al. 2005, J. Lee et al. 2005).

Quercetin, a PP abundant in many fruits, has demonstrated an ability to reduce cancer cell proliferation (Chang et al. 2013). Its anticancer effect has been studied as a sole agent and in combination with other anticancer drugs, such as vincristine, and with other PPs like berry-derived epicatechin (Forbes-Hernández 2020).

Catechins, another class of flavonoids that are present in green tea, apples, and plums, are highly bioavailable and proved to reduce the risk of cancers, including bile duct, bladder, breast, and colon cancers (Karin et al. 2002, Karin and Greten 2005, Aggarwal and Shishodia 2006; Guo et al. 2009). They were found to sensitize prostate cancer cells to radiological treatment, reduce the amount of the aldehyde enzyme dehydrogenase, and reduce stemness parameters (Castro Nava et al. 2016).

Green tea extract is rich in epigallocatechin gallate, a catechin that possesses chemopreventive properties (Li et al. 2015). Additionally, tannic acid from green tea and other plant sources has exhibited antioxidant, homeostatic, antiviral, antimutagenic, and antitumor properties (Kaczmarek 2020). Tannic acid has recently been formulated into hydrogels or cross-linked with chitosan to form films that exhibit antimicrobial properties, which can be used in therapeutic bandages. These bandages were also found useful in controlling the release of the cytotoxic anticancer drug doxorubicin HCl (Kaczmarek 2020).

Green tea is linked to glucagon (GCG)-like peptide 1 (GLP-1) for obesity, metabolic syndrome as well as diabetes treatment (Al-Noshokaty et al. 2025; Mostafa et al. 2015). This could be repurposed for various cancer types based on its effect on GLP-1.

PPs are also abundant in berries, which are rich in anthocyanins (pelargonidin, cyanidin, malvidin, peonidin, and delphinidin); phenolic acids (caffeic, chlorogenic, benzoic, coumaric, and ferulic acids); flavonols (quercetin and kaempferol); flavanols (catechins); hydrolyzable tannins (ellagitannins and gallotannins); and condensed tannins (proanthocyanidins), all contributing to their potent antioxidant effects (Forbes-Hernández 2020). Resveratrol (3,4,5-trihydroxy-trans-stilbene), an important nonflavonoid PP occurring in grapes and berries, has shown anticancer potential (Rizvi and Saleh, 2018).

Other notable PPs include **sulforaphane**, present in vegetables such as broccoli and bean sprouts, and is efficient in targeting breast cancer cells (Li et al. 2010). Curcumin has attracted attention in the last few decades due to its diverse biological anticancer activities. Recent reports investigated the effect of curcumin, present in turmeric and piperine (Rizvi and Saleh, 2018) on the proliferation of MCF7 breast cancer cell lines. This activity was achieved via inhibition of the *Wnt* signaling mechanism (Kakarala et al. 2010).

Genistein, an isoflavone by-product of soya beans, proved to possess beneficial biological protective effects against breast cancer (Fan et al. 2013).

Table 8.3 summarizes in vitro and in vivo studies, offering a detailed overview of the role of PPs in cancer therapy.

Significance of PPs as Multimodal Agents Targeting Multiple Cancer Hallmarks

PPs, owing to their diverse chemical structures and biological activities, have demonstrated the ability to interfere with multiple cancer-hallmark pathways as follows:

Table 8.3 Key PPs studied in vivo and in vitro for cancer therapy

Polyphenol name	Chemical structure	Natural source	Study type	Major findings	Mechanism of action	References
Resveratrol	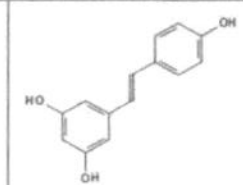	Peanuts, grapes, and berry fruits	In vitro	Resveratrol offers significant therapeutic advantages in combating diverse types of cancers through different molecular mechanisms. These include inhibiting cell migration and invasion, inducing oxidative stress and cytotoxicity, promoting autophagy, arresting the S phase of the cell cycle, and exhibiting apoptotic, antiangiogenic, and antiproliferative effects	Regulating various molecular pathways, such as p38/MAPK/ERK, PI3K/AKT, and NGFR-AMPK-mTOR	Bhuia et al. (2024)

(continued)

Table 8.3 (continued)

Polyphenol name	Chemical structure	Natural source	Study type	Major findings	Mechanism of action	References
Curcumin	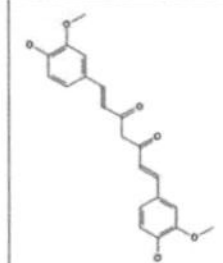	Turmeric	Network pharmacology and molecular docking techniques in vivo	Curcumin provides therapeutic effects on bladder cancer by targeting multiple pathways. Research suggests that curcumin may have anticancer effects by disrupting the cell cycle in bladder cancer cells. Curcumin disrupts bladder tumor cells and inhibits prostate tumor growth	Increasing the expression of genes such as ALB, TNF, ESR1, and CASP3 and reducing the expression of genes like EGFR, AKT1, BCL2, HSP90AA1, STAT3, and SRC, inhibiting MAPK and PI3K-AKT signaling pathways, EGFR tyrosine kinase inhibitor resistance, and IL-17 signaling pathway	Li et al. (2024) and Wang et al. (2024b)
Honokiol	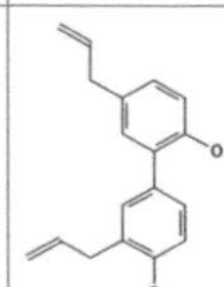	*Magnolia officinalis* bark and leaf, *Magnolia obovata*	In vitro	Inhibits the viability and proliferation of ovarian cancer cells, induces apoptosis, and reduces cell migration and invasion	Downregulating the YAP/TAZ pathway	Liu et al. (2024)
Oleuropein	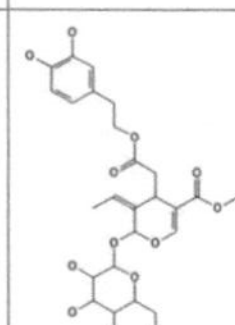	Premature olive fruit and leaves	In vitro and in vivo	Prevents cell proliferation, decreases cell viability, causes cell cycle arrest, and exhibits antioxidant properties	Targeting the AKT/NF-κB pathway and interacting with COX2, MMP, proliferator-activated receptors, and cannabinoid receptors	Nsairat et al. (2024)

Pterostilbene	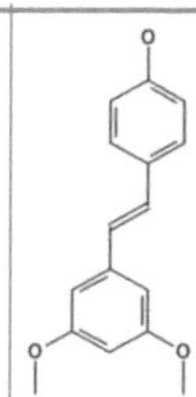	Grapes, blueberries, and other fruits	In vitro and in vivo	Inhibits the proliferation and metastasis of gastric cancer cells, suppresses the growth of cervical cancer cell lines, induces apoptosis in these cells in a dose-dependent manner, increases the expression of MICA/B, enhances NK cells, and disrupts cancer immune escape	JAK2/STAT3 pathway as the main mechanism of its anticancer activity, PI3K/AKT signaling pathway	He et al. (2024) and Li et al. (2024c)
Cyanidin-3-glucoside	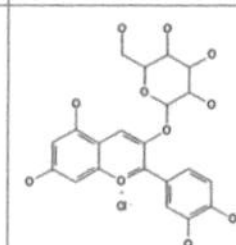	Blueberry, mulberry, and blackberry	In vivo and in vitro	Inhibits the proliferation of human colorectal cancer cells by promoting apoptosis and inhibiting angiogenesis Suppressive actions against DNA damage and enhanced DNA damage response pathways block the cancer cell cycle, limit migration of human retinal microvascular endothelial cells, and limit VEGF-mediated migration of human umbilical vein endothelial cells	Inducing caspase-mediated cell death	Yang et al. (2023)

(continued)

Table 8.3 (continued)

Polyphenol name	Chemical structure	Natural source	Study type	Major findings	Mechanism of action	References
Theaflavin-3,3′-di-gallate	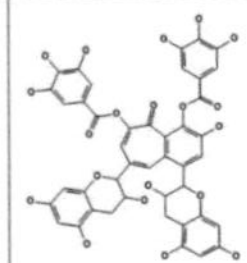	Black tea leaves	In vivo and in vitro	Reduces tumor volume in tumor xenografts	Upregulating the phosphorylation of PKCδ and the expression of aSMase	Sun et al. (2022)
Epigallocatechin gallate	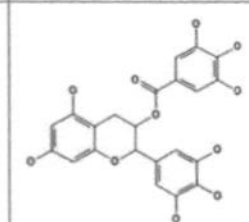	Green tea	In vivo and in vitro	Antitumor properties by inducing apoptosis in breast, leukemia, colon, brain cancers, and kidney, upregulating extrinsic apoptosis-related proteins, and activating OR2AT4 in the K562 leukemia cell line	Downregulating growth-related proteins: p38 MAPK and AKT	Choi et al. (2024)
Quercetin	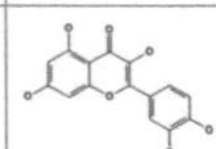	Various fruits, vegetables, and plants, such as onions, apples, berries, and green tea	In vivo and in vitro	Upregulates key mitophagy regulators, enhances colocalization of mitochondria with lysosomes, and increases mitophagy. Increases the expression of mitochondrial fusion genes (MFN1 and MFN2) and decreased the expression of fission genes (FIS1 and DRP1) in Hep3B and Huh7 cells, and reduced anticancer effects in hepatocellular carcinoma	Upregulating PINK1 and PARK2, targeting mitophagy	Chen (2024)

Genistein	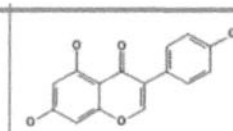	Soy-derived foods	In vitro	Reduces self-renewal capacity of cancer stem cells in MCF-7 breast cancer cells. Combined treatment of myokines and genistein has an additive anticancer effect, inhibiting cancer stemness in MCF-7 cells	Reducing the expression of cancer stem cell markers OCT4 and SOX2	Kwon et al. (2024)
Ellagic acid	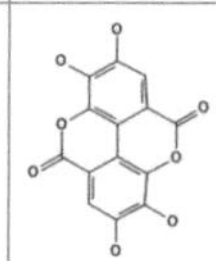	In the skin of muscadine grapes	In vitro	Prevents DNA damage caused by free radicals, induces apoptosis, and inhibits cancer cell growth	Suppressing tumor growth and angiogenesis	Otun et al. (2024)
Apigenin	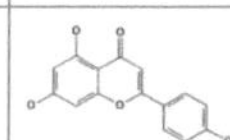	Dried parsley, dried chamomile flower, celery seed, vine spinach, kumquats, dried oregano, artichoke, juniper berries, and peppermint	In vitro and in vivo	Inhibits invasion and metastasis without causing significant toxicity to normal cells and enhances the effectiveness of conventional anticancer drugs Reduces ovarian cancer risk in a prospective study of nearly 70,000 women, while another study of around 40,000 women did not support a reduced risk of cancers (breast, ovarian, colorectal, endometrial, and lung)	–	Naponelli et al. (2024)

(continued)

Table 8.3 (continued)

Polyphenol name	Chemical structure	Natural source	Study type	Major findings	Mechanism of action	References
Luteolin	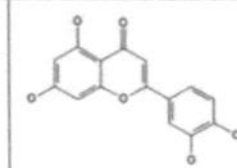	Radicchio, Chinese celery, dried parsley	In vitro and in vivo	Promotes apoptosis in lung cancer cells, inhibits tumor cell proliferation, invasion, and metastasis, supports chemo-radiotherapy, improves cancer complications, modulates the immune response, and inhibits the tumor cell proliferation. Effectively inhibits tumor epithelial-mesenchymal transition and migration, regulating migration-related factors such as AIM2, CCL2, and AR	Inhibiting the G1, S, G2, and M phases of the lung cancer cell cycle and regulating the expression of proliferation-related factors such as PCNA, LIMK, AR, claudin-2, and TAM RTKs, suppressing the EGFR-PI3K-AKT and integrin β1-FAK/SRC signaling pathways, and activating the MEK/ERK signaling pathway	Zhang et al. (2024)
Anthocyanins	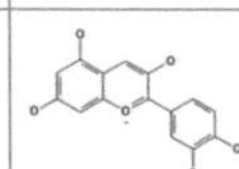	Edible flowers such as *Petunia hybrida, Begonia* sp., *Bellis perennis, Campanula* sp., *Calendula* sp., *Dahlia* sp., *Dianthus caryophyllus, Fuchsia* sp., and *Glechoma hederracea*	In vitro	Inhibits development and progression of urethane-induced lung tumors in C57BL/6 J mice	Modulating AMPK/mTOR signaling pathways and a decreasing intracellular lipid accumulation	Luo et al. (2024a)

Kaempferol	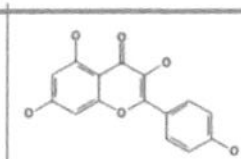	*Alpinia officinarum*	In vitro	Cell cycle arrest in hepatocellular carcinoma cells, SETD7 is the most effective target protein	Modulating the ATM/CHEK2/KNL1 signaling pathway	Li et al. (2024b) and Priyamvada et al. (2024)
Fisetin	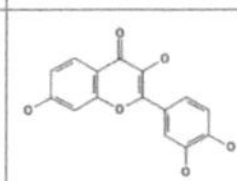	Strawberries, apples, onions	In vitro and in vivo	Eradicate pancreatic ductal adenocarcinoma, prostate cancer, lung cancer, melanoma, and breast and colon cancer	Targeting the CDK1–STAT3 axis	Xu et al. (2023)
Naringenin	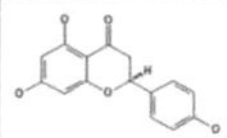		In vitro and in vivo	Naringenin, combined with 5-fluorouracil, enhances colorectal cancer inhibition through AMPK signaling and mitochondrial regulation. It also shows potential as an adjuvant in cervical cancer treatment, especially when paired with low doses of cisplatin	Naringenin, combined with 5-fluorouracil, enhances colorectal cancer inhibition and shows potential as an adjuvant with cisplatin for treating cervical cancer	Wang et al. (2024a)

(continued)

Table 8.3 (continued)

Polyphenol name	Chemical structure	Natural source	Study type	Major findings	Mechanism of action	References
Isorhamnetin	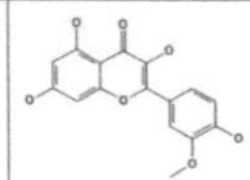	Fruits, vegetables, and medicinal herbs such as *Persicaria thunbergii H.*, *Elaeagnus rhamnoides*	In vitro and in vivo	Isorhamnetin has demonstrated anticancer effects in several in vitro and in vivo studies. It inhibits malignant cell growth by arresting the cell cycle at the G2/M and S phases, downregulating key proteins and signaling pathways, and promoting apoptosis through mitochondrial-dependent pathways. Isorhamnetin also enhances programmed cell death, suppresses NF-kB activation, and decreases tumor cell viability. In a gastric cancer xenograft model, isorhamnetin (1 mg/kg) effectively inhibited tumor growth both as a standalone treatment and when combined with capecitabine	Isorhamnetin has shown significant anticancer effects in various in vitro and in vivo studies. It inhibits malignant cells by arresting the cell cycle at the G2/M and S phases, down-regulating critical proteins and signaling pathways, and promoting apoptosis through mitochondrial pathways. Isorhamnetin also enhances the effects of capecitabine, suppresses NF-kB activation, and reduces tumor cell viability. In a gastric cancer xenograft model, isorhamnetin effectively inhibited tumor growth both as a standalone treatment and in combination with capecitabine	Biswas et al. (2024)

Rosmarinic acid	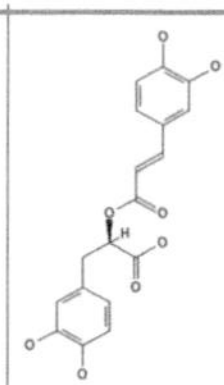	Rosemary, *Perilla frutescens* (L.) Britt (Zisu) (*Boraginaceae*, *Lamiaceae families*).	In vitro and in vivo	Rosmarinic acid shows strong antitumor effects in colorectal cancer, especially in colitis-associated cases. It reduces inflammation and tumor growth by modulating TLR4-mediated NF-κB and STAT3 activation. At 20 mg/kg, it shrinks tumors by enhancing oxidative stress, slowing cell proliferation, and inducing apoptosis through increased caspase-3 and TUNEL-positive cells	Rosmarinic acid has strong antitumor effects in colorectal cancer, especially in colitis-associated cases. It reduces colitis severity, inflammation, and tumor growth by modulating TLR4-mediated NF-κB and STAT3 activation. At 20 mg/kg, rosmarinic acid prevents tumor progression, shrinks tumor volume, enhances oxidative stress, slows cell proliferation, and induces apoptosis by increasing caspase-3 and TUNEL-positive cells	Luo et al. (2024b) and Khaksar et al. (2024)
Caffeic acid	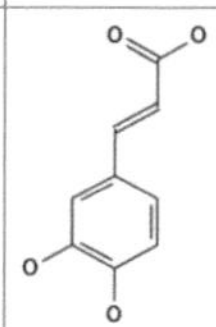	*Ligularia sagitta*	In vitro	Caffeic acid targets drug-resistant cancer cells in acidic environments by inhibiting proliferation, enhancing apoptosis, and modulating signaling pathways	Caffeic acid shows promise in overcoming drug resistance in cancer by targeting resistant cells in acidic environments, inhibiting proliferation, enhancing apoptosis, and modulating signaling pathways	Lim et al. (2024) and Wang et al. (2024c)

(continued)

Table 8.3 (continued)

Polyphenol name	Chemical structure	Natural source	Study type	Major findings	Mechanism of action	References
Allicin	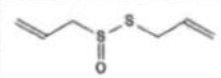	Garlic	In vitro	Allicin inhibits cervical cancer cell activity by targeting circEIF4G2 and downregulating the HOXA1/AKT/mTOR pathway. It also enhances the effectiveness of doxorubicin in resistant breast cancer cells, promoting cell apoptosis and reducing levels of Nrf2, HO-1, and p-AKT proteins, suggesting its role in overcoming doxorubicin resistance through inhibition of Nrf2/HO-1 and AKT pathways	Allicin inhibits cervical cancer cell activity by targeting circEIF4G2 and downregulating the HOXA1/AKT/mTOR pathway. Combined with doxorubicin, allicin effectively suppresses cell growth and promotes apoptosis in doxorubicin-resistant breast cancer cells, reducing Nrf2, HO-1, and p-AKT levels. This suggests that allicin may help overcome doxorubicin resistance by inhibiting the Nrf2/HO-1 and AKT pathways	Yifan et al. (2024) and Shi et al. (2024)

Myricetin	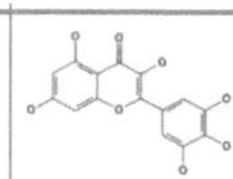	*Hovenia acerba* seed, myricetin leaves	In vitro	Myricetin induces ferroptosis in gastric cancer cells by inhibiting NOX4 degradation, leading to ROS production via the NOX4/NRF2/GPX4 pathway. It also promotes apoptosis in triple-negative breast cancer cells through ROS generation and causes oxidative stress and cell death in A549 lung cancer cells	Myricetin induces ferroptosis in gastric cancer cells by stabilizing NADPH oxidase 4 (NOX4) and promoting ROS production through the NOX4/NRF2/GPX4 pathway. In triple-negative breast cancer cells, it triggers apoptosis by enhancing iron-dependent ROS generation from hydrogen peroxide. Causes oxidative stress and cell death in A549 lung cancer cells	Lu et al. (2024)
Cinnamic acid	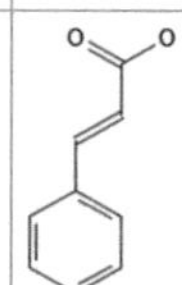	Cinnamon, Propolis	In vitro	Combining network pharmacology with cell culture analysis, it was concluded that cinnamic acid may effectively suppress ovarian cancer cell proliferation. Its anticancer effects are primarily through regulating core target proteins and modulating various signaling pathways	Using network pharmacology and cell culture analysis, it was concluded that cinnamic acid has potential as a therapeutic agent for ovarian cancer. Its anticancer effects are attributed to the regulation of core target proteins and modulation of multiple signaling pathways	Fan et al. (2024)

(continued)

Table 8.3 (continued)

Polyphenol name	Chemical structure	Natural source	Study type	Major findings	Mechanism of action	References
Hesperidin	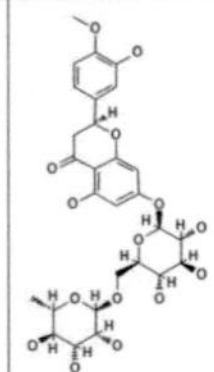	Tangerine peel, citrus fruits	In vitro and in silico	Hesperidin has cytotoxic and anti-inflammatory effects on oral cancer cells. Hesperidin nanoparticles are more effective and biocompatible than free hesperidin, suggesting their potential as adjunctive therapy for prostate cancer	Hesperidin shows cytotoxic and anti-inflammatory effects on oral cancer cells. Hesperidin nanoparticles are more effective and biocompatible than free hesperidin, suggesting their potential as adjunctive therapy for prostate cancer	Jayaraman et al. (2024) and Bakhshan et al. (2024)
Piceatannol	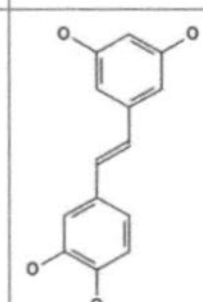	Grapes, passion fruit, and itadori leaf	In vitro and in vivo	Piceatannol shows potent antitumor effects by binding to Beclin-1, inhibiting Ser-295 phosphorylation, and promoting autophagy. It induces apoptosis in pancreatic cancer cells via mitochondrial and extrinsic pathways, increasing caspase activities and proapoptotic gene expression while also reducing cell invasion and migration	Piceatannol shows strong antitumor effects by binding to Beclin-1, inhibiting Ser-295 phosphorylation, and inducing autophagy in gastric cancer cells. It promotes apoptosis in pancreatic cancer cells through mitochondrial and extrinsic pathways and reduces cell invasion and migration	Çınar Ayan et al. (2022) and Huangfu et al. (2023)

Ferulic acid	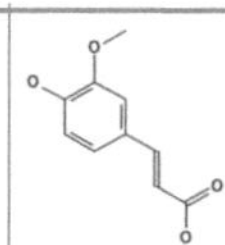	Medicinal plants, fruits, vegetables, and whole grains	In vivo	Ferulic acid exhibited anticancer effects by reducing the severity of NDEA-CCl_4-induced hepatic cancer (HCAR) in rats, a model for primary liver cancer	Ferulic acid reduced NDEA-CCl_4-induced hepatic cancer (HCAR) in rats, demonstrating its anticancer potential	Somade et al. (2024)
Punicalagin	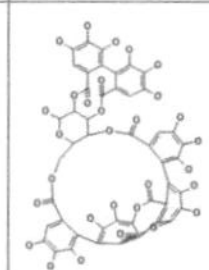	Pomegranate peel (*Punica granatum* L.)	In vitro	Punicalagin inhibits osteosarcoma cell proliferation, invasion, and angiogenesis by blocking NF-κB signaling. It also suppresses colorectal cancer cell invasion by targeting MMP-2, MMP-9, Snail, and Slug In colon cancer ($CaCo_2$) cells, punicalagin enhances oxaliplatin-induced apoptosis but does not significantly affect autophagy Additionally, punicalagin restricts proliferation and invasion in gastric cancer cells (AGS, HGC-27, 23,132/87) while sparing normal gastric epithelial cells, suggesting its potential as a novel chemotherapy option for gastric cancer	Punicalagin inhibits proliferation, invasion, and angiogenesis in osteosarcoma cells by blocking NF-κB signaling. It also reduces colorectal cancer cell invasion by targeting MMP-2, MMP-9, Snail, and Slug In colon cancer cells (Caco2), punicalagin enhances the apoptotic effects of oxaliplatin but does not affect autophagy Additionally, punicalagin limits proliferation and invasion in gastric cancer cells (AGS, HGC-27, 23,132/87) while sparing normal cells, highlighting its potential as a novel chemotherapy agent for gastric cancer	Sun et al. (2023), Alsufiani (2024), and Sun et al. (2024)

(continued)

Table 8.3 (continued)

Polyphenol name	Chemical structure	Natural source	Study type	Major findings	Mechanism of action	References
Silibinin	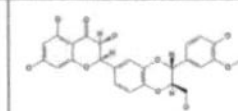	Seeds of milk thistle (*Silybum marianum*)	In vitro	Silibinin alters cell morphology and decreases F-actin fluorescence in MCF-7 and MDA-MB-231 cells through the YAP/TAZ pathway. Combining silibinin with vinblastine significantly enhances cytotoxicity and apoptosis by affecting the Bcl-2/Bax ratio Additionally, silibinin and N-acetylcysteine reduce ROS and ox-mtDNA in triple-negative breast cancer cells, potentially counteracting tumor progression and aging through unclear mechanisms	Silibinin affects MCF-7 and MDA-MB-231 cells by increasing cell protrusion and aspect ratio while decreasing cell area and F-actin fluorescence via the YAP/TAZ pathway. Inhibiting this pathway contributes to cell damage Combining silibinin with vinblastine shows enhanced cytotoxicity and apoptosis in MDA-MB-231 cells, affecting the Bcl-2/Bax ratio Additionally, silibinin and N-acetylcysteine reduce ROS and ox-mtDNA in triple-negative breast cancer cells, potentially impacting tumor progression and aging through unclear mechanisms	Fu et al. (2024a), Dariushnejad et al. (2024), and Roca et al. (2024)

Chemical structures retrieved from https://pubchem.ncbi.nlm.nih.gov//edit3/index.html. Accessed September 2024

Autophagy-Apoptosis-Epigenetics Axis

PPs can trigger autophagy and start cancer cell death (Benvenuto et al. 2020). Autophagy cooperates with apoptosis to prevent tumor growth. The autophagic process involves the contribution of cancer-associated fibroblasts, adipocytes, and immune cells. Alteration in autophagy occurs during tumorigenesis, which affects the tumor microenvironment (TME) and tumor immune microenvironment (TIME), resulting in changes in both drug absorption and metabolism, which in turn affects the bioavailability of treatments (Elanany et al. 2023). Such changes also affect the immune cell infiltration at the tumor site, reversing immunotherapy and, unfortunately, worsening the prognosis (Benvenuto et al. 2020).

Another cancer hallmark is resisting cell death, specifically apoptosis, the programmed cell death that typically eliminates damaged or abnormal cells. Cancer cells often upregulate antiapoptotic proteins such as B-cell lymphoma-2 (Bcl-2) or downregulate proapoptotic factors, enabling them to survive in hostile environments. These cells also achieve replicative immortality by maintaining the length of their telomeres—protective caps on the ends of chromosomes—through the activation of the enzyme telomerase, allowing for endless cell division.

One of the PPs' targets is the insulin-like growth factor 1 (IGF-1) signaling cascade. This plays a significant role in the growth and development of various cancers, such as breast and pancreatic cancer, via proliferative and antiapoptotic milieus, respectively (Sair and Liu 2022; El-Mesallamy et al. 2013).

Non-protein coding RNAs (ncRNAs) such as microRNAs and long ncRNAs (lncRNAs) have been extensively studied in the context of various cancers' pathophysiology (Hamdy et al. 2024; Abaza et al. 2023), being future recommended targets exploited to achieve effective cancer-related drug development.

Some PPs, including epigallocatechin-3-gallate, genistein, resveratrol, and curcumin, act as microRNA modulators in endometrial cancer (Liu and Liu 2025), prostate cancer (Dong and Zhang 2025), hypertension and cardiovascular diseases (Gareev et al. 2024), and cervical cancer (Wang et al. 2023). Therefore, targeting cancer-associated miRNAs using PPs constitutes a promising, precise approach for next-generation cancer chemotherapy.

One completed clinical trial among human breast and ovarian cancer syndrome (HBOC) that was characterized by mutations in BRCA1 and BRCA2 suppressor genes, which increase the carrier's risk of developing breast and ovarian cancer, especially before 40 years old. The inductions of epigenetic changes by a nutritional intervention with specific distribution of macronutrients, micronutrients, and PPs were done in the current clinical trial to ensure a decrease in oxidative stress, and therefore, in DNA damage https://clinicaltrials.gov/study/NCT05306002?cond=Cancer&term=Epigenetics&intr=Polyphenols&rank=1.

PPs are epigenetic regulators that function through autophagy-dependent or independent pathways (Patra et al. 2020). PPs can cause a shift in DNA methylation, histone modifications, and microRNA expression. They can also cause deregulation of oncogenic or tumor-suppressing microRNA(s) involved in neoplastic cell transformation, representing a potential anticancer mechanism that can be utilized either with PPs alone or in combination with conventional anticancer chemotherapeutics (Sathya and Devi 2017).

Epigallocatechin gallate in green tea showed an ability to activate apoptosis and restrict carcinogenic cell growth by arresting the synthetic (S) phase of the cell cycle. These derivatives also inhibit cell signaling and interfere with cancer cell metabolic pathways (Yang et al. 2000).

Oleuropein, the major PP in olive oil, demonstrated interesting biological activities, including anticancer, anti-inflammatory, and antiangiogenic effects. The antiproliferative effect of oleuropein PPs is based on suppression of NF-κB as well as targeting cyclin D1 and cyclooxygenase 2 (COX-2) in the MDA-MB-231 breast cancer cell line. Oleuropein additionally induces phosphorylation of ERK and Akt—responsible for cancer dissemination and invasion—and increases the ratio of Bax/Bcl-2, favoring the apoptotic pathway in pancreatic cancer cell lines (Gou et al. 2017). In glioma cells in vitro, oleuropein demonstrated anticancer mechanisms via regulation of Bax, Bcl-2, and matrix metalloproteinases 2 and 9 (MMP-2, MMP-9), pathways important for apoptosis (Liu et al. 2016a, b, c).

Regulation of Cancer-Related ncRNAs by Key Polyphenols

The anticancer properties of many polyphenols are mediated through the regulation of noncoding RNAs (ncRNAs). The key polyphenols involved in ncRNA regulation are discussed and summarized below, focusing on their ncRNA targets, cancer models, and biological outcomes (Table 8.4).

Resveratrol

Resveratrol has been shown to exert broad anticancer effects by modulating oncogenic and tumor-suppressive ncRNAs. In prostate cancer cells, for example, resveratrol (and its analog pterostilbene) downregulates oncomiRs of the miR-17 family, thereby relieving repression of the tumor suppressor PTEN (Dhar et al. 2015). Consistent with this, PTEN protein levels rose in resveratrol-treated cells (Dhar et al. 2015). Resveratrol also reactivates the lncRNA *PCAT29* in prostate cancer: it reverses IL-6/STAT3-mediated suppression of *PCAT29*, potently stimulating its expression and enhancing its growth-inhibitory effects (Al Aameri et al. 2017). In a prostate cancer model, knockdown of *PCAT29* increased cell viability, whereas resveratrol-induced upregulation of *PCAT29* suppressed cell growth (Al Aameri et al. 2017). underscoring *PCAT29*'s tumor-suppressive role.

In lung/bronchial epithelial cells, resveratrol similarly targets oncogenic signals via miRNAs. Han et al. (2012) showed that resveratrol treatment of transformed human bronchial (16HBE-T) cells markedly upregulates miR-622. miR-622 directly

Table 8.4 Key polyphenols involved in ncRNA regulation, focusing on their ncRNA targets, cancer models, and biological outcomes

Polyphenol	Target ncRNA	Outcome	Cancer model	References
Resveratrol	↓miR-17 miR-20a, miR-106a, and miR-106b ↑PCAT29	Suppressed proliferation (↑PTEN tumor suppressor protein)	prostate cancer cells (DU145)	Dhar et al. (2015) and Al Aameri et al. (2017)
	↑miR-34a	Inhibition of proliferation	Colon cancer (DLD-1 and SW480)	Yang et al. (2015a, b)
	↓ miR-21	Induction of apoptosis (↓regulation of BCL-2	Pancreatic cancer (PANC-1, CFPAC-1, and MIAPaca-2)	Liu et al. (2013)
	↓ NEAT1	Reduced proliferation, migration, and invasion	Multiple myeloma	Geng et al. (2018)
	↓ MALAT1 (lncRNA)	Inhibition of nuclear β-catenin and reduced migration	CRC (LoVo cells)	Ji et al. (2013)
	↓ AK001796 ↑miR-622	Reduced progression and induction of cell cycle arrest	NSCLC (A549, H446, and 16HBE-T)	Han et al. (2012) and Yang et al. (2015)
	↑miR-345, ↑miR-498	Induction of cell cycle arrest. Increased sensitivity to gefitinib	NSCLC (H1299 and PC-9/GR)	Lu et al. (2019)

(continued)

Table 8.4 (continued)

Polyphenol	Target ncRNA	Outcome	Cancer model	References
Curcumin	↑miR-181b, ↑miR-15a, ↑miR-16, ↓HOTAIR	Enhanced apoptosis and reduced metastasis	Breast (MDA-MB-231)	Phuah and Nagoor 2014
	↑miR-378	Impaired proliferation via p38 signaling	Glioblastoma (U87)	Li et al. (2017)
	↓ H19	P53 downregulation and reduced proliferation	GCa (SGC-7901)	Liu et al. (2016)
	↓ miR-214, ↑NBR2, ↑PANDAR	Induction of MEG3 and reduced chemoresistance, progression, proliferation, and apoptosis	Ovarian cancer	Zhang et al. (2017)
	↓PVT1, ↑miR-143	Decreased metastasis and invasion; impaired expression of CXCL1/2; increased apoptosis ↑ miR-15a, miR-16	prostate cancer (DU145)	Liu et al. (2019)
	low HOTAIR	Increased sensitivity to adriamycin; decreased metastasis	AML (HL-60); renal carcinoma	Liu et al. (2021a, b)
	↑miR-7	Enhanced apoptosis	Pancreatic cancer (PANC-1)	Ma et al. (2014)

	↑miR-330-5p, ↑miR-30c, ↑miR-34a/c-5p, ↑miR-302b-3p, ↑miR-98, ↑miR-192-5p, ↑miR-206↑, ↑miR-874, ↑let-7c, ↑miR-101, ↑miR-384, ↓miR-186, ↓miR-21, ↑ miR-192-5p/215, ↓ UCA1	Antiproliferative, antimigration, EMT inhibition, chemosensitizer, and increased apoptosis	NSCLC (A549)	Wang et al. (2018) and Liu et al. (2019)
	↑MEG3	Increased apoptosis, regulation of PGK1, and suppressed proliferation	BCa, NSCLC	Lu et al. (2013) and Liu et al. (2018)
	↓LINC00691	Increased apoptosis	Thyroid cancer	Li et al. (2022)
	↓KCNQ1OT1	Increased apoptosis	CRC	Duan et al. (2020)
Honokiol	↑miR-148a-5p, ↑miR-148a-3p	Induction of apoptosis	NSCLC	Zhang et al. (2019)
Pterostilbene	↓miR-106a-5p & miR-17-5p	Preserved PTEN levels and smaller tumors in vivo	Prostate cancer cells (DU145 line)	Dhar et al. (2015)
Epigallocatechin gallate	↓miR-98-5p, ↑NEAT 1	Enhanced tumor suppression and increased cisplatin sensitivity	NSCLC (A549, H460, and H1299)	Jiang et al. (2016) and Hayakawa et al. (2022)
	↑mir-155, ↑HIF-1a, ↑miR-210, ↓miR-212, ↑miR-485-5p	Reduced proliferation and migration (through MAPK) and induction of apoptosis	NSCLC (A549)	Wang et al. (2011) and Dharshini and Mandal (2024)
	↑miR-16	Apoptosis induction (by ↓Bcl-2)	Hepatocellular carcinoma (HepG2)	Tsang and Kwok (2010)
	↑miR-1	Suppressed growth	Osteosarcoma (MG-63 and U2OS)	Zhu and Wang (2016)

(continued)

Table 8.4 (continued)

Polyphenol	Target ncRNA	Outcome	Cancer model	References
Quercetin	↓miR-27a	Reduced drug resistance	Renal cancer	Li et al. (2014)
	↑miR-let-7a	Induction of apoptosis and reduced viability	Pancreatic cancer (PDA)	Appari et al. (2014)
	↑ miR-122	Reduced migration and invasion	Liver carcinoma (mouse model)	Boesch-Saadatmandi et al. (2012)
	↑miR-16	Inhibited migration and invasion	NSCLC (A549)	Sonoki et al. (2015)
	↓MALAT1	Inhibition of migration	Prostate cancer (PC-3)	Lu et al. (2020)
	↓UCA1	Induction of apoptosis	Urothelial carcinoma	Hayakawa et al. (2022)
Genistein	↓miR-27a	reduced proliferation and migration	Ovarian cancer (SKOV3)	Javed et al. (2021)
	↓HOTAIR	Reduced proliferation and migration	Prostate (PC3 and DU145); renal cancer (RCC)	Chiyomaru et al. (2013) and Imai-Sumida et al. (2020)
	↑miR-27a, ↓MET	Increased apoptosis	NSCLC (A549)	Yang et al. (2016) and
	↓TTTY18	Increased apoptosis	CRC (SW480)	Chen et al. (2020)
Apigenin	↑miR-34a-5p, ↓SNAI 1	Induce apoptosis	NSCLC (A549)	Aida et al. (2021)
Kaempferol	↑mir-340	Reduced proliferation and increased apoptosis	NSCLC (A549)	Zhang et al. (2021)
Berberine	↑CASC2	Suppressed oncogenesis	CRC (HT29 and HCT116)	Dai et al. (2019b)
	↓HOTAIR	enhanced chemosensitivity via the miR-34a-5p axis	NSCLC (A549 and H1975)	Zheng et al. (2020)
Silibinin	↓HOTAIR and ZFAST	Decreased oncogenesis	Bladder cancer (T24 and UM-UC-3)	Imai-Sumida et al. (2017)

targets KRAS mRNA, and its overexpression in these cells induced cell cycle arrest in G0 and inhibited proliferation. Thus, resveratrol-induced miR-622 acts as a tumor suppressor by silencing KRAS and contributing to growth arrest. High-content studies suggest that resveratrol also alters other KRAS-targeting miRNAs, but miR-622 has been most clearly validated (Han et al. 2012).

In colorectal cancer (CRC) models, resveratrol upregulates the tumor suppressor miR-34a. For example, Yang et al. reported that resveratrol inhibited growth and induced apoptosis in CRC cells via miR-34a induction. miR-34a is a p53-regulated oncosuppressor that targets multiple proliferation and survival genes; its upregulation by resveratrol contributes to cell cycle arrest and apoptotic death in colon carcinoma lines (Yang et al. 2015).

Conversely, in pancreatic cancer, resveratrol downregulates the oncomiR miR-21. Liu et al. (2013) found that resveratrol treatment of pancreatic cancer cell lines (PANC-1, CFPAC-1, MIAPaCa-2) significantly reduced miR-21 levels. Since miR-21 can reverse the downregulation of the antiapoptotic protein BCL-2, its inhibition by resveratrol leads to BCL-2 downregulation and increased apoptosis. Thus, resveratrol's proapoptotic effect in pancreatic cancer is mediated in part by suppression of miR-21/BCL-2 signaling (Liu et al. 2013).

Resveratrol also targets oncogenic lncRNAs. In multiple myeloma (MM), the lncRNA *NEAT1* is highly expressed and drives proliferation and invasion. Resveratrol treatment significantly decreases *NEAT1* levels in MM cells. Geng et al. showed that *NEAT1* downregulation by resveratrol impairs Wnt/β-catenin signaling, resulting in reduced MM cell proliferation, migration, and invasion (Geng et al. 2018). Similarly, in CRC cells, resveratrol potently suppresses the metastasis-associated lncRNA *MALAT1*. This in turn inhibits CRC cell invasion and metastasis by decreasing the nuclear localization of β-catenin and attenuating the Wnt/β--catenin pathway (Ji et al. 2013).

The oncogenic lncRNA *AK001796* is overexpressed in lung cancer, and its knockdown induces G0/G1 arrest. In a study on the A549, H446, and 16HBE-T lung tumor models, resveratrol treatment lowered *AK001796* levels and slowed tumor growth, indicating that suppression of this lncRNA is part of resveratrol's anticancer mechanism (Yang et al. 2015). Another study showed that treatment with the resveratrol derivative *trans*-3,5,4-trimethoxystilbene reduced resistance of the H1299 and PC-9/GR NSCLC cell lines to gefitinib by upregulating miR-345 and miR-498. This effect was mediated by the suppression of the MAPK/Akt/Bcl-2 pathway. Additionally, overexpression of miR-345 and miR-498 induced cell cycle arrest in the G1/0 phase (Lu et al. 2019).

Curcumin

Curcumin exerts broad anti-tumor effects by modulating noncoding RNAs across diverse cancer types. For example, in breast cancer (MDA-MB-231 cells), curcumin upregulates tumor-suppressive miR-181b (and miR-15a/16) to induce apoptosis and inhibit metastasis (via downregulation of CXCL1/2) (Phuah and Nagoor 2014). In glioblastoma (U87) cells, miR-378 overexpression enhances curcumin's antiproliferative effect via targeting p38 MAPK signaling (Li et al. 2017).

In ovarian cancer models, curcumin induces demethylation and restoration of lncRNA MEG3 and concomitant downregulation of miR-214, thereby attenuating extracellular-vesicle-mediated cisplatin resistance and reducing tumor viability (Zhang et al. 2017).

In prostate carcinoma (DU145), curcumin induces tumor suppressor miR-143 (among others), which downregulates PGK1, resulting in decreased proliferation and invasion (Liu et al. 2019).

Similarly, in acute myeloid leukemia (HL-60) and renal cell carcinoma, curcumin-mediated HOTAIR inhibition increases Adriamycin sensitivity and suppresses metastatic traits, respectively (Liu et al. 2021a, b).

In gastric cancer (SGC-7901), curcumin suppresses oncogenic lncRNA H19, leading to increased p53 activity and reduced proliferation (Liu et al. 2016a, b, c). Meanwhile, in other GI cancers like pancreatic and colorectal cancers, curcumin suppresses cell growth and promotes apoptosis by upregulating miR-7 and by the knockdown of KCNQ1OT1, respectively (Duan et al. 2020; Ma et al. 2014).

In NSCLC, curcumin downregulates lncRNA UCA1 and broadly reprograms miRNA expression—for example, it upregulates miR-192-5p/215, miR-330-5p, let-7c, and miR-101 and downregulates miR-186 and miR-21—collectively inhibiting proliferation and migration/EMT and enhancing apoptosis (Liu et al. 2019; Wang et al. 2018). Restoration of tumor-suppressive lncRNA MEG3 similarly inhibits proliferation and induces apoptosis in NSCLC and bladder cancer cells (Liu et al. 2018; Lu et al. 2013).

Curcumin decreases LINC00691 expression to promote apoptosis in thyroid carcinoma (B-CPAP cells) (Li et al. 2022). Together, these findings show that downregulation of oncogenic ncRNAs (e.g., HOTAIR, H19, and UCA1) and/or upregulation of tumor-suppressive ncRNAs (e.g., miR-181b, miR-143, and MEG3) by natural agents can enhance apoptosis and chemosensitivity and impair metastasis across multiple malignancies.

Honokiol

Honokiol exerts its anti-NSCLC effects through microRNA regulation. Zhang et al. reported that in A549 NSCLC cells, honokiol treatment markedly increases the levels of miR-148a-3p and miR-148a-5p (Zhang et al. 2019). These miRNAs target components of the PI3K/Akt signaling pathway. Honokiol-induced upregulation of miR-148a-3p/5p (and let-7c-5p) is associated with significant inhibition of cell proliferation and induction of apoptosis. Importantly, chemical inhibition of miR-148a-3p (and let-7c-5p) largely reverses honokiol's cytotoxic effects, indicating that miR-148a-3p/5p mediate honokiol-induced tumor cell death (Zhang et al. 2019).

Pterostilbene

Pterostilbene, a dimethylated analog of resveratrol, exerts antitumor effects in prostate cancer by downregulating oncogenic miR-17-5p and miR-106a-5p, thereby restoring PTEN expression. In tumor-bearing mice, this resulted in increased PTEN mRNA/protein levels and significantly reduced tumor growth, highlighting pterostilbene's potential as an miRNA-targeting therapeutic (Dhar et al. 2015).

Epigallocatechin Gallate (EGCG)

EGCG exerts anticancer effects by altering oncogenic ncRNA networks. In NSCLC, EGCG suppresses oncogenic miR-98-5p via p53 activation, relieving repression on copper transporter 1 (CTR1) to enhance cisplatin uptake (Jiang et al. 2016). Concurrently, EGCG promotes the lncRNA NEAT1, which sponges miR-98-5p and further upregulates CTR1, thereby sensitizing resistant lung tumor cells to cisplatin (Hayakawa et al. 2022; Jiang et al. 2016). In parallel, EGCG induces tumor-suppressive miR-485-5p in NSCLC cells, which targets and downregulates RXRα, leading to apoptosis and loss of cancer stem-like properties (Jiang et al. 2018). EGCG also stabilizes HIF-1α to upregulate miR-210 in A549 lung cancer cells, suppressing proliferation and modulating MAPK-driven miRNAs (e.g., miR-155↑, miR-212↓) to inhibit migration (Dharshini and Mandal 2024; Wang et al. 2011).

In other tumor types, EGCG similarly rewires ncRNA-mediated survival pathways. In hepatocellular carcinoma, EGCG markedly increases miR-16 levels, which directly target BCL-2 mRNA; this decreases BCL-2 protein and triggers mitochondrial apoptosis (Tsang and Kwok 2010). Likewise, in osteosarcoma cell lines (MG-63, U-2OS), EGCG upregulates miR-1, a suppressor of the oncogene c-MET. Loss of miR-1 abrogates the growth-inhibitory effect of EGCG, confirming that the EGCG–miR-1–c-MET axis is crucial for inducing apoptosis in these bone cancer cells (Zhu and Wang 2016). Together, these data indicate that EGCG's anticancer action involves upregulation of tumor-suppressive miRNAs and downregulation of oncomiRs/lncRNAs, leading to apoptotic and chemosensitizing effects in multiple cancers.

Quercetin

Quercetin targets oncogenic miRNAs and lncRNAs to overcome resistance and invasiveness in various cancers. In renal carcinoma cells, quercetin (often combined with hyperoside) downregulates Sp1-driven miR-27a, inducing its repressor ZBTB10. This shifts the balance toward reduced Sp transcription factor activity and lower survivin, thus sensitizing tumor cells to therapy (Li et al. 2014). In lung adenocarcinoma (A549), quercetin upregulates miR-16, which destabilizes claudin-2 mRNA—a tight-junction protein that promotes proliferation/migration. The miR-16 inhibitor rescues claudin-2 levels, confirming that quercetin's increase of miR-16 (and consequent claudin-2 suppression) underlies its antimigratory effect (Sonoki et al. 2015).

Quercetin also downregulates oncogenic lncRNAs that drive metastasis. In prostate cancer cells (PC-3), quercetin suppresses lncRNA MALAT1, reversing epithelial-to-mesenchymal transition and reducing invasion and migration (Lu et al. 2020). Quercetin has similarly been shown to downregulate UCA1 (urothelial carcinoma-associated 1) in cancer cells, which sensitizes cells to apoptosis and chemotherapeutics (Hayakawa et al. 2022).

Quercetin also exhibits anticancer effects in various gastrointestinal cancers. In liver cancer mouse models, quercetin-enriched diets restored the levels of the tumor suppressor gene miR-122, reducing cell migration and invasion (Boesch-Saadatmandi et al. 2012). Additionally, miR-let-7a was overexpressed by the combination of quercetin and catechins. In turn, this inhibited the resultant k-ras gene in pancreatic ductal adenocarcinoma, reducing cell viability and inducing apoptosis (Appari et al. 2014).

Genistein

Genistein reprograms ncRNA circuits across diverse cancers to halt proliferation. In ovarian cancer, genistein has been reported to downregulate oncogenic miR-27a, increasing the protein Sprouty2, which decreases growth and migration (Javed et al. 2021).

Crucially, genistein suppresses the oncogenic lncRNA HOTAIR in both prostate and renal carcinomas. In prostate cancer models, genistein downregulates HOTAIR (partly via miR-34a upregulation), leading to reduced proliferation, migration, and invasion (Chiyomaru et al. 2013). In kidney cancer cells, genistein impairs HOTAIR's interaction with the PRC2 complex, thereby derepressing tumor suppressor genes and inhibiting PI3K/AKT signaling to curb renal tumor growth (Imai-Sumida et al. 2020).

In lung adenocarcinoma (A549), genistein triggers tumor cell apoptosis by affecting the miR-27a/MET axis. Specifically, genistein treatment paradoxically upregulates miR-27a, which in turn downregulates MET protein; this relieves MET-driven survival signaling and activates caspase-3/9 to induce apoptosis (Yang et al. 2016). Finally, in colorectal cancer, genistein has been shown to downregulate lncRNA TTTY18 (which is elevated in CRC), resulting in reduced TGF-β1 expression and suppression of SGK1/Akt and p38 MAPK pathways. This epigenetic modulation promotes apoptosis and decreases CRC cell invasiveness (Chen et al. 2020).

Apigenin

In A549 lung cancer cells, apigenin induces apoptosis through a miR-34a-5p–SNAI1 axis. Transcriptomic profiling showed that apigenin markedly increases miR-34a-5p levels, which targets and downregulates the transcriptional repressor SNAI1. Since SNAI1 suppresses E-cadherin and blocks apoptotic programs, its inhibition leads to activation of caspases and cell death. Thus, apigenin's proapoptotic effect is attributed to miR-34a-5p induction that silences SNAI1, reversing EMT and promoting apoptosis in NSCLC (Aida et al. 2021).

Kaempferol

Kaempferol has been shown to activate tumor-suppressive miR-340 in lung cancer, with downstream effects on the PTEN/PI3K/AKT pathway. In A549 cells, kaempferol treatment upregulates miR-340 expression, which inhibits AKT signaling by relieving PTEN inhibition. This double hit—dampened PI3K/AKT survival signaling coupled with miR-340-driven proapoptotic programs—leads to suppressed proliferation and enhanced apoptosis. The miR-340–PTEN axis thus provides a mechanistic basis for kaempferol's antiproliferative effects in lung cancer (Zhang et al. 2021).

Berberine

Berberine modulates lncRNA networks to promote cancer cell death and chemosensitivity. In CRC cells, berberine sharply induces lncRNA CASC2 expression, which in turn sequesters the AUF1 protein away from BCL-2 mRNA. This prevents AUF1-mediated BCL-2 mRNA stabilization, leading to BCL-2 downregulation and robust apoptosis (Dai et al. 2019a, b). In NSCLC, berberine (particularly in combination with gefitinib) downregulates the oncogenic lncRNA HOTAIR and relieves miR-34a-5p suppression. Induction of miR-34a-5p and loss of HOTAIR synergistically reverse EMT and sensitize tumors to therapy (Zheng et al. 2020).

Silibinin

In bladder cancer models, silibinin exerts tumor-suppressive effects by epigenetically silencing oncogenic lncRNAs. Silibinin treatment of T24 and UM-UC-3 urothelial carcinoma cells led to marked downregulation of lncRNAs HOTAIR and

ZFAS1, which are associated with metastasis and proliferation (Imai-Sumida et al. 2017). This suppression coincides with decreased PI3K/AKT and actin-cytoskeleton signaling. By targeting HOTAIR and ZFAS1, silibinin dismantles lncRNA-driven oncogenic pathways and thereby inhibits bladder cancer growth and invasion (Imai-Sumida et al. 2017).

Cell Signaling Events: Cell Cycle Regulation—Cell Proliferation Axis

Disruptions in signaling pathways, such as hyperactivation of proliferative signals or inhibition of apoptotic pathways, can lead to uncontrolled cell division and tumor growth.

Curcumin is multifunctional, being capable of modulating various cancer targets and signaling pathways (Benvenuto et al. 2020). Curcumin PP compounds were reported to exert suppressive effects on the proliferation and growth of medulloblastoma cells via some molecular pathways, namely, suppressing Wnt/β-catenin, NF-κB, and SHH signaling pathways and triggering the apoptosis-related genetic effectors (Bax, Bcl-2, PARP, caspase-3, and caspase-9), therefore decreasing histone deacetylase 4 (HDAC4) function (Elahi et al. 2024). Thence, targeting the differential signaling pathways in medulloblastoma either by chemotherapy in nano-formula or in combination with PPs targeting nc-epigenetics to improve current and future therapeutics (Sokolov et al. 2024).

Chrysin (5,7-dihydroxyflavone) is a flavonoid in honey or propolis (Mani and Natesan 2018).

Genistein (5,7-dihydroxy-3-(4-hydroxyphenyl) chromen-4-one) is present in soy (Ganai and Farooqi 2015).

Quercetin is present in broccoli and onions (Anand David et al. 2016). Besides, in vitro and in vivo studies demonstrated that berry PPs can inhibit tumor cell proliferation by regulating the cell cycle, stimulating apoptosis, and activating signaling pathways including phosphatidyl inositol triphosphate (PI3K)/AKT/PKB/mTOR, NF-κB, Wnt/Beta-catenin, and ERK/MAPK (Forbes-Hernández 2020).

Sustaining proliferative signaling, wherein cancer cells continuously stimulate their growth by activating growth-promoting pathways, often through the overexpression of growth factors or mutations in growth factor receptors like epidermal growth factor receptor (EGFR). Additionally, cancer cells evade growth suppressors, such as the tumor suppressor proteins p53 and retinoblastoma (Rb), which normally act as brakes on cell division. By inactivating these pathways, cancer cells can bypass cell cycle checkpoints and proliferate uncontrollably (Hamdy et al. 2024).

Inflammation-Angiogenesis-Metastasis Axis

The chemopreventive effect of some PPs, such as anthocyanins, tannins, EGCG, oleuropein-dihydroxyphenyl, punicalagin, quercetin, resveratrol, and theaflavin, was evidenced for melanoma, breast cancer, and other metastatic cancers (Kotecha et al. 2016). This was achieved via the modulation of inflammation, angiogenesis, and metastasis via being linked to apoptosis, proliferation, and autophagic pathways such as EGFR/MAPK, mammalian target of rapamycin (mTOR)/PI3K/Akt, Janus kinase (JAK)/STAT, PGE-2/vascular endothelial growth factor (VEGF), and PGE-1/ERK/HIIF-1α (Benvenuto et al. 2020), targeted by several of the mentioned PPs.

Challenges in Utilizing PPs for Cancer Therapy

Despite the widespread recognition of the potential of PP in cancer therapy, several challenges continue to hinder its effective application. According to the established therapeutic protocols for cancer treatment, PPs cannot be relied upon alone. Therefore, PPs are used in combination with standardized chemotherapy (Ramos 2008) to maintain and improve health in human beings by initiating/stimulating the body's antioxidant, antiapoptotic, antiangiogenic, and anti-inflammatory activities (Liu et al. 2016a, b, c). Besides, dietary PPs administered orally travel through the intestinal gut without being absorbed, affecting the gut microbiota (Kim et al. 2016). PPs' absorption involves glucose transporters in the bowel (Atiq and Parhar, 2020). Moreover, PPs' extensive metabolism, rapid elimination, low water solubility, chemical instability, and poor bioavailability due to their instability at low pH are all challenging and limiting factors for the clinical use of PPs (Enaru et al. 2021). Despite PPs' ability to pass the blood-brain barrier (BBB) through carrier-based transport, for treating brain tumors, PPs are often refluxed from the brain, resulting in their poor bioavailability. This challenge must be addressed if PPs are to be used efficiently for brain tumor treatment.

Breakthroughs in Utilizing PPs for Cancer Therapy

To overcome the PPs' poor bioavailability, different strategies were employed to modify, engineer, or fabricate PPs to improve their stability. One approach involved coating PPs with special materials to be administered as prodrugs, which helps overcome excessive metabolism. The development of novel pharmaceutical formulations, such as nanoparticles (NPs) or liposomes as carriers, has significantly improved the bioavailability and efficacy of PPs while reducing systemic toxicity.

Moreover, NPs were conjugated with antibodies to achieve selective targeting of tumor cells. **Nano-systems** are considered a solution for the poor pharmacokinetics and the low bioavailability of PPs (Rambaran 2020). Innovative fabricated packaging or carrier(s) are required to benefit from PPs' therapeutic advantages while overcoming their limitations, particularly regarding bioavailability.

Several interesting entities formulated as NPs were recently marketed to screen, diagnose, prevent, or treat cancer (Stylianopoulos and Jain 2015). The use of such chemicals at a nano-level proved to be effective in preclinical studies (in vitro and ex vivo) as well as initial cancer treatment trials (Eetezadi et al. 2015; Fernandes et al. 2015; Johnstone et al. 2016).

Nano-formulas facilitated drug delivery to different sites in the body, decreasing toxicity and providing better control over the drug release rate, with an improvement to pharmacokinetic and pharmacodynamic properties (ADME) (Kumari et al. 2016). The highly permeable nature of NPs and their effective retention capability facilitate the distribution of chemical therapeutics in the nano-form through the body. Thus, administering homogenous therapeutics in nano-form would be a preferable choice for treatment (Fernandes et al. 2018).

The idea of developing ***edible PPs in their nano-form*** offers a promising outlook to enhance PPs' medicinal performance and effectiveness, especially in the "*Drug Repurposing*" approach (Sun 2017). The new NP formulas allow drug delivery of chemotherapeutics, as well as anti-inflammatory, antioxidant, and anticancer therapies, to site-specific targets.

Lipid-based nanocarriers (LBNC), such as liposomes, oil dispersions (micelles), and lipid-NPs, are major classes used in anticancer therapy due to their high biocompatibility properties. The first two announced FDA-approved liposomal formulations were **DOXIL**, a liposomal doxorubicin injection for ovarian cancer, and **Marqibo**, a liposomal vincristine sulfate injection for lymphoblastic leukemia.

Biodegradable polymeric nanoparticles (BDPNPs) synthesized from poly(lactic acid) (PLA), poly(lactic-co-glycolic) acid (PLGA), human serum albumin or bovine serum albumin (HSA or BSA), and chitosan polymers have gained popularity because of their controlled and sustained release properties, subcellular sizes, and biocompatibility (Costa et al. 2022). Polyethylene glycol (PEG)-modified PLA NPs loaded with anticancer drugs represent a new advancement in cancer nanomedicine drug delivery.

Hybrid polymer nanoparticles (HPNPs) are an example of **combined PPs** composed of PLGA and chitosan, demonstrating enhanced tumor uptake and potent cytotoxicity compared with unmodified NPs (Banik, Fattahi and Brown, 2016).

Novel hyaluronic acid-based hydrogels (HAHGs), like chitosan hydrogels and agarose or starch-based materials cross-linked by PPs (Kaczmarek 2020), showed an antioxidant effect that enhanced their antitumor mechanism of action.

Hyaluronic acid-polyvinyl alcohol hydrogels incorporating curcumin consist of hyaluronic acid and polyvinyl alcohol cross-linked with curcumin. Curcumin acts as a cross-linker and an active ingredient, providing structural integrity and direct antitumor effects. It helps to suppress the proliferation of cancer cells, induce apoptosis, and inhibit angiogenesis. The in vivo antitumor studies demonstrated that HA-CHEMS-Cur-TPGS NPs significantly inhibited tumor growth with no toxicity or side effects (Huang et al. 2020).

HAHGs are prepared by cross-linking hyaluronic acid and gelatin with PP, like tannic acid, a hydrolyzable amphiphilic tannin found in various natural sources such as gallnuts, grapes, tea, and coffee (Koopmann et al. 2020). Tannic acid acts as a natural cross-linker, providing antioxidant properties that enhance the stability and bioactivity of the hydrogels. These antioxidant properties reduce oxidative stress in the tumor microenvironment, supporting the antitumor activity of the hydrogels by promoting apoptosis in cancer cells (Baer-Dubowska et al. 2020).

A chrysin-curcumin complex formulated with PEG was used to target breast cancer cell lines (Harshul Batra et al. 2019). The chrysin-curcumin combination encapsulated in PEGylated PLGA NPs showed a potent effect suppressing the expression of the telomerase gene in the SW480 colorectal cancer cell line and demonstrated dose-dependent cytotoxicity against Caco-2 cells (Lotfi-Attari et al. 2017).

Genistein-quercetin or **genistein-epigallocatechin gallate** was reported to be highly effective in the treatment of glioma (Atiq and Parhar 2020).

Next-Generation NP-Based Cancer Treatment

Biocompatible Drug Delivery System (BDDS)

BDDS is a recent innovative biotechnological solution employed to find small, hydrophobic, inert carriers with a large surface area for one or more drugs (García-Fernández et al. 2020). NPs fulfill these requirements by being small with a large surface area and possessing the ability to carry various drugs. Therefore, NPs are considered an excellent drug delivery system. Nano-vehicles to be used for such cargos are polymeric-based or lipid-based conjugate-NPs (El-Readi and Althubiti, 2019). Several ongoing research proposals embrace the development of a chitosan/PLA hybrid-based NP system for the combined delivery of toxic chemotherapeutic alkylating agents or anti-inflammatory alkaloids with PPs for primary and metastatic cancers in vivo and in vitro testing.

Plant-derived NPs demonstrated potent activity against cancer cells. For example, ZnO NPs produced from *Cassia auriculata* leaf extract showed activity against MCF-7 breast cancer cells (Anjum et al. 2021). Gold NPs from *Trachyspermum ammi* seed extract suppressed HepG2 cancer cell lines through ROS-driven apoptosis (Perveen et al. 2021).

Chemotherapy's normal toxicity and multidrug resistance (MDR) occur due to an increased efflux of membrane transport proteins and overexpression of P-glycoprotein (P-gp; MDR1). The hallmarks of MDR include reduced drug activation, potentiation of drug metabolism, or alteration in drug targeting (Gottesman et al. 2002). Therefore, several mechanisms were proposed to improve chemotherapy and overcome MDR using lower doses. The combination therapy of chemotherapeutics with natural phytochemicals that possess anticancer, anti-inflammatory, antioxidant, and potential antiautophagic or epigenetic-modulating properties is a promising solution. The inhibition of P-gp function using natural products is another solution proposed to reverse MDR and thus enhance the overall anticancer therapeutic efficacy (El-Readi and Althubiti, 2019).

Two-by-Two PP Combination with Different Nano-formulation(s)

Curcumin and resveratrol displayed synergistic antiproliferative and apoptotic effects, demonstrating better efficacy in colon and hepatocellular carcinoma (Singh et al. 2013). Resveratrol was reported to enhance curcumin's in vivo and in vitro antitumoral potential through poly(ADP-ribose) polymerase 1 (PARP-1) cleavage, increasing the Bax/Bcl-2 ratio, dephosphorylating ERK1 and ERK2, and modulating the expression of the autophagic protein microtubule-associated protein 1A/1B-light chain 3 (LC3; LC3 II) as well as promoting autophagic vacuole formation (Tabrez et al. 2020).

Green Nanotechnology (GNT) as the Future for PP Formulation

GNT is proposed as an alternative protocol for NP synthesis to reduce the dose of hazardous chemotherapeutics and minimize the harsh reaction conditions used for NP synthesis. In this endeavor, this approach helps to fulfill sustainable development goals (SDGs) for the 2030 vision. Investigators used plants and plant-derived materials instead of organic compounds to reduce the environmental burden of NP-chemical production (Hamdy et al. 2022a).

The Good Face of Plant-Based Green Nanotechnology: Green NP Synthesis

Plant parts, including leaves, fruits, roots, stems, and seeds, were utilized to synthesize different NPs. Plant extracts containing compounds like polysaccharides, proteins, amino acids, organic acids, and phytochemicals like PPs, flavonoids, terpenoids, alkaloids, tannins, and alcohols were used for stable NP synthesis (Ali et al. 2020). Green NP formulations face technical and economical limitations associated with the source, type, and concentration of plant extracts; reagent stoichiometric ratios; optimal experimental conditions (such as temperature, pH, and time); yield; and the characterization/application of the final product (Rijuta Ganesh Saratale et al. 2018).

Chemical alkylating drugs, alkaloids from plant sources, and cancer antibiotics are used in the treatment of various types of cancers (Amreddy et al. 2018).

Recent advancements in metal-organic frameworks (MOFs) showed promising applications in cancer research, particularly for drug delivery and imaging applications. With highly organized, tunable porous structures, MOFs can encapsulate a wide range of anticancer agents, allowing for both controlled release and targeted delivery (Elmehrath et al. 2023; Dai et al. 2019a, b). Recent evidence suggested that stable MOFs, formed by natural herb-derived agents and metal ions, could significantly enhance therapeutic efficacy by mimicking the coordination structures of natural metalloenzymes where metals act as active centers (Alkaabi et al. 2024; Kostova and Balkansky 2013).

Metal-polyphenol networks (MPNs) have emerged as a complementary approach, leveraging the coordination between metal ions and polyphenolic compounds to form flexible, amorphous networks. MPNs not only offer superior biocompatibility but also integrate the intrinsic therapeutic benefits of PP, such as antioxidant and anticancer activities (Liu et al. 2021). These networks provide a robust platform for multifunctional nano-formulas in cancer theranostics, offering bio-adhesiveness, biocompatibility, versatile drug loading, and stimulus-response profiles (Liu et al. 2021; Dai et al. 2019a, b). The primary challenge in MPN research revolves around optimizing polyphenol-metal interactions to maximize therapeutic efficacy while maintaining structural integrity (Dai et al. 2019a, b). Although both MOFs and MPNs hold significant promise individually, future research should explore hybrid systems that combine the structural advantages of MOFs with the bioactivity of MPNs.

Summary

This chapter explored the potential of PPs in cancer therapy, emphasizing their role as multimodal agents that target various cancer hallmarks (Fig. 8.2). Key clinical and preclinical studies were reviewed demonstrating the efficacy of PPs in inhibiting tumor growth, inducing apoptosis, affecting epigenetics, and reducing metastasis. The mechanisms of action of some PPs were addressed, highlighting the ability of PPs to modulate signaling pathways, oxidative stress, and key inflammatory factors in cancer progression. Combining targeted therapies could be the solution to overcome conventional chemotherapy toxicity and/or resistance, as well as to counteract PPs' low bioavailability. PPs can target the molecular hallmarks of cancer, making them ideal anticancer agents. They can be used either as a single polyphenolic compound with conventional chemotherapeutic drugs or in combination as two PPs and antineoplastic agents that are carried or modified within NPs to increase

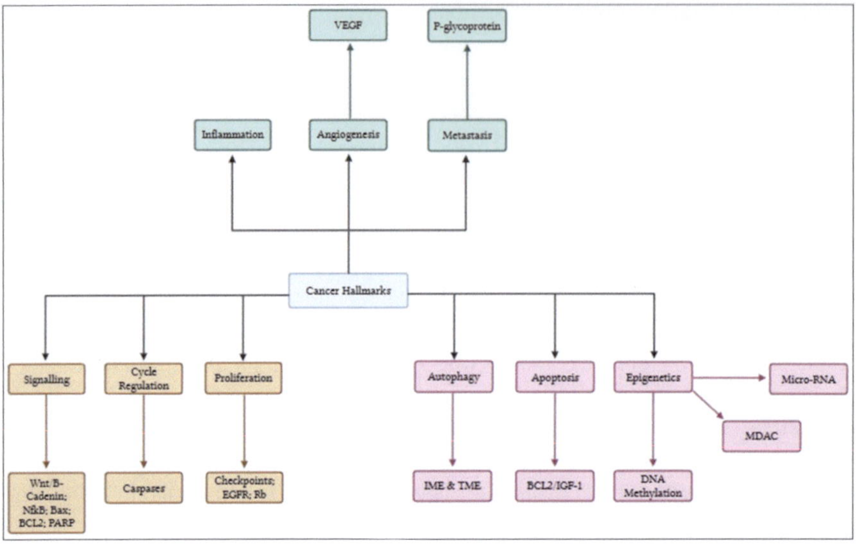

Fig. 8.2 Polyphenols' potential in cancer therapy, emphasizing their role as multimodal agents that target various cancer hallmarks

their activity toward several morbid tumors with reduced side effects. PPs can be used for green nano-formulation of known anticancer agents to produce novel pharmaceutical products. These products represent a revolution in drug delivery with expected and proven synergistic anticancer and anti-inflammatory activity and are believed to help limit cancer progression and metastasis. The combination therapy can be administered to cancer patients in an emulsion or through nano-encapsulation, providing an overall inhibitory effect against oncogenic factors, excessive oxidative stress, and chronic inflammation (Hamdy et al. 2022b). Overcoming or bypassing the MDR effect encountered by tumor cells can be achieved by combining PP polymers and NPs.

Conclusion

PPs offer significant promise in cancer therapy by targeting multiple molecular cancer hallmarks, including epigenetics, yet their clinical potential is limited by challenges like poor bioavailability. Advances in nano-formulation and combination therapies, if given with conventional chemotherapy, have shown the potential to overcome these barriers, enhancing efficacy and reducing side effects. Continued research is crucial to fully realize their role in cancer treatment.

Future Prospectives

Addressing new trends combining synthetic drugs and natural PPs targeting various cancer hallmarks and cancer-related enzymes, such as the 20S proteasome complex correlated with several cancer types, has become of great necessity (Atta et al. 2023).

Combination therapy using PPs and natural bioactive compounds such as the tropolone derivative hinokitiol, which inhibits polyphenol oxidase and inhibits stemness progression and self-renewal via the CD44/Nanog/SOX2/Oct4 pathway, could open new avenues for targeting cancer stem cells and modulating key pathways of apoptosis and autophagy modulation (Chiang et al. 2024).

Additionally, combining PPs and plant alkaloid derivatives like camptothecin, known to inhibit topoisomerase I activity, may improve therapeutic efficacy (Khodair et al. 2024).

The coadministration of PPs with vitamins, like vitamins D and E or the red pigment prodigiosin, that have a known disease/cancer-modulatory effect, with anti-inflammatory, endothelial protective effect, and antioxidative activities, presents another promising strategy to enhance treatment outcomes (Chen et al. 2023; Hamdy et al. 2009; Anwar et al. 2022). Future studies should prioritize these combination therapies, innovative drug formulations, and clinical trials to fully exploit the synergistic potential of PPs in cancer treatment.

References

Abaza, T., El-Aziz, M.K.A., Daniel, K.A., ... Kontos, C.K., Youness, R.A. (2023). Emerging Role of Circular RNAs in Hepatocellular Carcinoma Immunotherapy. International Journal of Molecular Sciences, 2023, 24(22), 16484

Aggarwal, B.B. & Shishodia, (2006). S. Molecular Targets of Dietary Agents for Prevention and Therapy of Cancer. Biochemical Pharmacology, 71, 1397–1421.

Aida, R., Hagiwara, K., Okano, K., Nakata, K., Obata, Y., Yamashita, T.,…Hagiwara, H. (2021). miR-34a-5p might have an important role for inducing apoptosis by down-regulation of SNAI1 in apigenin-treated lung cancer cells. Molecular Biology Reports, 48, 2291–2297.

Ajaz Ahmad Ganai, Humaira Farooqi, (2015). Bioactivity of genistein: A review of in vitro and in vivo studies. Biomedicine & Pharmacotherapy 76:30–38.

Al Aameri, R. F., Sheth, S., Alanisi, E. M., Borse, V., Mukherjea, D., Rybak, L. P., & Ramkumar, V. (2017). Tonic suppression of PCAT29 by the IL-6 signaling pathway in prostate cancer: Reversal by resveratrol. PLoS One, 12(5), e0177198.

Alkaabi, A., Ahsan, K., Munawar, N., Samadi, A., El-Maghraby, H., Amin, A., Greish, Y. (2024). Synthesis, and Characterization of Safranal@ MIL-88B (Fe) Nanostructures and Their Preliminary Anticancer and Antibacterial Characteristics. Materials Advances

Al-Noshokaty, T. M., Abdelhamid, R., Abdelmaksoud, N. M., Khaled, A., Hossam, M., Ahmed, R., Saber, T., Khaled, S., Elshaer, S. S., & Abulsoud, A. I. (2025). Unlocking the multifaceted roles of GLP-1: Physiological functions and therapeutic potential. Toxicology reports, 14, 101895.

Alsufiani HM. The synergistic effect of oxaliplatin and punicalagin on colon cancer cells Caco-2 death. Int J Health Sci (Qassim). 2024 Mar-Apr;18(2):33–37.

Alba, García-Fernández Elena, Aznar Ramón, Martínez-Máñez Félix, Sancenón (2020) New Advances in In Vivo Applications of Gated Mesoporous Silica as Drug Delivery

Nanocarriers Abstract Small 16(3). https://doi.org/10.1002/smll.v16.3 https://doi.org/10.1002/smll.201902242

Amreddy, N. et al., (2018). Recent Advances in Nanoparticle-Based Cancer Drug and Gene Delivery. Advances in cancer research, 137, 115–170.

Anand David AV, Arulmoli R, Parasuraman S (2016). Overviews of Biological Importance of Quercetin: A Bioactive Flavonoid. Pharmacogn Rev 10:84–89.

Ando, K., Hu, Q., Kasagi, Y., Oki, E., & Mori, M. (2021). Recent developments in cancer research: Expectations for a new remedy. Annals of Gastroenterological Surgery, 5(4), 419–426.

Anjum S., Hashim M., Malik S.A., Khan M., Lorenzo J.M., Abbasi B.H., Hano C (2021). Recent Advances in Zinc Oxide Nanoparticles (ZnO NPs) for Cancer Diagnosis, Target Drug Delivery, and Treatment. Cancers. 2021; 13:4570

Anwar, M.M., Albanese, C., Hamdy, N.M., Sultan, A.S, (2022). Rise of the natural red pigment 'prodigiosin' as an immunomodulator in cancer. Cancer Cell Int. Dec 28;22(1):419.

Appari, M., Babu, K. R., Kaczorowski, A., Gross, W., & Herr, I. (2014). Sulforaphane, quercetin and catechins complement each other in elimination of advanced pancreatic cancer by miR-let-7 induction and K-ras inhibition. International journal of oncology, 45(4), 1391–1400.

Ashley G. Muller et al. (2019). Delivery of natural phenolic compounds for the potential treatment of lung cancer. Daru: journal of Faculty of Pharmacy, Tehran University of Medical Sciences, 27(1).

Atta H, Alzahaby N, Hamdy NM, Emam SH, Sonousi A, Ziko L (2023). New trends in synthetic drugs and natural products targeting 20S proteasomes in cancers. Bioorg Chem. Apr;133:106427. doi: https://doi.org/10.1016/j.bioorg.2023.106427.

Ayesha, Atiq Ishwar, Parhar (2020) Anti-neoplastic Potential of Flavonoids and Polysaccharide Phytochemicals in Glioblastoma Molecules 25(21) 4895. https://doi.org/10.3390/molecules25214895

Bakhshan MA, Sheikhzadeh S, Delirezh N (2024). Hesperidin nanoparticles for prostate cancer therapy: preparation, characterization and cytotoxic activity. Biomed Mater. May 3;19(3).

Benvenuto, M., Albonici, L., Focaccetti, C., Ciuffa, S., Fazi, S., Cifaldi, L., Miele, M. T., De Maio, F., Tresoldi, I., Manzari, V., et al. (2020). Polyphenol-Mediated Autophagy in Cancer: Evidence of In Vitro and In Vivo Studies. Int. J. Mol. Sci. 2020, 21, 6635. https://doi.org/10.3390/ijms21186635

Bhuia MS, Chowdhury R, Akter MA, Ali MA, Afroz M, Akbor MS, Sonia FA, Mubarak MS, Islam MT (2024). A mechanistic insight into the anticancer potentials of resveratrol: Current perspectives. Phytother Res. May 20.

Biswas P, Kaium MA, Islam Tareq MM, Tauhida SJ, Hossain MR, Siam LS, Parvez A, Bibi S, Hasan MH, Rahman MM, Hosen D, Islam Siddiquee MA, Ahmed N, Sohel M, Azad SA, Alhadrami AH, Kamel M, Alamoudi MK, Hasan MN, Abdel-Daim MM (2024). The experimental significance of isorhamnetin as an effective therapeutic option for cancer: A comprehensive analysis. Biomed Pharmacother. 2024 Jul;176:116860.

Boesch-Saadatmandi, C., Wagner, A. E., Wolffram, S., & Rimbach, G. (2012). Effect of quercetin on inflammatory gene expression in mice liver in vivo–role of redox factor 1, miRNA-122 and miRNA-125b. Pharmacological Research, 65(5), 523–530.

Brittany L., Banik Pouria, Fattahi Justin L., Brown (2016) Polymeric nanoparticles: the future of nanomedicine WIREs Nanomedicine and Nanobiotechnology 8(2) 271-299. https://doi.org/10.1002/wnan.2016.8.issue-2 https://doi.org/10.1002/wnan.1364

Brown, J. M., & Giaccia, A. J. (1998). The unique physiology of solid tumors: opportunities (and problems) for cancer therapy. Cancer research, 58(7), 1408–1416.

Carrera-Quintanar, L., Lopez Roa, R. I., Quintero-Fabián, S., Sánchez-Sánchez, M. A., Vizmanos, B., & Ortuño-Sahagún, D. (2018). Phytochemicals that influence gut microbiota as prophylactics and for the treatment of obesity and inflammatory diseases. Mediators of inflammation, 2018.

Castro Nava A, et al. (2016). Development of novel radiochemotherapy approaches targeting prostate tumor progenitor cells using nanohybrids. Int J Cancer 137:2492–503.

Chang WW, Hu FW, Yu CC, et al. (2013). Quercetin in elimination of tumor initiating stem-like and mesenchymal transformation property in head and neck cancer. Head Neck 35:413–19.

Chen F. Inhibiting Pink1/Parkin-mediated mitophagy enhances the anticancer effects of quercetin in hepatocellular carcinomaf. Biochem Biophys Res Commun. 2024 Jun 18;712–713:149899.

Chen Y-C, Chiang Y-F, Lin Y-J, Huang K-C, Chen H-Y, Hamdy NM, Huang T-C, Chang H-Y, Shieh T-M, Huang Y-J, et al. (2023). Effect of Vitamin D Supplementation on Primary Dysmenorrhea: A Systematic Review and Meta-Analysis of Randomized Clinical Trials. Nutrients; 15(13):2830.

Chen, C. & Kong, (2004). A.N.T. Dietary Chemopreventive Compounds and ARE/Epre Signaling. Free Radical Biology and Medicine, 36, 1505–1516.

Chen, X., Wu, Y., Gu, J., Liang, P., Shen, M., Xi, J., & Qin, J. (2020). Anti-invasive effect and pharmacological mechanism of genistein against colorectal cancer. Biofactors, 46(4), 620–628.

Chiang YF, Huang KC, Chen HY, Hamdy NM, Huang TC, Chang HY, Shieh TM, Huang YJ, Hsia SM (2024). Hinokitiol Inhibits Breast Cancer Cells In Vitro Stemness-Progression and Self-Renewal with Apoptosis and Autophagy Modulation via the CD44/Nanog/SOX2/Oct4 Pathway. Int J Mol Sci. Mar 31;25(7):3904.

Chiyomaru, T., Yamamura, S., Fukuhara, S., Yoshino, H., Kinoshita, T., Majid, S.,…Enokida, H. (2013). Genistein inhibits prostate cancer cell growth by targeting miR-34a and oncogenic HOTAIR. PloS one, 8(8), e70372.

Choi YR, Na HJ, Lee JA, Kim Y, Kim YS, Kim MJ. Discovery of (-)-epigallocatechin gallate, a novel olfactory receptor 2AT4 agonist that regulates proliferation and apoptosis in leukemia cells. Heliyon. 2024 May 5;10(10):e30298.

Chong Huang, Fen Chen, Ling Zhang, Yue Yang, Xinggang Yang, Weisan Pan. (2020). 99mTc Radiolabeled HA/TPGS-Based Curcumin-Loaded Nanoparticle for Breast Cancer Synergistic Theranostics: Design, in vitro and in vivo Evaluation. Int J Nanomedicine. 2020; 15: 2987–2998)

Çınar Ayan İ, Güçlü E, Vural H, Dursun HG. Piceatannol induces apoptotic cell death through activation of caspase-dependent pathway and upregulation of ROS-mediated mitochondrial dysfunction in pancreatic cancer cells. Mol Biol Rep. 2022 Dec;49(12):11947–11957.

Cirillo G, Curcio M, Vittorio O, et al. (2016). Polyphenol conjugates and human health: a perspective review. Crit Rev Food Sci Nutr., 56:326–37.

ClinicalTrials.gov, Natural PP in cancer treatment, U.S National Library of Medicine, url https://clinicaltrials.gov/ct2/show/NCT01912820?term=Natural+PP+in+cancer+treatment&draw=2&rank=1

Dai W, Dai YG, Ren DF, Zhu DW. Dieckol, a natural polyphenolic drug, inhibits the proliferation and migration of colon cancer cells by inhibiting PI3K, AKT, and mTOR phosphorylation. J Biochem Mol Toxicol. 2023 May;37(5):e23313.

Dai, Q., Geng, H., Yu, Q., Hao, J., Cui, J. (2019a). Polyphenol-Based Particles for Theranostics. Theranostics, 9(11), 3170–3190.

Dai, W., Mu, L., Cui, Y., Li, Y., Chen, P., Xie, H., & Wang, X. (2019b). Berberine promotes apoptosis of colorectal cancer via regulation of the long non-coding RNA (lncRNA) cancer susceptibility candidate 2 (CASC2)/AU-binding factor 1 (AUF1)/B-cell CLL/lymphoma 2 (Bcl-2) axis. Medical Science Monitor: International Medical Journal of Experimental and Clinical Research, 25, 730.

Dariushnejad H, Roshanravan N, Wasman HM, Cheraghi M, Pirzeh L, Ghorbanzadeh V. Silibinin, Synergistically Enhances Vinblastine-Mediated Apoptosis in Triple Negative Breast Cancer Cell Line: Involvement of Bcl2/Bax and Caspase-3 Pathway. Int J Hematol Oncol Stem Cell Res. 2024 Apr 1;18(2):174–182.

de Lima Cherubim, D.J., Buzanello Martins, C.V., Oliveira Fariña, L., da Silva de Lucca, R.A., (2020). PP as natural antioxidants in cosmetics applications. J Cos Derm, 19:33–37.

Desai, K., Baralo, B., Kulkarni, A., Keshava, V. E., Iqbal, S., Ali, H., ... & Thirumaran, R. (2024). Cancer statistics: The United States vs. worldwide.

Dhar, S., Kumar, A., Rimando, A. M., Zhang, X., & Levenson, A. S. (2015). Resveratrol and pterostilbene epigenetically restore PTEN expression by targeting oncomiRs of the miR-17 family in prostate cancer. Oncotarget, 6(29), 27214.

Dharshini, L. C. P., & Mandal, A. K. A. (2024). Regulation of gene expression by modulating microRNAs through Epigallocatechin-3-gallate in cancer. Molecular Biology Reports, 51(1), 230.

Dong, H., Zhang, C. (2025) Epigenetic and biogenetic regulation by polyphenols in prostate cancer in the context of 3P medicine. *EPMA Journal* **16**, 113–125.

Duan, Q., Cai, L., Zheng, K., Cui, C., Huang, R., Zheng, Z.,… Yu, J. (2020). lncRNA KCNQ1OT1 knockdown inhibits colorectal cancer cell proliferation, migration and invasiveness via the PI3K/AKT pathway. Oncology letters, 20(1), 601–610.

Eetezadi S, Ekdawi SN, Allen C. (2015). The challenges facing block copolymer micelles for cancer therapy: in vivo barriers and clinical translation. Adv Drug Deliv. Rev 91:7–22.

Elahi M, Arefnezhad R, Sattar-Shamsabadi M, et al. (2024). Protective Effects of Curcumin Against Medulloblastoma: A Review. Natural Product Communications;19(4).

Elanany, M. M., Mostafa, D., & Hamdy, N. M. (2023). Remodeled tumor immune microenvironment (TIME) parade via natural killer cells reprogramming in breast cancer. Life sciences, 330, 121997.

Elmehrath, S., Nguyen, H. L., Karam, S. M., Amin, A., Greish, Y.E. (2023). BioMOF-Based Anti-Cancer Drug Delivery Systems. Nanomaterials; 13 (5), 953,

El-Mesallamy, H. O., Hamdy, N. M., Zaghloul, A. S., & Sallam, A. M. (2013). Clinical value of circulating lipocalins and insulin-like growth factor axis in pancreatic cancer diagnosis. Pancreas, 42(1), 149–154.

Enaru B, Socaci S, Farcas A, Socaciu C, Danciu C, Stanila A, Diaconeasa Z. Novel Delivery Systems of PP and Their Potential Health Benefits. Pharmaceuticals (Basel). 2021 Sep 22;14(10):946.

Fan P, Fan S, Wang H, et al. (2013). Genistein decreases the breast cancer stem-like cell population through Hedgehog pathway. Stem Cell Res Therapy 4:146.

Fan R, Liang Z, Wang Q, Chen S, Huang S, Liu J, Huang R, Chen J, Zhao F, Huang W. Beneficial action of cinnamic acid against ovarian cancer via network pharmacology analysis and the pharmacological activity assessment. Naunyn Schmiedebergs Arch Pharmacol. 2024 May;397(5):2987–2994.

Fayez, S., Ayoub, I. M., Mostafa, N. M., Moussa, A. Y., Gamal ElDin, M. I., & El-Shazly, M. (2022). Nutraceuticals in Cancer Therapy: Challenges and Opportunities. Handbook of Oxidative Stress in Cancer: Therapeutic Aspects, 189–208.

Feng Y, Lu J, Jiang J, Wang M, Guo K, Lin S. Berberine: Potential preventive and therapeutic strategies for human colorectal cancer. Cell Biochem Funct. 2024 Jun;42(4):e4033.

Fernandes E, Ferreira JA, Andreia P, et al. (2015). New trends in guided nanotherapies for digestive cancers: a systematic review. J Control Release 209:288–307.

Fernandes, C. et al., (2018). Tumor Microenvironment Targeted Nanotherapy. Frontiers in pharmacology, 9, 1230.

Flohé, L. et al., (1997). Redox Regulation of NF-Kappa B Activation. Free Radical Biology and Medicine, 22, 1115–1126.

Tamara Y. Forbes-Hernández (2020). Berries PP: Nano-delivery systems to improve their potential in cancer therapy Journal of Berry Research 10; 45–60 45

Fresco P et al., (2006). New insights on the anticancer properties of dietary PP. Med Res Rev, 26:747–66.

Fu J, Liu W, Liu S, Zhao R, Hayashi T, Zhao H, Xiang Y, Mizuno K, Hattori S, Fujisaki H, Ikejima T. Inhibition of YAP/TAZ pathway contributes to the cytotoxicity of silibinin in MCF-7 and MDA-MB-231 human breast cancer cells. Cell Signal. 2024a Jul;119:111186.

Fu W, Liu L, Tong S. Berberine inhibits the progression of breast cancer by regulating METTL3-mediated m6A modification of FGF7 mRNA. Thorac Cancer. 2024b Jun;15(17):1357–1368. doi: https://doi.org/10.1111/1759-7714.15321.

Gao, D., & Ge, G. (2023). Exploring the underlying biology of cancer and potential therapeutic strategies: a special issue focused on mechanism-based studies. *Acta Biochimica et Biophysica Sinica*, *55*(6), 891–892.

Gareev, I., Beylerli, O., Ilyasova, T., & Wang, C. (2024). Polyphenol-Mediated Modulation of Non-Coding RNAs: A New Therapeutic Approach for Hypertension - A Review. Current hypertension reviews, 20(3), 127–140.

Geng, W., Guo, X., Zhang, L., Ma, Y., Wang, L., Liu, Z.,…Xiong, Y. (2018). Resveratrol inhibits proliferation, migration and invasion of multiple myeloma cells via NEAT1-mediated Wnt/β--catenin signaling pathway. Biomedicine & Pharmacotherapy, 107, 484–494.

Gharpure KM, Wu SY, Li C, et al. (2015). Nanotechnology: future of oncotherapy. Clin Cancer Res 21:3121–30.

Gopalakrishnan, A. et al., (2008). Anti-carcinogenesis by Dietary Phytochemicals: Cytoprotection by Nrf2 in Normal Cells and Cytotoxicity by Modulation of Transcription Factors NF-κB and AP-1 in Abnormal Cancer Cells. Food and Chemical Toxicology, 46, 1257–1270.

Gou Q., Gong X., Jin J., Shi J., Hou Y. (2017). Peroxisome proliferator-activated receptors (PPARs) are potential drug targets for cancer therapy. Oncotarget. 2017;8:60704

Guo, W. et al., (2009). Dietary PP, Inflammation, and Cancer. Nutrition and Cancer, 61, 807–810.

Hail N Jr. et al., (2008). Cancer chemoprevention: a radical perspective. Free Radic Biol Med 45:97–110.

Hamdy N.M., Eskander G., Basalious E.B. (2022b). Insights on the Dynamic Innovative Tumor Targeted-Nanoparticles-Based Drug Delivery Systems Activation Techniques. International Journal of Nanomedicine, 17, pp. 6131–6155

Hamdy NM, Boseila AA, Ramadan A, Basalious EB (2022a). Iron Oxide Nanoparticles-Plant Insignia Synthesis with Favorable Biomedical Activities and Less Toxicity, in the "Era of the-Green": A Systematic Review. Pharmaceutics; 14(4):844.

Hamdy NM, Suwailem SM, El-Mesallamy HO (2009). Influence of vitamin E supplementation on endothelial complications in type 2 diabetes mellitus patients who underwent coronary artery bypass graft. J Diabetes Complications. May-Jun;23(3):167–73.

Hamdy, N. M., Zaki, M. B., Rizk, N. I., Abdelmaksoud, N. M., Abd-Elmawla, M. A., Ismail, R. A., & Abulsoud, A. I. (2024). Unraveling the ncRNA landscape that governs colorectal cancer: A roadmap to personalized therapeutics. *Life sciences*, *354*, 122946.

Han, Z., Yang, Q., Liu, B., Wu, J., Li, Y., Yang, C., & Jiang, Y. (2012). MicroRNA-622 functions as a tumor suppressor by targeting K-Ras and enhancing the anticarcinogenic effect of resveratrol. Carcinogenesis, 33(1), 131–139.

Harshul Batra, Shrikant Pawar, Dherya Bahl (2019), Curcumin in combination with anti-cancer drugs: A nanomedicine review. Pharmacological Research 139:91–105.

Havalli Bommegowda Rashmi, Pradeep Singh Negi, (2020). Phenolic acids from vegetables: A review on processing stability and health benefits. Food Research International 136:109298.

Hayakawa, S., Ohishi, T., Oishi, Y., Isemura, M., & Miyoshi, N. (2022). Contribution of non-coding RNAs to anticancer effects of dietary polyphenols: chlorogenic acid, curcumin, epigallocatechin-3-gallate, genistein, quercetin and resveratrol. Antioxidants, 11(12), 2352.

He P, Li Y, Hu J, Deng B, Tan Z, Chen Y, Yu B, Dong W. Pterostilbene suppresses gastric cancer proliferation and metastasis by inhibiting oncogenic JAK2/STAT3 signaling: In vitro and in vivo therapeutic intervention. Phytomedicine. 2024 Jun;128:155316.

Hoelder S et al., (2012). Discovery of small molecule cancer drugs: successes, challenges and opportunities. Mol Oncol. 6(2):155–176.

Hou Z, Sang S, You H, et al. (2005). Mechanism of action of (−)-epigallocatechin-3-gallate: auto-oxidation-dependent inactivation of epidermal growth factor receptor and direct effects on growth inhibition in human esophageal cancer KYSE 150 cells. Cancer Res 65:8049–56.

Huangfu L, Wang X, Tian S, Chen J, Wang X, Fan B, Yao Q, Wang G, Chen C, Han J, Xing X, Ji J. Piceatannol enhances Beclin-1 activity to suppress tumor progression and its combination therapy strategy with everolimus in gastric cancer. Sci China Life Sci. 2023 Feb;66(2):298–312.

Imai-Sumida, M., Chiyomaru, T., Majid, S., Saini, S., Nip, H., Dahiya, R.,…Yamamura, S. (2017). Silibinin suppresses bladder cancer through down-regulation of actin cytoskeleton and PI3K/Akt signaling pathways. Oncotarget, 8(54), 92032.

Imai-Sumida, M., Dasgupta, P., Kulkarni, P., Shiina, M., Hashimoto, Y., Shahryari, V.,…Yamamura, S. (2020). Genistein represses HOTAIR/chromatin remodeling pathways to suppress kidney cancer. Cellular physiology and biochemistry: international journal of experimental cellular physiology, biochemistry, and pharmacology, 54(1), 53.

Iqbal, M. J., Kabeer, A., Abbas, Z., Siddiqui, H. A., Calina, D., Sharifi-Rad, J., & Cho, W. C. (2024). Interplay of oxidative stress, cellular communication and signaling pathways in cancer. *Cell Communication and Signaling*, *22*(1), 7.

Isis S. Santos et al., (2013). Nanoencapsulation of PP for protective effect against colon–rectal cancer. Biotechnology Advances, 31(5), 514–523.

Javed, Z., Khan, K., Herrera-Bravo, J., Naeem, S., Iqbal, M. J., Sadia, H.,…Akbar, A. (2021). Genistein as a regulator of signaling pathways and microRNAs in different types of cancers. Cancer Cell International, 21, 1–12.

Javid Lotfi-Attari, Younes Pilehvar-Soltanahmadi, Mehdi Dadashpour, Shahriar Alipour, Raana Farajzadeh, Shahrzad Javidfar, Nosratollah Zarghami. (2017). Co-Delivery of Curcumin and Chrysin by Polymeric Nanoparticles Inhibit Synergistically Growth and hTERT Gene Expression in Human Colorectal Cancer Cells. Nutr Cancer. 2017 Nov-Dec;69(8):1290–1299).

Jayaraman S, Natararaj S, Veeraraghavan VP. Hesperidin Inhibits Oral Cancer Cell Growth via Apoptosis and Inflammatory Signaling-Mediated Mechanisms: Evidence From In Vitro and In Silico Analyses. Cureus. 2024 Feb 2;16(2):e53458.

Jemal A, Siegel R, Xu J, Ward E. (2010). Cancer statistics, 2010. CA Cancer J Clin 60:277–300.

Jensen DK et al., (2012) Design of an inhalable dry powder formulation of DOTAP-modified PLGA nanoparticles loaded with siRNA. J Control Release, 157:141–148.

Ji, Q., Liu, X., Fu, X., Zhang, L., Sui, H., Zhou, L.,…Ren, J. (2013). Resveratrol inhibits invasion and metastasis of colorectal cancer cells via MALAT1 mediated Wnt/β-catenin signal pathway. PloS one, 8(11), e78700.

Jiang, P., Wu, X., Wang, X., Huang, W., & Feng, Q. (2016). NEAT1 upregulates EGCG-induced CTR1 to enhance cisplatin sensitivity in lung cancer cells. Oncotarget, 7(28), 43337.

Jiang, P., Xu, C., Chen, L., Chen, A., Wu, X., Zhou, M.,…Feng, Q. (2018). Epigallocatechin-3-gallate inhibited cancer stem cell–like properties by targeting hsa-mir-485-5p/RXRα in lung cancer. Journal of cellular biochemistry, 119(10), 8623–8635.

Jo MJ, Jeong S, Yun HK, Kim DY, Kim BR, Kim JL, Na YJ, Park SH, Jeong YA, Kim BG, Ashktorab H, Smoot DT, Heo JY, Han J, Lee DH, Oh SC. Genipin induces mitochondrial dysfunction and apoptosis via downregulation of Stat3/mcl-1 pathway in gastric cancer. BMC Cancer. 2019 Jul 27;19(1):739.

Johnstone TC, et al. (2016). The next generation of platinum drugs: targeted Pt(II) agents, nanoparticle delivery, and Pt(IV) prodrugs. Chem Rev 116:3436–86.

Kaczmarek, B. (2020) Tannic Acid with Antiviral and Antibacterial Activity as A Promising Component of Biomaterials—A Minireview. Materials 2020, 13, 3224. https://doi.org/10.3390/ma13143224

Kahkashan Perveen, Fohad Mabood Husain, Faizan Abul Qais, Altaf Khan, Suhail Razak, Tayyaba Afsar, Pravej Alam, Ali M. Almajwal, Mahmoud M. A. Abulmeaty. (2021). Microwave-Assisted Rapid Green Synthesis of Gold Nanoparticles Using Seed Extract of Trachyspermum ammi: ROS Mediated Biofilm Inhibition and Anticancer Activity. (2021). Biomolecules. 2021;11:197.

Kakarala M, Brenner DE, Korkaya H, et al. (2010). Targeting breast stem cells with the cancer preventive compounds curcumin and piperine. Breast Cancer Res Treat 122:777–85.

Karin, M et al., (2002). NF-κB in Cancer: From Innocent Bystander to Major Culprit. Nature Reviews Cancer 2002, 2, 301–310.

Karin, M. & Greten, F.R. (2005). NF-κB: Linking Inflammation and Immunity to Cancer Development and Progression. Nature Reviews Immunology, 5, 749–759.

Karin, M. et al., (2004). The IKK NF-κB System: A Treasure Trove for Drug Development. Nature Reviews Drug Discovery, 3, 17–26.

Karuppaiya V, Annamalai A, Krishnamurthy S, Soundarapandian K. Dieckol prevents prostate cancer cell proliferation by transcriptionally attenuating JAK/STAT3 signaling pathway. Environ Toxicol. 2024 Mar;39(3):1187–1196.

Kathrin Pallauf, Gerald Rimbach, (2013). Autophagy, PP and healthy ageing. Ageing Res Rev 12:237–252.

Khaksar S, Kiarostami K, Ramdan M. Effect of Rosmarinic Acid on Cell Proliferation, Oxidative Stress, and Apoptosis Pathways in an Animal Model of Induced Glioblastoma Multiforme. Arch Med Res. 2024 Jun;55(4):103005.

Khan N, Afaq F, Saleem M, et al. (2006). Targeting multiple signaling pathways by green tea polyphenol (−)-epigallocatechin-3-gallate. Cancer Res 66:2500–5.

Khodair A.I. et al. (2024). Camptothecin structure simplification elaborated new imidazo[2,1-b]quinazoline derivative as a human topoisomerase I inhibitor with efficacy against bone cancer cells and colon adenocarcinoma. European Journal of Medicinal Chemistry, 265: 116049.

Kim BYS et al., (2010). Current concepts: nanomedicine. N Engl J Med 363:2434–43.

Kim, K et al., (2016). Greater Flavonoid Intake Is Associated with Improved CVD Risk Factors in US Adults. British Journal of Nutrition, 115(8), 1481–1488.

Koopmann, A.-K.; Schuster, C.; Torres-Rodríguez, J.; Kain, S.; Pertl-Obermeyer, H.; Petutschnigg, A.; Hüsing, N. (2020). Tannin-Based Hybrid Materials and Their Applications: A Review. Molecules 2020, 25, 4910

Kostova, I., & Balkansky, S. (2013). Metal complexes of biologically active ligands as potential antioxidants. Current medicinal chemistry, 20(36), 4508–4539.

Kotecha R, Takami A, Espinoza JL (2016). Dietary phytochemicals and cancer chemoprevention: a review of the clinical evidence. Oncotarget 7:52517–52529.

Ku, Y.-S., Ng, M.-S., Cheng, S.-S., Lo, A.W.-Y., Xiao, Z., Shin, T.-S., Chung, G., Lam, H.-M., (2020). Understanding the composition, biosynthesis, accumulation and transport of flavonoids in crops for the promotion of crops as healthy sources of flavonoids for human consumption. Nutrients, 12:1–23.

Kumari A. et al., (2010). Biodegradable polymeric nanoparticles-based drug delivery systems. Colloids Surf B: Biointerfaces,75:1–18.

Kumari P, Ghosh B, Biswas S. (2016). Nanocarriers for cancer-targeted drug delivery. J Drug Target 24:179–91.

Kwon H, Kim Y, Kim JH. A combination of myokines and genistein suppresses cancer stemness in MCF-7 human breast cancer cells. Nutr Res Pract. 2024 Jun;18(3):436–445.

Lambert JD, Elias RJ. (2010). The antioxidant and pro-oxidant activities of green tea PP: a role in cancer prevention. Arch Biochem Biophys. 501:65–72.

Lee, J. et al., (2005). Nrf2 As a Novel Molecular Target for Chemoprevention. Cancer Letters 2005, 224, 171–184.

León-González AJ. et al., (2015). Pro-oxidant activity of PP and its implication on cancer chemoprevention and chemotherapy. Biochem Pharmacol 98:371–80.

Li J, Feng S, Wang X, Zhang B, He Q. Exploring the Targets and Molecular Mechanisms of Curcumin for the Treatment of Bladder Cancer Based on Network Pharmacology, Molecular Docking and Molecular Dynamics. Mol Biotechnol. 2024a Jun 1.

Li X, Zhou M, Zhu Z, Wang Z, Zhang X, Lu L, Xie Z, Wang B, Pan Y, Zhang J, Xu J. Kaempferol from Alpinia officinarum hance induces G2/M cell cycle arrest in hepatocellular carcinoma cells by regulating the ATM/CHEK2/KNL1 pathway. J Ethnopharmacol. 2024b Jun 8;333:118430.

Li Y, Zhang T, Korkaya H, et al. (2010). Sulforaphane, a dietary component of broccoli/broccoli sprouts, inhibits breast cancer stem cells. Clin Cancer Res 16:2580–90.

Li YJ, Wu SL, Lu SM, et al. (2015). (−)-Epigallocatechin-3-gallate inhibits nasopharyngeal cancer stem cell self-renewal and migration and reverses the epithelial–mesenchymal transition via NF-κB p65 inactivation. Tumor Biol 36:2747–61.

Li Z, Zhang J, You S, Zhang J, Zhang Y, Akram Z, Sun S. Pterostilbene upregulates MICA/B via the PI3K/AKT signaling pathway to enhance the capability of natural killer cells to kill cervical cancer cells. Exp Cell Res. 2024c Feb 15;435(2):113933.

Li, W., Liu, M., Xu, Y.-F., Feng, Y., Che, J.-P., Wang, G.-C., & Zheng, J.-H. (2014). Combination of quercetin and hyperoside has anticancer effects on renal cancer cells through inhibition of oncogenic microRNA-27a. Oncology reports, 31(1), 117–124.

Li, W., Yang, W., Liu, Y., Chen, S., Chin, S., Qi, X., … Mei, X. (2017). MicroRNA-378 enhances inhibitory effect of curcumin on glioblastoma. Oncotarget, 8(43), 73938.

Li, Z., Gao, Y., Li, L., & Xie, S. (2022). Curcumin inhibits papillary thyroid cancer cell proliferation by regulating lncRNA LINC00691. Analytical Cellular Pathology, 2022(1), 5946670.

Lim SC, Lee TB, Han SI. Caffeic Acid Enhances Anticancer Drug-induced Apoptosis in Acid-adapted HCT116 Colon Cancer Cells. Anticancer Res. 2024 Jun;44(6):2587–2595.

Liu F, Zhang Y, Xia X, Han J, Cao L. Honokiol regulates ovarian cancer cell malignant behavior through YAP/TAZ pathway modulation. J Obstet Gynaecol Res. 2024 Jun;50(6):1010–1019.

Liu M., Wang J., Huang B., Chen A., Li X. (2016a). Oleuropein inhibits the proliferation and invasion of glioma cells via suppression of the AKT signaling pathway. Oncol. Rep. 2016; 36:2009–2016

Liu, D., & Liu, X. (2025). Polyphenols as microRNA modulator in endometrial cancer: implications for apoptosis induction. *Molecular genetics and genomics : MGG*, *300*(1), 34.

Liu, F., Ni, W., Yu, S., Li, F., Zhu, J., Chen, Z.,…Li, C. (2021a). Curcumin affects the prognosis of renal cell carcinoma through a negative feedback loop of H19/miR-675/HDAC/CTCF. Arch Med Res, 12, 1–11.

Liu, G., Xiang, T., Wu, Q. F., & Wang, W. X. (2016b). Curcumin suppresses the proliferation of gastric cancer cells by downregulating H19. Oncology letters, 12(6), 5156–5162.

Liu, G., Zhao, X., Zhou, J., Cheng, X., Ye, Z., & Ji, Z. (2018). Long non-coding RNA MEG3 suppresses the development of bladder urothelial carcinoma by regulating miR-96 and TPM1. Cancer biology & therapy, 19(11), 1039–1056.

Liu, J.-M., Li, M., Luo, W., & Sun, H.-B. (2021b). Curcumin attenuates Adriamycin-resistance of acute myeloid leukemia by inhibiting the lncRNA HOTAIR/miR-20a-5p/WT1 axis. Laboratory investigation, 101(10), 1308–1317.

Liu, P., Liang, H., Xia, Q., Li, P., Kong, H., Lei, P.,…Tu, Z. (2013). Resveratrol induces apoptosis of pancreatic cancers cells by inhibiting miR-21 regulation of BCL-2 expression. Clinical and Translational Oncology, 15, 741–746.

Liu, P., Shi, X., Zhong, S., Peng, Y., Qi, Y., Ding, J., & Zhou, W. (2021c). Metal-phenolic networks for cancer theranostics. Biomaterials science, 9(8), 2825–2849.

Liu, Y., Sun, H., Makabel, B., Cui, Q., Li, J., Su, C.,…Zhang, J. (2019). The targeting of non-coding RNAs by curcumin: Facts and hopes for cancer therapy. Oncology reports, 42(1), 20–34.

Liu, Z., Wang, W., Huang, G., Zhang, W., & Ni, L. (2016c). In vitro and in vivo evaluation of the prebiotic effect of raw and roasted almonds (Prunus amygdalus). Journal of the Science of Food and Agriculture, 96(5), 1836–1843.

Lu Y, Sun J, Yang M, Xing Y, Zhu W, Zhu J, Ma X, Wang Y, Wang L, Jia Y. Myricetin Induces Ferroptosis and Inhibits Gastric Cancer Progression by Targeting NOX4. J Agric Food Chem. 2024 Mar 27;72(12):6178–6188.

Lu, K.-H., Li, W., Liu, X.-H., Sun, M., Zhang, M.-L., Wu, W.-Q., … Hou, Y.-Y. (2013). Long non-coding RNA MEG3 inhibits NSCLC cells proliferation and induces apoptosis by affecting p53 expression. BMC cancer, 13, 1–11.

Lu, M., Liu, B., Xiong, H., Wu, F., Hu, C., & Liu, P. (2019). Trans-3, 5, 4′-trimethoxystilbene reduced gefitinib resistance in NSCLC s via suppressing MAPK/Akt/Bcl-2 pathway by upregulation of miR-345 and miR-498. Journal of Cellular and Molecular Medicine, 23(4), 2431–2441.

Lu, X., Chen, D., Yang, F., & Xing, N. (2020). Quercetin inhibits epithelial-to-mesenchymal transition (EMT) process and promotes apoptosis in prostate cancer via downregulating lncRNA MALAT1. Cancer management and research, 1741–1750.

Luo H, Gao M, Lu H, Chen Q, Lian X. Anthocyanins prevent the development and progression of urethane-induced lung cancer by regulating energy metabolism in mice. Food Nutr Res. 2024a Apr 17;68. doi: https://doi.org/10.29219/fnr.v68.10434 . PMID: 38716355; PMCID: PMC11075467.

Luo J, Solimini NL, Elledge SJ. (2009). Principles of cancer therapy: oncogene and non-oncogene addiction. Cell 136:823–37.

Luo Y, Prestwich GD. (2002). Cancer-targeted polymeric drugs. Curr Cancer Drug Targets 2:209–26.

Luo Y, Zhang G, Hu C, Huang L, Wang D, Chen Z, Wang Y. The Role of Natural Products from Herbal Medicine in TLR4 Signaling for Colorectal Cancer Treatment. Molecules. 2024b Jun 7;29(12):2727.

Ma, J., Fang, B., Zeng, F., Pang, H., Zhang, J., Shi, Y.,…Xia, J. (2014). Curcumin inhibits cell growth and invasion through up-regulation of miR-7 in pancreatic cancer cells. Toxicology letters, 231(1), 82–91.

Mahmoud Zaki, El-Readi Mohammad Ahmad, Althubiti (2019) Cancer Nanomedicine: A New Era of Successful Targeted Therapy Journal of Nanomaterials 20191-13. https://doi.org/10.1155/2019/4927312

Mandalari, G., Faulks, R. M., Bisignano, C., Waldron, K. W., Narbad, A., & Wickham, M. S. (2010). In vitro evaluation of the prebiotic properties of almond skins (Amygdalus communis L.). FEMS microbiology letters, 304(2), 116–122.

Martínez-Esquivias F, Guzmán-Flores JM, Chávez-Díaz IF, Iñiguez-Muñoz LE, Reyes-Chaparro A. Pharmacological network study on the effect of 6-gingerol on cervical cancer using computerized databases. J Biomol Struct Dyn. 2023 Sep 29:1–12.

Md. Arshad, Ali Temoor, Ahmed Wenge, Wu Afsana, Hossain Rahila, Hafeez Md. Mahidul, Islam Masum Yanli, Wang Qianli, An Guochang, Sun Bin, Li (2020) Advancements in Plant and Microbe-Based Synthesis of Metallic Nanoparticles and Their Antimicrobial Activity against Plant Pathogens Nanomaterials 10(6) 1146. https://doi.org/10.3390/nano10061146

Mejia, S. B., Kendall, C. W., Viguiliouk, E., Augustin, L. S., Ha, V., Cozma, A. I., ... & de Souza, R. J. (2014). Effect of tree nuts on metabolic syndrome criteria: a systematic review and meta-analysis of randomised controlled trials. BMJ open, 4(7).

Michael M., Gottesman Tito, Fojo Susan E., Bates (2002) Multidrug resistance in cancer: role of ATP–dependent transporters Nature Reviews Cancer 2(1) 48-58. https://doi.org/10.1038/nrc706

Mi Y, Shan H, Wang B, Tang H, Jia J, Liu X, Yang Q. Genipin inhibits proliferation of gastric cancer cells by inducing ferroptosis: an integrated study of network pharmacology and bioinformatics study. Med Oncol. 2024 Jan 4;41(2):46.

Miguel Asensi et al., (2011). Natural PP in cancer therapy. Journal of Critical Reviews in Clinical Laboratory Sciences, 48(5–6), 197–216.

Mostafa, A. M., Hamdy, N. M., El-Mesallamy, H. O., & Abdel-Rahman, S. Z. (2015). Glucagon-like peptide 1 (GLP-1)-based therapy upregulates LXR-ABCA1/ABCG1 cascade in adipocytes. Biochemical and biophysical research communications, 468(4), 900–905.

Motohashi, H. & Yamamoto, M. (2004). Nrf2–Keap1 Defines a Physiologically Important Stress Response Mechanism. Trends in Molecular Medicine, 10, 549–557.

Munawar Abbas et al., (2017). Natural PP: An overview. Journal of International Journal of Food Properties, 20(8), 1689–1699.

Naponelli V, Rocchetti MT, Mangieri D. Apigenin: Molecular Mechanisms and Therapeutic Potential against Cancer Spreading. Int J Mol Sci. 2024 May 20;25(10):5569.

Nsairat H, Jaber AM, Faddah H, Ahmad S. Oleuropein impact on colorectal cancer. Future Sci OA. 2024 May 23;10(1):FSO.

Orazio Vittorio et al., (2016). PP delivery by polymeric materials: challenges in cancer treatment. Journal of Drug Delivery, 24(1), 162–180.

Ortega AL, Mena S, Estrela JM. (2011). Glutathione in cancer cell death. Cancers 3:1285–310.
Otun S, Achilonu I, Odero-Marah V. Unveiling the potential of Muscadine grape Skin extract as an innovative therapeutic intervention in cancer treatment. J Funct Foods. 2024 May;116:106146.
Pacardo DB, Ligler FS, Gu Z. (2015). Programmable nanomedicine: synergistic and sequential drug delivery systems. Nanoscale 7:3381–91.
Patra, S., Praharaj, P. P., Panigrahi, D. P., Panda, B., Bhol, C. S., Mahapatra, K. K., Mishra, S. R., Behera, B. P., Jena, M., Sethi, G., Patil, S., Patra, S. K., & Bhutia, S. K. (2020). Bioactive compounds from marine invertebrates as potent anticancer drugs: the possible pharmacophores modulating cell death pathways. Molecular biology reports, 47(9), 7209–7228. https://doi.org/10.1007/s11033-020-05709-8
Paul SM. et al., (2010). How to improve R&D productivity: the pharmaceutical industry's grand challenge. Nat Rev Drug Discovery, 9(3):203–214.
Phuah, N. H., & Nagoor, N. H. (2014). Regulation of microRNAs by natural agents: new strategies in cancer therapies. BioMed research international, 2014(1), 804510.
Priyamvada P, Ashok G, Joshi T, Anbarasu S, Anbarasu A, Ramaiah S. Unravelling the molecular mechanistic pathway underlying the anticancer effects of kaempferol in colorectal cancer: a reverse pharmacology network approach. Mol Divers. 2024 May 25.
Qiong-Qiong Yang et al., (2019). Nanochemoprevention with therapeutic benefits: An updated review focused on epigallocatechin gallate delivery. Journal of Critical Reviews in Food Science and Nutrition, 60(8), 1243–1264.
Rambaran, T. F., Bergman, J., Nordström, P., & Nordström, A. (2020). Effect of Berry Polyphenols on Glucose Metabolism: A Systematic Review and Meta-Analysis of Randomized Controlled Trials. Current developments in nutrition, 4(7), nzaa100. https://doi.org/10.1093/cdn/nzaa100
Ramos, S. (2008), Cancer chemoprevention and chemotherapy: Dietary PP and signalling pathways. Mol. Nutr. Food Res., 52: 507–526.
Ramunë Jacobsen et al., (2009). Barriers to cancer pain management: A review of empirical research. Medicina, 45, 427.
Renuka Mani, Vijayakumar Natesan, (2018). Chrysin: Sources, beneficial pharmacological activities, and molecular mechanism of action. Phytochemistry 145:187–196.
Rijuta Ganesh Saratale, Indira Karuppusamy, Ganesh Dattatraya Saratale, Arivalagan Pugazhendhi, Gopalakrishanan Kumar, Yooheon Park, Gajanan S. Ghodake, Ram Naresh Bharagava, J. Rajesh Banu, Han Seung Shin (2018). A comprehensive review on green nanomaterials using biological systems: Recent perception and their future applications. Colloids and Surfaces B: Biointerfaces 170:20–35.
Rizvi, S. A. A., & Saleh, A. M. (2018). Applications of nanoparticle systems in drug delivery technology. Saudi pharmaceutical journal : SPJ : the official publication of the Saudi Pharmaceutical Society, 26(1), 64–70. https://doi.org/10.1016/j.jsps.2017.10.012
Roca E, Colloca G, Lombardo F, Bellieni A, Cucinella A, Madonia G, Martinelli L, Damiani ME, Zampieri I, Santo A. The importance of integrated therapies on cancer: Silibinin, an old and new molecule. Oncotarget. 2024 May 23;15:345–353.
Sabrina Bossio, Anna Perri, Rocco Malivindi, Francesca Giordano, Vittoria Rago, Maria Mirabelli, Alessandro Salatino, Antonio Brunetti, Emanuela Alessandra Greco, Antonio Aversa. (2022) Oleuropein Counteracts Both the Proliferation and Migration of Intra-and Extragonadal Seminoma Cells. Nutrients. 2022;14:2323
Sair, A. T., & Liu, R. H. (2022). Molecular regulation of phenolic compounds on IGF-1 signaling cascade in breast cancer. Food & function, 13(6), 3170–3184.
Sathya, S., Devi, K. P. (2017). The use of polyphenols for the treatment of alzheimer's disease Role of the Mediterranean Diet in the Brain and Neurodegenerative Diseases, 2017, pp. 239–252 (Book Chapter)
Shams Tabrez, Medha Priyadarshini, Maryam Urooj, Shazi Shakil, Ghulam Md Ashraf, Mohd Shahnawaz Khan, Mohammad Amjad Kamal, Qamre Alam, Nasimudeen R. Jabir, Adel Mohammad Abuzenadah, Adeel G. A. Chaudhary, Ghazi Abdullah Damanhouri (2013) Cancer

Chemoprevention by PP and Their Potential Application as Nanomedicine, J Env Sci Heal 31: 67–98.

Shi G, Li X, Wang W, Hou L, Yin L, Wang L. Allicin Overcomes Doxorubicin Resistance of Breast Cancer Cells by Targeting the Nrf2 Pathway. Cell Biochem Biophys. 2024 Feb 27.

Siddiqui IA, et al., (2012). Impact of nanotechnology in cancer: emphasis on nanochemoprevention. Int J Nanomed 7:591–605.

Side effects of cancer treatment. American Cancer Society. url https://www.cancer.org/treatment/treatments-and-side-effects/physical-side-effects.html

Singh CK, George J, Ahmad N (2013). Resveratrol-based combinatorial strategies for cancer management. Ann N Y Acad Sci 1290:113–121.

Singh, S., Behl, T., & Kumar, D. (2024). Molecular mechanism (s) of angiogenesis, inflammation, and oxidative stress in cancer. In *Targeting Angiogenesis, Inflammation, and Oxidative Stress in Chronic Diseases* (pp. 57–73). Academic Press.

Singh, S., Kaur, I., Kariyat, R., (2021). The multifunctional roles of PP in plant-herbivore interactions. Int J Mol Sci 22:1–19.

Sokolov, D., Sharda, N., Banerjee, A., Denisenko, K., Basalious, E. B., Shukla, H., Waddell, J., Hamdy, N. M., & Banerjee, A. (2024). Differential Signaling Pathways in Medulloblastoma: Nano-biomedicine Targeting Non-coding Epigenetics to Improve Current and Future Therapeutics. *Current pharmaceutical design*, *30*(1), 31–47.

Somade OT, Ajayi BO, Adeyi OE, Dada TA, Ayofe MA, Inalu DC, Ajiboye OI, Shonoiki OM, Adelabu AO, Onikola RT, Isiaka ID, Omotoso O, James AS, Olaniyan TO, Adegoke AM, Akamo AJ, Oyinloye BE, Adewole E. Ferulic acid interventions ameliorate NDEA-CCl4-induced hepatocellular carcinoma via Nrf2 and p53 upregulation and Akt/PKB-NF-κB-TNF-α pathway downregulation in male Wistar rats. Toxicol Rep. 2024 Jan 14;12:119–127.

Sonoki, H., Sato, T., Endo, S., Matsunaga, T., Yamaguchi, M., Yamazaki, Y.,…Ikari, A. (2015). Quercetin decreases claudin-2 expression mediated by up-regulation of microRNA miR-16 in lung adenocarcinoma A549 cells. Nutrients, 7(6), 4578–4592.

Soraya Sajadimajd et al., (2020). Advances on Natural PP as Anticancer Agents for Skin Cancer. Pharmacological Research, 151, 104584.

Sofia M., Costa Raul, Fangueiro Diana P., Ferreira (2022) Drug Delivery Systems for Photodynamic Therapy: The Potentiality and Versatility of Electrospun Nanofibers Abstract Macromolecular Bioscience 22(5) 10.1002/mabi.v22.5. https://doi.org/10.1002/mabi.202100512

Srini Srinivasan et al., (2005). V. Bioavailability of Ingredients in Dietary Supplements. In Pharmaceutical Dissolution Testing; Informa Healthcare: Cambridge, UK, 407–419.

Stylianopoulos T, Jain RK. (2015). Design considerations for nanotherapeutics in oncology. Nanomed: Nanotechnol Biol Med 11:1893–907.

Sun DP, Huang HY, Chou CL, Cheng LC, Wang WC, Tian YF, Fang CL, Lin KY. Punicalagin is cytotoxic to human colon cancer cells by modulating cell proliferation, apoptosis, and invasion. Hum Exp Toxicol. 2023 Jan-Dec;42:9603271231213979.

Sun DP, Uen YH, Kang NW, Chang CC, Tian YF, Fang CL, Lin KY. Punicalagin Restricts Growth, Promotes Apoptosis, and Reduces Invasion in Human Gastric Cancer Cells. Dose Response. 2024 Jun 19;22(2):15593258241264954.

Sun L, Wen S, Li Q, Lai X, Chen R, Zhang Z, Cao J, Sun S. Theaflavin-3,3'-di-gallate represses prostate cancer by activating the PKCδ/aSMase signaling pathway through a 67 kDa laminin receptor. Food Funct. 2022 Apr 20;13(8):4421–4431.

Sun, J., et.al., (2017). A systematic analysis of FDA-approved anticancer drugs. BMC systems biology, 11(Suppl 5), 87.

Surh, Y.-J., Na, H.-K. (2007). NF-κB and Nrf2 as Prime Molecular Targets for Chemoprevention and Cytoprotection with Anti-Inflammatory and Antioxidant Phytochemicals. Genes & Nutrition, 2, 313–317.

Tasnima Khushnud & Shaker A. Mousa. (2013). Potential Role of Naturally Derived PP and Their Nanotechnology Delivery in Cancer. Molecular Biotechnology, 55, 78–86.

Thakur VS, Gupta K, Gupta S. (2012). Green tea PP increase p53 transcriptional activity and acetylation by suppressing class I histone deacetylases. Int J Oncol 41:353–61.

Tomás-Barberán, F. A., Selma, M. V., & Espín, J. C. (2016). Interactions of gut microbiota with dietary PP and consequences to human health. Current opinion in clinical nutrition and metabolic care, 19(6), 471–476.

Tsang, W. P., & Kwok, T. T. (2010). Epigallocatechin gallate up-regulation of miR-16 and induction of apoptosis in human cancer cells. The Journal of nutritional biochemistry, 21(2), 140–146.

Tusi, S.K. et al., (2010). Attenuation of NF-κB and Activation of Nrf2 Signaling by 1,2,4-Triazine Derivatives, Protects Neuron-Like PC12 Cells Against Apoptosis. Apoptosis, 15, 738–751.

Valen, G. (2004). Signal Transduction Through Nuclear Factor Kappa B in Ischemia-Reperfusion and Heart Failure. Basic Research in Cardiology, 99, 1–7.

Wanda Baer-Dubowska, Hanna Szaefer, Aleksandra Majchrzak-Celińska & Violetta Krajka-Kuźniak . (2020). Tannic Acid: Specific Form of Tannins in Cancer Chemoprevention and Therapy-Old and New Applications. Current Pharmacology Reports (2020) 6:28–37).

Wang AZ, Langer R, Farokhzad OC. (2012). Nanoparticle delivery of cancer drugs. Annu Rev Med 63:185–98.

Wang D, Zhou Y, Hua L, Hu M, Zhu N, Liu Y, Zhou Y. The role of the natural compound naringenin in AMPK-mitochondria modulation and colorectal cancer inhibition. Phytomedicine. 2024a Aug;131:155786.

Wang S, Wang C, Zhang J, Jiang K, Nian F. Allelopathy and potential allelochemicals of Ligularia sagitta as an invasive plant. Plant Signal Behav. 2024b Dec 31;19(1):2335025.

Wang S, Zhang F, Chen J. Application and potential value of curcumin in prostate cancer: a meta-analysis based on animal models. Front Pharmacol. 2024c May 9;15:1379389.

Wang, G., Wang, J. & Momeni, M.R. (2023). Epigallocatechin-3-gallate and its nanoformulation in cervical cancer therapy: the role of genes, MicroRNA and DNA methylation patterns. *Cancer Cell Int* **23**, 335.

Wang, H., Bian, S., & Yang, C. S. (2011). Green tea polyphenol EGCG suppresses lung cancer cell growth through upregulating miR-210 expression caused by stabilizing HIF-1α. Carcinogenesis, 32(12), 1881–1889.

Wang, W.-H., Chen, J., Zhang, B.-R., Lu, S.-J., Wang, F., Peng, L., … Sun, Y.-Z. (2018). Curcumin inhibits proliferation and enhances apoptosis in A549 cells by downregulating lncRNA UCA1. Die Pharmazie-An International Journal of Pharmaceutical Sciences, 73(7), 402–407.

Williams RJ, et al., (2004). Flavonoids: antioxidants or signalling molecules? Free Radic Biol Med, 36:838–49.

Xu JW, Yan Y, Wang L, Wu D, Ye NK, Chen SH, Li F. Marine bioactive compound dieckol induces apoptosis and inhibits the growth of human pancreatic cancer cells PANC-1. J Biochem Mol Toxicol. 2021 Feb;35(2):e22648.

Xu X, Ding Y, Jin J, Xu C, Hu W, Wu S, Ding G, Cheng R, Cao L, Jia S. Post-translational modification of CDK1-STAT3 signaling by fisetin suppresses pancreatic cancer stem cell properties. Cell Biosci. 2023 Sep 24;13(1):176.

Yang M, Abdullah, Ahmad N, Hussain M, Lu X, Xu J, Zhong H, Guan R. A review of recent advances on cyanidin-3-glucoside: the biotransformation, absorption, bioactivity and applications of nano-encapsulation. Food Funct. 2023 Jul 17;14(14):6320–6345.

Yang, Q., Xu, E., Dai, J., Liu, B., Han, Z., Wu, J.,…Jiang, Y. (2015a). A novel long noncoding RNA AK001796 acts as an oncogene and is involved in cell growth inhibition by resveratrol in lung cancer. Toxicology and applied pharmacology, 285(2), 79–88.

Yang, R.-C. et al., (2000). Potential Protective Effect of NF-κB Activity on the Polymicrobial Sepsis of Rats Preconditioning Heat Shock Treatment. Clinica Chimica Acta, 302, 11–22.

Yang, S., Li, W., Sun, H., Wu, B., Ji, F., Sun, T.,…Zhou, D. (2015b). Resveratrol elicits anti-colorectal cancer effect by activating miR-34c-KITLG in vitro and in vivo. BMC cancer, 15, 1–11.

Yang, Y., Zang, A., Jia, Y., Shang, Y., Zhang, Z., Ge, K.,…Wang, B. (2016). Genistein inhibits A549 human lung cancer cell proliferation via miR-27a and MET signaling. Oncology letters, 12(3), 2189–2193.

Yao LH, Jiang YM, Shi J, et al. (2004). Flavonoids in food and their health benefits. Plant Foods Hum Nutr. 59:113–22.

Yifan M, Rui X, Yuan L, Feiyun J. Allicin inhibits the biological activities of cervical cancer cells by suppressing circEIF4G2. Food Sci Nutr. 2024 Jan 4;12(4):2523–2536.

Zhang J, Ma Y. Luteolin as a potential therapeutic candidate for lung cancer: Emerging preclinical evidence. Biomed Pharmacother. 2024 Jul;176:116909.

Zhang, J., Liu, J., Xu, X., & Li, L. (2017). Curcumin suppresses cisplatin resistance development partly via modulating extracellular vesicle-mediated transfer of MEG3 and miR-214 in ovarian cancer. Cancer chemotherapy and pharmacology, 79, 479–487.

Zhang, J., Zhang, Y., Shen, W., Fu, R., Ding, Z., Zhen, Y., & Wan, Y. (2019). Cytological effects of honokiol treatment and its potential mechanism of action in non-small cell lung cancer. Biomedicine & Pharmacotherapy, 117, 109058.

Zhang, Z., Shi, J., Nice, E. C., Huang, C., & Shi, Z. (2021). The multifaceted role of flavonoids in cancer therapy: leveraging autophagy with a double-edged sword. Antioxidants, 10(7), 1138.

Zheng, F., Li, J., Ma, C., Tang, X., Tang, Q., Wu, J.,…Hann, S. S. (2020). Novel regulation of miR-34a-5p and HOTAIR by the combination of berberine and gefitinib leading to inhibition of EMT in human lung cancer. Journal of cellular and molecular medicine, 24(10), 5578–5592.

Zhu, K., & Wang, W. (2016). Green tea polyphenol EGCG suppresses osteosarcoma cell growth through upregulating miR-1. Tumor Biology, 37, 4373–4382.

Correction to: From (Epi)genetics-to-Precision (E2P): Non-protein Coding RNAs (ncRNAs) a Step-toward Precision Medicine

Nadia M. Hamdy, Ahmed S. Sultan, and John Frederick Kennedy

Correction to:
Chapters 6 and 7 in: N. M. Hamdy et al. (eds.), *From (Epi)genetics-to-Precision (E2P)*: Non-protein Coding RNAs (ncRNAs) a Step-toward Precision Medicine, Subcellular Biochemistry 114, https://doi.org/10.1007/978-3-032-08530-6

The original version of Chapter 6 and Chapter 7 were inadvertently published with the incorrect name of the co-author. The name "Tohada Nashoaty" has now been corrected to "Tohada M. AL-Noshokaty" in both these chapters.

The updated versions of these chapters can be found at
https://doi.org/10.1007/978-3-032-08530-6_6
https://doi.org/10.1007/978-3-032-08530-6_7

N. M. Hamdy et al. (eds.), *From (Epi)genetics-to-Precision (E2P)*, Subcellular Biochemistry 114, https://doi.org/10.1007/978-3-032-08530-6_9

MIX
Papier aus verantwortungsvollen Quellen
Paper from responsible sources
FSC® C105338

If you have any concerns about our products,
you can contact us on
ProductSafety@springernature.com

In case Publisher is established outside the EU,
the EU authorized representative is:
Springer Nature Customer Service Center GmbH
Europaplatz 3, 69115 Heidelberg, Germany

Printed by Libri Plureos GmbH
in Hamburg, Germany